Springer Series in Materials Science

Volume 352

The Springer Series in Materials Science covers the complete spectrum of materials research and technology, including fundamental principles, physical properties, materials theory and design. Recognizing the increasing importance of materials science in future device technologies, the book titles in this series reflect the state-of-the-art in understanding and controlling the structure and properties of all important classes of materials.

Dorian A. H. Hanaor
Editor

Sustainable Materials for the Built Environment

Springer

Editor
Dorian A. H. Hanaor
Faculty of Science and Engineering
Southern Cross University
East Lismore, NSW, Australia

Materialeyes, Materials Innovation Consulting
Berlin, Germany

ISSN 0933-033X ISSN 2196-2812 (electronic)
Springer Series in Materials Science
ISBN 978-3-031-97817-3 ISBN 978-3-031-97818-0 (eBook)
https://doi.org/10.1007/978-3-031-97818-0

This Springer imprint is published by the registered company Springer Nature Switzerland AG
The registered company address is: Gewerbestrasse 11, 6330 Cham, Switzerland

Foreword

The built environment is a mirror of our values and a site of profound consequence. Every act of construction is a negotiation with the earth—an intervention in its ecosystems, an imprint on its material flows, and an influence on human and non-human lives. In its immense materiality, the built environment shapes our worlds, embedding within it our cultural, social, and ecological priorities. And so, this shaping is never neutral. Material selection for a project in one place always influences (and may come at the expense of) other environments and people. As Jane Hutton writes in her book: "*this global material flow plays out in highly differentiated ways on the ground: deeper quarries in one place, material improvements in another; toxic deposition in one place, refined minerals put to use in another*."[1]

The choices we make as engineers, architects, and builders are indeed imbued with responsibility, especially in a time of increasing environmental precarity. This book stands as both a reflection and a challenge to those who work within this critical terrain, offering an exploration of materials, their resources, and performances.

Philosophically speaking, the body of work in this book may be interpreted as one that draws upon a lineage of thought that recognizes the profound contextuality and entanglement between the making of our buildings and the agencies involved in the formation of materials and products. Two masterminds, Bruno Latour's insistence on a "*parliament of things*,[2]" where human and non-human actants must be recognized as co-constructors of reality, and Donna Haraway's proposition of "*staying with the trouble*,[3]" which asks us to engage deeply and ethically with the complexities of our entangled world—are examples of such important and pivotal scholarship. These frameworks should be acknowledged when reading the body of work in this book because they resonate profoundly in asking what the role of design and science is in materializing our world through building.

---

[1] Hutton, Jane Elizabeth. Reciprocal Landscapes: Tracing Materials Between New York City and Beyond. Routledge, 2018.

[2] Latour, Bruno. *Politics of nature*. Harvard University Press, 2004.

[3] Haraway, Donna J. "Staying with the trouble: Making kin in the Chthulucene." In *Staying with the Trouble*. Duke University Press, 2016.

To add my stance, in the pages that follow, "sustainable materials", as the title of the book implies, should not be framed as an isolated problem to be solved but as an ongoing dialogue—a *sympoiesis*—that demands innovation, humility, and a willingness to embrace material variability as a critical mode of resistance to industrial standardization and extraction.

This volume's organization reflects a deliberate effort to balance historical practices, emerging innovations, and systemic analyses of material impacts. Part I, *Earth-Based Construction*, is a testament to the enduring relevance of earthen materials in architecture. It brings forth traditional practices like cob and rammed earth, reframing them as active participants in a contemporary discourse that spans digital fabrication, structural performance, and cultural knowledge. Part II, *Alternative Materials and Processes*, includes explorations of additional nonconventional materials to challenge the dominance of steel and concrete. Part III, *Life Cycle Impacts and Circularity of Materials in the Built Environment*, situates material choices within their broader systems of production, use, and eventual decay. Here, the book takes a systems-thinking approach, underscoring the critical importance of life cycle assessment and circular material flows aiming to tie these methodologies to a synthetic whole.

My own research has often sought to understand the role of natural materials—clay-soils, plant fibers, and biological composites—not only as low-tech solutions but as expressive and embodied mediators. These materials, with their inherent variability and texture, are often inherently non-commodified, thus inviting an engagement with site-specificities of resources, labor, and craft. For example, in exploring earth-fiber composites, my intention is not merely to innovate new methods of construction but to highlight how these materials embody a dialogue with the environment—a dialogue that is based on embodied interaction with materials that are physiologically healthy and historically known to the body. This book aligns deeply with this ethos, calling for a material practice that is at once technical and ethical, precise, and open to the complexities of the world.

The contributions gathered here illustrate a much-required in how we approach materials in construction—one that moves beyond the constraints of a purely STEM-driven framework. While acknowledging the pragmatic requirements of industrial efficiency, standardization, scale, and pace, they also embrace the heterogeneity, porosity, and adaptability that natural and alternative materials offer. This convergence requires not only different technologies and analyses but also resilient mindsets—one that sees material variability not as a limitation but as an inherent and valuable aspect of the design process.

As we navigate uncertainties of accelerated digitalism and resource scarcity, the stakes of our work as engineers, architects, and educators have never been higher. This book sets forth knowledge that will be needed for the next generation of resourceful building, converging technology, low-tech, and site-responsive design in meaningful ways. It is also a reminder to design not for permanence but for adaptability, not for extraction but for regeneration, and not for isolation but for interdependence. It is a privilege to introduce this work, which I believe will serve as both a provocation and a guide to building professionals, navigating the complexities of sustainability in an engineered world. Let it remind us that every act of construction is an opportunity to honor our shared future, one material at a time.

Lola Ben-Alon
Graduate School of Architecture,
Planning and Preservation
Columbia University
New York, NY, USA

Contents

Contributors

Fernando Ávila Department of Continuum Mechanics and Theory of Structures, Polytechnic University of Valencia, Valencia, Spain

Sinan Acikgoz Department of Engineering Science, University of Oxford, Oxford, UK

Rosa Agliata Dipartimento di Ingegneria, Università della Campania "Vanvitelli", Aversa (CE), Italy

Mª Desirée Alba-Rodríguez Departamento de Ingeniería del Diseño, Escuela Politécnica Superior, Universidad de Sevilla, Sevilla, España

Ali Alsalman Department of Civil Engineering, College of Engineering, University of Basrah, Basrah, Iraq

Lateef N. Assi A Terracon Consultants Inc., Tulsa, OK, USA

David Avilés Sustainable Minerals Institute, The University of Queensland, Brisbane, Australia

Lola Ben-Alon Graduate School of Architecture, Planning and Preservation, Columbia University, New York, NY, USA

Nicolas Bertoldo Department of Chemistry, Materials and Chemical Engineering "Giulio Natta", Politecnico Di Milano, Milan, Italy

Sean Blackley School of Engineering, Swinburne University or Technology, Hawthorn, VIC, Australia

Álvaro Blanca-Hoyos Department of Structural Mechanics and Hydraulic Engineering, University of Granada, Granada, Spain

Elena Boerman Institute for Building Design and Technology, Karlsruhe Institute of Technology (KIT), Karlsruhe, Germany

Ilija Bošković Faculty of Architecture, University of Belgrade, Belgrade, Serbia

Rosa Caponetto Department of Civil Engineering and Architecture, University of Catania, Catania, Italy

Jake Cardillo School of Engineering, Swinburne University or Technology, Hawthorn, VIC, Australia

Alexandra Carr School of Art, Design and Architecture, University of Plymouth, Plymouth, UK

Kealy Carter Darla Moore School of Business, University of South Carolina, Columbia, SC, USA

Anya Montserrat Pulido Cavada Department of Architecture and Design, Higher Polytechnic School, San Pablo CEU University, Alcorcón, Madrid, Spain

Ana Cervilla-Maldonado Departamento de Construcciones Arquitectónicas, Escuela Técnica Superior de Arquitectura, Universidad de Granda, Granada, Spain

Conglin Chen School of Civil Engineering, Nanjing Tech University, Nanjing, China

Yujie Chen School of Civil Engineering, Nanjing Tech University, Nanjing, China

Federica Cucchiella Department of Industrial and Information Engineering and Economics, University of L'Aquila, L'Aquila, Italy

Joseph Dahmen University of British Columbia School of Architecture and Landscape Architecture, Vancouver, BC, Canada

Federico de Isidro Gordejuela Department of Architecture and Design, Higher Polytechnic School, San Pablo CEU University, Alcorcón, Madrid, Spain

Giovanni Dotelli Department of Chemistry, Materials and Chemical Engineering "Giulio Natta", Politecnico Di Milano, Milan, Italy

Moritz Dörstelmann Institute for Building Design and Technology, Karlsruhe Institute of Technology (KIT), Karlsruhe, Germany

Flora Faleschini Department of Civil, Environmental and Architectural Engineering, University of Padova, Padova, Italy

Andrei Farcasiu Department of Civil and Environmental Engineering, Carleton University, Ottawa, Canada

Paulina Faria Department of Civil Engineering, CERIS, NOVA University Lisbon, Caparica, Portugal

Ken Farnes RMIT University, Melbourne, Australia

Antonio Ferreira-Sánchez Departamento de Arquitectura, Universidad de Alcalá, Alcalá de Henares, Madrid, España

Raquel Maria Fuentes-García Departamento de Construcciones Arquitectónicas, Escuela Técnica Superior de Arquitectura, Universidad de Granda, Granada, Spain

Rafael Gallego Department of Structural Mechanics and Hydraulic Engineering, University of Granada, Granada, Spain

Zhiwei Gao School of Civil Engineering, Nanjing Tech University, Nanjing, China

Koorosh Gharehbaghi RMIT University, Melbourne, Australia

Giada Giuffrida Department of Civil Engineering and Architecture, University of Catania, Catania, Italy

Mohamed Gomaa School of Engineering, Swinburne University of Technology, Melbourne, Australia;
School of Design and Architecture, Swinburne University of Technology, Melbourne, Australia

Roberto Alonso González-Lezcano Department of Architecture and Design, Higher Polytechnic School, San Pablo CEU University, Alcorcón, Madrid, Spain

Naiyla Goverdhan RMIT University, Melbourne, Australia

Peter Griesbach Department of Civil and Environmental Engineering, Carleton University, Ottawa, Canada

Simon Guihéneuf UMR CNRS 6027, Université Bretagne-Sud, Lorient, France

Zhanggen Guo School of Civil Engineering, Nanjing Tech University, Nanjing, China

Tushar Gupta Department of Mining Engineering, National Institute of Technology Rourkela, Rourkela, Odisha, India

Mohammad Hajmohammadian Baghban Department of Manufacturing and Civil Engineering, Norwegian University of Science and Technology (NTNU), Gjøvik, Norway

Dorian A. H. Hanaor Faculty of Science and Engineering, Southern Cross University, East Lismore, NSW, Australia;
Materialeyes, Materials Innovation Consulting, Berlin, Germany

Theodore Hanein earth4Earth Technology Ltd., London, UK;
School of Civil Engineering, University of Leeds, Leeds, UK

Viktors Haritonovs Institute of Civil Engineering, Faculty of Civil and Mechanical Engineering, Riga Technical University, Riga, Latvia

Lex Hartnett School of Engineering, Swinburne University or Technology, Hawthorn, VIC, Australia

Matthew Horan School of Engineering, Swinburne University or Technology, Hawthorn, VIC, Australia

Alejandro Jiménez Rios Department of Architecture & Civil Engineering, University of Bath, Bath, UK

Rahman S. Kareem Department of Structure, Shatrah Technical Institute, Southern Technical University, Shatrah, Dhi Qar, Iraq

Nouman Khadim Vrije Universiteit Amsterdam, Amsterdam, Netherlands

Łukasz Klapiszewski Institute of Chemical Technology and Engineering, Faculty of Chemical Technology, Poznan University of Technology, Poznan, Poland

Johannes Konnerth Department of Natural Sciences and Sustainable Resources, Institute of Wood Technology and Renewable Materials, University of Natural Resources and Life Sciences Vienna, Austria (BOKU University), Tulln a.d. Donau, Austria

João A. Labrincha Department of Materials and Ceramic Engineering/ CICECO-Aveiro Institute of Materials, University of Aveiro, Campus Universitário de Santiago, Aveiro, Portugal

Dinh Linh Le University School for Advanced Studies IUSS Pavia, Pavia, Italy

Louise Lemesre UMR CNRS 6027, Université Bretagne-Sud, Lorient, France

Valeriano Lucas-Ruiz Departamento de Construcciones Arquitectónicas II, Escuela Técnica Superior de Ingeniería de Edificación, Universidad de Sevilla, Sevilla, España

Elena Lucchi Università di Pavia, Pavia, Italy

Hamidreza Malekian School of Design and Architecture, Swinburne University of Technology, Melbourne, Australia

Chiara Marchionni Department of Civil, Construction-Architectural and Environmental Engineering, University of L'Aquila, L'Aquila, Italy

Madelyn Marrero Departamento de Construcciones Arquitectónicas II, Escuela Técnica Superior de Ingeniería de Edificación, Universidad de Sevilla, Sevilla, España

Francisco Martín-Villegas Departamento de Construcciones Arquitectónicas, Escuela Técnica Superior de Arquitectura, Universidad de Granda, Granada, Spain

Ana Mafalda Matos CONSTRUCT, Faculty of Engineering, University of Porto, Porto, Portugal

Paul Mayencourt Department of Architecture, College of Environmental Design, Berkeley, USA;
Department of Environmental Science, Policy, and Management, Rausser College of Natural Resources, University of California Berkeley, Berkeley, CA, USA

Hamideh Mehdizadeh Department of Manufacturing and Civil Engineering, Norwegian University of Science and Technology (NTNU), Gjøvik, Norway

Qipei Mei Department of Civil and Environmental Engineering, University of Alberta, Edmonton, Canada

Remo Merijs-Meri Institute of Chemistry and Chemical Technology, Faculty of Natural Sciences and Technology, Riga Technical University, Riga, Latvia

Qinglong Miao School of Civil Engineering, Nanjing Tech University, Nanjing, China

Paula Milheiro-Oliveira CONSTRUCT, Faculty of Engineering, University of Porto, Porto, Portugal;
Center of Mathematics, University of Porto, Porto, Portugal

Alec Miller Department of Civil and Environmental Engineering, Monash University, Clayton, Australia

Melissa Neave Centre for Urban Research, School of Global Urban and Social Studies, RMIT University, Melbourne, VIC, Australia

Quan T. Nguyen Faculty of Construction Economics and Management, Hanoi University of Civil Engineering, Hanoi, Vietnam

Rui M. Novais Department of Materials and Ceramic Engineering/ CICECO-Aveiro Institute of Materials, University of Aveiro, Campus Universitário de Santiago, Aveiro, Portugal

John Ochsendorf Massachusetts Institute of Technology School of Architecture and Planning, Cambridge, MA, USA

Duhan Olmez School of Art, Design and Architecture, University of Plymouth, Plymouth, UK

Arnaud Perrot UMR CNRS 6027, Université Bretagne-Sud, Lorient, France

Cecilia Pesce earth4Earth Technology Ltd., London, UK

Maximilian Pramreiter Department of Natural Sciences and Sustainable Resources, Institute of Wood Technology and Renewable Materials, University of Natural Resources and Life Sciences Vienna, Austria (BOKU University), Tulln a.d. Donau, Austria;
Department of Architecture, College of Environmental Design, Berkeley, USA;
Department of Environmental Science, Policy, and Management, Rausser College of Natural Resources, University of California Berkeley, Berkeley, CA, USA

Esther Puertas Department of Structural Mechanics and Hydraulic Engineering, University of Granada, Granada, Spain

Bora Pulatsu Department of Civil and Environmental Engineering, Carleton University, Ottawa, Canada

Damien Rangeard Research and Development Department, Laboratoire CBTP, Noyal-sur-Vilaine, France

Scott Rayburg School of Engineering, Swinburne University or Technology, Hawthorn, VIC, Australia

Miakonirainy Vololontsoa Razafitrimo UMR CNRS 6027, Université Bretagne-Sud, Lorient, France

Alejandro Veliz Reyes School of Art, Design and Architecture, University of Plymouth, Plymouth, UK

Arturs Riekstins Institute of Civil Engineering, Faculty of Civil and Mechanical Engineering, Riga Technical University, Riga, Latvia

Marianna Rotilio Department of Civil, Construction-Architectural and Environmental Engineering, University of L'Aquila, L'Aquila, Italy

Gianluca Ruggieri Dipartimento Di Scienze Teoriche E Applicate, Università Dell'Insubria, Varese, Italy

Sergio Renato Maria Sabbadini A.N Studio Di Architettura Disstudio.It, a.N.a.B., Milan, Italy

Roberta Salomone Department of Economics, University of Messina, Messina, Italy

Marco Simoni earth4Earth Technology Ltd., London, UK

Mayara S. Siverio Lima Unit of Construction Management and Tunneling, University of Innsbruck, Innsbruck, Austria

Agnieszka Ślosarczyk Institute of Building Engineering, Faculty of Civil and Transport Engineering, Poznan University of Technology, Poznan, Poland

Veronica Soebarto School of Architecture and Built Environment, Adelaide University, Adelaide, Australia

Simon Steffl Institute for Industrial Production (IIP), Karlsruhe Institute of Technology (KIT), Karlsruhe, Germany

Qiansen Sun School of Civil Engineering, Nanjing Tech University, Nanjing, China

Kong Fah Tee Department of Civil and Environmental Engineering, King Fahd University of Petroleum and Minerals, Dhahran, Saudi Arabia; Interdisciplinary Research Center for Construction and Building Materials, KFUPM, Dhahran, Saudi Arabia

Liseane P. Thives Department of Civil Engineering, Federal University of Santa Catarina, Florianópolis, Brazil

Daniel Trento Department of Civil, Environmental and Architectural Engineering, University of Padova, Padova, Italy

Carene Umubyeyi Princeton University Civil and Environmental Engineering, Princeton, NJ, USA

Ignacio Valverde-Palacios Departamento de Construcciones Arquitectónicas, Escuela Técnica Superior de Arquitectura, Universidad de Granda, Granada, Spain

Humberto Varum CONSTRUCT, Faculty of Engineering, University of Porto, Porto, Portugal

Alejandro Veliz Reyes School of Art, Design and Architecture, University of Plymouth, Plymouth, UK

Rebekka Volk Department for Sustainable Systems Engineering (INATECH), University of Freiburg, Freiburg, Germany

Karissa Wenger Arup, Boston, MA, USA

Tim T. Werner Department of Geography, Earth and Atmospheric Sciences, University of Melbourne, Carlton, VIC, Australia

Xiaoling Wu Department of Civil and Environmental Engineering, Monash University, Clayton, Australia

Mohan Yellishetty Department of Civil and Environmental Engineering, Monash University, Clayton, Australia

Mariano Angelo Zanini Department of Civil, Environmental and Architectural Engineering, University of Padova, Padova, Italy

Ling Zhou School of Civil Engineering, Nanjing Tech University, Nanjing, China

Chapter 1
Introduction

Dorian A. H. Hanaor

1.1 Introducing the Collection: Sustainable Materials for the Built Environment

The built environment is the foundation of human civilization, comprising the spaces where we live, work, and interact. It is both a reflection of our cultural and technological achievements and a critical determinant of our environmental impact. As the global population continues to grow relentlessly, the demand for safe, comfortable, and resilient housing and infrastructure has never been greater. Yet, the construction sector, including its entire value chain, is broadly considered to be responsible for around 40% of global greenhouse gas emissions. The built environment therefore stands at a crossroads as the need to simultaneously address climate change and artificial housing scarcities worldwide becomes a priority for societies. Going forward, the way we design, build, and maintain our structures will profoundly influence our ability to address climate change, resource scarcity, and biodiversity loss while providing citizens with access to an adequate and affordable housing within their countries, recognized as a fundamental human right [1]. This book, **Sustainable Materials for the Built Environment**, seeks to provide a comprehensive exploration of the materials, methods, and philosophies that can guide us toward a more sustainable future.

Intended for a diverse audience including architects, materials scientists, sustainability advocates, and anyone interested in greener construction practices, this book brings together insights from leading thinkers and practitioners from diverse fields. It offers multidisciplinary perspectives on the challenges and opportunities

D. A. H. Hanaor (✉)
Materialeyes, Materials Innovation Consulting, Berlin, Germany
e-mail: dorian@materialeyes.net
URL: https://www.materialeyes.net

Faculty of Science and Engineering, Southern Cross University, East Lismore, NSW 2480, Australia

D. A. H. Hanaor (ed.), *Sustainable Materials for the Built Environment*, Springer Series in Materials Science 352, https://doi.org/10.1007/978-3-031-97818-0_1

of sustainable construction, presenting innovative solutions that balance environmental stewardship, social equity, and economic viability. By examining both traditional and emerging materials, this volume aims to inspire a reimagining of the built environment, one that harmonizes human needs with the natural world.

At the heart of this work lies a recognition that the solutions to today's crises of climate change, resource depletion, and social inequity are often rooted in the past. Ancient building practices, refined over millennia, are being revitalized through modern scientific rigor and digital technologies. Simultaneously, advancements in materials science and circular design principles are redefining what it means to build responsibly. This underscores a central theme: sustainability thrives at the intersection of respect for ecological limits and bold technological experimentation.

The narratives presented here go beyond the technical aspects of materials and their applications, inviting readers to consider the broader implications of material choices. Buildings are more than assemblages of matter; they are the stage upon which human life unfolds. By prioritizing materials that harmonize with natural systems, we can not only reduce carbon footprints of construction but also cultivate spaces that nurture well-being, cultural identity, and community resilience. The case studies and methodologies detailed in these pages demonstrate that sustainable construction materials are not a compromise but rather a means to achieve architectural beauty, structural integrity, and environmental stewardship.

1.2 What Do We Seek to Sustain?

Before examining methodologies for the processing and deployment of sustainable materials in the built environment, it is imperative to examine the foundational concepts of sustainability and their specific implications for construction materials. The Brundtland Commission's definition, "Sustainable development is development that meets the needs of the present without compromising the ability of future generations to meet their own needs" [2], provides a critical framework. This invites a deeper inquiry: What, precisely, constitutes the "needs" we seek to fulfil, and which systems or resources require preservation to ensure intergenerational equity?

Defining what we seek to sustain underpins goal formation, strategic prioritization, and long-term societal planning. In an era characterized by exponential technological growth, competing ideological frameworks, and fragmented policy landscapes, clarity of individual and collective purpose has become a scarce resource. Within neoclassical economic paradigms, the concept of "utility" has long served as a proxy for value, quantifying human satisfaction through metrics of production and consumption. However, this reductionist approach, which conflates well-being with monetary exchange, neglects the qualitative dimensions of lived experience. The built environment, as the physical manifestation of societal priorities, exemplifies this dissonance [3]. While humanity has honed advanced architectural and engineering capabilities over millennia, contemporary systems disproportionately prioritize the proliferation of consumer goods over the creation of dignified, resilient habitats.

This misalignment has precipitated a global housing crisis, a paradox of progress where unprecedented technical abilities coexist with systemic failure to address fundamental human needs. This imbalance has resulted in the erosion of craftsmanship, community cohesion, and ecological stewardship in favour of transient economic gains. As we confront these challenges, the imperative to redefine "progress" becomes inseparable from reimagining the materials, processes, and values that shape our built world.

Central to the discourse on human well-being and identifying what we seek to sustain is the recognition that the primary utility derived from economic activity lies not in material accumulation but in the experiential dimensions of life. Experiences, rooted in daily existence, social interaction, and cultural participation, are profoundly shaped by the built environment, where individuals spend the majority of their lives. Dwellings, workplaces, and communal spaces constitute the physical substrate of society, where human agency intersects with ecological systems. The built environment functions as both a mediator and a manifestation of the anthroposphere, the aggregate of human activities, and its dynamic interplay with Earth's spheres: the hydrosphere, lithosphere, atmosphere, and biosphere. Sustainable construction materials and construction practices therefore underpin our ability to sustain and improve human experiences on our planet.

1.3 Materials for Sustainability

Efforts to advance sustainable construction face a certain paradox: while the discourse often fixates on high-tech innovation, many of the most ecologically and socially viable solutions derive not from novel technologies but from the reinterpretation of ancient building practices. Techniques such as rammed earth, stonemasonry, and timber framing, which have been refined over millennia, offer low-carbon, culturally resonant alternatives to energy-intensive materials like concrete and steel. Similarly, agricultural by-products and bio-based composites demonstrate how circularity can be literally woven into the fabric of modern construction. The barrier to their adoption is seldom technological feasibility but rather systemic inertia. Unlike cutting-edge fields such as quantum computing or space exploration, sustainable built environments rarely demand radical scientific breakthroughs; instead, they require raising commercial readiness levels (CRL), addressing challenges of market structures that favour industrialized materials, regulatory frameworks ill-adapted to nature-based and vernacular practices, and financial models prioritizing short-term returns over long-term resilience.

This volume confronts these obstacles, examining how to align the construction sector with the urgent dual imperatives of addressing the global housing crisis and achieving net-zero targets. Through interdisciplinary analysis, it explores materials and methods that harmonize human needs with planetary boundaries, from revitalized traditional techniques to emerging digital fabrication processes. The chapters, authored by leading scholars and practitioners across disciplines, demonstrate that sustainability in the built environment is not a question of scarcity of ideas, resources, or knowledge, but rather a question of prioritization. By integrating various fields relating to architecture and engineering and merging ancient techniques with contemporary tools, this work illuminates pathways to transform construction from a driver of ecological degradation into a catalyst for regeneration. In doing so, it reframes sustainability not as a technical challenge but as a collective reimagining of value—one that places human well-being and ecological stewardship at the centre of how we build.

1.4 Themes

The book is organized into three main thematic sections, each addressing a critical aspect of sustainable construction.

Part 1: Earth-Based Construction

This section explores the resurgence of earth-based materials, which have been used for millennia but are now being reimagined through modern technologies and scientific advancements [4–8] and examines regulatory and technical hurdles that hinder the wide scale uptake of this sustainable construction approach [9]. Earthen materials, characterised by the textural triangle in terms of components of sand, silt and clay, can serve as the basis for wall elements subjected to mild compressive stresses. As we will see in the chapters of this part, structural performance of earthen materials can be further enhanced with the addition of fibrous biomass or cementitious stabilisation [10, 11]. The large mass of the walls and the porosity of the earthen materials may provide hygroscopic temperature regulation and thermal inertia [12, 13]. Chapters in this section explore the life cycles of earth buildings, from traditional techniques to cutting-edge digital fabrication methods such as 3D printing [14, 15]. Researchers from around the world present insights into the thermal, mechanical, and structural properties of earthen materials, highlighting their potential to reduce carbon emissions, enhance energy efficiency, and preserve cultural heritage. The section also addresses the challenges of scaling up earth-based construction in contemporary contexts, offering recommendations for industry adoption and regulatory frameworks.

Part 2: Alternative Materials and Processes

Focusing on replacing conventional materials such as Portland Cement based concrete, and steel in the built environment, the second part of the book shifts focus to innovative materials and processes that offer sustainable alternatives to conventional construction practices. Chapters examine the use of bio-based materials, such as fast-growing natural fibers and hemp-lime composites as alternatives to steel reinforced concrete [16], as well as alternative binders including biocementation, geopolymers other alkali activated materials [17, 18]. In this section the opportunities for circular economies, waste materials, residues and agricultural byproducts are discussed as pathways towards sustainable alternatives to versatile construction materials. By bridging the gap between traditional knowledge and modern innovation, this section highlights the potential of alternative materials to replace conventional carbon-intensive construction materials and thereby decarbonize the built environment while maintaining structural integrity and performance.

Part 3: Life Cycle Impacts and Circularity of Materials

The final section of the book takes a holistic view of sustainability, focusing on the lifecycle impacts of construction materials and building practices, identifying opportunities for improved practices, industrial symbiosis and circularity. Chapters in this section present findings from life cycle assessments (LCA) of various materials, including engineered wood products, bio-based insulation, and recycled concrete. The section also explores strategies for reducing embodied carbon, enhancing material reuse, and integrating circular economy principles into construction and renovation practices. By examining the economic, social, and environmental dimensions of sustainability, this section provides actionable insights for policymakers, architects, and engineers seeking to create a more sustainable built environment.

This book brings together diverse perspectives from leading researchers, practitioners, and policymakers, offering a multidisciplinary approach to sustainable construction. Through their analysis of both traditional and emerging materials, the chapters in this book, authored by esteemed experts from around the world, paint a picture of optimism. They highlight the immense potential of sustainable materials and practices to address the most pressing challenges of our time. From rediscovering ancient techniques to embracing cutting-edge technologies, the solutions presented here demonstrate that a sustainable built environment is not only possible but also within our reach. By working in harmony with the natural world, we can create structures that enhance human experience while safeguarding the planet for future generations.

The chapters of this book serve as a call to action and a reminder that the most important aspect of sustainability is the quality of life we create within the spaces we inhabit. It is an invitation to rethink our priorities, to align our efforts with the goal of fostering resilient, equitable, and environmentally responsible communities. Through its comprehensive examination of sustainable materials and practices, this volume aims to inspire innovation, collaboration, and a renewed commitment to building a better world.

References

1. G. Bristol, in *Handbook of Human Rights* (Routledge, 2012), p. 659
2. S.W.S. Wced, Our Common Future **17**, 1 (1987)
3. A. V. Reyes, A. Carr, and T. Crabtree, Co-making and prototyping community housing futures. The De Gruyter Handbook of Automated Futures: Imaginaries, Interactions and Impact, **2**, 339 (2024)

4. L. Ben-Alon, V. Loftness, K.A. Harries, E.C. Hameen, M. Bridges, J. Green Build. **15**, 87 (2020)
5. M. Gomaa, S. Schade, D.W. Bao, Y.M. Xie, J. Clean. Prod. **398**, 136569 (2023)
6. G. Giuffrida, R. Caponetto, F. Nocera, M. Cuomo, Sustainability **13**, 11948 (2021)
7. G. Giuffrida, R. Caponetto, M. Cuomo, in *IOP Conference Series: Earth and Environmental Science* (IOP Publishing, 2019), p. 012018
8. G. Giuffrida, L. Dipasquale, R.M. Pulselli, R. Caponetto, Sustainability **16**, 1367 (2024)
9. K. Harries, L. Ben Alon, B. Sharma, Characterisation, properties and applications (2019)
10. M. Gomaa, W. Jabi, A.V. Reyes, V. Soebarto, Autom. Constr.. Constr. **124**, 103577 (2021)
11. G. Giuffrida, V. Costanzo, F. Nocera, M. Cuomo, R. Caponetto, in *International Conference on Sustainability in Energy and Buildings* (Springer, 2022), p. 164
12. L. Ben-Alon, A.R. Rempel, Build. Environ. **238**, 110339 (2023)
13. M. Gomaa, J. Carfrae, S. Goodhew, W. Jabi, A. Veliz Reyes, Architect. Sci. Rev. **62**, 230 (2019)
14. F. Faleschini, D. Trento, M. Masoomi, C. Pellegrino, M.A. Zanini, Constr. Build. Mater. **398**, 132496 (2023)
15. D. Trento, F. Faleschini, M. Masoomi, C. Pellegrino, M.A. Zanini, in *International Conference on Earthen Construction* (Springer, 2024), p. 3
16. J. Abed, S. Rayburg, J. Rodwell, M. Neave, Sustainability **14**, 5570 (2022)
17. R.M. Novais, G. Ascensão, M. Seabra, J. Labrincha, Waste Manage. **52**, 245 (2016)
18. R.M. Novais, G. Ascensao, D.M. Tobaldi, M.P. Seabra, J.A. Labrincha, J. Clean. Prod. **171**, 783 (2018)

Part I
Earth Based Construction

From Ancient Wisdom to Modern Mastery: Earth's Resurgence in Sustainable Building

Around the globe, earth-based construction has been a major feature of the built environment for most of human history. Even today, a significant portion of dwellings, particularly in the global south, feature some element of earthen construction, whether it is rammed earth walls, clay-based plaster, or adobe bricks. Earthen materials, characterised by the textural triangle according to components of clay, silt, and sand, stand out as a logical choice for construction owing to their abundance and versatility. In nearly every natural environment, earthen materials can be sourced that provide sufficient compressive strength for wall-elements in low-rise dwellings alongside sustainable outcomes in construction and structure use.

With the industrialisation of societies and the transition away from vernacular construction to centralised production and distribution of building materials, earthen materials fell out of usage, their place being taken by mass produced masonry units and castable concretes; materials that offer convenience, uniformity, and scalability at the expense of significant environmental footprints. As we begin to come to terms with the consequences of environmental degradation, climate change, and technological dominance of nature, the beauty, utility, and natural compatibility of earthen materials are sparking imaginations across many sectors. As housing access and human employment become increasingly concerning in societies around the world, an opportunity presents itself whereby earthen and stone construction, rather than being seen as manually intensive and possibly expensive labour, can be reinterpreted and reframed as meaningful skill-driven work that connects us to heritage, craft, and local materials while meeting our housing needs with a positive environmental and social impacts. A contrast to the screen-bound routines dominating modern life.

While hands-on work is still a necessary and rewarding aspect of any construction endeavour, the efforts of scientists, engineers, architects, and thinkers and doers from across all disciplines are bringing knowledge to bear towards finding optimal ways to efficiently utilise the naturally occurring soils that surround us to create high-value materials, structures, and communities. Vernacular earth-based construction is therefore an exciting an exciting material nexus point uniting past, present and future—a bridge between ancestral wisdom and cutting-edge innovation, between local economies and global sustainability goals.

This first part of the book **Sustainable Materials for the Built Environment** explores this renaissance and brings insights from prominent practitioners and scholars around the world relating the many facets of earthen-construction. In the chapters of this part, the chemistry, physics, mineralogy, materials engineering, and construction methods relating to earth-based materials and their performance are examined. Further attention is given to the practical aspects of implementing earth-based materials in the twenty-first century in light of market requirements, circularity, barriers to adoption, and social and environmental impacts.

Dorian A. H. Hanaor

Chapter 2
The Life Cycles of Natural Buildings—From Traditional to Digital

Lola Ben-Alon

Abstract Material design research that looks at natural earth- and bio-based materials necessitates critical insights from a life cycle perspective. From traditional architecture to digital fabrication and additive manufacturing, understanding the environmental and social impacts is crucial for field advancements, pedagogy, and novel research development. This chapter catalogues, demystifies, and categorizes the terminology and attributes of natural materials through a life cycle approach. From the system boundaries definition to the thermal performance in the operational phase and the end-of-life recyclability, health, and biophilia, each building mix design in earth-fiber construction can illuminate the projected required development, and recommendations for the industry, driven by the need to incentivize an embodied approach to mandatory energy requirements.

2.1 Impacts and the Building Share—Unveiling Raw Data

When we talk about life cycle systems in material design research, we refer to the complete phases of a building—not just the use phase, but also its past iterations and potential future "reincarnations." The term "life" can be misleading here; unlike living organisms, buildings don't have a finite life. Instead, they transition through various forms—from landscapes to raw matter, to materials, to waste, and ideally, back to soil as nutrients.

A relevant question arises: *why focus on construction systems?* Construction is the expression of dynamic energy flows, from the molecular to the territorial scale, activating and reshaping the geography of the site. It is also the organization of matter. And what is a building's matter if not energy? Matter is but captured energy. As Kiel Moe explains [15]: *An object, such as a rock formation, is often thought as a fixed, inert object or shape. However, it is actually an expression of dynamic energy systems and processes that range from the molecular to the territorial. Energy*

L. Ben-Alon (✉)
Graduate School of Architecture, Planning and Preservation, Columbia University, New York, NY 10027, USA
e-mail: rlb2211@columbia.edu

D. A. H. Hanaor (ed.), *Sustainable Materials for the Built Environment*, Springer Series in Materials Science 352, https://doi.org/10.1007/978-3-031-97818-0_2

of many processes are embedded or embodied in materials. Energy is what binds matter, what activates matter, what transforms matter, and what organizes matter."

Construction systems are also dictated by material powers and can influence material sources. Sod structures provide an interesting case: in the US prairie, colonial settlers used earth materials, weaving strong dry root systems into blocks of mud. However, instead of adopting Indigenous domed earth construction methods, the new settlers imposed European rectilinear designs, often resulting in structural failures and collapse. Material practices tied to cultural knowledge can therefore be disrupted by power systems, leading to erroneous imitations.

While there are numerous studies documenting embodied energy and emissions in buildings, a true life cycle analysis of building materials should encompass the entire cycle, from a range of perspectives and entangled narratives: from the extraction of raw materials, land use and land grabs, through processing, regional health, transportation to the construction site, production, maintenance, cultural impacts, to the end of life via demolition, circular economy, recycling, upcycling, or reuse.

A designer's work, therefore, should encompass more than just the technical aspects of material selection and carbon or energy accounting—but rather engage in the synthesis of trade routes procurement procedures, health impacts evaluation, and equity analysis, all while balancing embodied carbon with the realities of transportation, chemical and thermal processing, storage, and assembly lines. How can we better visualize the origins of materials to capture exchanges of energy, labor, emissions, and ideas?

Critical analysis of data is an essential step to start with. In order to engaged in a fuller acknowledgment of life cycles, it is necessary to reinterpret primary data to challenge conventional narratives. Figure 2.1 underscores this by showing how the building sector's carbon footprint can vary depending on how data is allocated. The leftmost pie chart, based on EPA data, attributes 13% of U.S. greenhouse gas (GHG) emissions to the building sector. When we reallocate emissions from the electric power sector (since it is not a final end-use sector), the building sector's contribution rises to 32%. The rightmost chart, which accounts for building materials within the industrial sector, pushes this figure to nearly 38%. This kind of critical inquiry into energy and emissions data is crucial, as it informs both funding priorities and policy decisions. By interrogating and refining how data is presented, we can push for more accurate and equitable environmental policies.

We must continue to interrogate institutional data critically, peeling back so to speak surface-level interpretations to examine the underlying raw data that shape our understanding of energy in relation to material systems. In Fig. 2.2, three pie charts illustrate the significant, yet often overlooked, "other" category of energy consumption in both residential and commercial buildings. The residential sector shows miscellaneous electric loads such as dishwashers, pool heaters, and spas, with 2.50 quads of energy use attributed to "Other (Electric)." In commercial buildings, "Other (All Fuels)" accounts for 2.28 quads, with notable contributors including street lights, kitchen ventilation, and dry transformers. Together, these charts highlight the hidden yet substantial energy demands outside of standard categories like heating, cooling, and lighting. Categories of 'others' represent a recursive echo of

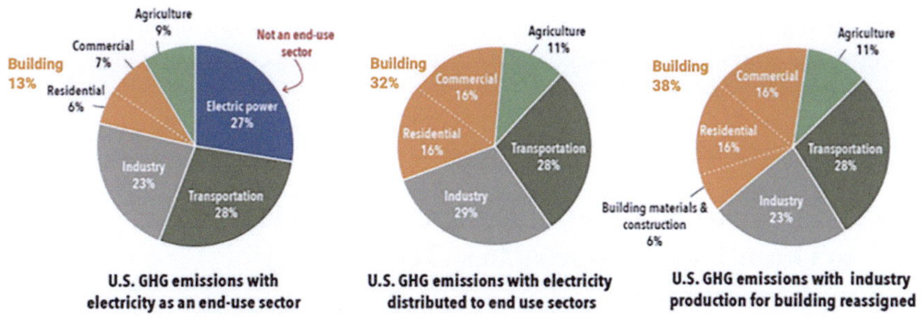

Fig. 2.1 Greenhouse gas emissions by sector and reallocation for buildings. A series of pie charts illustrating how the building sector's GHG emissions vary with different data allocations: direct allocation (left), including emissions from electricity (center), and including emissions from building materials production (right). Image by the author

overlooked data. This demands from us not just observation, but an active dissection and reevaluation of the life cycle inventory. Through this mining and hacking of data, we should aim to construct a more comprehensive narrative that acknowledges and integrates these hidden streams of energy, thus refining our approach to sustainability and design.

Many investments in the past decades are focused on improving energy generation and electrification. Nonetheless, these costly investments ultimately rely on extracting rare minerals for energy storage devices. However, building related-activities should be highlighted, especially the need to address the repetitive and

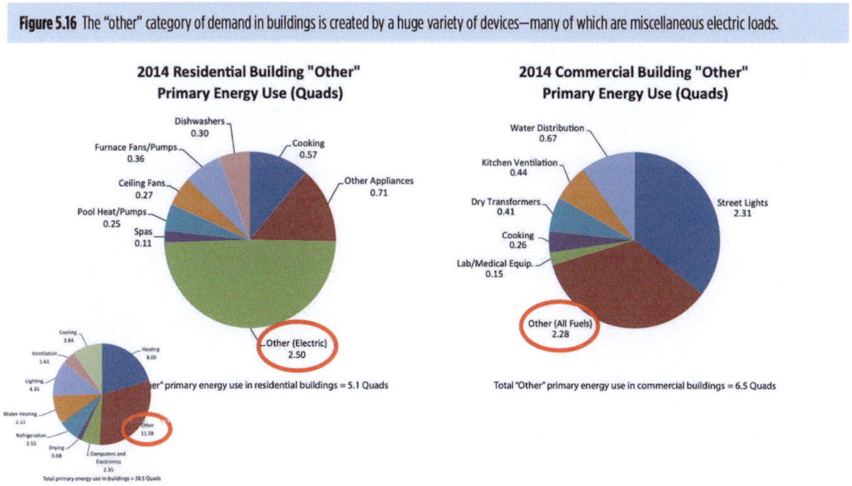

Fig. 2.2 Breakdown of "Other" energy use in residential and commercial buildings. *Annotated image, source* DOE, Quadrennial Technology Review: An Assessment of Energy Technologies and Research Opportunities, Chapter 5: Increasing Efficiency of Building Systems and Technologies, September 2015

accumulating carbon emissions caused by the building envelope and its properties. Building envelope's abatement potentials are not only less costly but can also yield in profitable investments.

2.2 Life Cycles in Material Design—From Qualitative to Quantitative Approaches: Terminology and Assumptions

2.2.1 Life Cycle Thinking

Epistemologically, life cycle—our point of departure, is originally observed and acknowledged through looking at the natural world. In the biological world, every plant or living element has a *cycle of life*—from sprouting as a seed to germinating out to land, air, and water.

Not only living beings but also living bodies, like water, have life cycles. The diagram in Fig. 2.3 illustrates the water cycle, beginning and ending at the ocean. From the ocean, water evaporates and condensates into clouds. Water then becomes precipitation that flows back to the ocean or to the land, and from the land, water will percolate, infiltrate, or flow back into the ocean.

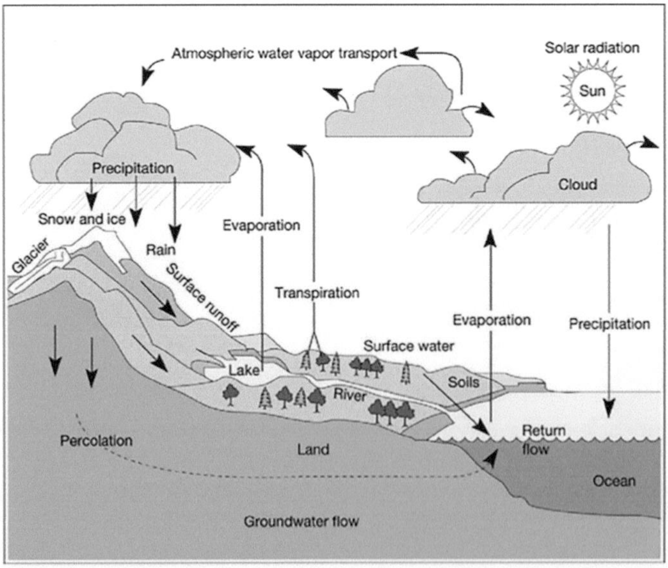

Fig. 2.3 The water cycle. Diagram courtesy of Dr. Fred Mackenzie, University of Hawai'i at Manoa

Emissions to air, water, and land can disturb this natural cycle by intruding on the movement of water throughout this system or harming the livelihood and health of water throughout these systems; acidification, for instance, makes the water more acidic due to increased SO_2 absorption within the water, as shown in Fig. 2.4. And, eutrophication causes a harmful algae bloom due to the excess use of fertilizers that run off into water bodies, thus blocking sunlight. Both interfere with marine life and the ecology of water.

But how do we measure these impacts? Environmental Life Cycle Assessment (LCA) is a rigorous and systematic methodology that can sometimes feel like a never-ending rabbit hole. In its non-applied theoretical state, LCA is an iterative, endlessly looping calculation of resource and material flows. The recursive nature of material flows, with inputs branching out into further upstream processes, stem from the constant mining, refining, and energy operations required from everything materials. LCA "tiers" represent these endless cycles, but since it is impossible to account for all possible iterations, LCA models typically halt the recursion at the 4th or 5th tier.

LCA examines the product or building element through its various life cycle phases—like a biological machine—enumerating potential impacts at each stage. A general rule of thumb is that if a product or material requires minimal processing and is produced locally, we can reasonably assume that its environmental and social impacts will be more favorable compared to processed or imported alternatives. However, intuition alone is insufficient for a thorough analysis. This is where LCA becomes critical, as it allows for the comparative evaluation of products, pushing forward policy and informed decision-making. While it is a valuable tool for tracking

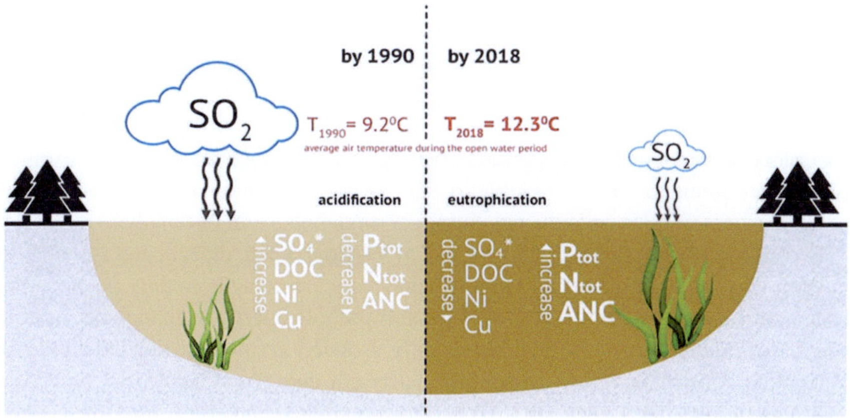

Fig. 2.4 Acidification and eutrophication of water bodies. Diagram courtesy of Dr. Fred Mackenzie, University of Hawai'i at Manoa

performance, LCA is not a cure-all for the built environment's environmental challenges. Despite how scientific literature often positions it, LCA cannot be viewed as an "exact" science with a single definitive theory or axiom. Instead, it is a method that aids in understanding complexity but does not offer absolute solutions.

To conduct a "proper" LCA we should follow four basic steps: (1) Define the goal, scope, system boundaries, and functional unit of the studied system, (2) Conduct an inventory analysis, or, if using software, check the inventory used "behind the scenes", (3) Analyze the resulting impacts, and (4) Interpret the results.

The life cycle inventory (LCI) is of utmost importance. It is the raw data used to create the impact assessment, and with which accuracy, or lack of, should be assumed for the LCA. Essentially, the LCI involves investigating which raw materials go into the system of making a certain product or material, and, as a result, what are the outputs and emissions to air, land, and water. The life cycle impact assessment (LCIA) is the conversion of LCI inventory raw data of energy, raw materials, and emissions, into meaningful metrics for the LCA audience. LCIA categories of such can be global warming potential and ozone depletion potential. Therefore, if the inventory is incorrect—then the impacts are incorrect and misleading. Figure 2.5 shows a general depiction of a building material system boundaries and the schematic inputs and outputs. Generally speaking, what comes out of the system—the thermal and emission outputs, are often harder to assess than the input energy and raw materials.

2.2.2 System Boundaries

To begin with, we need to establish a system for any life cycle analysis (LCA), and every system must have a boundary. But what exactly is a system boundary? A system encompasses anything we choose to study, measure, or engage within the world, focusing on a particular object of study while necessarily excluding others. This boundary delineates what is inside the system and what belongs to the 'surroundings.' It's crucial to recognize that whenever we study something, we inevitably exclude many other elements. It's impossible to analyze the entire universe with each study, so we must externalize or leave out certain components from our analysis.

Broadly speaking, there are three types of systems: open, closed, and isolated. In an open system, both matter and energy can freely cross the boundary between the system and its surroundings. A closed system permits energy exchange but not the transfer of matter across its boundary. Lastly, isolated systems, which exist only as conceptual constructs, allow for neither matter nor energy exchange. So, where do buildings fit into this framework? All buildings and their materials function as open systems. In fact, all materials in our surroundings originate from geological sources before they become architectural. They are inherently open systems, embodying inputs from the past and outputs for the future, along with unpredictable impacts and extensive planetary processes.

The challenge becomes: how do we capture the full extent of processing required for a specific material system? As an example, consider the system boundary diagram

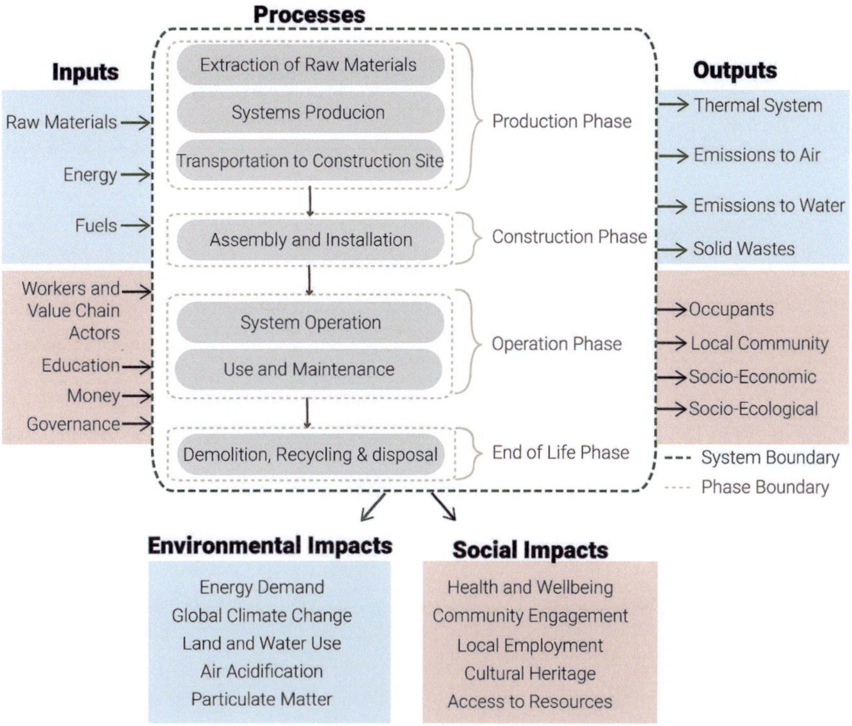

Fig. 2.5 System inputs and outputs for social and environmental assessment. Image by the author

in Fig. 2.6, which is derived from a study of earth materials, which will be further elaborated in this chapter (the full study and methodology can be found in [6]). The green dashed line indicates the boundary of the analysis—everything outside of this line is excluded from the study. The input streams on the left include both energy and materials, while the outputs on the right consist of the desired product—in this case, a wall—as well as undesirable emissions. In this study, the system boundaries are structured to focus on the cultivation phase of the material while acknowledging that certain elements are always excluded from the system.

System boundaries in LCA are often classified into categories like "cradle to cradle" (from production back to production), "cradle to grave" (from production to disposal), or "cradle to gate." The latter, depicted in Fig. 2.6, represents processes from the extraction of raw materials up to the point where the product is ready for construction—what we call "shovel ready."

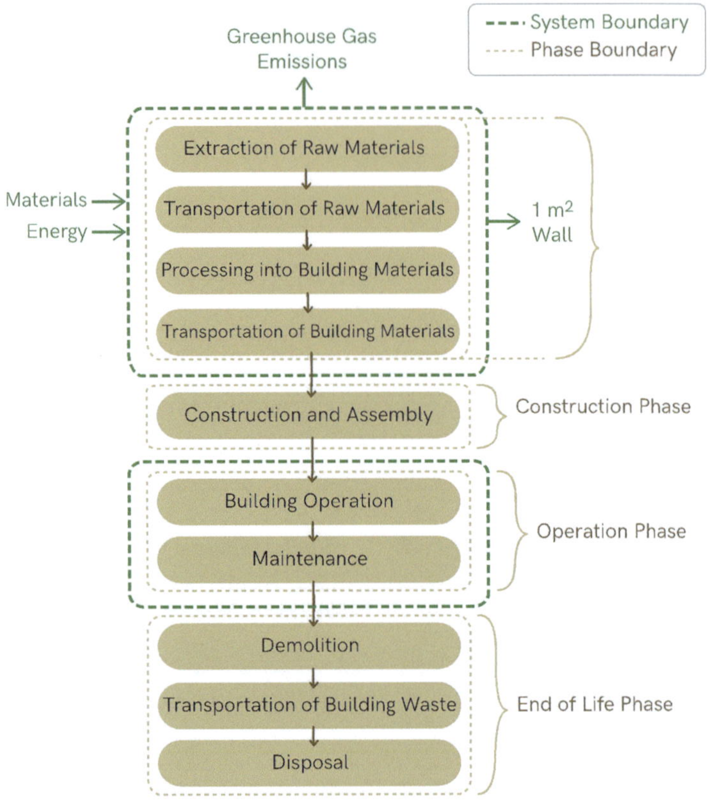

Fig. 2.6 System boundaries example for building materials

2.2.3 Matter

Materials can be classified into "families" or "groups," a method often rooted in an engineered, mechanical view of materiality. In his seminal work on material selection for mechanical design, Ashby cataloged six material families: metals, ceramics, glasses, polymers, elastomers, and hybrids—a categorization based on various parameters, including atomic structure, formological behavior, and how a material is born into the world [2]. For example, both metals and ceramics are mined from the earth, but they emerge from very different geological processes. Metals are formed through magmatic and hydrothermal processes, resulting in a crystalline atomic structure, whereas ceramics are produced through sedimentation processes and are characterized by their rigidity and brittleness.

While Ashby's categorization offers a clear, mechanical perspective, it is very much a product of twentieth century material science—one that has heavily influenced today's building material industry. The legacy of this engineering-focused approach is evident in the prevalence of transparent façade elements (from the glass

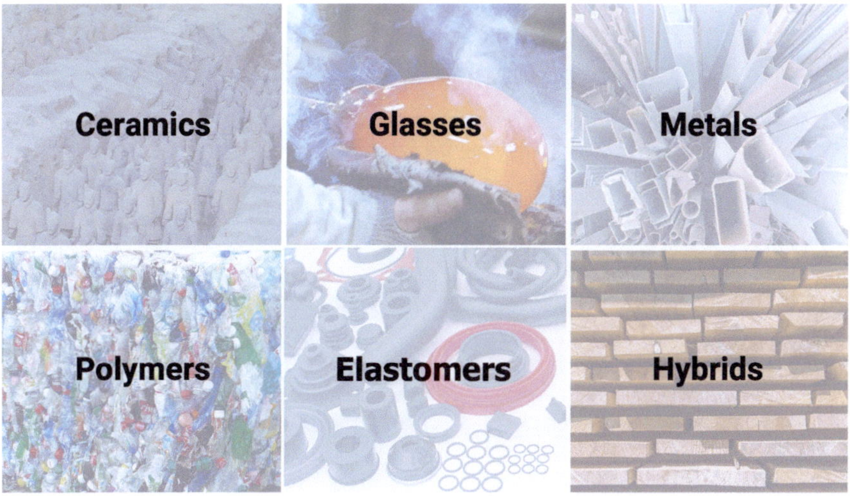

Fig. 2.7 Categorization of engineered system families. Image by the author

family), steel and aluminum structures (from the metals family), luxurious vinyl flooring (from the polymers family), and silicone sealants (from the elastomers family) (Fig. 2.7).

But it's precisely this narrow view of materials that we must now challenge. In the face of current social, environmental, and technical urgencies, we must move beyond such rigid categorizations. It's time to shift focus toward naturally occurring, readily available, extensively variable, and minimally processed materials—material with shorter supply chains, reduced thermochemical processes, and healthier life cycles for both land and people. Figure 2.8 offers a counter-response to Ashby's diagram. It makes a step towards adaptation to the building industry by looking at a material's raw origin as its foundational attribute. Instead of focusing on properties after intense refinement, this approach classifies materials into groups like geological, plant-based, animal products, fungi, living bacteria, and micro-organisms. These categories offer a more expansive view of what materials are and how they can function in geogenic and biogenic architectural composites, a nod to Ashby's "hybrids" but with a more regenerative and biophilic lens.

In fact, nearly everything we refine and synthesize in modern material science can be found in nature in a less concentrated form. Take polymers, for example. In nature, molecular compounds made of organic matter are inherently polymeric, aiding in plasticity and flexibility, making them crucial for material design. Biopolymers—those still synthesized, but from bio-based sources—are gaining attention in contemporary architecture and design. These biopolymers, often called bioplastics, are derived from plant- and animal-based proteins and polysaccharides. But here lies a critical challenge: the term "bio-based" has become misleading. It often

Fig. 2.8 Categorization of natural material families. Image by the author

encompasses highly processed materials such as certain types of PVC or epoxy, which, although technically bio-based, are fossil fuel derivatives. To address this, I argue for clearer terminology—specifically, focusing on ***naturally occurring materials***, meaning substances produced directly by living organisms without intensive chemical modification.

Additionally, the scale at which materials are examined can inform not only their performance but also the scale of their production processes. In material research, we often talk about three key scales: macro, mezzo, and micro. The macro scale refers to the structural assemblage or architectural form; the mezzo scale focuses on material testing, such as for rheology or fabrication parameters; and the micro scale investigates the molecular or nano-level arrangements that influence the material's behavior at the mezzo and macro levels. This multi-scalar approach is exemplified in Fig. 2.9, which presents material design research conducted at the Columbia GSAPP Natural Materials Lab, where we enrich 3D printed earth with plant fibers. The study shows how the fibers, coated in clay paste and arranged according to the printing direction, behave thermally and structurally during extrusion from a 3D printer nozzle. This type of research highlights how materials can be understood and manipulated across scales, providing insights into their performance within architectural systems.

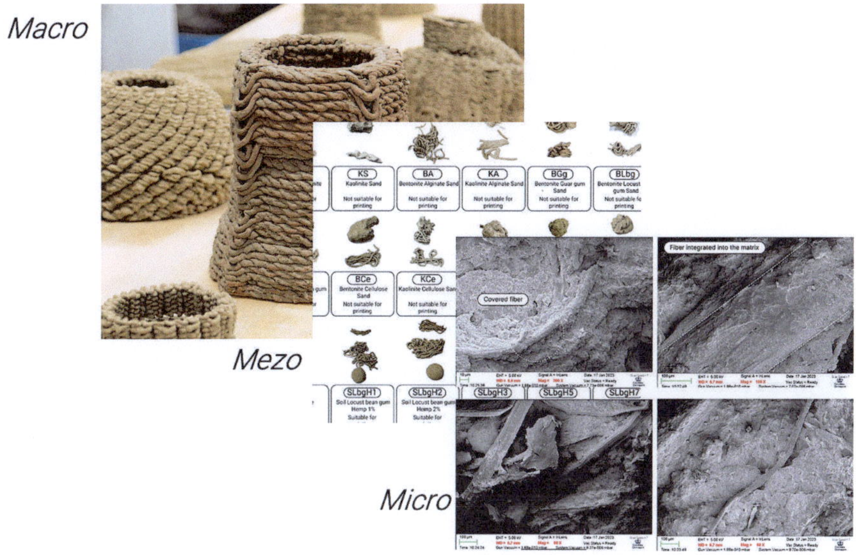

Fig. 2.9 Material design research for 3D printed earth-fiber, from the macro, to the mezzo, and micro scales. Image by the author following previous studies by the author: [1, 3, 11]

2.3 Environmental Cycles: Energy, Climate, Metabolism, and Survivability

When we consider the geological trade routes of building materials, we can imagine their histories, map their origins, and represent them in a system boundary diagram. But how do we quantify those journeys? How do we put a number on the thermo-chemical processing, the mechanical forces exerted, the surface treatments applied, or the miles of transportation and intermediate storage that each material endures? How can we move from storytelling and visualization to enumeration?

Take cob construction system as an example. Cob is an ancient and sustainable building system made from a mix of clay-rich soil, sand, straw, and water, forming a plastic wet-state composite that has been used for millennia in vernacular architecture. While there is extensive inventory research for LCA on conventional construction assemblies, only a few studies exist for earthen materials. Cob construction has often been overlooked, perhaps due to its variability or perceived "low-tech" status. Yet, these materials and their life cycles deserve just as much scrutiny. The diagram in Fig. 2.10 presents the various constituents of a cob building system and their processing stages. The material flows are broken down to the specific processes each undergoes; the straw component alone involves a series of movements from tilling and seeding to crop management, harvesting, and baling. The analysis includes the field

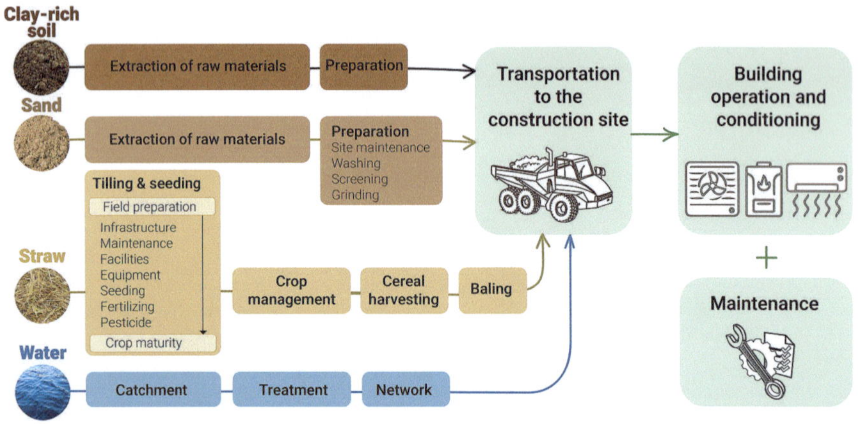

Fig. 2.10 Life cycle inventory diagram for cob—an earth-fiber construction system. Image by the author following a previous study [6]

preparation required, the infrastructure, the maintenance of equipment, the fertilizers and pesticides used, and the energy input throughout the process. Each of these stages adds another layer of complexity and a measurable contribution to the environmental footprint of cob as a construction system.

In a related study, we evaluated the life cycle inventory (LCI) of straw-bale construction versus fiberglass insulation. As shown in Fig. 2.11, the processes for strawbale production include wheat cultivation, from tilling and seeding to harvesting and baling. In contrast, the fiberglass production process requires melting and refining, adding binders, compressing, oven curing, cooling, and cutting. By diagramming alone, this exercise highlights the different levels of processing and thermochemical energy required for each material alternative, contributing to their overall environmental impact. Beyond this qualitative visual analysis, the inventory data was enumerated and synthesized into a life cycle impact analysis.

In this way, we begin to account for the often-unseen processes that go into building materials. Through diagramming, we can track not just the physical materials, but the processes and entanglements they possess. Then, we move from visualizing material flows to quantifying the true carbon or energy costs of these systems, shedding light on the broader ecological impacts they may have, as shown in Fig. 2.12. This transition accounts for the required LCA adherence to established regulations and standards, such as the ISO 14040 series, to ensure consistency and accuracy. Referring back to the LCA for cob, the inventory data for each stage was then assessed using models for impact enumeration. We specifically used the Cumulative Energy Demand (CED) and Tool for the Reduction and Assessment of Chemical and Other Environmental Impacts (TRACI), since they are particularly relevant within the US

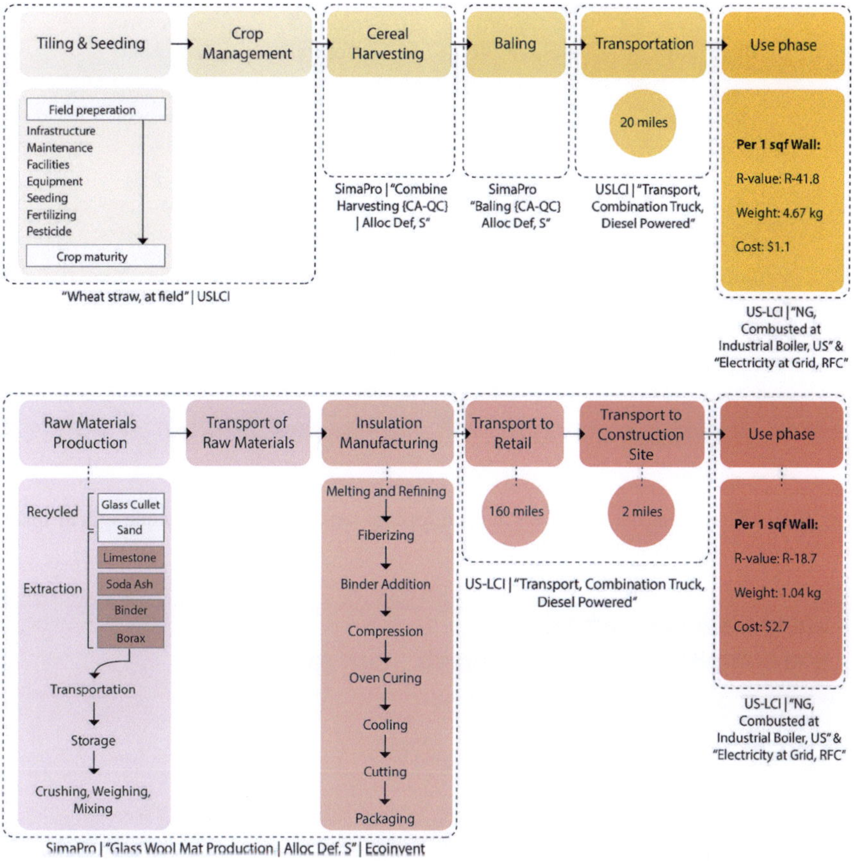

Fig. 2.11 Cradle to grave life cycle inventory (LCI) diagram for straw-bale versus fiberglass bat. Following a previous study by the author: [5]

context. We evaluated straw through financial-based allocations, which account for scenarios where agrowaste materials are considered valuable resources for construction, rather than just by-products of cereal production. In this case, wheat straw prices were drawn from both interviews with field builders and contractors, as well as existing literature.

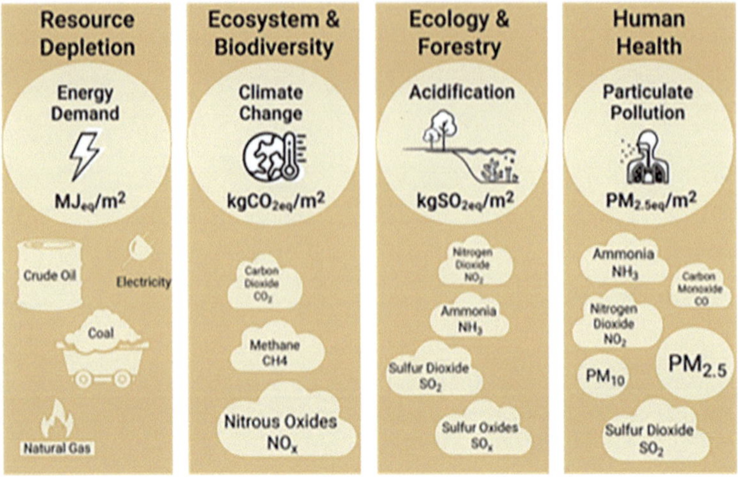

Fig. 2.12 Life cycle impact assessment diagram for cob—an earth-fiber construction system. Image by the author following a previous study: [6]

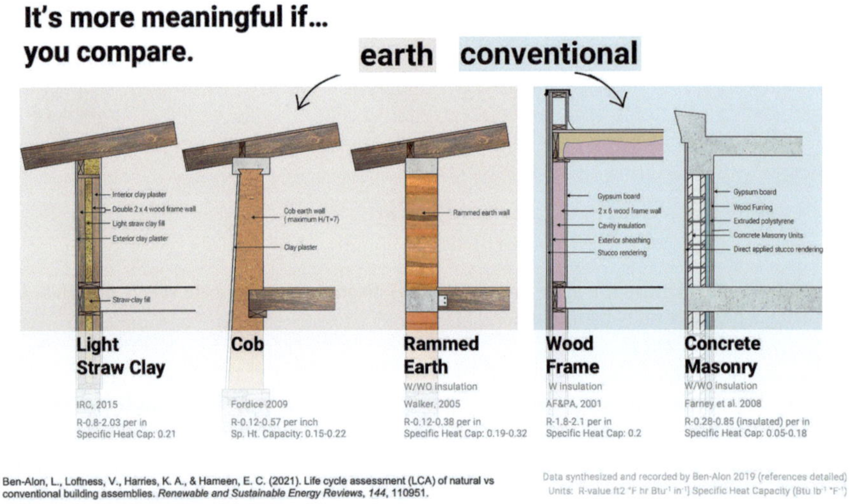

Fig. 2.13 The comparative analysis of wall assemblies: earth versus conventional. Image by the author following a previous study: [6–8]

As a final comparison, the earthen assemblies—light straw clay, cob, and rammed earth were compared to conventional wall assemblies—insulated wood frame and concrete masonry, using a functional unit of 1 m^2, for the comparison, to allow future comparison with other assemblies, and expansion to complete structures. In this sense, this study assumes that other roof and flooring systems are comparable (Fig. 2.13).

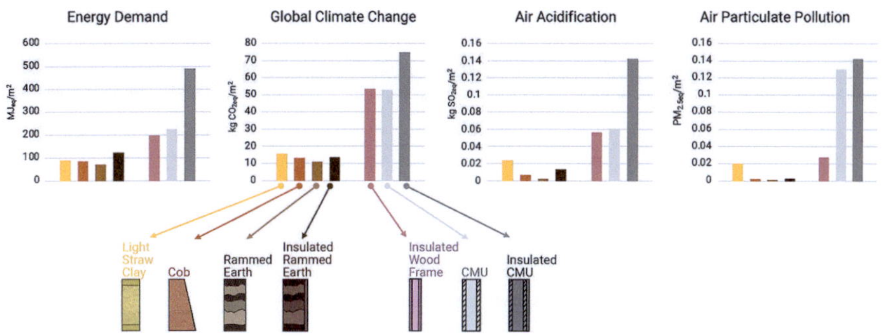

Fig. 2.14 Inventory flow diagrams of straw bale construction versus fiberglass insulation materials. Image by the author following a previous study: [8]

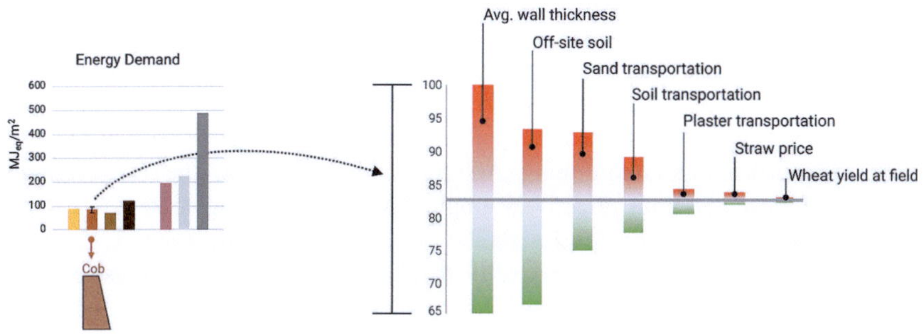

Fig. 2.15 Sensitivity Analysis of cob material construction system. Image by the author following a previous study: [8]

The visual summary of the life cycle impact assessment (LCIA) results for embodied energy is shown in Fig. 2.14. The results are clear—earth assemblies outperform conventional assemblies across all impact categories for the embodied phase, which includes production from cradle to construction site.

However, a sensitivity analysis of these results shows that even under the worst-case scenario for the earthen assemblies—such as increased transportation distances or greater wall thickness—the environmental performance remained significantly superior to that of conventional assemblies, as shown in Fig. 2.15.

But what about the operational phase? In this study, the operational ("use") phase was assessed using a hybrid methodology that combined field data gathered through stakeholder surveys, computational thermal simulations, and impact assessments. Specifically, a survey involving 35 homeowners of earth buildings was conducted to collect data on HVAC usage and comfort patterns within actual structures built with

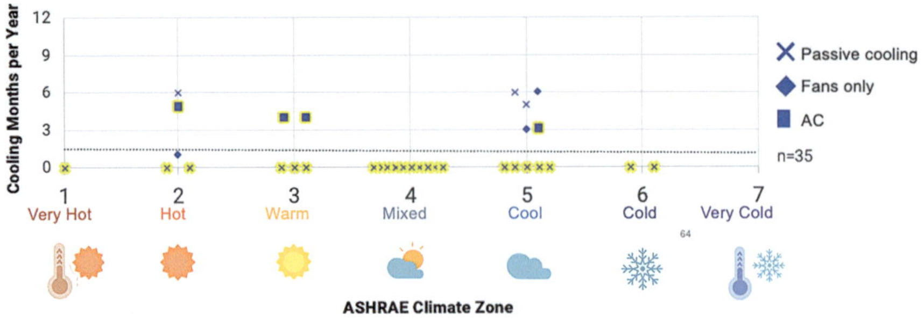

Fig. 2.16 Sensitivity analysis of cob material construction system. Image by the author following [4]

earth-fiber construction systems. The results revealed a striking trend: these earth-fiber assemblies exhibit remarkable energy efficiency, as they require minimal space conditioning for cooling throughout the year, regardless of climate conditions. This phenomenon can be attributed to their high thermal inertia and intrinsic evaporative cooling properties. As illustrated in Fig. 2.16, only four out of the 35 surveyed structures (11%) utilized air conditioning (AC) for cooling purposes. The overwhelming majority of these buildings effectively employed passive cooling strategies, even in the hotter and warmer climates designated as ASHRAE climate zones 1, 2, and 3. This actual field data shows that in reality, this is how earth-fiber buildings are used, highlighting the effectiveness of earth-fiber materials in providing thermal comfort while significantly reducing energy consumption and thereby contributing to more passive architectural regimes.

There are, therefore, multiple synergies possible between earthen wall construction and dynamic passive heating and cooling strategies like shading, natural ventilation, passive solar heating, and night insulation. For this purpose, thorough thermal and hygroscopic characterization is critically needed to pursue dramatically reduced space-heating and cooling needs, as well as passive thermal survivability in natural materials.

The subsequent analysis continues the examination of the four earthen wall assemblies—cob, light straw clay, rammed earth, and insulated rammed earth—alongside three conventional assemblies: concrete masonry unit (CMU), insulated CMU, and fiberglass insulated wood frame. By incorporating adaptive thermal comfort ranges across six distinct climates, the results analyzed thermal responses with and without well-operated passive heating and cooling systems, alongside energy balance and operational impacts as part of the life cycle thinking.

To further enhance our understanding, the next step involved acquiring better data regarding the actual thermal performance of each material through computational models and scenario simulations. This approach aimed to account for material porosity, thermal mass properties, and hygrothermal characteristics to fully capture the true and complex performance of the earth-fiber assemblies. The chosen climates represented dry, warm-hot conditions, as well as colder climates where earth materials have historically been utilized. An expected insight, shown in Fig. 2.17 illustrates

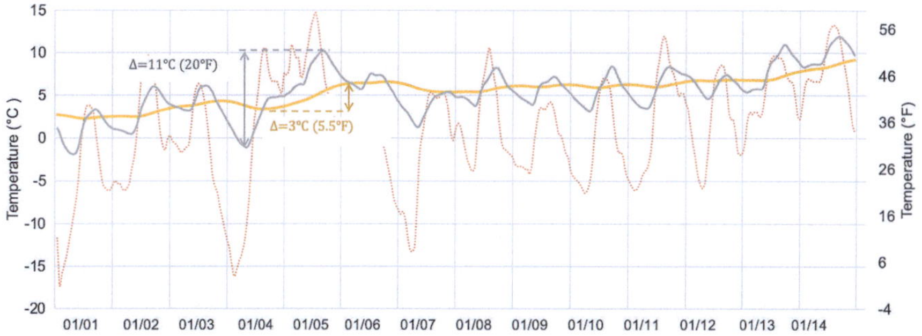

Fig. 2.17 Sensitivity analysis of cob material construction system in Tucson, Arizona contrasting fiberglass insulated woodframe structures (grey line) with light straw clay structures (bold yellow line) in response to outdoor temperatures. Image by the author following [4]

the temperature fluctuations between fiberglass-insulated wood frame structures and light straw clay assemblies, demonstrating that while the wood frame experiences significant temperature variations, the light straw clay maintains a remarkably stable indoor environment.

Nonetheless, the results regarding indoor temperatures within these dwellings revealed insightful phenomena concerning thermal comfort in earth-fiber buildings. Earthen structures with passive conditioning exhibited strikingly lower diurnal temperature swings across all climates and seasons compared to conventionally-walled structures. This notable finding highlights the consistently greater thermal stability achieved under passive conditioning regimes among earthen walls, indicating that substantial further effort is needed to integrate earthen buildings into national and regional building codes. The effectiveness of passive solar heating was particularly pronounced in the insulated earth-fiber assemblies—insulated rammed earth and light straw clay. In desert climates, passive cooling strategies effectively brought 50–80% of uncomfortable (overheated) hours into the adaptive comfort zone. Meanwhile, in steppe and cool-summer Mediterranean climates, the combination of passive heating and cooling eliminated 25–70% of uncomfortable hours.

Figure 2.18 illustrates the heat flux mechanisms underlying this performance, showcasing inherent patterns within insulated earth-fiber assemblies—particularly the light straw clay. Notably, light straw clay demonstrates superior hygroscopic performance, with its ability to uptake and evaporate moisture contributing to the thermal stability of the wall, reflected by the notable larger blue bars, showing the latent heat flux in the interior of the wall assembly. This characteristic not only enhances thermal comfort but also potentially contributes to the relative humidity management and a result, the intrinsic cooling effect, of the assembly. Therefore, light straw clay represents a promising assembly with significant potential to inform building codes concerning energy performance.

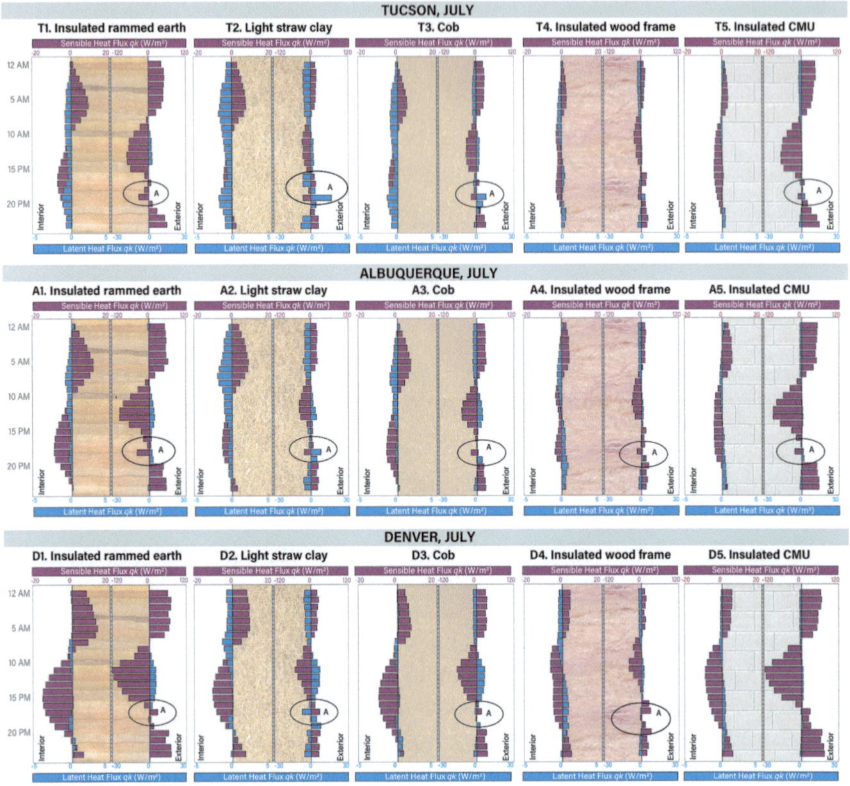

Fig. 2.18 Sensible heat flux (in purple) and latent heat flux (in blue) illustrated on the cross-section of each wall assembly, depicting heat transfer dynamics from both the interior and exterior of the walls. Image by the author following [9]

The potential energy and emissions through a comparative life cycle assessment (LCA), therefore, should account for the effects of the thermal performance on the operational conditioning expenses over the life cycle use phase, while examining the tradeoffs with the embodied stages of each wall assembly. The temperature and humidity levels not only assess the extent to which HVAC systems would be required but also the amount of energy demand in the building system. The final summary of the aggregated embodied and operational impacts, shown in Fig. 2.19, includes maintenance, over a 50 years life span. Notably, the embodied performance of earthen assemblies is significantly superior, effectively offsetting the operational heating and cooling demands, even in climates where uninsulated earthen mass may not perform optimally.

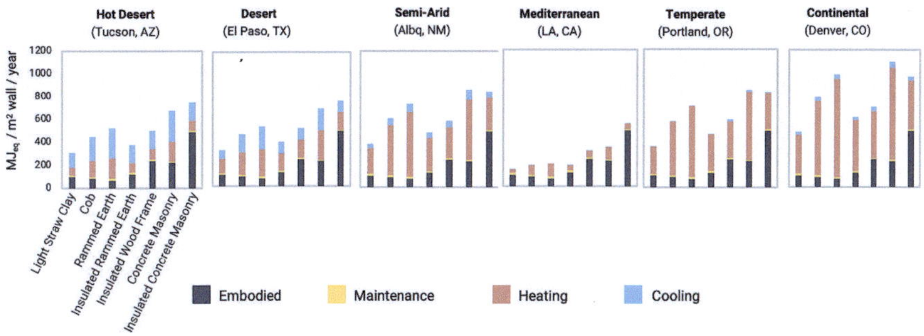

Fig. 2.19 The LCA results for cradle to grave phases. Image by the author following [8]

The results are further illustrated in Fig. 2.20, demonstrating that insulated rammed earth or compressed earth blocks (CEBs) require 65% less energy demand throughout the production and operation phases compared to insulated concrete masonry. Similarly, light straw clay infill within a wooden frame can reduce energy demand by up to 53% compared to fiberglass insulation. These findings are even more striking when considering emissions, which arise not only from energy generation but also from chemical reactions during material processing and fugitive emissions during quarry operations. Light straw clay outperforms all other earth-fiber and conventional assemblies across its cradle-to-grave life cycle. Its exceptional embodied and operational carbon performance is attributed to its low extractive constituent materials and the synergistic combination of thermal mass and thermal resistivity. This unique interplay effectively reduces both heating and cooling energy demand while enhancing hygroscopic performance.

Building on the compelling evidence of light straw clay's superior energy efficiency and minimal emissions, subsequent development of fabrication applications of this material in contemporary construction was pursued, with a digital fiber-rich earth mixture developed to facilitate an insulative infill method suitable for 3D printing. This approach not only draws inspiration from the traditional light straw clay but also embraces the potential of additive manufacturing with earth materials—a rapidly emerging field that aligns with sustainable design principles. Currently developed in recent research using a "cob"-like mix design, the following results look at novelly developing fiber-rich light-straw-clay-like 3D printable recipes. Increasing the content of plant fibers in 3D printed earth was here pursued with the goals of achieving higher thermal resistivity potential for code compliance, enhanced ductility and flexural strength, and lightweight assemblies that are more viable for construction economy and transportation viability. Figure 2.21 summarizes a continuous research that was geared towards the development and material design research of 3D printed light earth-fiber mixtures with a range of plant fibers from leaf, stem, and bast origins, including wheat straw, banana, hemp, kenaf, and sisal fiber.

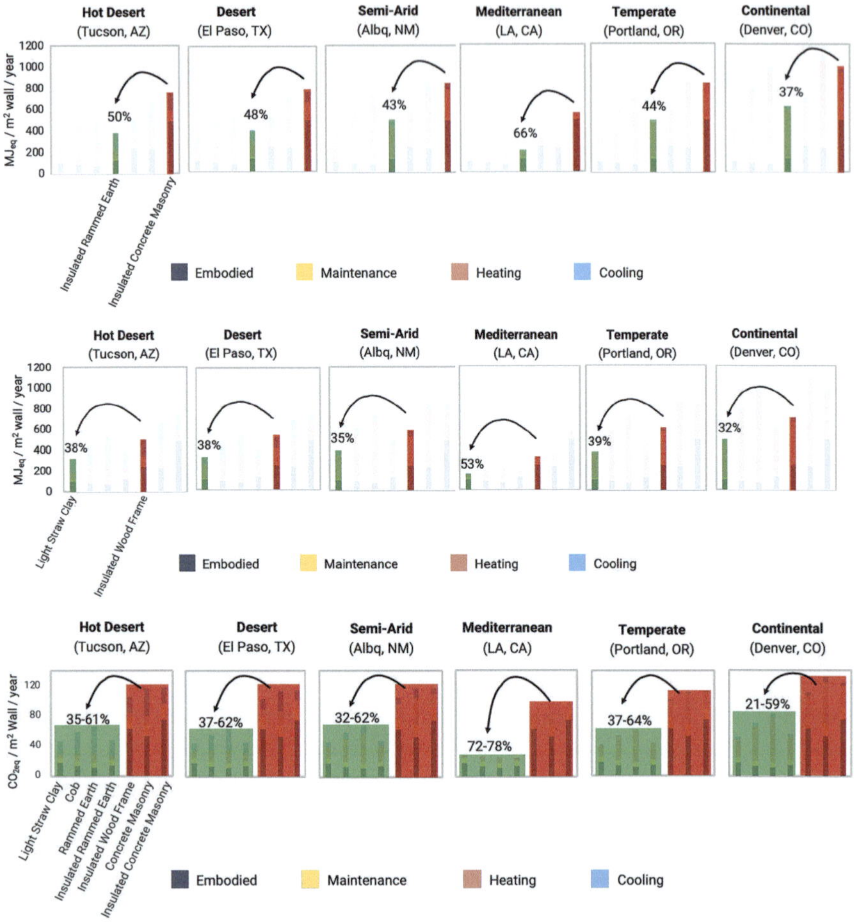

Fig. 2.20 LCA cradle-to-grave impact reductions by earth-fiber assemblies compared to conventional assemblies. Image by the author following [8]

Figure 2.22 shows the global warming potential (GWP) from a cradle-to-gate perspective for traditional earthen construction materials for a functional unit of the amount of material necessary to obtain an R = 1 m^2K/W and the related reference thickness that is required. Since fiber-rich material mixtures possess a bio-storage capability for carbon sequesteration. Whereas the GWP_{100} is the global warming potential over a use phase of 100 years, and the $GWP_{Bio\ index}$ is the carbon uptake over 60-year storage, so that:

$$GWP_{net} = GWP_{100} - GWP_{bio\ index} \tag{2.1}$$

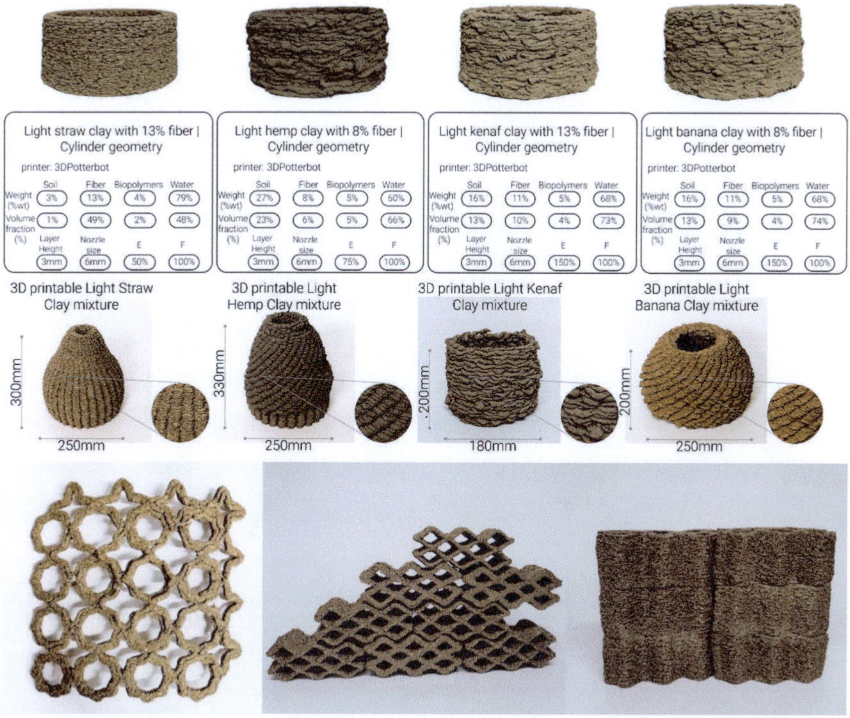

Fig. 2.21 Development of 3D printable fibrous earth mixtures, from a "recipe" approach and pattern experimentation for texture, tile elements, printed bricks, and vertical wall elements [3]. A demonstrtaion of 3D printed earth-fiber tower, by student Mohammad-Hossein Zowqi, at Columbia GSAPP, reflects on this material innovation

The results of this analysis, shown in Fig. 2.22, illustrate that among the traditional earth construction materials, light straw clay exhibits lower GWP_{net} emissions while providing a higher thermal resistivity. For the 3D printed earth-fiber mixtures, wall sections designed with air gaps were characterized by lower global warming potential and better thermal performances if compared to full 3D printed earth-fiber wall assemblies. Of the various plant fiber additives for 3D printed earth, the wheat straw fiber addition was shown to surpass all other fiber mixtures in terms of its smallest GWPnet (50.63 $kgCO_{2e}$/FU) and highest R-value (8.16 m^2K/W). In contrast, hemp-rich 3D printed earth mixtures were shown to exhibit higher GWP_{net} due to their higher density and assembly thickness (and thus, the amount of materials required) to achieve the required R-value.

Fig. 2.21 (continued)

2.4 Social Cycles, Biophilia, Accessibility, and Community Engagement

Social impacts and material geographies of material designs require further investigation; the environment is not just the air, ground, and water, but the people with whom we work and live. The metabolic flow of materials is linked through economic forces that reflect the history of place.

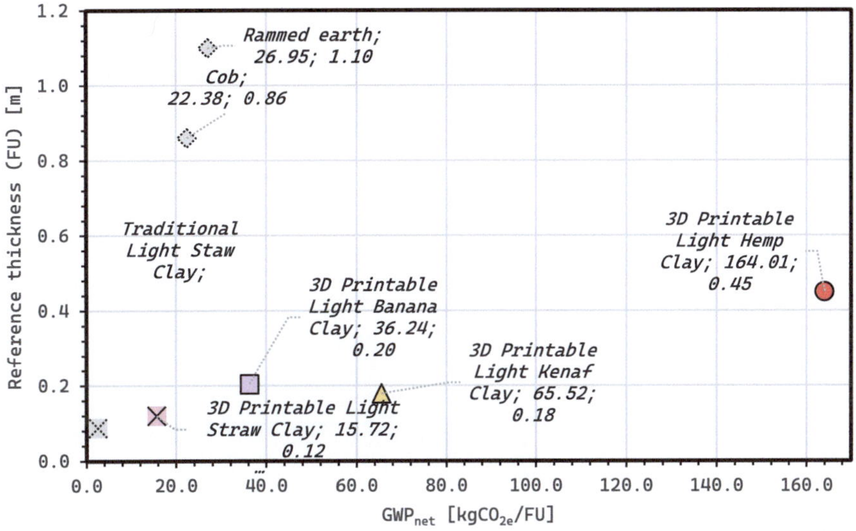

Fig. 2.22 GWP_{net} values and the reference thicknesses for the developed 3D printable earth-fiber composites versus traditional wall systems, for a FU of R = 1 m^2K/W [3]

Figure 2.23 shows the consumption of raw materials in the US over time, as we moved from organic biomass to industrial minerals, and the fluctuations of construction materials in response to societal events, wars, and economic turbulences. Materials and culture, reflected through major events in our society are constantly enmeshed and interweaved.

While there is a widespread consensus on the social benefits of earthen and bio-based materials, a systemic analysis that quantifies these benefits can shed light on the perceptual advantages and just encounters. From the extraction of its raw components to its end of life, a construction material's production, transportation, use, and disposal inherently engage a variety of stakeholders at different levels: from workers in production facilities to designers, builders, occupants, and demolition contractors. The social impact of these involvements can—and should—be evaluated through a number of categories defined by stakeholder groups and social/socio-economic criteria, such as health and wellbeing, local employment, and workers conditions.

Social life cycle assessment (S-LCA) is a quantitative and qualitative method that aims to assess the social concerns of human health, well-being, and social welfare. Similar to an environmental LCA, this S-LCA impact assessment is quantified using indicators across the entire life cycle of the building material while prioritizing areas of "hot spots" in the analysis, provided in terms of health and safety, worker conditions, regional impacts, and community engagements, and regional sustainability. S-LCA is still an evolving methodology that is informed by the challenge of quantifying societal attributes. Often used with a hot-spot analysis, S-LCA helps identify

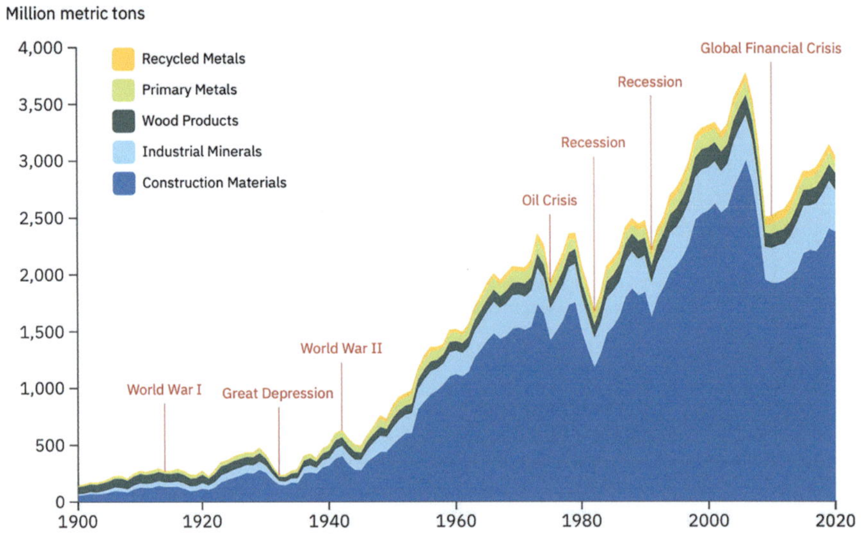

Fig. 2.23 Consumption of raw materials in the US over time and major societal events [12]

impacts perceived firsthand by a range of stakeholders: manufacturers, designers, contractors, researchers, and owners.

A comparative S-LCA was developed for earth, bio-based, and conventional (concrete and steel) materials following an online survey. Using the UNEP/SETAC guidelines, the survey was designed to collect social impact data from stakeholders in four categories: workers, occupants, local community, and society, and in response to eight impact categories: health and safety, human resource management, professional development and supplier relationship, use stage responsibility, community development, regional environmental and human health, broader societal commitments, and public interest. In order to align with the process of construction material production, this S-LCA also accounts for six additional categories that reflect the life cycle of building materials: extraction, production, transportation, construction, occupancy, and demolition/recycling. While most of the social impact categories affect all the life cycle stages of a material, not all apply to the occupancy phase. The distribution of the social impact categories based on the life cycle stages and stakeholders is shown in Fig. 2.24.

The final S-LCA results, shown in Fig. 2.25, highlight that earthen materials scored the highest across all social impact categories, followed by bio-based materials, particularly in health and safety, impacts on local communities, and regional sustainability. In contrast, concrete and steel performed the worst in most categories except worker conditions, where they excelled due to better professional training and social benefits, reflecting the more established industries existing for these materials. These findings raise questions about the role of industrialization in reducing the social benefits of building materials. While natural materials currently outperform

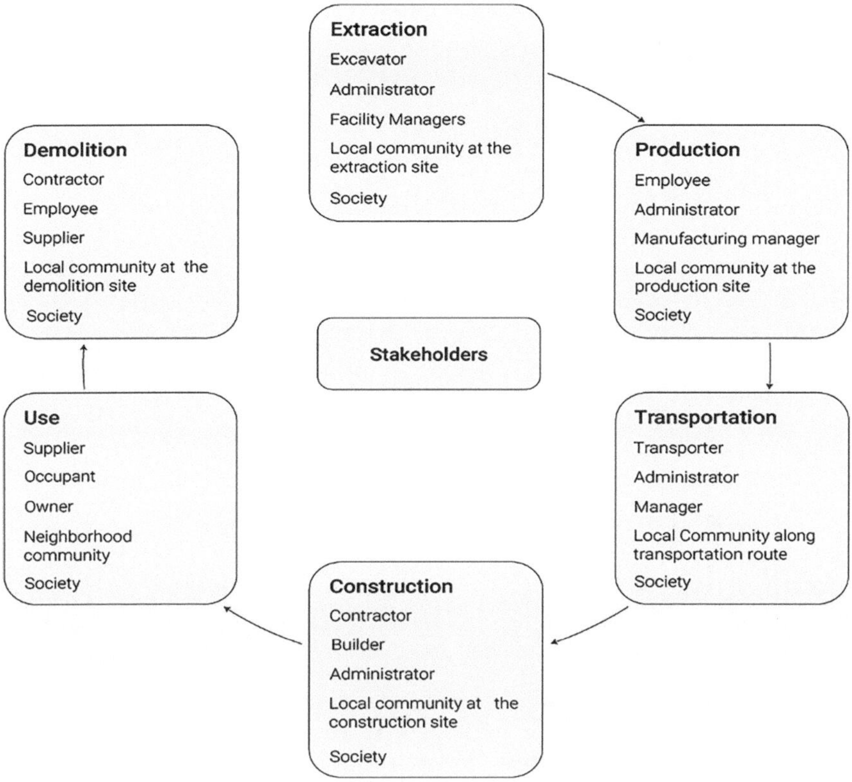

Fig. 2.24 The S-LCA stages, and stakeholders relevant to this study [10]

processed materials, the scores might exacerbate in the regional impact categories and be enhanced for workers' conditions as the natural material industries grow.

In terms of their health attributes over their use phase, natural materials were shown to create a biophilic effect, one that can cultivate positive evidence-based therapeutic effects. Specifically, when it comes to hospital and healthcare environments, natural materials like wood, bamboo, cork, and earth, were shown to affect occupant-related outcomes, such as increased recovery, reduced medication intake, and enhanced staff performance. Wood furniture was shown to reduce patients' systolic blood pressure, a physical manifestation of stress [18]. Similarly, medical staff reported preferring 60% wooden surfaces instead of rooms with no wood or full wood [16]. The effects of natural materials were shown to be important for balanced thermoregulatory, cardiovascular, and hormonal parameters while meeting regulatory requirements, as summarized in Fig. 2.26. For instance, patients in rooms with wood paneling had 20% less stress (indicated by measured plasma cortisol levels) than patients in rooms with concrete walls [17].

The location and placement of natural materials across a hospital room should follow visual and touch surface areas to provide patients with stimuli. Therefore,

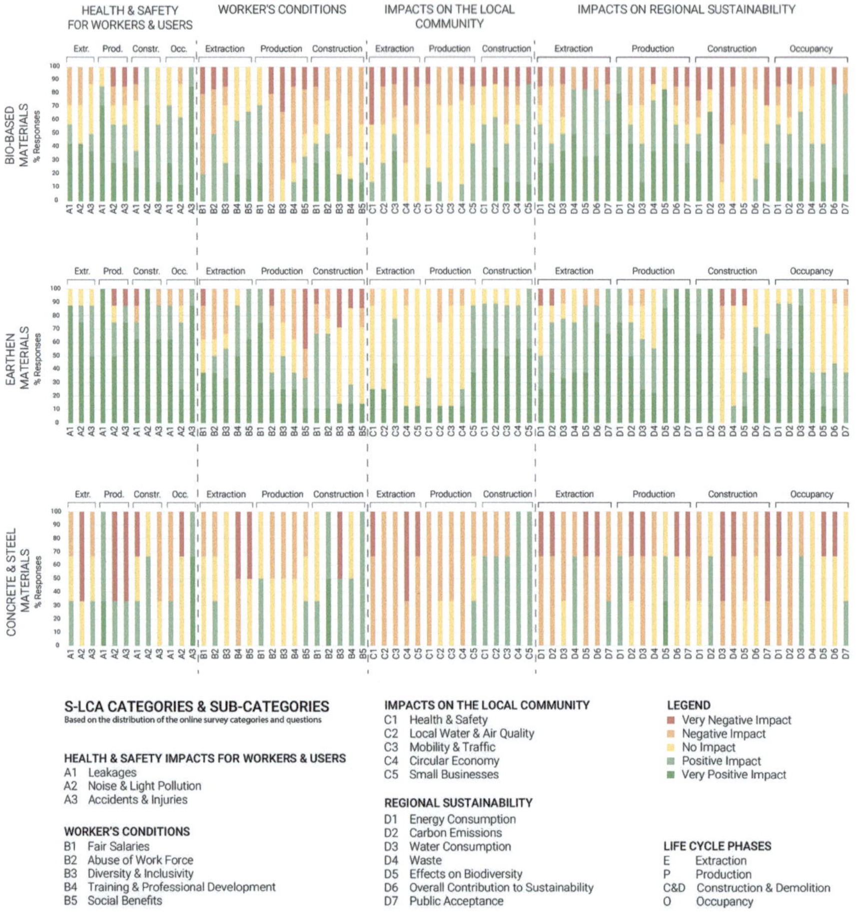

Fig. 2.25 Survey responses analysis—spread of responses for each life cycle phase [10]

locations in front and above a patient's bed and high touch points for the placement of materials should be preferred. Additional physical characteristics, such as temperature and humidity regulation may also be achieved by using natural materials, located in high touch points near and around a patient's bed.

2.5 Conclusions

The advancement and justification of natural earth- and bio-based materials through the lens of life cycle assessment (LCA) unveils critical insights in material design research. As we navigate the intersection of traditional knowledge and new fabrication techniques, the environmental and social implications that underpin materials are

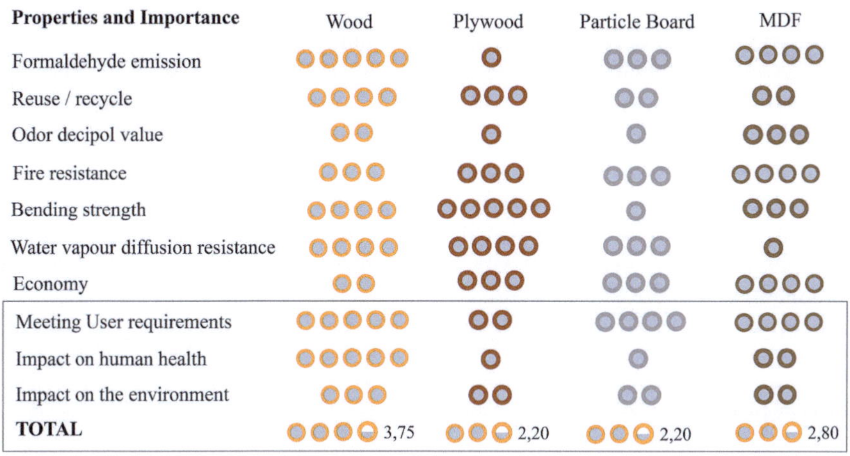

Fig. 2.26 Evaluation of natural versus artificial wood material. Summarized from (image by Maghlakelidze et al. [14], following Kokulu and Acun Ozgunler [13])

more important than ever for informed design decision-making. This chapter asks to catalog, demystify, and categorize the terminology and attributes of life cycle thinking and natural material systems. By establishing and diagramming system boundaries that encompass the entire life cycle—from material extraction and processing to operational performance and end-of-life recyclability—we can elucidate the complex relationships between material "recipes" so to speak and their impacts.

The designer's role, in this multifaceted analysis landscape, transcends mere technical material selection and carbon accounting. Instead, designers (and building professionals, in general) must engage in a synthesis of procurement procedures, health impact evaluations, and equity analyses, balancing embodied carbon with the realities of transportation and processing, thermal comfort, and the exchanges of labor, resources, and ideas.

Both environmental life cycle assessment (LCA) and social life cycle assessment (S-LCA) should be further used as qualitative and quantitative exercises for a fuller realization of the intricate relationships between material choices, environmental impacts, and social dynamics in architectural practice.

Specifically, for earth-fiber construction techniques—the study outlined in this chapter reveals a compelling narrative: even under adverse scenarios, such as increased transportation distances or adjustments in wall thickness, the LCA of earthen assemblies demonstrates significantly lower embodied carbon and energy footprints compared to conventional materials. Further diving into operational phases, the coupling of thermal resistivity with thermal heat capacity was shown to be superior, making assemblies like insulated rammed earth and light straw clay particularly beneficial. From a social and accessibility stand-point, aspects of labor practices, community health, and cultural heritage reveal that earth materials are perceived as advantageous by stakeholders, with the exception of professional training and working conditions—a sign of reflection of their non-commodified nature and required development within the building industry.

As the future of digital earth-fiber construction unfolds, embracing plant fiber and agrowaste materials within raw ("non-stabilized") earth pastes offers transformative opportunities for environmentally sound and socially just architectural practices. By synthesizing insights from LCA and SLCA, we can advocate for a built environment that honors both ecological integrity and the spirit of people and place. As we move forward, this chapter invites us to contemplate and commit to an architectural ethos that embraces variability, complexity, and the intricate interplay between materials, communities, and the natural world.

References

1. T. Akemah, L. Ben-Alon, Developing 3D-printed natural fiber-based mixtures, in *RILEM Bookseries*, vol. 45 (2023), pp. 555–572. https://doi.org/10.1007/978-3-031-33465-8_42
2. M.F. Ashbey, *Materials Selection in Mechanical Design*, 4th edn. (Elsevier, 2011). https://doi.org/10.1016/C2009-0-25539-5
3. O. Beatrice Carcassi, L. Ben-Alon. Maximizing fiber content in 3D-printed earth materials: printability and mechanical assessments, in *Construction and Building Materials* (2023)
4. L. Ben-alon, *Integrating Earthen Building Materials and Methods into Mainstream Housing ... Stages* (Carngie Mellon University, 2019). https://doi.org/10.1088/1755-1315/323/1/012139
5. L. Ben-Alon, M. Carre, H. Charles, Y. Liu, P. Tschofen, *A Comparative LCA of Fiberglass and Straw Bale Insulation* (Carnegie Mellon University, 2017). https://doi.org/10.2172/875800
6. L. Ben-Alon, V. Loftness, K.A. Harries, G. DiPietro, E.C. Hameen, Cradle to site life cycle assessment (LCA) of natural vs conventional building materials: a case study on cob earthen material. Build. Environ. **160**, 106150 (2019). https://doi.org/10.1016/j.buildenv.2019.05.028
7. L. Ben-alon, Lola, V. Loftness, K.A. Harries, E. Cochran, Integrating earthen building materials and methods using perception surveys and life cycle assessment (LCA), in *Lehm 2020: 8th International Conference on Building with Earth* (2020), pp. 1–10
8. L. Ben-Alon, V. Loftness, K.A. Harries, E.C. Hameen, Life cycle assessment (LCA) of natural vs conventional building assemblies. Renew. Sustain. Energy Rev. **144**, 110951 (2021). https://doi.org/10.1016/j.rser.2021.110951
9. L. Ben-Alon, A.R. Rempel, Thermal comfort and passive survivability in Earthen buildings. Build. Environ. **238**, 110339 (2023). https://doi.org/10.1016/J.BUILDENV.2023.110339
10. Z. Berrada, L. Ben-Alon, What about the materials' social impact?, in *ACSA/AIA Intersections Research Conference: Material Economies* (2023), pp. 1–8
11. Z.E. Bryson, W.V. Srubar, S. Kawashima, L. Ben-Alon, Towards 3D printed earth-and bio-based insulation materials: a case study on light straw clay. In *18th International Conference on Non-Conventional Materials and Technologies* (2022). https://par.nsf.gov/biblio/10386939
12. Center for Sustainable Systems, *U.S. Material Use Factsheet* (University of Michigan, 2024)
13. N. Kokulu, S. Acun Ozgunler, Evaluation of the effects of building materials on human health and healthy material selection. Gazi Univ. J. Sci. **32**(1), 14–25 (2019). https://dergipark.org.tr/en/pub/gujs/issue/43586/422072
14. M. Maghlakelidze, A. Mesa, A.O. Sawyer, M.G. Morad, L. Ben-Alon, Biophilic design and natural materials in healthcare environments. J. Green Build. (2023)
15. K. Moe, Matter is but captured energy, in *Matter: Material Processes in Architectural Production*, ed. by G.P. Borden, M. Meredith, 1st edn. (Routledge, London, 2011)

16. A.Q. Nyrud, T. Bringslimark, K. Bysheim, Benefits from wood interior in a hospital room: a preference study. Archit. Sci. Rev. **57**(2), 125–131 (2014). https://doi.org/10.1080/00038628.2013.816933
17. H. Ohta, M. Maruyama, Y. Tanabe, T. Hara, Y. Nishino, Y. Tsujino, E. Morita, S. Kobayashi, O. Shido, Effects of redecoration of a hospital isolation room with natural materials on stress levels of Denizens in cold season. Int. J. Biometeorol. **52**(5), 331–340 (2008). https://doi.org/10.1007/s00484-007-0125-4
18. X. Zhang, Z. Lian, Q. Ding, Investigation variance in human psychological responses to wooden indoor environments. Build. Environ. **109**, 58–67 (2016). https://doi.org/10.1016/J.BUILDENV.2016.09.014

Chapter 3
The Geographies of Sustainable Bio-Based Materials for Modern Methods of Construction

Alejandro Veliz Reyes, Alexandra Carr, Duhan Olmez, David Avilés, and Mohamed Gomaa

Abstract This chapter outlines a research agenda around material flows and geographies involved in the adoption of bio-based materials by modern methods of construction (MMC). In doing so, we argue that supply chains for bio-based materials are not only a matter of technical specification and procurement, but are influenced by broader societal, economic, and environmental forces that detach digital innovation from its physical, place-based manifestations. This challenge is particularly acute in post-industrial communities aiming to transition from localised manufacturing industries to knowledge- and creativity-based economies. The research is contextualised within a stream of research projects, of which four are presented as illustrative case studies. These projects have focused on local bio-based materials and communities often considered peripheral to mainstream digital innovation and MMC discourse: automation in earthen construction, participatory housing design in a rural community, the development of digital infrastructures for timber manufacturing, and MMC housing procurement in a government-funded rural housing programme. These projects are framed by the investigation on geographies of making set forth by Carr and Gibson. The Chapter concludes by identifying further work opportunities for this research agenda and suggests a wider and multidisciplinary approach to investigating the adoption of bio-based materials in construction.

A. Veliz Reyes (✉) · A. Carr · D. Olmez
School of Art, Design and Architecture, University of Plymouth, Plymouth, UK
e-mail: alejandro.velizreyes@plymouth.ac.uk

A. Carr
e-mail: alexandra.carr@plymouth.ac.uk

D. Olmez
e-mail: duhan.olmez@plymouth.ac.uk

D. Avilés
Sustainable Minerals Institute, The University of Queensland, Brisbane, Australia
e-mail: d.aviles@uq.edu.au

M. Gomaa
School of Engineering, Swinburne University of Technology, Melbourne, Australia
e-mail: mgomaa@swin.edu.au

D. A. H. Hanaor (ed.), *Sustainable Materials for the Built Environment*, Springer Series in Materials Science 352, https://doi.org/10.1007/978-3-031-97818-0_3

3.1 Introduction

Bridport, in rural south-west England, was the cordage and sailcloth capital of the United Kingdom (UK) from medieval times—at least from 1213 [1]—until the mid-eighteenth century [2]. Its industrial ecosystem was inextricably linked to its natural and urban productive landscapes: it has a good natural environment to grow hemp and flax, and the resulting cordage industry shaped the morphology of the town itself with (still visible today) long urban plots developed to house rope-making facilities [3] (Fig. 3.1). Bridport's industry was recognised with a Royal Charter to avoid local competition that restricted the production of cordage to Bridport's inhabitants, 'and restricting hemp sales within five miles of Bridport' [4]. The town reached national acclaim from its key role on the development of the maritime and naval prowess of the British Empire, only to later decline due to competition from imported cordage (Russia being a main supplier of hemp and cheaper labour from the 1550s), and the difficulties with supplying enough cordage for an ever-increasing maritime sector. A second wave of development is reported from the 1820s with the expansion of its port in Bridport Harbour (today called West Bay), with commercial traffic coming to an end in the 1980s after a slow decline [5]. Despite beneficial environmental, urban, and historical conditions for industrial growth, the town rapidly de-industrialised and became one of many post-industrial UK coastal towns, struggling to move away from manufacturing and material trade economic models and, today, showing signs of economic and social stagnation: problematic health outcomes [6], youth emigration in search for higher education and career opportunities, and low proportion of adults in employment, to name a few [7].

While this post-industrial decline may seem historical, processes of 'de-industrialisation' still take place around the world. Rural places, typically associated to commodified aesthetics, materials, or productive systems [8], struggle to 'shift' into knowledge economies and their reconversion is often seen as a matter of knowledge, geographic, entrepreneurial, or digital marginalisation. Works such as McGranahan et al.'s [9] suggest an injection of entrepreneurship and creative professionals as a way to prompt knowledge-based economic development in rural spaces. Wider studies across rural Europe show encouraging signs of rural spaces as incubators of social innovation resulting from close-knit communities and feelings of marginalisation from mainstream governmental policies and industrial strategies [10]. However, the binarism of material and industrial production vs knowledge or creative economic models is challenged by Gibson et al. [11], arguing that both approaches are often—mistakenly—seen as opposed as a result of 'overlapping conflations, assumptions and, in some cases, political and definitional manoeuvrings' (p. 87).

Today, Bridport communities are successfully resuming local hemp and flax industries and benefitting from those as break crops in local farms. Without a cordage maritime market to service, local uses of hemp include insulation bio-based material for local housing projects. The town has, however, lost some of the infrastructural

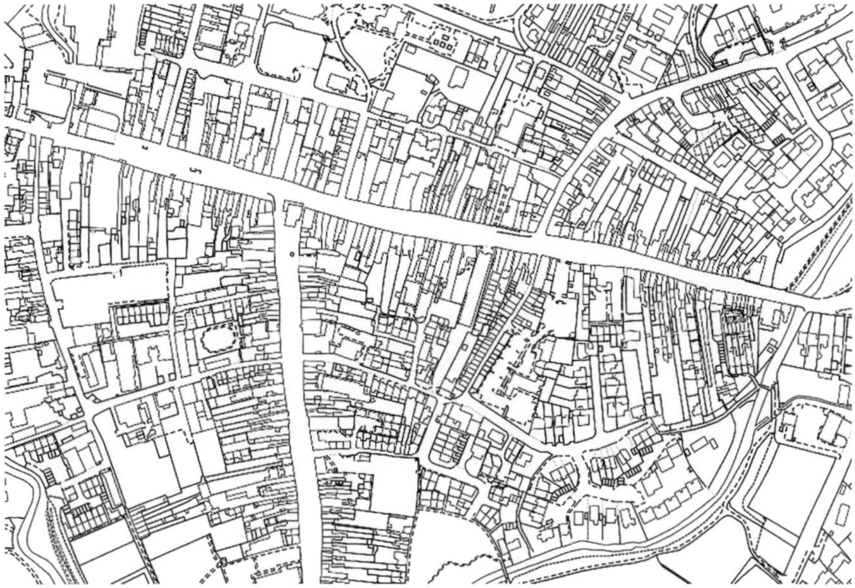

Fig. 3.1 Plan view of Bridport town centre, showing the density of long, slim plots originally allocated for ropemaking facilities. *Source* EDINA Digimap, 2024

'connective tissue' that brings together its local material flow ecosystem, with material providers, farmers, housing organisations, and builders not always clustered around local supply chains for housing construction, lack of educational facilities to develop appropriate skills, and a lack of a localised productive ecosystem. Wessex Community Assets (WCA), a local community housing organisation, approached our research team at the University of Plymouth to scope the adoption of local materials in the production of community housing, while concurrently injecting modern methods of construction (MMC) within their supply chain. By doing so, their expectation is to generate new building innovation around bio-based materials, as well as to grow the social infrastructures necessary to stimulate the local economy such as the creation of high-value skills and employment opportunities for its community—a vision for a new community-led economy [12] for the revitalisation of rural coastal towns in the UK that considers MMC for social good.

In doing so, we note the shortcomings of a reductionist discourse placing digital, creative and industrial innovation in contrast to place-based forms of physical material production. Instead of positioning this Chapter within this binary research space, we argue that digital and material flows are part of more complex and critical forms of knowledge production [13, 14] that can result in place-based societal benefits and impacts. In more detail, the Chapter outlines this theoretical position under the conceptual lens of these different 'geographies of making' [11, 15, 16]—with a particular focus on the geographies of low-carbon sustainable forms of material production and digitally-enabled forms of construction, here illustrated by four concatenated research projects delivered between 2018–2024. This research trajectory was initiated with the feasibility studies "Computing craft: Manufacturing cob structures using robotically controlled 3D printing", and "Participatory Housing Manufacturing"—both funded through the Engineering and Physical Science Research Council (EPSRC) network "Connected Everything: Accelerating Digital Manufacturing Research Collaboration and Innovation". Both projects explored the use of bio-based, low-carbon materials in the context of digital forms of architectural production, and implemented methodologies grounded on making, hands-on prototyping, and material experimentation. Then, the Chapter outlines the project "100 Factories" funded by the UK Forestry Commission which focused on the delivery of the digital infrastructures required for SMEs to join the housebuilding sector, creating regional networks able to deliver low-carbon housing under a Design for Distributed Manufacturing approach. Last, the Chapter concludes with an overview of an ongoing investigation on regional networks of digital manufacturing in the context of the Rural Housing Programme delivered by the Chile Ministry of Housing and Urbanism, funded by the Arts and Humanities Research Council (AHRC).

Methodologically, our work embodies material, disciplinary, and knowledge frictions which speak to broader industrial divides—such as the difficulties of incorporating local bio-based materials into construction procurement processes, and the problematic divides between the creative thinking of the built environment, and its production [11]. Following a "research in residence" approach [17], our 'embodied interventions' [15], while providing further insight into issues of building production, additionally shed light on how participatory and hands-on methods can generate broader societal benefit, and how hands-on methods can provide unique lived experience insights into place-based material innovation [18] that engages with local communities and geographies of making (Fig. 3.2). This aligns with institutional expertise on "research in residence", embedding our research teams within communities and material cultures in order to better understand the complexities of technology adoption, local development, and digital equity needs. Concurrently, working "in residence" allows for results that align with the needs of community groups, generating impact pathways and mobilising knowledge throughout a project delivery instead of a follow-on, later stage of the research process [19, 20]. In addition to the projects introduced in this chapter, this approach has been implemented in the challenge areas of ageing and access to heritage [21], technology adoption in design studios [22], or care home design [24], to name a few.

This chapter is structured as a case study report. Following a section outlining the theoretical position on 'geographies of making' by Gibson and Carr [11, 15, 16], we present four case study projects that embody critical approaches to intersect bio-based material flows with MMCs. As mentioned above, these include: (1) automation in earthen construction, (2) participatory mapping of timber and hemp material flows in Bridport, (3) the creation of digital infrastructures needed to grow local industrial ecosystems of SMEs for timber manufacturing in housing projects, and (4) a case

Fig. 3.2 Fieldwork in Bridport has included collaborative mapping of material flows (left), and creative making in pop-up Fablab facilities (right) among other methods. *Source* A. Carr, 2023

study on the adoption of MMC in the context of a government-funded rural housing programme. Finally, the Conclusion section summarises this chapter and sets out a research agenda for the adoption of MMC for social good.

3.2 Theoretical Framework

The pressing need to reduce carbon emissions has led into the discussion about how countries can decarbonise their economies in a time of climate crisis. In this sense, the agenda has been dominated by the "big discussion" around "mission-oriented" strategies, in which governments and economic elites would play a significant role [25, 26]. On the other hand, there have been calls to re-orient the analysis of production networks and value chains into global decarbonisation efforts in the economy [27]. However, those debates tend to focus on global and national scales (often, in the Global North), with less attention being paid to local and regional experiences of economic transformation [28]—or at least, how local resources could be connected to those global value chains by means of raw materials, labour power, or strategic locations. In consequence, the question of how local place-based making practices can find their space in productive processes to respond to their own challenges (and at the same time contribute to decarbonise the economy) has been overlooked. Furthermore, it is relevant to understand how a more local scale of enquiry can bridge the knowledge gap between traditional making practices embedded in their landscape, with new technologies that could provide an opportunity to break into contemporary production processes and economic benefit. In this way, this chapter seeks to scholarly re-engage with bio-based materials in the context of local communities, and understand how the geographies of making could "come into being" [29] when intersected with MMC.

In the context of our research, Bridport and other communities dealing with digital inequity have transited through different economic waves, which reflect the uneven spatial character of capitalist development in which different places are connected to global production processes and others are disconnected depending on the flows of capital [30]. In this sense, the Anthropocene present new challenges for research and development and, as Carr, et al. [31] discuss, the academic community needs to critically engage with the current economic processes and understand how particular and often devalued making practices could be most relevant to address some of the challenges of the Anthropocene. We identify bio-based construction and MMC as a relevant area to visibilise these challenge spaces—for example, MMCs have typically driven an agenda of optimisation, modularisation and industrial linkages with areas like manufacturing and computer sciences. Building upon existing global industrial manufacturing (such as 3D printing or robotic production) and data (such as building information modelling) frameworks, MMC do however struggle to scale within a construction industry of high fragmentation, low skill, and low technological adoption [32]. These systemic barriers have been recently under scrutiny in the UK Parliament following the enquiry "what's gone wrong?" [33]. Back to Bridport, we can illustrate this struggle with the example of a relatively simple timber panel construction. The local community is concerned about their local environment and has expressed their willingness to find ways of working with local ash timber. Ash has been affected by the dieback fungal disease, likely to decimate a majority of the ash forests across the UK [34]. Digital modes of production, such as computer-controlled milling, however, operate based on flat sheet materials such as plywood panels—plywood is not produced in the UK and is mostly imported from Brazil, China, or Latvia to a great carbon expense. Ash timber—as any other timber!—however, bends, cracks, it has knots and natural warps, a defined grain direction and defects from the dieback infection (Fig. 3.3). It's a complex, heterogeneous material that struggles to fit within digitally-enabled modes of production despite locally established supply chains, availability of local workforce, and urgent local environmental concerns.

We can hypothesise, then, that the following gaps can be behind the frictions we have encountered between bio-based sustainable materials, and industrial MMC approaches:

- Disciplinary: The separation of physical making from the mainstream creative and knowledge-based 'design economy'.
- Economic: The separation of local supply chains and skills from global industrial reconfigurations.
- Geographic: The separation of rurality and place-based research from digital innovation discourse and wider industrial policies.

Here we identify the work of Chantel Carr and Chris Gibson as instrumental to understand these conflating gaps [11, 16]. Their conceptualisation of "geographies if making" helps us question how specification of local materials is not only a matter of embodied carbon quantification or construction procurement challenges,

Fig. 3.3 Experimentation with local ash timber in a panel design, testing out the feasibility of local materials within digital forms of production. *Source* Carr, 2022

but a broader concatenation of social, economic, and geographic issues encompassing industrial policy and deliberate decisions to place creativity as an economic regeneration opportunity "replacing" post-manufacturing local economies, rather than connected to pre-existing place-based forms of physical making [11]. Only recently, outputs from design-driven research have attempted to establish a discussion around local supply chains and their broader economic benefit (for instance see [35]). Although rarely explored in the context of construction, this challenge area compounds with a commodification of craft as a "rural" activity taking place outside core centres of capital and labour (e.g. cities) and thus, peripheral to policies such as industrial strategies (which, in the UK, comprised a "Transforming Construction" challenge fund), levelling-up [36], and innovation funding opportunity [37].

Last, we stress how our 'research in residence' approach resonates with Carr and Gibson's work around 'slow scholarship' [15] which, although aimed at a geography audience, speaks of participatory methods and embeddedness as a way to further investigate the relationships between body and material work. We note, however, that beyond the understanding of labour our research has implemented hands-on and participatory methods to explore the frictions between low-carbon materials and digital forms of building production—thus, looking at the intersections of labour, technology, and material cultures over extended periods of time. Here we have dealt with a variety of forms of knowledge production and representation, including collectively produced creative media, digital fabrication data and technologies, material studies, prototypes (including building-size prototypes) and interviews that capture lived experiences of a wide array of participants.

3.3 Case Studies

3.3.1 Automation in Earthen Construction

Explorations into the use of automation and digital manufacturing technologies for earthen construction, such as 3D printing, began around 2011 but initially remained at the prototypical stage due to a lack of definitive knowledge on the correlation between materials, machines, and craftspeople [38]. The first significant attempt to holistically examine the feasibility of incorporating automation technologies into traditional earth construction emerged in 2018 through an EPSRC-funded collaborative project between Cardiff University and the University of Plymouth. This project explored the potential of combining cob, a low-cost material typical from vernacular constructions in the south-west of England, with robotic 3D printing to support the local building industry. By integrating traditional construction techniques with emerging technologies, the research aimed to establish a new ethos that embraces vernacular knowledge and local skills for digital innovation. The development of a robotic 3D printing cob framework followed a standard innovation process, progressing from concept to proof of concept and prototyping. The project had three primary goals:

1. Developing a descriptive framework and situating robotic fabrication within a theoretical view of tectonics and craft practice.
2. Establishing empirical knowledge on material performance aspects and technical correlations within the man–machine workflow.
3. Involving prospective community members, represented by students and professionals, in hands-on explorations of digital earth construction to create a transitional roadmap from scientific explorations to practical application.

A discussion on the aforementioned multidisciplinary goals was further developed in [13], which introduced a framework integrating nuanced craft elements and extended Pye and Stein's definitions to include the issues of 'risk' and 'certainty' involved in the negotiation between material behaviour, technology, and traditional forms of making—a negotiated space of architectural production tensioned by the frictions between localised forms of production, and digital innovation. This framework highlighted the need for multidisciplinary collaboration and suggested further research in knowledge representation, robotic material culture, and human–robot collaboration. To explore the practical and industrial aspects of 3D-printed cob, several studies investigated its workability, design, material processing, structural and thermal performance, and life cycle analysis (LCA) which included elements of material innovation and characterisation, such as the formulation of bespoke cob mixes for 3D printing extrusion. Discussions with local experts and systematic testing, led to a revised cob mixture with increased moisture content, which facilitated 3D printing through overcoming issues like slow extrusion rates and inconsistency. This learning process concluded into developing of a unique extrusion system capable

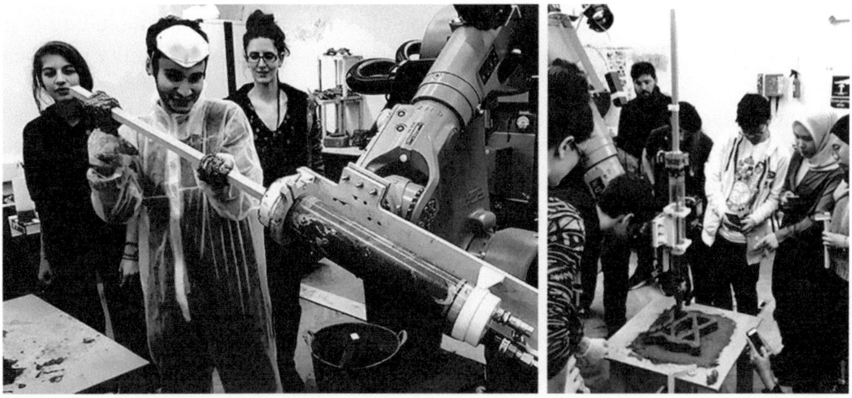

Fig. 3.4 Hands-on workshops at Cardiff University on robotic 3D printing of cob in 2017 (left) and 2018 (right). *Source* Veliz Reyes et al. 2019, Gomaa 2018 (personal archive).

of improving printing speed and quality [39, 40]. On the other hand, performance-based studies showed improved thermal conductivity for 3D-printed cob walls [41], where the structural performance studies confirmed the feasibility of 3D-printed cob as load-bearing walls in multi-storey houses [42]. Finally, a follow-on LCA study indicated favourable environmental benefits [43] when compared to concrete.

Methodologically, each study was integrated into a series of hands-on workshops involving students and professionals at the Welsh School of Architecture, Cardiff University (Fig. 3.4). Engaging students early in their academic careers helps create a continuous roadmap from learning to research and ultimately to practical application. This hands-on experience enhances their understanding and skills while fostering innovation and fresh perspectives that drive the field forward. Overall, this work bridges the gap between traditional earth construction and contemporary digital practices, offering a practical framework for both new builds and the renovation of historic cob buildings. It emphasises the importance of interdisciplinary approaches, involving material science, design, architecture, robotics, and mechanical engineering, in modern architectural design and production. Additionally, it contributes to robotic material culture and human–robot collaboration, advocating for the continuous pursuit of innovative hybrid solutions in the evolving field of construction materials and methods.

3.3.2 Participatory Design in Bridport

This project is funded by the Arts and Humanities Research Council and focuses on investigating material futures for local housing in Bridport, including the implementation of participatory methods and digital technologies. A residential fieldwork plan has been delivered through a collection of design interventions including pop-up makerspaces, interviews and hands-on workshops with 63 community members over a year.

More notably, this project supported the creation of a local "Living Lab" (Fig. 3.5)—a disused barn in Denhay Farm which served as a space for construction cocreation, material workshops, prototyping, and installation of different local materials in the context of housing prototypes and local hemp and timber supply chains. This participatory approach not only allowed for a long-term embeddedness within the local community [17], but more profoundly with ethical forms of investigation which require participation, diversity, and the inclusion of community members often ignored by developers of digital innovations [44]. As a result, the project helped us collaboratively develop an understanding of the landscape of low-carbon, bio-based materials for construction within a localised context through place-based practices. In addition to prototyping, methods such as collaborative mapping allowed for broader discussions around material flows within the region, and how the movement of materials is a cross-sectional area of enquiry which overlaps with community concerns in areas such as:

- The socio-technical infrastructures necessary to adopt MMC with local and sustainable materials, such as skills programmes and availability of equipment.
- The identification of specific areas of intervention around housing, such as retrofit, new building development, and the compartmentalisation of existing housing units.
- The need to replace carbon-intensive supply chains by low-carbon alternatives (i.e. materials such as cement, steel, and plastic-based insulation materials).

By facilitating the cocreation of a vision for a desirable future [45], local construction industry, community members were able to benefit from design thinking and creative methods (such as the production of maps and visual media materials), engaging in more aspirational speculation not otherwise available through standard methods of public consultation such as polls. Upon completion of this research project, the Living Lab in Denhay Farm was utilised to deliver a skills programme, allowing community members to participate in the construction of the Open House, a community facility designed by Common Practice in Bridport (Fig. 3.6).

3.3.3 Digital Infrastructures for Timber Innovation

On scaling this line of research into a wider set of scenarios, this project aimed to develop the digital systems necessary to allow small and medium-sized enterprises (SMEs) to join the MMC housebuilding industry. By partnering with Open Systems Lab, we are particularly focused on the open-source building system Wikihouse, consisting of volumetric prefabricated timber blocks (Fig. 3.7). By employing a participatory co-design methodology, the project enables the active involvement of industry stakeholders from the early stages, addressing both the technical and

Fig. 3.5 Local Living Lab for housing prototyping in Denhay Farm, Bridport. *Source* [44]

Fig. 3.6 Prefabrication of building elements for the Common House in Denhay Farm, Bridport. *Source* [46]

Fig. 3.7 Assembly of Wikihouse blocks in a residential project. *Source* [48]

socio-cultural challenges associated with digital adoption in the construction sector. This has included on-site collaboration with design companies, manufacturers, and community housing organisations. The outputs of the project (described below) are being trialled with local SMEs involved in the timber and manufacturing sectors: Woodlab Studio Ltd (Devon) and Maahee's Way Ltd (Leicester). In terms of local and economic development, this project expects to allow SMEs to join a sector currently dominated by large-scale, capital intensive MMC companies that struggle to cover the UK limited housing demand. Instead of high-risk investments, a more distributed network of manufacturers would spread the delivery risks, and create employability opportunity in the SME sector—in the UK, SMEs account for 60% of employment, and over 50% of the total industry revenue [47].

In detail, the project is currently developing 2 outputs. A manufacturing control platform, which will set a series of standards and processes to set quality checks, tolerances in manufacturing and material specifications, and other environmental specifications in order to improve the final quality and accuracy of the building blocks. These will be built using the open-source database of cutting and assembly files created by Open Systems Lab, and comprise testing and evaluation of the control plan by industry partners, and the Digital Fabrication Lab at the University of Plymouth. The second output is the database architecture and development for a material passport (MP). An MP will help "track and trace" building blocks throughout their lifecycle, facilitating traceability in space, provenance, ownership, transport (e.g. embodied carbon metrics), and condition of each block, supporting the reusability of building elements over time (Fig. 3.8). By considering "buildings as material banks", material passports have a limited implementation in the UK and recent initiatives are developing scoping activity in partnership with industry stakeholders [49]. An ongoing

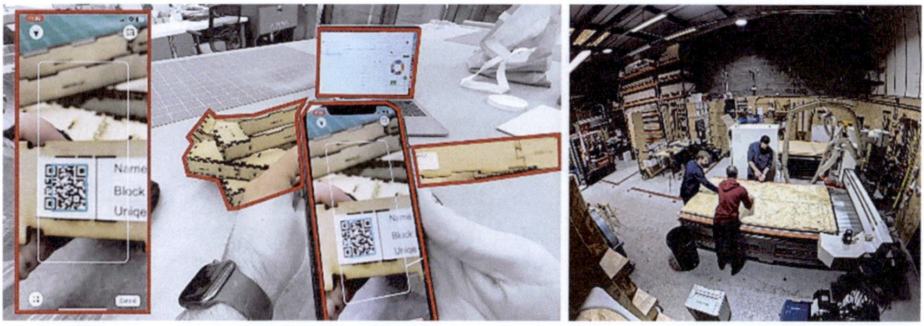

Fig. 3.8 Prototyping a material passport for volumetric timber components, including a user interface (left) and on-site codesign activity with industry partners (right). *Source* Olmez, 2024 (personal archive)

scoping review of MPs by our research team is to be published separately, and has reviewed a database of publications reporting on MPs prototypes and developments across different construction scenarios, regulatory contexts, and city-regional implementation case studies. Relevantly for this Chapter, the review has additionally identified community benefits of MPs, by supporting housing groups on making decisions about their own building and material stock, putting data about their material banks in the hands of local communities and decision-makers (for example, see [50]) and promoting the creation of new professional and data management employment opportunity [51, 52]. In summary, the review has identified a series of areas of research needed to facilitate their implementation in industry, namely:

1. Implementation and scalability.
2. Systems and databases' integration and coexistence.
3. MPs data architecture and development.
4. Socio-cultural industry barriers for adoption.
5. Trust for implementation.
6. User support and expertise to implement MPs.
7. Policy and legal frameworks for MP implementation.

3.3.4 MMC Adoption in a Rural Housing Programme

Last, this ongoing project is a pilot internationalisation initiative, in partnership with the award-winning Chile's Rural Housing Programme by the Ministry of Housing and Urbanism, the Chile Economic Development Agency, and the local Universidad Autonoma de Temuco and their research team expertise in local bio-based materials [53–55] and Wikihouse [56].

This project will prototype a small-scale innovation scenario of MMC adoption within a local rural ecosystem in southern Chile to help us translate our findings into an international policy context. This will include, for instance, the identification of needs for inter-stakeholder communication, and a rethinking of the role of innovation within industrial partners involved in the delivery of rural housing construction (an environment with very limited on-site "quality checks" due to accessibility challenges in Chilean rural settings). Relevantly for our research trajectory, the project will take place in Temuco—an area of the country with high deprivation rates, environmental issues originated in problematic forestry practices and low-quality housing, and conflict around land and land ownership with local indigenous mapuche communities [57]. These frame the environmental and social aspects of this project in similar ways our work has done in England, including references to sustainable forestry, regenerative landscapes, and social issues underpinning the decline of rural communities.

As the project develops, we are working directly with local innovation stakeholders on MMC models being used as probes to further identify challenges, knowledge gaps and design opportunities for MMC adoption in the local Temuco ecosystem. These are expected to be aligned to the goals of the Rural Housing Programme [58] which provides funding for either additional rooms and improvements in existing properties (such as refurbishment for ageing or disabled people, or building performance improvements), or the creation of new productive spaces in existing constructions (e.g. storage, animal shelter, crafts, cheesemaking spaces, etc.).

3.4 Conclusion

This Chapter outlines a research trajectory that explores the intersections between bio-based materials and MMCs. The ultimate goal is not necessarily one of performance optimisation, but social good—how the material flows and methods we use can contribute to broader societal benefits and address issues of digital inequity and marginalisation in different social and geographic contexts. These projects are eminently multi-disciplinary and composed of multiple partnership arrangements, comprising government bodies, community organisations, technology companies, manufacturers, or other academic institutions.

Although not the emphasis of this Chapter, we additionally note the wide disciplinary basis of this line of research. This helps tackle the disciplinary challenges identified in our theoretical framework, allowing for multiple viewpoints, types of knowledge, and their representations, to influence the elicitation of results through both quantitative and qualitative data. This includes, so far, organisations and researchers in our network are involved in policy, architecture, computer science, construction, product design, community work, digital arts, and heritage studies. By doing so, we propose a wider multi-disciplinary approach to the understanding of productive and material supply chains for sustainable construction which require consideration of economic, social and environmental forces influencing these new, physical and digital geographies of making. Here, the framework set out by Carr, Gibson, and other coauthors frames our enquiry by providing a conceptual lens into geographies of making. Although their work arguably sits within an economic geography remit, we do note the transferability of their findings into our research—with some differences that we have noted in the theoretical framework section.

Last, we note that this is a selection of major projects, however our research is currently engaged in a series of initiatives that test the growth of MMC opportunities in communities suffering from digital inequity, including an ongoing proposal to test out MMC adoption by community housing organisations in a urban context, and international partnership-building activity in Turkiye.

Acknowledgements This chapter reports on the research projects "Computing craft" (Cardiff University and University of Plymouth, EPSRC Connected Everything Network Plus: Industrial Systems in the Digital Age, grant ref P/P001246/1), "Building with/for Bridport" (Wessex Community Assets and University of Plymouth, AHRC Design Exchange Partnerships Round 2, grant ref AH/Y00003X/1), "One Hundred Factories: Unlocking the Collective Capacity of local timber MMC manufacturers" (Open Systems Lab, Woodlab Ltd, Maahee's Way Ltd, and University of Plymouth, Forestry Commission Timber in Construction Innovation Fund, grant ref TIC_19), and "Modern methods of construction in a world-leading rural housing programme: An implementation case study in Temuco, Chile" (University of Plymouth, AHRC Impact Acceleration Account Fellowship 2024–25).

References

1. R. Lane-Poole, A medieval cordage account. Mariner's Mirror **43**, 1 (1956)
2. H.T. Wood, Industrial England in 1754. J. R. Soc. Arts **58**, 2996 (1910)
3. English Heritage, *Bridport and West Bay: The Buildings of the Flax and Hemp Industry* (English Heritage, Swindon, 2006)
4. M. Morris, Naval cordage procurement in early modern England. Int. J. Marit. Hist. **11**, 1 (1999)
5. P. Thomson, Bridport harbour: the rise and decline of a coastal port. Mariner's Mirror **99**, 3 (2013)
6. S. Asthana, A. Gibson, Averting a public health crisis in England's coastal communities: a call for public health research and policy. J. Publ. Health **44**, 3 (2021)
7. Bridport Town Council. *Bridport area census profile—population and age profile*, June 2024. Available from: https://www.bridport-tc.gov.uk/bridport-area-census-profile-population-and-age-profile/
8. S. Kordel, Selling ruralities: how tourist entrepreneurs commodify traditional and alternative ways of conceiving the countryside. Rural. Soc. **25**, 3 (2016)
9. D.A. McGranahan, T.R. Wojan, D.M. Lambert, The rural growth trifecta: outdoor amenities, creative class and entrepreneurial context. J. Econ. Geogr. **11**, 3 (2011)
10. B. Slee, R. Lukesch, E. Ravazzoli, Social innovation: the promise and the reality in marginalised rural areas in Europe. World **3** (2022)
11. C. Gibson, C. Carr, A. Warren, Making things: Beyond the binary of manufacturing and creativity, in *The Routledge Companion to the Cultural Industries*. ed. by K. Oakley, J. O'Connor (Routledge, Abingdon, 2015), pp. 86–96
12. J.K. Gibson-Graham, Diverse economies: performative practices for 'other worlds.' Prog. Hum. Geogr. **32**, 5 (2008)
13. A. Veliz Reyes, et al., *Negotiated matter: a robotic exploration of craft-driven innovation.* Arch. Sci. Rev. **62**, 5 (Special Issue "Means, Methods, Machines and Making in Architecture") (2019)

14. P. Loh, J. Burry, M. Wagenfeld, Reconsidering Pye's theory of making through digital craft practice: a theoretical framework towards continuous designing. Craft Res. **7**, 2 (2016)
15. C. Carr, C. Gibson, Animating geographies of making: embodied slow scholarship for participant-researchers of maker cultures and material work. Geogr. Compass **11**, 6 (2017)
16. C. Carr, C. Gibson, Geographies of making: Rethinking materials and skills for volatile futures. Prog. Hum. Geogr. **40**, 3 (2016)
17. A. Veliz Reyes, A. Carr, T. Crabtree, *Co-making and imagining community-built housing*, in *The De Gruyter Handbook of Automated Futures*, ed. by V. Fors, M. Berg, M. Brodersen (De Gruyter, Berlin, Boston, 2024), pp. 339–358
18. H. Goulden, E. Morrison, *An Equitable Future for Research and Innovation: Building Sustained Community Involvement in Knowledge Production* (The Young Foundation's Institute for Community Studies, Commissioned by UK Research and Innovation, 2022).
19. M. Marshall, Researchers-in-Residence: a solution to the challenge of evidence-informed improvement? Prim. Health Care Res. Dev. **15**, 4 (2014)
20. F. Gradinger et al., Reflections on the researcher-in-residence model co-producing knowledge for action in an integrated care organisation: a mixed methods case study using an impact survey and field notes. Evid. Policy **15**, 2 (2019)
21. A. Veliz Reyes, et al., Codesign principles for the effective development of digital heritage extended reality systems for health interventions in rural communities. CoDesign (2024)
22. A. Veliz Reyes, A multimodal study of augmented reality in the architectural design studio. Int. J. Technol. Des. Educ. (2024)
23. K.S. Willis, Making a 'place' for ICTs in rural communities: The role of village halls in digital inclusion, in *C&T '19: 9th International Conference on Communities & Technologies—Transforming Communities*. ed. by H. Tellioglu, M. Teli, L. Nathan (ACM, Vienna, Austria, 2019), pp. 136–142
24. R. Burke, A. Veliz Reyes, Socio-spatial relationships in design of residential care homes for people living with dementia diagnoses: a grounded theory approach. Arch. Sci. Rev. **66** (Special Issue: Design for Dementia) (2021)
25. M. Mazzucato, Mission-oriented innovation policies: challenges and opportunities. Ind. Corp. Chang. **27**, 5 (2018)
26. M. Mazzucato, *Mission Economy: A Moonshot Guide to Changing Capitalism*, 2nd edn. (Penguin, 2022)
27. N. Coe, C. Gibson, Progress in economic geography? Decarbonising global production networks (GPNs). Progr. Econ. Geogr. **1**, 1 (2023)
28. S. Barrientos, Regional value chains in the global South: governance implications for producers and workers? Camb. J. Reg. Econ. Soc. **15**, 2 (2022)
29. H. Hawkins, L. Price, Towards the geographies of making: an introduction, in *Geographies of Making, Craft and Creativity*, ed. by L. Price, H. Hawkins (Routledge, London, 2018)
30. N. Smith, *Uneven Development: Nature, Capital, and the Production of Space*, 3rd edn. (The University of Georgia Press, Athens, Georgia, 2008)
31. C. Carr, et al., Work, value and space: three key questions of making for the anthropocene, in *Geographies of Making, Craft and Creativity*, ed. by L. Price, H. Hawkins (Routledge, London, 2018)
32. Ž Turk, Structured analysis of ICT adoption in the European construction industry. Int. J. Constr. Manag. **23**, 5 (2023)
33. UK Parliament. *MMC Sector May Continue to Struggle without a Fresh Approach from the Government*, June 2024. Available from: https://committees.parliament.uk/committee/518/built-environment-committee/news/199612/mmc-sector-may-continue-to-struggle-without-a-fresh-approach-from-the-government/
34. T.L.R. Coker et al., Estimating mortality rates of European ash (*Fraxinus excelsior*) under the ash dieback (*Hymenoscyphus fraxineus*) epidemic. Plants People Planet **1**, 1 (2018)
35. S. Islam, et al., *Circular Biobased Construction in the North East and Yorkshire* (Material Cultures, ARUP, 2021), p. 175

36. M. Fransham, et al., Level best? The levelling up agenda and UK regional inequality. Reg. Stud. **57**, 11 (2023)
37. D. Bell, M. Jayne, The creative countryside: policy and practice in the UK rural cultural economy. J. Rural. Stud. **26**, 3 (2010)
38. M. Gomaa, et al., Digital manufacturing for earth construction: a critical review. J. Clean. Prod. **338** (2022)
39. A. Veliz Reyes, et al., Computing craft: early development of a robotically-supported cob 3D printing system, in *36th International Conference on Education and Research on Computer-Aided Architectural Design in Europe (eCAADe) 2018: Computing for a Better Tomorrow*, ed. by A. Kępczyńska-Walczak (Lodz, Poland, 2018)
40. M. Gomaa, et al., 3D printing system for earth-based construction: case study of cob. Automat. Constr. **124** (2021)
41. M. Gomaa et al., Thermal performance of 3D printed cob. Archit. Sci. Rev. **62**, 3 (2019)
42. M. Gomaa, et al., Feasibility of 3DP cob walls under compression loads in low-rise construction. Constr. Build. Mater. **301** (2021)
43. H. Alhumayani, et al., Environmental assessment of large-scale 3D printing in construction: a comparative study between cob and concrete. J. Clean. Prod. **270** (2020)
44. A. Carr, A. Veliz Reyes, Codesigning material futures: a case study of housing codesign in a rural community, in *Careful Collaborations: Ethics and Care in Cultural Knowledge Exchange and Trans-Disciplinary Research*, ed. by S. Leighton, E. Barrett (National Centre for Academic and Cultural Exchange, 2023)
45. M. Mejía et al., From researching to making futures: a design mindset for transdisciplinary collaboration. Interdisc. Sci. Rev. **48**, 1 (2022)
46. Common Practice, *Company Website*, July 2024. Available from: https://commonpractice.io/
47. L. Turner, *The Role of SMEs in the UK Construction Industry*, July 2024. Available from: https://ww3.rics.org/uk/en/journals/construction-journal/the-role-of-smes-in-the-uk-construction-industry.html
48. Open Systems Lab, *Company Website*, July 2024. Available from: https://www.opensystemslab.io/
49. A. Rute Costa, R. Hoolahan, *Material Passports: Accelerating Material Reuse in Construction* (Lancaster University and Orms Architects Ltd., London, 2024)
50. E. Goselink, Construction material passports and databanks—circular approach for neighbourhood renovation, in *Cityloops* (Cityloops, 2021), p. 66
51. M. Honic, I. Kovacic, H. Rechberger, Improving the recycling potential of buildings through material passports (MP): an Austrian case study. J. Clean. Prod. **217** (2019)
52. M. Honic, et al. Data- and stakeholder management framework for the implementation of BIM-based material passports. J. Build. Eng. **23** (2019)
53. C. Rojas, et al., Thermal insulation materials based on agricultural residual wheat straw and corn husk biomass, for application in sustainable buildings. Sustain. Mater. Technol. **20** (2019)
54. C. Rojas-Herrera, et al., Characterization and utilization of sawdust waste generated from advanced manufacture for its application as a thermal insulation in sustainable buildings using the blowing technique. J. Build. Eng. **88** (2024)
55. M. Valenzuela, et al., Towards the development of performance-efficient compressed earth blocks from industrial and agro-industrial by-products. Renew. Sustain. Energy Rev. **194** (2024)
56. G. Romero Quidel, et al., Assessment of modular construction system made with low environmental impact construction materials for achieving sustainable housing projects. Sustainability **15**, 10 (2023)
57. F. Rojo Mendoza, C. Salinas-Silva, V. Alvarado-Peterson, The end of indigenous territory? Projected counterurbanization in rural protected indigenous areas in Temuco, Chile. Geoforum **133** (2022)
58. UN Habitat, *Government Rural Habitability Programme Wins Award for Improving Livelihoods in Remote Chilean Communities* (2024)

Chapter 4
Are We There Yet? Progress Review of Digital Manufacturing for Earth Construction

Mohamed Gomaa, Hamidreza Malekian, Veronica Soebarto, and Alejandro Veliz Reyes

Abstract The construction industry is currently transforming under the dual imperatives of digitisation and decarbonisation. Digital Manufacturing of Earth Construction (DMEC) is emerging as a sustainable alternative to traditional building methods, offering solutions to environmental challenges while improving efficiency. This chapter provides a progress review of DMEC from 2021 to 2024, focusing on three key domains: manufacturing technologies, material and performance aspects, and research reliability and visibility. Serving as a "Checkpoint 2.0," this chapter builds on previous review by authors, aiming to consolidate critical information to support the broader adoption of digital earth construction techniques. The review highlights significant advancements, such as the development of hybrid systems and enhanced earthen mixtures. It also examines the increasing interest from industry and government stakeholders in scaling up DMEC for industrial applications. The digital earth construction framework has evolved from initial proofs-of-concept to full-scale prototypes, marking a significant step toward commercial viability.

M. Gomaa (✉)
School of Engineering, Swinburne University of Technology, Melbourne, Australia
e-mail: mgomaa@swin.edu.au

M. Gomaa · H. Malekian
School of Design and Architecture, Swinburne University of Technology, Melbourne, Australia
e-mail: hmalekian@swin.edu.au

V. Soebarto
School of Architecture and Built Environment, Adelaide University, Adelaide, Australia
e-mail: veronica.soebarto@adelaide.edu.au

A. V. Reyes
School of Art, Design and Architecture, University of Plymouth, Plymouth, UK
e-mail: alejandro.velizreyes@plymouth.ac.uk

D. A. H. Hanaor (ed.), *Sustainable Materials for the Built Environment*, Springer Series in Materials Science 352, https://doi.org/10.1007/978-3-031-97818-0_4

4.1 Introduction

The construction industry is undergoing a profound transformation driven by two interrelated and critical trends: digitisation and decarbonisation. Digitisation involves the integration of advanced digital technologies, such as Building Information Modelling (BIM), 3D printing, robotics, and the Internet of Things (IoT), to enhance efficiency, accuracy, and productivity throughout construction processes. These technologies enable better design, real-time data collection, improved project management, and more precise construction techniques. On the other hand, decarbonisation focuses on reducing the carbon footprint of construction activities, aiming to mitigate their environmental impact by adopting low-carbon materials, improving energy efficiency, and implementing sustainable design practices [1].

Together, these trends are paving the way for a more efficient, sustainable, and environmentally responsible construction industry. As the pursuit of fully automated construction processes gains momentum, companies and institutions around the globe are rapidly advancing towards constructing full-scale structures and houses using digital and robotic manufacturing methods. With growing awareness of global warming and resource depletion, researchers and stakeholders in the building and construction industries are increasingly prioritising the decarbonisation of construction processes and materials. Given that the construction industry consumes nearly 50% of the world's raw materials, it has become imperative to reconsider conventional materials and explore more sustainable alternatives. In this context, earth-based materials are strongly re-emerging as viable substitutes for cement-based counterparts, offering affordable and eco-friendly solutions for various sectors, including residential projects [2, 3].

Over the past decade, research on automating earth construction has made significant strides. Early prototypes and projects have demonstrated the great potential of earth construction as a modern and renewed method, providing numerous environmental and economic benefits. However, the transition from prototype stages to industrial-scale applications has been slower than anticipated due to challenges on multiple fronts. Additionally, despite the growing number of studies, research efforts around the world have remained fragmented, with a considerable lack of communication and collaboration between teams [4, 5].

The need for a checkpoint to review, evaluate, and reflect on the progress in this field has become apparent. In 2022, Gomaa et al. [6] published a critical review that served as the first comprehensive assessment of the major advancements in Digital Manufacturing of Earth Construction (DMEC), focusing on its potential and the challenges of industrial-scale adoption. This study, referred to as "Checkpoint 1.0," revealed that while the majority of research between 2011 and 2021 focused on exploring the geometrical and design aspects of DMEC, there was a lack of rigorous information on constructability and performance aspects, such as manufacturing systems, material standards, and structural and thermal performance. This deficiency in replicable data has hindered the construction industry's adoption of new techniques and delayed the approval of these methods by regulatory authorities, making the goal of industrialising DMEC more challenging to achieve.

Despite these challenges, the study reported a noticeable increase in the adoption rate of DMEC between 2018 and 2021, forecasting that future efforts would evolve to address the current challenges. The analysis of reported projects and research identified three key domains requiring further improvements: manufacturing technologies, performance aspects, and research quality and visibility. This chapter, serving as "Checkpoint 2.0," aims to provide an updated review of the progress in the digital earth construction field since 2021, building on the foundation laid by the previous study. The chapter explores trending research directions and improvements in the aforementioned domains, ultimately enhancing the future utilisation of digital techniques in earth construction.

4.2 Theoretical Framework

This chapter serves as an extension to the critical review article by Gomaa et al. [6] in 2022, which established selection and classification criteria for studies on Digital Manufacturing of Earth Construction (DMEC). Following the same criteria, this review is based on literature primarily sourced from ScienceDirect and Google Scholar, with priority given to officially published journal articles and conference proceedings. The inclusion criteria for this review are as follows:

- Published between 2022 and 2024
- Focus on digitally manufactured earth constructions (DMEC)
- Emphasis on work leading to an actual or near-building scale
- Inclusion of only soil/earth-based materials
- Preference for experimental work over theoretical approaches.

Search terms included combinations related to digital fabrication and earth construction, such as "digital earth construction" and "additive manufacturing of earth." The search yielded 43 relevant works that met the inclusion criteria, comprising 12 studies from 2022, 17 studies from 2023, and 14 studies up to August 2024 (Fig. 4.1). It is important to note that this book is expected to include approximately five chapters on digitising earth construction; however, these are not included in the 2024 count. Given the field's rapid development and growing academic interest, it is anticipated that the number of publications will have increased further by the time of this book's release.

The aim of this chapter is not to provide an exhaustive review, but rather to highlight key developments in the recent literature that align with the forecasted target domains identified in the previous critical review. It will offer an overview of the current trends, focusing on how close we are to achieving commercial scale adoption. Therefore, this chapter will be structured around the progress made in the target domains of improvement, with a particular emphasis on the challenges, potential, and future directions.

Fig. 4.1 Number of recorded studies on digital earth construction between 2011 and 2021 (blue), and recorded studies between 2022 and 2024 (red)

4.3 Progress Precedents

Since 2022, the literature has shown significant progress across all target domains of Digital Manufacturing of Earth Construction (DMEC). There is a growing adoption of DMEC, indicating that ongoing efforts are leading to improved solutions for commercial applications. This new analysis of recent DMEC studies reveals the following progress ratios per each target domain:

- Manufacturing technologies: 35%
- Performance aspects: 50%
- Research quality and visibility: 15%.

The ratio of the last domain, research quality, has been assessed based on the number of review studies that have systematically collected and evaluated other work on DMEC, as well as the number of newly launched initiatives and industry-led projects focused on digitising earth construction. From the authors' perspective, this distribution indicates a healthy and balanced approach, suggesting steady progress towards the commercialisation of DMEC in the construction industry, potentially within the next 3–5 years, depending on the location. This optimism is further reinforced by the growing number of newly funded projects and initiatives from industrial stakeholders worldwide, with project timelines ranging from 18 months to 5 years.

Each of the following three sections is designed to cover one target domain. Each section will include a brief overview of the pre-2021 state from the relevant domain in the previous review study [6], which serves as "Checkpoint 1.0.". This will be followed by highlights from selected studies conducted after 2021, serving as the corresponding "Checkpoint 2.0." This structure is intended to clearly illustrate the progress made in each domain over time. Studies are presented in a chronological order.

4.3.1 Manufacturing Technologies Domain

The advancement of digital tools and machinery is pivotal for integrating earth materials into digital construction processes. Digital construction frameworks for earth materials primarily rely on manufacturing systems that combine the moving device (e.g., robotic arm) and the material deposition device (e.g., extruder). Up until 2021, most research in DMEC concentrated on additive manufacturing (AM), with other techniques like subtractive, formative, or hybrid methods remaining relatively unexplored. The majority of studies focused on wet earth methods (e.g., cob) due to their favourable compatibility with 3D printing processes, while rammed earth and other earthen techniques remained underexplored. Although notable advancements were made in 3D printing technologies for earth, such as those developed by producers like WASP and 3D Potter, challenges persisted, particularly in integrating fibrous earth mixtures into feeding systems. Similarly, free-formed rammed earth components continued to heavily rely on manual processes, thus not qualifying as fully automated.

Since 2022, there have been significant advancements in manufacturing systems for both wet and dry-based earth construction. New hybrid systems have been developed to deliver stronger earthen elements, and more studies have been conducted on automating rammed earth, wattle and daub, and sand. In addition, innovative approaches for implementing reinforcement and reusable light formwork have been introduced. For example, two studies in 2023 and 2024 [7, 8] explored the use of willow, a fast-renewable and locally abundant material, in novel circular material systems for construction. They developed a custom machine that expertly weaves willow fibres into a continuous macro-fibre, which is then threaded through a bed of wooden poles to create cellular voids filled with earth via pneumatic extrusion. This innovative integration of material, structure, and form sets the stage for a new era of sustainable construction. The "InterTwig" structure, a 4-m-tall freestanding demonstrator built using this system, showcases the potential of combining willow and earth in a seamless, eco-friendly construction process (Fig. 4.2).

Similarly, Jauk et al. [9] explored the enhancement of 3D-printed clay objects through filament reinforcement, which significantly improves mechanical properties. The study involved the continuous reinforcement of clay during the printing process, resulting in stronger and more versatile structures. This approach offers a viable

Fig. 4.2 Left: full-scale demonstrator, right: computational structural analysis model [7]

Fig. 4.3 Additive manufacturing of ceramics using reinforced filaments [9]

solution for creating robust architectural and construction applications, with future work focusing on optimising reinforcement materials and techniques for large-scale implementations (Fig. 4.3).

On the other hand, Zhu et al. [10] developed an automated robotic system for constructing rammed earth walls with precision and customisable designs. Their methodology integrates digital design with robotic fabrication, using a parametric model to generate graphic patterns based on landscape painting concepts. These patterns are then converted into execution codes for a KUKA robotic arm, which extrudes earthen material in specific colour proportions. The study highlights the potential of digital fabrication in bridging the gap between design and construction, reducing the loss of design intent during the translation to physical structures. Further research is needed to evaluate the system's performance in large-scale projects and its impact on the structural properties of the constructed walls (Fig. 4.4).

Abdelaal et al. [11] presented another innovative approach that combines modular design with digital fabrication to enhance rammed earth construction. By using modular rammed earth (RE) tiles with intricate curvatures, their method allows for mass customisation and production, facilitating the creation of continuous free-form RE walls. This modular system improves efficiency, quality control, and reduces construction waste, making it a sustainable alternative to traditional rammed earth methods. The flexibility of the tiles allows architects to create diverse and visually appealing structures through simple adjustments like rotation and flipping. The study underscores the potential of modular, prefabricated RE wall sections as a scalable, eco-friendly solution for modern construction practices (Fig. 4.5).

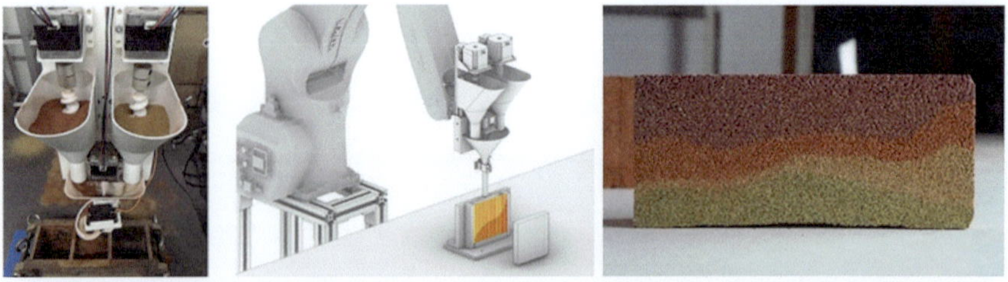

Fig. 4.4 End-effector design for earth deposition and colour mixture (left), material distribution in using the robotic fabrication technique (center), a produced prototype (right) [10]

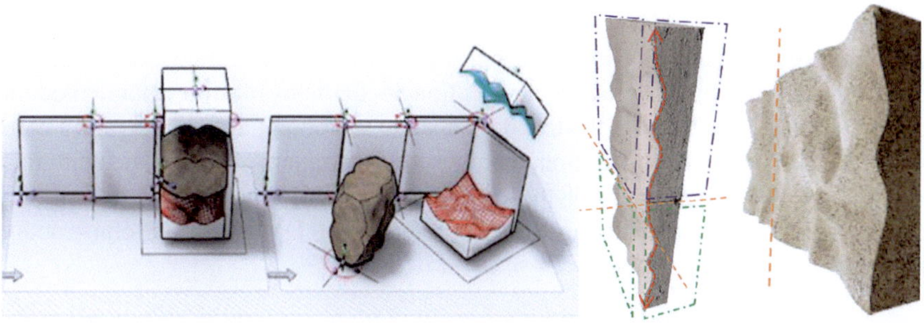

Fig. 4.5 Utilising 3-axis CNC carving to create form liners (left), produced rammed earth tiles (right) [11]

Terrene 2.0, a study by Lasting et al. [12], introduces ChitoSand, a novel composite made from sand, chitosan biopolymer, flax fibres, citric acid, and burlap fabric. This combination enhances the material's compressive and tensile properties while remaining biodegradable. The research employs reusable pneumatic formwork to support the ChitoSand composite during hardening, allowing for the creation of complex, optimised geometries using anticlastic shellular structures. ChitoSand exhibits strong mechanical properties, with improved flexural strength, making it a viable alternative to traditional construction materials. Using polyhedral graphic statics, the researchers efficiently designed complex geometries, ensuring precise control over force distribution. The use of reusable pneumatic formwork further reduced waste and guaranteed accurate construction, with the structures maintaining stability over time. ChitoSand and the associated construction methods offer a sustainable alternative to conventional practices. This research provides a comprehensive roadmap for advancing sustainable construction, highlighting its potential for broader application. The interdisciplinary collaboration in this study underscores the importance of combining expertise in materials science, engineering, and architecture to achieve sustainable development goals Fig. 4.6.

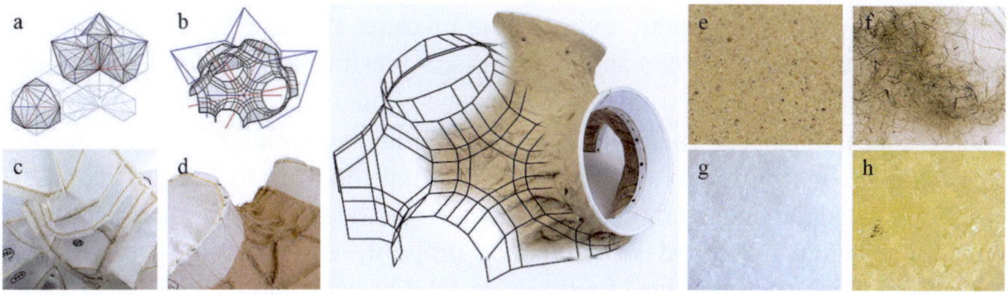

Fig. 4.6 Terrene 2.0 summary diagram. Structural design and construction method including; **a** form's polyhedral force diagram, **b** 'hybrid' geometry, **c** pneumatic formwork, **d** burlap sleeve [12]

In conclusion, the advancements in manufacturing technologies for DMEC since 2022 have significantly contributed to the field's progress. These innovations, ranging from hybrid systems and robotic fabrication to filament reinforcement, modular design, and sustainable composites, demonstrate the potential for integrating earth materials into modern construction processes on a commercial scale. The continued exploration and development of these technologies will be crucial in overcoming current challenges and realising the full potential of DMEC in the construction industry.

4.3.2 Material and Performance Domain.

Material and building performance analysis provides stakeholders with crucial information to enhance building safety and efficiency for both occupants and the environment. Architectural performance encompasses a wide range of technical domains, including structural and thermal performance [13]. Consequently, numerous studies on digital earth construction have assessed mechanical, structural, thermal, and environmental properties. Up until 2021, research into the mechanical, structural, and environmental properties of Digital Manufacturing of Earth Construction (DMEC) highlighted its potential for low-rise buildings, showcasing higher material efficiency and design flexibility [14–17]. Large-scale prototypes of earth constructions suggested that digitally manufactured earth walls could serve as load-bearing elements, with examples including the integration of timber systems into earth walls to support floors and staircases [18, 19]. However, full-scale structural testing remains essential to standardise these constructions and further exploration of earthen mixture properties is necessary.

Despite the promise shown by these early studies, many relied on small-scale prototypes or simulations, underscoring the need for more comprehensive investigations. In particular, research has called for a deeper understanding of the optimal water content in earth mixtures for 3D printing, a key design parameter that can mitigate issues like shrinkage and cracking during hardening. Since 2022, substantial research has shifted towards developing revised earth mixtures that can withstand industrial applications, utilising rigorous and near full-scale testing conditions. These studies have addressed a broad spectrum of performance-related aspects, such as mechanical strength, thermal performance, rheology, workability, and printability.

A notable contribution to this body of research is the work by Ferretti et al. [20], which advanced the understanding of 3D-printed earthen buildings' material performance. Their study examined the mechanical properties of 3D-printed wall segments made from an earthen mixture, focusing on load-bearing capacities (Fig. 4.7). By combining soil with a rice husk-lime bio-composite, the study leveraged the silica content of rice husk and the bio-cementation effects of lime, resulting in compressive strength comparable to traditional rammed earth structures. The study also highlighted that good cohesion between 3D-printed layers contributed to mechanical

Fig. 4.7 Tested 3d printed wall specimen for compressive strength [20]

strength similar to that of hand-compacted specimens, while the infill within these walls effectively prevented crack propagation, enhancing stability.

In a similar vein, Silva et al. [21] explored the use of potato starch gel and sisal fibres to stabilise earthen-based matrices for 3D printing. Their research aimed to create alternative mixtures from soil and organic materials that would improve printability and structural properties. Through extensive testing, including pumpability, extrudability, and compressive strength assessments, the study found that these natural additives significantly enhanced the workability and mechanical strength of the mixtures. The findings suggest that these eco-friendly materials could offer a sustainable alternative to traditional construction methods. Building on that, Zavaleta et al. [22] investigated the use of chitosan and sisal fibres to enhance the properties of 3D-printed earthen composites. Their research addressed the limitations of traditional earthen materials, such as poor mechanical strength and low water durability. The study's findings revealed significant improvements in extrusion quality and resistance to water erosion, highlighting the potential of natural additives in advancing sustainable construction practices.

Alqenaee and Memari [23] focused on conducting further verifications on adapting traditional cob construction techniques for modern 3D printing. Their research involved modifying cob mixture proportions to balance printability, buildability, and mechanical properties. The study provided a recommended ratio for a printable cob mixture and explored various printing techniques to optimise structural integrity. While the research highlights the potential of 3D-printed cob for small-scale construction, it also identifies challenges such as shrinkage and deformation that warrant further investigation. Further expanding the exploration of earthen mixtures, Asaf et al. [24] developed and tested twelve sand and clay mixtures to assess their suitability for 3D printing. Their findings emphasised the importance of rheological properties, particularly the thixotropic effect, in maintaining stability after printing. This study underscores the critical role of particle size distribution and clay mineralogy in optimising the performance of 3D-printed mixtures, providing valuable guidelines for future construction practices.

Rückrich et al. [25] proposed a field-oriented methodology to ensure consistent and high-quality printing results. They developed three characterisation tests to assess the flowability, cohesiveness, and deformation properties of earth mixtures. The research identified two potential indices, flowability and green strength, as standards for evaluating the printability of local earth-based mixtures. These indices could facilitate the standardisation of industrially pre-processed materials, though further refinement is necessary.

The study by Maierdan et al. [26] took a different approach by focusing on the potential of xanthan gum (XG) as a biopolymer binder to improve the rheological, mechanical, and 3D printing properties of earth-based concrete. The research demonstrates that XG can effectively mitigate viscosity, yield stress, and storage modulus over time to enhances the material's buildability for 3D printing. The study also reveals that XG significantly improves compressive strength. Despite not forming strong bonds with clay in the fresh state, XG may create a robust polymer matrix during the dehydration process, reinforcing the clay and sand particles and improving compressive strength. Ultimately, the study concludes that XG is a promising dual-function agent, acting as both a binder and a rheology modifier, making it suitable for sustainable 3DP construction applications.

The methods for maximizing the fibre Content in 3D-printed earth mixtures were also explored in a study by Carcassi et al. [27]. The research develops various earth-fibre composites using bio-based additives and agrowaste fibres such as wheat straw, hemp, banana, and kenaf. These fibres were incorporated into the mix designs at varying content levels, with the study achieving significant increases in fibre content up to 13% by weight in some cases. The key findings indicate that higher fibre content significantly improves the compressive strength of the materials, with an increase of up to 125% compared to non-fibre-reinforced mixtures. Additionally, the thermal conductivity of the fibre-reinforced mixtures was notably lower, improving the material's insulating properties. The study concludes that optimizing fibre content in 3D-printed earth materials can lead to more sustainable, high-performance construction materials, though challenges remain in balancing printability with mechanical and thermal performance.

Lastly, Schmitz et al. [28] conducted a comprehensive study to understand the interrelationship between process and material parameters in rammed earth construction, with the goal of establishing a robust process for automated robotic rammed earth. Their research provided key recommendations for optimising compaction processes, which could inform the further development of robotic systems for rammed earth construction. Overall, these recent studies have made significant advancements in the field of 3D-printed earth construction by improving material properties, enhancing structural integrity, and exploring innovative design strategies. Research has demonstrated that integrating natural additives such as rice husk-lime bio-composites, potato starch gel, and sisal fibres can significantly enhance the compressive strength, workability, and durability of earthen mixtures. Additionally, studies have identified optimal ratios and methods for 3D printing materials like cob and clay, while also developing new standards for evaluating printability and structural performance. Collectively, these efforts have brought 3D-printed earth construction closer to mainstream application, offering sustainable alternatives to traditional building methods and paving the way for future innovations in the field.

4.3.3 Research Reliability and Visibility Domain

The adoption of Digital Manufacturing of Earth Construction (DMEC) has seen a noticeable increase in recent years. However, up until 2021, these efforts were still in their infancy compared to other well-established digital construction methods like 3D concrete printing (3DCP). While many early experiments in DMEC claimed improvements in productivity, quality, and performance (thermal, structural, and environmental), they often lacked systematic assessment and verification. Most recorded studies were student projects, proofs of concept, or commercial prototypes, with few being published in peer-reviewed journals. This lack of definitive information raised concerns about the validity of the results and hindered industrial adoption and research replication. Additionally, research in DMEC was fragmented, with many groups working in isolation, leading to redundancy and slow collective progress.

Since 2022, however, significant efforts have been made to address these issues. As seen in the previous two domains, numerous studies have emerged to explore, verify, and validate the performance aspects of earthen mixtures, while also introducing improved production techniques. Additionally, several review studies have collectively analysed the potential and challenges of adopting DMEC on a larger scale [2, 29–32]. These efforts have not gone unnoticed by industry and governments. Globally, several firms and organisations are now directing their resources and attention toward scaling digital earth construction for industrial applications. Companies like WASP 3D in Italy and 3D Potter in the US have long been leaders in promoting 3D-printed earth construction. Newer entrants like Luyten 3D in Australia are also making significant strides, particularly in sustainable construction. In 2023, Luyten 3D secured a \$2.9 million AUD grant from the Australian Government's CRC-P (Cooperative Research Centres Projects [33]) to develop sustainable 3D-printed houses using local materials. Through such initiatives, Luyten 3D is helping to revolutionise the construction industry, promoting rapid, sustainable, and scalable building solutions. Building on this momentum, Swinburne University of Technology and Luyten 3D have recently secured over A\$600,000 in competitive funding from the Australian Research Council (ARC) Linkage Program and the Innovative Planet Research Institute (IPRI) in late 2024 [34]. These grants aim to comprehensively explore advanced engineering methodologies to enhance the industrial readiness of 3D-printed earth construction. This reflects a growing institutional and governmental commitment to transitioning digital earth construction from experimental prototypes to viable industrial solutions.

In the UK, the Earth Build Project (EBP) was established in 2023 to harness 3D-printing technology for scalable, decarbonised construction across Europe [35]. EBP is currently focused on adapting 3D-printing techniques to suit the UK climate and ensuring compliance with building regulations. The project has built a robust research network, secured Network Grant funding from the EPSRC, and formalised partnerships with companies like WASP and institutions such as IAAC. By bringing together top researchers in digital earth construction from the UK, Europe, and Australia, the Earth Build Project represents a significant step forward in commercialising sustainable 3D-printed construction. The Earth Building Association of Australia (EBAA), an NGO dedicated to promoting earth construction techniques since 1981, has also shown a strong interest in modernising earth construction through digital techniques [36]. Since 2022, EBAA has sponsored workshops on 3D printing of cob and supported the inclusion of digital earth construction researchers in their annual conferences. The organisation advocates for the recognition and integration of earth building techniques within the broader construction industry and regulatory frameworks in Australia.

Globally, initiatives like Mud Futures, launched in 2021 by Sushant University in India, have further highlighted the growing interest in earth-based construction. As part of the International Prize for Contemporary Architecture (IPAC) 2021–2022, Mud Futures explored the potential of mud as a key material in sustainable architecture [37]. The year-long global symposium brought together architects, researchers, and practitioners to discuss how traditional mud-building techniques can be reimagined and integrated with contemporary design and construction practices. Through workshops, lectures, and discussions, Mud Futures has inspired innovation in earth-based construction and continues to promote the importance of sustainable building practices.

4.4 Conclusion

Over the past decade, the development of the digital earth construction framework has evolved from initial proof-of-concept stages to full-scale physical prototyping, demonstrating the significant potential of earth-based construction. This progression has expanded the scope of digital manufacturing in construction beyond environmentally harmful cement-based materials, underscoring the ecological advantages of earth-based alternatives. Collaborative efforts have laid a robust foundation for advancing digital earth construction toward industrial scalability, bridging the gap between traditional earth construction methods and modern digital practices.

This chapter has provided a comprehensive review of the current state of digital manufacturing in earth construction, highlighting both the achievements and the pathways for further improvement. By examining diverse research threads, the chapter aimed to build knowledge bridges that could accelerate the adoption of digital earth construction techniques in modern architecture. Despite the promising advancements, it is evident that additive manufacturing alone will not suffice for the broader application of earth-based construction methods. There is a critical need to move beyond 3D printing prototypes designed solely for research purposes. Approaches such as dynamic molds, hybrid construction methods, and the industrialisation of machinery and processes from a market perspective are necessary to achieve practical implementation on a larger scale.

The field of digital manufacturing in earth construction is transitioning from the early to intermediate stages, marked by accumulating performance data and growing involvement from research teams, companies, and funding bodies. This increasing interest and investment are crucial for pushing the boundaries of what is possible in earth-based construction, paving the way for innovative, sustainable building practices that can be widely adopted.

References

1. B. Awuzie, A. Ngowi, D. Aghimien, Towards built environment decarbonisation: a review of the role of artificial intelligence in improving energy and materials' circularity performance. Energy Build. **319** (2024). https://doi.org/10.1016/j.enbuild.2024.114491.
2. D. Rocha, P. Faria, S.S. Lucas, Additive manufacturing of earth-based materials: a literature review on mortar composition, extrusion, and processing earth. Materials 17 (2024). https://doi.org/10.3390/ma17010202
3. M. Gomaa, S. Schade, D.W. Bao, Y.M. Xie, Automation in rammed earth construction for industry 4.0: precedent work, current progress and future prospect. J. Clean Prod. **398**, 136569 (2023). https://doi.org/10.1016/j.jclepro.2023.136569
4. Y. Gin, D.U. Shah, M.H. Ramage, An assembly-oriented novel low-carbon masonry building method with unfired 3D printed earthen blocks, in *Proceedings of the IASS Annual Symposium 2023 Integration of Design and Fabrication* (Melbourne, 2023)
5. A. Veliz Reyes, W. Jabi, M. Gomaa, A. Chatzivasileiadi, L. Ahmad, N.M. Wardhana, Negotiated matter: a robotic exploration of craft-driven innovation. Archit. Sci. Rev. **62** (2019). https://doi.org/10.1080/00038628.2019.1651688
6. M. Gomaa, W. Jabi, V. Soebarto, Y.M. Xie, Digital manufacturing for earth construction: a critical review. J. Clean. Prod. **338**, 130630 (2022). https://doi.org/10.1016/j.jclepro.2022.130630
7. G. Casalnuovo, E. Zanetti, T. Haußer, M. Dörstelmann, M.R. La, Digital structural design for natural composites a case study of willow-earth hybrid construction habits of the Anthropocene, in *Habits of the Anthropocene: Scarcity and Abundance in a Post-Material Economy*, pp. 282–292 (2023)
8. E. Zanetti, E. Olah, T. Haußer, G. Casalnuovo, R. La Magna, M. Dörstelmann, InterTwig—willow and earth composites for digital circular construction, in *Sustainable Development Goals Series*, vol. Part F2790 (Springer, 2024), pp. 491–511. https://doi.org/10.1007/978-3-031-36554-6_32
9. J. Jauk, L. Gosch, H. Vašatko, M. Königsberger, J. Schlusche, M. Stavric, Filament-reinforced 3D printing of clay. Materials **16** (2023). https://doi.org/10.3390/ma16186253.
10. G. Zhu, X. Zhou, J. Zhang, G. Liu, S. Hao, D. Luo, Automatic robotic construction for customisable rammed earth walls, in *Human-Centric, Proceedings of the 28th International Conference of the Association for Computer-Aided Architectural Design Research in Asia (CAADRIA)*, pp. 110–118 (2023)
11. A. Abdelaal, M. Gomaa, Y.M. Xie, A novel modular technique for manufacturing free-form rammed earth walls, in *Proceedings of the IASS Annual Symposium 2023 Integration of Design and Fabrication* (IASS, Mlebourne, 2023)
12. L. Lasting, M. Akbari, D. Keuchel, N.K. Lee, S. Pradeep, S. Chawla et al., Terrene 2.0: biomaterial composites design & shellular structures development for augmented earthen construction. Mater Des. **239** (2024). https://doi.org/10.1016/j.matdes.2024.112792

13. B. Kolarevic, A. Malkawi, *Performative Architecture: Beyond Instrumentality*, 1st edn. (Spon Press, New York, 2005)
14. M. Gomaa, W. Jabi, A. Veliz Reyes, V. Soebarto, 3D printing system for earth-based construction: case study of cob walls. Autom. Constr. **124**, 103577 (2021). https://doi.org/10.1016/j.autcon.2021.103577
15. M. Gomaa, J. Carfrae, S. Goodhew, W. Jabi, R.A. Veliz, Thermal performance exploration of 3D printed cob. Archit. Sci. Rev. **62**, 1–8 (2019). https://doi.org/10.1080/00038628.2019.1606776
16. M. Gomaa, J. Vaculik, V. Soebarto, M. Griffith, W. Jabi, Feasibility of 3DP cob walls under compression loads in low-rise construction. Constr. Build. Mater. **301**, 124079 (2021). https://doi.org/10.1016/j.conbuildmat.2021.124079
17. A. Perrot, D. Rangeard, E. Courteille, 3D printing of earth-based materials: processing aspects. Constr. Build. Mater. **172**, 670–676 (2018). https://doi.org/10.1016/j.conbuildmat.2018.04.017
18. A. Chiusoli, *3D Printed Earth Wall with Embedded Staircase*, pp. 2–3 (2019)
19. Y.-C. Chang, D. Fiore, F. Sevostianov, G.L. van Stirum, G. Refalian, Q. Li et al., *Digital adobe—IAAC* (Institute for Advanced Architecture of Catalonia, 2018)
20. E. Ferretti, M. Moretti, A. Chiusoli, L. Naldoni, F. De Fabritiis, M. Visonà, Mechanical properties of a 3D-printed wall segment made with an earthen mixture. Materials **15** (2022). https://doi.org/10.3390/ma15020438
21. G. Silva, R. Ñañez, D. Zavaleta, V. Burgos, S. Kim, G. Ruiz et al.: Eco-friendly additive construction: analysis of the printability of earthen-based matrices stabilized with potato starch gel and sisal fibers. Constr. Build. Mater. **347** (2022). https://doi.org/10.1016/j.conbuildmat.2022.128556
22. D. Zavaleta, R. Ñañez, G. Silva, G. Ruiz, M.A. Pando, R. Aguilar et al., Additive construction using enhanced earthen-based composites: improvement of the mechanical strength and water durability using chitosan and agave fibers. Constr. Build. Mater. **411** (2024). https://doi.org/10.1016/j.conbuildmat.2023.134159.
23. A. Alqenaee, A. Memari, Experimental study of 3D printable cob mixtures. Constr. Build. Mater. **324** (2022). https://doi.org/10.1016/j.conbuildmat.2022.126574
24. O. Asaf, A. Bentur, P. Larianovsky, A. Sprecher, From soil to printed structures: a systematic approach to designing clay-based materials for 3D printing in construction and architecture. Constr. Build. Mater. **408** (2023). https://doi.org/10.1016/j.conbuildmat.2023.133783
25. S. Rückrich, G. Agranati, Y.J. Grobman, Earth-based additive manufacturing: a field-oriented methodology for evaluating material printability. Archit. Sci. Rev. **66**, 133–143 (2023). https://doi.org/10.1080/00038628.2022.2154739
26. Y. Maierdan, D. Zhao, P.H. Chokshi, M. Garmonina, K.S. Rheology, 3D printing, and particle interactions of xanthan gum-clay binder for earth concrete. Cem. Concr. Res. **182** (2024). https://doi.org/10.1016/j.cemconres.2024.107551
27. O.B. Carcassi, Y. Maierdan, T. Akemah, S. Kawashima, L. Ben-Alon, Maximizing fiber content in 3D-printed earth materials: printability, mechanical, thermal and environmental assessments. Constr. Build. Mater. **425** (2024). https://doi.org/10.1016/j.conbuildmat.2024.135891
28. L.P. Schmitz, J. Gosslar, E. Dorresteijn, D. Lowke, H. Kloft, Experimental investigations on the compaction energy for a robotic rammed earth process. Front. Built. Environ. **10** (2024). https://doi.org/10.3389/fbuil.2024.1363804
29. K. Haddad, S. Lannon, E. Latif, Investigation of cob construction: review of mix designs, structural characteristics, and hygrothermal behaviour. J. Build. Eng. **87** (2024). https://doi.org/10.1016/j.jobe.2024.108959
30. X. Yin, C. Guo, B. Sun, H. Chen, H. Wang, A. Li, The state of the art in digital construction of clay buildings: reviews of existing practices and recommendations for future development. Buildings **13** (2023). https://doi.org/10.3390/buildings13092381
31. E. Binega Yemesegen, A.M. Memari, A review of experimental studies on cob, hempcrete, and bamboo components and the call for transition towards sustainable home building with 3D printing. Constr. Build. Mater. **399** (2023). https://doi.org/10.1016/j.conbuildmat.2023.132603

32. R. Robayo-Salazar, R. Mejía de Gutiérrez, M.A. Villaquirán-Caicedo, S. Delvasto Arjona, 3D printing with cementitious materials: challenges and opportunities for the construction sector. Autom. Constr. **146** (2023). https://doi.org/10.1016/j.autcon.2022.104693
33. CRC-P, *Successful CRC Projects Round 15* (2024)
34. Swinburne University of Technology, (2024). https://www.swinburne.edu.au/news/2024/11/close-to-1m-in-funding-for-swinburne-industry-linked-project-3d-printing-houses-and-conserving-digital-artefacts/
35. EBP, *The Earth Build Project* (2024)
36. EBAA, *Earth Building Association of Australia* (2024)
37. IPAC, *Mud-Futures* (2021)

Chapter 5
Earth-Based Materials in Contemporary Construction: Challenges and Opportunities

Giada Giuffrida and Rosa Caponetto

Abstract The growing concern regarding the worsening of environmental conditions and the energy crisis has led, in recent decades, to a strong push to rethink and renew the construction and production sector. This need has brought the advent of design approaches for sustainability that are based on the circular economy paradigm, the reduction of resources and energy consumptions, and the minimization of climate-changing emissions. Following these premises, some bio and geo-based materials such as raw earth, have received great attention by virtue of their availability, low economic and environmental cost, and good material performances. This contribution deals with framing contemporary earthen construction within the broader challenges of the transition path to a more sustainable built environment. Furthermore, the main limitations that earth construction must face today to establish itself as a valid alternative to conventional construction are exposed and commented on. Finally, the main challenges and opportunities for raw earth materials are outlined by analyzing the last advances in research and industrial production.

5.1 Introduction

Building with raw earth represents an effective and sustainable solution to many of the current challenges related to the natural and built environments, as well as society. By combining tradition and innovation, this material offers significant benefits in terms of environmental sustainability, energy efficiency, health and well-being; at the same time, using raw earth as a building material promotes social inclusion, economic accessibility and cultural heritage preservation, contributing to a greener, fairer, and more sustainable future [1].

In light of the above, raw earth construction fully addresses the goals of recent international policies aimed at creating a more sustainable and healthy future for forthcoming generations: from the 2015 Paris Agreement to the 2030 Agenda for

G. Giuffrida (✉) · R. Caponetto
Department of Civil Engineering and Architecture, University of Catania, Catania, Italy
e-mail: giada.giuffrida@unict.it

D. A. H. Hanaor (ed.), *Sustainable Materials for the Built Environment*, Springer Series in Materials Science 352, https://doi.org/10.1007/978-3-031-97818-0_5

Sustainable Development, particularly addressing the following SDGs: SDG 3 (Good Health and Well-Being), SDG 7 (Affordable and Clean Energy), SDG 11 (Sustainable Cities and Communities), SDG 12 (Responsible Consumption and Production), and SDG 13 (Climate Action).

But how exactly does raw earth construction meet these previously mentioned needs? Its environmental sustainability primarily derives from its easy availability, recyclability, low-carbon processing methods, and reduced waste production compared to other construction technologies [2]. As a renewable and abundant material, raw earth alleviates the pressure on non-renewable resources and offers a low-carbon alternative to traditional materials like concrete and steel. Its hygrothermal properties help maintain stable indoor temperatures, reducing the need for artificial heating and cooling, and thus energy consumption [3]. Additionally, raw earth construction processes are less energy-intensive, contributing to a lower overall carbon footprint of buildings [2].

From a biodiversity perspective, raw earth construction helps mitigate biodiversity loss by limiting the extraction and transportation of materials over long distances and promoting sustainable building practices that preserve natural habitats.

The quality of raw earth-built living spaces is high due to the absence of toxic substances, the material's ability to regulate temperature and humidity, provide acoustic insulation, and create a visually pleasing and healthy environment, thereby promoting the psychological well-being of occupants through its naturalness and sustainability [4].

Socially, raw earth construction enhances economic accessibility to housing due to the local availability of the material, reduced transportation costs, and the potential for self-building. This technology also stimulates community involvement by encouraging active participation in self-construction, creating opportunities for learning and collaboration. This strengthens cultural identity by valuing traditional techniques (and thus preserving cultural heritage) and promoting support for the local economy [5].

In recent years, raw earth construction has made significant progresses: the use of natural stabilizers and modern additives has improved its mechanical properties and durability. Additionally, the development of hybrid systems combining raw earth with wood, bamboo, and metal, along with the introduction of new technologies and machineries, has enhanced its performance, making it more suitable for contemporary housing needs. Technological experimentation and innovation have thus modernized raw earth architecture, resulting in aesthetically appealing buildings that integrate harmoniously with the natural environment and enhance local resources, respecting the principles of ecophilia [5].

Figure 5.1 reports a state of the art of raw earth construction, with reference to its background, its advantages, the main construction technologies, the current limits, and the main challenges. These aspects will be further discussed in the following paragraphs.

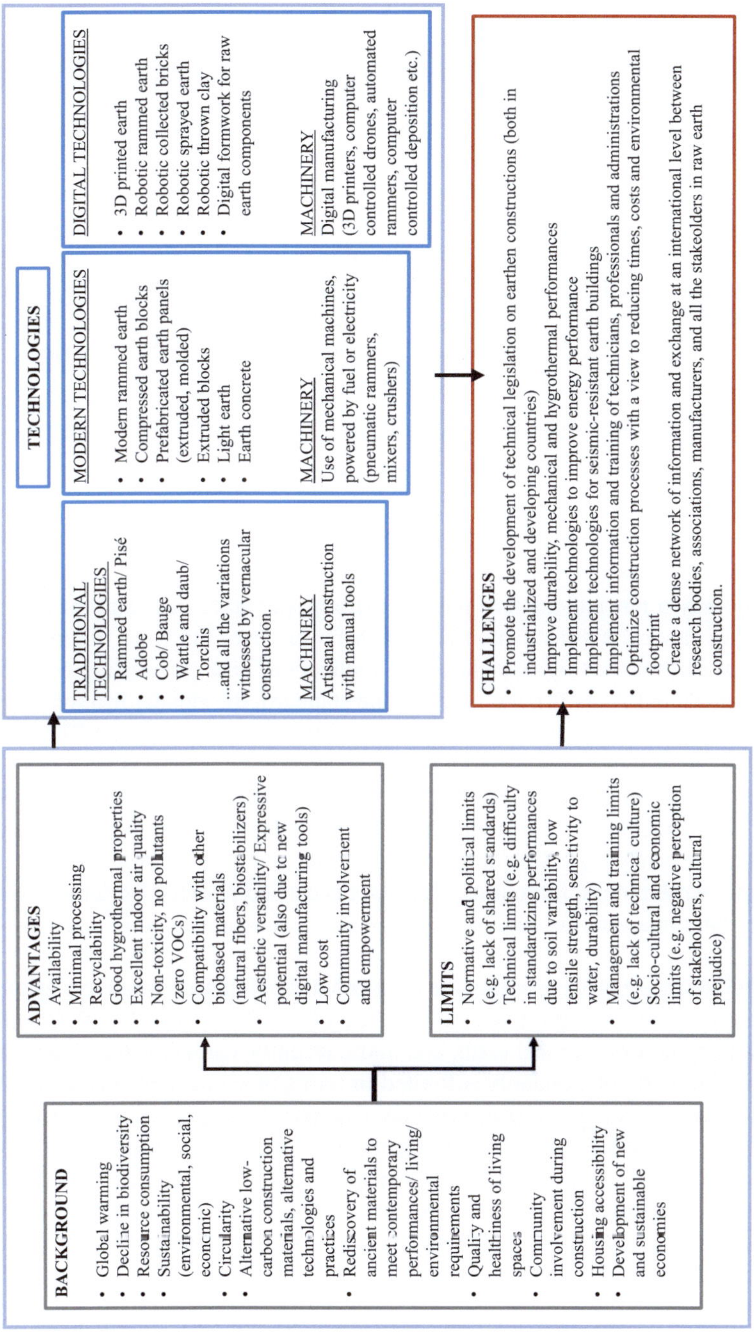

Fig. 5.1 Overview on the state-of-the-art of earth-based materials for contemporary construction

5.2 Limits

Despite the progressive increase of scientific literature and the excellent examples of earth-based architectures designed in recent years, raw earth struggles to establish itself as a material for widespread construction in industrialized countries. A number of recent studies have identified the main obstacles to using raw earth through literature surveys and interviews/questionnaires [6, 7]. The main limitations concern normative, technical, management and training aspects, but also socio-cultural and economic ones. In the following paragraphs, each of these issues will be scrutinized and analyzed, while in the next section, the latest developments and research/practice horizons that could help in solving the previous limitations will be exposed.

5.2.1 Normative and Political Limits

Lack of shared technical standards. The development and diffusion of raw earth construction are hindered by the lack of widely accepted guidelines and standards. Indeed, only a few nations are endowed with proper standards or normative documents for raw earth construction, and even fewer cover the wide variety of available raw earth technologies or fields of application. In an interesting contribution by Achenza and Agus [8], the name of the standard, the standardization body, the Nations, and the main contents of several raw earth standards are listed. The analyzed standards refer to those of Brazil, Colombia, France, Germany, India, Kenya, New Zealand, New Mexico, Peru, Spain, Sri Lanka, Tunisia, USA, and Zimbabwe. Other drafts of standards and guidelines have been produced in the meantime (e.g. the Californian standard for Cob construction, or the new version of the German Standards for earth blocks, masonries, mortars, panels and plasters). Reference is made to the main analyzed technologies and to the guidelines on soil selection procedures, product expected performances, laboratory tests, design, manufacturing and construction aspects. Concerning the analyzed technologies, it is interesting to point out that compressed earth blocks (CEB) are the subject of most of the standards, followed by adobe, rammed earth, and, more recently, earth concrete. Nonetheless, the existing body of standards is limited in terms of scope, contents, and covered technologies. Existing standards therefore risk having a limited impact both in terms of scope and status of regulatory text. On the basis of the abovementioned review [8], the authors report an analysis of the historical and political dynamics that led Italy not to have a raw earth standard and consequently propose a structure for further investigations aimed at drafting a new regulation for raw earth construction (in particular adobe) in Italy.

Lack of policies encouraging raw earth construction. Side by side to the lack of shared technical standards, it emerges the lack of national and international policies encouraging the diffusion and production of raw earth buildings and components. As we will see in the following pages, there are a few isolated cases where individual initiative has led to the creation of companies that base their assets on the production of raw earth building components. Several companies working on raw earth materials and products are present in France, Spain, Austria, Canada, Germany, Italy, United States, Australia, and New Zealand.

Only in rare cases these enterprises have been assisted in their growth by local municipalities, while the launch of their activities required the cooperation of various partners and the use of mechanisms of national and international fundings (as UE ones). For instance, for the French company *Cycle Terre*, a cooperation of several parties aimed at the creation of a waste recovery chain resulting from the soils excavated for the construction of the new *Grand Paris* transport network [9]. The cooperative includes the *Grand Paris* Society, in charge of the construction of the transport network in *Île-de-France* region, but also a local municipality –in which two new *Grand Paris* stations are being built-, several research centers, a real estate developer, an architectural firm and a training center [9].

Rather, in most cases, it appears evident that these commercial realities have developed where priorly existed a raw earth technical knowledge, usually due to a well-spread existing earthen heritage (e.g. in France and Spain) [10], or to a relatively renewed earth building practice, as the one stimulated in Germany in between the two World Wars [11]. This is for instance the case of several French, Spanish, German and Austrian companies.

Another major limitation of current earthen construction is the difficulty of intertwining with other supply chains in a circular economy perspective. In recent years, a copious scientific literature has developed on the possibilities offered by the stabilization of earth-based materials through the use of natural materials [12, 13], deriving from by-products and processing waste from agriculture [14], or from other production chains [15]. Despite the good technical performance of such alternative stabilization strategies [16], the methods and flows through which the integration between various production chains could be achieved remain to be studied. In one of the authors' latest contributions [17], the environmental convenience of stabilizing rammed earth mixtures by using natural sisal fibers and marble-cutting waste is exposed. The technical and environmental performances of these mixtures appear to be particularly competitive in relation to more conventionally stabilized mixtures (through the addition of sand and lime) [17]. At the same time, the authors highlight the difficulties in terms of production chain, mainly due to the need to activate collaborative chains for the recovery of by-products and waste materials which could be repurposed in new earth-based materials, while distributing waste management centers throughout the territories to minimize transport-related environmental impacts.

5.2.2 Technical Limits

In 2022, an interesting contribution by Fabbri et al. [18] describes the remaining challenges of the technical committee Rilem 274-TCE, by highlighting the main points that should be addressed in the future to explain the technical behavior of raw earth materials and improve it. In recent years there has been an exponential increase in scientific publications on the technical performances of earth-based materials [7], nevertheless a review of such contents is outside the scope of this brief contribution. However, it can be said that the technical limitations relate to some issues in particular, which are examined in the following.

Soil selection and characterization for raw earth construction. Professionals and practitioners would like to follow easy-to-apply rules and methodologies to assess if a soil is suitable or not for earth construction. This approach has rarely been investigated in the literature (see for instance [19]). Latest researches report that particle size distribution is not a key parameter for predicting the strength of raw earth materials, while the amount and type of present clays is a factor of major interest [18].

Hygrothermal properties and methodologies for their assessment. Hygrothermal properties of earth-based materials have a direct influence on the energy performance of buildings and the comfort of inhabitants [20]. That is the reason why several contributions have focused on the assessment of the parameters governing heat transport and storage, moisture storage, vapor and liquid transport (see [18] for details) in raw earth materials and components. As reported by Fabbri et al. [18], even if earth-based materials are intrinsically variable, the main issue is given by the lack of standardized methodologies for the assessment of these parameters, which causes a dramatic dispersion of their assessed performances.

Mechanical properties and structural behavior. Similarly to the previous point, there are no shared methodologies for the assessment of static and dynamic mechanical properties of earth-based materials, and consequently, the structural behavior of earth buildings, especially in seismic-prone zones, needs to be further studied to avoid any risks related to the safety of earthen structures. In recent years, also thanks to the issuing of some standards for raw earth construction in seismic areas [21], there has been an increasing awareness of the need for structural reinforcements for raw earth buildings. On the other hand, if it is true that raw earth is a material with a low tensile strength—even lower than the one of concrete-, it seems obvious that it should be reinforced to increase its structural performance. The experimental evidence (discussed by the authors in another contribution [22]) has demonstrated that some structural reinforcement strategies (such as those based on the use of timber reinforcements) are more suitable to be combined with raw earth massive walls. On the other hand, lightweight earth structures (such as wattle and daub ones) appear to be more suitable for earthen construction in seismic areas [11].

Reliability. Closely related to the previous limitation is the reliability one. Unlikely other construction materials, raw earth quality control during the construction process is usually problematic [23, 24]. There is a need to develop easy-to-apply methodologies that establish unified criteria for production, for the monitoring of in-situ drying, shrinkage-related issues, and quality control over time [18, 25].

Durability. A common criticism raised against earthen construction is its low durability over time. In 2020, Beckett et al. [26] carried out a review on the types of exposure and risks to which earthen buildings are subject and reported a series of durability tests on earth-based materials. These tests seek to assess the material's resistance to direct, indirect or dripping rain, pooling water, wetting and drying cycles, and freeze–thaw cycles. These tests, although standardized (see for instance the capillary water absorption test, the accelerated erosion test, the Geelong test or the wire brush test) have been relatively little performed in the literature and rarely performed on full-scale raw earth elements. Furthermore, they are difficult to carry out on unstabilized raw earth samples (i.e. not stabilized with chemical binders).

Health and hygiene quality and control over raw materials. It is necessary to verify if the soil extraction site has hosted polluting activities. If the soil is to be mixed with others extracted elsewhere, this certainly becomes more complex. In this sense, some companies working with raw earth have a very stringent analysis protocol, which aims to identify any polluting agents present in the masses of soils arriving at the plant, for each load of material received. Some authors [6] have stated that it would be more appropriate to work with materials excavated on the construction site to more easily reconstruct their history and lower the environmental impacts related to transport. However, this is not always possible, due to the possible non-adaptability to construction purposes of the soil present on site or its insufficient quantity. The claimed healthiness and eco-compatibility of raw earth materials could be a way to impose themself on more conventional building materials, but this requires further tests that investigate the relationship between the raw earth elements and the absorption capacity of bacteria or fungi [27], aldehydes, volatile organic compounds (VOCs) and fine dusts. Furthermore, the healthiness assessment of raw materials (excavated soil) and secondary raw materials (which, as we already said, can be used for the stabilization of earth-based materials) should always be verified in the pre-production phase.

5.2.3 *Management and Training Limits*

Raw earth construction still has to overcome a series of management limits, which postulate the improvement of planning, design, construction, use and end-of-life phases of earth buildings.

Production and integration with other supply chains. Concerning soil stabilization for building purposes, we already discussed the potential of using waste, by-products and secondary raw materials (obtained from other production chains). In this sense, it is necessary to identify clear paths for the reuse procedures of such materials, encouraging industrial symbiosis.

Construction process. Raw earth construction processes are often too artisanal and unsystematized. As already pointed out in the past [6], there is a problem of cooperation between parties, of exchange and reliability of information, of accountability, and eventually of safety. The issue of the exchange of information and communication concerns the various stakeholders.

Services maintenance over time. Unlike other construction systems, the study of electrical and plumbing systems is, for raw earth construction, fundamental and must be carried out in a fully integrated manner at the earliest design phase. As known, the damaging of a water pipe can dramatically lower the mechanical resistance of earth masonry. Furthermore, as Morel et al. [6] point out, sometimes fewer finishing works are used in earthen buildings, leaving the structural walls exposed. This necessarily postulates the need for greater attention during the design and implementation phases.

Ordinary maintenance by end-users. Closely linked to the previous point, the end-users must be made aware of the methods of use and maintenance of raw earth walls and exposed raw earth facades.

Lack of technical culture. Raw earth construction requires education, training, and specific skills for all professionals in the sector. Furthermore, to avoid the onset of disputes, there is a need to agree on the responsibilities and contract the mutual obligations of each interested party. Poor training and lack of technical culture of raw earth construction can lead to errors in the design and execution phases, with dangerous effects on the quality of earthen building components [6].

5.2.4 Socio-Cultural and Economic Limits

Costs. The nascent raw earth construction market is still a niche, and although there are manufacturers of raw earth materials and products (especially premixed products for earth mortars and renders, but also compressed earth blocks and extruded earth panels), it is complex to quantify the costs of a raw earth project. If this complexity is, in a certain sense, limited for construction with prefabricated raw earth components, this is even greater for on-site construction, where there is a real difficulty in quantifying the materials, rentals, and labor required.

Users' lack of confidence (cultural prejudice). Alongside the other issues, there is the barrier represented by the social perception of raw earth construction, which is perhaps the most complex to overcome. There is a lack of awareness of the benefits of earth-based technologies, not only by potential users but also by technical designers. Raw earth construction is often perceived as more expensive and time-consuming (also due to the lack of automated technologies), and this is particularly true in countries like Italy where raw earth technical culture is not widespread. The lack of technical information also leads to prejudices in potential customers.

5.3 Turning Challenges into Opportunities

5.3.1 *Transversal Standards and Wide-Ranging Policies*

About the lack of shared standards and normative documents, there is a need to create a common technical front aiming at underlining the similarities, rather than the differences, between existing raw earth standards, focusing on the normalization of raw earth construction (concerning its most widespread techniques). This regulatory advancement can only be produced by technicians specialized in raw earth construction who, gathered in specific technical committees, can lay the foundations for the definition of shared guidelines and standards. These documents should relate to the characterization procedures for soils to be used in earth construction, the methodologies for the assessment of the expected performances of earth products (both in the laboratory and on site), the production and construction procedures, and the correct structural and hygrothermal design principles. To facilitate the development of appropriate standards, from 2022 three RILEM Technical Committees are working on several aspects related to raw earth construction: the TCs PEM, MAE and BEC [28].

About political issues, incentive measures must be established to encourage local administrations to develop their territory according to the principles of ecological transition, encouraging the use of circular and sustainable materials. In particular, concerning the extracted soils and secondary raw materials (obtained from other production chains) to be used for raw earth construction, it is necessary to identify clear and unambiguous political paths that aim to simplify the procedures of adoption, and possibly certification, of such materials [6]. Furthermore, these procedures should reward the ecological choice of using natural materials with low environmental impact, such as raw earth or other recycled materials, through the application of tax relief for all the stakeholders involved in the production and building processes.

5.3.2 *Technical Research Converges Around a Few Key Points*

Current scientific literature focuses on the innovation of raw earth materials and the improvement of their performances [13, 14], on the refinement of existing construction technologies and the implementation of new ones [29], also thanks to the use of digital manufacturing methods [30, 31]. It is right to remember here that the technical solutions proposed by traditional raw earth construction can act as a "living database" to carry forward innovation of Architecture, Engineering and Construction sector. It also emerges the need to adopt a performance-based approach to raw earth materials and building components, which must fulfill specific performance targets confirmed by clear checks [32].

As reported in previous contributions [18, 28], the new global standards on raw earth construction should address several items regarding:

- Material selection, analysis, and design (mainly composition and grading);
- Production and manufacturing including molds, machinery to be used, and reference construction techniques;
- Testing methodologies for performance assessments and quality control;
- Structural design, including aspects related to seismic-resistant design;
- Thermal design, including indication about thermal, hygrometric and moisture buffering performances;
- Lifetime behavior, durability, and maintenance guidelines.

For many of these aspects, scientific literature and professional practice have already given numerous and valid answers. In many cases, these responses are based on the material resources and characteristics of the places, confirming the validity of a local approach to contemporary raw earth construction, often linked to the specific geographical characteristics and risks connected. It is outside the scope of this chapter to report the latest technical achievements relating to earthen construction; in this regard, please refer to a recently published volume on the subject [33].

5.3.3 The Importance of Education and Training to Overcome Management Issues

The issues concerning the exchange of information and communication could be solved by the involvement of builders and designers (structural and installation engineers) from the early stages of the architectural project. Their involvement could guide and frame the architectural project to the characteristics and limits of raw earth materials and technologies adopted, in particular as regards external constraints (seismic or flooding risk, areas with particularly severe climates). Many other management issues reported above are intrinsically linked to the lack of training and education in the field of raw earth construction. This lack affects both the design and management aspects of the built environment (technical culture), and the problems relating to the maintenance of systems (by technicians, workers etc.) or of raw earth building components (by end-users).

In the authors' opinion, technical training (aimed at professionals in the construction sector) should be acquired voluntarily (an aspect which is closely tied to socio-cultural aspects which will be discussed in the next point). Above all, such training should be provided by organizations as professional associations, consulting companies or manufacturers of raw earth products, as is already the case for various European companies.

Another opportunity to improve several aspects connected to technical culture, and to the execution of raw earth constructions could be provided by the use of digital manufacturing technologies. Indeed, innovative digital construction processes for earthen buildings and components include 3D printed earth, robotic rammed earth, and robotic sprayed earth [30], but several other techniques are being developed, always drawing from the knowledge of traditional construction techniques and hybridizing them with today's technical and operative possibilities [29, 34]. Contemporary research on the use of digitally manufactured earth construction has begun to be in vogue since 2011 [31], but still has wide margins for development, both in the field of in-situ digital fabrication of entire buildings or in the production of prefabricated components, to be assembled on site. Digital fabrication technologies can provide improved efficiency of the production process (for instance, it has been proved that additive manufacturing can provide a 25–60% in material reduction and 30% time-saving as compared to traditional manufacturing techniques [31]). They also provide a higher level of quality in detailing, accuracy, finishing, and reproducibility and increased safety during the manufacturing process [31]. An extension of the lifetime behavior of the product and improved compliance with modern building regulations for structural, thermal, and environmental performances are also expected. Lastly, digital manufacturing increases satisfaction of the contemporary needs of occupants, in terms of design, comfort, and well-being [30].

Ultimately, raw earth construction can benefit from the push for digital transition that now permeates the contemporary construction sector. Methods and techniques for systematizing information relating to the performance of raw earth materials can be developed to catalog previous studies and predict (through AI-based approaches) mechanical, thermal, and physical performances of novel materials based on the raw materials inputs or design procedures used. Methods for optimizing the production process of raw earth materials and components can be refined through life cycle analysis (LCA) and life cycle cost (LCC), in close collaboration with entrepreneurs operating in the sector, pursuing the objectives of economic and environmental sustainability. The need to improve project management, systematize the information and the technical characteristics of raw earth components can benefit from a BIM-based approach. As it happens for conventional construction, BIM tools concentrate information and channel it with a view to interoperability between softwares and communication between all the actors involved in the various phases of the construction process (including the building site). In particular, the systematization of information related to the performances of innovative raw earth materials and components can aim at the creation of technical data sheets and databases that can feed the BIM libraries, including information about their environmental impacts, to encourage comparison of different "green" design options. Through these tools, it will be possible to take into consideration performance, environmental and economic aspects.

The challenge of the twin ecological and digital transition therefore perfectly suits raw earth construction, which, through the digitalization of information and workflows, could raise the quality of design, execution and management of earthen buildings [6].

5.3.4 A Widespread Raw Earth Culture to Overcome Socio-Cultural and Economics Issues

During the last decades, several companies specialized in raw earth construction have been born, especially for those building sites where specific equipment and machinery were required (as in rammed earth construction). The cost of new earthen construction in Europe and in the world is still quite high, also because of its insufficient use and spread. To reduce costs, a great opportunity can be given by automated or digital constructive techniques or industrialized production processes [35, 36].

Concerning the cultural prejudice, the dissemination of raw earth culture, and its expressive and environmental potential is fundamental for its development. Raw earth construction will increasingly be able to gain a foothold in contemporary construction only if it becomes known and recognizable. In this sense, the following initiatives are producing the trigger for the creation of an ever-increasing technical culture on raw earth construction:

- Dedicated master's degrees
- Undergraduate university courses
- Professional training courses
- Associations and networks of professionals
- International architectural competitions
- Research projects
- Industrial research.

Indeed, innovation coming from industry is also giving new inputs to the development of raw earth construction, according to a bottom-up approach.

As previously stated [6], at the same time it is necessary to eradicate some false beliefs, and to do this it is essential to spread exemplary projects throughout the territories (the Orangerie of Lyon in France, the Ricola Herb Center and the Swiss Ornithological Institute in Switzerland, the Oaxaca School in Mexico, the Amangiri Resort and Spa in Utah, the Schlins Workshop in Austria, are just some examples) to show how raw earth can be adopted to create modern, efficient and environmentally compatible architecture and living spaces.

Furthermore, it is the authors' opinion that these "exemplary" construction sites should also be "exploratory", with a view to opening up to innovative solutions and to the hybridization of technical and non-technical knowledge. Moreover, they could be conceived as "educational", aiming at the transmission of technical knowledge and emphasizing the pedagogical role of the construction site, thus encouraging its openness to designers and operators of the construction sector, even if not directly involved in the work.

5.4 Conclusions and Final Remarks

Raw earth construction has undoubted advantages compared to conventional construction. However, it also presents limits, which in this contribution have been divided into regulatory and political, technical, management and training, socio-cultural and economic. These limits must not be considered as insurmountable, in fact, the second part of the contribution presents insights into what may be the challenges to overcome and the opportunities to be seized.

The reported insights refer to the regulations and policies that could favor the development of raw earth construction, the key points around which technical research is focusing, the importance of education and specific training for overcoming the management issues linked to earth construction, and the need to spread a raw earth culture to be able to overcome socio-cultural and economic issues. The contribution also highlights how digital manufacturing techniques and digitalization in general can help in creating methods and techniques that can simplify, rationalize, and optimize the production and management of raw earth construction, favoring its adoption in contemporary Architecture, Engineering, and Construction sector.

The return to raw earth construction today means returning to the principles underlying the origins of the history of buildings. It means taking into account the importance of raw materials present in the territories, and being aware of their limitations. It also means designing circularly, with materials that can be regenerated and reused at their end-of-life without harm to the environment. Finally, it means being able to rely on the quality of buildings in which we live, both in terms of comfort, safety and reliability requirements.

Today, it is possible to innovate raw earth construction and make it competitive, attractive, and reliable even in comparison with other sustainable construction technologies. To achieve this, it is necessary to overcome the aforementioned limitations by placing the values of sustainability, reuse, circular economy, along with the challenges of the ecological and digital transition, at the center of future building and design practices.

References

1. L. Ben-Alon, Down to earth: using natural building materials for community resiliency, in *Curriculum for Climate Agency: Design in Action* (Columbia University, ACSA Press, 2021), pp. 44–51
2. L. Ben-Alon, V. Loftness, K.A. Harries, E. Cochran Hameen, Life cycle assessment (LCA) of natural vs conventional building assemblies. Renew. Sustain. Energy Rev. **144**, 110951 (2021). https://doi.org/10.1016/j.rser.2021.110951
3. B. Marques, H. Varum, H. Corvacho, M.C. Guedes, L. Baptista, Using raw earth construction systems on contemporary buildings: reflections on sustainability and thermal efficiency. Renew. Energy Environ. Sustain. **6**, 46 (2021). https://doi.org/10.1051/rees/2021041

4. F. Pacheco-Torgal, S. Jalali, Earth construction: lessons from the past for future eco-efficient construction. Constr. Build. Mater. **29**, 512–519 (2012). https://doi.org/10.1016/j.conbuildmat.2011.10.054
5. Z. Berrada, L. Ben-Alon, *The Development of a Social Life Cycle Assessment (S-LCA) Framework to Compare the Social Impact of Different Building Materials* (2023)
6. J.-C. Morel, R. Charef, E. Hamard, A. Fabbri, C. Beckett, Q.-B. Bui, Earth as construction material in the circular economy context: practitioner perspectives on barriers to overcome. Phil. Trans. R. Soc. B **376**, 20200182 (2021). https://doi.org/10.1098/rstb.2020.0182
7. J. Morel, R. Charef, What are the barriers affecting the use of earth as a modern construction material in the context of circular economy? IOP Conf. Ser. Earth Environ. Sci. **225**, 012053 (2019). https://doi.org/10.1088/1755-1315/225/1/012053
8. M. Achenza, A. Agus, Verso la normalizzazione del materiale terra cruda in Italia, in *New Horizons for Sustainable Architecture, Proceedings from Colloqui.AT.e* (Edicom Edizioni, 2020)
9. Cycle Terre, *Dossier-Presse-Cycle Terre-2018* (2018)
10. H. Guillaud, H. Houben, J. Dethier, CRAterre, in *Traité de construction en terre* (Parenthèses Editions, 2006)
11. G. Minke, *Building with Earth* (Birkhäuser, Basel, 2006)
12. A. Laborel-Préneron, J.E. Aubert, C. Magniont, C. Tribout, A. Bertron, Plant aggregates and fibers in earth construction materials: a review. Constr. Build. Mater. **111**, 719–734 (2016). https://doi.org/10.1016/j.conbuildmat.2016.02.119
13. C. Turco, A.C. Paula Junior, E.R. Teixeira, R. Mateus, Optimisation of compressed earth blocks (CEBs) using natural origin materials: a systematic literature review. Constr. Build. Mater. **309**, 125140 (2021). https://doi.org/10.1016/j.conbuildmat.2021.125140
14. A.E. Losini, A.C. Grillet, M. Bellotto, M. Woloszyn, G. Dotelli, Natural additives and biopolymers for raw earth construction stabilization—a review. Constr. Build. Mater. **304**, 124507 (2021). https://doi.org/10.1016/j.conbuildmat.2021.124507
15. N. Jannat, A. Hussien, B. Abdullah, A. Cotgrave, Application of agro and non-agro waste materials for unfired earth blocks construction: a review. Constr. Build. Mater. **254**, 119346 (2020). https://doi.org/10.1016/j.conbuildmat.2020.119346
16. G. Giuffrida, V. Costanzo, F. Nocera, M. Cuomo, R. Caponetto, Natural and recycled stabilizers for rammed earth material optimization, in *Sustainability in Energy and Buildings*, ed. by J. Littlewood, R.J. Howlett, L.C. Jain (Springer Nature Singapore, Singapore, 2023), pp. 164–174
17. G. Giuffrida, L. Dipasquale, R.M. Pulselli, R. Caponetto, Sustainability of earth-based materials incorporating marble cutting waste, in *Second RILEM International Conference on Earthen Construction*, eds. by C. Beckett, A. Bras, A. Fabbri, E. Keita, C. Perlot, A. Perrot. ICEC 2024. RILEM Bookseries, vol. 52 (Springer, Cham, 2024). https://doi.org/10.1007/978-3-031-62690-6_32
18. A. Fabbri, J.C. Morel, J.-E. Aubert, Q.-B. Bui, D. Gallipoli, A. Ventura, V.B.V. Reddy, E. Hamard, A. Pelé-Peltier, H.N. Abhilash, An overview of the remaining challenges of the RILEM TC 274-TCE, testing and characterisation of earth-based building materials and elements. RILEM Tech. Lett. **6**, 150–157 (2022). https://doi.org/10.21809/rilemtechlett.2021.149
19. S. Burroughs, Recommendations for the selection, stabilization, and compaction of soil for rammed earth wall construction. J. Green Build. **5**, 101–114 (2010). https://doi.org/10.3992/jgb.5.1.101
20. G. Giuffrida, M. Detommaso, F. Nocera, R. Caponetto, Design optimisation strategies for solid rammed earth walls in Mediterranean climates. Energies **14**, 325 (2021). https://doi.org/10.3390/en14020325
21. NTE E 080, *Diseño y Construcción con Tierra Reforzada* (Ministerio de Vivienda, Construcción y Saneamiento, Perú, 2017)
22. G. Giuffrida, R. Caponetto, F. Nocera, M. Cuomo, Prototyping of a novel rammed earth technology. Sustainability **13**, 11948 (2021). https://doi.org/10.3390/su132111948

23. J. Canivell, J.J. Martín-del-Río, R.M. Falcón, C. Rubio-Bellido, Rammed earth construction: a proposal for a statistical quality control in the execution process. Sustainability **12**, 2830 (2020). https://doi.org/10.3390/su12072830
24. A. Azil, M. Le Guern, K. Touati, N. Sebaibi, M. Boutouil, F. Streiff, S. Goodhew, M. Gomina, Earth construction: Field variabilities and laboratory reproducibility. Constr. Build. Mater. **314**, 125591 (2022). https://doi.org/10.1016/j.conbuildmat.2021.125591
25. K. Touati, M. Le Guern, Y. El Mendili, A. Azil, F. Streiff, J. Carfrae, M. Fox, S. Goodhew, M. Boutouil, Earthen-based building: In-situ drying kinetics and shrinkage. Constr. Build. Mater. **369**, 130544 (2023). https://doi.org/10.1016/j.conbuildmat.2023.130544
26. C.T.S. Beckett, P.A. Jaquin, J.-C. Morel, Weathering the storm: a framework to assess the resistance of earthen structures to water damage. Constr. Build. Mater. **242**, 118098 (2020). https://doi.org/10.1016/j.conbuildmat.2020.118098
27. A. Laborel-Préneron, K. Ouédraogo, A. Simons, M. Labat, A. Bertron, C. Magniont, C. Roques, C. Roux, J.-E. Aubert, Laboratory test to assess sensitivity of bio-based earth materials to fungal growth. Build. Environ. **142**, 11–21 (2018). https://doi.org/10.1016/j.buildenv.2018.06.003
28. P. Faria, C.T.S. Beckett, A. Fabbri, E. Keita, J.-C. Morel, C. Perlot, A. Perrot, RILEM contribution to Earthen building, in *Second RILEM International Conference on Earthen Construction*, ed. by C. Beckett, A. Bras, A. Fabbri, E. Keita, C. Perlot, A. Perrot (Springer Nature Switzerland, Cham, 2024), pp. 194–205
29. H. Kloft, A. Salamatian, J. Gosslar, E. Dorresteijn, D. Lowke, Robotic rammed earth-concrete (RREC): a novel additive manufacturing technology to strengthen Rammed earth structures by integrated Rammed concrete parts, in *Second RILEM International Conference on Earthen Construction*, ed. by C. Beckett, A. Bras, A. Fabbri, E. Keita, C. Perlot, A. Perrot (Springer Nature Switzerland, Cham, 2024), pp. 60–70
30. M. Schweiker, E. Endres, J. Gosslar, N. Hack, L. Hildebrand, M. Creutz, A. Klinge, H. Kloft, U. Knaack, J. Mehnert, E. Roswag-Klinge, Ten questions concerning the potential of digital production and new technologies for contemporary earthen constructions. Build. Environ. **206**, 108240 (2021). https://doi.org/10.1016/j.buildenv.2021.108240
31. M. Gomaa, W. Jabi, V. Soebarto, Y.M. Xie, Digital manufacturing for earth construction: a critical review. J. Clean. Prod. **338**, 130630 (2022). https://doi.org/10.1016/j.jclepro.2022.130630
32. C. Perlot, A.W. Bruno, M. Posani, G. Habert, S. Vucetic, Challenges for bio-stabilised earth-based construction, in *Second RILEM International Conference on Earthen Construction. ICEC 2024. RILEM Bookseries*, ed. by C. Beckett, A. Bras, A. Fabbri, E. Keita, C. Perlot, A. Perrot, vol. 52 (Springer, 2024)
33. C. Beckett, A. Bras, A. Fabbri, E. Keita, C. Perlot, A. Perrot, *Second RILEM International Conference on Earthen Construction: ICEC 2024* (Springer Nature Switzerland, Cham, 2024)
34. P. Ayres, M.R. Thomsen, B. Sheil, M. Skavara, *Fabricate 2024: Creating Resourceful Futures* (UCL Press, 2024)
35. A. Von Mag, M. Rauch, Paredes de tapial y su industrialización (encofrados y sistemas de compactación). Inf. Constr. **63**, 35–40 (2011). https://doi.org/10.3989/ic.10.013
36. M. Gomaa, S. Schade, D.W. Bao, Y.M. Xie, Automation in rammed earth construction for industry 4.0: precedent work, current progress and future prospect. J. Clean. Prod. **398**, 136569 (2023). https://doi.org/10.1016/j.jclepro.2023.136569

Chapter 6
Cob Vernacular Architecture Conservation and Its Contribution Towards a Sustainable Built Environment

Alejandro Jiménez Rios

Abstract Cob is a low-impact building material with a minimal carbon footprint that doesn't require energy-intensive manufacturing processes, uses locally available materials, is non-toxic and biodegradable. Moreover, cob vernacular architecture is often deeply rooted in local cultures and traditions. By conserving and maintaining these structures, cultural heritage is preserved and a connection to traditional building methods, which can foster a sense of identity and pride in local communities is maintained. In this chapter, the importance of cob vernacular heritage conservation and how this can contribute towards a sustainable built environment with an emphasis on the sustainable properties of this material is discussed. The chapter content is in alignment with the United Nations Sustainability Development Goals and with well stablished international conservation methodologies and philosophies, such as the Venice Charter, the Nara Document on authenticity, among others.

6.1 Vernacular Architecture

The word "vernacular" originates from the Latin term "vernaculus", which means local, domestic or indigenous [1]. Hence, vernacular architecture is commonly known as folk or popular architecture [2]. This architectural style can be found throughout the world and is characterized by local environmental factors. Vernacular buildings are created using native materials by the inhabitants of a particular place to meet their housing needs [3]. Vernacular local adaptability has allowed for centuries the creation of what is presently referred to as "nature-based solutions" [4]. It is a true representation of the world's cultural diversity as it is the result of complex societal mechanisms through the intergenerational refinement of building knowledge and construction techniques. As clearly described by Piesik [5], vernacular skills and know-how are customarily passed from experienced individuals to novice learners of

A. Jiménez Rios (✉)
Department of Architecture & Civil Engineering, University of Bath, Bath, UK
e-mail: ajr225@bath.ac.uk

D. A. H. Hanaor (ed.), *Sustainable Materials for the Built Environment*, Springer Series in Materials Science 352, https://doi.org/10.1007/978-3-031-97818-0_6

Fig. 6.1 **a** Vernacular houses in the picturesque Irish town of Skerries, **b** turf houses in the Glaumbær Farm & Museum, Iceland, **c** Chefchaouen, Morocco, and its distinctive blue vernacular architecture, and **d** the colorful landscape created by the vernacular architecture of the city of Guanajuato, Mexico

the trade. As vernacular buildings are often constructed and inhabited by the owners themselves, vernacular architecture has also been paradoxically defined as "architecture without architects" [6]. Figure 6.1 presents some examples of vernacular architecture found in different countries and climates.

Vernacular architecture embodies a series of economic benefits. For centuries, it has been the engine of local economies by using local craftsmanship and materials to satisfy those communities' demands for shelter. According to Santos et al. [7], vernacular architecture solutions bring about cost-effective solutions by enhancing human comfort, which ultimately results in low energy consumption. This observation is on the contrary contested by Bustan-Gaona et al. [8], who determined after conducting a multidisciplinary study on vernacular buildings in Equator, the inadequacy of light conditions inside the analyzed local study cases. Such lighting deficiency is caused by the small windows dimensions of this vernacular building typology. This results into higher energy consumption levels and an increased operational cost for the inhabitants. From another perspective, further economic benefits are attached to vernacular architecture, particularly to the one built using earthen materials and techniques, by Zami and Lee [9]. These authors have performed a thorough review on the topic and present such typology as a low-cost alternative to cope with increased urban housing demand.

Furthermore, intangible values and traditional knowledge are deeply embodied in vernacular architecture as it gives a unique character to a region or culture and helps preserve cultural heritage and traditional knowledge [7]. A recent study case conducted by Alnaim [10], who studied the vernacular courtyard house typology of Iranian houses, concluded that such building style can meet the cultural, social and climatic demands of local inhabitants by creating a comfortable and private household atmosphere. The loss of uniqueness character and values caused by the global implementation of standardized materials and construction techniques, is exposed for instance by Tarrad and Ahmad [11], who described Jordanian cities as being severely affected by this phenomena. What once was a "*vibrant town with identified aesthetic architectural style*", is now perceived as an "*anonymous and monotonous unit of concrete*". Sentiments of nostalgia are also evocated by Sharami and Hosseini [12], when they compared the declining vernacular Iranian neighborhood structure, known as "Mahalla", in front of the "*cruelly drawing and widening lines for the modernization of the urban structure*". Fortunately, universal recognition of values inherent to vernacular architecture and the necessity of preserving them, has been adopted by the United Nations (UN) [13] (see Sect. 6.3).

Studying and comprehending vernacular architecture not only aids in preserving its historical, cultural, and architectural significance, but it can also guide stakeholders toward contemporary sustainable solutions [14]. Such outcomes can be achieved through a bioclimatic design, approach that obtains valuable insights from vernacular practices. As recognized by Huang et al. [15], vernacular buildings are specifically tailored to efficiently cope with local climatic and socio-cultural conditions. These authors emphasize the importance of empirical work to better understand and adopt vernacular bioclimatic design strategies, which may considerably differ from general tools, i.e. Mahoney Table, Olgyay Bioclimatic Chart, Givoni's Building Bioclimatic Chart, etc. [16]. On another study case, Sharami and Hosseini [12] highlighted the positive environmental characteristics of vernacular Isfahani architecture. They have concluded that such typology provides optimal conditions for semi-arid and hot climates due to their energy efficient and low-impact components, both at housing and urban development levels. Numerous authors have presented other examples of how utilizing vernacular materials, construction techniques, and typologies can enhance the sustainability of modern buildings [17–19].

Of paramount importance to adequately address contemporary issues, are culture and tradition integration with environment and nature aspects, if vernacular principles are to be adopted. To foster the implementation of vernacular skills and know-how, Santos et al. [7] have identified the need for increased awareness of the potential benefits obtained and the value of the vernacular practice, as well as the necessity of providing the required financial aid. Such measurements need to be accompanied by an aligned sustainable urban development strategy and adequate preservation of existing heritage measurements.

6.2 Rediscovering Cob

For hundreds of years, humans across the globe have harnessed the earth beneath their feet to create shelter. Earthen architecture stands as a testament to humanity's ingenuity, utilizing the abundant and accessible resources found in local environments. Today, the heritage of earthen architecture is rich and varied, encompassing archaeological sites, living monuments, groups of buildings, and even entire historic towns and cultural landscapes.

The most common earthen construction classifications are based on either the type of method implemented (dry or wet), the structural function of the earthen elements (load bearing or non-load bearing, or by the type of element produced (monolithic, masonry, or infill). Among the many methods of earthen construction, cob, which is classified as a wet, load bearing, monolithic technique, stands out for its simplicity and versatility. This chapter explores the characteristics, preparation, construction processes, and mechanical properties of cob, shedding light on why this ancient technique remains relevant and how it could be beneficial in contemporary architecture.

6.2.1 The Unique Characteristics of Cob

Cob is a mixture of soil, water, and natural fibers such as straw. The simplicity of its ingredients contradicts the complexity of its behavior as a building material. The quality of the soil, the amount of water (main parameter determining the mixture consistency) [20], and the type and quantity of fibers, all significantly influence the final product's strength and durability. The natural fibers added to the cob mixture serve multiple purposes. They improve workability by keeping the wet soil cohesive during mixing and placement [21], while enhancing the material's tensile strength and ductility [22]. This means that cob can deform under stress without cracking excessively, an essential property for withstanding environmental stresses and low magnitude seismic events. Additionally, fibers help manage soil shrinkage as it dries, reducing the likelihood of cracks or limiting their opening [23].

Fibers, typically between 15 and 50 cm in length, account for 1–3% by mass of the mixture, which shows an important variability in the characteristics of the fibers reported by different authors. This influences not only cob mechanical properties, but also its thermal performance [24]. The fibers decrease the material's density, improving its insulating properties, although unfortunately reducing its compressive strength at a certain threshold. The drying of the material during the construction process is also accelerated by the fibers. On the other hand, fibers increase cob humidity absorption rate during operation, which is a desired characteristic that helps keeping healthy indoor living conditions. Moreover, the presence of fibers can enhance resistance to erosion and frost, making cob a resilient choice for various climates [25].

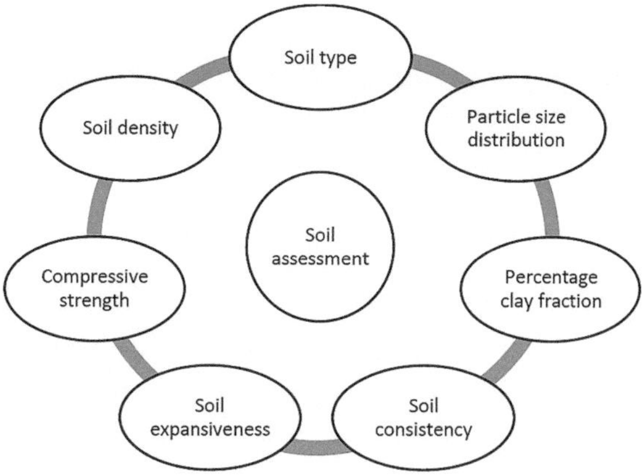

Fig. 6.2 Parameters influencing the suitability of a soil to be used in Cob

Several components presented in Fig. 6.2 play an important role determining the suitability of a given soil to be used on cob. Soil type depends on its underlying geology, mineralogy, stratigraphy, and color, which may also be of interest for aesthetic purposes. A soil granulometric composition, also known as particle size distribution, depends on its particle's sizes. Soil particles are classified as cobbles (greater than 63 mm), gravels (63–2 mm), sand (2.0–0.063 mm), silt (0.063–0.002 mm) or clay (smaller than 0.002 mm) [26]. Harries et al. [27] suggested the use of soils with a clay content between 10 and 25%, silt content between 10 and 20%, and sand/gravel content of about 55% to 70%. Watson and McCabe [28] proposed similar ranges for silts and clays, whereas they recommended having about 25–30% of sand and 30–40% of gravel.

Clay content is crucial for soil behavior in cob because clay binds the coarser particles together, creating a cohesive and uniform mixture. This cohesion arises due to the water suction between clay particles [29]. Nevertheless, clay content must be limited to no more than 30% as excessive amounts of clay would lead to considerable expansion and contraction, which can destabilize the structure. Soil consistency is determined by its plasticity and cohesion. In simple terms, "*Plasticity is the ability of a material to be molded without fracturing, while cohesion is the ability to maintain that deformed shape*" [30]. Based on its water content, the state of a soil can be classified as liquid, plastic, semi-solid or solid, and the limits between such stages are referred to as Atterberg limits (see Fig. 6.3).

The amount of swelling/shrinkage of a soil, also known as soil expansiveness, will be mainly determined by clay content and type. Montmorillonites will suffer bigger volumetric changes in comparison to kaolinites or illites [23]. While mixing of the cob constitutive components, it is also important to use the minimum amount of water necessary for an adequate workability as, by reducing the amount of water that will evaporate when cob dries, the amount of volume shrinkage and the opening of cracks can be controlled. Soil's particle type, size distribution, and compaction degree also affect its density, property that is mainly used to compute the soil's

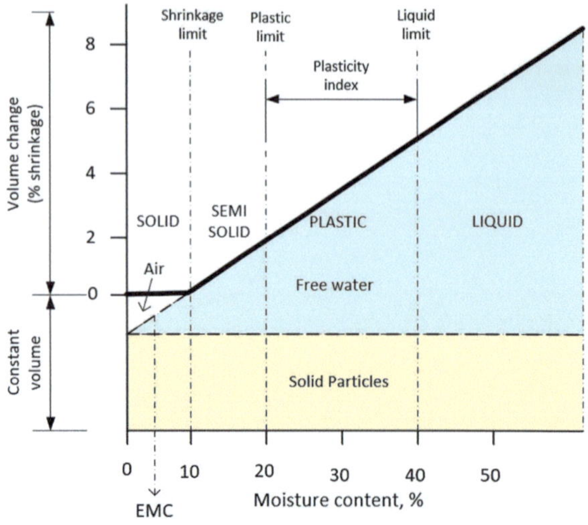

Fig. 6.3 Diagram showing the different soil classification phases and the corresponding Atterberg limits. EMC = Equilibrium Moisture Content

porosity and thermal performance. Well compacted soils would be stronger than poorly compacted materials [25].

Finally, cob's load bearing capacity is also influenced by the compressive strength of its soil. Soil strength comes from the friction forces between its particles which are typically assessed on function of the soil macroscopic angle of friction, ϕ. The shear strength of a soil, τ is computed as $\tau = \sigma' \tan\phi' + c$, where σ', represents the soil effective stress, c its cohesion, and ϕ' represents the effective macroscopic angle of friction. The effective stress of a soil can be calculated as $\sigma' = \sigma - u$, where the total stress is represented by σ, and u denotes the pore water pressure. Thus, if pore water pressure increases, caused by an increase in the moisture content, the effective stress of the soil will decrease, consequently reducing its shear strength. These equations are derived assuming full soil saturation and effective stress parameters. Nevertheless, cob, being a manufactured unsaturated soil, presents a structural behavior not entirely described with such assumptions. The field of unsaturated soil mechanics, which provides equations with a higher degree of complexity, needs to be consulted if a more refined description of the macroscopic unsaturated behavior of soils is required [31].

6.2.2 Preparation of Cob and Construction Process

Preparing the mixture for cob construction is a less industrialized process than that of modern materials like concrete or steel. Yet, it requires careful attention to detail. Ideally, the soil should be sourced locally or recycled from other projects to minimize environmental impact. When necessary, the soil's properties can be adjusted through stabilization, either by modifying its grading or by adding substances like pure sand, clay, or, in extreme conditions, cement or lime to make the structure stronger and resistant to external hazards.

Traditionally, loam preparation was a labor-intensive process performed by hand or animal power. Today, modern machinery can expedite this process, but the fundamental steps remain the same. The soil is first dried and sieved to remove large particles. Water is then added, and the mixture is kneaded and trampled until it reaches a plastic consistency. At this stage, fibers are incorporated, and the mixture is further worked until it regains a plastic consistency that allows for its placement on the wall (see Fig. 6.4).

Building with cob begins on a solid foundation, traditionally stone masonry, but often reinforced concrete is used in modern constructions which increases the environmental impact of the whole construction. The cob mixture is placed directly on top of the foundation, normally in the form of small lumps that allow manual handling by a person, and that accumulate to form layers along the footprint of the projected walls. The construction grows taller layer by layer. Each layer, or lift, is about 30–60 cm high and requires a drying period of approximately 5–7 days before the next layer can be added. As each lift dries, the lateral surface of the walls is trimmed and smoothed to ensure evenness and prepare it for subsequent layers. This step is crucial for achieving a monolithic wall with no weak planes, as the layers bond together during the drying process (see Fig. 6.5).

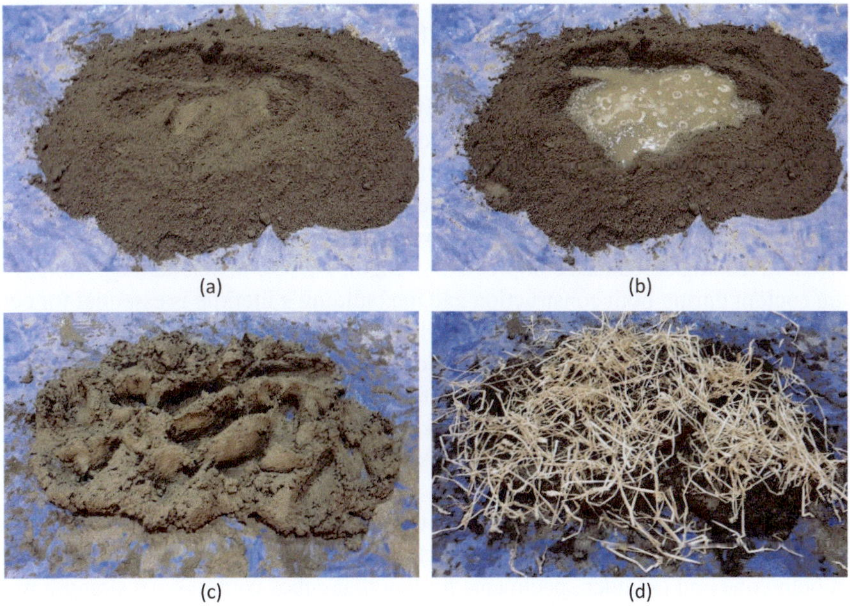

Fig. 6.4 Cob mixing process: **a** soil is dried and sieved, **b** a hole is made in the middle of the soil and water is added, **c** the water/soil combination is mixed until it is brought to a viscous consistency, **d** straw is added and the mixture is further worked until it is ready for use

(a) (b)

Fig. 6.5 **a** Cob wallettes being built at the Laboratory of Civil, Structural, and Environmental Engineering, Trinity College Dublin and **b** cob wall being built at the 2018 BASEHabitat International Summer School, Austria

6.2.3 Mechanical Properties and Structural Behavior of Cob

Cob's mechanical properties are primarily determined by its clay content and the presence of fibers. These elements provide the material with its cohesive strength and relatively high ductility. A tabular summary of some of cob's mechanical properties as reported in the literature is presented in Table 6.1. Moisture content also plays a significant role in cob's strength, with higher moisture after a certain optimal threshold leading to reduced compressive strength. Therefore, maintaining low moisture content during both construction and the building's lifetime is essential for cob structural integrity. Cob also exhibits good shear strength compared to other earthen materials. Unlike modular earthen techniques such as adobe, cob walls are monolithic, which means they lack the inherent weakness planes found in brick or block constructions. These properties, combined with its ability to regulate indoor humidity and its thermal insulating capabilities, makes cob an attractive option for sustainable building practices [32].

As cob is a non-standardized material, the value of the mechanical properties can vary considerably [39]. Keefe [20] studied the variability of cob compressive strength on both cube and cylinder specimens, finding that cubes compressive strength was 20–25% higher than that of the cylinders. This variation in percentage is also common when the compressive strength of concrete is obtained from specimens with similar ratios. Furthermore, according to this author, grading, compaction, clay content and clay mineralogy, are all properties that can affect cob compressive strength. More

Table 6.1 Cob mechanical properties as reported in the literature

Property/References	[33]	[22, 34]	[35]	[36–38]
Compressive strength (MPa)	0.45–1.40	1.59	1.12–1.35	0.697–0.900
Tensile strength (MPa)	NR*	0.159–0.254	NR*	NR*
Young's modulus (MPa)	170–335	651	16.9–23.13	103.5–203
Poisson's ratio (−)	NR*	0.15	0.12–0.09	0.28

*Not reported

importantly, Keefe has reported that there is a negative linear relation between moisture content and cob unconfined compressive strength, i.e., the higher the moisture content, the lower the unconfined compressive strength of cob. This observation has reinforced the data described by Saxton [40] and has as well been confirmed by Akinkurolere et al. [41].

6.2.4 Numerical Modelling of Cob

Given the numerous uncertainties in studying existing earthen structures, various numerical strategies often depend on experimental campaigns. These campaigns are essential for determining the necessary numerical input parameters, model calibration, and for validating the numerical results obtained [42]. Only a few authors have conducted numerical modelling studies of cob, all of them implementing the so called continuum homogeneous models approach [43].

One notable study by Miccoli et al. [34] utilized a finite element method macro-modeling approach in the software DIANA. Their approach assumed plane stress, material homogeneity, and material continuity. By employing a total strain rotating crack constitutive model with a multi-linear definition of the stress–strain relationship for compressive behavior, they successfully calibrated their simulations and replicated the shear stress-shear strain relationship observed experimentally [22]. However, the model failed to replicate the diagonal failure pattern observed in experiments, likely due to small non-symmetric conditions in the samples and testing setup. Quagliarini and Maracchini [35] also explored numerical simulations of cob, but this time using the finite element software MIDAS FEA. They applied a smeared-crack macro-modeling approach with a 2D plane stress simplification. Their stress–strain curves matched experimental results up to the peak strength of the material, although the post-peak behavior, which is crucial for understanding the material's energy dissipation capacity, was not replicated by the numerical model. The authors noted that the damage patterns, interpreted through principal stresses, aligned with typical crack patterns observed in compressive tests. Further expanding the understanding of cob numerical modeling, Jiménez Rios and O'Dwyer conducted a study on cob's nonlinear monotonic behavior using the software tools ANSYS and ABAQUS [44]. They examined the strengths and weaknesses of various constitutive models. Their findings indicated that the Concrete and Concrete Damaged Plasticity models were the most effective at replicating cob's nonlinear monotonic behavior in ANSYS and ABAQUS, respectively.

6.3 Conservation of Earthen Vernacular Architecture

In 2015, the General Assembly of the UN adopted the 2030 Agenda for Sustainable Development. This resolution aimed to transform our world through 17 interrelated Sustainable Development Goals (SDG) and 169 targets based on the economic, social, and environmental dimensions of sustainable development. In particular, SDG 11 focuses on making cities and human settlements inclusive, safe, resilient and sustainable. Within this framework, Target 11.4 emphasizes the importance of protecting and safeguarding the world's heritage (https://sdgs.un.org/goals). The UN Educational, Scientific and Cultural Organization (UNESCO) plays a crucial role in promoting cultural heritage conservation worldwide. By adopting the 1972 World Heritage Convention, UNESCO member states committed to identifying, protecting, conserving, and presenting the world's heritage.

UNESCO member states are advised on terms of conservation philosophy and techniques by the International Council on Monuments and Sites (ICOMOS). ICOMOS also evaluates properties inscription on the UENSCO World Heritage List, conducts comparative studies, and assessments on the conservation estate of inscribed properties. Moreover, a series of International Scientific Committees (ISC) have been established within ICOMOS to gather, study, and disseminate information related to principles, techniques, and policies on conservation to enhance the management and protection of cultural heritage. Among the 31 ISCs, the ISC on Earthen Architectural Heritage (ISCEAH) and the ISC on Vernacular Architecture (CIAV) are particularly relevant to this chapter. As a result of the efforts done by these institutions, the conservation practice of earthen vernacular architecture is nowadays based on a set of principles and guidelines broadly accepted and applied by the international community contained within the following charters: the Venice Charter [45], the Nara Document on Authenticity [46], the Charter on the Built Vernacular Heritage [47], and the Burra Charter: the Australia ICOMOS Charter for Places of Cultural Significance [48] (see Fig. 6.6).

6.3.1 The Venice Charter

The Venice Charter, approved in 1964, establishes the fundamental principles for preserving and restoring historical monuments. In this document the definition of a historical monument encompasses any architectural work, urban or rural setting in which is found evidence of a particular civilization, a significant development, or a historic event. It emphasizes the use of all available sciences and techniques to

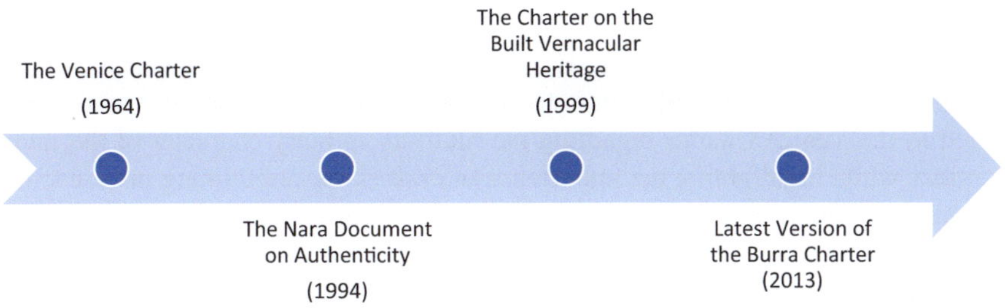

Fig. 6.6 Simplified timeline of key documents dealing with the conservation of earthen vernacular architecture

conserve monuments, both as works of art and as historical evidence. Key guidelines include maintaining monuments in use, preserving aesthetic and historical values, and documenting all interventions comprehensively. New constructions, demolitions, modifications or moving of all or part of a monument must be restricted and only on cases of absolute necessity for the safeguarding of the monument itself or justified by national or international interest, should they be allowed.

6.3.2 The Nara Document on Authenticity

Adopted in 1994, the Nara Document on Authenticity extends and complements the Venice Charter by addressing cultural and heritage diversity. It highlights the importance of authenticity in conservation, acknowledging that different cultures attribute value based on various criteria. Form and design, materials, and substance, use and function, traditions and techniques, location and setting, spirit and feeling, as well as other external and internal factors must be examined based on the local context of every cultural heritage asset to determine its authenticity. Defining authenticity helps to "*clarify and illuminate the collective memory of humanity*". The values upon which authenticity is based, are attributed by local cultures and their socio-cultural and historical contexts. Respect of local culture and its people must be of paramount importance for conservation professionals and other stakeholders involved in the conservation of the built cultural heritage environment.

6.3.3 The Charter on the Built Vernacular Heritage

The 1999 Charter on Built Vernacular Heritage bestow vernacular architecture with the representation of the culture of a community and of its ties with its surrounding environment. The set of built vernacular architectural assets represents the world's cultural diversity and its conservation is promoted on this document. The Charter recognizes the threats posed by economic and cultural homogenization and underscores the need for community, government, and specialist involvement in preserving vernacular buildings. Moreover, this charter reinforces the principles of conservation stated in the Venice Charter regarding the multidisciplinary character of the interventions while highlighting the importance of craft skills and training in vernacular architecture conservation efforts. The Charter advocates for governments, authorities, and organizations to develop educational, training, and informational initiatives aimed at raising public awareness about vernacular heritage, supporting communities in conserving traditional systems, materials, and skills, and promoting the principles of vernacular conservation. Additionally, it encourages the establishment of regional networks to share expertise and experiences related to vernacular heritage. In accordance with this vision, vernacular architecture is better preserved if its assets are treated and studied as a regional group or at settlement level. The landscape surrounding them as well as the intangible values attached to these architectural assets play an important role to better understand the way in which they are used by local communities, and design better and more sympathetic preservation interventions.

6.3.4 The Burra Charter

The Burra Charter (latest version adopted in 2013) provides guidelines for conserving places of cultural significance, stressing the principle of inter-generational equity, which requires that places of cultural significance to be preserved for present and future generations. It advises a three-tiered process that begins with understanding the significance of the place, followed by the development of a conservation policy, and finalizes with conducting the managing of the site accordingly to such policy. This charter is deeply interlinked with the Nara document, as the significance of a cultural site is highly dependent on its authenticity.

6.3.5 Conservation Standards and Specialized Guidance

The conservation of earthen vernacular architecture and cultural heritage requires in addition a multidisciplinary approach guided by international standards and principles. By understanding and applying these frameworks, researchers and practitioners can ensure the preservation of our shared heritage for future generations. Other than the ICOMOS charters, various standards and specialized guidance have been developed for the conservation of earthen cultural heritage. Internationally, the International Organization for Standardization (ISO) developed ISO 13822 Bases for Design of Structures—Assessment of Existing Structures [49], which addresses the assessment of existing structures, emphasizing their economic and cultural value. Regionally, the European Committee for Standardization (CEN) published EN 16853 [50], which defines the conservation process and provides guidelines on decision making, planning and implementation of conservation of the built cultural heritage environment.

While specialized standards for conserving earthen structures are limited, some guidance documents exist specifically for cob. Recently, The Cob Research Institute (CRI) developed a cob construction appendix, which has been approved for inclusion in the International Residential for building codes in the United States [51]. Similarly, the Devon Earth Building Association (DEBA) has proved that cob dwellings comply with United Kingdom building regulations and the suitability of cob to be used to build new buildings in accordance with such regulation [52]. Finally, the Getty Conservation Institute has also contributed guidelines on conserving earthen buildings, particularly through its Seismic Retrofitting Project (SRP) within the Earthen Architecture Initiative (EAI) framework (https://www.getty.edu/conservation/our_projects/field_projects/earthen/).

6.4 Contributions of Cob Vernacular Architecture Conservation Towards a Sustainable Built Environment

Rediscovering cob offers a bridge between ancient wisdom and modern sustainability. Its simplicity, versatility, and environmental benefits make it an attractive choice for contemporary construction. By understanding and applying the principles of cob building, we can create structures that are not only resilient and efficient but also deeply connected to the natural world. Moreover, by improving our understanding of cob's behavior under different conditions and refining our modeling techniques, we can better preserve and innovate within this ancient building method, contributing to sustainable and resilient construction practices for the future. These conditions can be fulfilled only if the existing rich heritage on cob vernacular architecture is properly preserved and the corresponding wisdom and know-how is adequately documented, communicated, and disseminated. Thus, in 2001 UNESCO launched the World Heritage Earthen Architecture Programme (WHEAP) aiming at developing policies for the conservation, revitalization, and increased awareness of earthen architecture. One of the notable outcomes of this project was the Inventory of World Heritage Earthen Architecture [53], which identified the use of earth on 150 UNESCO sites. The inventory also highlighted the use of cob in 29% of the sites surveyed. This indicated that cob was present in nearly 20% of the World Heritage List earthen architecture sites across four continents (see Fig. 6.7).

Unfortunately, these invaluable assets are to a large extent vulnerable to different socio economic and natural hazards, especially in front of climate change and its effects [55]. The deterioration or disappearance of cultural heritage represents a significant loss, both economic and cultural. Preservation efforts promote social cohesion, create collective identity, encourage cultural participation, and support economic development. They also contribute to urban planning, placemaking, and climate change resilience. These benefits have been recognized by the UN in Target

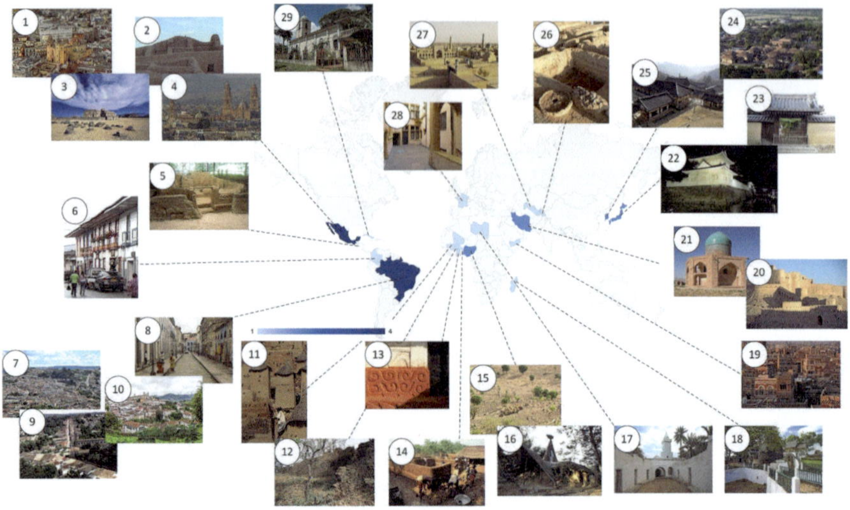

Fig. 6.7 UNESCO World Heritage List sites where the use of cob has been reported [53]: 1 Historic Town of Guanajuato and Adjacent Mines, 2 Archaeological Zone of Paquimé, Casas Grandes, 3 Historic Centre of Oaxaca and Archaeological Zone of Monte Alban, 4 Historic Centre of Morelia, 5 Joya de Cerén, Archaeological Site, 6 Coffe Cultural Landscape of Colombia, 7 Historic Centre of the Town of Diamantina, 8 Historic Centre of Sao Luís, 9 Historic Centre of the Town of Goiás, 10 Historic Town of Ouro Preto, 11 Cliff of Bandiagara, 12 Ruins of Loropéni, 13 Asante Traditional Buildings, 14 Royal Palaces of Abomey, 15 Sukur Cultural Landscape, 16 Osun-Osogbo Sacred Grove, 17 Old Town of Ghadamès, 18 Royal Hill of Ambohimanga, 19 Old City of Sana'a, 20 Bam and its Cultural Landscape, 21 Soltaniyeh, 22 Historic Monuments of Ancient Kyoto, 23 Historic Monuments of Ancient Nara, 24 Historic Villages of Korea: Hahoe and Yangdong, 25 Haeinsa Temple Janggyeong Panjeon, the Depositories, 26 Proto-urban Site of Sarazm, 27 Itchan Kala, 28 Historic Site of Lyons, 29 Viñales Valley

11.4 of the SDGs which states that safeguarding cultural heritage contributes to making cities and human settlements inclusive, safe, resilient, and sustainable [56]. Moreover, by safeguarding cultural traditions, the revitalization of local economies and the creation of job opportunities in rural settlements can be fostered [57]. Thus, endeavors to preserve the built cultural heritage environment are both required and desired.

As highlighted by Sharami and Hosseini [12]: "*It is possible to achieve sustainable contemporary architecture and urban design using contextual knowledge of vernacular architecture and combining it with new scientific achievements*". In that regard, the VerSus research project [54] identified principles and strategies inherent in vernacular architecture that can be applied to contemporary building projects. These principles organized around environmental, socio-cultural, and socio-economic sustainability, and intrinsic to most types of vernacular architecture, demonstrate how traditional knowledge can inform modern sustainable practices. Furthermore, specific strategies were proposed to achieve the successful adoption of these principles as graphically presented in Fig. 6.8.

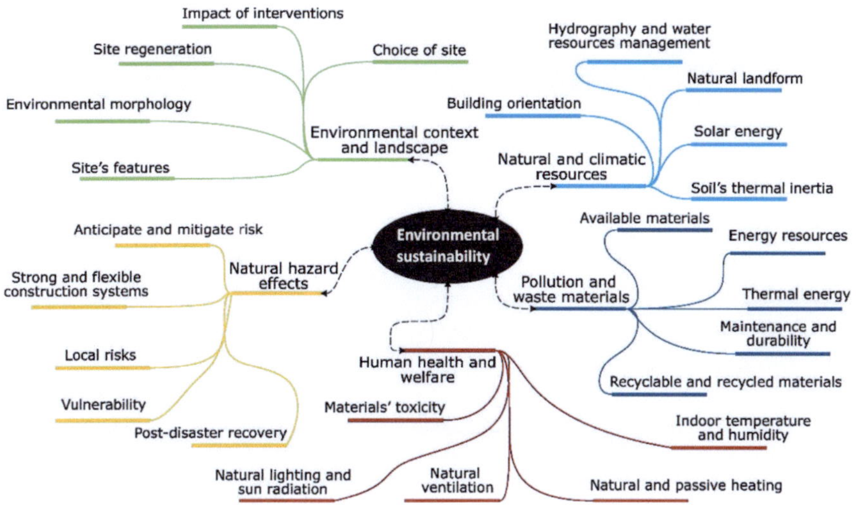

Fig. 6.8 Principles (first level items) and strategies (branched items) towards achieving: **a** environmental sustainability, **b** socio-cultural sustainability, and **c** socio-economic sustainability (developed based on [54])

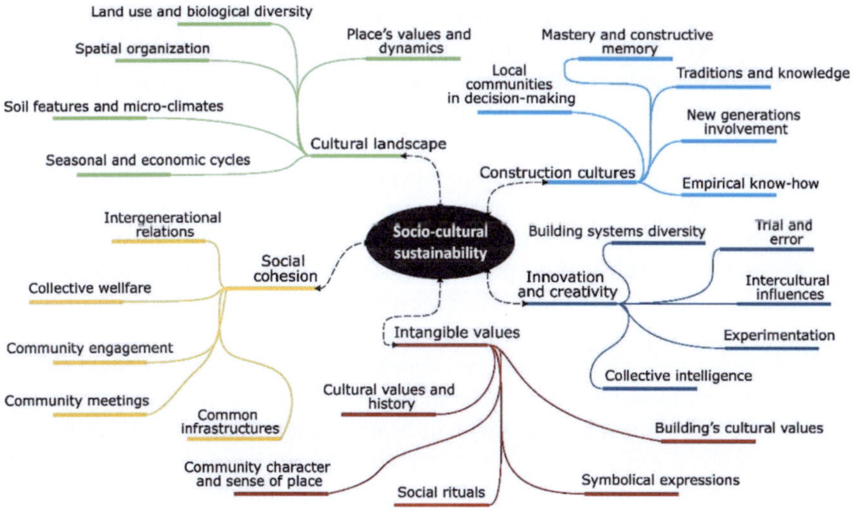

Fig. 6.8 (continued)

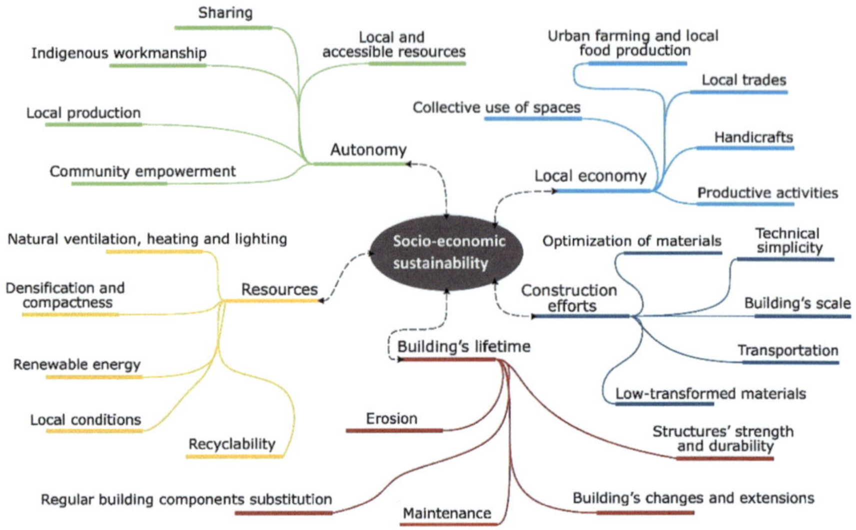

Fig. 6.8 (continued)

While Industry 4.0 has significantly driven digitalization and boosted productivity, it has been found insufficient to meet sustainability targets by 2030 [58]. This shortfall is due to its shortcomings addressing the environmental impacts of increased industrial activity and the challenges posed by automation to the human component of the socio-economic system. Consequently, a new paradigm based on three core principles: human-centrism, resilience, and sustainability, Industry 5.0, has emerged [59]. Within this transformative vision, social media has been recognized as a potential tool to foster and enhance collaboration among various stakeholders. Liang et al. [60] investigated the potential of social media to promote sustainability within the context of Built Cultural Heritage and Environment (BCHE) conservation, concluding that if utilized as a civic engagement tool, especially within the framework of the Historic Urban Landscape approach [61], social media could significantly contribute to sustainability. Tourism plays a vital role in the economy of areas with rich BCHE, attracting visitors with its beauty, history, and cultural significance. However, without proper management, overtourism can cause severe problems for both cultural heritage sites, such as anthropological degradation, and local communities, such as gentrification. Aiming at developing a sustainable tourism management model, de la Plata et al. [62] digitized cultural sites and presented the information to tourists in an engaging and accessible way, ensuring broad accessibility while preserving the heritage sites in Merida, Spain.

Further efforts to enhance sustainability practices within the conservation of the BCHE have focused on participatory approaches. Bizzarri et al. [63], using a multi-agent participatory methodology involving conservation professionals, a national government agency, local workers, and university students, managed to increase local community interest and encouraged residents of nearby villages to actively participate in conserving their cultural heritage. Their project, based on traditional skills and sustainable local materials, ensured the long-term maintenance of a case study site in Oman. Similarly, Ekici et al. [64] involved local communities, visitors, non-governmental organizations, and governmental bodies in developing policies for sustainable conservation practices in a rural village in Turkey. This approach, heavily influenced by social sustainability, effectively preserved the village's integrity. In both cases, it was observed that a crucial factor in the successful implementation of any participatory approach was "stakeholder consciousness," which refers to the awareness achieved through the active involvement of various agents in the co-design and co-implementation of heritage revitalization initiatives and cultural heritage policies.

Many researchers have emphasized sustainability by examining the circularity and adaptability of the BCHE. Redden and Crawford's extensive literature review supports the idea that historic buildings are highly sustainable and environmentally efficient [65]. They argued that some studies inadequately assess these valuable assets due to the implementation of limited methodologies that often overlook critical indicators like resource depletion, material waste, and pollution. Pintossi et al. [66] conducted a local case study in Salerno, Italy, addressing challenges posed by Adaptive Reuse of Cultural Heritage (ARCH) [67] through a participatory approach and proposed solutions for policymakers to promote ARCH for sustainable development based on BCHE conservation.

The achievement of the UN SDGs is seriously compromised as noted in the last progress SDG's report [68], and in particular Target 11.4, which aims to protect and safeguard the world's heritage, as there is no data to measure Indicator 11.4.1. This means that progress on Target 11.4 is neither being tracked nor discussed in progress reports. To achieve Target 11.4, member states need to enhance collaboration, allocate budgets, and develop infrastructure to support cultural heritage conservation. Local governments should implement policies that localize the global agenda, promote policy initiatives, and develop infrastructure towards heritage conservation, and especially towards earthen vernacular architectures, which is particularly vulnerable and often neglected.

Ensuring the continuous protection of earthen vernacular architecture requires a significant effort which could be compensated with a more sustainable future. Key activities to achieve this include: the training of conservation professionals, essential to preserving vernacular traditions and the associated irreplaceable knowledge and skills, which should be taught and researched in a modern context; wide-spread communication and dissemination activities to raise public awareness about the importance of conserving remaining earthen vernacular buildings; and further development and adoption of specialized standards for cob conservation would greatly help protect and preserve this rich cultural heritage.

References

1. C. Fadillah, C. Benedikta, Traditional heritage: vernacular architecture to vernacular typography with Rumah Gadang as the object. AIP Conf. Proc. **2710**(1), 020021 (2024). https://doi.org/10.1063/5.0144428
2. A. Rapoport, *House Form and Culture* (Prentice-Hall Inc., USA, 1969)
3. A. Jiménez Rios, D. O'Dwyer, *Earthen buildings in Ireland*, in *6th International Congress on Construction History*, ed. by I. Wouters et al. (CRC Press, Brussels, Belgium, 2018), pp. 787–794. https://www.taylorfrancis.com/chapters/edit/10.1201/9780429446719-12/earthen-buildings-ireland-jim%C3%A9nez-rios-alejandro-dwyer-dermot
4. M. Bostenaru Dan et al., *Architectural Heritage and Archetypal Landscape Approaches Facing Environmental Hazards.* Sustainability **16**(4) (2024). https://doi.org/10.3390/su16041505.
5. S. Piesik, *Habitat: Vernacular Architecture for a Changing Planet* (Thames & Hudson Ltd., UK, 2017)
6. B. Rudofsky, *Architecture Without Architects: A Short Introduction to Non-pedigreed Architecture* (The Museum of Modern Art, 1964). https://www.moma.org/documents/moma_catalogue_3459_300062280.pdf
7. M.M. Santos, A.V. Ferreira, J.C.G. Lanzinha, Sustainable vernacular architecture to improve thermal comfort in African Countries, in *Proceedings of CIRMARE 2023*, ed. by J.C.G. Lanzinha, E.L. Qualharini (Springer Nature Switzerland, Cham, 2024), pp. 564–575. https://doi.org/10.1007/978-3-031-48461-2_48
8. D. Bustán-Gaona et al., Natural lighting performance of vernacular architecture, case study oldtown Pasa, Ecuador. Energy Convers. Manag. X **20**, 100494 (2023). https://doi.org/10.1016/j.ecmx.2023.100494
9. M.S. Zami, A. Lee, Economic benefits of contemporary earth construction in low-cost urban housing–state-of-the-art review. J. Build. Apprais. **5**(3), 259–271 (2010). https://doi.org/10.1057/jba.2009.32
10. M.M. Alnaim, The typology of courtyard space in Najdi architecture, Saudi Arabia: a response to human needs, culture, and the environment. J. Asian Archit. Build. Eng. **23**(1), 91–110 (2024). https://doi.org/10.1080/13467581.2023.2229399.
11. M. Tarrad, U. Ahmad, Factors affecting the formation of architectural space and construction systems in the Jordanian vernacular architecture, in *Proceedings of the 3rd International Civil Engineering and Architecture Conference*, ed. by M. Casini (Springer Nature Singapore, Singapore, 2024), pp. 781–791. https://doi.org/10.1007/978-981-99-6368-3_63
12. H. Jafari Sharami, S.J. Hosseini, Theoretical framework of the Isfahani style: inspiring sustainable aspects of a vernacular urban development. Front. Archit. Res. (2024). https://doi.org/10.1016/j.foar.2023.12.008
13. UNESCO. *The 2005 Convention on the Protection and Promotion of the Diversity of Cultural Expressions* (2005). https://www.unesco.org/creativity/en/2005-convention
14. A. Jiménez Rios, Learning from the past: parametrical analysis of cob walls. Appl. Sci. **13**(15), 9045 (2023). https://doi.org/10.3390/app13159045
15. H. Huang et al., Bioclimatic design strategy of vernacular architecture in the south-east of China: a case study in Fujian, China. Int. J. Low-Carbon Technol. **19**, 1–17 (2024). https://doi.org/10.1093/ijlct/ctad079
16. A.A.M. Sayigh, *Bioclimatic Approach to Building Design* (1995). https://indico.ictp.it/event/a02304/contribution/23/material/0/0.pdf
17. P. Oliver, *Built to Meet Needs: Cultural Issues in Vernacular Architecture* (Routledge, 2007)
18. F. Pacheco-Torgal, S. Jalali, Earth construction: lessons from the past for future eco-efficient construction. Constr. Build. Mater. **29**, 512–519 (2012). https://doi.org/10.1016/j.conbuildmat.2011.10.054
19. M. Correia, G. Carlos, S. Rocha, *Vernacular Heritage and Earthen Architecture* (CRC Press, 2013)
20. L. Keefe, *Earth Building: Methods and Materials, Repair and Conservation* (Routledge, 2012)

21. E. Hamard et al., Cob, a vernacular earth constuction process in the context of modern sustainable building. Build. Environ. **106**, 103–119 (2016). https://doi.org/10.1016/j.buildenv.2016.06.009
22. L. Miccoli, U. Müller, P. Fontana, Mechanical behaviour of earthen materials: a comparison between earth block masonry, rammed earth and cob. Constr. Build. Mater. **61**, 327–339 (2014). https://doi.org/10.1016/j.conbuildmat.2014.03.009
23. G. Minke, *Building with Earth: Design and Technology of a Sustainable Architecture* (Walter de Gruyter, 2012). https://doi.org/10.1007/3-7643-7873-5
24. A. Laborel-Preneron et al., Plant aggregates and fibers in earth construction materials: a review. Constr. Build. Mater. **111**, 719–734 (2016). https://doi.org/10.1016/j.conbuildmat.2016.02.119
25. Historic England, *Practical Building Conservation, Earth, Brick and Terracotta* (2015)
26. British Standard, *BS 1377-1:2016 Methods of Test for Soil for Civil Engineering Purposes—Part 1: General Requirements and Sample Preparation* (British Standard, United Kingdom, 2016), p. 48
27. R. Harries, D. Clark, L. Watson, *A rational return to earth as a contemporary building material*, in *Terra 2000: 8th International Conference on the Study and Conservation of Earthen Architecture* (Devon, UK, 2000), p. 452. http://primo.getty.edu/GRI:GETTY_ALMA21136207270001551
28. L. Watson, K. McCabe, The cob building technique. Past, present and future. Inf. Constr. **63**(523), 59–70 (2011). https://doi.org/10.3989/ic.10.018
29. P. Jaquin, C. Augrade, *Earth Building: History, Science and Conservation*, vol. viii (IHS BRE Press, Bracknell, 2012), p. 87
30. M. Carter, S.P. Bentley, *Soil Properties and Their Correlations* (Wiley, 2016). https://doi.org/10.1002/9781119130888
31. C.W.W. Ng, B. Menzies, *Advanced Unsaturated Soil Mechanics and Engineering* (CRC Press, 2007). https://doi.org/10.1201/9781482266122
32. R.J. Ansaa-Asare et al., Impact of earth variability and implementation processes on hydromechanical and microstructure properties of sustainable construction materials. Constr. Build. Mater. **417**, 135303 (2024). https://doi.org/10.1016/j.conbuildmat.2024.135303
33. C. Ziegert, *Lehmwellerbau: Konstruktion, Schäden und Sanierung* (Fraunhofer-IRB-Verlag, 2003). https://books.google.ie/books?id=NzVRAAAACAAJ
34. L. Miccoli et al., Static behavior of cob: experimental testing and finite-element modeling. J. Mater. Civ. Eng. **31**(4), 04019021 (2019). https://doi.org/10.1061/(ASCE)MT.1943-5533.0002638
35. E. Quagliarini, G. Maracchini, Experimental and FEM investigation of cob walls under compression. Adv. Civ. Eng. (2018). https://doi.org/10.1155/2018/7027432
36. A. Jiménez Rios, D. O'Dwyer, Experimental validation for the application of the flat jack test in cob walls. Constr. Build. Mater. **254**, 119148 (2020). https://doi.org/10.1016/j.conbuildmat.2020.119148
37. A. Jiménez Rios, D. O'Dwyer, Data collected from the application of the single flat jack test in cob walls. Data Brief. **31**, 105764 (2020). https://doi.org/10.1016/j.dib.2020.105764
38. A. Jiménez Rios, D. O'Dwyer, Adaptations to the flat jack test for its application in cob walls. MethodsX **7**, 101003 (2020). https://doi.org/10.1016/j.mex.2020.101003
39. P. Poullain, M. Barnaure, S. Bonnet, Mechanical properties of cob-earth composites: variability and focus on the different calculation methods of Young's modulus. J. Build. Eng. **72**, 106622 (2023). https://doi.org/10.1016/j.jobe.2023.106622
40. R.H. Saxton, The performance of cob as a building material. Struct. Eng. **73**(7), 111–115 (1995)
41. O. Akinkurolere et al., Engineering properties of cob as a building material. J. Appl. Sci. **6**(8), 1882–1885 (2006). https://doi.org/10.3923/jas.2006.1882.1885
42. A. Jiménez Rios et al., Computational methods applied to earthen historical structures. Front. Built Environ. **9** (2023). https://doi.org/10.3389/fbuil.2023.1219108
43. A.M. D'Altri et al., Modeling strategies for the computational analysis of unreinforced masonry structures: review and classification. Arch. Comput. Meth. Eng. **27**(4), 1153–1185 (2020). https://doi.org/10.1007/s11831-019-09351-x

44. A. Jiménez Rios, D.W. O'Dwyer, Numerical modelling of cob's non-linear monotonic structural behaviour. Int. J. Comput. Meth. **17**(5), 1940013 (2019). https://doi.org/10.1142/S0219876219400139
45. ICOMOS, *International Charter for the Conservation and Restoration of Monuments and Sites* (ICOMOS, Venice, 1964). https://www.icomos.org/images/DOCUMENTS/Charters/venice_e.pdf
46. ICOMOS, *The Nara Document on Authenticity* (ICOMOS, Nara, Japan, 1994). https://whc.unesco.org/archive/nara94.htm
47. ICOMOS, *Charter on the Built Vernacular Heritage* (ICOMOS, Mexico, 1999). https://www.icomos.org/images/DOCUMENTS/Charters/vernacular_e.pdf
48. ICOMOS, *The Australia ICOMOS Charter for Places of Cultural Significance* (ICOMOS, Burra, Australia, 2013). https://openarchive.icomos.org/id/eprint/2145/
49. ISO, *ISO 13822:2010 Bases for Design of Structures—Assessment of Existing Structures* (ISO, Switzerland, 2010). https://www.iso.org/standard/46556.html
50. CEN, *EN 16853:2017 Conservation of Cultural Heritage—Conservation Process—Decision Making, Planning and Implementation* (British Standard, 2017). https://www.nen.nl/en/nen-en-16853-2017-en-233380
51. ICC, *2021 International Residential Code (IRC)* (2021). https://codes.iccsafe.org/content/IRC2021P2
52. A. Stokes, *Cob Dwellings, Compliance with the Building Regulations 2000 (As Amended)* (2008). https://www.devonearthbuilding.com/leaflets/building_regs_pamphlet_08.pdf
53. D. Gandreau, L. Delboy, *World Heritage Inventory of Earthen Architecture* (2012). https://whc.unesco.org/document/116577
54. M. Correia, L. Dipasquale, S. Mecca, *VERSUS: Heritage for Tomorrow* (Firenze University Press, 2014). https://library.oapen.org/handle/20.500.12657/55140
55. UNESCO, *Policy Document on the Impacts of Climate Change on World Heritage Properties* (2006). https://whc.unesco.org/en/CC-policy-document/
56. United Nations, *Transforming Our World: The 2030 Agenda for Sustainable Development* (2015). https://sdgs.un.org/sites/default/files/publications/21252030%20Agenda%20for%20Sustainable%20Development%20web.pdf
57. United Nations, *Tracking Progress Towards Inclusive, Safe, Resilient and Sustainable Cities and Human Settlements, SDG 11 Synthesis Report* (2018). https://uis.unesco.org/sites/default/files/documents/sdg11-synthesis-report-2018-en.pdf
58. European Commission, *Industry 5.0: A Transformative Vision for Europe* (2022). https://research-and-innovation.ec.europa.eu/knowledge-publications-tools-and-data/publications/all-publications/industry-50-transformative-vision-europe_en
59. European Commission, *Industry 5.0: Towards a Sustainable, Human-Centric and Resilient European Industry* (2021). https://data.europa.eu/doi/10.2777/308407
60. X. Liang, Y. Lu, J. Martin, A review of the role of social media for the cultural heritage sustainability. Sustainability (Switz.) **13**(3), 1–17 (2021). https://doi.org/10.3390/su13031055
61. UNESCO, *Recommendation on the Historic Urban Landscape* (2011). https://unesdoc.unesco.org/ark:/48223/pf0000215084.page=55
62. A. Rueda Márquez de la Plata, P.A. Cruz Franco, J.A. Ramos Sánchez, Architectural Survey, Diagnostic, and Constructive Analysis Strategies for Monumental Preservation of Cultural Heritage and Sustainable Management of Tourism. Buildings **12**(8) (2022). https://doi.org/10.3390/buildings12081156
63. S. Bizzarri et al., A multidisciplinary approach to the conservation of Salūt, Sultanate of Oman. Stud. Conserv. **67**(S1), 23–31 (2022). https://doi.org/10.1080/00393630.2022.2084965
64. S.C. Ekici, Ö. Özçakır, A.G. Bilgin Altinöz, Sustainability of historic rural settlements based on participatory conservation approach: Kemer Village in Turkey. J. Cult. Herit. Manag. Sustain. Dev. (2022). https://doi.org/10.1108/JCHMSD-05-2021-0087
65. R. Redden, R.H. Crawford, Valuing the environmental performance of historic buildings. Aust. J. Environ. Manag. **28**(1), 59–71 (2021). https://doi.org/10.1080/14486563.2020.1772133

66. N. Pintossi, D. Ikiz Kaya, A. Pereira Roders, Cultural heritage adaptive reuse in Salerno: challenges and solutions. City Cult. Soc. **33** (2023). https://doi.org/10.1016/j.ccs.2023.100505
67. G. Foster, Circular economy strategies for adaptive reuse of cultural heritage buildings to reduce environmental impacts. Resour. Conserv. Recycl. **152**, 104507 (2020). https://doi.org/10.1016/j.resconrec.2019.104507
68. United Nations, *The Sustainable Development Goals Report 2023* (2023). https://unstats.un.org/sdgs/report/2023/The-Sustainable-Development-Goals-Report-2023.pdf

Chapter 7
Earth as a Building Material: From Traditional Building Techniques to Additive Manufacturing

Flora Faleschini, Daniel Trento, and Mariano Angelo Zanini

Abstract The research interest for earth-based building materials has increased extremely in recent years because of a tremendous rise of housing-demand worldwide, together with the necessity of fostering environmentally friendly solutions within the construction industry. Nevertheless, the diffusion of earthen construction is still limited due to many reasons among which it is worth mentioning the poor durability of these materials (e.g., their sensitivity to moisture) and the lack of Codes and regulation for their design. In the meanwhile, new construction methods disrupt the market, among them, additive manufacturing (AM) has demonstrated its potential to transform the construction industry, especially through the 3D printing technology by extrusion. This technique has been widely studied for concrete production but only recently has been applied for earth-based materials, being only validated under laboratory conditions with few real-scale applications. In this chapter, a review on earth-based materials for the building industry is presented, paying attention to the characteristics needed by these materials to be suitably applied in modern AM processes. Advantages, current limits and open challenges in 3D printing such geomaterials will be presented. Lastly, some case studies will be shown.

7.1 Introduction

As the world has been experiencing exponential population growth over recent years [1], the need for new buildings and infrastructure, especially in developing countries, is a highly relevant issue [2, 3]. The construction sector remains one of the key economic industries [4], and its impact on global economy is significant, as about 110 million people are employed directly in this sector [5], which corresponds to 7–10% of the actual workforce [4]. Furthermore, it plays an important role in determining the quality of life [6], nonetheless this sector is responsible for 35% of all carbon

F. Faleschini (✉) · D. Trento · M. A. Zanini
Department of Civil, Environmental and Architectural Engineering, University of Padova, Padova, Italy
e-mail: flora.faleschini@dicea.unipd.it

D. A. H. Hanaor (ed.), *Sustainable Materials for the Built Environment*, Springer Series in Materials Science 352, https://doi.org/10.1007/978-3-031-97818-0_7

emissions and produces between 45 and 65% of the waste currently landfilled [7]. In this context, the 2021 global construction industry report highlights the absence of a significant green revolution within the sector as construction industry emissions are projected to increase, further contributing to global warming [8, 9]. The relevance of sustainability in the building industry is clear and evident since stakeholders are increasingly aware of respecting the environment and ethical values as collective responsibility [10].

The implementation of sustainable construction materials and techniques significantly enhances the eco-efficiency of the construction sector, thereby fostering a more sustainable form of development. Concrete is still the most-used construction material and the second most-consumed substance in the world, just after water [11]. Considering all stages of production, cementitious concrete is responsible for 4–8% of the world CO_2 emissions [12]. Hence, the need to replace cementitious conglomerates with more sustainable materials to improve constructions environmental footprint has led to consider earth as a viable alternative [13]. At the same time, developed countries are recovering and valorizing ancient construction systems, which involve locally available materials like earth [14]. Earth materials typically consist of mixtures of soil and water, often incorporating fibers like straw [13]. These materials are commonly utilized for external walls, offering significant thermal mass that enhances passive thermal design [15]. Their minimal environmental footprint is attributed to the low energy consumption during construction and the high recyclability of nearly all building materials used, which are frequently entirely natural and compatible [16]. In this way, earthen construction has been encouraged as an economically attractive solution joining with the value of vernacular architecture. In fact, a new sensibility is spreading, embraced by cultural heritage values, moving towards a functional philosophy that views architectural assets not merely for their historic and artistic significance but also as representation of a traditional building art that requires comprehension rather than rejection, therefore, the idea of preservation should encompass building methods as well.

Parallelly, the spread of technology and automation processes to accelerate industry production is hard to reconcile with traditional and old valuable manufacture techniques. Over the last few decades, all economic sectors have been migrating progressively to digital processes to achieve higher efficiency. This has happened to the construction industry as well, which now offers greater complexity of shapes and methods thanks to digital manufacturing [17]. Numerous studies have shown that digital/automated processes can offer significant advantages to the construction sector, including increased design flexibility and improved productivity [18]. As construction techniques with concrete, timber and masonry are undergoing a transformation due to the integration of digital manufacturing methods like 3D printing and robotic fabrication, earth construction remains one of the least explored building types that has not been directly involved by this renovation [19, 20]. Nonetheless, some studies highlighted that integrating digital manufacturing with earth vernacular construction methods can lead significant sustainability potentials as compared to other materials [15, 21].

This chapter aims to critically review the state of state-of-the-art on earth constructions. The chapter primarily explores the properties of earth as a building material, showing the traditional techniques and the available standards. In the second part, additive manufacture is analyzed as an innovative construction process for earth; the insights include processing methods and properties, also making comparison with other printable construction materials. Finally, some real-scale buildings realized with such a technology are presented and open challenges are analyzed.

7.2 Earth Constructions

7.2.1 Historical Overview

Earth has been used as a construction material for millennia, however, there is no consensus about when mankind began to use this method. Some adobe-based dwellings discovered in Turkmenistan and dated on the period between 8000 and 6000 BC may suggest that earth construction technique was adopted more than 9000 years ago [22, 23]. According to another source [24], the most ancient adobe blocks were found along the Tigris River basin dating back to 7500 BC, while in Mesopotamia some evidence on the use of earth for construction purposes dates from the period between 5000 and 4000 BC [25]. Summarizing, it is likely that earth constructions appeared before 9000 years ago, with the start of early agricultural societies, a period comprised between 12,000 and 7000 BC [23]. However, many examples of ancient earth constructions have endured over the years, making it to now. In China, during the Tang dynasty (618–907), fortified settlements as Jiahoe, Gaochang and Xi'an were encircled with large, rammed earth walls, whereas the city of Kashgar was realized in adobe [26, 27]. The Ziggurat of Ur, erected about 2100 BC, featured a core made of adobe bricks and was covered with fired bricks held in place by bitumen (Fig. 7.1a). In North Africa, the Phoenicians expanded from eastern Mediterranean establishing settlements along the African north coast. Among them, their capital Carthage (now in Tunisia) was founded in 814 BC, and archeologists demonstrated earth walls were used here [28]. Rammed earth technology may have been carried to Europe by this civilization when Phoenicians expanded in the Iberic Peninsula and Mediterranean islands. Rammed earth towers built by Hannibal are described by Roman historian Pliny, and architect Vitruvius describes rammed earth used in the French city of Marseilles and the Greek city of Athens being constructed entirely from adobe. [26]. When Islam civilization arrived in Southern Europe in 711 AD, conflicts at this time led to build many rammed earth and adobe fortifications [28]. The fortifications of Calatayud and Plad'Almalain in Spain, dating back to 884 AD, were uncovered through excavations [28]. Additionally, the Muslim defensive walls surrounding the historic cities of Cordoba, Seville, and Granada were constructed using rammed earth techniques. The Alhambra Palace in Granada (Fig. 7.1b), which is now a World Heritage site, was built from rammed earth around

1238 [29]. Through time, earth has continued to be used as a construction material, but it was progressively replaced by the increasing use of fired bricks. Earth-based materials were used by Native Americans in modern southern United States. The Pueblo peoples in contemporary New Mexico built adobe constructions that housed multiple families and rose several stories high. Among these structures, the Taos Pueblo stands out as the most famous, dating back to around 1000 AD (Fig. 7.1c). In South America, the former recorded earth bricks date back to the Moche culture, which developed from 100 AD [26]. The city of Cerro Blanco in Peru was the center of this civilization, where two adobe pyramids dedicated to the sun and the moon were built, named Huaca del Sol (Fig. 7.1d) and Huaca de la Luna [26]. Historical evidence has shown that earth building techniques have developed separately in several parts of the world. A common aspect can be observed among them, i.e. the soil and the subsoil for earth constructions were employed when timber or stone were not available. Indeed, if stone or timber were available, they would be preferred by local populations.

The global significance of earthen architecture is evident and deserves acknowledgment, safeguarding, and preservation by the international community. It is worth mentioning that about 10% of World Heritage sites comprise earthen monuments or sites, being an important part of mankind's history [34].

Today, earth is not considered a common building material, but the number of households living in earthen dwellings worldwide ranges between 8 and 10% [35],

Fig. 7.1 Earth-based constructions from the past: **a** the Ziggurat of Ur [30]; **b** the Alcazaba fortress of the Alhambra Palace in Granada [31]; **c** Taos Pueblo in New Mexico [32]; **d** Huaca del Sol in Peru [33]

while almost 30% of the world constructions are made from earth [13]. These architectures can also be found in France, Germany and the UK that has more than 500,000 earth-based manufacts [23]. However, the importance of this construction method has been increasing substantially largely due to the sustainable agenda, in which the construction sector plays an important role.

7.2.2 Earth Construction Techniques and Properties

Earth architecture is a construction technique based on local materials and local needs, where local builders' know-how plays an important part. These buildings are typically realized with a mixture of soil, water, and fibers with different techniques according to the mixture proportions and the building process characteristics [36]. There are several methods to classify earth construction techniques for producing and shaping the geometries, but the most affordable is based on the distinction between dry and wet methods [15] (Fig. 7.2).

In dry methods, the earth mixture is placed at its optimum water content (e.g., evaluated by Proctor test), and mechanical strength is achieved through compaction densification [15]. Rammed earth and compressed earth/sand blocks are included in this category, which are all load bearing [15, 36]. Instead, wet methods refer to techniques involving mixtures with relatively high-water content that achieve a plastic state, the mechanical strength is developed through densification and shrinkage during the drying process until reaching an optimum moisture content. Among them, adobe, cob, wattle and daub, and earth plaster are the most well-known, but just adobe and cob have load-bearing capacity [13, 15, 36].

Typically, topsoil is not used in earth mixtures to avoid any contamination with organic substance (minimum excavation depth is one meter). Subsoil is usually preferred and mixed with water, fibers and stabilizers when required.

The process of making adobe bricks involves filling wooden molds with damp earth, which are subsequently left out in the sun to dry. When the blocks dry, shrinkage

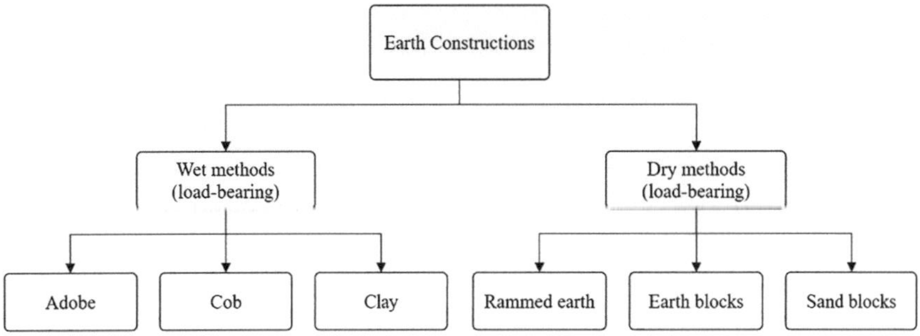

Fig. 7.2 Classification of load-bearing earth construction methods adapted from [12, 14, 31]

cracks may appear on the surface, for this reason straw or other natural fibers are used to avoid this phenomenon [37] (Fig. 7.3a). Cob technique consists in building masonry walls layer by layer with a mixture of earth, water, and straw [16] (Fig. 7.3b). Clay is a sub ingredient of earth constructions, it can be used mixed and manufactured, or as it is found in nature for its excellent cohesive properties.

Rammed earth consists of walls formed by compacting soil between temporary formworks [38] (Fig. 7.3c). Earth blocks and sand blocks on a smaller scale can be realized in the same way. Earth projection technique is often included in rammed earth category [23], it consists in compacting and stabilizing the soil to projecting it inside molds as for shotcrete.

Earth mixtures are employed even for non-load bearing applications, among them wattle and daub architecture entail erecting walls using frames made of interwoven or lashed wooden elements (wattles). These frames are then coated with a thick layer of puddled, clayey soil (daub) [38] (Fig. 7.3d). Earth plasters are made from clayey soils mixed with a certain amount of sand; they are appreciated for being permeable to water vapor [39].

Fig. 7.3 Earth-based construction methods: **a** Adobe bricks wall [40]; **b** Cob wall [41]; **c** Rammed earth [42]; **d** Wattle and daub [43]

7.2.3 Current Standards

The earthen architecture is highly variable depending on the territory where it is diffused, implying heterogeneity in terms of quality of the available resources, environmental conditions, and needs of the users. Hence, the need to regulate these buildings in different geographical areas has risen in many countries according to the expected societal needs and constraints. Thus, the regulatory framework is very fragmented. Most documents have not the characteristics of standard, typically referring to specific aspects of earth constructions.

Table 7.1 summarizes some of the most relevant available standards worldwide, the instructions are provided for: object, requirement of the product, methodology, quality control and anti-seismic criteria. In this context, the regulations that have transferred the concepts and best practices in standard tools recognized by the national standard bodies or equivalent authorities are considered exclusively for earth-based materials without the use of stabilizers.

7.3 Additive Manufacturing for Earth-Based Materials

7.3.1 Additive Manufacturing as a Construction Technique

Additive manufacturing (AM) is a production process consisting in adding thin layers of material consequently on top of each other [44]. It finds applications in several industrial fields, as it can be used with a wide variety of materials.

In this context, the construction sector has been introducing digitally based technologies, e.g. additive manufacturing, to speed up the production rate, reduce the labor costs and afford design freedom [45–50]. Adopting AM methods in modern constructions provides three significant benefits [13, 51–53]. First, it reduces the materials/resources consumption and increases the efficiency, with an estimated 25–60% material consumption reduction and 30% time saving compared to traditional manufacturing techniques [45, 46, 54]. This advantage is considered of great importance for remote areas and those affected by natural disasters and war conflicts, where AM would allow the quick realization of new houses and infrastructures [49, 55]. Second, AM extends the product life, making the repair and retrofitting processes easier [13]. Third, it enhances value chains by offering shorter, and more localized production and supply [17, 56]. Another great potential from AM is the reduction of formworks use and related provisional structures during construction. Traditionally, formworks are used to cast concrete using a subtractive process to shape the fresh concrete as it hardens. Their cost linked to labor time and materials impacts from 35 to 60% of the overall concrete cost [56, 57].

In the construction industry, substantial research has been developing over the last few years on AM for concrete and cement products. As a result, a great development in 3D printing for cement-based materials has been reached [58, 59]. In terms of

Table 7.1 Standards for earth-based materials and buildings

Area/ country	Standard	Object	Requirements of the product	Methodology	Quality control criteria/ anti-seismic criteria
Africa	ARS 670-683	Compressed earth blocks and earth mortars		Standard for classification and identification of the material, mechanical tests	
France	AFNOR XP P13-901	Compressed earth blocks	Geometric, physical, and mechanical characteristics of the blocks. Characteristics of the constituents	Geometric characteristics, dry and wet compressive strength, capillarity, abrasion resistance	
Germany	DIN 18945	Earth blocks	Geometry, mechanical performance, deformability, density, content of harmful compounds, resistance to water, reaction to fire, thermal conductivity	Shape and dimensions, compressive strength, density, resistance to water vapor diffusion, behavior under humidity and frost	Certification of compliance. External monitoring and certification by a recognized body. Declaration of conformity Product data sheet
	DIN 18946	Earth masonry mortar	Compressive strength, density, bond, shear strength, resistance to water, resistance to fire	Requirements refer to on-site processing, properties of the fresh and hardened mortar	Inspection and certification by a recognized body. Certificate of compliance. Declaration of conformity, Product data sheet
	DIN 18947	Earth plasters	Flexural and compressive strength, density, bond, resistance to water, resistance to fire	Requirements refer to on-site processing	Factory production control by the manufacturer. Declaration of conformity

(continued)

Table 7.1 (continued)

Area/ country	Standard	Object	Requirements of the product	Methodology	Quality control criteria/ anti-seismic criteria
India	IS 13827	Earthen buildings			Seismic strengthening of bearing walls. Internal bracing systems. Holdfasts to the walls
New Zealand	NZS 4297	Engineering design of earth buildings	Design requirements, design criteria	The material must contain clay and silt to achieve satisfactory results with or without chemical stabilization	Earthquake loads are analyzed
	NZS 4298	Earth buildings	Requirements for the materials and workmanship requirements for the unfired earth	Requirements for rammed earth, adobe bricks, pressed bricks, CINVA-ram bricks, poured bricks	Mode of walls at large deformation
	NZS 4299	Earth buildings not requiring specific design	Performance criteria, design requirements		Minimum bracing or bond beams with structural diaphragms depending on the seismic zone
Peru	NTE 080	Earth buildings in simple or stabilized adobe	Features, performance, and design		Seismic behavior of adobe buildings

(continued)

Table 7.1 (continued)

Area/ country	Standard	Object	Requirements of the product	Methodology	Quality control criteria/ anti-seismic criteria
Spain	UNE 41410	Compressed earth blocks	Geometry, mechanical performance, durability, resistance to water, resistance to fire	Size and tolerance, density, constituents, compressive strength, resistance to wet/ dry cycles, resistance to freeze/thaw cycles, water vapor permeability, reaction to fire	
USA	ASTM E2392/ E2392M	Earthen wall building systems	Technical requirements and considerations for sustainable development	Requirements for adobe, cob, rammed earth, and other techniques used as structural and non-structural systems	
Zimbabwe	SAZS 724			Compressive strength test, drop test, roll test, formwork deformation test, wall density test, erosion test, render adhesion test	

mechanical properties, one of the most significant features of 3D printed concrete is its anisotropy. Indeed, due to layer deposition the compressive strength in the parallel printing direction is higher than those of the other two directions, whereas flexural strength in direction of stacked layers is higher than that in direction of parallel printing paths [58, 60]. Further, interlayer interfaces are inevitably created due to the layer-upon-layer deposition. In this region, increased porosity can be found, which affects mechanical strength, this being one of the main reasons for anisotropy [61].

Besides technical aspects, the introduction of AM in construction projects would change the way construction works are performed, with non-negligible social consequences. Security is a major concern in the building industry because injuries and fatalities from construction works cause great losses to individuals and societies. In AM process, most dangerous works are performed by printers with automated tasks, resulting in reduced injuries [46, 56, 62]. Further, social consequences can be seen in a larger spectrum, as the printing process is highly automated, manpower required can be drastically reduced, and such a reduction will introduce new type of construction workers who need intellectual skills rather than manual ones, thus improving gender equality [45, 54].

On the other side, these benefits highly depend on the context, size, and complexity of the construction project. Regarding the use of robots on-site and off-site, some challenges include the interaction of them with people, and the interoperability of more machines contemporarily, dealing with possible overlapping working areas. Some solutions have already been proposed in literature, including programming strategies, design of new robotic platforms, and the use of perceptional systems [53, 63].

7.3.2 3D Printing with Earth

The integration of vernacular construction techniques with digital fabrication can lead to several social and sustainability advantages. However, the introduction of a new technology, e.g. additive manufacturing, should maintain and represent the local identity, avoiding standardizing the product. When earth is employed for digital fabrication, the potential of realizing sustainable constructions with vernacular knowledge generates responsible innovation by leveraging local industries and technologies, utilizing local resources [64, 65].

7.3.2.1 Processing Aspects

Material control in AM is one of the most challenging areas, the rheological properties of printable materials at the fresh state significantly differs from those of conventional ones [66, 67]. Among the developed techniques for additive manufacturing, extrusion-based methods are the most used and studied for construction purposes [68, 69]. In this kind of process, layers are extruded with a robot by pression and screw-driven mechanisms. The efficiency of this process is linked to the buildability of the material and the speed rate [69]. This construction process can be resumed into three stages:

1. *Pumping and extrusion.* During this step, the mix must be fluid enough to flow through the pump and exit smoothly from the nozzle avoiding segregation, phase separation or blockages [70]. To ease this phase, the yield stress and viscosity of the material should be maintained at minimum, to limit the stresses inside the extruder equipment and the pressure.
2. *Placement and early-stage.* Here the material must maintain its shape developing sufficient early strength to support the succeeding layers without excessive deformations. To ensure such a behavior, the yield stress and viscosity should reach a minimum level, and this contrasts with the previous phase where lower values

of these parameters are preferred. When deposited, the material to be printed must show a homogeneous surface quality without visible cracks, tears, splits, or pores [71]. To deal with these issues, ensuring a balance between the nozzle travel speed and extrusion rate is essential [72, 73]. Indeed, an excessive extrusion rate can lead to issues such as compression stresses on the filament near the nozzle; at this regime, the extruded material expands around the nozzle and localized buckling may occur [74]. Instead, high travel speed leads to tensile stresses inside the filament, and the printed filament may show tearing or cracking. Other aspects of great importance deal with the nozzle, in terms of dimension, shape, height, and movement path: all these parameters affect the printing quality. When the nozzle opening is excessively small, friction increases between the nozzle and the extruded material, which generates tangential stresses on the filaments, causing tears and deformations. Furtherly, a small nozzle is more prone to face blockage [75]. The nozzle shape also affects intercepting shock, which is typically higher in rectangular dies due to the presence of corners [72, 75].

3. *Hardening*. This stage occurs when the material hardens and reaches its final mechanical properties. The deposited material needs to be hard enough to support the increasing load of consequently deposited layers.

To ensure fast production and structural stability, the mixture rheology must be carefully optimized to satisfy the requirements of each of the above steps. For any printing material, a good balance between fluidity for pumping and structural stability should be achieved by mixture optimization: this applies to earth, too. When dealing with such a material, it should be recalled that the strength evolution over time is highly non-linear. Without stabilizers, drying is the only mechanism that allows for strength development, and due to its lengthy this may represent a hard limit. To deal with it, fast-setting stabilizers, e.g. alginate, are often added to allow rapid strength evolution. According to Perrot et al. [19], earth with alginate needs 0.1 h to develop sufficient strength to sustain the weight of 1 m wall, whereas the same strength without alginate is reached in more than 24 h. Apart from strength development, elastic properties evolution is also important to avoid the structure collapse by buckling. Indeed, slender printed structures may collapse due to self-buckling [76], this phenomenon is linked to insufficient elastic modulus profile along the printed element. As for the strength, the evolution of elastic modulus can be accelerated by adding proper stabilizers. Anyhow, when the setting time is not enough, cold joints between extruded layers may appear, this indicates that the time interval between each successive deposit was too long to ensure good adhesion between the layers. Hence, a compromise should be attained in terms of fresh open time and buildability to ensure good adhesion between layers and stability.

7.3.2.2 Properties

Earth is a unique material for AM in construction being recyclable and climate friendly, with minimal carbon impact and potential passive thermal performance

[77]. Generally, the mechanical properties of earth-based mixtures depend on several factors, i.e. soil properties, moisture content, presence and quality of fibers, workmanship expertise. Compressive strength in printed earth elements is comparable to that of traditional ones, however the nozzle feature can affect final properties. In the case of circular nozzles, some voids might remain among layers due to the circular cross section of the extruded material, whereas a rectangular nozzle allows fulfilling the cross-section [78]. Perrot et al. [19] experienced a compressive strength of about 1.2 MPa for earth blocks with circular nozzle, whereas the same blocks printed with rectangular section die exhibited compressive strength of about 1.7 MPa. Compressive strength often has a close relationship with bulk density, as higher density values result in higher compressive strength.

A relationship can also be observed between density and conductivity, but in this case walls with lower density result in better thermal insulation. Gomaa et al. [79] performed thermal conductivity tests on 3D printed walls at 1:4 scale with four possible wall-section variations. Results demonstrate that the thermal conductivity of 3D oriented earth walls can reach 0.32 W/mK, which is lower than most conventional brick and masonry walls. Such a good insulating performance is due to the incorporated inner voids in the wall-section.

Concerning environmental impacts, 3D printing for monolithic constructions and prefabrication has lower energy consumption compared to traditional techniques [80]. Further, if earth is used as printing material the global warming potential is even lower than for concrete printing [81].

The feasibility of the use of 3D printed earth to act as load-bearing walls in low-rise constructions was also proved by previous works [82], offering a correlation between 3D cross section design and the overall design of the building.

7.3.2.3 Fibers in 3D Printing of Earth-Based Materials

Differently from earth for traditional building techniques, 3D printable mixtures usually have a high-water content, i.e. about 20–25%, and a reduced fiber content to reduce viscosity and ease the extrusion, thus resulting in high shrinkage during the drying process with possible extensive cracking [13], and low thermal properties [83]. However, fiber addition in 3D printing should be considered as a viable solution to improve flexural strength and ductility while also increasing thermal resistivity and lightness, contributing to construction economy and transportation viability [84]. Looking at traditional earth constructions, fiber reinforcement has demonstrated to be effective to give stability for shape and water retention [85]. Most fibers used in traditional earth constructions are derived from plants, and the fiber types may vary notably according to local availability and vegetation [84]. Fibers may also limit the lateral strains under compression behaving as a diffuse confinement to increase the energy absorption capacity, tensile strength, ductility and toughness [86, 87]. Furtherly, the thermal resistivity increases for higher fiber content and length that may grant thermal comfort indoor [83, 84, 88].

Gomaa et al. [79] proposed a 3D printable earth mixture with straw at 2% by weight, reaching values of thermal conductivity comparable to those of traditional cob. Beyond the fiber content of 2% by weight, the authors highlighted extrusion problems due to clogging and increased viscosity [79]. In a second work, the same authors [20] optimized the mixture slightly adjusting the water content but maintaining a straw content at 2% by weight. Lately, Binega Yemesegen [89] confirmed through literature review that printable earth mixtures should contain 72–73% subsoil, 25–26% water and 2% fiber by weight. However, Faleschini et al. [65] proposed several mixtures with comparable mechanical properties employing jute, coconut fiber, sisal, and goat hair at 0.5% fiber content by weight, optimizing the printability without affecting mechanical properties. Recently, Carcassi et al. [84] presented 3D printable earth-mixtures maximizing natural fiber content from commonplace "cob" with 2% fibers by weight to "light fiber clay" containing 13% fibers.

7.3.3 Standards on Additive Manufacturing for Constructions

Additive manufacturing has emerged as a revolutionary tool in the construction industry with the potential to significantly enhance sustainability in the building industry and speed up construction processes [90]. Although this technology holds huge potential in the future of construction industry, the number of codes and standards considering the use of 3D printing in construction is currently few. For 3D printing to be widely adopted in construction, there should be compliance with existing codes and standards, which are designed for traditional building methods. This highlights the importance of developing new codes for 3D printing to aid in transforming traditional building materials into printable ones. This transformation is crucial for moving from formwork-based technologies to more automated additive manufacturing approaches [91].

Future standards for 3D printing in construction should address several key aspects to ensure its efficient application. These include guidelines for geometric complexity, ensuring the safety and reliability of complex structures enabled by 3D printing. Standards should specify the printed object size, minimum number of layers, and procedures for compressive and flexural strength tests for mechanical properties testing. Additionally, guidelines for integrating steel bar reinforcement within printed layers are necessary to enhance structural load-carrying capacity, reliability and durability. Finally, specifications for nozzle size and particles dimension are crucial, as they significantly impact the quality and properties of 3D printed structures [92]. As the field of AM gained the attraction of researchers and continues to grow, these standards will likely be updated and expanded to keep pace with technological advancements and industry needs.

Some efforts to establish standards for 3D printing in constructions are underway, with notable initiatives such as those by the firm Winsun. They worked with China's national construction department to update traditional building codes to adapt standards to 3D printing technologies [93]. Recently, ISO/ASTM 52939:2023 was published, this document sets the requirements for additively manufactured structural and infrastructure elements to ensure safety and efficiency in 3D printing construction industry. The purpose of this document is to outline the essential requirements for the production and delivery of high-quality additively manufactured structures, whether for residential or infrastructure purposes, in the construction sector. Indeed, important steps are defined which will be controlled and monitored to ensure high quality printed constructions. However, this document is not technology-, or material-specific, and consequently more detailed regulations should be applied depending on the adopted approach. Nonetheless, printed elements must be approved by a locally certified engineer and comply with both local and regional specifications and requirements. Further, ISO/ASTM 52900 and ISO/ASTM 52950:2021 are about general principles in AM, the former focuses on fundamentals and vocabulary, the latter shows an overview of data processing.

7.3.4 Real-Scale Examples of 3D Printed Constructions

After the development of contour-crafting [94], many 3D-printed large-scale structures were built [95]. Most of these printed buildings are made from concrete [96], but there are examples of earth-based 3D-printed houses which will be discussed in this section. In 2017 a 20 m^2 pavilion was printed in 15 days at an international construction fair to show the combination of robotic construction and natural material [97]. Some key limits identified in this project regarded the maximum 3D printing height, here limited to 20 cm/day due to slow hardening of adobe, and also to the connections among printed walls and the other building components, such as foundation piles, window framing, roof structures, etc. [97]. WASP (World's Advanced Saving Project) is an Italian company which is one of the pioneering companies in onsite 3D printing with earthen materials and renowned for its innovations in 3D printing technology. One of their first project is GAIA [98], a 30 m^2 house made of locally available material soil (30% sand, 30% clay, and 40% silt), natural waste (25% rice husk, 40% straw-chopped rice), and 10% hydraulic lime showcasing the potential of 3D printing in sustainable constructions (Fig. 7.4a). The other one is TECLA [99], printed with Crane, i.e. WASP's large-scale 3D printer, in 2021. TECLA was formed of two connected dome-shaped volumes with an outer wall comprising 350 stacked layers of 3D-printed clay with a thickness of 12 mm (Fig. 7.4b). WASP in collaboration with IAAC (Institute for Advanced Architecture of Catalonia) also printed a wall portion 40 cm thick. The mix for this wall is the mixture of clay and rice fibers which was used in the previous projects by WASP. The process of printing this wall took 40 h [100].

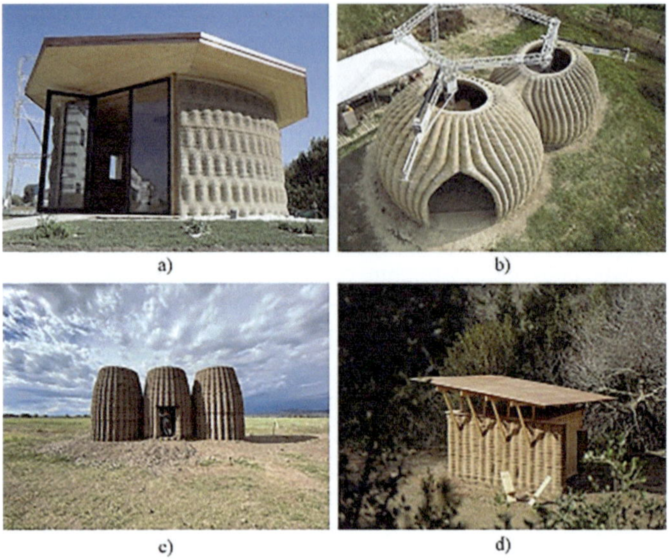

Fig. 7.4 Real-scale examples of 3D printed constructions: **a** GAIA prototype [98]; TECLA prototype [99]; Casa Covida [101]; TOVA prototype [102]

Another example of 3D printed house is Casa Covida (Fig. 7.4c) which is in Colorado's San Luis Valley. The name of the house underscores the project response to the COVID-19 pandemic and its impact on how people live and design spaces. This building is made of a mixture of sand, silt, clay, water, and straw, all are locally available. This house comprises 3 separate spaces which are for co-habitation during the time of pandemic [101].

Finally, TOVA (Fig. 7.4d) is the first 3D printed house in Spain which is made of locally available material soil, aloe, egg white and enzymes [102].

7.3.5 Future Developments and Open Challenges

Although AM in the building industry leads to several advantages, its use and advancement are still in their beginning and face numerous obstacles.

Typically, printed surfaces are rougher than conventional cast surfaces, especially when a circular-section nozzle is employed. Hence, such a manufacturing process may provide structures which do not fit with designers' and end users' requirements, as per the current technology development. Nevertheless, some stakeholders appreciate the printed surface texture, especially for earth-based materials, as it gives a raw appearance that matches with soil.

Another issue is related to the compatibility of raw materials for printing. The constituents, especially the soil, need to fulfill certain requirements to be compatible with AM in terms of printability and buildability. The greatest advantage of AM with earth-mixtures lies in the use of local soil, especially when the building site is in remote areas. Although, if the local soil is not suitable for printing, the transportation required to bring the desired material would increase the costs and jeopardize the economic and environmental benefits of choosing 3D printing. The durability and effective use of local soils in the aforementioned projects is a topic that need to be still fully understood.

Further, other challenges are still unsolved to implement AM with most current construction site layouts. The printer adaptability to work at different levels even with nonlinear geometry also arises issues related to the transportation and the location of the equipment in the building site [56]. Additionally, a continuous feeding system is required including material preparation, mixing, and delivery to the nozzle [103]. Hence, new construction plans should be selected, as most printing activities work in continuous at almost constant rate, whereas discrete activities typically distinguish traditional building techniques [104].

Due to brittleness and discontinuity between layers, strength-related issues arise when printing load-bearing elements [105]. Printed structures are expected to be anisotropic due to the printing path which creates linearity at the same level, but discontinuity between subsequent layers. Further, voids may appear between filaments [106], creating a weaker interface where cracks are more likely to appear. Hence, the structural capacity of the hardened element strongly depends on the effectiveness of the printing process. If a good bond between layers is created, the final element is more prone to behave as a structural element. A good balance between mixture open time, machine speed rate and hardening are required, but according to some authors [45], the current technology might not be suitable for use in large-scale models or buildings due to the stability and strength limitations of printed goods made with existing printing materials. Hardening is also a matter of paramount importance for earth-based printed materials, as it is highly influenced by environmental conditions, such as temperature and relative humidity.

Similar to structural capacity, durability is also affected by pores volume and distribution. Referring to results carried out on 3D printed concrete, Zhang et al. [107] employed X-ray computed tomography (X-CT) to analyze the distribution of voids and the difference in the macro-behavior between 3D mix and traditionally cast concrete. It was highlighted that the voids in printed concrete are well-interconnected or even continuous within the filament, which could lead higher weaknesses based on freeze–thaw tests, deeper water penetration according to rapid chloride ion migration tests, and continuous enhancement in the early stages based on sulphate attack tests [107]. Similarly, interfacial defects may have an important effect on the carbonation depth and permeability of printed materials. Sun et al. [108] experienced higher chloride penetration depth and larger carbonation depth along the filament interfaces than that in the matrix, but the chloride ion penetration and the carbonation depth gradually decreases in the gravity direction. No specific indications are provided for earth materials, but similar considerations may apply regarding the influence of increased local porosity and defects at filament borders on material durability.

Furthermore, earth shrinkage may also be increased due to the higher drying speed in 3D printing process than in ordinary casting procedures.

7.4 Concluding Remarks

This chapter has reviewed the current state-of-the-art regarding the use of earth in construction, specifically providing some insights regarding its use as a printable material. Despite some real-scale applications were already realized in some well-known projects, the current research is still limited, and many open issues need to be investigated. Mostly, the current research gaps are in common with concrete 3D printing: technological limits, normative barriers and materials optimization are the most important aspects on which the research needs to focus and advance significantly. In terms of specific problems of earth materials, instead, durability and shrinkage aspects are worth mentioning.

3D printing of earth and soils may represent one of the most sustainable alternatives to current construction methods in the next future, as it adopts natural and locally-available materials, with minimum environmental impacts. It can even work synergistically with other traditional construction methods. This makes 3D printing of earth very attractive also for those regions in the world that are less accessible to the concrete market, paving the way for a soft-transition to an automatized construction industry that valorizes ancient construction materials and methods in new valuable processes.

References

1. United Nations, *World Population Prospects 2022* (2022). https://population.un.org/wpp/. Accessed 29 Apr 2024
2. R. Asghar, M.A. Khan, R. Alyousef, M.F. Javed, M. Ali, Promoting the green construction: scientometric review on the mechanical and structural performance of geopolymer concrete. Constr. Build. Mater. **368**, 130502 (2023)
3. E.A. de Souza, P.H.R. Borges, T. Stengel, B. Nematollahi, F.P. Bos, 3D printed sustainable low-cost materials for construction of affordable social housing in Brazil: potential, challenges, and research needs. J. Build. Eng. **87**, 108985 (2024)
4. M. Santamouris, K. Vasilakopoulou, Present and future energy consumption of buildings: challenges and opportunities towards decarbonisation. e-Prime Adv. Electr. Eng. Electron. Energy **1**, 100002 (2021)
5. M. Santamouris, Green commercial buildings save energy. Nat. Sustain. **1**, 613–614 (2018)
6. D.T. Doan, A. Ghaffarianhoseini, N. Naismith, T. Zhang, A. Ghaffarianhoseini, J. Tookey, A critical comparison of green building rating systems. Build. Environ. **123**, 243–260 (2017)
7. L. Lima, E. Trindade, L. Alencar, M. Alencar, L. Silva, Sustainability in the construction industry: a systematic review of the literature. J. Clean. Prod. **289**, 125730 (2021)
8. United Nations Environment Programme, *2021 Global Status Report for Buildings and Construction* (2021)

9. X. Li, X. Liu, Y. Huang, J. Li, J. He, J. Dai, Evolutionary mechanism of green innovation behavior in construction enterprises: evidence from the construction industry. Eng. Constr. Archit. Manag. **31**, 159–178 (2024). https://doi.org/10.1108/ECAM-02-2022-0186
10. M.H. Alencar, L. Priori, L.H. Alencar, Structuring objectives based on value-focused thinking methodology: creating alternatives for sustainability in the built environment. J. Clean. Prod. **156**, 62–73 (2017). https://doi.org/10.1016/j.jclepro.2017.03.221
11. Z. Winn, *Cleaning Up One of the World's Most Commonly Used Substances* (MIT News, 2023)
12. J. Watts, *Concrete: The Most Destructive Material on Earth* (The Guardian, 2019)
13. M. Gomaa, W. Jabi, V. Soebarto, Y.M. Xie, Digital manufacturing for earth construction: a critical review. J. Clean. Prod. **338**, 130630 (2022)
14. Q. Angulo-Ibáñez, Á. Mas-Tomás, V. Galvañ-Llopis, J.L. Sántolaria-Montesinos, Traditional braces of earth constructions. Constr. Build. Mater. **30**, 389–399 (2012). https://doi.org/10.1016/j.conbuildmat.2011.11.024
15. E. Hamard, B. Cazacliu, A. Razakamanantsoa, J.C. Morel, Cob, a vernacular earth construction process in the context of modern sustainable building. Build. Environ. **106**, 103–119 (2016)
16. E. Quagliarini, A. Stazi, E. Pasqualini, E. Fratalocchi, Cob construction in Italy: some lessons from the past. Sustainability **2**, 3291–3308 (2010). https://doi.org/10.3390/su2103291
17. B. García de Soto, I. Agustí-Juan, J. Hunhevicz, S. Joss, K. Graser, G. Habert, B.T. Adey, Productivity of digital fabrication in construction: cost and time analysis of a robotically built wall. Autom. Constr. **92**, 297–311 (2018). https://doi.org/10.1016/j.autcon.2018.04.004
18. B. Zareiyan, B. Khoshnevis, Interlayer adhesion and strength of structures in contour crafting—effects of aggregate size, extrusion rate, and layer thickness. Autom. Constr. **81**, 112–121 (2017). https://doi.org/10.1016/j.autcon.2017.06.013
19. A. Perrot, D. Rangeard, E. Courteille, 3D printing of earth-based materials: processing aspects. Constr. Build. Mater. **172**, 670–676 (2018). https://doi.org/10.1016/j.conbuildmat.2018.04.017
20. M. Gomaa, W. Jabi, A. Veliz Reyes, V. Soebarto, 3D printing system for earth-based construction: case study of cob. Autom. Constr. **124**, 103577 (2021). https://doi.org/10.1016/j.autcon.2021.103577
21. I. Agustí-Juan, F. Müller, N. Hack, T. Wangler, G. Habert, Potential benefits of digital fabrication for complex structures: environmental assessment of a robotically fabricated concrete wall. J. Clean. Prod. **154**, 330–340 (2017). https://doi.org/10.1016/j.jclepro.2017.04.002
22. G. Minke, *Building with Earth* (2021)
23. F. Pacheco-Torgal, S. Jalali, Earth construction: lessons from the past for future eco-efficient construction. Constr. Build. Mater. **29**, 512–519 (2012)
24. B. Berge, *The Ecology of Building Materials* (2009)
25. B.R. Foster, Ancient mesopotamia. By Susan Pollock. Am. J. Archaeol. **104** (2000). https://doi.org/10.2307/507460
26. P. Jaquin, History of earth building techniques, in *Modern Earth Buildings: Materials, Engineering, Constructions and Applications* (2012)
27. H. Jiyao, J. Weitung, Earth—culture—architecture: the protection and development of rammed earth and adobe architecture in China, in *6th International Conference on the Conservation of Earthen Architecture: Adobe 90 preprints: Las Cruces, New Mexico, U.S.A.*, 14–19 Oct 1990
28. P.A. Jaquin, C.E. Augarde, C.M. Gerrard, Chronological description of the spatial development of rammed Earth techniques. Int. J. Archit. Herit. **2**, 377–400 (2008). https://doi.org/10.1080/15583050801958826
29. F.J. Gallego Roca, I. Valverde Espinosa, J.M. López Osorio, A.M.D.L. Santos Fernández, J. Martin Barranco, E. Ontiveros Ortega, R. Fajardo Martinez, The city walls of Granada (Spain): use, conservation and restoration, in *7a Conferencia Internacional sobre e Estudo e Conservação da Arquitectura de Terra = 7th International Conference of the Study and Conservation of Earthen Architecture. Silves, Portugal*, 24–29 Oct 1993

30. Ziggurat at Ur. https://www.britannica.com/topic/ziggurat-at-Ur#/media/1/618946/157749. Accessed 27 May 2024
31. Alhambra de Granada. https://www.alhambradegranada.org/en/info/alcazaba.asp. Accessed 27 May 2024
32. E. Gnerre, *Pueblo de Taos* (2008). http://whc.unesco.org/en/documents/110335. Accessed 27 May 2024
33. Huacas del Sol y la Luna: todo lo que debes saber. https://www.denomades.com/blog/huacas-sol-luna/. Accessed 27 May 2024
34. N. Barbacci, Earthen architecture—Valorisation and underestimation, in *International Archives of the Photogrammetry, Remote Sensing and Spatial Information Sciences—ISPRS Archives* (2020)
35. A.T.M. Marsh, Y. Kulshreshtha, The state of earthen housing worldwide: how development affects attitudes and adoption. Build. Res. Inf. **50**, 485–501 (2022). https://doi.org/10.1080/09613218.2021.1953369
36. L. Keefe, *Earth Building: Methods and Materials, Repair and Conservation* (2012)
37. E. Quagliarini, S. Lenci, The influence of natural stabilizers and natural fibres on the mechanical properties of ancient Roman adobe bricks. J. Cult. Herit. **11**, 309–314 (2010). https://doi.org/10.1016/j.culher.2009.11.012
38. G.D. Shaffer, An archaeomagnetic study of a wattle and daub building collapse. J. Field Archaeol. **20**, 59–75 (1993). https://doi.org/10.1179/009346993791974334
39. E. Hamard, J.C. Morel, F. Salgado, A. Marcom, N. Meunier, A procedure to assess the suitability of plaster to protect vernacular earthen architecture. J. Cult. Herit. **14**, 109–115 (2013). https://doi.org/10.1016/j.culher.2012.04.005
40. Adobe technique. https://www.urbunhut.com/adobe_building.php. Accessed 27 May 2024
41. The beauty of Cob. https://www.ecoearthbuilds.com/the-beauty-of-cob. Accessed 27 May 2024
42. Diary of a Rambling Antiquarian. https://www.babelstone.co.uk/BabelDiary/2017/10/liao-superior-capital-revisited.html. Accessed 27 May 2024
43. Wattle and Daub. https://bushcraftbuddy.com/wattle-and-daub-the-number-1-iron-age-building-material/. Accessed 27 May 2024
44. N. Labonnote, A. Rønnquist, B. Manum, P. Rüther, Additive construction: state-of-the-art, challenges and opportunities. Autom. Constr. **72**, 347–366 (2016)
45. P. Wu, J. Wang, X. Wang, A critical review of the use of 3-D printing in the construction industry. Autom. Constr. **68**, 21–31 (2016)
46. I. Hager, A. Golonka, R. Putanowicz, 3D printing of buildings and building components as the future of sustainable construction? Proc. Eng. **151**, 292–299 (2016)
47. R. Maskuriy, A. Selamat, P. Maresova, O. Krejcar, O.O. David, Industry 4.0 for the construction industry: review of management perspective. Economies **7**, 68 (2019)
48. X. Zhang, M. Li, J.H. Lim, Y. Weng, Y.W.D. Tay, H. Pham, Q.C. Pham, Large-scale 3D printing by a team of mobile robots. Autom. Constr. **95**, 98–106 (2018). https://doi.org/10.1016/j.autcon.2018.08.004
49. R. Robayo-Salazar, R. Mejía de Gutiérrez, M.A. Villaquirán-Caicedo, S. Delvasto Arjona, 3D printing with cementitious materials: challenges and opportunities for the construction sector. Autom. Constr. **146**, 104693 (2023)
50. S.A. Khan, H. Ilcan, R. Imran, E. Aminipour, O. Şahin, A. Al Rashid, M. Şahmaran, M. Koç, The impact of nozzle diameter and printing speed on geopolymer-based 3D-printed concrete structures: numerical modeling and experimental validation. Res. Eng. **21**, 101864 (2024). https://doi.org/10.1016/j.rineng.2024.101864
51. I. Agustí-Juan, G. Habert, Environmental design guidelines for digital fabrication. J. Clean. Prod. **142**, 2780–2791 (2017). https://doi.org/10.1016/j.jclepro.2016.10.190
52. S. Ford, M. Despeisse, Additive manufacturing and sustainability: an exploratory study of the advantages and challenges. J. Clean. Prod. **137**, 1573–1587 (2016). https://doi.org/10.1016/j.jclepro.2016.04.150

53. S. El-Sayegh, L. Romdhane, S. Manjikian, A critical review of 3D printing in construction: benefits, challenges, and risks. Arch. Civ. Mech. Eng. **20**, 1–25 (2020)
54. S.H. Ghaffar, J. Corker, M. Fan, Additive manufacturing technology and its implementation in construction as an eco-innovative solution. Autom. Constr. **93**, 1–11 (2018)
55. M. Bazli, H. Ashrafi, A. Rajabipour, C. Kutay, 3D printing for remote housing: benefits and challenges. Autom. Constr. **148**, 104772 (2023)
56. D. Delgado Camacho, P. Clayton, W.J. O'Brien, C. Seepersad, M. Juenger, R. Ferron, S. Salamone, Applications of additive manufacturing in the construction industry—a forward-looking review. Autom. Constr. **89**, 110–119 (2018). https://doi.org/10.1016/j.autcon.2017.12.031
57. E. Lloret, A.R. Shahab, M. Linus, R.J. Flatt, F. Gramazio, M. Kohler, S. Langenberg, Complex concrete structures: merging existing casting techniques with digital fabrication. CAD Comput. Aid. Des. **60**, 40–49 (2015). https://doi.org/10.1016/j.cad.2014.02.011
58. D. Liu, Z. Zhang, X. Zhang, Z. Chen, 3D printing concrete structures: state of the art, challenges, and opportunities. Constr. Build. Mater. **405**, 133364 (2023)
59. J. Zhang, J. Wang, S. Dong, X. Yu, B. Han, A review of the current progress and application of 3D printed concrete. Compos. Part A Appl. Sci. Manuf. **125**, 105533 (2019)
60. P. Feng, X. Meng, J.F. Chen, L. Ye, Mechanical properties of structures 3D printed with cementitious powders. Constr. Build. Mater. **93**, 181–209 (2015). https://doi.org/10.1016/j.conbuildmat.2015.05.132
61. M. van den Heever, A. du Plessis, J. Kruger, G. van Zijl, Evaluating the effects of porosity on the mechanical properties of extrusion-based 3D printed concrete. Cem. Concr. Res. **153**, 106695 (2022). https://doi.org/10.1016/j.cemconres.2021.106695
62. O. Zaid, M.H. El Ouni, Advancements in 3D printing of cementitious materials: a review of mineral additives, properties, and systematic developments. Constr. Build. Mater. **427**, 136254 (2024). https://doi.org/10.1016/j.conbuildmat.2024.136254
63. N. Makul, Advanced smart concrete—a review of current progress, benefits and challenges. J. Clean. Prod. **274**, 122899 (2020)
64. B. Garrett, 3D printing: new economic paradigms and strategic shifts. Glob. Policy **5**, 70–75 (2014). https://doi.org/10.1111/1758-5899.12119
65. F. Faleschini, D. Trento, M. Masoomi, C. Pellegrino, M.A. Zanini, Sustainable mixes for 3D printing of earth-based constructions. Constr. Build. Mater. **398**, 132496 (2023). https://doi.org/10.1016/j.conbuildmat.2023.132496
66. H. Hassan, E. Rodriguez-Ubinas, A. Al Tamimi, E. Trepci, A. Mansouri, K. Almehairbi, Towards innovative and sustainable buildings: a comprehensive review of 3D printing in construction. Autom. Constr. **163**, 105417 (2024). https://doi.org/10.1016/j.autcon.2024.105417
67. J. Xiao, G. Ji, Y. Zhang, G. Ma, V. Mechtcherine, J. Pan, L. Wang, T. Ding, Z. Duan, S. Du, Large-scale 3D printing concrete technology: current status and future opportunities. Cem. Concr. Compos. **122**, 104115 (2021)
68. S. Paritala, K.K. Singaram, I. Bathina, M.A. Khan, S.K.R. Jyosyula, Rheology and pumpability of mix suitable for extrusion-based concrete 3D printing—a review. Constr. Build. Mater. **402**, 132962 (2023)
69. A. Perrot, D. Rangeard, A. Pierre, Structural built-up of cement-based materials used for 3D-printing extrusion techniques. Mater. Struct. Mater. Constr. **49**, 1213–1220 (2016). https://doi.org/10.1617/s11527-015-0571-0
70. A. Perrot, D. Rangeard, Y. Mélinge, P. Estellé, C. Lanos, Extrusion criterion for firm cement-based materials. Appl. Rheol. **19**, 53042–53051 (2009). https://doi.org/10.3933/ApplRheol-19-53042
71. A.U. Rehman, J.H. Kim, 3d concrete printing: a systematic review of rheology, mix designs, mechanical, microstructural, and durability characteristics. Materials **14**, 3800 (2021)
72. L. Yang, S.M.E. Sepasgozar, S. Shirowzhan, A. Kashani, D. Edwards, Nozzle criteria for enhancing extrudability, buildability and interlayer bonding in 3D printing concrete. Autom. Constr. **146**, 104671 (2023). https://doi.org/10.1016/j.autcon.2022.104671

73. N. Roussel, J. Spangenberg, J. Wallevik, R. Wolfs, Numerical simulations of concrete processing: From standard formative casting to additive manufacturing. Cem. Concr. Res. **135**, 106075 (2020)
74. W. Lao, M. Li, T.N. Wong, M.J. Tan, T. Tjahjowidodo, Improving surface finish quality in extrusion-based 3D concrete printing using machine learning-based extrudate geometry control. Virtual Phys. Prototyp. **15**, 178–193 (2020). https://doi.org/10.1080/17452759.2020.1713580
75. P. Shakor, S. Nejadi, G. Paul, A study into the effect of different nozzles shapes and fibre-reinforcement in 3D printed mortar. Materials **12**, 1708 (2019). https://doi.org/10.3390/MA12101708
76. R.J.M. Wolfs, F.P. Bos, T.A.M. Salet, Early age mechanical behaviour of 3D printed concrete: numerical modelling and experimental testing. Cem. Concr. Res. **106**, 103–116 (2018). https://doi.org/10.1016/j.cemconres.2018.02.001
77. A. Curth, N. Pearl, A. Castro-Salazar, C. Mueller, L. Sass, 3D printing earth: local, circular material processing, fabrication methods, and life cycle assessment. Constr. Build. Mater. **421**, 135714 (2024). https://doi.org/10.1016/j.conbuildmat.2024.135714
78. D. Rocha, P. Faria, S.S. Lucas, Additive manufacturing of earth-based materials: a literature review on mortar composition, extrusion, and processing earth. Materials **17**, 202 (2024). https://doi.org/10.3390/ma17010202
79. M. Gomaa, J. Carfrae, S. Goodhew, W. Jabi, A. Veliz Reyes, Thermal performance exploration of 3D printed cob. Archit. Sci. Rev. **62**, 230–237 (2019). https://doi.org/10.1080/00038628.2019.1606776
80. I. Bianchi, S. Volpe, F. Fiorito, A. Forcellese, V. Sangiorgio, Life cycle assessment of building envelopes manufactured through different 3D printing technologies. J. Clean. Prod. **440**, 140905 (2024). https://doi.org/10.1016/j.jclepro.2024.140905
81. H. Alhumayani, M. Gomaa, V. Soebarto, W. Jabi, Environmental assessment of large-scale 3D printing in construction: a comparative study between cob and concrete. J. Clean. Prod. **270**, 122463 (2020). https://doi.org/10.1016/j.jclepro.2020.122463
82. M. Gomaa, J. Vaculik, V. Soebarto, M. Griffith, W. Jabi, Feasibility of 3DP cob walls under compression loads in low-rise construction. Constr. Build. Mater. **301**, 124079 (2021). https://doi.org/10.1016/j.conbuildmat.2021.124079
83. K. Zeghari, A. Gounni, H. Louahlia, M. Marion, M. Boutouil, S. Goodhew, F. Streif, Novel dual walling cob building: dynamic thermal performance. Energies (Basel) **14**, 7663 (2021). https://doi.org/10.3390/en14227663
84. O.B. Carcassi, Y. Maierdan, T. Akemah, S. Kawashima, L. Ben-Alon, Maximizing fiber content in 3D-printed earth materials: printability, mechanical, thermal and environmental assessments. Constr. Build. Mater. **425**, 135891 (2024). https://doi.org/10.1016/j.conbuildmat.2024.135891
85. S. Paul, M.S. Islam, T.E. Elahi, Comparative effectiveness of fibers in enhancing engineering properties of Earth as a building material: a review. Constr. Build. Mater. **332**, 127366 (2022). https://doi.org/10.1016/j.conbuildmat.2022.127366
86. A. Koutous, E. Hilali, Reinforcing rammed earth with plant fibers: a case study. Case Stud. Constr. Mater. **14**, e00514 (2021). https://doi.org/10.1016/j.cscm.2021.e00514
87. P. Zak, T. Ashour, A. Korjenic, S. Korjenic, W. Wu, The influence of natural reinforcement fibers, gypsum and cement on compressive strength of earth bricks materials. Constr. Build. Mater. **106**, 179–188 (2016). https://doi.org/10.1016/j.conbuildmat.2015.12.031
88. M. Ouedraogo, H. Bamogo, I. Sanou, V. Mazars, J.E. Aubert, Y. Millogo, Microstructural, physical and mechanical characteristics of adobes reinforced with sugarcane bagasse. Buildings **13**, 117 (2023). https://doi.org/10.3390/buildings13010117
89. E. Binega Yemesegen, A.M. Memari, A review of experimental studies on Cob, Hempcrete, and bamboo components and the call for transition towards sustainable home building with 3D printing. Constr. Build. Mater. **399**, 132603 (2023)
90. G.H. Ahmed, A review of "3D concrete printing": materials and process characterization, economic considerations and environmental sustainability. J. Build. Eng. **66**, 105863 (2023)

91. G. De Schutter, K. Lesage, V. Mechtcherine, V.N. Nerella, G. Habert, I. Agusti-Juan, Vision of 3D printing with concrete—technical, economic and environmental potentials. Cem. Concr. Res. **112**, 25–36 (2018)
92. S.C. Paul, G.P.A.G. van Zijl, I. Gibson, A review of 3D concrete printing systems and materials properties: current status and future research prospects. Rapid Prototyp. J. **24**, 784–798 (2018)
93. J. Bacon-Shone, *WinSun: Demonstrating the Viability of 3D Printing at Construction Scale* (World Economic Forum, 2016)
94. B. Khoshnevis, Automated construction by contour crafting—related robotics and information technologies, in *Automation in Construction* (2004)
95. J. Zhang, B. Khoshnevis, Optimal machine operation planning for construction by Contour crafting. Autom. Constr. **29**, 50–67 (2013). https://doi.org/10.1016/j.autcon.2012.08.006
96. R. Wolfs, D. Bos, T. Salet, Lessons learned of project Milestone: the first 3D printed concrete house in the Netherlands. Mater. Today Proc. (2023). https://doi.org/10.1016/j.matpr.2023.06.183
97. A. Dubor, J.-B. Izard, E. Cabay, A. Sollazzo, A. Markopoulou, M. Rodriguez, On-site robotics for sustainable construction, in *Robotic Fabrication in Architecture, Art and Design 2018* (2019)
98. A. Chiusoli, The first 3D printed House with earth | Gaia, in *In House, Press Releases* (2018). https://www.3dwasp.com/en/3d-printed-house-gaia/. Accessed 6 Jun 2024
99. Tecla. https://www.3dwasp.com/en/3d-printed-house-tecla/?_gl=1*9acrhk*_up*MQ..*_ga*MjAyNjg5MjM1MS4xNzE3Njg2NzUz*_ga_T6FJKDX4T8*MTcxNzY4Njc1Mi4xLjEuMTcxNzY4Njg2OC4wLjAuMA. Accessed 6 Jun 2024
100. *IAAC and WASP: New 3D Printing Strategies Towards the Realisation of Load-Bearing Earthen Structures*. https://iaac.net/iaac-wasp-new-3d-printing-strategies-towards-realisation-load-bearing-earthen-structures/. Accessed 6 Jun 2024
101. Rael San Fratello. *Casa Covida*. https://www.rael-sanfratello.com/made/casa-covida. Accessed 6 Jun 2024
102. A. Severi, TOVA is Spain's first 3D printed building with Crane WASP, in *In House, Press Releases* (2022). https://www.3dwasp.com/en/tova-3d-printed-building-with-crane-wasp/?_gl=1*t9kays*_up*MQ..*_ga*MjA3OTAyNTEyMC4xNzE3Njg3NjEy*_ga_T6FJKDX4T8*MTcxNzY4NzYxMS4xLjAuMTcxNzY4NzYxMS4wLjAuMA. Accessed 6 Jun 2024
103. O. Davtalab, A. Kazemian, B. Khoshnevis, Perspectives on a BIM-integrated software platform for robotic construction through Contour crafting. Autom. Constr. **89**, 13–23 (2018). https://doi.org/10.1016/j.autcon.2018.01.006
104. F. Bos, R. Wolfs, Z. Ahmed, T. Salet, Additive manufacturing of concrete in construction: potentials and challenges of 3D concrete printing. Virtual Phys. Prototyp. **11**, 209–225 (2016). https://doi.org/10.1080/17452759.2016.1209867
105. B. Berman, 3-D printing: the new industrial revolution. Bus. Horiz. **55**, 155–162 (2012). https://doi.org/10.1016/j.bushor.2011.11.003
106. R.A. Buswell, W.R. Leal de Silva, S.Z. Jones, J. Dirrenberger, 3D printing using concrete extrusion: a roadmap for research. Cem. Concr. Res. **112**, 37–49 (2018)
107. Y. Zhang, Y. Zhang, L. Yang, G. Liu, Y. Chen, S. Yu, H. Du, Hardened properties and durability of large-scale 3D printed cement-based materials. Mater. Struct. Mater. Constr. **54**, 1–14 (2021). https://doi.org/10.1617/s11527-021-01632-x
108. X. Sun, J. Zhou, Q. Wang, J. Shi, H. Wang, PVA fibre reinforced high-strength cementitious composite for 3D printing: mechanical properties and durability. Addit. Manuf. **49**, 102500 (2022). https://doi.org/10.1016/j.addma.2021.102500

Chapter 8
Clay-Based Plasters: Back to the Future

Paulina Faria

Abstract The use of clay-based mortars is almost as old as mankind and they continue to be applied nowadays, namely indoors as plasters. That is because the environmental and technological advantages of clay-based plastering mortars clearly surpass their barriers and drawbacks. Selected aspects of the formulation, application, protection and use of clay-based plasters are presented and discussed here based on recent research. Ways of further increasing its environmental and technological performance are highlighted, with a view towards facilitating a broader use.

8.1 Introduction

Most probably, the first building product Mankind produced was a mortar, made by mixing local earth with water. It was used to fill the space between branches to build shelter and protect from rain, wind and sun. With time, requirements concerning buildings increased and, much later, mortars started to be used to layer masonry units and coat externally (by renders) and internally (by plasters) [1] different types of building walls and ceilings. Renders and plasters can be applied on the substrate in one or more layer and their main function is to protect the masonry or monolithic walls and ceilings they coat.

As before, the first mortar used to plaster walls and ceilings was clay-based as it could be made with a local, available, unexpensive material: earth, excavated from the local soil, composed by different fraction and types of clay, silt and sand, after the removal of coarser particles. To produce the plastering mortar, the earth was mixed with water. Therefore, in clay plasters (also called earth plasters), the sand acts as aggregate, the silt as filler and the clay as binder.

In terms of materials, the binder is the main change in comparison to other common plastering mortars, known by the name of their binder, namely cement plasters, lime plasters and gypsum plasters: while the others are produced after thermal treatment, the clay is used as it is in nature.

P. Faria (✉)
Department of Civil Engineering, CERIS, NOVA University Lisbon, Caparica, Portugal
e-mail: paulina.faria@fct.unl.pt

D. A. H. Hanaor (ed.), *Sustainable Materials for the Built Environment*, Springer Series in Materials Science 352, https://doi.org/10.1007/978-3-031-97818-0_8

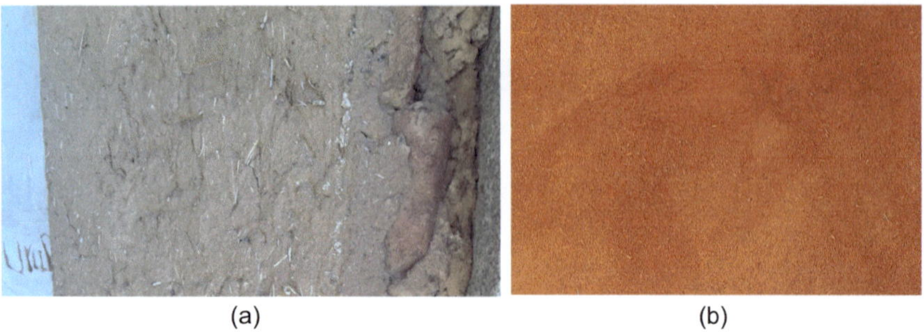

Fig. 8.1 Old wall coated with a clay plaster with fibres, finished with a limewash, partially removed (**a**) and new wall coated with another clay plaster, produced with a different earth and without supplementary finishing (**b**)

Therefore, for millennia, people produced and applied clay plasters as they were the more affordable and available. Frequently the plasters were limewashed not to show they were clay plasters, and also to provide a whitish surface and more luminous indoor space, together with some durability to an eventual contact with water (Fig. 8.1a).

Nowadays, clay plasters are technically recognized mainly due to the contribution to comfort they can provide but also to their low environmental impact and very positive life cycle assessment [2, 3]. Therefore, some people concerned with the environment and healthy buildings require them for their buildings. Others also require this type of plasters because the natural building products are nowadays fashionable. Mainly in the latter case, it is often a request that the clay plasters are not painted (Fig. 8.1b).

Therefore, it is important to know the potentiality of clay-based plasters and how they can fulfil buildings and users´ requirements. For that, the formulation of clay plasters is important.

8.2 Clay Plasters Formulation and Main Requirements

Clay plasters are commonly produced with one layer of a clay-based mortar (also called earth mortar), made by mixing a sieved earth, composed by clay, silt and sand particles, eventually with supplementary sand or other aggregate, with water. In turn, clay-based mortars can be produced locally, with local earth, or sold pre-dosed. In this latter case they are commonly produced with an earth local to the producing company, the earth and eventually additions (supplementary sand or similar size aggregate, natural fibres) are dosed by precise mass, the dry materials are packed and transported to a commercial company or directly to the building site. In the pre-dosed clay plasters producing company, the formulation can be better controlled than when the composition is formulated in situ. In big working sites with many square meters to be plastered, some producing companies have movable equipment and can also prepare the pre-dosed mortars on site. In the building site, both the local materials used to formulate the mortar or the pre-dosed materials are mixed with water and applied, manually or by mechanical projection. The projection equipment used for clay-based plasters is similar to the one used to apply gypsum plasters. Commonly, they are regularized with a 2 m ruler and then the surface is finished with a trowel or a sponge.

Each earth is composed by several different types and fractions of clay, silt, sand and coarser aggregates. As mentioned before, the clay fraction acts as binder providing important properties to the plaster [4, 5], namely controlled drying shrinkage, good adhesion to the wall, good cohesion between the particles, adequate mechanical performance, but also good hygroscopicity and a naturally pigmented colour (Fig. 8.2).

The clay fraction of an earth is commonly composed by one main type of clay minerals, while other clay minerals can also co-exist. In literature clay plasters with the kaolinitic [4, 6] and illitic [4, 7] clay minerals are common; with montmorilonitic clay minerals it is not recommended because of high shrinkage [4].

Additional and very important properties are also being studied on clay plasters, such as performance in case of fire [8, 9] and their ability to capture pollutants, and release them when ventilation is provided [10–12], further contributing to safety and indoor air quality.

There are no specific international standards for clay plastering mortars. The European standard for rendering and plastering mortars is the EN 998-1 [1], that

Fig. 8.2 Samples of clay plasters with diversity of colours

defines 0.4 N/mm^2 as the minimum compressive strength these mortars should have. This standard refers to the EN 1015 (composed by several parts) standards to define how to produce and test, for different properties, plaster samples. However, the mortar production, curing and some test methods are much more appropriated for cement-based mortars and not so much for clay-based ones. However, since 2013 there is a German national standard specific for clay plastering mortars: DIN 18947 [13], defining specific requirements, labelling and test methods, such as the adsorption and the dry abrasion tests, while other methods are based on some EN 1015 parts, such as the mortars production in a laboratory [14], flow table consistence [15], bulk density of the fresh [16] and hardened [17] mortars, flexural and compressive strengths [18] and adhesion to a substrate [19].

To produce clay-based plastering mortars, the earth is crushed to roughly disaggregate the different fractions and sieved to remove the coarser aggregates, bigger than sand. In case the volumetric proportions clay-silt-sand in the raw earth are considered adequate, that means that the clay plastering mortar are adhesive and does not crack after drying when coating a substrate. If the earth is too clayish, to avoid drying shrinkage and cracking when applied on a wall, supplementary sand or other aggregate with similar particle size distribution or natural fibres should be added, before mixing with the water content necessary to provide workability [20]. For that, the volume of added aggregate or fibres should be optimized: increased added volumes of extra aggregate or/and fibres are tested on small plaster samples on a substrate, and in general the minimum volume that reduces visible shrinkage is adopted (Fig. 8.3).

The reason why the volume of additional aggregate to include in the mortar formulation is commonly minimized is because it reduces the adhesion to the substrate and also because frequently it is a sand that has to be specifically extracted, while the earth used to produce the mortar can be a local construction and demolition waste. Therefore, that optimizes the plaster performance and minimizes environmental impact and costs. However, some recycled aggregates with similar to sand particle size can be used instead of natural sand, as well as fibres.

Natural fibres, resulting from agriculture wastes, can also be added to the clay plastering mortar formulation. Depending on the type (for instance thin like typha fibres, or thicker as oat straw), size and volume of fibres added, different properties can be changed [21], namely drying shrinkage, density, flexural and compressive strength.

After application on a wall or ceiling, the plastering mortars just dries, without any chemical reaction. Particularly when the plastering mortars contains plant fibres, the initial drying should have good ventilation; otherwise, biological colonization can occur during the initial drying as it takes long.

The clay minerals and the hardening just by drying allows clay plasters to have a high hygroscopicity when compared to other plasters and other indoor coatings [22, 23], acting as passive moisture buffers. Furthermore, they are reversible, easily repairable with a humid sponge and reusable just by mixing the dry mortar again with water. However, contact with water used indoors or, by accident, entering by an open window when it rains may damage the clay plasters.

Fig. 8.3 From down to up, examples of testing increased volume and type of aggregate added to two different clay mortars, to define the minimum content that controls cracking

While the contribution of vegetal fibres and other biomass can be important to increase the hygroscopicity of other plasters, namely gypsum-based [24, 25], the influence they may have on the highly hygroscopic clay plasters is not significant. However, the addition of vegetal fibres can also contribute to decrease the mortars life cycle analysis, due to the contribution of the plant to CO_2 sequestration, and also to provide a more natural aesthetics, as fibres are seen in the surface of unpainted clay plasters (Fig. 8.1b).

To provide some durability when in contact with water, when that is a requirement for the plaster, different architectural protections and finishings can be design and applied on specific areas, to avoid being a barrier to the use of clay plasters [26]. In general, the contact of plasters with water only occurs by accident (cleaning with water, a window left open when it rains), except when they are applied close to taps and showers. Of course, on those areas specific protection, formulation or, more commonly, a water resistant coating should be applied, such as ceramic tiling.

Reactive additions can be added in the mortar formulation, although that may turn the plaster not reversible and reusable; only recyclable. However, both the finishings and the reactive additions [27] should not reduce significatively the moisture buffer capacity of the clay plasters.

8.3 Clay Plasters Reactive Additions

The main reason to add reactive addition to clay plasters formulation is to provide some durability when in contact with water. However, plasters, applied indoors, may not need that type of durability. Therefore, eventual additions should not increase too much the environmental impact of the clay plasters, keep their high adsorption and the reversibility/reusability of the mortars.

In different parts of the world vernacular additions were added to clay plasters, and particularly to clay-based renders, to provide that durability. An example is the addition of cow dung, that includes small fibres together with compounds resulting from the animal digestion [28–30]. This addition does not implicate bad smell or toxicity because it is added after drying. Furthermore, it allows reducing the supplementary sand or other aggregates added because it also contains natural fibres, already cut by the animal. Also, it does not reduce the previously mentioned advantages of clay plasters: adsorption, low embodied energy and reusability. Finally, it improves the resistance when in contact with water for mortars with different types of clays and cow dung type and preparation. However, the intensity of the effect may change with the type of earth and processing of the cow dung [31, 32]. Another vernacular example is the addition of cactus mucilage and many other natural materials, namely biopolymers [33] and linseed oil [26]. Many other materials are being tested [34] and some can be agro-industry wastes. Therefore, the biostabilization is nowadays an important line of research, supported by the work of RILEM Technical Committee 318—BEC, Bio-stabilised earth-based construction: performance-approach for better resilience [35].

The main advantages of these bio-based additions are that they do not significantly reduce the moisture buffer capacity of the plasters and the reversibility and reusability of the mortars, while they may not increase significantly its embodied energy and, therefore, environmental impact. Another advantage is that in general they do not change much the natural colour of the clay plaster.

Another type of reactive additions are mineral binders. Remembering that the main function of plasters is to protect the substrates where they are applied, physic-mechanical and chemical compatibility is required. Therefore, although clay-based plasters should fulfil requirements defined in the EN 998-1 [1], namely a minimal compressive strength of 0.4 N/mm^2, high strength and low water vapour permeability mortars are not required. As for other plasters, on corners or other areas more prone to mechanical impacts, protections can be applied [26]. Furthermore, past literature has shown the durable use of air lime and gypsum-based binders in historic clay-based building elements [36, 37]. So, these are the more appropriate binders when the choice is to add a binder to a clay plastering mortar, and other mineral binders, such as cement, should be avoided. To avoid increasing too much the embodied energy of binder-added clay plasters, recycled lime and gypsum may be used.

However, although the resistance to water increases, adding low content of air lime may decrease the mechanical performance of the clay mortars [6, 38], and adding a high content turns the mortar into a lime-based mortar and no longer a

Fig. 8.4 From right to left, slight colour change of a dark chocolate coloured clay plaster, depending on supplementary sand added, and impacting colour changes when adding low contents of air lime and natural hydraulic lime

clay-based one. Furthermore, the addition of a low content of air lime promotes a higher reduction of moisture buffering capacity in comparison to the same addition of hemihydrate gypsum, while the latter promotes a slight increase in mechanical properties [26]. Therefore, it seems that, if the choice is to add a mineral binder, hemihydrate gypsum may be the best choice [39]. However, the addition of both binders turns the clay plaster no longer reversible and reusable, as it turns into an artificial stone, only recyclable. Furthermore, adding mineral binders to naturally coloured earth will most probably significantly change the plaster natural colour, which should also be taken in consideration before the decision (Fig. 8.4).

8.4 Clay Plasters Surface Protection

Another alternative to increase durability of clay plasters, not only to water, is to apply a surface finishing or protective treatment. These systems can be applied by different ways, such as by brush, roller sponge or spray. Surface finishings can be varnish or paint systems. However, these systems may significantly reduce the moisture buffering capacity of the clay plasters [40]. In the case of varnishes, they do not change the colour but the texture (Fig. 8.5); in the case of common paints or limewash, they also change the colour. The advantage of the limewash, in comparison to common paints, is that the moisture buffering reduction is very low [26]. Clay-based paints can also be applied. They are produced with coloured clay with very low particle size [41].

Surface protective treatments can be obtained by the application of different materials, also by brush, roller or spray, namely some of the ones that can be added in the formulation of the clay plasters, such as linseed oil or cactus mucilage, and others such as beeswax, cellulose water or other bio-based products [42, 43]. Some do not

Fig. 8.5 Decorative clay plaster with a varnish finishing in a toilet of a restaurant in Lisbon, Portugal (*Credits* Vera Schmidberger)

form a surface film, impregnating the external millimetres of the plasters and consolidating the surface; others, such as beeswax, form a film and reduce more significantly the moisture buffer capacity. Some slightly change the colour while with others the colour change is more visible. The effect on both water adsorption and damage by water dripping also change with different treatments. The same with the increase of environmental impact, although in general is not highly affected when using natural materials.

8.5 Final Remarks

Although earth as a building material has many applications, one of the oldest is to produce clay-based mortars. The structural use of earthen walls has many advantages but some limitations, namely in high rise buildings due to the high mass of those walls. However, the use of earth to produce plastering mortars does not seem to have limitations, as those plasters can be applied on many different walls. Those walls can be made with vernacular, traditional and innovative materials and products, such as masonries with adobe, compressed earth blocks, ceramic bricks, concrete blocks and straw bales, monolithic rammed earth and cob walls, and even 3D printed walls. Therefore, clay plasters are compatible both with new, vernacular and historic building walls, and can be both applied as first plaster and as renovation plaster. They can even be applied as sacrificial plasters to protect archaeological structures.

The advantages of clay plasters are broad. In past times the main issue was to have availability and low cost of the material to produce the plastering mortars. Nowadays many other requirements are added to the previous. The low environmental impact can be kept, once the addition of mineral binders with high embodied energy is not recommended nor needed. As plasters, applied indoors, do not have contact with rain, the clay plasters can be left visible without any treatment or protection. Alternatively, some protective finishing can be applied on the surface, or low content of low embodied energy materials can be added to the formulation. Bio-based materials, namely resulting from agro-industrial wastes, can be a possibility. The choices should depend on the requirements defined for the plasters on different areas of different buildings. In general, it is important that the surface protection or the additions to the formulation do not reduce the moisture buffering capacity, which in turn may also provide some pollutants capture, and a significative contribution to comfort and environmental indoor quality, the reversibility, facilitating small repairs and changes in texture if required, and the reusability of the mortar at the end of its life cycle.

References

1. EN 998-1, *Specification for Mortar for Masonry—Part 1: Rendering and Plastering Mortar* (CEN, Brussels, 2016)
2. T. Santos, J. Almeida, J.D. Silvestre, F. Faria, Life cycle assessment of mortars: a review on technical potential and drawbacks. Constr. Build. Mater. **288**, 123069 (2021)
3. P. Melià, G. Ruggieri, S. Sabbadini, G. Dotelli, Environmental impacts of natural and conventional building materials: a case study on earth plasters. J. Clean. Prod. **80**, 179186 (2014)
4. J. Lima, P. Faria, A. Santos Silva, Earth plasters: the influence of clay mineralogy in the plasters' properties. Int. J. Architect. Heritage **14**(7), 948–963 (2020)
5. M. Lagouin, J.-E. Aubert, A. Laborel-Préneron, C. Magniont, Influence of chemical, mineralogical and geotechnical characteristics of soil on earthen plaster properties. Constr. Build. Mater. **304**, 124339 (2021)
6. M.I. Gomes, P. Faria, T.D. Gonçalves, Earth-based mortars for repair and protection of rammed earth walls. Stabilization with mineral binders and fibers. J. Clean. Prod. **172**, 2401–2414 (2018)
7. P. Faria, T. Santos, J.-E. Aubert, Experimental characterization of an earth eco-efficient plastering mortar. J. Mater. Civ. Eng. **28**(1), 04015085 (2016)
8. Y. Jiang, A. Phelipot-Mardele, F. Collet, C. Lanos, M. Lemke, M. Ansell, A. Hussain, M. Lawrence, Moisture buffer, fire resistance and insulation potential of novel bio-clay plaster. Constr. Build. Mater. **244**, 118353 (2020)
9. J. Tadonbou, P. Pliya, A.-L. Beaucour, P. Faria, A. Noumowe, Behavior of earth mortars at high temperature. Constr. Build. Mater. **425**, 136093 (2024)
10. S. Arris-Roucan, A. Fabbri, F. McGregor, C. Perlot, J.-C. MOREL, Sorption mechanisms of CO_2 in earthen materials under variable hygric states. Mater. Struct. **57**, 86 (2024)
11. A. Ranesi, P. Faria, R. Veiga, E. Gall, Eco-efficient coatings for healthy indoors: ozone deposition velocities, primary and secondary emissions, Build. Environ. **254**, 111306 (2024)
12. E.K. Darling, C.J. Cros, P. Wargocki, J. Kolarik, G.C. Morrison, R.L. Corsi, Impacts of a clay plaster on indoor air quality assessed using chemical and sensory measurements. Build. Environ. **57**, 370–376 (2012)
13. DIN 18947, *Earth Plasters—Requirements, Test and Labelling (in German)* (German Institute for Standardization, Berlin, 2024)
14. EN 1015-2, + *A1:2006, Methods of Test for Mortar for Masonry—Part 2: Bulk Sampling of Mortars and Preparation of Test Mortars* (CEN, Brussels, 1998)

15. EN 1015-3, *Methods of Test for Mortar for Masonry—Part 3: Determination of Consistence of Fresh Mortar (By Flow Table)* (CEN, Brussels, 1999)
16. EN 1015-6, *Methods of Test for Mortar for Masonry—Part 6: Determination of Bulk Density of Fresh Mortar* (CEN, Brussels, 1998)
17. EN 1015-10:1999, Methods of Test for Mortar for Masonry—Part 10: Determination of Dry Bulk Density of Hardened Mortar. CEN, Brussels.
18. EN 1015-11, *Methods of Test for Mortar for Masonry—Part 11: Determination of Flexural and Compressive Strength of Hardened Mortar* (CEN, Brussels, 2019)
19. EN 1015-12, *Methods of Test for Mortar for Masonry—Part 12: Determination of Adhesive Strength of Hardened Rendering and Plastering Mortars on Substrates* (CEN, Brussels, 2016)
20. M. Emiroğlu, A. Yalama, Y. Erdoğdu, Performance of ready-mixed clay plasters produced with different clay/sand ratios. Appl. Clay Sci. **115**, 221–229 (2015)
21. J. Lima, P. Faria, Eco-efficient earthen plasters. The influence of the addition of natural fibers, in *Natural Fibres: Advances in Science and Technology Towards Industrial Applications. From Science to Markets*, vol. 12, ed. By R. Fangueiro, S. Rana, Springer (RILEM Book Series, 2016), 315–327
22. T. Santos, M.I. Gomes, A. Santos Silva, E. Ferraz, P. Faria, Comparison of mineralogical, mechanical and hygroscopic characteristic of earthen, gypsum and cement-based plasters. Constr. Build. Mater. **254**, 119222 (2020)
23. J. Lima, P. Faria, R. Veiga, Comparison of an earth mortar and common binder mortars for indoor plastering, in *ICSEFCM2021—Proceedings of the 2nd International Conference on Sustainable, Environmentally Friendly Construction Materials*, Szczecin, Poland, Aug–Sept 2021, ed. by E. Horszczaruk, P. Brzozowski (2021), pp. 71–76
24. A. Ranesi, P. Faria, M.T. Freire, M. Gonçalves, R. Veiga, Gypsum plastering mortars with Acacia dealbata biowaste additions: effect of different fractions and contents on the relative humidity dependent properties. Constr. Build. Mater. **404**, 133283 (2023)
25. A. Ranesi, P. Faria, M.T. Freire, M. Gonçalves, R. Veiga, Eco-efficient plastering mortars for improved indoor comfort—The influence of A. dealbata bark addition, Constr. Build. Mater. **421**, 135572 (2024)
26. P. Faria, J. Lima, Earth plasters (in Portuguese), in *Cadernos de Construção com Terra* vol. 3, 1ª edn. (Argumentum, Lisbon, 2018)
27. F. Stazi, A. Nacci, F. Tittarelli, E. Pasqualini, P. Munafò, An experimental study on earth plasters for earthen building protection: the effects of different admixtures and surface treatments. J. Cult. Herit.Herit. **17**, 27–41 (2016)
28. Y. Kulshreshtha, P.J. Vardon, G.M.H. Meesters, M.C.M. Van Loosdrecht, N. Mota, H.M. Jonkers, What makes cow-dung stabilized earthen block water-resistant? Constr. Technol. Architect. **1**, 540–548 (2022)
29. S.M. Rao, R. Sukumar, R.E. Joshua, M. Rekapalli, Bio-stabilization of soft soil using cattle manure. Proc. Inst. Civil Eng. Ground Improve. **176**(1), 41–53 (2023)
30. S.M. Rao, *Stabilization of Earthen Building Materials by Select Native Binders* 1st edn. (Springer, Switzerland, 2023).
31. R.N. Pachamama, P. Faria, M.A.P. Rezende, P.M. Moraes, Cow dung biostabilized earth mortars: reusability and influence of processing, humid curing and cow diet. Buildings **14**(11), 3414 (2024)
32. R.N. Pachamama, P. Faria, M.A.P. Rezende, A. Santos Silva, Effect of cow dung additions on tropical and Mediterranean earth mortars—Mechanical performance and water resistance. Materials **17**, 2885 (2024)
33. A. Vissac, A. Bourgès, D. Gandreau, R. Anger, L. Fontaine, *Argiles et biopolymères - les stabilisants naturels pour la construction en terre* (CRATerre Éditions, Villefontaine, 2017)
34. J.L. Parracha, A.S. Pereira, R. Velez da Silva, V. Silva, P. Faria, Effect of innovative bioproducts on the performance of bioformulated earthen plasters. Constr. Build. Mater. **277**, 122261 (2021)
35. P. Faria, C. Becket, A. Fabbri, E. Keita, J.-C. Morel, C. Perlot, A. Perrot, RILEM contribution to earthen building, in *Second RILEM International Conference on Earthen Construction. ICEC 2024. RILEM Bookseries* vol. 52, ed. by C. Beckett, A. Bras, A. Fabbri, E. Keita, C. Perlot, A. Perrot (Springer, Cham, 2024), 194–205

36. M. Mattone, S. Rescic, F. Fratini, R.M. del Fà, Experimentation of earth-gypsum plasters for the conservation of earthen constructions. Int. J. Archit. Herit. **11**(6), 763–772 (2017)
37. J.L. Parracha, A. Santos Silva, M. Cotrim, P. Faria, Mineralogical and microstructural characterisation of rammed earth and earthen mortars from 12th century Paderne Castle. J. Cult. Heritage **42**, 226–239 (2020)
38. T. Santos, L. Nunes, P. Faria, Production of eco-efficient earth-based plasters: influence of composition on physical performance and bio-susceptibility. J. Clean. Prod. **167**, 55–67 (2017)
39. S. Rescic, M. Mattone, F. Fratini, L. Luvidi, Earthen plasters stabilized through sustainable additives: an experimental campaign. Sustainability **13**(3), 1090 (2021)
40. A. Ranesi, R. Veiga, P. Rodrigues, P. Faria, The different effect of two paint systems on moisture buffer capacity of traditional and modern plasters. J. Build. Eng. **91**, 109701 (2024)
41. A.F. Carvalho, F.P. Cardoso, *Manual cores da terra: Paint Production with Soil Pigments*, 2nd edn. (Instituto da Pedra, São Paulo, 2023)
42. J.L. Parracha, A.S. Pereira, R. Velez da Silva, N. Almeida, P. Faria, Efficacy of iron-based bioproducts as surface biotreatment for earth-based plastering mortars. J. Clean. Prod. **237**, 117803 (2019)
43. T. Santos, R.I. Gomes, P. Faria, Assessment of natural aging and ecological surface treatments in clay-based renderings (in Portuguese). Conservar Património **35**, 31–44 (2020)

Chapter 9
Rammed Earth in Modern Construction: Physical and Mechanical Properties

Fernando Ávila, Esther Puertas, Álvaro Blanca-Hoyos, and Rafael Gallego

Abstract Rammed earth is a traditional construction technique widely used all over the world, and which is currently experiencing a resurgence of attention as an eco-friendly building solution. Nevertheless, the general absence of national and international standards based on the structural comprehension of such constructions poses a challenge for architects and builders looking to incorporate this technique into modern building projects. In this context, the present chapter develops a thorough analysis of the state-of-the-art of the knowledge regarding the mechanical and physical properties of rammed earth, modern testing techniques applied to this traditional material and main results, values of consensus, and uncertainties. However, the mechanical properties of unstabilized rammed earth are sometimes not enough to meet the requirements of existing building codes, so stabilization and reinforcement techniques are applied to enhance its mechanical behavior. The second part of the chapter focuses, therefore, on the stabilizing and reinforcing techniques for rammed earth.

9.1 Introduction and Historical Background

Rammed earth (RE) traditional construction technique is experiencing a significant revival as a sustainable and eco-friendly building solution. Used historically across several cultures and geographic regions, this construction method involves compacting layers of moist earth between temporary formwork to create walls. Today, as the construction industry increasingly prioritizes environmental sustainability, RE is gaining renewed interest for its minimal environmental impact, use of local materials, and potential for reducing the carbon footprint of buildings.

F. Ávila (✉)
Department of Continuum Mechanics and Theory of Structures, Polytechnic University of Valencia, Valencia, Spain
e-mail: favila@mes.upv.es

E. Puertas · Á. Blanca-Hoyos · R. Gallego
Department of Structural Mechanics and Hydraulic Engineering, University of Granada, Granada, Spain

D. A. H. Hanaor (ed.), *Sustainable Materials for the Built Environment*, Springer Series in Materials Science 352, https://doi.org/10.1007/978-3-031-97818-0_9

From ancient times, humans have employed earth as a construction material. Its low cost, versatility, and favorable mechanical and insulating properties have made it an excellent choice throughout history [1, 2]. Cultures across the globe have devised various building techniques using earth as the primary material, adapting these methods to local conditions and improving construction practices. Many examples of earth-built structures from different civilizations have survived to the present day.

The earliest examples of earth architecture preserved today can be found in the Near and Middle East. Some of the oldest structures have been discovered in the Turkestan Region (Kazakhstan), where archaeological sites dating from 8000 to 6000 BCE reveal houses made with adobe [3]. Regarding rammed earth, some of the oldest examples in the same zone are the building foundations found in the area corresponding to the ancient Assyrian civilization, dating back to around 5000 BCE [1].

In the African continent we can also find some of the oldest examples of earthen constructions. In Upper and Middle Egypt, some earth block constructions have been estimated to be over 4000 years old. Notable examples, such as the fortification of Medinet Habu and the Temple of Ramses II at Gourna, still stand today. In Sub-Saharan Africa, numerous examples of centuries-old earth construction techniques are present and still in use, sustained by the oral transmission of knowledge. The widespread use of earthen construction across the African continent is primarily due to its excellent thermal performance, which keeps interiors cool during the day and warm at night, as well as its low cost and high potential for material reuse [4, 5].

In Europe, with some examples of earth construction (earth-filled timber-framed houses and wattle-and-daub building technique) date from the Bronze Age, but the greatest expansion of rammed earth construction was between the 15th and the nineteenth centuries, mainly in Central Europe [1]. Earlier examples (built from between the thirteenth and sixteenth centuries) are found in Southern Spain, including some UNESCO heritage sites like the Alhambra of Granada (Fig. 9.1a, the Royal Alcazars of Seville, or the Desmochada Tower, part of the Almohad enclosure of the city of Caceres [6].

The case of America, particularly Central and South America, is unique due to its historical context. Pre-Columbian America had a rich tradition of earth construction, which later blended significantly with techniques introduced by the Spaniards after 1492 [8]. Notable pre-Hispanic examples include the Sacred City of Caral-Supe in Peru (3000 - 1800 BCE), the North Platform of Monte Albán in Mexico (first century BE) or the burial chambers of Tierradentro in Colombia (sixth - tenth century). Numerous historic urban centers in Latin America, such as Potosí (Bolivia), Quito (Ecuador), Antigua Guatemala (Guatemala), Cartagena de Indias (Colombia), Valparaíso (Chile), Puebla (Mexico) or Old Havana (Cuba), showcase a mixture of indigenous and Spanish earth construction techniques, featuring notable adobe and rammed earth structures [6, 9].

Today, earthen construction is gaining renewed attention due to two fundamental characteristics. Firstly, its reduced cost makes it an ideal solution for countries and regions with limited resources, and secondly, its low environmental impact aligns

(a) (b)

Fig. 9.1 Traditional rammed earth walls in the Alhambra of Granada (Spain) (**a**) and modern rammed earth building of L'Orangerie in Lyon (France) [7] (**b**)

with the imperative shift towards more sustainable construction practices. These factors position earthen construction as a crucial ally in achieving the UN Sustainable Development Goals [10] and as a compelling alternative to conventional building materials, thus promoting environmental sustainability in the construction industry. In fact, both earth construction in general and rammed earth construction in particular have seen increasing relevance in contemporary building practices, with numerous modern constructions employing these techniques in recent years (Fig. 9.1b).

However, and despite its benefits, the widespread adoption of RE in contemporary construction faces substantial challenges. One of the primary obstacles is the lack of comprehensive national and international standards tailored to the structural properties and performance of rammed earth constructions. This gap in standardization presents difficulties for designers and builders who seek to integrate this traditional technique into modern projects while ensuring compliance with current building codes and regulations.

The present chapter aims to address this gap by providing an in-depth analysis of the current knowledge on the mechanical and physical properties of rammed earth. It reviews modern testing techniques applied to this material, highlighting key results, consensus values, and existing uncertainties. Understanding these properties is crucial for developing reliable structural assessments and for advancing the application of rammed earth in contemporary buildings. There are several cases, however, where unstabilized rammed earth (URE) does not meet the mechanical performance requirements set by the existing building codes. To overcome this limitation, stabilization and reinforcement techniques are employed to enhance its strength and durability [11]. This topic is addressed in the second part of this chapter, evaluating the existing techniques and their benefits and drawbacks.

9.2 Characterization of the Material Properties

The resurgence of interest in rammed earth as a sustainable building material in recent years, has prompted extensive research into its mechanical and physical properties. The mechanical behavior of RE, including its compressive, tensile, and shear strengths, is critical for evaluating its performance in modern construction. By examining the results from existing literature, this section aims to establish consensus values while addressing the uncertainties that still persist in the field.

Beyond its mechanical properties, the section also explores the thermal and acoustic insulation capabilities of rammed earth, which are crucial for ensuring user comfort in buildings. The durability and degradation of RE are also discussed, emphasizing the importance of protecting these structures from environmental factors to prolong their lifespan.

9.2.1 Compressive Behavior

Rammed earth structural elements are typically designed to work primarily under compression. Consequently, their unconfined compressive strength (UCS) has become the most common parameter for defining their mechanical performance, similarly to other quasi-brittle building materials, such as concrete. The procedure commonly used to obtain the UCS of rammed earth materials involves conducting uniaxial compression tests. In the absence of specific standards for testing the UCS of rammed earth samples, researchers have adapted ASTM standards like D1633 [12] or D2166 [13], designed for the evaluation of the compressive strength of soil–cement cylinders and cohesive soil, respectively, or proposed alternative procedures based on existing standards for cement mortars and masonry design rules [14–16]. Although the variability in UCS results for RE in the literature is partly due to the material inherent heterogeneity, establishing a standardized test procedure would be essential for making results from different studies truly comparable.

It is widely recognized that UCS of RE is affected by the geometrical characteristics of the specimens (size and shape) and the manufacturing procedures and conditions (moisture content and compaction energy). However, the exact relationship between these parameters and the mechanical properties of rammed earth is still not fully known. Previous studies [15] have observed significant variations in the compressive strength of RE samples with different shapes and sizes made from the same material. In this case, the reduction in the UCS for the larger samples was attributed to variations in the material grading, but other authors [17] evaluating specimens of different scales have also found that UCS for smaller samples was higher than for larger ones. This underscores the need to approach the results of compression strength tests on conventional (small) RE specimens with caution, as they may not directly correspond to the mechanical capacity of full-scale walls.

Regarding the shape of the samples, studies in the literature have reported substantial differences in UCS results for prismatic versus cylindrical samples [15, 18]. One possible explanation is that the friction between the earth and the formwork during ramming is greater in prismatic specimens, particularly at the corners, allowing cylindrical specimens to be more effectively compacted and therefore exhibit better mechanical performance. Additionally, differences in load distribution patterns between prismatic and cylindrical specimens may also contribute to the observed variations in UCS results [19]. The procedures and results for some of the main studies regarding the compressive behavior of URE using prismatic and cylindrical samples are reported in Tables 9.1 and 9.2, respectively.

Another key aspect for the compressive strength of RE materials is the compaction energy applied to the soil during manufacturing, in fact, it is well-known that increasing the material density tends to result in higher UCS. However, it is very uncommon that existing studies measure or report the manufacturing compaction energy. In this regard, a recent study [16] proposed a new manufacturing methodology based on the widely used Proctor test, including the use of a standardized Proctor mold and the control of the compaction energy. The results showed a very significant reduction of the dispersion of the UCS of the samples (coefficient of variation lower than 2%) with respect to other studies in literature about RE.

Table 9.1 Results from uniaxial compression tests of URE cylindric samples in literature

Refs.	Diam. (cm)	Height (cm)	Density (t/m^3)	Moisture (%wt)	Testing standard	UCS (MPa)	E (MPa)
[20]	4.0	8.0	1.65	21	–	1.04	103
[21]	7.1	14.2	1.92	13	–	3.79	342
[22]	7.5	15.0	2.04	12	ASTM D2166	1.77	
[23]	7.5	15.0	2.14	7	ASTM D2166	1.85	34
[24]	7.5	15.0	1.95	12	–	2.23	143
[16]	10.1	11.5	1.94	13	ASTM D1633	1.40	–
[25]	10.0	20.0	2.19	8	AS 5101.4	0.70	147
[26]	10.0	20.0	2.10	10	–	1.26	1034
[27]	10.0	20.0	2.08	8	–	1.40	–
[28]	10.0	20.0	1.91	14	EN 1239 4	1.95	348
[29]	10.0	20.0	1.79	12	–	2.00	763
[15]	10.0	20.0	1.85	13	–	2.46	160
[30]	16.0	30.0	1.92	11	–	1.90	500
[15]	30.0	60.0	1.85	13	–	1.90	–

Table 9.2 Results from uniaxial compression tests of URE prismatic samples in literature

Refs.	Base (cm)	Height (cm)	Density (t/m^3)	Moisture (%wt)	Testing standard	UCS (MPa)	E (MPa)
[31]	10 × 10	10	1.66	–	IS 4332	1.10	1050
[32]	10 × 10	10	2.20	6	EN 12390-3	1.48	65
[33]	15 × 15	15	2.02	–	–	1.90	–
[34]	20 × 20	20	–	9	–	2.80	
[35]	20 × 20	20	–	11	–	3.04	–
[30]	20 × 20	40	–	10	–	1.00	–
[29]	25 × 25	50	1.88	12	–	1.15	365
[36]	10 × 40	50	1.82	–	–	0.71	60
[37]	11 × 50	50	2.19	10	DIN 18945	3.73	4143
[15]	30 × 30	60	1.92	13	–	0.81	65
[30]	40 × 40	65	1.85	13	–	1.90	–
[38]	30 × 100	100	2.00	–	–	1.30	500
[36]	50 × 200	250	1.92	–	–	0.33	80

Regarding the elastic modulus (E) of the RE materials, the dispersion in the results is even more evident, with values from less than 100 MPa to more than 1000 MPa. This considerable variability is related to the intrinsic heterogeneity of the materials, but also to the significant differences in the manufacturing and testing procedures, as observed for the UCS. Additionally, the studies in literature show differing definitions of elastic modulus that contribute to this dispersion. Some authors [23, 24] propose the calculation of a secant stiffness modulus in the elastic region of the compressive stress–strain curve of the material, following the expression indicated in the standard ASTM C469 [39]:

$$E = (\sigma_2 - \sigma_1)/(\varepsilon_2 - \varepsilon_1) \tag{9.1}$$

where ε_1 is a longitudinal strain equal to 5×10^{-5}, σ_1 is the stress corresponding to the longitudinal strain ε_1, σ_2 is the stress corresponding to 40% of the ultimate load and ε_2 is the strain produced by σ_2. As Eq. (9.1) was designed for concrete samples, other authors have proposed modification to adapt this expression to the elastic behavior of RE, considering σ_1 equal to 35% maximum stress and σ_2 equal to 75% of the maximum stress, with ε_1 and ε_2 as their corresponding strains [32].

However, this is not the only option to address the stiffness of RE, and other authors prefer using the tangent modulus, defined as the slope of the stress–strain curve in its elastic area [21, 25, 40, 41]. In addition, some studies [28, 42] calculate the secant modulus as the ratio between the maximum stress and its corresponding

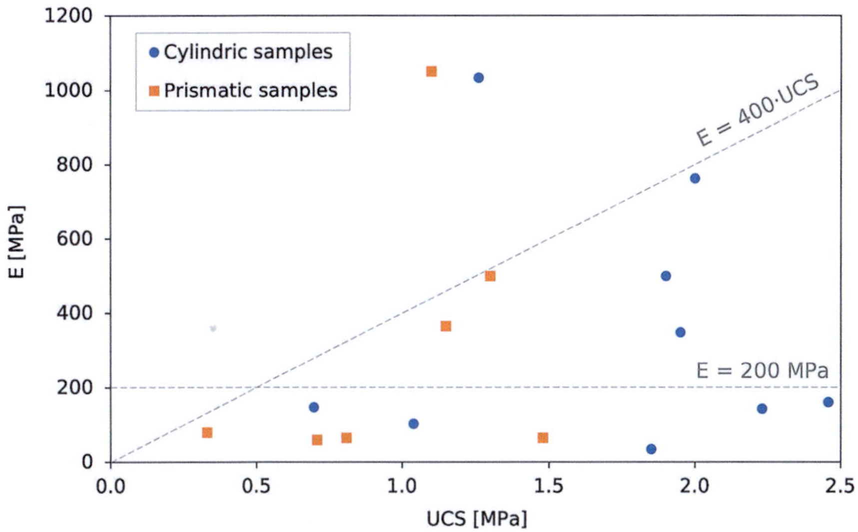

Fig. 9.2 Elastic modulus as a function of the uniaxial compressive strength of URE samples. Values from references indicated in Tables 9.1 and 9.2

strain, and in one investigation [43] a cyclic triaxial test was used to define the elastic modulus.

There is a general agreement that there is a relationship between the compressive strength of RE and its elastic modulus, but the exact correlation between these two parameters is still unclear. The Australian Earth Building Handbook [44] recommends the use of an elastic modulus of 500 MPa for all RE materials, but according to the results in literature, this value seems to be too high for URE samples. Another well-known standard, NZS 4297 [45], indicates that E can be calculated as $300 \cdot f_c\prime$, where $f_c\prime$ is the characteristic compressive strength, calculated as follows:

$$f_c\prime = 1 - 1.5 \cdot UCS_1 \cdot CV_{UCS} \tag{9.2}$$

where UCS_1 is the lowest UCS value of the series and CV_{UCS} is the coefficient of variation of the UCS series. Beyond these recommendations, as it can be observed in Fig. 9.2, the majority of the existing studies about URE obtain an E-value lower than 400 times the compressive strength and, in many cases, the elastic modulus is under 200 MPa regardless of the UCS of the material.

9.2.2 Tensile and Shear Behavior

Rammed earth elements are generally designed to work mostly under compressive loads, as they are known to have very low tensile and shear strength. However, these two parameters have proven to be essential to define (and prevent) failure of RE structures subjected to extreme loads, such as an earthquake [30, 46].

According to the results in literature [24, 30, 37], the tensile strength of RE is equal to ca. 10% of the compressive strength of the material. This relationship, obtained by means of pull-off and Brazilian tests, results into approximate tensile strength values between 0.10 and 0.35 MPa, in agreement with other studies in literature [19].

Similar values, with some few exceptions, are obtained for the shear strength, as shown in Table 9.3. Although some authors [23, 47] have performed direct shear tests to characterize the shear strength of RE, the most common procedure is to calculate it indirectly carrying out diagonal compression tests (DCT) [16, 26, 36, 37] on relatively large samples, in most cases following the procedure stablished in ASTM E519 [48] for masonry assemblies.

The shear strength of RE samples, highly influenced by its moisture content [51, 52], is directly related to the cohesion and friction angle of the material. Applying the Mohr–Coulomb theory, these parameters can be obtained in laboratory through shear box tests and triaxial tests [49, 53], or calibrated through numerical models [16, 30, 37, 38]. As it can be observed in Table 9.3, values of cohesion are typically in the range between 50 and 200 kPa, while the friction angle has values from 37° to 50°, excluding some very few exceptions.

Table 9.3 Characterization of the shear behavior of URE in literature

Refs	Dimensions (cm)	ρ (t/m^3)	MC (%wt)	Test (standard)	f_s (MPa)	c (kPa)	φ (°)
[49]	⌀7.6, h14.7	2.00	11	TCT	–	13	41
[30]	⌀16, h30	1.90	10	BT	0.18	170	51
[26]	20 × 55 × 55	2.10	–	DCT	0.15	189	37
[47]	2 × 6 × 6	2.13	11	DST (EN 1377-7)	–	68	44
[29]	3.5 × 10 × 10	–	–	DST	–	198	45
[23]	15 × 15 × 18	2.14	12	DST (ASTM D3080)	–	50	65
[35]	20 × 20 × 20	–	11	DST	0.30	260	51
[36]	10 × 40 × 50	1.82	–	DCT	0.06	–	–
[16]	10 × 50 × 50	1.92	13	DCT (ASTM E519)	0.14	–	–
[37]	11 × 50 × 50	2.19	10	DCT (ASTM E519)	0.75	–	39
[50]	45 × 50 × 50	–	12	DST	–	31	37
[36]	50 × 200 × 250	1.92	–	DCT	0.04	–	–

ρ Density; *MC* moisture content; f_t Tensile strength; f_s Shear strength; c Cohesion; φ Friction angle; *BT* Brazilian test; *DCT* Diagonal compression test; *TCT* Triaxial compression test; *DST* direct shear test

9.2.3 Hygrothermal and Acoustic Properties

Ensuring the thermal and acoustic comfort inside a building is essential in any architectural project. These factors are related to the general configuration and situation of the building, but also to insulating properties of the materials used.

The main parameters to measure the thermal properties are the thermal conductivity and the thermal resistance. In the case of unstabilized rammed earth, it has been observed that the compacted material has a thermal conductivity ranging from 1.0 to 1.4 W/(m K) [7, 19], which is very similar to that of brick masonry and better than the one measured in concrete or stone. The value measured for RE, nevertheless, is highly dependent on the material density and moisture content [1, 7, 54, 55].

Not only the conductivity of the material itself should be considered, however, but also the thickness of the structural element. For this purpose, the thermal resistance is defined as the ratio between the thickness of the element and its thermal conductivity. Therefore, the thermal resistance of a 50-cm thick wall would be of around 0.4 m^2K/W, quite far from the minimum value of 1.3 m^2K/W indicated in some earth building standards [44]. This issue could be addressed by adding certain polymers or phase-change materials to the soil, or by installing thermal insulating panels attached to the walls, similar to how conventional walls are commonly insulated [56–58].

Additionally, RE walls have the capacity to buffer temperature variations inside the building, due to their thermal inertia, reducing thermal amplitude by more than 70% [57, 59, 60]. This property, together with the ability of RE to balance air humidity (significantly reducing the relative humidity fluctuations) [1, 60, 61], has a very positive impact on the thermal comfort.

Acoustic insulation, on the other hand, is commonly measured evaluating the decibel reduction provided by a certain structural element, so it depends on the material and the thickness of the element. This acoustic insulating capacity is typically represented by the sound reduction index [62] or the sound transmission class (STC) [63, 64]. Most studies regarding URE show values of both parameters of about 57 dB [19], proving a good acoustic performance of the material (for reference, a STC of 60 dB means that amplified sounds are hardly audible [65]). In the absence of direct tests to measure the sound reduction index, an approximate value can be obtained using the expression indicated in BS 8233 [66]:

$$R = 21.65 \log_{10}(\rho \cdot t) - 2.3 \tag{9.3}$$

where R is the sound reduction index in decibels, ρ is the material density in kg/m^3 and t is the thickness of the wall in meters.

In addition, rammed earth, due to its porosity, can reduce sound reverberation inside the building [67, 68], although further studies about this topic would be required to correctly assess this property.

9.2.4 *Durability and Degradation*

The sensitivity of rammed earth to the weathering action is a relevant matter of concern for designers, building companies and users. Unprotected RE walls are potentially vulnerable to rain and wind erosion, according to the degradation observed in several existing structures and the laboratory tests carried out in different studies. These tests include accelerated erosion and wire brush tests [40, 69–71], performed according to ASTM D559 [72] (designed for compacted soil–cement samples), water erosion tests [73] (according to NZS 4298 [74]), and rainfall simulation and drip tests [75].

Based on the results of these studies, cement stabilization can help reduce erosion damage to RE materials. However, it is generally recommended to externally protect RE elements, whether stabilized or not, using waterproofing paints and plasters. Additionally, reducing direct exposure to rain and wind through the use of sloping roof systems is advised [75–77].

External protection and stabilization are also useful to enhance the durability of RE under aggressive environments. A significant reduction of URE mechanical properties has been observed when exposed to sulfate, alkaline and acid environments, which can be minimized if cement is added as a stabilizer to the mixture [76, 78]. However, some precautions should be taken when using cement to avoid secondary alkali-aggregate reactions and sulfate induced swelling induced by the presence of the stabilizer [69].

Another factor that can lead to the degradation of rammed earth materials is the exposure to dry–wet cycles and frost-defrost cycles. Although further research on this topic is still needed, some studies [73] suggest that the inclusion of cement and the presence of gravel in the soil can significantly enhance the ability of RE to withstand frost-defrost cycles without substantial degradation.

9.3 Stabilization and Reinforcement

When employing the RE technique in new constructions, its mechanical performance frequently does not meet the standards defined by building codes. To enhance these mechanical properties, as well as improve its thermal and acoustic behavior or durability, various additives can be added to the soil, resulting in what is known as stabilized rammed earth (SRE). These stabilizers improve the performance of rammed earth through physical and chemical interactions with the soil particles and water present in the mixture. Some of these additives, such as lime or natural fibers, have been used since ancient times, while others, like cement, sub-products from other industries (e.g. coal combustion residues), artificial fibers, and advanced materials, have been introduced more recently but are having a high impact today.

The use of stabilizers in rammed earth is becoming increasingly common, significantly enhancing its mechanical properties and allowing for its application in a wider range of construction projects. However, if additives are used systematically and without careful consideration of the specific requirements that URE cannot fulfill, there is a risk that RE constructions could lose some of their most valuable characteristics that make this construction technique particularly attractive: low cost and minimal environmental impact.

9.3.1 Additives and Stabilizers

Additives and stabilizers can be classified based on whether their use in the stabilization of RE is through chemical reactions (such as cement, lime, and gypsum) or, in contrast, based on their shapes (natural and synthetic fibers).

9.3.1.1 Cement, Lime and Other Chemical Stabilizers

These stabilizers are used to chemically enhance the mechanical properties of RE and strengthen its matrix. The most commonly used stabilizer is Portland cement, as it allows to achieve the highest compressive strength and durability [11, 31]. Soils with a lower percentage of clay are more suitable for cement stabilization since higher clay contents are more likely to exhibit more significant shrinkage [52]. Additionally, there are other requirements that the soils used for cement stabilization should meet, such as not containing organic material.

According to several studies, the unconfined compressive strength improves as a higher percentage of cement is added, as does the tensile strength [11, 24]. Furthermore, durability parameters such as erosion resistance are also improved [76, 78]. Another advantage, in terms of technical and legal certainty, is that cement stabilization is considered in most modern rammed earth standards in countries such as Australia, New Zealand, and the United States, making it easier for designers and builders to use cement in RE constructions.

However, despite its mechanical advantages and the improved durability provided, cement stabilization could compromise the sustainability benefits of RE. The proximity of resources (potential use of local soils) and the positive impact on local economies are affected by using cement as a stabilizer. Cement stabilization also impacts the environmental cost of RE by increasing the global warming potential over 100 years by 70%, while increasing by 35% the acidification potential of land and water compared to soil without cement stabilization [69]. The low cost of this building material could also be compromised, as, for example, 10% cement stabilized RE cost per unit weight can be between two and four times the one of URE [22, 23].

Hence, the use of natural stabilizers has gained growing interest with the aim of improving the mechanical and durability properties of RE without compromising its environmental and economic benefits. Among these, lime stabilization is the most commonly used, due to its historically use to stabilize earthen materials in general and RE in particular, both as an additive to the mixture or as an external protection

[32]. Historical heritage sites made of lime-stabilized rammed earth (LSRE) can be found in several regions, such as the Alhambra of Granada (Spain), the Fujian Tulou (China) or numerous traditional houses in Southern Portugal or France [23, 79–82].

Lime stabilization results from the hydration reaction of lime, which increases the concentration of $Ca2+$ and $OH-$ [31, 32]. This reaction improves the mechanical performance and durability by reducing the porosity of RE, which is closely related to the UCS [83]. Lime is not only used as part of the soil and water mixture, but it is also sometimes applied at the wall surface to reduce the effects of rain and wind erosion. However, using cement or lime as surface plasters can affect the hydric behavior of RE, causing disadvantages related to increased porosity when applied after the initial weight stabilization of RE instead of being part of the initial mixture [80].

To reduce the footprint and environmental costs of cement and increase the benefits of lime stabilization, the combination of these two stabilizers with subproducts such as fly or bottom ash, recycled concrete aggregates or brick waste material has been studied by some authors [41, 61, 84]. The use of these materials combined with cement can provide similar improvements, halving the percentage of cement needed to achieve the mechanical requirements established by building standards [61]. However, the stabilization of RE materials always increase their environmental impact, in terms of carbon emissions and embodied energy, among others, as shown in Table 9.4.

Table 9.4 Carbon emissions and embodied energy per cubic meter of RE

Additive(s)	CO_2 emissions (kg)	Refs	Embodied energy (MJ)	Refs
None (URE)	3–9	[22, 85]	49	[69]
2.5% cement	42	[22]	–	
4% cement	–		280	[86]
5% cement	86	[22]	–	
6% cement	–		400	[86]
7.5–8% cement	131	[22, 23]	500	[86]
10% cement	179	[22]	630	[86]
12% cement	–		750	[86]
5% cement + 2.5% FA	129	[23]	–	
5% cement + 5% FA	–		155	[69]
2.5% cement + 5% FA	120	[23]	–	
7.5% FA	106	[23]	–	
25% FA + 6% CCR	–		68	[69]
20%vol EPS	18	[22]	–	
10% PCM	1630	[22]	–	

FA Fly ash; *CCR* Calcium carbide residue; *EPS* Expanded polystyrene; *PCM* Phase-change materials.
Source [11]

9.3.1.2 Natural and Synthetic Fibers

To enhance the mechanical properties of rammed earth, particularly those where it fails to meet construction standards, such as tensile and shear strength, additives that enhance the material properties through their physical shape, rather than their chemical properties, are often used. These additives are typically presented in the form of fibers, and their use for RE enhancement is more accurately referred to as reinforcement (or additivation) rather than stabilization, although the term "fiber stabilization" is not rare in literature [11, 42].

Fiber reinforcement can be performed by adding short pieces of fiber directly to the earth mixture, in which case the fibers act as an additive, or by using fibers in the form of fabric for internal or external structural reinforcement [87–89]. This section focuses on fibers used as an additive to the RE mixture, while internal or external reinforcement using fibers will be covered in the following section.

Fibers used as additives in RE mixtures can be classified as natural or synthetic based on their origin. Examples of natural fibers include cereal straw, palm tree fibers, leaf fibers, and sheep wool [11, 90]. Recent research has also explored the use of synthetic fibers, such as waste tire fibers, fiberglass and diverse plastic fibers [21, 24]. While synthetic fibers offer better control over dimensions and characteristics, natural fibers maintain the low environmental cost of RE, which is one of its primary advantages. Further research is steel needed to properly evaluate the effect of the diverse types of natural and synthetic fibers on the mechanical performance of RE materials.

9.3.2 Reinforcement Techniques

The use of fibers as short pieces can impact the workability of rammed earth in terms of compactability, especially when a high percentage of fibers is added. Additionally, mixing the material with fibers can be challenging due to their presence. To address these issues, fiber reinforcement has also been explored as internal or external reinforcement. In these cases, instead of individual pieces, a grid of fibers is placed internally and externally to enhance the tensile and flexural strength of RE.

As an alternative to fabrics, bars made from various materials have been used to reinforce RE. The main aim of these bars is to enhance the flexural strength of RE and are sometimes employed to reinforce critical points such as corners, joints, or wall openings. Similar to fibers, bars can be made of natural materials (e.g. wood and bamboo) or from more modern materials like steel [91], which have been the subject of recent studies investigating their impact on RE.

9.3.2.1 Internal Reinforcement

When the fiber grid or fabric is placed between layers of rammed earth (Fig. 9.3a), it is considered internal fiber reinforcement. This fabric is embedded within the RE matrix to enhance flexural strength and strain energy dissipation. Variables such as the diameter of the fibers and the spacing between them are crucial for assessing flexural strength [88].

However, similar to using short fiber pieces, the presence of interlayer fabric can affect the compactability of the material. Another significant challenge with this reinforcement technique is the limited knowledge about the effect of a fabric in the interlayer on the isotropy of the RE element. Flexural strength tests of internally reinforced RE have shown that the layer above the fabric could be expelled with crack propagation [88]. Further research is needed to understand potential negative effects, particularly on compressive strength, to avoid compromising some mechanical properties while enhancing others.

Given that RE is often compared to concrete as a quasi-brittle material, reinforcing RE walls with bars (Fig. 9.3b) to improve flexural strength is a logical research focus. However, including bars in RE presents concerns about ensuring a proper and uniform compaction to the mixture [93]. The inclusion of bars can hinder the compaction of the material nearby due to the difficulty in using the rammer close to the bars, as vibration techniques used for concrete are obviously not applicable for RE. Hence, the bond performance between the earth mixture and the bars is a critical factor, as the enhancement of flexural and tensile strength depends on the effective interaction between these elements.

This kind of internal reinforcement technique can be implemented with steel bars, like conventional reinforced concrete, which may include ribs for better grip. Alternatively, natural wooden or bamboo bars can be used with the aim of maintain the ecological benefits of RE and avoid some steel-related issues such as bar corrosion.

(a) (b)

Fig. 9.3 RE with internal reinforcement: specimen with interlayer synthetic fabric [88] (**a**) and representation of a wall with internal insulation and steel bars [92] (**b**)

9.3.2.2 External Reinforcement

To avoid issues resulting from poor compaction and increasing the anisotropy of RE elements caused by internal reinforcement with bars or fabrics, recent research has explored alternative external reinforcement techniques. These methods involving the application of external reinforcements are often carried out after the RE elements have reached their maximum strength [80, 88], frequently with the aim of increasing the structural performance of an existing building that does not meet the current standards or that has been damaged (for instance due to a seismic event).

While internal reinforcement techniques can compromise the workability and compactability of the mixture, external reinforcement techniques face the challenge of ensuring that the RE structure and the reinforcement work effectively together. After the RE wall has been built, the reinforcement should be applied with materials and procedures that ensure strong adhesion to the RE surface, making it work as a single composite structure.

This is particularly relevant for external fabric reinforcements (Fig. 9.4), which can be made from natural or synthetic materials, as it happened with the fibers used in internal reinforcement. The selection and application of the material used to bond the fabric to the RE surface presents its own challenges, as differences in moisture content between the RE and the added material could affect the hygroscopic characteristics of the earthen structure. Alternatively, wire mesh can be used in a similar way to fabric, typically placed at both sides of the RE element and attached to the it by wire connectors linking both sides [36, 94].

Despite these challenges, external reinforcement with fabric has proven useful according to several studies, particularly for reinforcing heritage buildings made of rammed earth to improve their seismic performance and prevent collapses in previously damaged areas [89]. Unlike internal reinforcement, which must be incorporated during construction, external reinforcement allows for the enhancement of critical structural points with superficial treatments, preventing collapse mechanisms such as wall overturning, which are common in historical buildings subjected to seismic forces [89, 91].

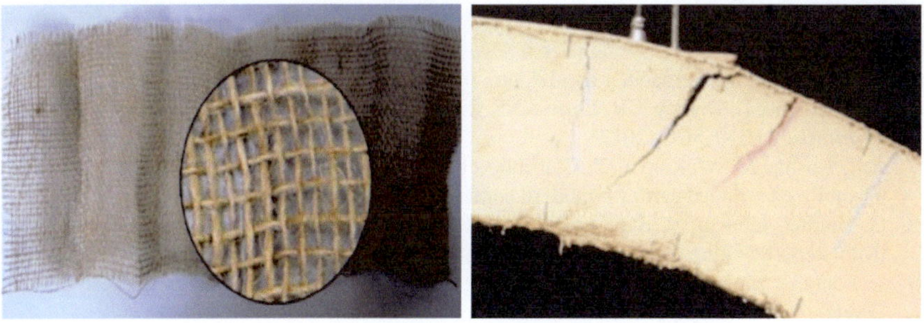

(a) (b)

Fig. 9.4 Jute fabric (**a**) and application as an external reinforcement for a RE arch (**b**) [87]

Other solutions to enhance their structural and seismic performance of RE structures are the inclusion of vertical rigid elements (columns) and horizontal confining elements. Steel and concrete columns can be incorporated to the RE structure during construction to increase the loading capacity and the seismic behavior of the assembly [95]. Cast-in-situ or precast reinforced concrete columns can be also combined with horizontal reinforcements (concrete tie beams or wire ties) to increase energy dissipation and bearing capacity [96]. Concrete bond beams, commonly placed at the top or half-height of walls, have proven useful to prevent wall overturning and reduce out-of-plane drift, enhancing the seismic behavior of RE buildings.

Another reinforcement method involves installing external wooden elements to act as a confining structure for the rammed earth building. These reinforcements help the walls work monolithically, minimize maximum displacements during seismic events, and enhance the energy dissipation capacity of the structure [79, 94]. The wooden elements should be positioned on both sides of the wall and securely connected with steel rods passing through [94]. This technique is especially beneficial for the rehabilitation of existing RE buildings.

Acknowledgements This research is part of the project *Evaluación de la vulnerabilidad estructural de patrimonio construido en tapia – VulTapia* (Evaluation of the structural vulnerability of heritage built in rammed earth), C-ING-134-UGR23, funded by Consejería de Universidad, Investigación e Innovación and by ERDF Andalusia Program 2021-2027. This study has been partially supported by a doctoral to Álvaro Blanca-Hoyos within the program "*Formación de Investigadores. Contratos Predoctorales (FPU) Universidad de Granada-Banco Santander*".

References

1. G. Minke, *Building with Earth: Design and Technology of a Sustainable Architecture* (Birkhauser—Publishers for Architecture, Basel, Switzerland, 2006)
2. H. Niroumand, M.F.M. Zain, M. Jamil, Various types of earth buildings. Procedia Soc. Behav. Sci. **89**, 226–230 (2013)
3. R. Pumpelly, *Explorations in Turkestan: Expedition of 1904* (Carnegie Institution of Washington, Washington, D.C., USA, 1908), p. 1.
4. J. Kennedy, *Building Without Borders* (New Society Publishers, Gabriola Island, BC, Canada, 2004)
5. M. Dabaieh, Earth vernacular architecture in the Western Desert of Egypt (2013), p. 24–30.
6. D. Gandreau, L. Delboy, CRATerre-ENSAG, *UNESCO World Heritage Inventory of Earthen Architecture* (UNESCO, 2012)
7. A.E. Losini, M. Woloszyn, T. Chitimbo et al., Extended hygrothermal characterization of unstabilized rammed earth for modern construction. Constr. Build. Mater. **409**, 133904 (2023)
8. G. Bauluz del Río, P. Bárcena Barrios, *Bases para el diseño y construcción con tapial* (Ministerio de Obras Públicas y Transportes (MOPT), Madrid, Spain, 1992)
9. L. Fontana, Molding earth outline: typology, technology and morphology of earth building (Exeter, U.K., 2006), pp. 1137–1154
10. United Nations, *The Sustainable Development Goals Report* (2022).
11. F. Ávila, E. Puertas, R. Gallego, Characterization of the mechanical and physical properties of stabilized rammed earth: a review. Constr. Build. Mater. **325**, 126693 (2022)

12. ASTM, *D1633-17 Standard Test Methods for Compressive Strength of Molded Soil-Cement Cylinders* (2017)
13. ASTM, *D2166/D2166M-16 Standard Test Method for Unconfined Compressive Strength of Cohesive Soil* (2016).
14. E. Erdogmus, E. Garcia, in *12th North American Masonry Conference*, Denver, Colorado, USA (2015)
15. V. Maniatidis, P. Walker, Structural capacity of rammed earth in compression. J. Mater. Civ. Eng. **20**, 230–238 (2008)
16. F. Ávila, M. Fagone, R. Gallego, E. Puertas, G. Ranocchiai, Experimental and numerical evaluation of the compressive and shear behavior of unstabilized rammed earth. Mater Struct 56 (2023)
17. Q.B. Bui, J.C. Morel, S. Hans, N. Meunier, Compression behaviour of non-industrial materials in civil engineering by three scale experiments: the case of rammed earth. Mater. Struct. **42**, 1101–1116 (2009)
18. R. El Nabouch Ranime, Q.B. Bui, Olivier, P., Perrotin, Assessing the in-plane seismic performance of rammed earth walls by using horizontal loading tests, Eng. Struct. **145**, 153–161 (2017)
19. F. Ávila, E. Puertas, R. Gallego, Characterization of the mechanical and physical properties of unstabilized rammed earth: a review. Constr. Build. Mater. **270**, 121435 (2021)
20. Q. Liu, L. Tong, Engineering properties of unstabilized rammed earth with different clay contents, J. Wuhan Univ. Technol. Mater. Sci. Ed. **32**, 914–920 (2017)
21. P. Zare, S.N. Shayan, M. Abbaspour, A. Fahimifar, M.M.H.S. Majdeddin, P. Zare, Experimental investigation of non-stabilized and cement-stabilized rammed earth reinforcement by Waste Tire Textile Fibers (WTTFs), Constr. Build. Mater. **260**, 120432 (2020)
22. M. Pakand, V. Toufigh, A multi-criteria study on rammed earth for low carbon buildings using a novel ANP-GA approach. Energy Build. **150**, 466–476 (2017)
23. M. Kosarimovahhed, V. Toufigh, Sustainable usage of waste materials as stabilizer in rammed earth structures. J. Clean. Prod. **277**, 123279 (2020)
24. V. Toufigh, E. Kianfar, The effects of stabilizers on the thermal and the mechanical properties of rammed earth at various humidities and their environ-mental impacts. Constr. Build. Mater. **200**, 616–629 (2019)
25. D. Ciancio, C.T.S. Beckett, J.A.H. Carraro, Optimum lime content identification for lime-stabilised rammed earth. Constr. Build. Mater. **53**, 59 65 (2014)
26. R. A. Silva, D. V. Oliveira, L. Schueremans, P.B. Lourencco, and T. Miranda, Modelling the structural behaviour of rammed earth components. In *Civil-Comp Proceedings*, vol. 106 (2014)
27. A. Arrigoni, R. Pelosato, G. Dotelli, C.T.S. Beckett, D. Ciancio, Weathering's beneficial effect on waste-stabilised rammed earth: a chemical and microstructural investigation. Constr. Build. Mater. **140**, 157–166 (2017)
28. A. Koutous, E. Hilali, Compression stress-strain curve of rammed earth: measuring and modelling. Results Eng. **18**, 101012 (2023)
29. E. N. Ranime, Mechanical behavior of rammed earth walls under Pushover tests, Université Grenoble Alpes (2017).
30. T.T. Bui, Q.B. Bui, A. Limam, S. Maximilien, Failure of rammed earth walls: from observations to quantifications. Constr. Build. Mater. **51**, 295–302 (2014)
31. D.D. Tripura, K.D. Singh, Characteristic properties of cement-stabilized rammed earth blocks. J. Mater. Civ. Eng. **27**, 04014214 (2015)
32. F. Ávila, E. Puertas, R. Gallego, Mechanical characterization of lime-stabilized rammed earth: Lime content and strength development, Constr. Build. Mater. **350**, 128871 (2022)
33. D.M. Lilley, J. Robinson, Ultimate strength of rammed earth walls with openings. Proc. Inst. Civ. Eng. Build. **110**, 278–287 (1995)
34. L. Zhang, T. Zhou, Z. Zhang, H. Tao, W. Tan, Z. Liang, Study on the bond-slip behavior between steel bar and rammed earth. J. Build. Eng. **90** (2024)
35. T. Zhou, H. Zhang, Z. Zhang, L. Zhang, W. Tan, Investigation of intralayer and interlayer shear properties of stabilized rammed earth by direct shear tests. Constr. Build. Mater. **367**, 130320 (2023)

36. L.E. Yamin, C.A. Phillips, J.C. Reyes, D.M. Ruiz, Seismic behavior and rehabilitation alternatives for adobe and rammed earth buildings, *In: 13th World Conference on Earthquake Engineering*, vol. 10 (2004)
37. L. Miccoli, U. Muller, P. Fontana, Mechanical behaviour of earthen materials: a comparison between earth block masonry, rammed earth and cob, Constr. Build. Mater. **61**, 327–339 (2014)
38. T. Bui, Q.B. Bui, A. Limam, J.C. Morel, Modeling rammed earth wall using discrete element method. Contin. Mech. Thermodyn. **28**, 523–538 (2016)
39. ASTM, *C469/C469M Standard Test Method for Static Modulus of Elasticity and Poisson's Ratio of Concrete in Compression* (2014).
40. M.M. Hallal, S. Sadek, S.S. Najjar, Evaluation of engineering characteristics of stabilized rammed-earth material sourced from natural fines-rich soil. J. Mater. Civ. Eng. **30**, 04018273 (2018)
41. S. Raj, A.K. Sharma, K.B. Anand, Performance appraisal of coal ash stabilized rammed earth. J. Build. Eng. **18**, 51–57 (2018)
42. A. Koutous, E. Hilali, Reinforcing rammed earth with plant fibers: a case study. Case Stud. Constr. Mater. **14**, e00514 (2021)
43. L. Xu, K.K. Wong, A. Fabbri, F. Champiré, D. Branque, Loading-unloading shear behavior of rammed earth upon varying clay content and relative humidity conditions. Soils Found. **58**, 1001–1015 (2018)
44. P. Walker, *Standards Australia, HB 195-2002 The Australian Earth Building Handbook* (Standards Australia International Ltd, Sydney, Australia, 2002)
45. N. Z. Standard, *NZS 4297:1998. Engineering Design of Earth Buildings* (1998)
46. L. Miccoli, D.V. Oliveira, R.A. Silva, U. Muller, L. Schueremans, Static behaviour of rammed earth: experimental testing and finite element modelling. Mater. Struct. Constr. **48**, 3443–3456 (2015)
47. A. Corbin, C. Augarde, in *First International Conference on Bio-based Building Materials*, Clermont-Ferrand, France (2015), p. 93.
48. ASTM, *E519/E519M-15 Standard Test Method for Diagonal Tension (Shear) in Masonry Assemblages* (2015).
49. H. Nowamooz, C. Chazallon, Finite element modelling of a rammed earth wall. Constr. Build. Mater. **25**, 2112–2121 (2011)
50. R. El-Nabouch, Q.B. Bui, O. Plé, P. Perrotin, Characterizing the shear parameters of rammed earth material by using a full-scale direct shear box. Const. Build. Mater. **171**, 414–420 (2018)
51. A.H. Narayanaswamy, *Mechanical Testing Procedure for Local Building Materials: Rammed Earth and Laterite Building Stones* (Ecole Nationale des Travaux Publics de l'Etat, France, 2016)
52. V. Maniatidis, P. Walker, A review of rammed earth construction, in *Developing Rammed Earth for UK Housing*, vol. 109
53. R. El Nabouch, J. Pastor, Q.B. Bui, O. Plé, Limit analysis, rammed earth material and Casagrande test. Comptes Rendus - Mec. **346**, 98–109 (2018)
54. M. Hall, D. Allinson, Assessing the effects of soil grading on the moisture content-dependent thermal conductivity of stabilised rammed earth materials. Appl. Therm. Eng. **29**, 740–747 (2009)
55. L. Soudani, *Modelling and Experimental Validation of the Hygrothermal Performances of Earth as a Building Material* (Université de Lyon, France, 2016)
56. S. Samadianfard, V. Toufigh, Energy use and thermal performance of rammed-earth materials. J. Mater. Civ. Eng. **32**, 04020276 (2020)
57. S. Serrano, A. de Gracia, L.F. Cabeza, Adaptation of rammed earth to modern construction systems: Comparative study of thermal behavior under summer conditions. Appl. Energy **175**, 180–188 (2016)
58. G. Cuitiño, A. Esteves, G. Maldonado, R. Rotondaro, Análisis de la transmitancia térmica y resistencia al impacto de los muros de quincha, Inf. la Constr. **67** (2015)
59. S. Serrano, L. Rincón, B. González, A. Navarro, M. Bosch, L.F. Cabeza, Rammed earth walls in Mediterranean climate: material characterization and thermal behaviour. Int. J. Low-Carbon Technol. 281–288 (2016)

60. B. Jiang, R. Lu, M. Jiang, L. Wang, L. Chun, L. Wan, Experimental study on thermal and humidity properties of modified rammed earth buildings in winter, Build. Environ. **258** (2024)
61. A. Arrigoni, C.T.S. Beckett, D. Ciancio, R. Pelosato, G. Dotelli, A.C. Grillet, rammed earth incorporating recycled concrete aggregate: a sustainable, resistant and breathable construction solution, Resour. Conserv. Recycl. **137**, 11–20 (2018)
62. ISO 10140-2:2021. *Acoustics—Laboratory Measurement of Sound Insulation of Building Elements. Part 2: Measurement of Airborne Sound Insulation* (2021)
63. ASTM, *E413-16 Classification for Rating Sound Insulation* (2016)
64. S. Australia, *AS 1276-1979. Methods for Determination of Sound Transmission Class and Noise Isolation Class of Building Partitions* (1979)
65. B. Schnitta, Defining the acoustic requirements for a building and community. Tech Science Press, Henderson **49**, 8 (2015)
66. B. S. Institution, *BS 8233 Guidance on Sound Insulation and Noise Reduction for Buildings* (2014)
67. A. Niampira Daza, E. Zambrano, J. Alcides Ruiz, in *EuroRegio2016*, Porto, Portugal (2016)
68. A.W. Bruno, *Hygro-Mechanical Characterisation of Hypercompacted Earth for Building Construction* (Université de Pau et des Pays de l'Adour, 2016)
69. A. Arrigoni, C. Beckett, D. Ciancio, G. Dotelli, Life cycle analysis of environmental impact vs. durability of stabilised rammed earth. Constr. Build. Mater. **142**, 128–136 (2017)
70. K.K.G.K.D. Kariyawasam, C. Jayasinghe, Cement stabilized rammed earth as a sustainable construction material. Constr. Build. Mater. **105**, 519–527 (2016)
71. D. Ciancio, P. Jaquin, P. Walker, Advances on the assessment of soil suitability for rammed earth. Constr. Build. Mater. **42**, 40–47 (2013)
72. ASTM, *D559/D559M-15 Standard Test Methods for Wetting and Drying Compacted Soil-Cement Mixtures* (2015), pp. 1–8
73. P. Narloch, P. Woyciechowski, Assessing cement stabilized rammed earth durability in a humid continental climate. Buildings **10**, 26 (2020)
74. N. Z. Standard, *NZS 4298:1998. Materials and Workmanship for Earth Buildings* (1998)
75. Y. Luo, M. Yang, P. Ni, X. Peng, X. Yuan, Degradation of rammed earth under wind-driven rain: the case of Fujian Tulou, China. Constr. Build. Mater. **261**, 119989 (2020)
76. S. Ghasemalizadeh, V. Toufigh, Durability of rammed earth materials. Int. J. Geomech. **20**, 04020201 (2020)
77. K.A. Heathcote, Durability of earthwall buildings. Constr. Build. Mater. **9**, 185–189 (1995)
78. Y. Luo, P. Zhou, P. Ni, X. Peng, J. Ye, Degradation of rammed earth under soluble salts attack and drying-wetting cycles: the case of Fujian Tulou, China. Appl. Clay Sci. **212**, 106202 (2021)
79. M.J. de la Torre López, P.E. Sebastán, G.J. Rodríguez, A study of the wall material in the Alhambra (Granada, Spain), Cem. Concr. Res. **26**, 825–839 (1996)
80. M.I. Gomes, T.D. Gonçalves, P. Faria, Hydric behavior of earth materials and the effects of their stabilization with cement or lime: study on repair mortars for historical rammed earth structures. J. Mater. Civ. Eng. **28**, 04016041 (2016)
81. Y. Luo, H. Zhong, F. Bao, Z. Guo, P. Ni, Insights into natural and carbonation curing of ancient Chinese rammed earth mixed with brown sugar. Constr. Build. Mater. **317**, 125969 (2022)
82. D. Alex, Recognition of a heritage in danger: Rammed-earth architecture in Lyon city, France, IOP Conf. Ser. Earth Environ. Sci. **143** (2018)
83. A.B. Jesús, R.L. dos Santos, S.I. Luis Ronaldo, B.M. Eclesielter, L.R. Juliana, Optimizing the evolution of strength for lime-stabilized rammed soil, J. Rock Mech. Geotech. Eng. **11**, 882–891 (2019)
84. M. Shaaban, Sustainability of excavation soil and red brick waste in rammed earth. Civ. Eng. Archit. **9**, 789–798 (2021)
85. S. Bestraten, E. Hormías, A. Altemir, Construcción con tierra en el siglo XXI, Inf. la Constr. **63**, 5–20 (2011)
86. B.V. Reddy, P.P. Kumar, Embodied energy in cement stabilised rammed earth walls. Energy Build. **42**, 380–385 (2010)

87. F. Loccarini, G. Ranocchiai, T. Rotunno, M. Fagone, Experimental and numerical analyses of strengthened rammed earth masonry arches, Comput. Struct. **239**, 106329 (2020)
88. E. Bernat-Maso, L. Gil, C. Escrig, Textile-reinforced rammed earth: experimental characterisation of flexural strength and thoughness, Constr. Build. Mater. **106**, 470–479 (2016)
89. M. Fagone, H. Kloft, F. Loccarini, G. Ranocchiai, Jute fabric as a reinforcement for rammed earth structures. Compos. Part B Eng. **175**, 107064 (2019)
90. A. Laborel-Preneron, J.E. Aubert, C. Magniont, C. Tribout, A. Bertron, Plant aggregates and fibers in earth construction materials: a review. Constr. Build. Mater. **111**, 719–734.
91. L. Zhang, T. Zhou, K. Hu, W. Tan, Z. Liang, Seismic performance of rammed earth walls with embedded steel wire mesh or bamboo reinforcement mesh. Structures (Oxford) **65** (2024)
92. L. Birznieks, *Designing and Building with Compressed Earth* (Delft University of Technology, 2013)
93. L. Zhang, T. Zhou, Z. Zhang, H. Tao, W. Tan, Z. Liang, Study on the bond-slip behavior between steel bar and rammed earth. J. Build. Eng. **90**, 109441 (2024)
94. Asociación Colombiana de Ingeniería Sísmica, Manual para la rehabilitación de viviendas construidas en adobe y tapia pisada (2004)
95. T. Zhou, B. Liu, Experimental study on the shaking table tests of a modern inner-reinforced rammed earth structure. Constr. Build. Mater. **203**, 567–578 (2019)
96. X. Yang, H. Wang, Seismic behavior of rammed earth walls with pre-cast concrete tie columns. Adv. Mater. Sci. Eng. **2018**, 1–10 (2018)

Chapter 10
Raw Rammed Earth Construction in Contemporary Contexts

Joseph Dahmen, Carene Umubyeyi, Karissa Wenger, and John Ochsendorf

Abstract Replacing energy-intensive building materials with low carbon sustainable alternatives is a key aspect of addressing global climate change. The built environment produces approximately 37% of anthropogenic carbon emissions (Building Materials and the Climate in Constructing a New Future, United Nations Environment Programme and Yale Center for Ecosystems + Architecture, 2024). A significant share of these emissions is due to fossil fuels consumed to manufacture and distribute building materials. The need for low carbon materials has led to a renewed interest in traditional earthen construction methods. These inherently circular materials provided humans with shelters for thousands for years prior to the emergence of energy-intensive materials. Recent estimates indicate that approximately 8–10% of the global population live in some form of earthen housing (Marsh and Kulshreshtha in Build Res Inf 50:485, 2022), which range from provisional shelters to architecturally significant buildings. Rammed earth is one form of earth construction, in which soil is compacted in layers into formwork to produce monolithic walls. Traditional rammed earth construction practices relied on naturally-occurring clays in soil for adhesion, a method known as "unstabilized" or "raw" rammed earth. This method was used to construct buildings that have lasted for hundreds of years. Cement stabilizers are often added to rammed earth to adapt its performance to contemporary standards, a method known as "stabilized rammed earth." Although stabilizers can

Portions of this chapter were adapted from an article published by the authors in *Construction and Building Materials*. The original article can be found at [36].

J. Dahmen (✉)
University of British Columbia School of Architecture and Landscape Architecture, Vancouver V6T 1Z2, BC, Canada
e-mail: jdahmen@sala.ubc.ca

C. Umubyeyi
Princeton University Civil and Environmental Engineering, Princeton, NJ, USA

K. Wenger
Arup, Boston, MA, USA

J. Ochsendorf
Massachusetts Institute of Technology School of Architecture and Planning, Cambridge, MA, USA

D. A. H. Hanaor (ed.), *Sustainable Materials for the Built Environment*, Springer Series in Materials Science 352, https://doi.org/10.1007/978-3-031-97818-0_10

improve strength and durability, they do so at an environmental cost. This chapter discusses the environmental advantages and durability considerations governing the use of raw rammed earth construction in contemporary architectural contexts.

10.1 Earthen Construction Overview

Buildings account for as much as 37% of global anthropogenic greenhouse gas (GHG) emissions [1]. A significant subset of these emissions, known as embodied energy, is due to manufacturing building materials and construction activity. Embodied energy accounts for 8–10% of global GHG emissions. Replacing energy intensive construction materials with environmentally sustainable low-carbon alternatives will be a key part of achieving ambitious GHG reduction targets laid out by the Intergovernmental Panel on Climate Change (IPCC) and others. Traditional building practices such as earthen construction offer valuable insights in this regard. Earthen dwellings have provided comfortable shelter to humans for thousands of years, long before the widespread availability of non-renewable fossil fuels and energy-intensive supply chains. The legacy of this approach is still evident in the approximately 650–700 million people, approximately 9% of the global population, who live in some form of earthen housing [2]. Earthen buildings range from simple shelters to architecturally significant buildings represented on the World Heritage List, such as the famous Alhambra in Granada, Spain shown in Fig. 10.1. [3, 4]. Rammed earth construction, in which monolithic walls are produced by compacting multiple layers of loose soil into formwork, is among the oldest earth construction technologies, extending back approximately 10,000 years [5, 6]. Traditionally, builders selected soils that contained sufficient amounts of naturally-occurring clays to provide the necessary adhesion of soil particles required for the technique [7]. More recently, ordinary Portland cement (hereafter referred to as 'cement') and other chemical binders are added to soil mixtures to provide higher design strengths and ensure predictable performance. However, these techniques add to the overall environmental impacts of rammed earth. The urgent need for low embodied energy materials to reduce climate change impacts gives traditional raw earth techniques new relevance in contemporary contexts. This chapter explores the use of rammed earth as a low-impact material in contemporary architecture.

10.2 Stabilized Versus Raw Rammed Earth

A growing number of contemporary builders have adopted rammed earth and compressed earth block construction in Europe, Australia, New Zealand and North America [8], responding to the demand for sustainable construction materials and methods. Contemporary architectural design using rammed earth (Figs. 10.2, 10.3) is documented by [9, 10]. These efforts are supported by a growing body of knowledge

Fig. 10.1 Unstabilized rammed earth has been used to construct buildings such as the Alhambra in Spain that have lasted for hundreds of years. Photo by Jebulon, Image available for use under the CC BY 4.0 license: https://en.wikipedia.org/wiki/Alhambra#/media/File:Dawn_Charles_V_Palace_Alhambra_Granada_Andalus

about rammed earth construction methods and performance. Normative guidelines for rammed earth applications in contemporary construction contexts include [8, 11, 12]. Technical guidance is offered by [13–17]. Rammed earth methods and materials have been the topic of international conferences convened in the past decade [18]. Commercial activities and academic research point to the potential of rammed earth as a sustainable construction method for contemporary contexts, highlighting the beneficial effects of a minimally processed material whose thermal mass can reduce heating and cooling loads when designed appropriately. However, the majority of the building codes and technical advice guiding the use of rammed earth in industrialized regions mandate the incorporation of cement stabilizers, driven by a desire to bring its structural performance and durability into compliance with contemporary expectations. Research shows that adding cement stabilizers to rammed earth increases its compressive strength [19]. Stabilization may also reduce initial surface erosion due to precipitation [20], although compressive strength does not correlate precisely with erosion resistance. Rather, erosion resistance is a factor of material consolidation due to compaction [7]. What is clear is that the incorporation of stabilizers drives up the embodied energy of rammed earth substantially, raising questions about its appropriateness as a material for sustainable architecture.

Embodied energy (EE) refers to the total amount of energy required to produce a product, including raw material acquisition, manufacturing, transportation and installation. As a measure of the environmental impact of materials, it correlates closely to

Fig. 10.2 Contemporary design using rammed earth—windhover Contemplative centre at Stanford university in Palo Alto, California (2015). Architect: Aidlin darling design. Builder: rammed earth works. Photo © Matthew Millman. Used with permission

CO_2 emissions. Each kilogram of cement requires approximately 4.5 MJ of energy per ton to produce [21], releasing approximately 0.9 kg of CO_2 into the atmosphere [22]. Cement stabilizers account for the greatest share of the EE of stabilized rammed earth due to the energy content of cement, which substantially outweighs other energy inputs [23]. Both the EE and climate impacts of stabilized rammed earth vary linearly with stabilizer use [24], which influences overall environmental more than the energy required to compact soils [19, 24, 25]. The incorporation of cement stabilizers in rammed earth reduces its environmental benefits, but by how much?

10.3 Assessing the Environmental Costs of Stabilization

Nine research projects characterizing the EE of rammed earth construction are summarized in Table 10.1. The different studies have been converted into MJ/m^3 of finished rammed earth for the purpose of comparison. Based on these results, a table of likely ranges of EE for stabilized and raw (unstabilized) rammed earth have been derived. Unlike building products manufactured in highly controlled facilities, rammed earth is typically constructed in situ, making consistent comparisons across projects challenging due to the differences in construction practices across individual building sites. However, the available data can be used to estimate the

Fig. 10.3 Contemporary design using rammed earth—studio rick joy, Tucson, AZ architect: rick joy. Builder: Quentin branch. Used with permission

expected range of EE required for both raw and stabilized rammed earth, which is presented in Table 10.2.

Assessing the full scope of building material impacts would require a "cradle to grave" life cycle assessment (LCA) approach that would track the full scope of embodied and operating impacts over the lifetime of the building, from raw material acquisition, through manufacturing, assembly, operating life, and eventual demolition. However, the most current research on the impacts of building materials are limited in scope to what is known as a "cradle to gate" approach. This approach accounts for all environmental impacts incurred up to the point that the product leaves the manufacturing facility, corresponding to stages A1–A3 below. Because rammed earth is produced at the building site in most (though not all) cases, cradle to gate accounting of rammed earth ends with the material installed at the building site. The different life cycle stages from material acquisition to construction, are outlined below, with aspects particular to rammed earth indicated below each heading.

- **A1: Raw material acquisition**

  Excavation of soil and any amendments required using excavating equipment. Energy required for the production of cement or other stabilizing agents.

Table 10.1 Comparison of the embodied energy of stabilized and raw rammed earth walls

Construction type	Data source	Method	Stabilization type (% wt)	Mixing and compaction method	Embodied energy (MJ/m^3)	Wall thickness, (m)	Embodied energy of Wall, (MJ/m^2)
Site Cast 25 MPa concrete reference*	Hammond and Jones [21]	LCA	12% OPC	None	2153	0.25	538
Cement stabilized rammed earth	Reddy and Kumar [23]	Empirical	6–8% OPC	Manual	450	0.3	135
	Reddy et al. [47]	Empirical	8% OPC	Manual	517	0.30	155
	Treloar et al. [25]	Estimation	8% OPC	Mechanical	561	0.30	168
	Lax [24]	LCA	6–9% OPC	Mechanical	EE varies linearly w/ cement content	0.30	–
	Arrigoni et al. [7]	LCA	10% OPC	Excluded from consideration	750	0.30	225
	Hammond and Jones [21]	Estimation	5% OPC	Mechanical	1292	0.30	388
	Hammond and Jones [21]	Estimation	6% OPC; 2% Lime	Mechanical	1577	0.3	473
Lime stabilized rammed earth	Fernandes et al.* [48]	LCA	3% Hydraulic lime	Mechanical	596	0.30	179
	Mateus et al. [49]	LCA	Lime	Mechanical	630	0.30	189
Unstabilized rammed earth	Reddy and Kumar [23]	Empirical	None	Manual	90	0.45	41
	OKOBaudat [50]	Unknown	None	Unknown	125	0.45	56

(continued)

Table 10.1 (continued)

Construction type	Data source	Method	Stabilization type (% wt)	Mixing and compaction method	Embodied energy (MJ/m^3)	Wall thickness, (m)	Embodied energy of Wall, (MJ/m^2)
	Arrigoni et al. [7]	LCA	None	Not included	163	0.45	73
	Lax [24]	LCA	None	Mechanical	GHG 2.5 times lower in raw than stabilized RE	0.45	–
	Hammond and Jones [21]	Estimation	None	Mechanical	810	0.45	365
	Nanz et al. [27]	Empirical	None	Mechanical	1170	0.45	527
	Nanz et al. [27]	Empirical	None	Mechanical	9628	0.45	4333

*Assumes a density of concrete of 2400kg/M3 with embodied energy value of.78MJ/Kg reported by [21] for cradle to gate stages A1–A3, increased by 15% to account for stages A4–A5 per [28]

Table 10.2 Aggregated data ranges: embodied energy ranges for stabilized and raw rammed earth

	Range	Embodied energy (MJ/m^3)	Typical thickness (m)	Embodied energy (MJ/m^2)	Comparison of RE EE/ m^2 to reference (%)	Comparison of EE to reference, including concrete foundation (%)*
25 MPa Concrete reference		2200	0.25	550	100	100
Cement stabilized rammed earth	Upper: OPC stabilization ~ 8%	1500	0.30	450	82	91
	Lower: OPC stabilization ~ 5%	1200	0.30	360	65	79
Lime stabilized rammed earth	Lime stabilization above 3%	800	0.30	240	44	63
	Lime stabilization 3% or below	600	0.30	180	33	55
Raw rammed earth	High: large amount of soil processing and amendment required/ soil transported	800	0.45	360	65	94
	Mid: moderate amount of soil processing and amendment required; soil provided at or near building site	500	0.45	225	41	76
	Lower: little to no amount of soil processing and amendment required; soil directly from foundation excavation	200	0.45	90	16	57

**Assumes a rammed earth wall 3 m above grade, supported by continuous 25 MPa concrete foundation 1m below grade at same thickness as RE wall above, compared to a. 25 m thick reference wall of concrete on a concrete foundation 1 m below grade at the same thickness (.25 m)

- **A2: Transportation of raw material to the manufacturing facility**

  Energy required to transport soil, amendments and stabilizers to the building site.

- **A3: Manufacturing energy.**

  Any soil processing required, such as blending, sieving, or other processing to achieve desired soil gradation and binding.
  Erecting formwork.
  Placing soil into formwork.
  Compacting soil blends in formwork.
  Stripping formwork.

- **A4: Transportation to building site**

  None if the walls are produced in situ; considerable if rammed earth is prefabricated offsite.

- **A5: Energy required for building assembly**

  None if the walls are produced in situ; can be a factor for rammed earth prefabricated offsite.

Since most rammed earth is produced at the building site, cradle to gate studies comparing rammed earth to other materials such as site-cast concrete must include the energy required by these materials at stages A4 (transport to construction site) and A5 (assembly and installation).

10.3.1 Method of Comparison

The results of the nine research projects consulted were converted to EE per cubic meter [MJ/m^3] where necessary to enable clear comparison. The following assumptions were used when converting:

- Density of RE assumed to be 1900 kg/m^3 except where otherwise noted
- Density of concrete is assumed to be 2400 kg/m^3
- Energy has been converted to primary energy

The results have been listed in Table 10.1. Cement stabilized variants are listed at the top. Lime stabilized blends, which are fired at a lower temperature than cement and require less energy to produce, are listed in the middle. Results for raw rammed earth are listed at the bottom. The research source and date of the research results are provided. Within each group, EE results are arranged from low to high.

The life cycle assessment framework requires a functional unit, defined as a quantity of a product required to provide a specific service, in order to compare building materials [26]. Thermal mass is a common rationale for specifying rammed earth in buildings, so a functional unit of 1 m^2 of wall of delivering a delivering a heat capacity of 500 kJ/K is used to enable comparison. Rammed earth has a specific heat capacity of approximately 850 J/kg K [27], comparable to that of concrete. One square meter of rammed earth at a thickness of 300 mm provides 500 kJ/K of heat capacity, compared to a concrete wall 250 mm thick providing the same performance due to its slightly greater density. In practice however, raw rammed earth is often constructed at a thickness of 450 mm to provide larger bearing area due to its lower compressive strength. This produces more thermal mass than is required for many buildings. Accordingly, Table 10.1 thus compares the EE/m^2 of wall area of a 300 mm thick stabilized rammed earth wall to a 450 mm thick raw rammed earth wall. Both of these walls are compared to a 250 mm thick concrete wall for reference. The EE of the concrete wall is provided using data taken from cradle-to-gate life cycle analysis data (stages A1–A3) provided by [21], with 15% additional EE reported to account for energy incurred during stages A4 (transporting ready-mix concrete to the building site), and A5 (setting formwork, lifting, placing materials, and stripping forms once the concrete has cured) [28]. Reinforcing steel has not been considered in the rammed earth or reference walls.

Five studies provide information on the EE of raw rammed earth construction. Disregarding the values reported at either extreme, which were due to manual production at the low end [23], and unusually high transportation of soils at the high end [27], suggests a range of approximately 125–1200 MJ/M^3 for raw rammed earth. In contrast, the seven studies reporting EE for stabilized rammed earth range between 400 and 1577 MJ/m^3. The lowest results in this range were reported where manual labour was used for excavation, mixing, placement and compaction. However, these findings are less relevant to industrialized regions, where these activities are generally accomplished with the assistance of machinery powered by electricity or distillate fuels. For industrialized regions, it is likely that stabilized rammed earth ranges between 1200 and 1600 MJ/m^3.

Table 10.2 provides aggregated ranges for stabilized and raw rammed earth. The ranges were determined by a qualitative analysis of the data reported in Table 10.1, and assumes mechanical extraction, transportation, processing and ramming. The wall thicknesses are the same as those indicated in Table 10.1, and include a concrete reference case for comparison. Whether rammed earth can be expected to land at the higher or lower end of the range provided depends on particular soil requirements, whether soil amendments such as additional sand or clay are required, and whether stabilizers are required. The aggregated ranges in Table 10.2 assume that the soil is supplied at the building site, with excavation energy considered except in the most minimal case, where it is assumed the soil used is provided by the foundation excavation, and can thus be considered an externality of that process.

The final column of Table 10.2 compares the impacts of different types of rammed earth walls with foundations included. Rammed earth is typically constructed on a continuous concrete foundation at the same thickness as the wall above to protect it from moisture below grade. This requires large quantities of concrete below grade, particularly in the case of raw rammed earth walls, which are nearly twice the thickness of concrete walls, and often thicker than stabilized rammed earth walls. The impacts of the concrete used in foundations are often overlooked when assessing the impacts of rammed earth. The final column thus assumes a rammed earth wall constructed to a height of 3 m above grade, resting on a concrete foundation of the same thickness as the wall above, constructed to a depth of 1 m below grade. Reinforcing steel has not been included in the analysis of either the wall or the foundation.

Limitations

The data reported above has several important limitations. EE is only one of six major impacts measured in an LCA study, which include impacts to human health, resource depletion, and other categories. Limiting the scope to EE may leave out important aspects of comparison. For example, while EE correlates closely with CO_2 equivalent emissions in most cases, it does not account for the CO_2 emissions caused during the production of cement due to the chemical reaction that occurs during calcination. This reaction, which releases approximately half of the CO_2 attributable to cement, is not reflected in the EE totals provided. This would significantly increase the climate impacts of cement stabilized rammed earth. It should also be noted that the cradle to gate approach employed is an effective means of limiting the scope of investigation, but it does not account for a number of potential benefits of rammed earth during the operational and end of life phases of the building. These include potential reductions to operating energy consumption during the occupied phase of the building, indoor air quality benefits caused by hygrothermal regulation, and ease of recycling and reuse during building demolition. One final consideration is that in practice, rammed earth builders sometimes add more stabilizers than required by mix design specifications to ensure that design strengths are met. Most mix designs are provided by weight, which is often converted to volume on the building site using approximate means. The risk of structural failure leads builders to err on the side of excessive use of cement stabilizers. Interviews conducted with North American builders suggested that actual rammed earth stabilization rates of 12–14% were not uncommon there, exceeding the intended specification by as much as 40–50% [29]. Overuse of stabilizers, whether intentional or not, can result in a stabilized rammed earth wall whose EE significantly exceeds that of a concrete wall serving a comparable purpose.

10.4 Strength of Raw Earth Construction

The impacts of cement stabilizers indicated in Tables 10.2 and 10.3, and the existence of many historic rammed earth buildings constructed without them, raises the question of how these traditional techniques could be adapted to contemporary contexts. Raw rammed earth construction requires special considerations, which starts with the selection of the soil, and more specifically, the clay constituent. Broadly, clays can be categorized as either expansive (typically montmorillonite) or nonexpansive (kaolinite or illite) [30]. Expansive clays have properties that enable them to incorporate water molecules in their microscopic structure, resulting in visible volumetric changes at the macroscopic scale [31]. The higher the content of expansive clays

Table 10.3 Erosion averages over 16 years for an raw wall constructed in the eastern united reproduced from [36]

Exposure catergory	Maximum erosion average (mm)	Maximum erosion average rate mm/year	Erosion per mm of precipitation (mm)
High	53	3.5	0.003
Medium	25	1.5	0.0012
Low	8	0.5	0.0004

present in the wall, the more shrinkage and swelling can be expected in the earthen material. Shrinkage cracks also facilitate water infiltration into earthen walls during subsequent wetting events [32]. To avoid excessive deterioration due to expansion and shrinkage, contemporary raw earth construction guidelines stipulate that only non-expansive clays be used as binders in the construction of rammed earth walls and that the overall clay content falls within a range of 15–30% by weight [8, 11].

The strength of raw rammed earth is a function of the following interrelated factors: type and percentage of clay, the soil type and gradation, optimal moisture content, and compaction energy applied. Soil is a highly variable resource, and a full discussion is beyond the scope of this chapter. However, numerous references exist that outline the use of natural soils to produce raw rammed earth walls, such as [8, 11, 13, 33]. Geotechnical evaluation of site soils is often advisable when developing mix designs and determining optimal moisture content, as well as creating samples for testing. With proper selection of soils and adequate compaction at optimal moisture levels, raw earth can be expected to achieved unconfined compressive strengths of approximately 1–2 MPa [34]. As compared to 2–9 MPa for stabilized rammed earth, depending on the amount of cement stabilization and compaction density [35]. The tensile strength of unreinforced rammed earth (whether stabilized or not) is approximately 10% of the compressive strength [34] which can be considered negligible. The strength of raw earth is derived from bonds between clay platelets. The ingress of moisture weakens these bonds, with clays reverting to plastic behaviour. This makes it essential to protect structural walls constructed of raw rammed earth from moisture due to precipitation or rising damp to retain design strengths [36].

It is clear from the data indicated above that the compressive strength of raw rammed earth is low by the standards of most conventional construction materials. However, the relatively low strength of raw rammed earth walls can be accommodated by designing thick walls with conservative slenderness ratios, which reduce compressive loading per unit area to acceptable levels. Tensile forces should be avoided altogether. In some cases, it may be advisable to provide a load bearing frame using a higher strength material in conjunction with rammed earth walls, as was done in the commercial building documented in [27]. Although the strength of unreinforced rammed earth is low, with careful design it is possible to use it effectively for a range of applications (Figs. 10.4 and 10.5).

Fig. 10.4 Unstabilized earth architecture: conference center in Alpbach, Austria (1998). Architecture: Din A 4 Arkitektur. Builder: Lehm Ton Erde Baukunst GmbH. Photo © Bruno Klomfar. Used with permission

10.5 Durability of Raw Rammed Earth Walls

The susceptibility of raw earth walls to weathering presents special challenges for contemporary construction. A full discussion of the factors involved can be found in [36]. The durability of a building material is a measure of its resistance to deterioration over time [37]. Raw earth walls are susceptible to weathering and deterioration due to environmental factors even when non-expansive clays are used as binders. As mentioned above, moisture ingress has an adverse effect on the bonds between clay platelets, causing a loss of compressive strength. The loss of mechanical strength also causes erosion at the wall face. This can be due to water infiltration and evaporation, surface erosion caused by rainfall, wind or snow, and damage caused by repeated freeze–thaw cycles [32], or rising damp due to capillary action from below. At the material scale, suction is an important element in the strength development of raw rammed earth [6]. As soils dry and suction increases, strength and internal cohesion between soil particles also increases. However, this increase has limits. Toll and Ong [38] have demonstrated that strength contributions from suction in a sandy clay decrease as the level of saturation increases. A typical exterior rammed earth wall in temperate climates has an estimated moisture content of 3–10% due to the relative

Fig. 10.5 Unreinforced earth in contemporary architecture: Atelier AHA Männedorf, Switzerland, (2022). Architecture: atelier.krecl GmbH: Lehm Ton Erde Baukunst GmbH Martin Rauch. Photo © Bruno Helbling. Used with permission

humidity absorbed. Since cohesion decreases in earthen materials with increases in internal water content, connections can be made between environmental conditions resulting in prolonged moisture ingress and consequent weathering and deterioration. The evaporation of water can also cause deterioration in rammed earth walls. As water evaporates from the soil, it carries dissolved ions with it, which migrate towards the wall surface and form salt crystals. These salt crystals reduce cohesion between soil particles and can consequently lead to weathering [39].

Entry of water into earthen walls is primarily caused by incident rainfall and capillary suction from the surrounding ground. In addition to water penetration, the angle and speed at which droplets impact a wall's surface and the formation of continuous streams of water can create pitting or localized regions of deep erosion [40]. This is particularly common when water accumulates and drips down from roof overhangs and other higher surfaces [41]. Careful detailing can reduce exposure to these deterioration mechanisms, therefore reducing the durability demands placed on the earthen materials themselves. The short-and long-term durability of raw rammed earth walls can be evaluated by employing laboratory and in-situ testing.

10.6 Laboratory Durability Assessments

Laboratory tests exist to assess the durability of earthen materials. These include the wire brush test which consists of a cyclical procedure of water immersion, oven drying, and firm brushing of soil–cement earthen samples to assess material loss due to repeated wetting and drying [42]. Also common is the accelerated erosion test, which involves spraying a specimen sample with a jet of water at a specified pressure and determining the maximum erosion depth and extent of moisture penetration in the material [13, 37]. Richards [43] used this test to investigate earthen heritage deterioration at an archeological site in northwest China. The drip test, developed by Yttrup [46] attempts to simulate rain droplets on an earthen surface by releasing water droplets from a wet cloth wick onto an inclined specimen [40]. These tests, which reproduce the effects of environmental conditions in a controlled laboratory setting, can provide some insight into comparative durability performance of different soils.

Although the tests outlined above may be helpful comparative assessments, their accuracy in predicting long term durability is the subject of debate. Research by [20, 40, 44, 45] indicates that laboratory testing often exaggerates the intensity of environmental factors responsible for deterioration, resulting in overly conservative performance criteria. Furthermore, the combined effects of multiple climatic conditions are difficult to simulate accurately in the laboratory, making the tests less effective in forecasting long-term performance [45]. These challenges demonstrate the value of full-scale experimental studies of rammed earth structures subjected to actual climatic conditions over multiple years or decades.

10.7 Long-Term Durability Assessments

Several long-term durability studies have been conducted on structures utilizing earthen materials to date. Guettala et al. [45] carried out a 4-year study to test the durability of eight small walls constructed using stabilized earthen bricks in the region of Biskra, Algeria. Prior to constructing the walls, brick specimens containing the same soil mix were subjected to laboratory tests, including an ASTM D559 wire brush test, a spray test, and a freeze–thaw test. The design of the experiment made it possible to compare the performance predicted by laboratory tests to actual observed weathering in the field. Results confirmed that the outcomes of accelerated weathering tests conducted in the laboratory were significantly more severe than observed weathering produced by the climatic conditions of Biskra after the 4-year test period. Similar results were found in a study by [46] where the erosion rates of raw compressed soil blocks subjected to a laboratory spray test were compared to the erosion rate of a test wall constructed using the same soil blocks and exposed to natural rain precipitation for three years. The erosion rate was quantified by comparing volume of soil removed from the specimens to the measured volume of water impinged, both in the laboratory and in the field. Results showed the quantified field erosion rate was approximately one third that of the laboratory rate suggesting that relying only on laboratory testing would result in overly conservative guidelines.

Two research projects in investigate the long term weathering of rammed earth walls, providing a more accurate quantification of the erosion of raw rammed earth walls in actual climatic conditions. Bui et al. [20] conducted a study on the durability of rammed earth wall specimens exposed to 20 years of natural weathering in France. The specimens were constructed on a concrete foundation and covered by an asbestos cement roof to protect them from water penetration. Two types of soil mixes were used, with lime stabilization added in several specimens. Erosion was quantified using stereo-photogrammetry, a method that involves estimating distances using photographic images taken from various locations. Results showed that raw rammed earth walls achieved a maximum erosion of approximately 18–21 mm over the course of the study, which amounts to a rate of approximately 1 mm/year [20] Similar rates of erosion are reported in a 16 year study of erosion mechanisms of a free standing raw rammed earth test wall constructed on the campus of the Massachusetts Institute of Technology in the United States in 2005 [36] (Fig. 10.6). The research documents the erosion of the 450 mm thick wall and identifies the environmental factors responsible for the observed deterioration. While erosion levels are highly dependent on localized environmental factors, results show that regions of the test wall with high levels of environmental exposure experienced an erosion rate of 3.5 mm/year, regions with medium exposure have an erosion rate of 1.5 mm/year, and regions with low exposure have an erosion rate of 0.5 mm/year, as reported in Table 10.3. The results presented are for a test wall with a steel cap overhanging the wall by 3.5 cm on each side. These values can be considered conservative, since they rely on erosion rates for a wall with minimal protection from precipitation that is exposed to weathering on both sides.

Fig. 10.6 Free standing unstabilized rammed earth test wall constructed on the campus of the Massachusetts Institute of Technology in 2005 [36]. Photo by Joseph Dahmen. Used with permission.

The durability studies mentioned above are useful in quantifying the weathering of raw earthen walls constructed with at least minimal protection from vertical precipitation. Without protection from vertical precipitation, raw rammed earth will deteriorate more rapidly. In contrast, providing ample roof overhangs will significantly slow down erosion due to precipitation. Historic rammed earth structures provide ample evidence that when adequately protected, raw rammed earth can last for hundreds of years, suggesting that this low impact material could be used to replace more carbon intensive construction materials with significant environmental benefit.

10.8 Conclusion

There is a growing interest in rammed earth construction as a response to the climate crisis. Embodied energy of building materials currently accounts for 8–10% of global GHG emissions. Replacing energy intensive conventional construction materials with environmentally sustainable low-carbon alternatives will be a key part of achieving ambitious greenhouse gas reduction targets laid out by the IPCC and others. However, most rammed earth in industrialized regions relies on chemical stabilizing agents that adversely affect its environmental benefits. In contrast, raw rammed earth, which relies on clays occurring naturally in soil for adhesion, has been used for millennia to construct comfortable buildings that can last for hundreds of years. The following guidelines may be useful when specifying raw rammed earth as an ecologically sound material for contemporary contexts:

- Minimize or eliminate the use of cement stabilizers
- Minimize the transportation of soils, preferably within 50 km.
- Use pier and beam foundations to minimize the use of continuous thick concrete footings below grade.
- Plan for erosion of 1–2 mm/year in temperate climates, which can be reduced or eliminated through detailing that protects walls from precipitation.

Unstabilized rammed earth is an age-old material that has new relevance for sustainable construction. Soil is a nearly ubiquitous resource that requires minimal processing to function as a safe and effective construction material, provided that its unique characteristics with respect to strength and weathering are accommodated through careful design and detailing.

References

1. Building Materials and the Climate, Constructing a New Future, United Nations Environment Programme and Yale Center for Ecosystems+Architecture (2024)
2. A.T. Marsh, Y. Kulshreshtha, The state of earthen housing worldwide: how development affects attitudes and adoption. Build. Res. Inf. **50**, 485 (2022)

3. D. Gandreau, L. Delboy, WHEAP (2012)
4. CRATerre, ENSAG World Heritage of Earthen Architecture Programme (2012)
5. S. Dobson, Rammed earth in the modern world, in *Rammed Earth Construction.* ed. by D. Ciancio, C. Beckett (CRC Press, London, 2015), pp.3–10
6. P.A, Jaquin, C. Augarde, C.M. Gerrard. A chronological description of the spatial development of rammed earth techniques. Int. J. Arch. Heritage: Conservation, Anal. Restoration. **2**(4), 377–400 (2008)
7. C. Arrigoni, C. Beckett, D. Ciancio, G. Dotelli, Life cycle analysis of environmental impact vs. durability of stabilised rammed earth. Constr. Build. Mater. **142**, 128–136 (2017). https://doi.org/10.1016/j.conbuildmat.2017.03.066
8. D. Easton, *The Rammed Earth House* (Chelsea Green Publishing Company, White River Junction, Vermont, 1996)
9. R. Rael, *Earth Architecture* (Princeton Architectural Press, NY, 2010)
10. O. Kapfinger, M. Sauer, M. Rauch, *Refined Earth: Construction & Design with Rammed Earth,* ed by O. Kapfinger, M. Sauer. (Edition Detail, Munich, 2015)
11. P.G. McHenry Jr., *Adobe and Rammed Earth Buildings* (University of Arizona Press, Tucson, AZ, 1984)
12. H. Houben, H. Guillaud, *Earth Construction: A Comprehensive Guide* (Intermediate Technology Publications, 1994)
13. P. Walker, R. Keable, J. Martin, V. Maniatidis, *Rammed Earth: Design and Construction Guidelines* (BRE Bookshop, London UK, 2005)
14. M. Hall, R. Lindsay, M. Krayenhoff, *Modern Earth Buildings: Materials, Construction and Applications* (Woodhead Publishing, Cambridge, UK, 2012)
15. H. Schroeder, Modern earth building codes, standards and normative development, in *Modern Earth Buildings: Materials, Engineering, Constructions and Applications* (Woodhead Publishing, Cambridge, UK, 2012), pp. 72–109
16. B.V.V. Reddy, *Compressed Earth Block & Rammed Earth Structures* (Springer Transactions in Civil and Environmental Engineering, 2022), pp. 347–363
17. A. Fabbri, J. Morel, C. Aubert, Q.B. Bui, D. Gallipoli, B.V.V. Reddy, in *State-of-the-art report of the RILEM TC 274-TCE* (Springer, 2022)
18. D. Ciancio, C. Beckett, C. Augarde, C., P. Jaquin, in *Proceedings of the Institution of Civil Engineers-Construction Materials* (Perth, Australia, 2016)
19. B.V.V. Reddy, P. Kumar, Embodied energy in cement stabilized rammed earth walls. Energ. Build. **42**(3), 380–385 (2009). https://doi.org/10.1016/j.enbuild.2009.10.005
20. Q.B. Bui, J.C. Morel, B.V.V. Reddy, W. Ghayad, Durability of rammed earth walls exposed for 20 years to natural weathering. Build. Environ. **44**(5), 912–919 (2009). https://doi.org/10.1016/j.buildenv.2008.07.001
21. G. Hammond, C. Jones, E.F. Lowrie, P. Tse, Embodied carbon, in *The Inventory of Carbon and Energy (ICE).* Version 2.0 (2011)
22. S. Chaturvedi, J. Ochsendorf, Global environmental impacts due to cement and steel. Struct. Eng. Int. **14**(3), 198–200 (2004). https://doi.org/10.2749/101686604777963748
23. B.V.V. Reddy, P. Kumar, Embodied energy in cement stabilised rammed earth walls. Energ. Buildings **42**(3), 380–385 (2009). https://doi.org/10.1016/j.enbuild.2009.10.005
24. C. Lax, *Life Cycle Assessment of Rammed Earth,* Master's thesis, University of Bath, (2010)
25. G. Treloar, C. Owen, R. Fay, Environmental assessment of rammed earth construction systems. Struct. Surv. **19**(2), 99–106 (2001). https://doi.org/10.1108/02630800110393680
26. ISO, 2006b. ISO 14040, *Environmental Management—Life Cycle Assessment—Principles and Framework.* (International Standards Organization, Geneva, 2006); ISO, 2006c. ISO 14044: *Environmental Management—Life Cycle Assessment—Requirements and Guidelines.* (International Standards Organization, Geneva, 2006)
27. L. Nanz, M. Rauch, T. Honermann, T. Auer, Impacts on the embodied energy of rammed earth façades during production and construction stages. J. Facade Des. Eng. **7**, 75–88 (2019)
28. Votorantim Cimentos, Environmental Product Declaration for Ready Mixed Concrete SPECTRA 40. Industry Report. São Paulo, Brazil (2023), https://api.environdec.com/api/v1/EPDLibrary/Files/32185f38-2244-4248-909c-08db0d9b78e5/Data Accessed 30 Dec 2024

29. J. Dahmen, *Personal Communication with Quentin Branch*, ed by D. Easton (Phoenix, AZ, 2005) (Napa, CA 2006)
30. R.F. Craig, *Craig's Soil Mechanics* (Spon Press, NY, 2004)
31. B. Velde, A. Meunier. *The Origin of Clay Minerals in Soils and Weathered Rocks* (Berlin, Heidelberg, 2008)
32. J.C. Morel, Q.B. Bui, Weathering and durability of earthen material and structures, in *Modern Earth Buildings: Materials, Engineering, Construction and Applications* (Woodhead Publishing, Cambridge, UK 2012), pp. 282–303
33. G. Minke, *Building with Earth: Design and Technology of a Sustainable Architecture* (Walter de Gruyter GmbH, Berlin, 2006)
34. (F. Ávila E. Puertas, R. Gallego, Characterization of the mechanical and physical properties of stabilized rammed earth: A review. Constr. Build. Mater. **325**, 126693 (2022)
35. B.V.V. Reddy, *Compressed Earth Block & Rammed Earth Structures* (Springer Transactions in Civil and Environmental Engineering, 2022) pp. 347–363
36. C. Umubyeyi, K. Wenger, J. Dahmen, J. Ochsendorf, Durability of unstabilized rammed earth in temperate climates: A long term study. Constr. Build. Mater. **409**, 133953 (2023).
37. K.A. Heathcote, *An Investigation into the Erodibility of Earth Wall Units*, PhD Dissertation, University of Technology Sydney, 2002
38. D.G. Toll, B.H. Ong, Critical-state parameters for an unsaturated residual sandy clay. Geotechnique **53**(1), 93–103 (2003)
39. S. Yan, N. Xie, J. Liu, L. Li, L. Peng, S. Jiang, Salt weathering of sandstone under dehydration and moisture absorption cycles: an experimental study on the sandstone from Dazu rock carvings. Earth. Surf. Proc. Land. **47**(4), 977–993 (2022)
40. K.A. Heathcote, Durability of earthwall buildings. Constr. Build. Mater. **9**(3), 185–189 (1995)
41. C. Beckett, P.A. Jaquin, J.C. Morel, Weathering the storm: A framework to assess the resistance of earthen structures to water damage. Constr. Build. Mater. **242**, 118098 (2020)
42. ASTM International, *D559/D559M-15 Standard Test Methods for Wetting and Drying Compacted Soil-Cement Mixtures* (ASTM International, West Conshohocken, 2020)
43. J. Richards, G. Zhao, H. Zhang, H. Viles, A controlled field experiment to investigate the deterioration of earthen heritage by wind and rain. Heri. Sci. **7**(1) (2019)
44. F.O. Ogunye, H. Boussabaine, Diagnosis of assessment methods for weatherability of stabilised compressed soil blocks. Constr. Build. Mater. **16**(3), 163–172 (2002)
45. A. Guettala, A. Abibsi, H. Houari, Durability study of stabilized earth concrete under both laboratory and climatic conditions exposure. Build. Mater. **20**(3), 119–127 (2006)
46. B.V.V. Reddy, K.S. Jagadish, Spray erosion studies on pressed soil blocks. Build. Environ. **22**(2), 135–140 (1987)
47. B.V.V. Reddy, G. Leuzinger, V.S. Sreeram, Low embodied energy cement stabilised rammed earth building—a case study. Energ. Buildings **68**, 541–546 (2014)
48. J. Fernandes, M. Peixoto, R. Mateus, H. Gervásio, Life cycle analysis of environmental impacts of earthen materials in the Portuguese context: Rammed earth and compressed earth blocks. J. Clean. Prod. **241** (2019). https://doi.org/10.1016/j.jclepro.2019.118286
49. R. Mateus, J. Fernandes, E.R. Teixeira, *Encyclopedia of Renewable and Sustainable Materials*, ed. by S. Hashmi, I.A. Choudhury (Elsevier, 2020), pp. 63–68
50. ÖKOBAUDAT, Sustainable Construction Information Portal, https://www.oekobaudat.de/en.html Accessed 23 Dec 2024

Chapter 11
New Technologies for Earth Construction: 3D Printing, a Reality or a Mirage?

Ana Mafalda Matos, Paula Milheiro-Oliveira, and Humberto Varum

Abstract The construction industry is on the brink of a revolutionary transformation with the advent of new technologies designed to address the pressing demands of sustainability, efficiency, and affordability. Among these innovations, 3D printing using earth materials stands out as a particularly intriguing development. This technique, which leverages locally sourced soil, proposes a paradigm shift from traditional construction methods. However, as with any emerging technology, it prompts a critical question: is 3D printing for earth construction a tangible solution to be adopted, or is it merely a futuristic vision that remains out of reach? This discussion explores the current state of 3D printing with earth materials, examining its practical applications, technological advancements, and the challenges that must be overcome to determine whether it is a present-day reality or an elusive mirage. This study specifically sought to identify the key mixture variables that impact the properties of a local and Portuguese stabilised earth, potentially suitable for 3D printing, and to determine the appropriate ranges for further investigation. To achieve this, a 2^3 central composite design (CCD) approach was employed to develop statistical models capable of describing fresh properties such as slump and other engineering characteristics like mechanical performance, resistivity, and mass loss. As such, a central composite design (CCD) was followed to find statistical models capable of describing those key stabilised earth mixtures properties in function of mixture input parameters: water to solids volume ratio (Vw/Vp); superplasticiser to cement weight ratio (Sp/p); cement to soil weight ratio (c/s).

List of Abbreviations

CCD	Central composite design

A. M. Matos (✉) · P. Milheiro-Oliveira · H. Varum
CONSTRUCT, Faculty of Engineering, University of Porto, Porto, Portugal
e-mail: anamatos@fe.up.pt

P. Milheiro-Oliveira
Center of Mathematics, University of Porto, Porto, Portugal

D. A. H. Hanaor (ed.), *Sustainable Materials for the Built Environment*, Springer Series in Materials Science 352, https://doi.org/10.1007/978-3-031-97818-0_11

Ci	Central points of CCD
CCi	Axial points of CCD
d	Days
DOE	Design of experiments
F	Flexure strength (MPa)
Fi	Factorial points of CCD
LOI	Loss on ignition (%)
Mass loss_7d	Mass loss after 7 days (%)
Mass loss_21d	Mass loss after 21 days (%)
Mass loss_28d	Mass loss after 28 days (%)
PC	Portland cement
PSD	Particle size distribution
SCM	Supplementary cementitious materials
SDG	Sustainable development goals
Sp	Superplasticizer
Resist_7d	Electrical resistivity at 7 days (%)
Resist_21d	Electrical resistivity at 21 days (%)
Resist_28d	Electrical resistivity at 28 days (%)
RH	Relative humidity (%)
Vw/Vs	Water to solids volume ratio
Sp/p	Superplasticiser to powder weight ratio
Rc	Compressive strength at 28 days (MPa)

11.1 Introduction

Earth (soils) is a raw natural building material as old as humanity, with records from the prehistoric period and worldwide use, which gives it a universal character. Earth was used in the construction of not only houses but also the most valuable and important monuments and architectural ensembles in the world. Even today, the earth is widely used as the main building material in regions such as Africa (particularly in the North), where this material does not compete economically. The tradition still exists in Central and Latin America and Asia, especially outside the large cities. Thus, a high percentage of the world's population, estimated to be two billion people, lives or works in earthen constructions among 150 countries [1, 2]; however, with high variability around the world. As such, the Earth materials provides economic, ecological, and social benefits, including health and comfort, due to their natural properties. The thick layers of earth elements possess high thermal inertia, which helps regulate relative humidity through their porous structure and clay content, contributing to excellent indoor air quality.

Traditional techniques based on raw earth are vernacular because they use locally available materials on the construction site or nearby. Earth construction techniques are varied, from rammed earth, adobe, and cob, depending on the identity of cultures and places, societal contexts and times in which they were integrated. For this reason, Earth Construction techniques usually present marked variations from region to region, and there may even be significant differences between buildings located in the same areas [3]. The construction methods were based on empirical know-how and passed from generation to generation without any scientific tool or analysis.

Those earth construction features are incompatible with the current building materials sector, particularly in Western economies, which today is based on standards, quality control, performance and modelling. Besides, the vernacular earth construction method is slow, time-consuming and require a large and very skilled workforce [4]. Other construction methods employing materials such as reinforced concrete, timber, or masonry have advanced, while Earth construction upgrades were almost non-existent. Therefore, the scientific community needs to reinvent Earth Construction and revisit its potential to meet the Agenda 2030 challenges by strengthening confidence in Earth materials, adapting it to current technologies, and unlocking its potential to construction stakeholders. Studying new construction methods for earth construction is crucial for advancing these practices, enhancing their viability, and addressing contemporary environmental, economic, and social challenges.

One of the front's action to upgrade earth construction techniques has been based on cement-based materials technology, namely the earth materials production and casting using the similar methods employed in concrete. Some examples can be seen in the following references [5–11], as concrete earth, poured Earth, flowable Earth, and self-compacting earth, among others. This approach has demonstrated that plasticisers and superplasticisers commonly used in cement-based materials, such as polycarboxylate ethers, are compatible with Earth materials. Additionally, stabilising raw earth with hydraulic binders—such as lime (10–15% by soil mass), Portland cement, and/or blended mixtures (typically 5–10% by soil mass), or calcium sulfoaluminate cement—enhances the workability of Earth materials by reducing viscosity [5, 11–15]. It also facilitates quicker setting, improves mechanical strength, and increases water resistance and, consequently, durability.

The rise of digital fabrication techniques, like 3D printing of cementitious materials (3DPC), has shown that advanced digital and automated production processes can offer significant advantages, including creating freeform architectural designs without the need for formwork [21]. Numerous institutions, universities, and companies around the globe have presented digitally fabricated prototypes of structural elements, furniture, and full-scale demonstration buildings [22–26]. Nevertheless, these projects predominantly rely on cement-based materials.

Combining natural or based on natural materials, such as earth, with the emerging digital production processes enables the integration of cutting-edge technology with the vernacular construction as grounds for contemporary digital innovation. In fact, a major driver was the 3D printing of a full-size earth-based structure in Italy by WASP [24], unveiling the potential of earth construction, and expanding the scope of 3D printing beyond cement-based materials. Very last research works of Gomaa et al. [22], Veliz et al. [23], Alhumayani et al. [24] showed the potential positive environmental benefits on the 3D printing of earth-based materials might have due to the reduction of the carbon footprint related with materials extraction and process, transportation, and the waste generation.

The advancement of a 3D-printable framework for Earth or Earth-based materials remains limited, with current experiments and applications still in their early developmental stages [22, 25–29]. Most of the research focuses on the design possibilities of 3D Printing for Earth construction. There is a need to start and provide scientific information rationally for the design of earth printable materials, considering typical soil variability and different types of soils, the fabrication machines employed, and construction workflow and processes, among others.

From the material's point of view, several issues must be stressed out. Alqenaeea and Memari [22] focused on 3D printable cob mixtures design. They found that 49% clay, 24% water, 15% sand, 10% lime, and 1.5% straws achieved printability and buildability simultaneously and adequate mechanical properties for low-rise houses. Thus, mixture design study and small changes from traditional cob mix ratios provided a 3D printable cob mixture version manufactured using similar concrete mixer, pumped system, and printed by a computer-controlled industrial robotic arm [22].

Silva et al. [30] analysed several earth mixtures for 3D printing: raw earth mixes, fibre-reinforced earth mixtures and stabilised (with Portland cement) fibre-reinforced earth mixtures. The study on fresh and hardening properties, using a rheometer, revealed that earth mixed with 24.3% of water (wt. %) reinforced with 1% (wt. %) of sisal fibre and stabilised with 13% (wt. %) of PC was suitable for 3D printing.

Perrot et al. [27] employed a biopolymer, alginate, to reach earth printable mixture by improving green strength. Two earth composites with 45% water content were studied, and 0% and 3% of alginate. Both mixtures were printed. However, the construction ratio was higher with alginate due to the fast increase in green strength. Compressive tests in printed samples with the circular cross-section showed similar results for both earth mixes with and without alginate, around 1.20 MPa, typical values for cob.

11.2 Research Significance

As the construction industry faces increasing pressure to adopt more sustainable and efficient practices, the potential of integrating advanced technologies with traditional building methods has garnered significant attention. Among these innovations, 3D printing has emerged and historically rooted in techniques like adobe, cob, and rammed earth, earth construction can be renowned for its environmental benefits and cultural heritage. The advent of 3D printing offers the possibility of enhancing these age-old methods with precision, speed, and versatility. However, lacks to provide scientific information on the constructability aspect, such as a rational basis for the design of new and advanced Earth-based materials, considering typical soils variability and different types of soils, the fabrication machines employed, and construction workflow and processes.

The study focuses on stabilising Portuguese soil by incorporating low dosages of cement. This approach aims to enhance the bonding between soil particles, thereby improving the mechanical strength and durability of the material. By optimising these properties, the research seeks to soil-based material suitable for 3D printing applications, balancing structural performance with sustainability.

One of the first barriers is the mix design of earth- or earth-based materials that can be printable. The present study aimed to study a stabilised Portuguese soil for future application in 3D printing construction method. To develop statistically significant models for explaining the printability-related properties of earth-based mixtures and other engineering characteristics, a statistical factorial design approach was employed. This approach analysed how the input parameters of the mixtures influenced the outcomes. The mix proportions were determined using three key ratios (factors): the water-to-soil volume ratio (Vw/Vs), the superplasticiser-to-cement weight ratio (Sp/p), and the cement-to-soil weight ratio (c/s). The response variables included slump, mass loss, electrical resistivity, and mechanical strength. Regression models were developed for each response variable, highlighting the significant influence of specific factors.

11.3 Materials and Methods

11.3.1 Raw Materials

A Portuguese soil from South region was used in the current research study. The raw soil particles presented some gravel and lumps, which seem not appropriate for actual printing technology [26]. Therefore, the raw soil particles were sieved through sieve of 1 mm. The soil fraction passed in the sieve was used in the current study, and the fraction retained was rejected, as seen in Fig. 11.1. The Particle size distribution (PSD) of soil (passed in a sieve 1 mm) was performed according to ASTM E 100 and ASTM D422 by sieving and hydrometer methods, respectively, and it is depicted in Fig. 11.2. As can be perceived, this soil consisted of 12% clay, 8% silt and 80% sand.

A scanning electron microscopic available at University of Porto [31], FEI QUANTA 400 FEG ESEM with integrated spectroscopic characterisation with EDS (Energy Dispersive x-ray Spectroscopy), allows access to soil particle morphology, as depicted in Fig. 11.3. As can be seen, soil particles present roughly similar dimensions (corroborating PSD analysis) with particles mostly with round edges, which can positively affect the workability of the composite. Besides, the EDS analysis presented mainly silicium and aluminium, typically found in soils (Fig. 11.3c).

The soil stabilisation was performed using Port (CEM I 42.5 R, complying with EN 197–1) and limestone filler according to EN 13139 and EN 12620. The main chemical, physical and mechanical properties of cementitious materials are highlighted in Table 11.1. A polycarboxylate ether superplasticiser was used to ensure

Fig. 11.1 A sample of the Portuguese soil used in the current study

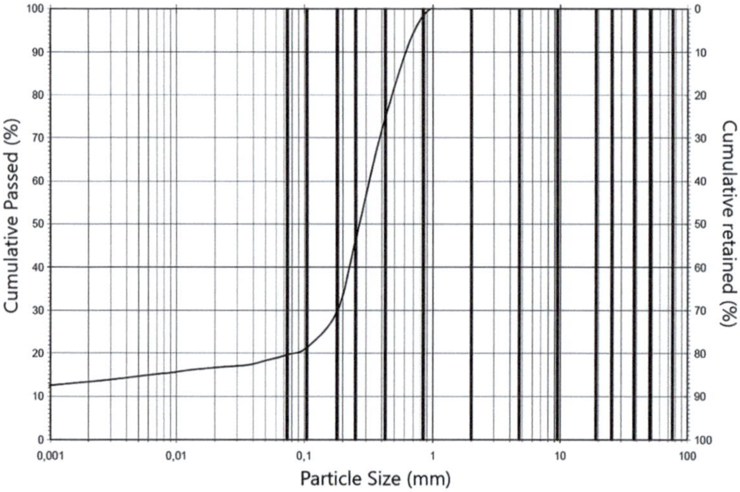

Fig. 11.2 Soil particle size distribution

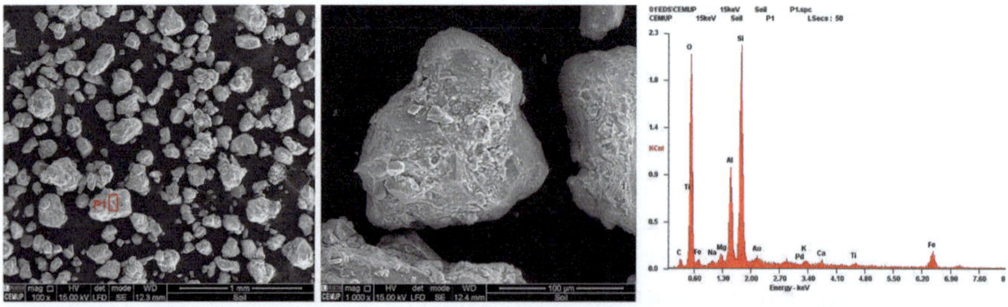

(a) (b) (c)

Fig. 11.3 Secondary electron mode SEM image of (**a**) soil particles (passed in sieve 1 mm) with amplification of 100; (**b**) soil particles (passed in sieve 1 mm) with amplification of 1000; EDS spectrum of P1

fine powder particle dispersion (with a density of 1070 kg/m^3 and a solid content of 29.5%). The water was obtained from Tap and used afterwards at room temperature.

11.3.2 Mix Design and Design of Experiments

The stabilised earth mix proportions design was based on the following three factors (mixture input parameters): water to soil volume ratio (Vw/Vs); superplasticiser to cement weight ratio (Sp/c); cement to soil mass ratio (c/s). Using the fundamental equation of formulation:

$$Vs + Vc + Vlf + Va + Vsp + Vv = 1.0\,m^3 \tag{11.1}$$

where, Vs is the volume of soil (earth), V is the volume of cement, Va is the volume of water, Vsp is the volume of superplasticiser, and Vv is the estimated void volume. The soil-to-mortar ratio volume was kept constant in all mixtures at 0.530. Considering the adopted values of Vs/Vs, Sp/c, c/s (see Table 11.3) and the raw materials density presented in Sect. 11.3.1, it is possible determine the quantity (kg/m^3) of each constituent material.

A Central Composite Design (CCD) was chosen to explore the possibility of developing statistically significant models that correlate the printability and other engineering properties of stabilised earth with the input parameters of the mixture previously described. The 2^3 Central Composite Design (CCD) included 8 factorial points (Fi), supplemented by 6 axial points (CCi) and 4 central runs. The influence of each factor -Vw/Vs, Sp/c, and c/s-was assessed at five levels: $-\,\alpha$, $-\,1$, 0, $+\,1$and $+\,\alpha$, as detailed in Table 11.2. To ensure the design's rotatability, the value of α was calculated as $nF^{1/4}$, where nF is the number of factorial combinations of the design, equal to 8 in this study. Accordingly, $-\,\alpha$, $-\,1$, 0, $+\,1$and $+\,\alpha$ were set to $-\,1.682$ and $+\,1.682$, respectively. Table 11.2 provides the correspondence between coded and absolute values of the input mixture parameters, while Table 11.3 displays the combinations of the three factors across all mixtures. The interval range for the design variables of stabilised Earth mixtures, as shown in Table 11.2, was determined based on preliminary tests, which verified the minimum workability and stability for both the most viscous (F1) and the most fluid (F8) stabilised Earth mixtures.

Several output properties were used as response models, namely, Slump, mass loss control between 7 and 28 days (Mass loss_7d, Mass loss_21d, Mass loss_28d, respectively), electrical resistivity evolution between 7 and 28 days (Resist_7d, Resist_21d, Resist _28d, respectively) and flexural and compressive strength at 28 days (F and Rc, respectively). The following section briefly describes the test methods employed.

Table 11.1 Cement and limestone filler's main oxide composition and physical properties

	Main oxide composition (%)											Physical properties	
	LOI	Insoluble residue	SiO_2	Al_2O_3	Fe_2O_3	CaO	MgO	Na_2O	K_2O	SO_3	Cl	Density (kg/m^3)	Specific surface (kg/m^2)
Cement	2.81	7.0	19.76	5.22	2.96	62.29	1.74	0.14	0.62	3.32	0.04	3080	373
Limestone filler					0.02	99.00	0.30			< 0.05	< 0.001	2720	550

Table 11.2 Correspondence of actual and coded values of Input mixture parameters

	Units	– 1682	– 1	0	+ 1	+ 1682
X_1: Vw/Vs		0.616	0.650	0.700	0.750	0.784
X_2: Sp/c	%	0.916	0.950	1.000	1.050	1.084
X_3: c/s		0.083	0.090	0.100	0.110	0.117

11.3.3 Response Properties

The stabilised earth mixtures (presented in Sect. 11.3.2) were produced using a mixer in accordance with EN 196–1, which operates at 62.5 and 125 rpm.

The flow diameter of fresh mixes was measured using a flow table in compliance with EN 12,350–5 (refer to Fig. 11.4). For the slump flow test, the fresh stabilised Earth mixtures were placed into a brass cone in two layers, with each layer compacted by tamping 25 times using a specialised rod. The excess mortar was levelled off using the straight edge of a trowel. The cone was then removed, and the table was dropped 25 times from a height of 12.5 mm within 15 s. Two perpendicular measurements of the spread flow diameter were recorded, and the average value was used to determine the slump flow.

Three prismatic specimens, with dimensions $40 \times 40 \times 160\ mm^3$ (see Fig. 11.5), for each mixture and used for both mass control and electrical resistivity (after demoulding, 7, 21 and 28 days) and mechanical strength (at 28 days). The specimens were demoulded after 48 h of the cast, and at this moment, the mass was accessed and used as reference mass of the following measurement at 7, 21 and 28 days. Afterwards, demoulding the specimens were kept in a controlled environment ($T = 20 \pm 2°C$ and $HR = 50 \pm 5\%$) until each testing time.

The electrical resistivity of the specimens was measured by the two-electrode method on prismatic specimens with two stainless steel meshes embedded at both extremities at 7, 21 and 28 days of curing. This method is a non-destructive test usually employed in cement-based materials [32–34] as measure of the hydration process, microstructure evolution and durability, according to their ability to carry electric charge [35]. Electrical resistivity is affected by the mixture design (such as water content, volume of voids, cement type, among others [36]) and curing conditions (such as temperature and moisture environment). Since specimens were exposed to the same conditions ($T = 20 \pm 2°C$ and $HR = 50 \pm 5\%$) [37], the hydration process of stabilised earth specimens was not compromised by the curing conditions, so the resistivity was only influenced by mixture design, and it is possible to compare results between mixtures. The electrical resistivity was calculated using Eq. 11.1, following Ohm's Law, to determine the relationship between the intensity of the current and the potential difference.

$$R = \frac{V}{I} <=> R = \rho \frac{L}{A} <=> \rho = \frac{V \times A}{L \times I} \tag{11.2}$$

Table 11.3 Level combinations considered in the experimental plan (coded values of design variables), mix proportions and GWP

		Vw/Vs	sp/c	c/s	Water	Superplasticizer	Cement	Limestone	Soil	GWP
					kg/m^3	kg/m^3	kg/m^3	kg/m^3	kg/m^3	(Kg CO_2/m^3)
Central points	Ci	0	0	0	368.98	287	142.04	145.05	1420.40	131.17
Factorial points (Fi)	F1	– 1	– 1	– 1	342.11	340	127.84	229.55	1420.40	125.67
	F2	1	– 1	– 1	396.07	203	127.84	8539	1420.40	116.46
	F3	– 1	1	– 1	341.85	375	127.84	229.55	1420.40	125.98
	F4	1	1	– 1	395.92	224	127.84	8539	1420.40	116.65
	F5	– 1	– 1	1	342.08	343	156.24	204.71	1420.40	145.62
	F6	1	– 1	1	396.05	206	156.24	6055	1420.40	136.41
	F7	– 1	1	1	341.83	379	156.24	204.71	1420.40	145.94
	F8	1	1	1	395.90	228	156.24	6055	1420.40	136.60
Axial points (CCi)	CC1	– 1.682	0	0	323.55	408	142.04	266.28	1420.40	138.96
	CC2	1.682	0	0	414.40	166	142.04	23.83	1420.40	123.37
	CC3	0	– 1.682	0	369.15	263	142.04	145.05	1420.40	130.96
	CC4	0	1.682	0	368.81	311	142.04	145.05	1420.40	131.38
	CC5	0	0	– 1.682	369.00	284	118.15	165.94	1420.40	114.39
	CC6	0	0	1.682	368.95	290	165.93	124.16	1420.40	147.95

Fig. 11.4 Example of a slump diameter measurement immediately after production

Fig. 11.5 Specimens of stabilised earth

where R is the electrical resistance (Ω), V is potential difference (Amps), I is current intensity (Volts), ρ is electrical resistivity (Ωm), A is the cross area of specimens in which the current passes (m^2), and L is the length between electrodes (m).

At the age of 28 days, after the mass loss and electrical resistivity measurements, the mechanical strength was evaluated according to EN 196–1.

11.3.4 Embodied CO_2

This study employed the global warming potential (GWP) as an environmental metric to evaluate the embodied carbon associated designed mixtures (see Table 11.3), required to produce 1 m^3 of each mixture in the CCD. By analysing the carbon footprint of key mixture components, including cement, soil, and admixtures, the

Table 11.4 GWP adopted for the constituent raw materials employed in the current work [39]

Constituent material	GWP (Kg CO_2/kg)
Cement	0.75
Limestone filler	0.0557
Superplasticizer	0.867
Water	0.000133
Soil	0.00987

GWP facilitates the comparison and optimisation of formulations to minimise their environmental impact at the material level.

The environmental data for the constituent materials utilised in this study were carefully selected based on comprehensive research of diverse datasets derived from previous scientific investigations into greenhouse gas emissions of concrete components and Environmental Product Declarations (EPDs). The specific GWP values adopted in this work are detailed in Table 11.4. Additionally, for the soil, it was considered the GWP of a natural sand [38]. The GWP computed for each mixture is presented in the last column of Table 11.3.

Furthermore, the correlation between each CCD mixture GWP (per 1 m^3) and compressive strength was analysed using the embodied carbon dioxide index (CI). This index was computed using the following equation, and it is an interesting way to compare carbon footprint in an integrated way.

$$CI = \frac{GWP\left(KgCO_2/m^3\right)}{Rc\_28d\,(MPa)} \tag{11.3}$$

11.4 Results and Discussion

11.4.1 Experimental Test Results

Experimental test results obtained for response variables slump (mm), Mass loss 7d, Mass loss 21d, Mass loss 28d (%), Resist_7d, Resist_21d, Resist_28d (ohm.m), F and Rc (MPa) are presented in Table 11.3. Those results were used for model fitting (Sect. 4.2). It follows some discussion here. A narrow range of slumps was obtained on CCD, from 100 and 150 mm, in which both printable and non-printable mixutres are expected (Table 11.5).

Figure 11.6 presents the mass loss evolution from 7 to 28 days of all earth mixtures of the CCD plan. As expected, the mass loss increases with time. However, the involvement was high between reference age (which was demoulding 48 h after production) and 7 days, on average 5% (all CCD), corroborating with previous results of the authors [11]. There is still some drying between 7 and 21 days (8.90% on

Table 11.5 Experimental test results obtained from the experimental plan to be used for model fitting

Mixture ID	Slump (mm)	Mass loss_ 7d (%)	Mass loss_ 21d (%)	Mass loss_ 28d (%)	Resist_7d (ohm.m)	Resist_21d (ohm.m)	Resist_28d (ohm.m)	F (MPa)	Rc (Mpa)	GWP (Kg CO_2/m^3)	CI (kgCO_2/ MPa)
C1	112	– 5.54	– 9.44	– 9.52	90.66	3134.59	4684.21	1.53	5	131.17	26.23
C2	110.5	– 2.53	– 6.39	– 6.59	82.45	2778.84	4695.73	1.3	4.43	131.17	29.59
C3	113	– 2.75	– 6.63	– 6.67	82.65	3584.77	4532.84	1.38	4.7	131.17	27.92
C4	111.5	– 2.68	– 7.79	– 7.85	63.13	3111.97	4435.6	1.51	4.85	131.17	27.03
F1	101.5	– 7.24	– 8.89	– 8.92	360.93	5447.76	6281.99	1.45	4.55	125.67	27.59
F2	138.5	– 6.85	– 11.38	– 11.46	52.97	3553.77	4953.27	1.06	2.73	116.46	42.69
F3	104	– 5.03	– 6.91	– 6.91	421.43	5141.6	5513.57	1.27	4.49	125.98	28.07
F4	139.5	– 6.24	– 11.37	– 11.74	85.65	3986.05	5186.35	1.13	3.04	116.65	38.38
F5	100.5	– 5.69	– 8.4	– 8.52	119.04	3219.23	4225.02	1.97	7.32	145.62	19.89
F6	136	– 6.82	– 10.11	– 10.31	84.3	3227.74	3917.4	1.17	3.8	136.41	35.94
F7	105	– 3.45	– 5.09	– 5.11	423.16	4269.28	4734.64	2.03	7.03	145.94	20.76
F8	140.5	– 5.92	– 9.76	– 9.99	120.63	3069.69	5714.82	1.24	4.13	136.6	33.04
CC1	100	– 8.57	– 10.43	– 10.26	380.39	4629.48	5669.61	2.02	6.68	138.96	20.8
CC2	150	– 6.17	– 12.98	– 13.34	45.27	2122.79	4584.38	0.83	2.37	123.37	52.04
CC3	112	– 3.9	– 7.81	– 7.88	109.79	3348.96	4593.9	1.45	4.71	130.96	27.78
CC4	106.5	– 5.65	– 11.23	– 11.33	87.06	4471.44	5642.3	1.35	4.35	131.38	30.23
CC5	110.5	– 4.6	– 9.28	– 9.42	76.37	3566.83	4961.39	1.08	3.26	114.39	35.13
CC6	113.5	– 4.42	– 6.39	– 6.49	209.63	3929.25	4983.05	1.81	6.09	147.95	24.31

Table 11.6 Results summary and main statistics for all CCD points and central points

	Slump (mm)	Mass loss_7d (%)	Mass loss_21d (%)	Mass loss_28d (%)	Resist_7d (ohm.m)	Resist_21d (ohm.m)	Resist_28d (ohm.m)	F (MPa)	Rc (Mpa)	GWP (Kg CO_2/m^3)	CI ($kgCO_2$/MPa)
All plan (18 mixtures)											
Max	100.0	− 2.53	− 5.09	− 5.11	45.27	2122.79	3917.4	0.83	2.37	131.17	29.59
Min	150.0	− 8.57	− 12.98	− 13.34	423.16	5447.76	6281.99	2.03	7.32	131.17	26.23
Mean	116.9	− 5.23	− 8.90	− 9.02	160.86	3699.67	4961.67	1.42	4.64	131.17	27.69
Std Dev	16.1	0.02	0.02	0.02	134.85	842.93	603.97	0.35	1.41		1.44
Cov (%)	13.7	32.2	23.9	23.5	83.8	22.8	12.2	24.3	30.4		5.49
Ci points (4 mixtures)											
Max	110.5	− 2.53	− 6.39	− 6.59	63.13	2778.84	4435.6	1.3	4.43	147.95	52.04
Min	113.00	− 5.54	− 9.44	− 9.52	90.66	3584.77	4695.73	1.53	5.00	114.39	19.89
Mean	111.8	− 3.38	− 7.56	− 7.66	79.72	3152.54	4587.10	1.43	4.75	131.17	30.41
Std Dev	1.0	0.02	0.01	0.01	11.70	330.88	125.34	0.11	0.24	0.25	0.26
Cov (%)	0.9	40.0	17.2	17.7	14.7	10.5	2.7	7.6	5.1	8.62	41.02

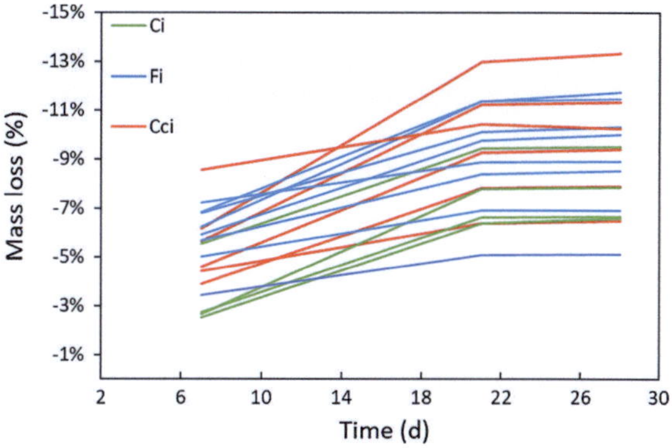

Fig. 11.6 Mass loss evolution from 7 to 28 days

average), but from 21 to 28 days. The higher mass loss was obtained for the CC2 mixture (13.3%), and the mixture F7 presented the lowest mass loss (5.1%), which are mixtures with lower and highest water volume, respectively, see Table 11.3.

The electrical resistivity involvement of stabilised earth mixtures is presented in Fig. 11.7. As can be perceived, this property has a substantial linear increase over time among all mixtures, and it seems that it still presents an increasing tendency at 28 days. From 7 to 21 days, the electrical resistivity presented a very high evolution, from 45–423 to 2122–5447 Ωm values, respectively. The highest electrical resistivity was reached by mixture F1, with a value of 6282 Ωm at 28 days.

Regarding compressive strength, the results values were between 0.53 and 7.07 MPa, which are in accordance with the literature survey [40], which stated that cement-stabilised earth materials average compressive strength is roughly 5 MPa, which may be double of raw earth materials, which ranges from 1 MPa to 2.5 MPa [41]. Mixture F5 reached the highest compressive strength and corresponds to 7.3 MPa.

As expected, the data shows the existence of significant linear correlations between Mass loss at different ages, as well as between resistivity at different ages (see

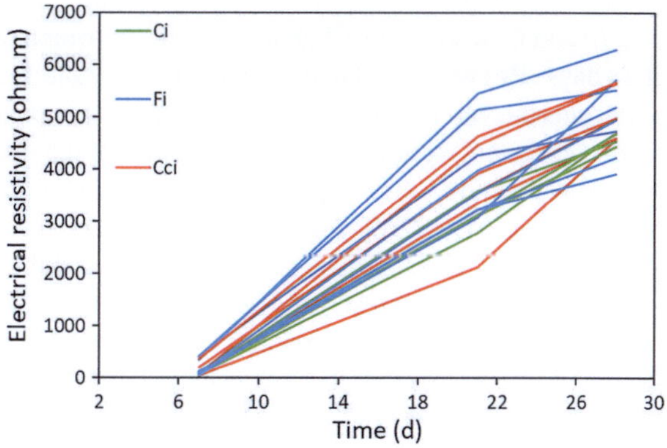

Fig. 11.7 Electrical resistivity evolution from 7 to 28 days

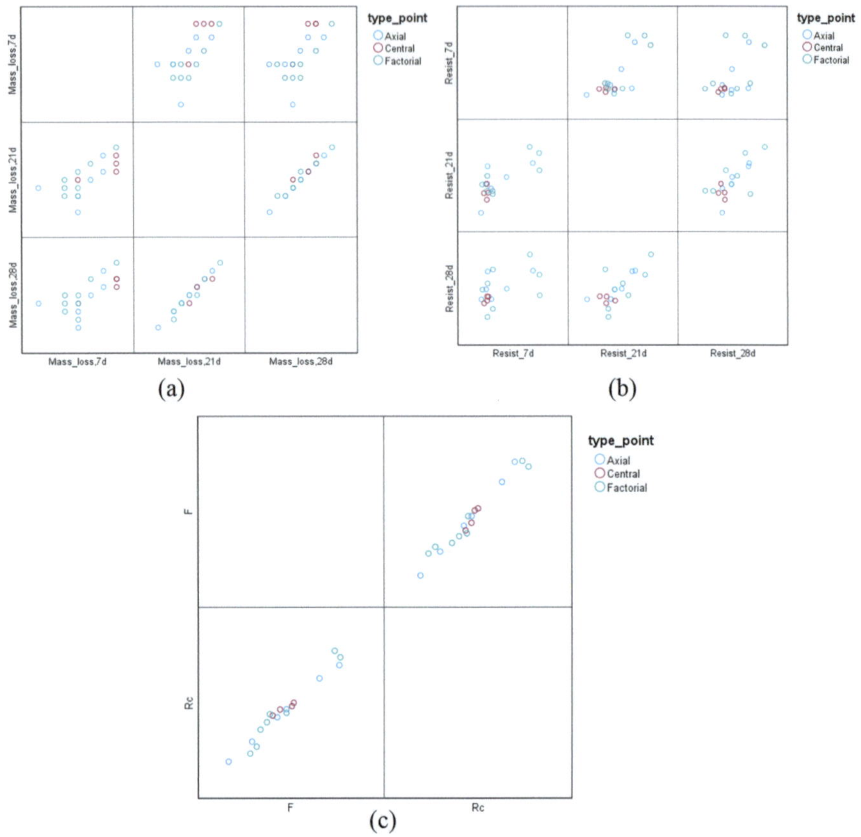

Fig. 11.8 Correlation between response properties: (**a**) Mass loss; (**b**) Resistivity; (**c**) F and Rc

Fig. 11.8). However, the strongest linear correlation is the one existing between F and Rc ($r = 0.981$, $p < 0.001\%$).

Figure 11.9 presents the influence of mixture parameters, where A = Vw/Vs, B = sp/p and C = s/c, in GWP and CI. The experimental compressive strength results at 28 days (Rc) were used to compute CI (see Table 11.5). Each plot line in the figure represents the variation of a single mixture parameter across its range. In contrast, the other parameters remain fixed at the reference point, located at the centre. The factors A = Vw/Vs and C = s/c significatively influence the GWP, see Fig. 11.7a, but in opposite ways. The GWP decreases by increasing the factor A = Vw/Vs from − 1 to + 1 (for coded values, see the correspondence in Table 11.2). This is mainly because when A = Vw/Vs increses, the Vw increases, the volume of the solid constituent materials is reduced, and the water presents the lowest GWP (see Table 11.4). On the other way, an increase in C = c/s increases GWP. This occurs mainly because the cement has a significant GWP (Table 11.4), and rising C = s/c, increases the cement content in the formulation.

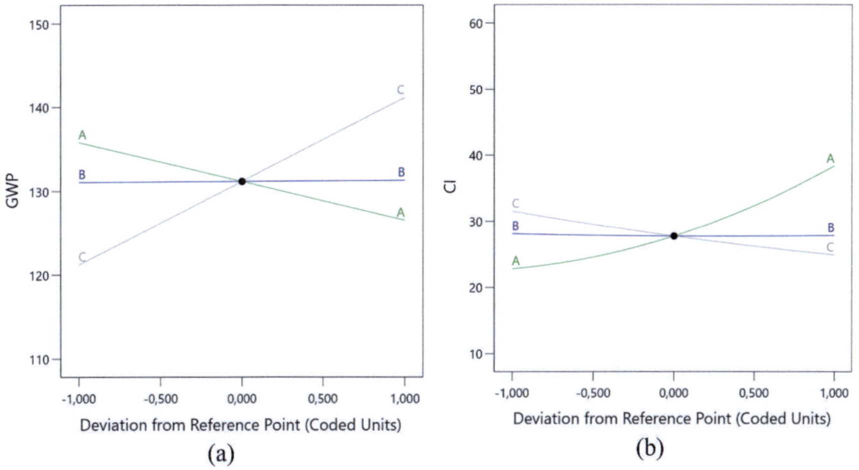

Fig. 11.9 Influence of each mixture parameter on: (**a**) GWP and (**b**) CI, where A = Vw/Vs, B = sp/p and C = c/s

The carbon dioxide index (CI) translates the relation between the CO_2 emission per each MPa of compressive strength. As such, the higher the CI, the higher the CO_2 for lower strength. In this case, CI increases with the increase of A = Vw/Vs and decreases with the increase of C = c/s, but the factor Vw/Vs has a higher influence, see Fig. 11.9b. It is evident that Vw/Vs and c/s are the most significant factors affecting environmental performance analysis. It can be also concluded that CI is higher for soils compared to concrete [32, 42].

11.4.2 Models Fitting

As aforementioned in Sect. 3.2, a 2-level 3-factor central composite design with 6 axial runs and 4 central points has been used. The full quadratic model (See Eq. 11.1) was estimated based on CCD; however, some terms of Eq. 11.1 may not be significant:

$$y = \beta_0 + \sum_{i=1}^{k} \beta_i x_1 + \sum_{i=1}^{k} \beta_{ii} x_i^2 + \sum_{i<j} \sum \beta_{ij} x_i x_j + \varepsilon \qquad (11.4)$$

this model, y represents the response variable, while xi denotes the design variables under consideration. The coefficients β are used to define the model, where β0 is the constant (intercept), βi represents the linear effect of xi, βii represents the quadratic effect of xi, and βij represents the interaction effect between xi and xj. The term ϵ accounts for the fitting error.

The state-of-the-art statistical procedure was applied to the modelling of the response, involving all steps of the process except validation with additional runs [43]. The steps include ANOVA testing, the diagnosis of residuals, checking for outliers and leverage points and analysing determination coefficients and model comparison statistics, as well as lack of fit. A detailed description of the aforementioned procedure can be found in a previous publication of the authors authors [44] or dedicated bibliography [43]. Some points are discussed as follows here.

Statistical modelling of the response surface for the 9 response variables was performed based on the DoE data. Table 11.7 exhibits the estimates of the model coefficients in coded values, as well as the associated values obtained for the fitting statistics R^2 and Adjusted R^2. The Coded coefficient estimates presented in Table 11.7, are an interesting way of analysis since they indicate the relative effect of the mixture parameter, Vw/Vp, Sp/p and c/s, on each response variable. Higher values indicate the stronger influence of the mixture parameter in the response variable. A positive coefficient means that the response (or transformed response in the case of Resist,7d) variable will increase if the given mixture parameter increases and vice-versa.

The two most relevant contributions of input factors are shown in bold face, and the most relevant of the two is also underlined. Non-significant contributions that are kept in the model equation are flagged with the word NS in upersript. Even though these contributions are non-significant, they are preserved in order to keep the hierarchy of the models and other statistical properties in general in the statistical modelling procedure (see [43]).

Analysis of Table 11.7 enables the conclusion that, with the exception of early age mass loss (7 days) and late resistivity (28 days), Vw/Vs is the main factor contributing to the response of the material. The mixture parameter sp/c only significantly contributes to resistivity at 28 days and is the second most important contribution to this response, the interaction with c/s being the most significant term in the response model. The ratio c/s comes second in the scale of the most significant contributions to the mechanical strength, flexural and compressive. Significant quadratic effects of Vw/Vs also exist on Slump and mass loss at all ages. In the case of early mass loss (7 days), this quadratic effect is even the most significant. The significant contribution of Vw/Vs to the responses is negative except for Slump, meaning that they will significantly decrease with an increase in Vw/Vs.

On the contrary, Slump will significantly increase with an increase in Vw/Vs. The influence of sp/c and its interaction with c/s on resistivity at 28 days is positive, meaning that late resistivity is increased as a response to an increase in Vw/Vs. The interaction between sp/c and c/s mostly influences late resistivity. Figures 11.8, 11.9, 11.10, 11.11, 11.12 show the contour plots of the surfaces that represent the response as a function of the factors considered in the study, keeping the less important factor value fixed.

The mixture parameter Vw/Vs clearly influences all variable responses, with the exception of early mass loss and late resistivity, and is key to finding good formulations for both fresh state properties and mechanical response. This is expectable because it translates the water volume in the mixture, as the higher is Vw/Vs, the higher the water content, which decreases mechanical strength. On the other way,

Table 11.7 Response fitted models' coefficients (models in coded values)

	Slump (mm)	Mass_loss,7d	Mass_loss,21d	Mass_loss,28d	Resist,7d $^{(-0.5)}$	Resist,21d	Resist,28d	F	Rc
Intercept	111.26	– 0.0415	– 0.0788	– 0.0816	0.0958	3699.7	4961.7	1.42	4.64
Vw/Vs	<u>**18.45**</u>	– 0.0053^{NS}	**– 0.0132**	**– 0.0132**	**0.0269**	**– 619.2**	– 205.7^{NS}	**– 0.3018**	**– 1.24**
sp/c	NS	NS	NS	0.00001^{NS}	NS	NS	**258.8**	NS	– 0.0231^{NS}
c/s	0.2596^{NS}	NS	0.0074	0.0074^{NS}	NS	NS	– 242.1	**0.1997**	**0.8955**
Vw/Vs x sp/s	– 0.1875^{NS}	NS	NS	– 0.0088^{NS}	NS	NS	286.2	NS	0.1238^{NS}
Vw/Vs x c/s	NS	NS	NS	NS	NS	NS	291.1	– 0.1325	– 0.3937
sp/c x c/s	NS	NS	NS	NS	NS	NS	**335.3**	NS	NS
$(Vw/Vs)^2$	**7.86**	**– 0.0188**	**– 0.0119**	**– 0.0111**	NS	NS	NS	NS	NS
$(c/s)^2$	0.6978^{NS}	NS	NS	NS	NS	NS	NS	NS	NS
R^2	0.978	0.560	0.671	0.713	0.628	0.434	0.748	0.953	0.988
Adjusted R^2	0.967	0.4976	0.601	0.593	0.604	0.398	0.610	0.943	0.983
Runs excluded (outliers)	CC2	CC2							

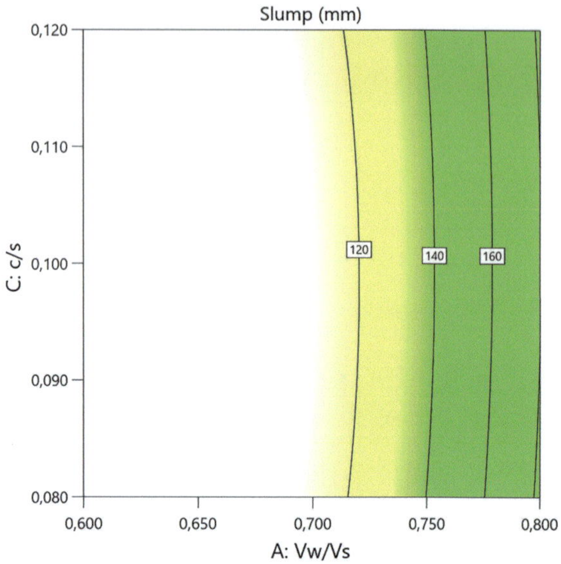

Fig. 11.10 Slump model curves of the response surface as depending on the significant factor Vw/Vs e c/s (sp/c = 1.0%)

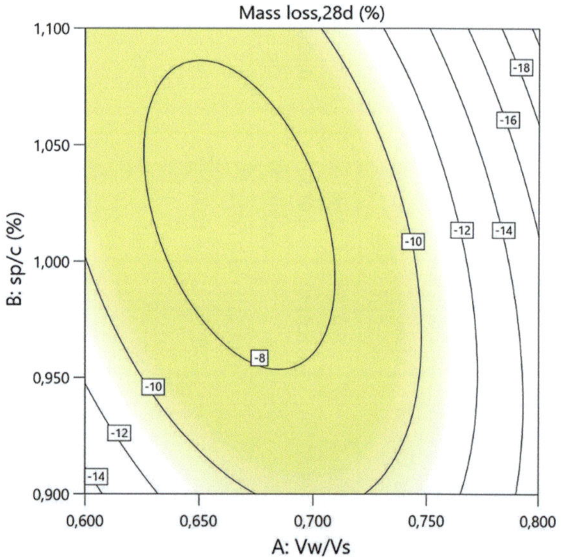

Fig. 11.11 Mass loss at 28 days model curves of the response surface depending on the significant factor Vw/Vs e c/s (sp/c = 1.0%)

it increases the flowability (higher slump). The mixture parameter c/s is the second most relevant in what concerns the mechanical performance of the stabilised earth material. Even though clay presents some similarities with cementitious materials in terms of colloidal interactions and adhesion forces [45], this is limited [46] compared to cement because there is no hydraulic reaction [10]. As such, when the earth is stabilised with cement, mechanical strength increases (higher c/s) (Fig. 11.13).

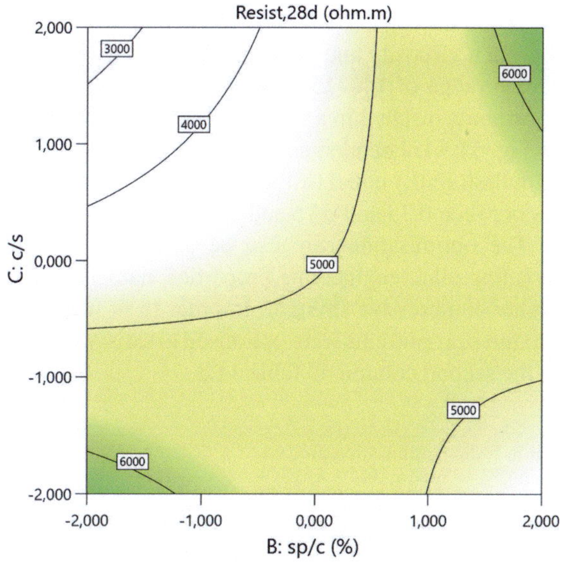

Fig. 11.12 Mass loss at 28 days model curves of the response surface depending on the significant factor sp/c e c/s (Vw/Vs = 0.70)

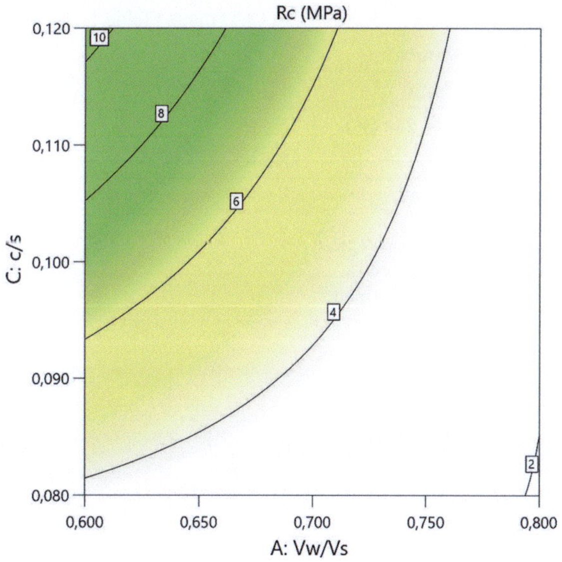

Fig. 11.13 Level curves of the response surface for compressive strength depend on the significant factor Vw/Vs and c/s (sp/c = 1.00%)

11.4.3 Material Optimisation Possibilities

Based on the obtained models, optimisation can be conducted to find the optimal earth mix ratio region using the desirability function of Design Expert software. An example of an optimisation process is indicated in Table 11.8, and the importance weights for all goals were set equal. The potential stabilised Earth mixtures must present a viscosity and deformability adequate for extrudability and buildability. A Slump range of 140–150 mm was simulated to satisfy those requirements. Those constraints resulted in a region belonging to the design space represented in yellow in Fig. 11.14. Founded solutions for optimised mixes can be found in this region with desirability equal to 100%. As can be perceived, solutions were found for Vw/Vs between 0.75 and 0.76 and for all c/s ranges of CCD.

The optimisation can also be performed to find the best multi-performance targeting other engineering properties, namely, low cracking risk at early ages and higher compressive strength. In order to fit those performance requirements, earth mixtures proportions were optimised to satisfy the optimisation constraints indicated in the second column of Table 11.8.

Table 11.8 Optimisation process

	Fresh state	Other engineering properties
Mixture parameters (factors)		
X_1: Vw/Vp	In CCD range (Table 11.2)	In CCD range (Table 11.2)
X_2: Sp/p	In CCD range (Table 11.2)	In CCD range (Table 11.2)
X_3:c/s	In CCD range (Table 11.2)	In CCD range (Table 11.2)
Response variables		
Y_1: Slump (mm)	140–150	140
Y_5: Mass loss_28d (%)	None	Minimise
Y_6: Rc (MPa)	None	Maximise

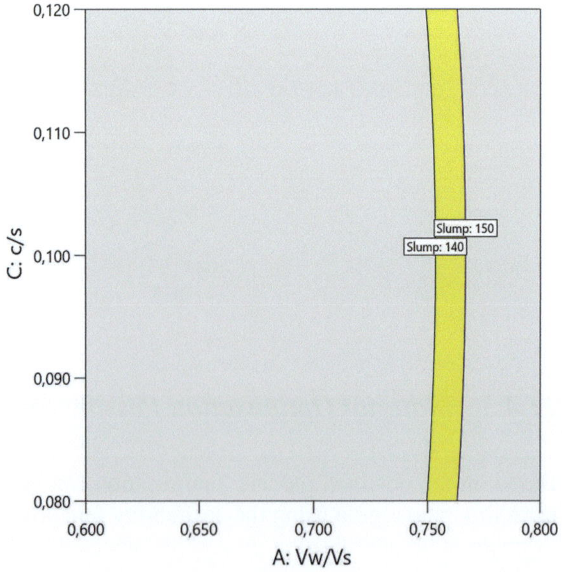

Fig. 11.14 Optimal region for fresh stabilised earth based on CCD performed

The Mass loss_28d was minimised to accomplish low cracking risk, and Compressive strength results at 28 days were maximised. The desirability function revealed that stabilised earth with mixture parameters Vw/Vs = 0.751; Sp/p = 1.100%; c/s = = 0.117 was the best solution. It must be stressed that, considering several performance requirements in both fresh state properties, mass loss and compressive strength, no solutions could be found with desirability of 100%. This is because there are no stabilised earth mixtures with a mass loss lower than 11% in the current CCD. Thus, additional studies are needed in a wider range to optimise mass loss reduction and increase mechanical strength simultaneously.

11.5 Conclusions

Studying new construction methods for earth construction is essential for addressing contemporary challenges related to environmental sustainability, economic constraints, cultural preservation, and technological innovation.

This research was a kick-off for an innovative technique for earth construction. The study focuses on stabilising Portuguese soil to explore its viability for extrusion 3D printing by addressing workability, structural integrity, and environmental performance.

This study aimed to investigate stabilised earth and select combinations and ranges of powder materials (cement + LF + soil), water and superplasticiser dosage, and understand how to reach potential printable earth mixture. Specifically, this study aimed to investigate which key mixture variable influences stabilised earth properties with adequate performance for printing and which are the appropriate ranges for deeper research. For this purpose, a 2^3CCD was followed to find statistical models capable of describing fresh properties (slump) and other engineering properties, such as mechanical behaviour, resistivity and mass loss. The main conclusions are the following:

- The mixture parameter $\frac{Vw}{Vs}$ was the main effect in most of the output properties evaluated (Slump, late mass loss, early resistivity and mechanical strength);
- The mixture parameter $\frac{sp}{c}$ and its interaction with c/s were the main contributions to late resistivity;
- The second most influencing factor contributing to mechanical strength is $\frac{c}{s}$.

This work stands as a preliminary investigation of the mixture parameters that most control fresh state as well as mechanical properties of stabilised soil materials. Modelling the slump response is particularly challenging. There is no doubt that one is facing a nonlinear model and the fitting shown in the present work dismissed one mixture as being a regression outlier. The lack of fit obtained is non-significant. However, more research would be advised, either replicating some of the points in the CCD or by, expanding the CCD to consider a 3rd order model or adding new factors.

In summary, this study established preliminary guidelines for designing mixtures and evaluating potential stabilised earth materials suitable for 3D printing, utilising locally sourced raw materials and standard production and testing techniques. Future research should explore innovative construction methods for earthen materials, leveraging knowledge from cement-based material technologies.

Acknowledgements This work was financially supported by: This work is financially supported by national funds through the FCT/MCTES (PIDDAC), under the MIT Portugal Programme under the project 2022.15478.MIT—DigiCrete—NEXT GENERATION OF DIGITAL CONCRETE": performance mix design and assessment of sustainable and circular cementitious composites; by Funding - UID/04708 of the CONSTRUCT - Instituto de I&D em Estruturas e Construções - funded by Fundação para a Ciência e a Tecnologia, I.P./ MCTES through the national funds and by FCT—Fundação para a Ciência e a Tecnologia through through the individual Scientific Employment Stimulus 2021.01765.CEECIND. Collaboration and materials supply by Secil, Omya Comital, Sika and Chryso Portugal is gratefully acknowledged.

The authors would also like to thank Structures Laboratory for technical support during the experimental work, namely, Engineer Paula Silva and Claudio Ferraz.

References

1. L. Keefe, in *Earth Building: Methods and Materials, Repair and Conservation* (Taylor & Francis, 2005)
2. CRAterre. Accessed: 21 Feb 2022. [Online]. Available: http://craterre.org/
3. A. Fabbri et al., An overview of the remaining challenges of the RILEM TC 274-TCE, testing and characterisation of earth-based building materials and elements. RILEM Tech. Lett. **6**, 150–157 (2021)
4. A. Fabbri, J.-C. Morel, J.-E. Aubert, Q.-B. Bui, D. Gallipoli, B.V. Venkatarama Reddy, *Testing and Characterisation of Earth-based Building Materials and Elements*, vol. 35 (State-of-the-Art Report of the RILEM TC 274-TCE, Springer International Publishing, 2022). https://doi.org/10.1007/978-3-030-83297-1
5. C.M. Ouellet-Plamondon, G. Habert, Self-compacted clay based concrete (SCCC): proof-of-concept. J. Clean. Prod. **117**, 160–168 (2016). https://doi.org/10.1016/J.JCLEPRO.2015.12.048
6. J. Romero Clausell, A. Quintana-Gallardo, C. Hidalgo Signes, B. Serrano-Lanzarote, Environmental evaluation of a self-compacted clay based concrete with natural superplasticizers. Mater. Struct./Mater. Constr. **54**(1), 1–16 (2021). https://doi.org/10.1617/S11527-020-01586-6/FIGURES/8
7. C. Ma, B. Chen, L. Chen, Variables controlling strength development of self-compacting earth-based construction. Constr. Build. Mater. **123**, 336–345 (2016). https://doi.org/10.1016/J.CONBUILDMAT.2016.07.017
8. C. Ouellet-Plamondon, G. Habert, Proof of concept of self-compacting clay concrete to scale-up earth construction, in *American Society of Agricultural and Biological Engineers Annual International Meeting 2014, ASABE 2014*, vol. 5, pp. 1 (2014). https://doi.org/10.13031/AIM.20141908765
9. C. Ma, L. Chen, B. Chen, Experimental study of effect of fly ash on self-compacting rammed earth construction stabilized with cement-based composites. J. Mater. Civ. Eng. **28**(7), 04016022 (2016). https://doi.org/10.1061/(ASCE)MT.1943-5533.0001518
10. D. Ardant, C. Brumaud, G. Habert, Influence of additives on poured earth strength development. Mater. Struct./Mater. Constr. **53**(5), 1–17 (2020). https://doi.org/10.1617/S11527-020-01564-Y/FIGURES/13

11. A.M. Matos, H. Varum, in Self-Compacting Earth-Based Composites: Mixture Design and Multi-Performance Characterisation, *Buildings 2022,* vol. 12(5), pp. 612 (2022). https://doi.org/10.3390/BUILDINGS12050612
12. H. Van Damme, H. Houben, Earth concrete. Stabilization revisited. Cem. Concr. Res. **114**, 90–102 (2018). https://doi.org/10.1016/j.cemconres.2017.02.035
13. S. Imanzadeh, A. Hibouche, A. Jarno, S. Taibi, Formulating and optimizing the compressive strength of a raw earth concrete by mixture design. Constr. Build. Mater. **163**, 149–159 (2018). https://doi.org/10.1016/J.CONBUILDMAT.2017.12.088
14. M.M. Barbero-Barrera, F. Jové-Sandoval, S. González Iglesias, Assessment of the effect of natural hydraulic lime on the stabilisation of compressed earth blocks. Constr. Build. Mater. **260**, 119877 (2020). https://doi.org/10.1016/J.CONBUILDMAT.2020.119877
15. J.D. Sitton, Y. Zeinali, W.H. Heidarian, B.A. Story, Effect of mix design on compressed earth block strength. Constr. Build. Mater. **158**, 124–131 (2018). https://doi.org/10.1016/J.CONBUILDMAT.2017.10.005
16. M.A. Hossain, A. Zhumabekova, S.C. Paul, J.R. Kim, A review of 3D printing in construction and its impact on the labor market. Sustain. (Switz.) **12**(20), 1–21 (2020). https://doi.org/10.3390/su12208492
17. J. Zhang, J. Wang, S. Dong, X. Yu, B. Han, in A Review of the Current Progress and Application of 3D Printed Concrete (Elsevier Ltd., 2019) https://doi.org/10.1016/j.compositesa.2019.105533
18. J.G. Sanjayan, B. Nematollahi, in 3D Concrete Printing for Construction Applications (Elsevier, 2019), pp. 1–11. https://doi.org/10.1016/b978-0-12-815481-6.00001-4
19. The first 3D printed House with earth|Gaia—3D Printers|WASP. Accessed: 21 Feb 2022. [Online]. Available: https://www.3dwasp.com/en/3d-printed-house-gaia/
20. The BOD—The first 3D printed building in Europe. Accessed: 21 Feb 2022. [Online]. Available: https://cobod.com/the-bod/
21. T.A.M. Salet, Z.Y. Ahmed, F.P. Bos, H.L.M. Laagland, Design of a 3D printed concrete bridge by testing. **13**(3), 222–236 (2018). https://doi.org/10.1080/17452759.2018.1476064
22. A. Alqenaee, A. Memari, Experimental study of 3D printable cob mixtures. Constr. Build. Mater. **324**, 126574 (2022). https://doi.org/10.1016/J.CONBUILDMAT.2022.126574
23. P. https, Computing craft: early development of a robotically-supported cob 3D printing system (2018). Accessed: 22 Jul 2022. [Online]. Available: http://hdl.handle.net/10026.1/12769
24. H. Alhumayani, M. Gomaa, V. Soebarto, W. Jabi, Environmental assessment of large-scale 3D printing in construction: a comparative study between cob and concrete. J. Clean. Prod. **270**, 122463 (2020). https://doi.org/10.1016/J.JCLEPRO.2020.122463
25. M. Schweiker et al., Ten questions concerning the potential of digital production and new technologies for contemporary earthen constructions. Build. Environ. **206**, 108240 (2021). https://doi.org/10.1016/J.BUILDENV.2021.108240
26. G. Silva, L. Quispe, S. Kim, J. Nakamatsu, R. Aguilar, Development of a stabilized natural fiber-reinforced earth composite for construction applications using 3D printing. IOP Conf. Ser. Mater. Sci. Eng. **706**(1) 2019. https://doi.org/10.1088/1757-899X/706/1/012015
27. A. Perrot, D. Rangeard, E. Courteille, 3D printing of earth-based materials: processing aspects. Constr. Build. Mater. **172**, 670–676 (2018). https://doi.org/10.1016/j.conbuildmat.2018.04.017
28. M. Gomaa, W. Jabi, A. Veliz Reyes, V. Soebarto, 3D printing system for earth-based construction: case study of cob. Autom. Constr. **124**, 103577 (2021). https://doi.org/10.1016/J.AUTCON.2021.103577
29. M. Gomaa, W. Jabi, V. Soebarto, Y.M. Xie, Digital manufacturing for earth construction: a critical review. J. Clean. Prod. **338**, 130630 (2022). https://doi.org/10.1016/J.JCLEPRO.2022.130630
30. G. Silva, L. Quispe, S. Kim, J. Nakamatsu, R. Aguilar, Development of a stabilized natural fiber-reinforced earth composite for construction applications using 3D printing. IOP Conf. Ser. Mater. Sci. Eng. **706**(1), 012015 (2019). https://doi.org/10.1088/1757-899X/706/1/012015
31. CEMUP|HOME. Accessed: 11 Jul 2024. [Online]. Available: https://www.cemup.up.pt/

32. A.M. Matos, P. Milheiro-Oliveira, M. Pimentel, Eco-efficient high performance white concrete incorporating waste glass powder. Constr. Build. Mater. **411**, 134556 (2024). https://doi.org/10.1016/J.CONBUILDMAT.2023.134556
33. S. Nunes, A. M. Matos, T. Duarte, H. Figueiras, J. Sousa-Coutinho, Mixture design of self-compacting glass mortar. Cem. Concr. Compos. **43** (2013). https://doi.org/10.1016/j.cemconcomp.2013.05.009
34. A.M. Matos, S. Nunes, C. Costa, J.L. Barroso-Aguiar, Spent equilibrium catalyst as internal curing agent in UHPFRC. Cem. Concr. Compos. **104** (2019). https://doi.org/10.1016/j.cemconcomp.2019.103362
35. H. Figueiras, S. Nunes, J.S. Coutinho, C. Andrade, Linking fresh and durability properties of paste to SCC mortar. Cem. Concr. Compos. **45**, 209–226 (2014). https://doi.org/10.1016/j.cemconcomp.2013.09.020
36. P. Azarsa, R. Gupta, Electrical resistivity of concrete for durability evaluation: a review. Adv. Mater. Sci. Eng. **2017**, 1–30 (2017). https://doi.org/10.1155/2017/8453095
37. P. Azarsa, R. Gupta, Electrical resistivity of concrete for durability evaluation: a review. Adv. Mater. Sci. Eng. **2017** (2017). https://doi.org/10.1155/2017/8453095
38. A. M. Gaspar De Oliveira Braga, Impacte ambiental comparado do ciclo de vida de betão com agregados grossos reciclados e naturais Engenharia Civil. MsCThesis, Instituto Superior Técnico, 2015
39. Portuguese 'grey' cement atic-associação técnica da indústria de cimento environmental product declaration DAPHabitat system (2023). Accessed 15 Oct 2024. [Online]. Available: www.daphabitat.pt
40. F. Ávila, E. Puertas, R. Gallego, Characterization of the mechanical and physical properties of stabilized rammed earth: a review. Constr. Build. Mater. **325**, 126693 (2022). https://doi.org/10.1016/J.CONBUILDMAT.2022.126693
41. F. Ávila, E. Puertas, R. Gallego, Characterization of the mechanical and physical properties of unstabilized rammed earth: a review. Constr. Build. Mater. **270**, 121435 (2021). https://doi.org/10.1016/J.CONBUILDMAT.2020.121435
42. A.M. Matos, P. Milheiro-Oliveira, M. Fonseca, M. Pimentel, Formulações ternárias para impressão 3D utilizando materiais nacionais, in *Encontro Nacional do Betão Estrutural 2024* (Porto, Portugal, 2024)
43. D.C. Montgomery, in *Design and Analysis of Experiments*, 5th edn (Arizona State University, 2005)
44. A.M. Matos, L. Maia, S. Nunes, P. Milheiro-Oliveira, Design of self-compacting high-performance concrete: study of mortar phase. Constr. Build. Mater. **167**, 617–630 (2018). https://doi.org/10.1016/J.CONBUILDMAT.2018.02.053
45. R.J.M. Pellenq, H. Van Damme, Why does concrete set? The nature of cohesion forces in hardened cement-based materials. MRS Bull. 2004 **29**(5), 319–323 (2011). https://doi.org/10.1557/MRS2004.97
46. A.P.S. Milani, C.I. Da Silva, Influence of superplasticiser on cement-stabilised self-compacting earth. **172**(2), 85–95 (2019), https://doi.org/10.1680/JGRIM.18.00025

Part II
Alternative Construction Materials and Processes

From Waste to Resilience: Innovating Beyond Industrial Materials

The enormous carbon footprint of established construction materials and the need for greater improvements in the lifecycle outcomes of the construction value chain in light of global challenges of climate change, environmental degradation, and housing scarcity have motivated a rethinking of the way structures are built and the materials we use in them.

The material landscape is ever changing. The abundant old-growth hardwood forests replete with oak, elm, cedars, and chestnut trees that once provided us with a beautiful palette with which to create comfortable built environments, have been depleted and are no longer available to meet our burgeoning needs. The industrial materials and processes that we have come to rely on to create huge swaths of built-up landscapes rely on brute-force manipulation of matter at planet-threatening scales.

These factors have motivated the search for new ways to identify and utilise alternative materials and processes to create successful structures and communities. In particular, innovators from diverse backgrounds have found ways to utilise waste products from industry, agriculture, and mass consumption as an asset in the production of sustainable materials, helping transition the built environment from our largest waste stream source to a vanguard of circular economies.

The chapter in this second part of the book Sustainable Materials for the Built Environment showcases inspiring work by discipline-bridging thinkers and practitioners, showing just how much can be achieved by identifying opportunities for circularity, looking for natural alternatives to industrial materials, and bringing together diverse elements to produce sustainable composite materials and compound structural elements that serve to enhance lifecycle outcomes across elements of the built environment.

Addressing modern demands, for load-bearing structures, retrofitting, energy-efficient insulation, resilient pavements or otherwise, requires more than revisiting historical practices. It demands audacity: leveraging AI to optimize material reuse, drawing inspiration from Indigenous land stewardship, or rethinking supply chains to bridge sectors like forestry and tech manufacturing. The projects profiled in this part of the book exemplify this ethos, demonstrating how cross-industry collaboration and systemic thinking can turn constraints into catalysts. By prioritizing synergies over silos, these innovators chart a path toward a built environment that thrives within planetary boundaries, proving sustainability is not a limitation but a launchpad for ingenuity.

Dorian A. H. Hanaor

Chapter 12
Using Marine Sediments to Improve the Mechanical Behaviour of Earthen Materials

Miakonirainy Vololontsoa Razafitrimo, Louise Lemesre, Simon Guihéneuf, Damien Rangeard, and Arnaud Perrot

Abstract Earth construction is emerging as a sustainable and local solution for reducing the environmental impact of the construction industry. Numerous research initiatives are currently revitalising this traditional construction material, gathering many experts. Earth is a mix of particles of various sizes, from clay and silt to sand and gravel. These materials can be used for structural or non-structural purposes, with compressive strengths ranging from 1 to 10 MPa. They also offer superior indoor hygrometric regulation properties. However, their water sensitivity poses durability issues. Recently, low-tech stabilisation solutions have emerged to enhance water resistance. Simultaneously, Europe dredges 50 million cubic meters of marine sediments annually, with France contributing significantly. The traditional solution of sea disposal raises environmental concerns, leading to stricter regulations favouring sediment valorisation. Like earth, sediments are mainly a mix of mineral particles but may also contain other pollutants. Although these sediments present technical and environmental challenges, their use reduces the need for natural resource extraction. Various solutions have been tested, including their application in earth construction. This proposed work evaluates the potential of incorporating sediments into earthen materials, examining the effects of salt and the possibility of stabilization with locally sourced seaweed extracts.

M. V. Razafitrimo · L. Lemesre · S. Guihéneuf · A. Perrot (✉)
UMR CNRS 6027, Université Bretagne-Sud, F-56100 Lorient, France
e-mail: arnaud.perrot@univ-ubs.fr

D. Rangeard
Research and Development Department, Laboratoire CBTP, 35532 Noyal-sur-Vilaine, France

D. A. H. Hanaor (ed.), *Sustainable Materials for the Built Environment*, Springer Series in Materials Science 352, https://doi.org/10.1007/978-3-031-97818-0_12

12.1 Introdution

Earth construction is emerging as a sustainable and local solution for reducing the environmental impact of the construction industry [1]. Many research initiatives are currently very dynamic and gather lot of experts to revive this vernacular construction material. The International Union of Laboratories and Experts in Construction, known as RILEM have launched three technical committees in 2022 on earthen construction [2] to follow-up the first earth construction technical committee dedicated to earthen construction in the twenty-first century. In France, a national project called "PN terre crue" bears witness to this, as does the rise of numerous projects in this field.

Earthen materials including components of clay, silt, sand and gravel, can be used either as structural or non-structural construction materials and can exhibit compressive strengths ranging from 1 to 10 MPa [3]. This material is also known to exhibit superior properties in terms of indoor hygric regulation [4, 5]. The major drawback of this construction material is its water sensitivity which can lead to durability and robustness issues. Nevertheless, some vernacularly inspired and low-tech stabilisation solutions have recently emerged to provide sufficient water resistance regarding national standards [6–9].

At the same time, in Europe, 50 million cubic meters of marine sediments are dredged each year in France [10–12]. The accumulation of sediments on the seabed and in harbours leads to obstructions that hamper marine activities and makes dredging operations mandatory. Dumping marine sediments at sea is the easiest way to manage the problem and has been considered the traditional solution. However, the disruption of seabed ecosystems arising from the deposit of sediments on the wild sea floor is raising environmental concerns [13–15], leading to stricter national regulation favouring sediment valorisation [16]. Therefore, port authorities are looking for sustainable solutions regarding sediments that cannot be disposed at sea [17] and are challenged to develop strategies to manage the volume of sediments extracted from the dredging operations.

This situation offers an opportunity to valorise these sediments while limiting the extraction of traditional natural resources. However, marine sediments can present technical, environmental and regulatory challenges. The presence of salt or pollutants from human activities can impact the valorisation of sediments in different applications. A literature review shows that many solutions have already been envisaged and tested with or without prior treatments such as supplementary cementitious materials [18, 19], fired clay bricks [20], filling material [21], artificial aggregates [12], pavement base materials [22] or even substrates for plant growth [23]. The possibility of using sediments in earthen construction have been limited to adobe bricks [20, 24] with only one case of incorporating sediments originated from harbours.

The proposed work assesses the potential of using sediments as a potential addition in earthen materials whatever the consistency of the material and the construction process. It proposes to address these challenges by first combining a sediment with two types of available earthen resources using a specific mix-design methodology

[25, 26], analysing the effect of salt and testing the possibility of stabilisation with locally-sourced seaweed extracts.

12.2 Materials and Methods

12.2.1 Materials: Sediments, Soil and Sands

To improve the distribution of earth particle size distribution, as demonstrated in the study by Perrot et al. [26], two reference earthen materials were designed from two local clayey and fine soils. The first was based on a quarry by-product originated from a granite rock quarry in Brittany. This by-product originates from a Breton granite quarry, composed of crushed rocks and by-products of crushing. It shows a grain size distribution ranging from 0 to 4 mm. This type of sand is widely used in the construction field, particularly in ready-mix concrete to reduce the use of natural sands.

The second reference soil is a mix of 20% in dry mass fraction of fine-grained clayey soil (0–80 μm), 30% of fine sand (63–200 μm), and 50% standardized sand (noted 0–4 mm). The sand is added to broaden the grain size distribution, and ensure high dry densities as suggested by [25], various sands of different grain sizes are added, including fine sand (0–200 μm) and standardized sand (0–4 mm). The studied sediments were collected from the Lorient harbour during an annual dredging operation. The sediment was then stored at ambient temperature and relative humidity for drying.

In total, six different materials were tested, and the particle size distributions (wet sieving for the coarse fractions and laser diffraction granulometry for the fraction of particles below 63 μm) were provided in Fig. 12.1. Geotechnical parameters such as Methylene Blue value, Atterberg limits (measured using the fall cone method) for the sediments and the clayey soil and organic content are provided in Table 12.1 and were determined following different methods of the French standard NF EN ISO 17892. It can be noted that the sediment and the clayey soil are relatively close in terms of clay activity (plasticity index and methylene blue value) and particle size distribution. However, the sediment contains significant amount of sodium chloride and organic matter.

12.2.2 Materials: Algae Extract

In order to stabilise the earthen materials and provide water resistance to these materials, four species of invasive marine algae from Brittany were studied in this article: Solieria Polysaccharide collected in Saint Gildas de Rhuys (latitude: 47° 30′ 0″ N, longitude: 2° 49′ 60″ W, France), Ulva Polysaccharide collected in Sarzeau (latitude:

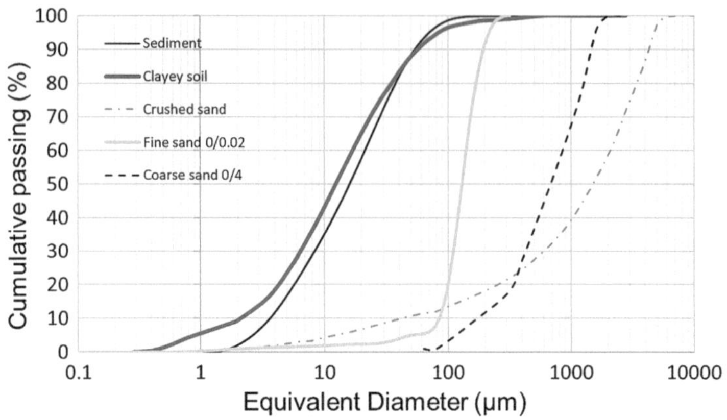

Fig. 12.1 Particle size distributions of the sediments, clayey soils and sands

Table 12.1 Physical and geotechnical properties of the clay, crushed sand and sediments

Measured parameter	Sediment	Clayey soils	Crushed sand	0/0.2 sand	0/4 sand
Specific density	2.48	2.61	2.65	2.70	2.70
Organic content	2.0%	–	–	–	–
NaCl content	2.2%	–	–	–	–
Liquid limit	72%	63%	–	–	–
Plastic limit	45%	39%	–	–	–
Plasticity index	27%	24%	–	–	–
Methylene blue value	1.7	1.25	0.15	–	–

47° 31′ 39″ N, longitude: 2° 46′ 09″ W, France). Sargassum either freeze-dried or in the form of precipitated polysaccharides collected in Saint Gildas de Rhuys (latitude: 47° 30′ 0″ N, longitude: 2° 49′ 60″ W, France).

In this study, the stabilisation solutions have been limited to sea-based products in order to adopt a circular economy approach enabled by the valorisation of local resources, coherently anchored in the territory of Southern Brittany.

12.2.3 Methods: Sample Preparation

Prismatic test specimens with dimensions of 40 × 40 × 160 mm^3 are fabricated for different formulations (Sediment + first reference soil and sediments + second reference soil). The targeted water content is the one that gives the mixture a plastic consistency, slightly above the plastic limit (w_p) in order to obtain a shapeable adobes-like consistency. Six water contents were tested for each soils in order to find the consistency that provides the highest dry density and compressive strength, in the same spirit of Proctor-like tests used in geotechnical engineering [26].

The casting is done manually by the operator (without the use of additional mechanical tools) in a $40 \times 40 \times 160\ mm^3$ mould. After moulding, the moulds are placed in a chamber with a stabilized atmosphere (20 °C and a relative humidity of 50%) for at least 3 days. Then, the specimens are demoulded and dried in an oven at 60 °C until their masses stabilise. The drying period at room temperature helps limit the water content gradient in the sample and the possible appearance of differential shrinkage.

Once drying is complete, the dry mass as well as the dimensions (length, width, and height) of each specimen are measured to determine the dry density (kg/m^3).

Three specimens are made with different compositions of the studied soils containing the sediment. For the present study, mixtures comprising 30%, 40%, and 50% sediment will be mixed respectively with 70%, 60%, and 50% of both reference soils. Samples made only with the reference soils are also prepared for comparison purposes.

12.2.4 Methods: Mechanical Testing

Once the specimens are considered dry (mass stabilization), they are broken in half by flexion. The half-specimens ($40 \times 40 \times 80\ mm^3$) will be tested in compression using a 300 kN loading machine equipped with a 50 kN load sensor. The compression tests are carried out at a constant loading velocity of 0.1 mm/s. The apparatus consists of a fixed plate, a movable plate, a drive mechanism with uniform movement at a controlled speed, and a force and deformation indicator.

The ends of the specimens must be parallel and perpendicular to the direction of the applied force. They are placed between the two plates of the press and properly centered. Two steel shims are used on the two flat faces of the specimens. The sample compressive strength is computed from the maximum recorded force divided by the sample initial section (perpendicular to the loading direction).

12.3 Results and Discussions

12.3.1 Compressive Strength of Earthen Materials Containing Sediments

In order to find the best water content to prepare moulded samples, the dry densities of the samples were plotted in function of the initial water content for the two reference soils. Examples of curves for an addition of 40% sediment for both reference soils are plotted in Fig. 12.2. It is worth noting that each curve shows a maximum dry density

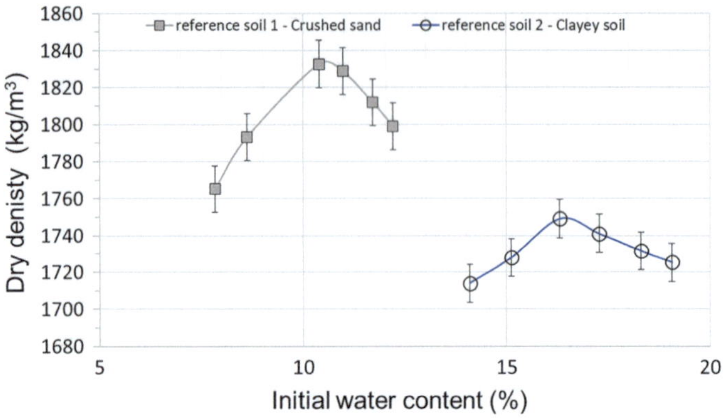

Fig. 12.2 Investigating the water content effect of the material compactibility: dry density versus initial water content for the two materials with a sediment content of 40%. The best water content is the one providing the maximal dry density

value for a specific water content value. Such specific water content allows the best compromise which provides the smallest porosity within the materials matrix. At dry state, the porosity originates from the initial water that evaporates during drying and the entrapped air which is close to 0 for the highest water content and may increase as the initial material gets drier and becomes hard to mould. The determination of the best water content for a specific process is mandatory for the optimization of the process in order to obtain the highest dry densities.

It is observed from Fig. 12.2 that the optimal water content is higher for the clayey soil (reference soil 2) than for the crushed sand (reference soil 1) because of the higher clay activity (and hence higher water demand) of the reference soil 2. This leads to smaller dry density for the materials formulated with this reference soil.

The compressive strengths of the different earthen materials are plotted in Fig. 12.3. It is worth noting that it has not been possible to obtain samples made only with reference soil 1 due to a lack of cohesion in the sample. Therefore, the unconfined compressive strength of this material can be considered equal to 0. It can be explained by the very low clay activity of the crushed sand as displayed in Table 12.1. It is also obvious that the compressive strength of the second reference soil is only about a quarter of the strength of the same soil combined with different proportions of sediments even if the obtained dry densities are slightly decreased. Figure 12.3 also seems to indicate that adding sediment can drastically increase the compressive strength of earthen materials. For reference soil 2, the sample exhibiting the highest compressive strength is a sediment content of 50% which is the highest tested dosage (6 MPa) while for reference soil 1, the highest compressive strength is obtained at the highest reached dry density for a sediment content of 30% (5.3 MPa). Ardant et al. [3] explain that the compressive strength of earthen materials depends on both the dry density of the sample which describes the sample porosity and the clay activity that can be described using the Methylene Blue Value (MBV). They have shown that a master curve linking the ratio between compressive strength at

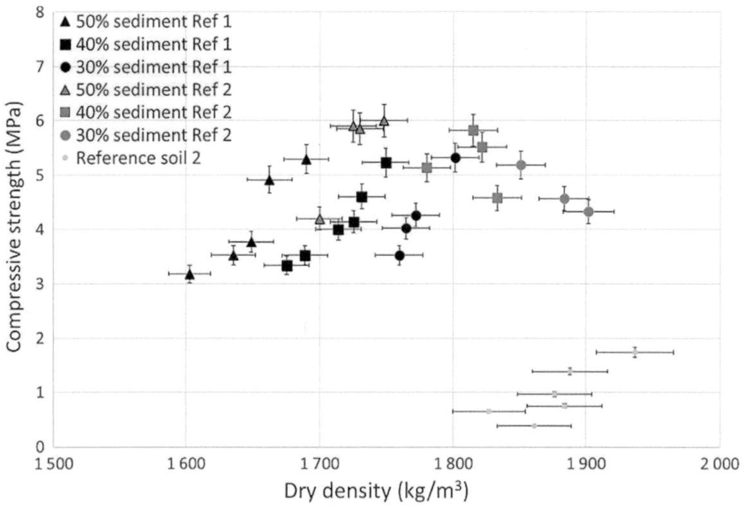

Fig. 12.3 Compressive strength versus dry density for the tested mix-design (Reference soils + various sediment contents)

dry state and MBV to the dry density of earthen materials exists and can be used to predict the strength of a sample. Such observation can explain the different results obtained with reference soils 1 and 2 as the sediments increase the clay activity and also change the dry density of the samples in different ways.

It is interesting to use the Ardant et al. model to study the effect of sediments on the mechanical behaviour of earthen materials. Therefore, the compressive strength to MBV ratio is plotted against dry density for the different tested mix designs (reference soils 1 and 2 at different sediment dosages) along with the prediction of Ardant's model in Fig. 12.4.

The observation of Fig. 12.4 clearly highlights the beneficial effects of sediment on the compressive strength of the earthen materials. It can be seen that the samples without sediment show compressive strengths that align with Ardant's model and behave as standard earthen materials. However, it is also clear that at a constant dry density, the compressive strength increases with the sediment contents. In Fig. 12.4, it is remarkable that each dot corresponding to samples having the same sediment content aligns and seems to be located on a curve almost parallel to the Ardant prediction curve. Moreover, the highest curve is the one corresponding to the highest tested sediment content (50%). Such results can be explained by the impurities contained in the sediment: organic matter and salt that can 1/stabilize the earthen matrix and 2/crystalize into solid particles. It is interesting to note that these compounds, often considered to be pollutants hindering the valorisation, here provide a significant reinforcement effect on dried hardened materials. Further research needs to be carried out to isolate and to quantify the effect of salts and organic matter on the mechanical behaviour of sediment-added earthen materials.

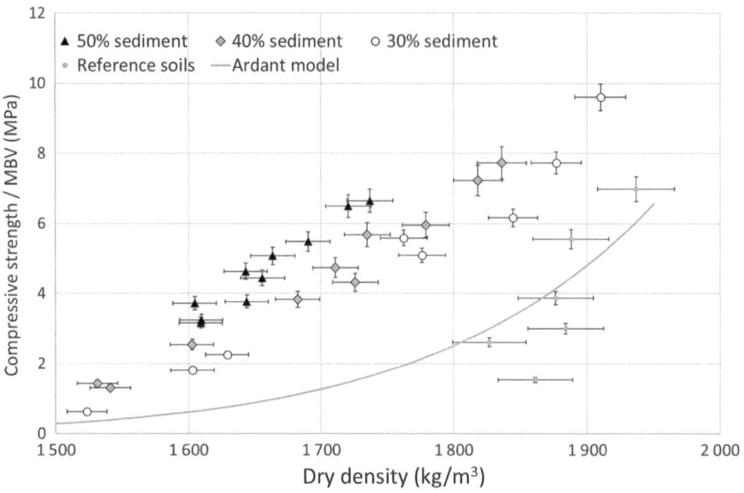

Fig. 12.4 Compressive strength/MBV versus dry density for the tested mix-design (Reference soils + various sediment contents)—comparison with the prediction of the Ardant model

12.3.2 Stabilization of Earthen Materials Containing Sediments with Algae Extracts

Water resistance tests have been performed on samples made of reference soil 2 with an addition of 40% sediment. The test consists in immersing the sample in tap water for 20 min, removing the sample and sealing it in a bag for 24 h before measuring the compressive strength. This test is inspired by the former French standard NF XP P 13–901 and is known to induce a severe solicitation on earthen materials.

The measured compressive strength of a given unstabilised sample is equal to 0 has the material without algae extract has been destroyed by the immersion period. However, it is interesting to note that the addition of the different extracts from invasive and locally-sourced algae have provided water resistance to the sediment-added earthen samples as the material shows mechanical strength ranging from 1.2 to 1.8 MPa depending on the algae extract type (Fig. 12.5).

These promising results show how the advantageous valorisation of locally-sourced waste such as marine dredge sediments and invasive algae can become a sustainable material for modern construction.

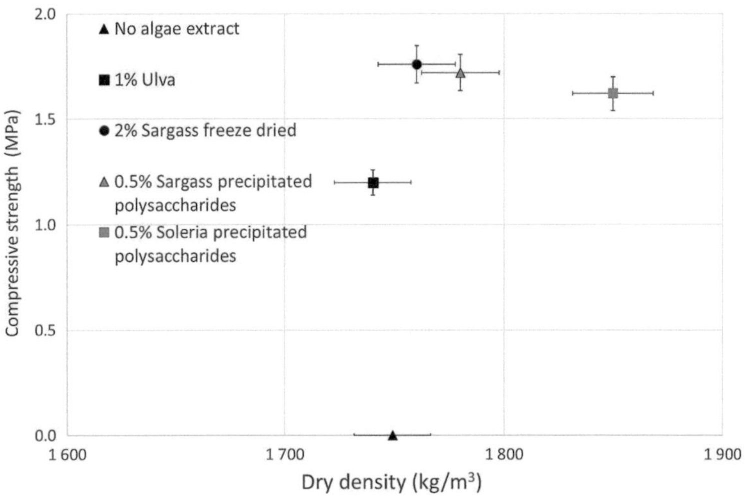

Fig. 12.5 Compressive strength of sediment-added earthen materials (reference soil 2 with a sediment content of 40%)

12.4 Conclusions

In conclusion, the use of sediment in earthen materials presents a promising avenue for sustainable and environmentally friendly building practices. The management of dredged marine sediments in Europe, particularly in France, presents an environmental challenge and an opportunity. The shift from traditional sea dumping to valorisation aligns with stricter environmental regulations and the need for sustainable solutions. This study demonstrates the potential for incorporating marine sediments into earthen construction materials, offering a dual benefit of resource efficiency and environmental conservation. The specific mix-design methodology developed in this work, combined with the use of locally-sourced seaweed extracts for stabilisation, provides a viable pathway for using sediments and earthen construction. Moreover, the study also demonstrates that the addition of sediment in earthen materials can enhance the compressive strength of the sample while seaweed extracts provide water resistance. Valorisation of marine-based waste can become a chance for earth construction.

Overall, the integration of marine sediments into earthen construction not only contributes to sustainable construction practices but also promotes the responsible management of dredged materials. Future research and practical applications should continue to explore and refine these methods, ensuring that earthen construction can fully realize its potential as a sustainable building solution. Quantification of the effect of salts and organic matter still needs to be done and the robustness of these strengthening and stabilizing solutions must be evaluated in relation to the resources variabilities.

Acknowledgements This research was partly funded by the ANR (Agence National de la Recherche) within the framework of the Labcom COLORE "Construction with local ressources" ANR-21-LCV3-0008 and by Lorient agglomeration through a partnership program. M.K Razafitrimo also thanks Campus France Madagascar for BGF grant.

References

1. A. Fabbri, J.-C. Morel, J.-E. Aubert et al., *Testing and Characterisation of Earth-Based Building Materials and Elements* (Springer, 2022)
2. E. Keita, A. Perrot, Processing of earth-based materials: current situation and challenges ahead. RILEM Tech. Lett. **8**, 141–149 (2023)
3. D. Ardant, C. Brumaud, A. Perrot, G. Habert, Robust clay binder for earth-based concrete. Cem. Concr. Res. **172**, 107207 (2023)
4. G. Giada, R. Caponetto, F. Nocera, Hygrothermal properties of raw earth materials: a literature review. Sustainability **11**, 5342 (2019)
5. H. Cagnon, J.-E. Aubert, M. Coutand, C. Magniont, Hygrothermal properties of earth bricks. Energy Build. **80**, 208–217 (2014)
6. Y. Du, G. Habert, C. Brumaud, Design of Tannin-based poured earth material via deflocculation-coagulation control: additive selection and Tannin variation. ACS Sustain. Chem. Eng. **10**, 14495–14502 (2022)
7. Y. Du, G. Habert, C. Brumaud, Influence of tannin and iron ions on the water resistance of clay materials. Constr. Build. Mater. **323**, 126571 (2022)
8. S. Guihéneuf, D. Rangeard, A. Perrot et al., Effect of bio-stabilizers on capillary absorption and water vapour transfer into raw earth. Mater. Struct. **53**, 138 (2020)
9. Y. Autem, N. Bourgougnon, S. Guihéneuf, A. Perrot, Comparative study of effects of various seaweed parietal polysaccharides on rheological, mechanical and water-durability properties of earth-based materials. Mater. Struct. **56**, 108 (2023). https://doi.org/10.1617/s11527-023-02195-9
10. G. Dorleon, S. Rigaud, I. Techer, Management of dredged marine sediments in Southern France: main keys to large-scale beneficial re-use. Environ. Sci. Pollut. Res. (2024). https://doi.org/10.1007/s11356-024-33129-9
11. J.S. Gray, M. Elliott, *Ecology of Marine Sediments: from Science to Management* (Oxford University Press, 2009)
12. F. Todaro, S. De Gisi, M. Notarnicola, Contaminated marine sediments: waste or resource? An overview of treatment technologies. Procedia Environ. Sci. Eng. Manage. **3**, 157–164 (2016)
13. K. Essink, Ecological effects of dumping of dredged sediments; options for management. J. Coast. Conserv. **5**, 69–80 (1999)
14. J. Stronkhorst, F. Ariese, B. van Hattum et al., Environmental impact and recovery at two dumping sites for dredged material in the North Sea. Environ. Pollut. **124**, 17–31 (2003). https://doi.org/10.1016/S0269-7491(02)00430-X
15. K. Gnandi, H.J. Tobschall, The pollution of marine sediments by trace elements in the coastal region of Togo caused by dumping of cadmium-rich phosphorite tailing into the sea. Environ. Geol. **38**, 13–24 (1999)
16. A.P. Lehoux, K. Petersen, M.T. Leppänen et al., Status of contaminated marine sediments in four Nordic countries: assessments, regulations, and remediation approaches. J. Soils Sediments **20**, 2619–2629 (2020)
17. A. Loudini, M. Ibnoussina, O. Witam et al., Valorisation of dredged marine sediments for use as road material. Case Stud. Constr. Mater. **13**, e00455 (2020). https://doi.org/10.1016/j.cscm.2020.e00455

18. Z. Zhao, M. Benzerzour, N.-E. Abriak et al., Use of uncontaminated marine sediments in mortar and concrete by partial substitution of cement. Cement Concr. Compos. **93**, 155–162 (2018). https://doi.org/10.1016/j.cemconcomp.2018.07.010
19. F. Agostini, F. Skoczylas, Z. Lafhaj, About a possible valorisation in cementitious materials of polluted sediments after treatment. Cement Concr. Compos. **29**, 270–278 (2007). https://doi.org/10.1016/j.cemconcomp.2006.11.012
20. K. Hamer, V. Karius, Brick production with dredged harbour sediments. An industrial-scale experiment. Waste Manage. **22**, 521–530 (2002). https://doi.org/10.1016/S0956-053X(01)00048-4
21. S. De Gisi, F. Todaro, E. Mesto et al., Recycling contaminated marine sediments as filling materials by pilot scale stabilization/solidification with lime, organoclay and activated carbon. J. Clean. Prod. **269**, 122416 (2020). https://doi.org/10.1016/j.jclepro.2020.122416
22. V. Dubois, N.E. Abriak, R. Zentar, G. Ballivy, The use of marine sediments as a pavement base material. Waste Manage. **29**, 774–782 (2009)
23. P. Macía, C. Fernández-Costas, E. Rodríguez et al., Technosols as a novel valorization strategy for an ecological management of dredged marine sediments. Ecol. Eng. **67**, 182–189 (2014). https://doi.org/10.1016/j.ecoleng.2014.03.020
24. M. Hussain, D. Levacher, N. Leblanc et al., Reuse of harbour and river dredged sediments in adobe bricks. Clean. Mater. **3**, 100046 (2022)
25. M. Audren, S. Guihéneuf, T. Le Borgne et al., Strategy for the mix design of building earthen materials made of quarry by-products. Buildings **13**, 2531 (2023). https://doi.org/10.3390/buildings13102531
26. A. Perrot, D. Rangeard, F. Menasria, S. Guihéneuf, Strategies for optimizing the mechanical strengths of raw earth-based mortars. Constr. Build. Mater. **167**, 496–504 (2018)

Chapter 13
Approaches for the Use of Rapidly Growing Willow Rods in Earth-Based Construction

Rebekka Volk, Simon Steffl, Moritz Dörstelmann, and Elena Boerman

Abstract This chapter investigates construction applications of natural fibres and fast-growing plant-based materials, which can be processed into load-bearing components using specially tailored digital manufacturing technology. The sourcing and production of the developed sustainable and recyclable freeform lightweight structures based on willow and earth are described in detail and material samples are shown. At the end-of-life, the components can be disjoined and the composite materials in the components can be fully separated mechanically for composting, recycling or incineration. Moreover, the developed structures were prototyped, implemented and tested in full-scale research demonstrators at the 2023 German National Garden Show in Mannheim (BUGA23). With this contribution we show the feasibility of such novel robotic construction methods of biogenic materials in structural applications for the first time. The envisioned application potential lies particularly in the new construction of residential buildings, hotels, office buildings and buildings with similar uses.

R. Volk (✉)
Department for Sustainable Systems Engineering (INATECH), University of Freiburg, Freiburg, Germany
e-mail: rebekka.volk@inatech.uni-freiburg.de

S. Steffl
Institute for Industrial Production (IIP), Karlsruhe Institute of Technology (KIT), Karlsruhe, Germany
e-mail: simon.steffl@kit.edu

M. Dörstelmann · E. Boerman
Institute for Building Design and Technology, Karlsruhe Institute of Technology (KIT), Karlsruhe, Germany
e-mail: moritz.doerstelmann@kit.edu

E. Boerman
e-mail: elena.boerman@kit.edu

D. A. H. Hanaor (ed.), *Sustainable Materials for the Built Environment*, Springer Series in Materials Science 352, https://doi.org/10.1007/978-3-031-97818-0_13

13.1 Introduction

The European Union is approximately 43.9% covered by forest and other wooded areas. Finland and Sweden have the largest share of this, whereas Germany ranges in the middle with a forested area of 30–45% [1]. At the same time, Germany was the largest roundwood producer in the EU in 2020 and shared the leading position in sawn timber production with Sweden [2]. Although wood is a renewable resource, with a rotation period of 30–140 years, it is not an unlimited resource. Despite high and overall increasing production figures, domestic wood quantities are not sufficient to meet European demand, which is why the material is imported into the EU from all over the world [2].

According to a WWF study, global wood consumption in 2020 was between 4.3 and 5.0 billion cubic meters, far exceeding the globally sustainably cultivated amount of around 3.0 billion cubic meters. The consumption trend here is increasing, although the risk limit of 4.2 billion cubic meters, which entails significant sustainability restrictions, has already been exceeded [3]. Moreover, timber is frequently utilized in engineered wood products which contain glues and adhesives. These high-performance materials are energy-intense to produce and are created from a combination of natural and fossil resources. Furthermore, energy recovery is the sole end-of-life option for these products (see Fig. 13.1).

Thus, other approaches are required either to reduce the demand or to develop other renewable materials for construction purposes in order to meet the demand. For this, willow as a fast-growing natural raw material seems suitable to replace wood that is not infinitely available or other construction materials that have a high environmental impact.

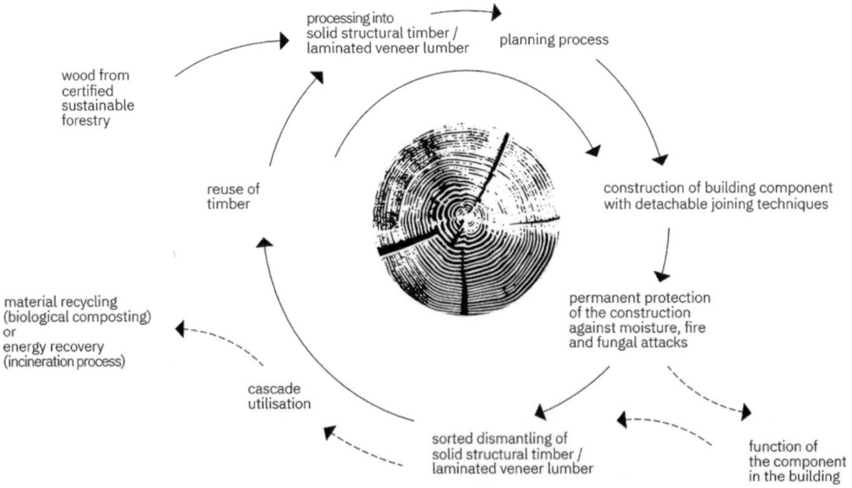

Fig. 13.1 "Circle" of wood. *Copyright* KIT Professorship of Sustainable Construction

Willow is a pioneer plant settling which colonize areas that are poor in nutrients and water and can quickly regenerate after strong environmental influences such as flooding. Their cultivation therefore does not (directly) compete with other agricultural products, which further underlines their importance for sustainable use of resources. Due to its rapid growth and short rotation periods of up to 30 years, willow is now mainly cultivated for energy production purposes [4] and harvested in 1–5 years short rotation cycles, depending on the intended use [5]. Construction purposes are not considered yet. Moreover, when willow is used in building components, the aim is either to reclaim it from the component and use it in other wooden application as a filling material or biodegrade it after it reaches the end of utilization cycle as a construction/product material.

Clay is a traditional construction material with high local availability [6]. It can be recycled infinitely [7] and can be processed and manufactured as a secondary building material in the same quality as the primary material [8]. To ensure optimal and efficient recyclability of the clay mixture, leaching processes can be used to separate mineral and plant additives. Alternatively, the organic fraction can remain in the material mixture (see Fig. 13.2).

A combination of both willow and clay is envisioned for innovative construction elements, that are less energy and emission-intense in production than conventional construction elements from concrete, faster regrowing than wood, potentially locally sourced, not harmful to nature, enabling rewetting of agricultural land and are fully recyclable. To achieve this, the research projects ReGrow [9], WillowWeave and ReSidence [10] developed a holistic experimental and digital construction approach within an integrative co-design process in order to produce such a construction component of a willow-clay composite on a 1:1 scale.

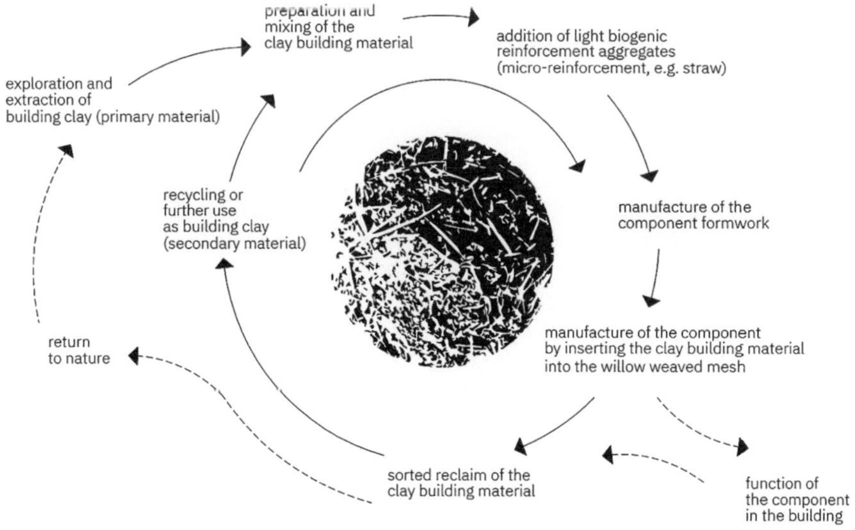

Fig. 13.2 "Circle" of clay. *Copyright* KIT Professorship of Sustainable Construction

13.2 Method

The aim was to develop circular construction methods based on regional materials and to create functionally integrated lightweight structures for use in construction. The design principles should make modular combination of the resulting components possible and special attention was paid to concepts for the subsequent use of the structures. Central elements that contribute to this goal are the following three design and manufacturing processes:

- Bespoke computational design and fabrication solutions to enable free-form wound lightweight structures based on the combination of willow and clay as natural sustainable materials.
- Robotic fabrication methodology for implementing weaved willow tree branches.
- Lightweight wood construction.

The aim was to investigate construction applications of fast-growing plant-based materials, which can be processed into load-bearing components using specially tailored digital manufacturing technology. For this, the whole life cycle of the building component was designed (see Fig. 13.3) for flexibility and circularity. When designing buildings and their components, maintaining material cycles through the detachability of materials and their joining techniques play a crucial role in order to be able to continue using the resources as well as maintaining their values and material quality as much as possible. Thus, the separability and recyclability of the used material were a critical aspect.

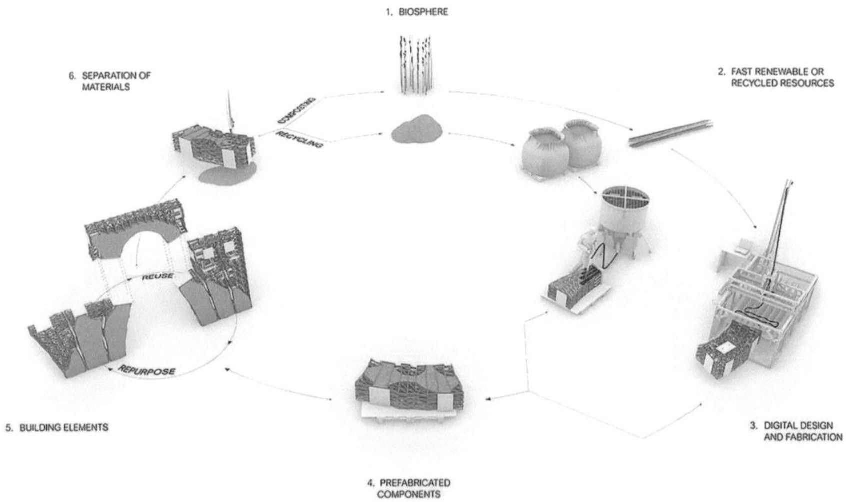

Fig. 13.3 Life cycle of willow-clay-composite from sourcing, production, use to end-of-life and a new cycle

Moreover, the resulting lightweight structures were assessed with respect to environmental impact categories, e.g. greenhouse gas emissions, and compared to conventional constructions of the same load bearing capacity and showed promising results.

The harvested willow rods (1. and 2.) are made flexible for weaving in a dampening or watering process. Then, they are coupled and additively fabricated by a custom machine to baskets (3.) that are then filled with moist earth. Finally, the prefabricated building components (4.) are drying for several weeks for hardening. The basket weaving and components can be built in any shape required for loadbearing structures (5.). For installation, the prefabricated elements are joined onsite with a crane and can be easily disjoined (6.). At the end-of-life, earth and willow can be separated mechanically, where the biogenic share can be utilized as biobased filling material in other applications, or alternatively composted or incinerated and the earth can be directly recycled.

The development was driven forward by interdisciplinary co-design workshops and supported by bespoke manufacturing technology for the production of the components. To achieve this, a comprehensive fabrication system was developed, starting with a two-axis machine capable of extruding and depositing continuous willow-jute composite strands made from spliced willow branches using additive techniques. Controlling the machine's extrusion, movement, and deposition path proves challenging due to factors such as the flexibility and varying thickness of willow branches, as well as the requirement for tensioning in the final components. Moreover, challenges due to low local availability of willows (unexpected high market demands), or variability of biogenic raw materials had to be faced and mastered.

13.3 Solutions

On the one hand, a bespoke design-to-manufacturing process [11] for the use of fast-growing willow tree branches was expanded and adapted in order to implement lightweight structures in the future construction industry not only in an material-efficient manner, but also on the basis of renewable raw materials. On the other hand, an approach to digital fabrication for structural applications was developed. This includes both the development of a new type of construction and design repertoire and the implementation of corresponding machine technologies for the production of load-bearing components. Both design ideas are combined in a complementary manner to create a modular concept for producing functionally integrated building components from renewable raw materials.

Figure 13.4 shows the necessary pre-processing of the willow rods to make them useable for robotic weaving processes. It includes planting and harvesting, damping or watering and the production of endless strains for the robotic weaving process. During the fabrication of the building components, a stabilisation by metal forms is provided which can be reused afterwards. Moreover, engineering wood products

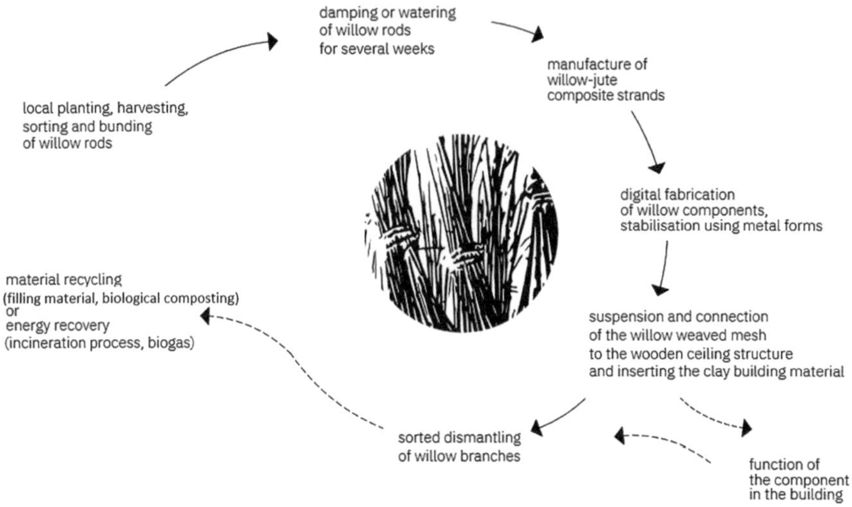

Fig. 13.4 "Circle" of willow. *Copyright* KIT Professorship of Sustainable Construction

are used to stabilize the whole component and to enable the clay filling. Then, the willow baskets are connected to each other and the demonstrator is manually filled with the clay building material including light biogenic reinforcement aggregates. Finally, the composite elements are dried for several weeks and then transported and erected with a crane onsite using steel ring screws, that can be detached after the modules reach their final position.

After use, a full recyclability of the elements is envisioned. The clay can be recycled and reprocessed to secondary building materials, but for the willow rods, after their recovery from the component one option is a post-treated utilization as organic filling material to keep the biomass in the technosphere. When reaching the last use cascade level, the end-of-life biological compositing turns willow into organic material and returns it to biosphere. The engineered wood products' recycling depends on the availability of local material recovery plants (crushing, sorting), the demand for this material as well as the used adhesives which would often lead to energy recovery as the most likely end-of-life scenario.

As part of the research, two sustainable and recyclable lightweight structures were developed and produced, erected and tested with full-scale research demonstrators at the 2023 German National Garden Show "Bundesgartenschau BUGA23" in Mannheim (see Fig. 13.5). Moreover, the willow-clay building demonstrator was expanded to investigate the research focus on local energy generation and microclimate adaptations. Information events, visitor surveys and an international symposium took place at the Federal Garden Show to give a large audience an insight into the research topic of sustainable and digital construction.

Fig. 13.5 Willow earth composite elements in detail and full-scale demonstrators at Bundesgartenschau Mannheim 2023 © KIT/IIP and KIT/IEB

13.4 Conclusions

Thanks to the current state of the art, robotic freeform winding can already be used to produce topology-optimized and low-waste lightweight structures based on natural and energetically low processed raw materials. This composite development represents a further step towards the sustainable construction of tomorrow. Future applications in residential buildings, hotels, office buildings and similar usages are envisioned and can cover a considerable share of new construction demand. By further developing the process technology to meet the special requirements of reinforcement made of biogenic sources like natural willow tree branches, sustainable load-bearing structures should be produced on a larger scale, something that has not been technically feasible to date. In addition, regionally available, rapidly renewable raw materials are being made usable for scaled applications. For this, the sustainable sourcing of willow and earth needs to be addressed with respect to an industrial scale production, since both willow short rotation coppice production facilities as well as clay mines need to be ventured. However, this is necessary support the transition to a circular economy in the construction sector and further reduce the CO_2 footprint of lightweight structures. At the same time, state-of-the-art digital design and manufacturing processes are being used, which guarantees high material efficiency and leads to a high potential for the future of the construction industry in terms of economic utilization.

References

1. L. Corselli-Nordblad, A. Johannson, (eds.), *Key Figures on Europe*, 2022 edn. (European Union, Luxemburg, 2022) (eurostat). ISBN 978-92-76-51643-9, ISSN 2315-201X. https://doi.org/10.2785/851035. https://ec.europa.eu/eurostat/documents/15216629/15358170/KS-EI-22-001-EN-N.pdf/273c56e9-a255-ebc6-fb8d-33f3bb7ca2cf?t=1669211457873. Access 21 Mar 2025
2. Joint Forest Sector Questionnaire (JFSQ), *Wood Products—Production and Trade* (European Union, Luxemburg, 2021) (eurostat). https://ec.europa.eu/eurostat/statistics-explained/index.php?title=Wood_products_-_production_and_trade. Access 21 Mar 2025
3. M. Beck-O'Brian, V. Egenolf, S. Winter, J. Zahnen, N. Griesshammer, S. Bringezu, Everything-from-wood: the resource of the future or the next crisis (2022). https://www.wwf.de/fileadmin/fm-wwf/Publikationen-PDF/Wald/WWF-Study-Everything-from-wood.pdf. Access 02 Aug 2024
4. Fachagentur Nachwachsende Rohstoffe e.V. (FNR), Anbauzahlen in Deutschland (2024). https://pflanzen.fnr.de/anbauzahlen. Access 02 Aug 2024
5. M. Bärwolff, H. Hansen, M. Hofmann, F. Setzer, Energieholz aus der Landwirtschaft. Rostock: Fachagentur Nachwachsende Rohstoffe e.V. (FNR) (2012). https://www.fnr.de/fileadmin/Projekte/2021/Mediathek/energieholz_web.pdf. Access 02 Aug 2024
6. C. Ballabio, P. Panagos, L. Monatanarella, Mapping topsoil physical properties at European scale using the LUCAS database. Geoderma **261**(2016), 110–123 (2015). https://doi.org/10.1016/j.geoderma.2015.07.006
7. W. Sobek, *Non nobis - Über das Bauen der Zukunft, Band 1: Ausgehen muss man von dem, was schon da ist* (avedition, Stuttgart, 2022). ISBN: 978-3-89986-369-7
8. H. Schroeder, *Lehmbau - Mit Lehm ökologisch planen und bauen* (Springer Vieweg, Wiesbaden, 2018). https://doi.org/10.1007/978-3-658-23121-7 (2018)
9. E. Zanetti, E. Olah, T. Haußer, D. Fischer, G. Casalnuovo, M. Zareian, R. La Magna, M. Dörstelmann, ReGrow willow—digital circular construction for Earth-willow hybrid structures, in *Fabricate 2024: Creating Resourceful Futures*, ed. By M.R. Thomsen, P. Ayres, B. Sheil, M. Skavara (UCL Press, 2024). https://doi.org/10.2307/jj.11374766
10. KIT, ReSidence, ieb (2024b). https://ddf.ieb.kit.edu/residence.php. Access 7 Aug 2024
11. G. Casalnuovo, E. Zanetti, T. Haußer, M. Dörstelmann, R. La Magna, Digital structural design for natural composites: a case study of willow-Earth hybrid construction, in *Habits of the Anthropocene: Scarcity and Abundance in a Post-Material Economy. Proceedings of the 43rd*

Annual Conference of the Association for Computer Aided Design in Architecture, ed. by A. Crawford, N. Diniz, R. Beckett, J. Vanucchi, M. Swackhamer (Association for Computer Aided Design in Architecture (ACADIA), 2023), 282–292

Chapter 14
Carbon Nanomaterial Additives in Concrete as a Pathway Towards High Performance and Sustainability

Dorian A. H. Hanaor

Abstract The addition of carbon nanomaterials at low concentrations has been shown to produce significant improvements to the strength and hardness of concrete. Nanostructured carbon based materials including graphene nanoparticles, graphene oxide and carbon nanotubes have been applied in various concrete types with interesting results, presenting diverse opportunities for the reduction of cement content and the improved performance and sustainability outcomes of cemented materials. Here we examine the mechanisms through which carbon nanomaterials enhance the performance of concrete and discuss the challenges and opportunities involved in the future development of these advanced materials. In particular the value of carbon nanomaterial modified concrete towards greater sustainability in the built environment is scrutinized. Graphene oxide appears to be the most widely used and effective method to improve cement performance in terms of strength and density of hydration products. The improvement of cement performance is found to occur through enhanced hydration kinetics and denser hydration products. However, the high environmental impact of their production and the lack of a clear framework to guide the implementation of such carbon nanomaterials as cement additives remain as obstacles towards the achievement of improved sustainability outcomes.

14.1 Introduction

Interactions of carbon with materials in the built environment underpin many concepts towards greater sustainability in materials production and usage including cement carbonation and the use of biomass as carbon-sinks in structures. Most carbon atoms in materials in the built environment are either in the form carbonate phases

D. A. H. Hanaor (✉)
Faculty of Science and Engineering, Southern Cross University, East Lismore, NSW 2480, Australia
e-mail: dorian@materialeyes.net
URL: https://www.materialeyes.net

Materialeyes, Materials Innovation Consulting, Berlin, Germany

D. A. H. Hanaor (ed.), *Sustainable Materials for the Built Environment*, Springer Series in Materials Science 352, https://doi.org/10.1007/978-3-031-97818-0_14

in masonry and concrete or as part of large organic macromolecules in polymers and wood. Carbon nanomaterials, and carbon fibres may seem at first glance not to be significantly relevant in the built environment. Carbon materials, that is those consisting primarily of carbon bonded to other carbon atoms, involve carbon–carbon bonds in either sp2 or sp3 hybridization. In sp3 hybridization a tetrahedral bonding is present, while sp2 bonding results in a hexagonal lattice.

Carbon in its sp2 bonded form is rather ubiquitous in nature, occurring in bulk form as graphite and readily forming small particles (including nanoparticles) in soot and ash. However, it has only been since the 1990s that carbon nanomaterials (CNMs) based on sp2 hybridized hexagonal lattices of carbon, including fullerenes, carbon nanotubes (technically a type of fullerene) and graphene have truly emerged as a highly promising class of materials. The emergence of carbon nanomaterials has taken place largely as the result of advances in the synthesis and characterisation of nanomaterials. Our ability to produce highly pure forms of carbon nanomaterials though physical and chemical processes has evolved hand-in-hand with our ability to characterise using various spectroscopic techniques, the structure and quality of these materials.

Methods have flourished over the last three decades for the production and characterisation of controlled carbon nanostructures including

- Highly ordered nanomaterials of sp2 bonded carbon: Fullerenes (so-called Buckyballs) having distinct numbers of carbon atoms in a closed sphere; single wall carbon nanotubes (SWCNT), in their zigzag, armchair or intermediate chiralities; and single-sheet graphene layers which still require meticulous care in creating, often on the surface of a single-crystal copper.
- Moderately-ordered nanomaterials or those made of nanobuilding blocks of predominant sp2 carbon: Carbon-nano onions consisting of numerous shells; multi-wall carbon nanotubes in varying degrees of defectiveness (in the form of sp3 bonding), length and shell number, carbon nanofibers; defective or multi-sheet graphenes, such as graphene-oxide, reduced graphene-oxide (rGO) and graphene nanoparticles (which are more-appropriately considered graphite nanoparticle).

Additionally, the largely disordered carbon black, which often is in the form of agglomerated nanoparticles comprising between 50 and 70% sp2 bonded carbon can be considered a nanomaterial, and indeed is widely used as a filler in various polymers and is widely used to impart durability in tyres.

The evolution of synthesis technologies has resulted in a growing market and a general downward trend in the cost of carbon nanomaterials, inspiring their gradual uptake in a range of niche applications. The market for multi-wall nanotubes and vapour grown carbon nanofibers has grown significantly as these have been taken up in applications of high performance composites and energy storage. Nevertheless the volume of sp2 carbon nanomaterials applied in industry is quite limited and near-pristine nanotubes and fullerenes are mostly confined to research and beyond the laboratory, their real-world use remains very limited.

The gap between the potential performance of nanomaterials and their uptake in real-world applications is nowhere starker than in the case of graphene. Since the early 2000s the popular science and engineering media has been awash with reports of outstanding mechanical, electrical and thermal performance achievable in single-layer sp2 bonded carbon and the astonishing engineering applications that will soon be enabled by this new wonder-material. And yet, the heralded graphene revolution has been slow to arrive. The challenges in harnessing the impressive theoretical properties of graphene arise from the difficulties in producing and applying this material in its single-layer and defect-free form. These obstacles have not prevented chemical suppliers from profiting from the sale a broad range of materials under the name "Graphene". However these are invariably in the form of aggregated particles consisting of multi-layer stacked graphene, and generally offer none of the performance characteristics of single sheet graphene.

Oddly enough, one area of application where graphene is already finding effective application, is in the improvement of concrete. It may seem counterintuitive that one of the first successful applications of a graphene carbon nanomaterials is as an additive to concrete–generally the most widely used and among the cheapest engineering materials. However upon closer inspection, it is evident that this application of graphene is not as bizarre as it may seem at first glance. Particularly small additions of graphene (graphene oxide most commonly) have been shown to result in profound changes to the behaviour of concrete, with implications towards pathways to using this material more sustainably. In particular the use of nanomaterial additives in concrete, are attractive as a method towards reducing the fraction of cement necessary to achieve required performance in structural use, allowing reductions to the carbon footprint of concrete structures.

14.2 Materials and Applications

14.2.1 Carbon Fibres

The earliest use of carbon materials as reinforcement in concrete involved carbon fibres. Carbon Fibre Reinforced Concrete (CFRC) was first reported in 1972, not long after the mass production of PAN and Pitch based carbon fibres became commonplace [1, 2]. Carbon fibre reinforcement may be in the form of short or continuous filaments. The rationale for reinforcement with carbon fibres, despite their generally higher cost, is their high stiffness and tensile strength. Values of flexural and compressive strength are enhanced by around 50% compared with unreinforced, concrete with additions of just a few percentages of carbon fibres, indicating mechanical performance levels comparable with steel reinforced concrete, with only a small fraction of the overall weight and fibre volume [3]. Additional value comes from their corrosion resistance, making them useful in situations where steel reinforcement may be prone to corrosion, such as in thin profiles or near marine environments, although the oxidation of carbon fibres is also a problem that is not negligible. Most carbon fibres are derived from polymers or other petroleum derived precursors (Commonly PAN). The energy used to produce carbon fibres and their fossil based precursors means that in terms of LCA outcomes, for a functional unit of 1 cubic metre of concrete, CFRC generally has a higher environmental footprint (in terms of Global Warming Potential, GWP) when compared with conventional steel reinforced concrete [4]. However this functional unit may be misleading, as the volume of fibres and concrete used varies with application requirements. Nevertheless a survey of the field shows that significant sustainability dividends are unlikely to be reached through the replacement of steel with carbon fibre in concrete. Nanomaterials however open up new avenues to exploit unique carbon structures to impart superior performance in concrete potentially allowing for a significant reduction in required cement fractions, with the addition of small fraction of nanoscale additives [5].

14.2.2 Graphene Oxide

Among the various carbon nanomaterials that may be used to improve the performance of concrete, graphene oxide (GO) stands out in terms of the feasibility of its application and the promise it offers towards effective enhancement in mechanical and physical properties of the materials.

Graphene oxide is less complex to produce than pristine single sheet graphene, and due to the functional groups present on the surfaces of graphene oxide this material is polar and thus tends to disperse well in water. The aqueous solubility of GO is key to its effective use as a cement additive. In many nanocomposite systems the aggregation of nanoparticles hinders the fulfilment of their potential, as they tend to clump together to reduce their surface energy, making GO a highly attractive candidate.

Graphene oxide is mainly produced from graphite through the use of oxidation and exfoliation agents. The Hummer's method is probably the most widely known, with variations of this method being used to produce GO towards research and real-world use. The production of GO through oxidation and exfoliation is relatively scalable, although the use of aggressive oxidation agents gives this material a considerable lifecycle impact, limiting sustainability outcomes that are achieved in such materials [6]. The functional groups present on the surface of graphene oxide may include hydroxyl, epoxy and carboxyls. The presence of these groups enables the aforementioned aqueous solubility and eliminates the need for organic solvents, reducing toxicity concerns and simplifying integration into concrete mixes [7].

Graphene oxide has been extensively researched as an additive in cement pastes, mortars and concrete since at least 2012 [8–10]. It is routinely found that the addition of graphene oxide to cement pastes at levels in the order of magnitude of 0.05 wt% produces enhancements to compressive strength typically around 30–60% and similar improvements to flexural strength [11, 12]. In concrete and mortar, similar low levels of addition produce slightly milder improvements relative to cement pastes, but the trend and the significant enhancements remain clear. Reports have further shown that when attention is given to the oxidation conditions of GO sheets, a doubling of flexural and compressive strength can be achieved [8]. Alongside a dramatically increased strength, small amounts of GO, as low as 0.02 wt%, produce a marked reduction in slump, indicating reduced workability, owing to the interactions of water molecules with the high level of surfaces brought into play by the GO additive [12]. This has implications for optimising the water ratio in GO enhanced concrete. The use of GO additives also reduces the permeability of concrete, through the formation of denser hydration product. This limits the ingress chloride and other ions and provides advantages in durability and corrosion resistance for steel reinforcement [13, 14].

14.2.3 Terminological Ambiguities and Material Realities

The integration of graphene into concrete represents an active and promising direction in material science, offering pathways to enhanced strength, durability, and sustainability. However, a critical challenge lies in the ambiguity surrounding the term "graphene" itself. While graphene oxide (GO) is widely studied for its compatibility with cementitious matrices, graphene's application in concrete is not limited to GO. The broader landscape is muddled by inconsistent terminology, where materials labelled as "graphene" or "graphene nanoparticles" often deviate significantly from the idealized monolayer carbon structure.

As discussed, Graphene, in its IUPAC definition, refers to a single atomic layer of sp2 bonded carbon atoms arranged in a hexagonal lattice. Yet, in practice, materials marketed as graphene frequently consist of multilayer graphite particles (often exceeding 10 layers) or disordered carbon structures, such as turbostratic carbon, which lacks the crystalline order of true graphene [15, 16]. This inconsistency stems from variations in synthesis methods, commercial labelling practices, and a lack of standardized definitions across industries [17]. For instance, "few-layer graphene" may describe stacks of 2–10 layers, while "graphene nanoplatelets" often comprise thicker, irregular graphite fragments [18, 19]. Such discrepancies complicate efforts to isolate graphene's specific contributions to concrete performance, as interactions between the additive and the cement will depend heavily on layer count, edge structure, and defect density, and particularly dispersion.

The ambiguity in material classification poses significant challenges for researchers. Studies reporting "graphene-enhanced concrete" may, in reality, involve graphite derivatives or carbon additives with vastly different behaviors. For example, turbostratic carbon, characterized by misaligned graphene layers, exhibits distinct interfacial interactions with cement hydrates compared to well-ordered GO. This variability undermines cross-study comparisons and obscures the mechanisms by which carbon nanomaterials influence concrete properties. Despite these hurdles, several investigations have explored non-GO graphene forms.

Multilayer graphene particles have been shown to improve compressive strength in cement pastes and concrete by filling nano-pores and bridging microcracks [18, 20–22]. Graphene nanoplatelets (GNPs) are commercially available and have been explored as cement additives. While, as mentioned, the terminology of GNP may be applied inconsistently, the LCA of GNPs remains quite favourable with equivalent emissions as low as 0.17 kg CO_2 resulting from the production of 1kg of GNPs [23]. This is in stark contrast to GO, which may have an equivalent CO_2 footprint over 1 tonne per kg of graphene oxide produced [24]. Such LCA outcomes are supportive of GNP usage. In comparison with the use of GO, the addition of GNPs to concrete tends to produce a more mild effect on compressive strength, in the order 10–20% and even a weaker enhancement of flexural strength, in the order of 0–10% [19–22].

Reports of graphene modified cements and mortars exhibiting greater enhancement of mechanical properties, in the order of doubling compressive strength, have been reported for the use of Graphene Nanoparticles (a term often used in the literature to describe widely differing materials) produced by chemical treatment with strong acids, followed by thermal shock and exfoliation [25]. It is however likely that many of the graphene nanoparticles reported in concrete are in fact more appropriately considered to be GO, due to the oxidising chemical treatments involved in their processing.

14.2.4 Carbon Nanotubes

Towards the enhancement of concrete, graphene oxide would appear to be the most promising and useful among carbon nanostructures. The ease of dispersion and the relatively facile fabrication of GO creates versatile avenues towards the modification of concrete to extract higher levels of performance. It is therefore not surprising that commercialisation efforts focus on the implementation of GO in concrete. However, the use of fibres, wires and elongated bars is a sensible approach towards the enhancement of flexural strength and ductility in concrete, and it is the use of such fibres that constitutes the most common approach towards the fabrication of reinforced concrete materials, and the use of carbon fibres (not nanoscale) is among the most attractive options towards the substitution of steel in concrete materials. We can therefore expect carbon nanotubes, a highly interesting 1D-nanostructure with outstanding axial properties, to be of interest towards their implementation in advanced and sustainable concrete formulations.

Various carbon nanotubes (CNTs) and nanofibers have been proposed as reinforcing phases in concrete. The logic behind such applications is clear, the high tensile strength of carbon nanotubes and nanofibres, and the formation of dense hydration products and carbonate phases at interfaces with the CNTs means that good strengthening effects should be possible in such composites [26]. In these applications, the general trend is the use multi-wall nanotubes or carbon nanofibers as alternative short-fibre reinforcing phases, as single wall nanotubes are prohibitively expensive. As with graphene, the key to the successful use of CNTs lies in their successful dispersion and appropriate surface functionalisation, with certain studies showing that the addition of CNTs may in fact decrease the mechanical performance of cements [26]. Commonly the observed CNT driven enhancement of compressive strength in hardened cement pastes is relatively mild and highly dependent on the chemical and physical methods used to disperse the nanoscale phase [27–29]. Enhancement is regarded to occur mainly through crack bridging.

Concretes made with CNT additives would generally fall under the category of High-Performance Fibre Reinforced Cementitious Composites (HPFRCCs). These are a subclass of Engineered Cementitious Composites (ECCs, also called bendable concrete). ECCs are high-performance, fibre-reinforced materials designed for exceptional ductility and crack control, achieving tensile strain capacities hundreds of times greater than unreinforced concrete [30]. In architecture, their durability and self-healing properties reduce maintenance needs and extend structural lifespans, aligning with sustainable goals by minimizing material waste and embodied carbon. The inclusion of CNTs in ECCs may enable thinner, lighter and more durable structures, reducing material use and prolonging service life. ECCs may benefit from the addition of CNTs not only because of enhanced mechanical performance but also chemical resistance, to acid and sulphate attack, which are particularly important in this type of material [31].

14.3 Mechanisms

The integration of carbon nanomaterials, such as graphene and graphene oxide (GO), into cementitious matrices has emerged as a transformative strategy for enhancing concrete performance. These materials exert their influence at the nanoscale, primarily by modulating the microstructure of hydrated cement phases during setting and hardening. A critical mechanism involves the morphological alteration of calcium silicate hydrate (C–S–H) binding phase in concrete. Studies indicate that graphene and GO act as nucleation templates, promoting the formation of densely packed, interlocked C–S–H structures with "flower-like" morphologies (Fig. 14.1) [8, 12]. These configurations enhance mechanical strength and durability by reducing porosity and improving interfacial bonding between hydration products and aggregate surfaces [32].

The nucleation of acicular, flower like or interlocking C–S–H hydration products is likely only part of the story. Carbon nanomaterials, including both graphene, GO and CNTs, enhance hydration kinetics of cement, forming denser hydration products surrounding the dispersed carbon nanoparticles and more extensive reactions of hydraulic phases [33–35]. A further important mechanism that has been discussed is the formation of interlocking carbonate phases including calcite and aragonite at interfaces with the cement gel [36, 37]. While not as widely reported as the C–S–H flower mechanism, the decomposition of graphene oxide into CO and CO_2 in the alkaline medium of the cement gel, and the carbonation of the surrounding hydrated phases of CH, ettringite and C–S–H to form needle-like $CaCO_3$ is likely an important factor in the often highly enhanced performance in C nanomaterial modified concretes [33, 37].

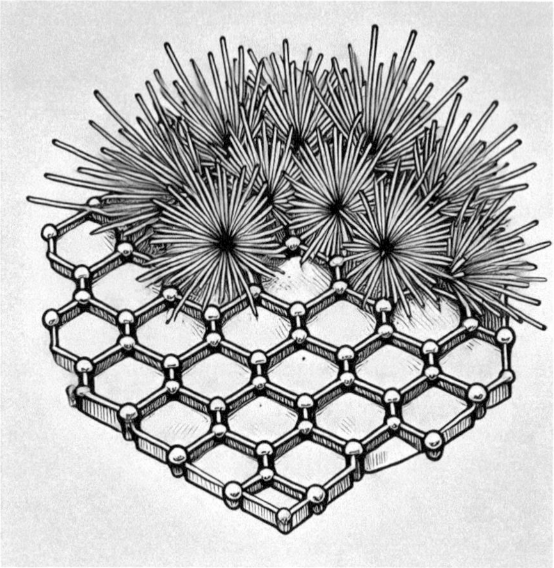

Fig. 14.1 Schematic illustration of graphene nucleation of flower-like C–S–H phases (generated using DALL E 3)

Further performance gains arise from synergistic interactions of graphene, graphene oxide or carbon nanotubes with supplementary nanoparticles. The combination of pozzolanic silica fume and graphene oxide stands out as an attractive combination of additives [38–41]. Silica fume helps create further C–S–H phases through reaction with hydroxide materials, which boost the strengthening effect brought about by the addition of graphene and GO. It has been widely shown that silica fume (nanoscale amorphous SiO_2) amplifies nucleation efficiency by helping disperse graphene and providing additional reactive sites for interconnected C–S–H growth [39, 40, 42]. Similar effects may be seen also with carbon nanotubes, whereby silica fume enhances dispersion and turns an otherwise mild enhancement by CNTs into a more significant strengthening effect in cementitious materials of various kinds [38].

Based on reported mechanisms for strength enhancement by carbon nanomaterials it is logical to expect that pozzolanic additives in general would provide synergy in developing further strength, through the formation and interlocking of greater levels of C–S–H, through the pozzolanic interactions. As GO is known to enhance hydration kinetics and extent of reactions, it is unsurprising that pozzolanic reactions and formations of further strength giving hydration products further benefit from the addition of graphene oxide, which acts as a catalyst for pozzolanic activity of additives. Further to silica fume, the addition of GO further works in synergy with nano-alumina (Al_2O_3) [43, 44], calcined clay [45], and Fly-ash [42, 46] to contribute to denser and stronger hydration products in cement. These synergistic effects highlight pathways towards extracting further utility from supplementary cementitious materials (SCMs), and reducing the need for clinker in concrete formulations.

Towards Commercialisation and Sustainable Built Environments

It has been demonstrated that the optimised addition of GO can produce economically enhanced concrete with higher levels of strength/cost ratio achieved [12], and it is likely that with a decreasing cost of GO production, these levels can be even further improved. Graphene enhanced concrete is mostly implemented in pilot stage and early commercialisation stage projects and is yet to become a mainstay feature of the construction landscape. The economic production potential of GO has inspired numerous commercial ventures into graphene enhanced concrete, making this one of the first sectors to see widespread uptake of graphene nanomaterials. A key drawcard driving the uptake of graphene enhanced concrete is the increased density, and the consequent improved resistance to corrosion, chloride ingress and freeze–thaw cycles.

Several key barriers to commercial uptake of carbon nanomaterial enhanced concrete are clearly identifiable. Firstly, the construction industry is well known as a highly conservative market, exhibiting great inertia and slow adoption rates of new materials, due to well-entrenched supply chains and rigid regulatory frameworks. Secondly, the high costs of graphene oxide additives means that there are few use-case scenarios where its inclusion makes significant economic sense to include this additive and thirdly, there remain technical challenges to using graphene or GO effectively, particularly relating to the dispersion of additives and prediction of long-term performance.

The innovation and commercialisation scene relating to the development and implantation of carbon nanomaterials in concrete is notably active. Venture arms of large construction materials producers and investor-funded startup companies are seeking to capitalise on the potential for new and improved materials that deliver superior sustainability outcomes and new markets for advanced materials. Currently most ventures provide carbon nanomaterials, mainly graphene oxide, as admixtures to be included in concrete mixes. Some selected ventures from around the world are outlined in Table 14.1. This list is by no means exhaustive, and the dynamic nature of this sector means that new ventures are constantly emerging and disappearing in the scene.

14.4 Towards Sustainable Built Environments

A key motivating factor driving the development of carbon nanomaterials for the enhancement of cementitious materials is the urgent need to reduce the enormous carbon footprint of the concrete sector, which is widely evaluated as being responsible for ~ 8% of GHG emissions. In this context the improvement of cement performance through the use of nanomaterials can reduce the environmental impact of the concrete industry through three main enabling pathways as outlined here.

Table 14.1 Commercial ventures developing carbon nanomaterials for concrete

Origin	Company	Focus	Stage
	Concretene	GO enhanced concrete	Early commercialisation in EU projects
	Graphenstone	GO enhanced impervious coatings	Commercialisation of sealing, paints and enhanced limes
	Blackleaf	Graphene inks and additives	Marketed as an additive to increase σ_c by 10% at concentrations from 0.02–0.04% [47]
	Edencrete	Carbon nanotube enhanced admixture for improved toughness and abrasion	Early stage commercialisation in pavement projects. Synergy with fly ash [48]
	Cemex	Carbon nanomaterials produced from CO_2 in flue gases	Pilot test study as part of Cemex's Smart Innovation platform
	Directa Plus	Suppliers of graphene and graphene inks	Trials ongoing in concrete and asphalt
	Nanoxplore	Supply and distribution of graphene	Suppliers to diverse materials producers, GO and GNPs
	Heidelberg Cement	Graphene infused concrete for electrical conduction	Development for heating and sensing surfaces, collaboration with Talga
	GrapheneUp	Suppliers of GUP graphene oxide admixtures for concrete	Range of commercially available products
	First Graphene	Suppliers of PureGRAPH admixtures reported to improve strength	Development with chemical manufacturer Fosroc International [49]
	Graphenea	Producers of graphene for various applications	Demonstration of materials as concrete admixtures

1. **Enabling the use of less concrete**

As we have seen, the integration of carbon nanomaterial additives such as GO and CNTs in concrete, together with other additives such as silica fume, nano-alumina or others, brings about significant improvements in compressive and flexural strengths. This, along with the improved density of hydration products and consequent betterment of corrosion resistance in steel reinforced concrete brought about by such additives can enable the design of slimmer, more material-efficient structural elements. The improvement in material performance may thus allow engineers to reduce cross-sectional dimensions in beams, columns, and slabs without compromising structural integrity, thereby lowering material consumption in typical applications [5, 50]. Typical reported enhancements to compressive and flexural strengths are in the region 30–50%, for cement pastes, mortars and concretes. However, a systematic approach to determine the extent to which these enhancements can be translated to reductions in materials usage remains to be developed. Industry adoption will depend not only on cost reductions in nanomaterial synthesis but also updated building codes and regulatory requirement to reflect these advanced material properties.

2. **Enabling concrete with a smaller cement fraction**

Another approach towards realising the benefits of nanomaterial additives in more sustainable concrete construction is to use a smaller cement fraction to achieve required mechanical performances in cemented materials, namely concrete. Again, the relationship is quite complex, and it remains challenging to translate reported mechanical performance enhancements to cement-savings in applied concrete. Based on standard mix designs, it would seem that the addition of small amounts of graphene oxide in the order of 0.02 to 0.05 wt% to concrete, may result in M25 concrete mixes exhibiting strengths comparable to typical M40 mixes [51, 52], which would imply a cement saving in the order of ~ 100 kg per cubic metre of concrete. Assuming ~ 1kg of CO_2 is emitted per kg of cement produced, achieving net carbon reductions through the use of GO to reduce the cement fraction in mix designs, assuming around 700g of GO is required per cubic metre of concrete, requires the GO to have embodied carbon below ~ 140 kg CO_2 equivalent per kg GO produced for any net CO_2 reductions to be realistic. As some reports put the embodied carbon of GO as high as 4800 kg CO_{2e} equivalent per kg GO [24, 53], it is relatively clear that the use of GO towards improving the embodied carbon of concrete necessitates synthesis approaches that are more environmentally friendly than the common Hummer's method. A more detailed LCA of this approach is necessary, and it is important to identify pathways that produce useful strength enhancements at particularly low additive levels.

3. **Enabling higher performance from alternative cement chemistries**

Carbon nanomaterial additives, while mainly studied in the context of Portland Cement (OPC) are also of value towards extracting higher levels of performance from alternative cementitious materials, and are particular interesting towards the modification of low-carbon cements. As we have already seen, synergy between carbon nanomaterials and other SCMs are widely studied, including enhancing the effect of pozzolanic additives such as fly ash, silica fume and nano-alumina. This approach sets the stage for greater versatility in the design of optimised cement mixtures. The use of GO, GNPs or CNTs has also been shown to affect the performance of more sustainable cement alternatives including geopolymers and natural hydraulic lime (NHL). Finding pathways to apply these additives in greener cements sets the stage to broadening the industry uptake of alkali-activated materials, carbonating cements and other binders not reliant on traditional clinkers.

Outlook

While a large volume of research has confirmed the outstanding ability of small levels of carbon nanomaterial additives to impart significant improvement to the mechanical and physical attributes of cementitious materials, translating these improvements to better and more sustainable built environments remains challenging. In particular, achieving net improvements to sustainability outcomes through the use of these materials necessitates progress in our ability to sustainably produce them, and implement them.

Sustainable Methods to Produce Additives

Lifecycle assessments (LCAs) have been conducted to scrutinize a range of graphene and graphene oxide materials. These differ in feedstock, processing, output and by-products. Towards use in concrete, it is evident that GO is of predominant interest, and the reported carbon dioxide equivalent of this material in terms of global warming potential varies widely from 10 to 1000s kg CO_2 equivalent per kg of produced carbon nanomaterial. It is clear that a long path remains to be trodden before we can reliably produce these materials in industrial quantitates in a sustainable manner. There are many factors to consider in developing more sustainable approaches to produce graphene related materials and carbon nanotubes. A wide range of waste and bio-based feedstock have been considered towards the production of graphene materials including lignin, waste tyres and agricultural waste. [24, 54, 55]. Carbon nanotubes are mostly produced from hydrocarbon precursors, but these too can be derived from waste feedstocks [56]. The establishment of scalable methods to produce carbon nanomaterials from sustainable feedstocks with low environmental impact are clearly needed in order for them to be of relevance in materials for the built environment, and various research efforts, under the umbrella of the EU's Graphene flagship project, are increasingly aligned with this [57].

Implementing Carbon Nanomaterials Towards Sustainable Built Environments

While the addition of very low levels of carbon nanomaterials has been widely demonstrated to enhance the strength and performance of cement pastes, mortars, and concretes, significant challenges must be addressed before these innovations can be effectively translated into real-world, sustainable concrete solutions. One approach that stands out towards the effective use of graphene oxide in the enhancement of concrete, is its implementation as a surface treatment agent. This my facilitate the nucleation of dense and hard hydration products on the outer surfaces of concrete, where they are most needed, while reducing the overall amount of nanomaterial consumed [14, 58]. The interplay between various carbon nanomaterials, cement types, aggregates, and supplementary additives creates a highly complex, multidimensional optimization problem that is difficult to navigate. Currently, there is no established industry framework or systematic methodology to guide the integration of carbon nanomaterials with low-carbon cements, supplementary cementitious materials (SCMs) such as silica fume or fly ash, recycled aggregates, and waste-based additives. Developing such a framework is essential to create optimized concrete mixes that not only meet structural performance requirements but also minimize environmental impact. Bridging this gap will require interdisciplinary collaboration, advanced computational modelling, and rigorous lifecycle assessments to ensure that the promise of nanomaterials can be realized in a way that aligns with global sustainability goals.

14.5 Concluding Remarks

The integration of carbon nanomaterials, such as graphene, graphene oxide and carbon nanotubes, into concrete represents an astonishingly effective way to extract higher performance from concrete, offering unparalleled improvements in mechanical strength, durability, and material efficiency. By enabling slimmer structural designs, reducing cement demand, and enhancing alternative binders like geopolymers, these nanomaterials may be able to play a role in addressing the urgent need to decarbonize an industry responsible for around 8% of global CO_2 emissions. However, their commercial viability hinges on overcoming entrenched industry inertia, prohibitive production costs, and unresolved technical challenges relating to the sustainable production and implementation of these materials. Collaborative efforts among researchers, policymakers, and industry leaders are critical to standardize material definitions, incentivize green production pathways, and update regulatory frameworks. As pilot projects and startups demonstrate early successes, the future of nano-carbon enhanced concrete lies in harmonizing its theoretical promise with lifecycle sustainability, ensuring it evolves from a laboratory marvel into a useful material in producing the sustainable concrete needed so badly in our built environments.

References

1. Y. Ohama, Carbon-cement composites. Carbon **27**(5), 729–737 (1989)
2. M. Ali, A. Majumdar, D. Rayment, Carbon fibre reinforcement of cement. Cem. Concr. Res.. Concr. Res. **2**(2), 201–212 (1972)
3. T. Muthukumarana et al., A review on the variation of mechanical properties of carbon fibre-reinforced concrete. Constr. Build. Mater. **366**, 130173 (2023)
4. J. Manso-Morato, et al., Fiber-reinforced concrete and its life cycle assessment: a systematic review. J. Build. Eng., p. 110062 (2024)
5. P. Akarsh et al., Graphene oxide as nano-material in developing sustainable concrete–a brief review. Mater. Today: Proc. **60**, 234–246 (2022)
6. L. Serrano-Lujan et al., Environmental impact of the production of graphene oxide and reduced graphene oxide. SN Appl. Sci. **1**(2), 179 (2019)
7. K. Kelly, W. Billups, Synthesis of soluble graphite and graphene. Acc. Chem. Res. **46**(1), 4–13 (2013)
8. S. Lv et al., Regulation of GO on cement hydration crystals and its toughening effect. Mag. Concr. Res.Concr. Res. **65**(20), 1246–1254 (2013)
9. S. Lv et al., Effect of graphene oxide nanosheets of microstructure and mechanical properties of cement composites. Constr. Build. Mater. **49**, 121–127 (2013)
10. H. Alkhateb et al., Materials genome for graphene-cement nanocomposites. J. Nanomechanics Micromech **3**(3), 67–77 (2013)
11. A. Gholampour et al., Revealing the dependence of the physiochemical and mechanical properties of cement composites on graphene oxide concentration. RSC Adv. **7**(87), 55148–55156 (2017)
12. S. Devi, R. Khan, Effect of graphene oxide on mechanical and durability performance of concrete. J. Build. Eng. **27**, 101007 (2020)

13. H. Zeng et al., Effect of graphene oxide on permeability of cement materials: an experimental and theoretical perspective. J. Build. Eng. **41**, 102326 (2021)
14. A.H. Korayem et al., Graphene oxide for surface treatment of concrete: A novel method to protect concrete. Constr. Build. Mater. **243**, 118229 (2020)
15. M. Inagaki, K. Takai, Terminology for graphene to graphite and for graphene oxide to graphite oxide. Carbon Rep. **1**(2), 59–69 (2022)
16. A.J. Pollard, C.A. Clifford, Terminology: the first step towards international standardisation of graphene and related 2D materials. J. Mater. Sci. **52**(24), 13685–13688 (2017)
17. A. Bianco, et al., *All in the Graphene Family—A Recommended Nomenclature for Two-Dimensional Carbon Materials* (Elsevier, 2013), pp. 1–6
18. A. Jiménez-Suárez, S.G. Prolongo, *Graphene Nanoplatelets* (MDPI, 2020), p. 1753
19. B. Wang, R. Jiang, Z. Wu, Investigation of the mechanical properties and microstructure of graphene nanoplatelet-cement composite. Nanomaterials **6**(11), 200 (2016)
20. G. Chen et al., Graphene nanoplatelets impact on concrete in improving freeze-thaw resistance. Appl. Sci. **9**(17), 3582 (2019)
21. L.J. Jaramillo, R. Kalfat, Fresh and hardened performance of concrete enhanced with graphene nanoplatelets (GNPs). J. Build. Eng. **75**, 106945 (2023)
22. Z. Jiang, O. Sevim, O.E. Ozbulut, Mechanical properties of graphene nanoplatelets-reinforced concrete prepared with different dispersion techniques. Constr. Build. Mater. **303**, 124472 (2021)
23. B.A. Salami et al., Graphene-based concrete: synthesis strategies and reinforcement mechanisms in graphene-based cementitious composites (part 1). Constr. Build. Mater. **396**, 132296 (2023)
24. F. Bahmei, et al., Comparison of environmental impacts in the production of graphene from biomass waste and the Hummers' method. J. Clean. Prod., p. 145145 (2025)
25. R.A. e Silva, et al., Enhanced properties of cement mortars with multilayer graphene nanoparticles. Constr. Build. Mater. **149**, 378–385 (2017)
26. S. Musso et al., Influence of carbon nanotubes structure on the mechanical behavior of cement composites. Compos. Sci. Technol. **69**(11–12), 1985–1990 (2009)
27. P. Zhang et al., Influence of carbon nanotube on properties of concrete: a review. Constr. Build. Mater. **369**, 130388 (2023)
28. A. Carriço et al., Durability of multi-walled carbon nanotube reinforced concrete. Constr. Build. Mater. **164**, 121–133 (2018)
29. S. Arrechea et al., Effect of additions of multiwall carbon nanotubes (MWCNT, MWCNT-COOH and MWCNT-Thiazol) in mechanical compression properties of a cement-based material. Materialia **11**, 100739 (2020)
30. V.C. Li, *Engineered Cementitious Composites (ECC) Material, Structural, and Durability Performance* (2008)
31. N. Bheel et al., Durability behaviours of engineered cementitious composites blended with carbon nanotubes against sulphate and acid attacks by applying RSM modelling and optimization. Buildings **13**(8), 2032 (2023)
32. F. Sanchez, K. Sobolev, Nanotechnology in concrete—a review. Constr. Build. Mater. **24**(11), 2060–2071 (2010)
33. J. Gong et al., Effect of graphene oxide on the properties of ternary limestone clay cement paste. Nanotechnol. Rev.. Rev. **13**(1), 20230222 (2024)
34. M. Murali et al., Utilizing graphene oxide in cementitious composites: a systematic review. Case Stud. Constr. Mater. **17**, e01359 (2022)
35. Z. Pan et al., Mechanical properties and microstructure of a graphene oxide–cement composite. Cement Concr. Compos.Concr. Compos. **58**, 140–147 (2015)
36. S. Chen et al., Carbon nanotube–cement composites: a retrospect. IES J. Part A: Civ. Struct. Eng. **4**(4), 254–265 (2011)
37. S.A. Yaseen, G.A. Yiseen, Z. Li, Elucidation of calcite structure of calcium carbonate formation based on hydrated cement mixed with graphene oxide and reduced graphene oxide. ACS Omega **4**(6), 10160–10170 (2019)

38. A. Chaipanich, R. Rianyoi, T. Nochaiya, The effect of carbon nanotubes and silica fume on compressive strength and flexural strength of cement mortars. Mater. Today: Proc. **4**(5), 6065–6071 (2017)
39. X. Li et al., Incorporation of graphene oxide and silica fume into cement paste: a study of dispersion and compressive strength. Constr. Build. Mater. **123**, 327–335 (2016)
40. D. Lu, X. Shi, J. Zhong, Interfacial nano-engineering by graphene oxide to enable better utilization of silica fume in cementitious composite. J. Clean. Prod. **354**, 131381 (2022)
41. Z. Lu et al., Comparative evaluation on the dispersion and stability of graphene oxide in water and cement pore solution by incorporating silica fume. Cement Concr. Compos.Concr. Compos. **94**, 33–42 (2018)
42. C.S.R. Indukuri, R. Nerella, S.R.C. Madduru, Effect of graphene oxide on microstructure and strengthened properties of fly ash and silica fume based cement composites. Constr. Build. Mater. **229**, 116863 (2019)
43. B. Arpitha, P. Parthasarathy, Effect of nano-alumina and graphene oxide on the behavior of geopolymer composites: a state of the art of review. Mater. Today: Proc. (2023)
44. W. Li et al., Effects of nanoalumina and graphene oxide on early-age hydration and mechanical properties of cement paste. J. Mater. Civ. Eng. **29**(9), 04017087 (2017)
45. C. Venkatesh et al., Synergistic effects of graphene oxide and limestone calcined clay cement on mechanical properties and durability of concrete. J. Build. Pathol. Rehabil. **9**(2), 116 (2024)
46. K. Chintalapudi, R.M.R. Pannem, Enhanced strength, microstructure, and thermal properties of Portland Pozzolana fly ash-based cement composites by reinforcing graphene oxide nanosheets. J. Build. Eng. **42**, 102521 (2021)
47. A. Scott, Improving concrete, one of the world's worst materials. Chem. Eng. News **102** (2024)
48. A. Magazine, *Innovations: Nanotube Concrete* (Concrete Pavement Progress, 2022)
49. G. Cement, *First Graphene Starts Deal with Fosroc to Develop Additives* (2022)
50. A. D'Alessandro et al., Innovative concretes for low-carbon constructions: a review. Int. J. Low-Carbon Technol. **12**(3), 289–309 (2017)
51. P. Walunjkar, M. Bajad, Review on effect of graphene oxide reinforcement on properties of cement concrete. Mater. Today: Proc. **115**, 256–264 (2024)
52. M. Garg, S. Singh, S. Singh, Comparative study of M35 and M40 grades of concrete by ACI, DOE, USBR and BIS methods of mix design using rounded aggregate. Int. Res. J. Eng. Technol. (IRJET) **2**(4), 782–785 (2015)
53. J. Munuera, et al., A review on sustainable production of graphene and related life cycle assessment. 2D Mater. **9**(1), 012002 (2021)
54. S. Temizel-Sekeryan, F. Wu, A.L. Hicks, Global scale life cycle environmental impacts of single-and multi-walled carbon nanotube synthesis processes. Int. J. Life Cycle Assess. **26**, 656–672 (2021)
55. S. Maroufi, M. Mayyas, V. Sahajwalla, Nano-carbons from waste tyre rubber: an insight into structure and morphology. Waste Manage. **69**, 110–116 (2017)
56. A. Bazargan, G. McKay, A review–synthesis of carbon nanotubes from plastic wastes. Chem. Eng. J. **195**, 377–391 (2012)
57. H. Lin et al., Environmental and health impacts of graphene and other two-dimensional materials: a graphene flagship perspective. ACS Nano **18**(8), 6038–6094 (2024)
58. A. Antolín-Rodríguez et al., Statistical study of the effectiveness of surface application of graphene oxide as a coating for concrete protection. Coatings **13**(1), 213 (2023)

Chapter 15
Advantages and Challenges of 3D Geopolymer-Based Concrete in Construction

Lateef N. Assi, Ali Alsalman, Rahman S. Kareem, and Kealy Carter

Abstract Rapid urbanization and industrialization have created significant challenges in the construction industry, including labor shortages, resource consumption, and safety concerns. Traditional concrete production, which relies heavily on Ordinary Portland Cement (OPC), is highly energy-intensive and a major contributor to carbon dioxide (CO_2) emissions, exacerbating climate change. Recently, the integration of three-dimensional printing (3DP) with geopolymer-based concrete has emerged as a promising solution for sustainable construction. This innovative approach offers economic, environmental, and productivity benefits over conventional methods. Studies have shown that 3D Geopolymer-Based Concrete significantly reduces construction costs by lowering labor and material expenses, primarily due to the elimination of formwork. The cost of manufacturing prefabricated bathroom units using 3DP is notably lower compared to traditional precast methods. Additionally, 3DP technology minimizes material waste, reduces construction time, and decreases labor costs, resulting in an overall more efficient and cost-effective process. Environmental assessments indicate that 3DP concrete construction has a better environmental performance than conventional methods, with notable reductions in CO_2 emissions and resource consumption. The use of geopolymer-based concrete further enhances these benefits by reducing the reliance on energy-intensive cement. Overall, the adoption of 3DP and geopolymer-based concrete in construction presents a viable path toward addressing the challenges posed by urbanization and industrialization, offering substantial improvements in cost efficiency, environmental impact, and productivity.

L. N. Assi (✉)
A Terracon Consultants Inc., Tulsa, OK, USA
e-mail: lateeftiger2000@gmail.com

A. Alsalman
Department of Civil Engineering, College of Engineering, University of Basrah, Basrah, Iraq

R. S. Kareem
Department of Structure, Shatrah Technical Institute, Southern Technical University, Shatrah, Dhi Qar, Iraq

K. Carter
Darla Moore School of Business, University of South Carolina, Columbia, SC, USA

D. A. H. Hanaor (ed.), *Sustainable Materials for the Built Environment*, Springer Series in Materials Science 352, https://doi.org/10.1007/978-3-031-97818-0_15

15.1 Introduction

3D printed geopolymer-based concrete (3DPGC) combines the benefits of geopolymer concrete with the advanced construction potential of 3D printing technology. This innovative approach has acquired attention due to its sustainability, proficiency, and ability to address challenges in developed construction. Geopolymer concrete is an eco-friendly alternative to conventional Portland cement concrete with 0% of Portland cement. It is made using industrial by-products like fly ash, metakaolin, and slag, which are activated using alkali solutions; mostly a combination of sodium silicate and sodium hydroxide. The use of geopolymer concrete results in several benefits such as lowering CO_2 emissions, and embodied energy, enhancing durability, and improving the mechanical properties of concrete including compressive and flexural strengths [1, 2]. 3D printing in construction utilizes additive manufacturing techniques to layer geopolymer concrete directly into desired shapes, eliminating the need for molds. This integration offers design flexibility; complex structures can be built with ease and material efficiency; reduced waste during construction, and labor reduction; automation reduces dependency on manual labor. This study investigates the advantages of 3DPGC, for instance, its cost-effectiveness, environmental benefits, and material competence, while discussing practical limitations, including interlayer bonding, anisotropy, and scalability. The findings highlight the potential of this transformative methodology in changing construction practices, emphasizing the need for further technical and regulatory developments.

15.2 Advantages of 3D Geopolymer-Based Concrete

The combination of 3DPC and geopolymer concrete can have several advantages. They are summarized in the next sections.

15.2.1 Economic Benefits

The economic benefits of 3DGPC are substantial, driven mainly by the reduction in labor and material costs. Conventional construction techniques depend primarily on formwork, which accounts for a significant portion of the costs in labor and materials. On the other hand, the 3D printing process eliminates the need for formwork. Typically, the labor cost accounts for more than half of the cost of conventional construction. The cost of materials and equipment accounts for about 20% [3]. On the other hand, the material cost for 3DPC is about half of the total cost, while the labor cost is reduced to about 35% and the equipment cost is about 20% [4]. 3DPC does not require formwork as used for conventional construction; the use of formwork increases the costs of labor, materials, and equipment. Figure 15.1 shows the cost

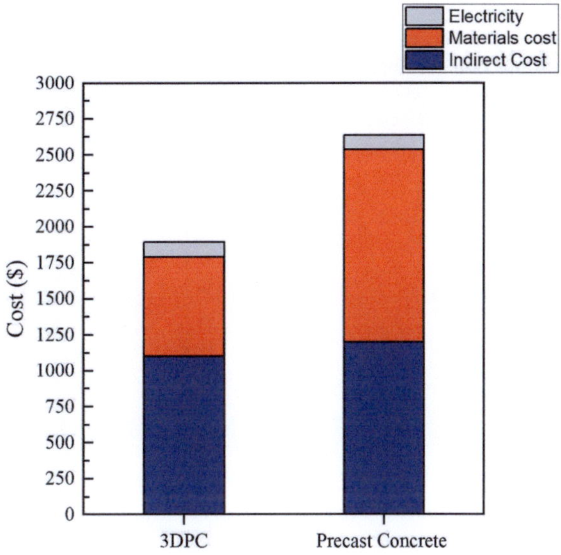

Fig. 15.1 Cost breakdown of 3DP and precast method

breakdown for manufacturing a prefabricated bathroom unit (PBU) by using t 3DPC and precast methods [5]. The indirect costs gained for the production of prefabricated bathroom units by the 3DPC and precast approaches were quite similar. However, the overall cost of the precast method of prefabricated bathroom units increases by 34.1% in comparison to using three-dimensional printing. Moreover, there was a reduction of 13.8% in labor cost by using the 3D printing. In terms of the material (counting raw materials and formwork), the cost of using a precast method is 86.6% higher. In addition to a significant reduction of material usage, the formworks are unnecessary in the fabrication of walls using 3DPC. As a result of reducing material wasted and no over-engineering, the cost of 3DPC may be lower than conventional concrete. The use of 3DPC reduces 30–60% of construction wastes, 50–80% of labor costs, and 50–70% of production time [6]. Due to the rapid development of the 3DP, the equipment cost of 3DP technology will be lower than that of the equipment cost of traditional construction when 3DP technology achieves its large-scale implementation [3]. Overall, the use of 3DP in construction relatively reduces the cost of construction compared to using traditional construction.

15.2.2 Environmental Benefits

3DPGC addresses essential ecological interests by reducing carbon emissions and supporting resource effectiveness. Contrasting conventional concrete, which relies on energy-intensive OPC, geopolymer concrete uses industrial by-products such as

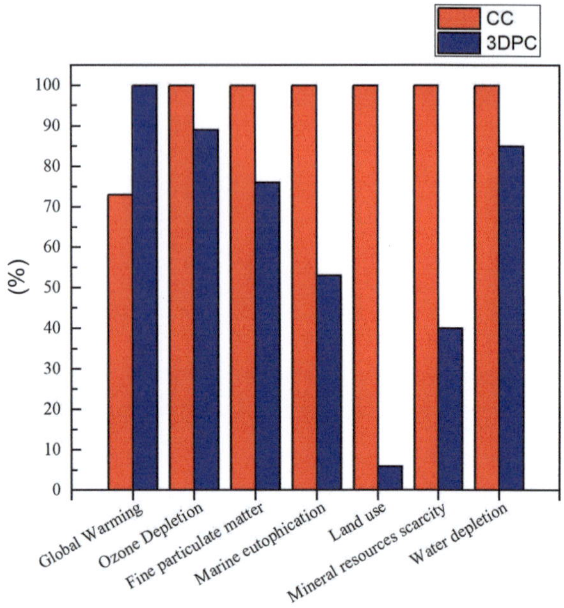

Fig. 15.2 The environmental effects of casting 1 m^2 load-bearing wall by using conventional concrete and 3DP concrete

fly ash, silica fume, and slag, resulting in a smaller environmental footprint. The environmental performance of using 3DP technologies is considerably better than that of conventional concrete construction. Figure 15.2 shows the comparison between the environmental effects of casting a 1 m^2 load-bearing wall by using conventional concrete and 3DPC [7]. The standard Life Cycle Assessment (LCA) method was used to assess the environmental impacts of the construction materials and processes. The figure shows that 3DPC has a general improvement in all categories collectively with 24%, except for the global warming category. This is primarily because of the used amount of cement. Therefore, the use of geopolymer-based concrete has attracted much attention in combination with 3DP techniques [8, 9]. One of the reasons for the poor performance of conventional concrete is the presence of reinforcing steel and cement is highly energy intensive and releases massive amounts of carbon dioxide (CO_2) into the atmosphere [10].

15.2.3 Productivity and Time Efficiency

The 3D printing process improves efficiency by reducing construction time and improving workflow productivity. Automated procedures reduce human error and streamline operations, reducing the duration of construction projects by up to 70% [5]. Figure 15.3 displays the working time and total weight of a prefabricated bathroom unit by using the 3D printed and precast (Weng et al. 2020). The representation exhibits an overall improvement of 48.1% in productivity when three-dimensional

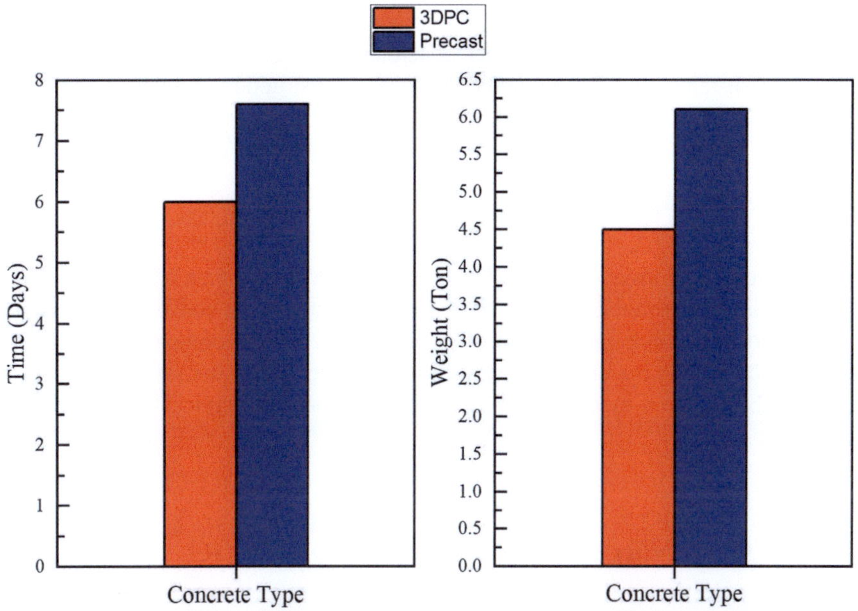

Fig. 15.3 Working time and total weight of a prefabricated bathroom unit by using the 3D printed and precast

printing (3DP) is used for prefabricating the bathroom unit. Moreover, the total weight of the bathroom unit was reduced by 26.2% compared with using the precast way.

15.2.4 Conclusions of Benefits

3DPGC presents an innovational advancement in sustainable construction, offering significant environmental, economic, and productivity advantages over conventional practices. Its use of industrial by-products, such as fly ash and slag, drastically reduces CO_2 emissions compared to OPC, while its precision in material deposition minimizes construction waste by 30–60%. Enhanced Life Cycle Assessment (LCA) metrics highlight its superior environmental performance, though challenges related to alkali activators persist. Economically, 3D printing eliminates the need for formwork and reduces labor costs by 50–80%, material costs by 50%, and production time by 50–70%. Moreover, productivity improvements include a 48.1% increase in efficiency for prefabricated components and a 26.2% reduction in structural weight, streamlining operations, and material handling. By addressing labor shortages and improving safety through reduced on-site manual work, 3D geopolymer concrete not only optimizes resource management but also improves workplace safety. With its ability to significantly cut costs, reduce waste, and deliver innovative architectural designs, 3D-printed geopolymer concrete stands as a pivotal technology for advancing sustainable, efficient, and eco-friendly construction practices.

15.3 Technical Foundations

The structural performance of 3DPGC elements is a crucial area of interest nowadays due to the potential enhancement of this technology in the construction industry and due to the ecological benefits of geopolymer concrete compared to conventional concrete. The major keys of this technology regarding structural performance are printability, fresh mechanical properties, hardened mechanical properties, and durability.

15.3.1 Printability and Fresh Properties of 3DPGC

Printability of the 3DPGC is associated with the recognized properties of layer-by-layer construction without the need for formwork. Generally, the printability of 3DPC refers to the extrusion of the concrete mixture easily through a nozzle and accumulated layer by layer while maintaining the shape and structural integrity of the pored elements (Le et al., 2012). The printability of 3DPC may include extrudability, buildability, and rheological characteristics. Extrudability of 3DPGC is the ease with which the concrete mixture can be extruded successfully without blocking. Studies have shown that there are several factors affecting the extrudability of 3DPGC such as rheological properties, particle size distribution, binder content, alkaline activating solution, and curing conditions [11, 12]. The extrusion process of any type of 3DPCs is required to be homogeneous without segregation or cracking [13, 14]; this process is significantly influenced the in case of 3DPGC due to the high viscosity of the activation solution. Based on the extrusion method of printed concrete, the fresh properties influence the printability of the pored elements. Typically, there are two common methods of the extrusion of 3D-printed concrete; screw and ram extrusions. Table 15.1 states the comparison between them [15–17]. On the other hand, buildability is the ability of the concrete to maintain its shape and support without failure or build up the desired element with tolerable deformation [18, 19]. In other words, buildability refers to the capability of the concrete to sustain its weight in addition to the weight of the concrete from the above layers without collapsing. Commonly, the thickness of a single layer of printable concrete is small; typically ranging from 1 to 10 cm to minimize the effect of the preliminary gravity stress [19]. Accordingly, the shear stress yields by gravity must be lower than the yield stress of cementitious materials to preserve the shape of the printed concrete. Hence it is essential to control the thickness of layers to prevent the risk of collapsing due to gravity. Studies have shown that there are several types of failure of 3DPC of multiple layers; yield failure of a

Table 15.1 Common extrusion methods of 3DPC

Extrusion	Characteristics	Advantages
Screw	Employs an Archimedean screw to push the concrete mixture through a nozzle, which allows for a steady flow of cementitious materials	• Appropriate for large-scale and continuous projects • Since the rotating screw constantly mixes the concrete the homogeneity and uniformity of the mixture consistency is ensured • Pressure is regulated which improves the precision and quality of the printed layers
Ram	Ram extrusion utilizes a piston (ram) to force the concrete through a nozzle, making it appropriate for batching processing	• It can apply higher pressures, which makes it appropriate for denser and more viscous mixtures • With fewer moving parts than screw extrusion, the maintenance might be minimal • Facilitates precise control over the quantity of material extruded per batch

single layer, yield failure of the bottom layer, failure triggered by geometrical nonconformity, and buckling failure [19]. It is well known that as the number of layers increases, the probability of failure of the bottom layer due to compression stress is higher [20]. Fresh properties of 3DPGC are critical for guaranteeing the quality and structural performance of the printed concrete elements. Rheology is the typical term used in the literature to describe fresh properties, which can be defined as the process of assessment of the viscosity, elasticity, and plasticity of cementitious materials under shear stress [21]. It was found that the rheological properties of 3DPGC were influenced significantly by the composition of the alkaline activating solution. It can be shown in Fig. 15.4 that the silicate-to-sodium ratio in the alkaline solution (Si/Na) ratio of the activator affected the rheological properties of the geopolymer; the yield stress and structural rebuilding ability enhanced with the reduction of the Si/Na ratio [22]. Reference [23] investigated the open time of 3DPGC which refers to the period during which the freshly mixed concrete stays workable. They stated that mixtures with potassium-based activators had shorter open time compared to the mixtures with sodium-based activating solutions.

15.3.2 Mechanical Properties

The mechanical characteristics of 3DPGC are significant as they affect directly its practical applications in the construction industry. These properties include compressive, tensile, and flexural strengths, elastic modulus, and durability. The early compressive strength is required to be developed shortly after the concrete has been printed which allows the structure to support loads. In addition, the capacity

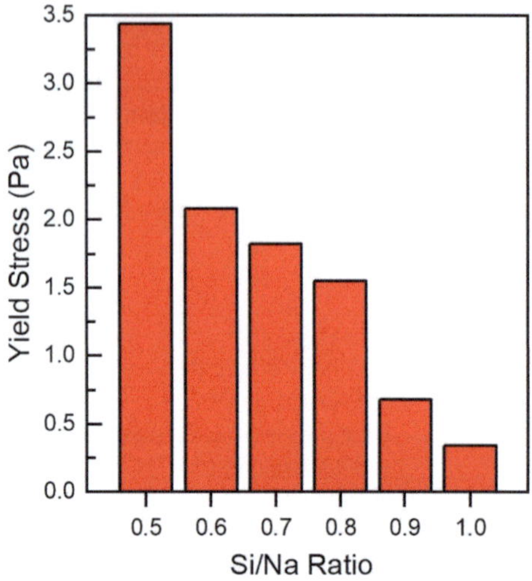

Fig. 15.4 Relationship between yield stress and Si/Na ratio

of the fresh concrete layers to adhere to each other and produce a cohesive structure is crucial for the integrity of the printed elements since they are directly related to the interlayer bond performance of the concrete [24]. Therefore, the development of these properties before the setting time requires more consideration compared to conventional concrete since this technology utilizes no formwork and depends on the strength of layer-by-layer construction; the early-age mechanical properties affect the buildability of the printed elements [15]. Reference [25] conducted a study on the mechanical properties of 3DPGC and they found the loading direction of the tested elements affected the compressive and flexural strength of the printed concrete. It was determined that 3DPGC samples exhibit greater compressive strength when loaded in the y direction compared to the x and z directions (Fig. 15.5). This behavior can be understood by comparing it to the micro-mechanics of composite materials. When loaded perpendicular to the layers, the load is directly transferred and distributed uniformly across the cross-section. However, when loaded in the transverse direction, the layers tend to slip or separate from each other, resulting in lower strength in the specimen. A typical value of the compressive of 3DPGC was reported in the range of 20–60 MPa depending on the mixture design and curing regimen. For tensile and flexural strengths, the values are in the range of 2 MPa to 4 MPa and 3 MPa to 10 MPa, respectively. However, elastic modulus values ranging from 10 to 30 GPa were reported in the literature [26, 27]. The flexural strength of printed concrete varies with testing orientation and can be higher than the mold-cast samples [15]. Previous studies have shown that the compressive strength of 3DPC can be higher when the application of the load is parallel to the printed layers. On the other hand, the tensile strength is lower when the load is perpendicular to the layers due to the weaker interlayer cohesiveness. Moreover, the flexural strength can vary substantially based on the loading directions that are relative to the orientation of the printed layers [17]. Overall, the anisotropy of 3DPC can be affected by several aspects such as the

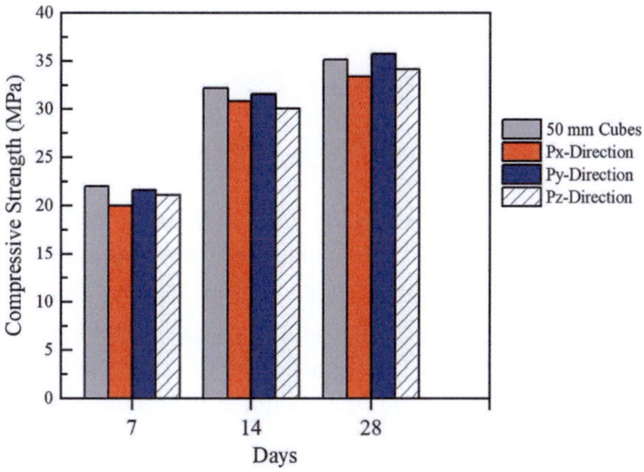

Fig. 15.5 Compressive strength of 3DPGC based on loading directions

printing process and devices, mixture proportions, and ambient conditions such as humidity and temperature. The major concerns of 3DPC regarding structural performance are the interlayer bond strength and anisotropy. The interlayer bond strength is influenced by several factors; which include the interval time, the speed of the printing head, and the height of the print nozzle, which have significant impacts on the interface bond strength of 3DPC [15]. The anisotropy of 3DPGC describes the variation of the mechanical properties of the concrete based on the layer-to-layer construction process. In contrast to conventional concrete, which is typically cast to construct a uniform material, 3DPC demonstrates distinct characteristics along different directions because of how each layer is placed and bonds with the previous one.

15.3.3 Mixture Proportions

To ensure the performance of 3DPGC, the mixture proportions require significant consideration. Mixture proportions include fly ash, ground granulated blast furnace slag, metakaolin, or a combination of them as the base materials for geopolymer concrete. An alkaline activating solution is required to initiate the polymerization process. High-range water-reducing admixtures (HRWRA) are fundamental to enhancing flowability and reducing water content while maintaining the desired workability. To control the viscosity of the concrete, viscosity modifier admixtures are essential; typically, hydroxypropyl methylcellulose (HPMC) is used. Aggregate is also required to increase the stiffness of the concrete; normally conventional sand, and recycled concrete sand, and manufactured with a well-graded particle size distribution to optimize packing density and rheology. On the other hand, coarse aggregate normally is excluded to ensure nozzle compatibility and avoid blocking [28, 29].

15.3.4 Reinforcement

Due to the brittle failure modes of 3DPGC, which result from the low tensile-to-compressive strength ratio, the incorporation of reinforcement is necessary to modify the structural performance of the printed concrete. However, incorporating reinforcement continuously into 3DPGC is challenging due to the layer-by-layer printing process [24]. Conventional steel reinforcement may be incorporated into 3DPC either during or after printing. On the other hand, fiber reinforcements (i.e. steel, polypropylene, glass, carbon) can be added to the mixture design as in fiber-reinforced concrete to enhance the ductility and reduce the propagation of cracks; it was found that steel fiber is not suitable for 3PDC because the bulging fibers can obstruct the adhesion between interface layers [11]. The typical fiber content ranges from 0.5–2% by volume fraction of the mixture. The integration of mesh or grid materials within the layers of 3DPC concrete can also increase the tensile strength of the concrete and ensure a uniform distribution of the applied loads. Moreover, employing post-tensioning practices in 3DPC concrete structures could improve their load-bearing capability and mitigate cracks [30–33].

15.3.5 Summary of Technical Challenges

Rheological properties, including yield stress and viscosity, are fundamental for asserting the shape of printed layers while accepting smooth bumps. However, achieving an optimum balance is challenging, as the high viscosity of geopolymer mixtures frequently indicates difficulties in extrusion and flow. Studies have shown that the Si/Na ratio; which is a critical ratio in the alkaline solution to ensure the performance of geopolymer concrete, significantly influences rheological properties; lower ratios enhance yield stress and structural rebuilding ability. Adjustments to rheology must also account for temperature and humidity variations, which can impact material behavior during printing. The use of high-viscosity alkali solution for geopolymer concrete and fine aggregates poses challenges in maintaining homogeneity and avoiding blockages. Screw-based extrusion systems have been shown to mitigate some of these issues by providing continuous mixing during the extrusion process, nevertheless, this requires accurate control of particle size distribution and binder content of the designed mixtures. The open time of 3DPGC, which is the interval during which the mixture stays workable, is an important challenge. Potassium-based alkali solutions tend to decrease the open time compared to sodium-based ones, leading to quicker settings and decreased workability during printing. This limitation limits the timeframe for achieving strong interlayer bonding and homogenous layer deposition. Buildability, which is the ability of the 3D-printed concrete to sustain the weight of sequential layers without collapsing, is a significant concern for 3DPGC. Studies suggest that inadequate yield stress in fresh conditions may lead to geometric alterations or collapse during the printing process. Additionally, maintaining a uniform mixture during the extrusion process is necessary for inhibiting segregation and guaranteeing consistent mechanical properties. However, the incorporation of high-viscosity activation solution and fine aggregates aggravates segregation risks, specifically during persistent printing periods or under variable environmental conditions. The inclusion of viscosity-modifying admixtures can improve the stability of fresh geopolymer concrete. Conversely, excessive use of such admixture can increase the mixture's viscosity, complicating extrusion and printability. Reinforcement incorporation, especially fibers is another challenge that can affect the nozzle compatibility, uniform distribution of the fibers, interference with layer bonding, and cost.

15.4 Reliability Process and Limitations

The reliability process of 3DPGC in general is still questionable since the application of this technology is contemporary and not standardized in codes and specifications. However, studies have proven the reliability of this technology such as accuracy and customization, reducing waste of materials, and the speed of the construction process. 3D printing techniques enable precise and sophisticated designs that are challenging to attain with conventional construction methods. The layer-by-layer methodology reduces waste compared to traditional poring procedures. In addition, structures can be printed more rapidly than they can be built using conventional approaches that involve formwork [30, 33]. However, studies have also stated the limitations of this technology. The printing procedure might lead to weaker interlayer bonds, yielding anisotropic mechanical properties and material inconsistencies since the rheological and setting characteristics of the concrete mixture are challenging. Other limitations can be the cost and the Size constraints; the preliminary investment for 3D printing equipment is substantial although these costs can be balanced out over time by savings in labor and materials. Difficulties also arise when attempting to expand 3D printing technology for large-scale construction because of restrictions in equipment size and logistical concerns [34, 35]. The durability of the 3DPGC still needs to be investigated entirely, such as the freezing and thawing cycles and exposure to deicing salt and sulfate attacks. Another limitation of 3DPGC is that alkali-activated concrete could have short open times and require careful handling to avoid early setting.

15.5 Regulatory Challenges

The regulatory challenges facing 3DPGC involve the development of new codes and standards that focus on these types of structures to achieve the desired safety and structural performance. This involves developing the aspects regarding rheology and workability, load-bearing capacity, type of reinforcement, and long-term serviceability under different environmental conditions. Quality control and certification of the materials are also required. Testing of the materials utilized in 3D printing procedures to satisfy established standards for strength, durability, and ecological impact is significant. Therefore; guaranteeing reliable material quality and performance for geopolymer concrete is essential for regulatory compliance [36]. Considering the possible risks related to material handling, equipment operation, and worker safety is another point that needs to be considered regarding the regulations of 3DPGC. In addition, environmental influences such as sources of materials, management of wastes (i.e., use of by-product materials), and CO_2 emissions generated from printing technology should be considered. However, geopolymer concrete has proven to be eco-friendly compared to conventional concrete; less energy is required and less CO_2 is released into the environment [1]. In short, these regulatory challenges emphasize that there is a need for cooperation between the industry sector, regulatory organizations, and research institutions such as universities to develop reliable schemes for the utilization of this innovative technology.

15.6 Conclusions

The integration of 3D printing technology with geopolymer-based concrete offers a transformative approach to addressing the challenges of modern construction. By leveraging the environmental, economic, and productivity benefits of this innovative technology, the construction industry can reduce its reliance on traditional, resource-intensive methods. The reduction in CO_2 emissions, construction waste, and material usage underpins the sustainability of 3D geopolymer concrete, while its ability to lower costs and improve construction efficiency highlights its economic viability. Despite these advantages, challenges such as ensuring structural integrity, addressing anisotropy, and developing standardized regulatory frameworks remain critical areas for further research and development. Advances in reinforcement techniques, optimization of mixture designs, and the refinement of printing processes will be instrumental in overcoming these obstacles. Furthermore, collaboration among industry stakeholders, regulatory bodies, and researchers will be essential to establishing comprehensive guidelines and promoting widespread adoption. In conclusion, 3D geopolymer-based concrete stands at the forefront of sustainable construction technologies, offering a promising path toward a more efficient, environmentally conscious, and innovative construction industry.

References

1. A. Alsalman, L.N. Assi, R.S. Kareem, K. Carter, P. Ziehl, Energy and CO_2 emission assessments of alkali-activated concrete and Ordinary Portland Cement concrete: a comparative analysis of

different grades of concrete. Clean. Environ. Syst. **3**, 100047 (2021). https://doi.org/10.1016/j.cesys.2021.100047
2. L.N. Assi, A. Alsalman, R.S. Kareem, K. Carter, P. Ziehl, Y. Alhamadani, Why sustainable concrete cannot penetrate concrete markets. J. Build. Eng. **91**, 109487 (2024). https://doi.org/10.1016/j.jobe.2024.109487
3. J. Zhang, J. Wang, S. Dong, X. Yu, B. Han, A review of the current progress and application of 3D printed concrete. Compos. A Appl. Sci. Manuf. **125**, 105533 (2019). https://doi.org/10.1016/j.compositesa.2019.105533
4. G. De Schutter, K. Lesage, V. Mechtcherine, V.N. Nerella, G. Habert, I. Agusti-Juan, Vision of 3D printing with concrete—technical, economic and environmental potentials. Cem. Concr. Res. **112**, 25–36 (2018). https://doi.org/10.1016/j.cemconres.2018.06.001
5. Y. Weng, M. Li, S. Ruan, T.N. Wong, M.J. Tan, K.L. Ow Yeong, S. Qian, Comparative economic, environmental and productivity assessment of a concrete bathroom unit fabricated through 3D printing and a precast approach. J. Clean. Prod. **261**, 121245 (2020). https://doi.org/10.1016/j.jclepro.2020.121245
6. D. Weinstein, P. Nawara, Determining the applicability of 3D concrete construction (contour crafting) of low income houses in select countries (2015). https://hdl.handle.net/1813/70743. Accessed 22 June 2024
7. H. Alhumayani, M. Gomaa, V. Soebarto, W. Jabi, Environmental assessment of large-scale 3D printing in construction: a comparative study between cob and concrete. J. Clean. Prod. **270**, 122463 (2020). https://doi.org/10.1016/j.jclepro.2020.122463
8. B. Panda, M.J. Tan, Experimental study on mix proportion and fresh properties of fly ash based geopolymer for 3D concrete printing. Ceram. Int. **44**, 10258–10265 (2018). https://doi.org/10.1016/j.ceramint.2018.03.031
9. M. Xia, J. Sanjayan, Method of formulating geopolymer for 3D printing for construction applications. Mater. Des. **110**, 382–390 (2016). https://doi.org/10.1016/j.matdes.2016.07.136
10. G. Habert, E. Denarié, A. Šajna, P. Rossi, Lowering the global warming impact of bridge rehabilitations by using ultra high performance fibre reinforced concretes. Cement Concr. Compos. **38**, 1–11 (2013). https://doi.org/10.1016/j.cemconcomp.2012.11.008
11. S. Al-Qutaifi, A. Nazari, A. Bagheri, Mechanical properties of layered geopolymer structures applicable in concrete 3D-printing. Constr. Build. Mater. **176**, 690–699 (2018). https://doi.org/10.1016/j.conbuildmat.2018.04.195
12. S. Muthukrishnan, S. Ramakrishnan, J. Sanjayan, Effect of alkali reactions on the rheology of one-part 3D printable geopolymer concrete. Cement Concr. Compos. **116**, 103899 (2021). https://doi.org/10.1016/j.cemconcomp.2020.103899
13. K. El Cheikh, S. Rémond, N. Khalil, G. Aouad, Numerical and experimental studies of aggregate blocking in mortar extrusion. Constr. Build. Mater. **145**, 452–463 (2017). https://doi.org/10.1016/j.conbuildmat.2017.04.032
14. A. Perrot, D. Rangeard, V.N. Nerella, V. Mechtcherine, Extrusion of cement-based materials—an overview. RILEM Tech. Lett. 3, 91–97 (2019). https://doi.org/10.21809/rilemtechlett.2018.75
15. T.T. Le, S.A. Austin, S. Lim, R.A. Buswell, A.G.F. Gibb, T. Thorpe, Mix design and fresh properties for high-performance printing concrete. Mater. Struct. **45**, 1221–1232 (2012). https://doi.org/10.1617/s11527-012-9828-z
16. Z. Malaeb, F. AlSakka, F. Hamzeh, 3D concrete printing, in *3D Concrete Printing Technology* (Elsevier, 2019), pp. 115–136. https://doi.org/10.1016/B978-0-12-815481-6.00006-3
17. R.J.M. Wolfs, F.P. Bos, T.A.M. Salet, Hardened properties of 3D printed concrete: the influence of process parameters on interlayer adhesion. Cem. Concr. Res. **119**, 132–140 (2019). https://doi.org/10.1016/j.cemconres.2019.02.017
18. V.N. Nerella, V. Mechtcherine, Studying the printability of fresh concrete for formwork-free concrete onsite 3D printing technology (CONPrint3D), in *3D Concrete Printing Technology* (Elsevier, 2019), pp. 333–347. https://doi.org/10.1016/B978-0-12-815481-6.00016-6
19. N. Roussel, Rheological requirements for printable concretes. Cem. Concr. Res. **112**, 76–85 (2018). https://doi.org/10.1016/j.cemconres.2018.04.005

20. N. Roussel, P. Coussot, "Fifty-cent rheometer" for yield stress measurements: from slump to spreading flow. J. Rheol. **49**, 705–718 (2005). https://doi.org/10.1122/1.1879041
21. J. Kruger, S. Zeranka, G. Van Zijl, An ab initio approach for thixotropy characterisation of (nanoparticle-infused) 3D printable concrete. Constr. Build. Mater. **224**, 372–386 (2019). https://doi.org/10.1016/j.conbuildmat.2019.07.078
22. D.-W. Zhang, D. Wang, X.-Q. Lin, T. Zhang, The study of the structure rebuilding and yield stress of 3D printing geopolymer pastes. Constr. Build. Mater. **184**, 575–580 (2018). https://doi.org/10.1016/j.conbuildmat.2018.06.233
23. S.H. Bong, B. Nematollahi, A. Nazari, M. Xia, J.G. Sanjayan, Fresh and hardened properties of 3D printable geopolymer cured in ambient temperature, in *First RILEM International Conference on Concrete and Digital Fabrication—Digital Concrete 2018*, ed. by T. Wangler, R.J. Flatt (Springer International Publishing, Cham, 2019), pp. 3–11. https://doi.org/10.1007/978-3-319-99519-9_1
24. S. Hou, Z. Duan, J. Xiao, J. Ye, A review of 3D printed concrete: performance requirements, testing measurements and mix design. Constr. Build. Mater. **273**, 121745 (2021). https://doi.org/10.1016/j.conbuildmat.2020.121745
25. B. Panda, S.C. Paul, L.J. Hui, Y.W.D. Tay, M.J. Tan, Additive manufacturing of geopolymer for sustainable built environment. J. Clean. Prod. **167**, 281–288 (2017). https://doi.org/10.1016/j.jclepro.2017.08.165
26. J.L. Provis, Geopolymers and related alkali-activated materials. Annu. Rev. Mater. Res. **44**, 299–327 (2014). https://doi.org/10.1146/annurev-matsci-070813-113515
27. J.S.J. Van Deventer, J.L. Provis, P. Duxson, Technical and commercial progress in the adoption of geopolymer cement. Miner. Eng. **29**, 89–104 (2012). https://doi.org/10.1016/j.mineng.2011.09.009
28. L.N. Assi, E. Eddie Deaver, P. Ziehl, Effect of source and particle size distribution on the mechanical and microstructural properties of fly ash-based geopolymer concrete. Constr. Build. Mater. **167**, 372–380 (2018). https://doi.org/10.1016/j.conbuildmat.2018.01.193
29. F. Bos, R. Wolfs, Z. Ahmed, T. Salet, Additive manufacturing of concrete in construction: potentials and challenges of 3D concrete printing. Virtual Phys. Prototyping **11**, 209–225 (2016). https://doi.org/10.1080/17452759.2016.1209867
30. R.A. Buswell, W.R. Leal De Silva, S.Z. Jones, J. Dirrenberger, 3D printing using concrete extrusion: a roadmap for research. Cem. Concr. Res. **112**, 37–49 (2018). https://doi.org/10.1016/j.cemconres.2018.05.006
31. S.A. Khan, M. Jassim, H. Ilcan, O. Sahin, İR. Bayer, M. Sahmaran, M. Koc, 3D printing of circular materials: comparative environmental analysis of materials and construction techniques. Case Stud. Constr. Mater. **18**, e02059 (2023). https://doi.org/10.1016/j.cscm.2023.e02059
32. T. Marchment, J. Sanjayan, Mesh reinforcing method for 3D concrete printing. Autom. Constr. **109**, 102992 (2020). https://doi.org/10.1016/j.autcon.2019.102992
33. T.A.M. Salet, Z.Y. Ahmed, F.P. Bos, H.L.M. Laagland, Design of a 3D printed concrete bridge by testing. Virtual Phys. Prototyping **13**, 222–236 (2018). https://doi.org/10.1080/17452759.2018.1476064
34. R.A. Buswell, R.C. Soar, A.G.F. Gibb, A. Thorpe, Freeform construction: mega-scale rapid manufacturing for construction. Autom. Constr. **16**, 224–231 (2007). https://doi.org/10.1016/j.autcon.2006.05.002
35. M.T. Souza, I.M. Ferreira, E. Guzi De Moraes, L. Senff, A.P. Novaes De Oliveira, 3D printed concrete for large-scale buildings: an overview of rheology, printing parameters, chemical admixtures, reinforcements, and economic and environmental prospects. J. Build. Eng. **32**, 101833 (2020). https://doi.org/10.1016/j.jobe.2020.101833
36. Y. Pan, Y. Zhang, D. Zhang, Y. Song, 3D printing in construction: state of the art and applications. Int. J. Adv. Manuf. Technol. **115**, 1329–1348 (2021). https://doi.org/10.1007/s00170-021-07213-0

Chapter 16
Sustainable Alternatives to Steel in Concrete Reinforcement

Jake Cardillo, Sean Blackley, Lex Hartnett, Matthew Horan, Scott Rayburg, and Melissa Neave

Abstract Due to environmental and global warming concerns, sustainable alternatives to traditional materials in the building industry are rapidly gaining in popularity. Historically, the construction industry has relied heavily on steel reinforcement in concrete, however the emissions associated with the production of steel are large, and steel production will need to be drastically reduced (or green steel technologies will need to become commercially available) to meet net zero goals. One way to reduce steel use in construction is to replace steel reinforcement in concrete with sustainable alternatives. While a number of concrete reinforcement alternatives have been proposed and researched, no study has explored which of these are most suitable as replacements for steel across a range of structural, cost and environmental indicators. This study aims to investigate and compare traditional steel reinforcement with emerging sustainable reinforcement types used in concrete, including recycled plastic reinforcement, carbon fibre reinforcement, and natural reinforcement types (bamboo and basalt). To accomplish this, a series of nested systematic literature reviews were conducted on each of these reinforcement types with a focus on strength, cracking control, cost, and carbon emissions. The results of this study show that all of the alternatives to steel for reinforcement have potential applications in industry. However, they all lack the flexural strength of steel, and so are unsuited to situations where flexural loads are limiting. In addition, carbon fibre was shown to be an unsustainable alternative to steel, as its manufacture requires higher carbon emissions than steel, while recycled plastic and natural reinforcement options were shown to be similar or more sustainable than steel. Finally, the study identified a key weakness in the existing literature, in that there are few studies that directly compare different reinforcement types against one another using the same sample characteristics. If alternative reinforcements are to gain popularity, this issue will need to be

J. Cardillo · S. Blackley · L. Hartnett · M. Horan · S. Rayburg
School of Engineering, Swinburne University or Technology, Hawthorn, VIC, Australia
e-mail: scott.rayburg@swinburne.edu.au

M. Neave (✉)
Centre for Urban Research, School of Global Urban and Social Studies, RMIT University, Melbourne, VIC, Australia
e-mail: melissa.neave@rmit.edu.au

D. A. H. Hanaor (ed.), *Sustainable Materials for the Built Environment*, Springer Series in Materials Science 352, https://doi.org/10.1007/978-3-031-97818-0_16

resolved to improve the clarity of the relative merits of each reinforcement type and what applications they are most suited to.

16.1 Introduction

Global warming, resulting from anthropogenic greenhouse gas emissions, has seen increases in atmospheric temperatures that are driving numerous global responses, including rising sea levels and more frequent natural disasters [1]. Although many human activities contribute to these increasing emissions, steel manufacturing has been identified as the most energy consuming and CO_2 emitting industry in the world, contributing to more than 7% of global annual carbon emissions [2, 3]. Steel has multiple uses but over half of global steel production is consumed by building and infrastructure, with one of its most common applications being to reinforce concrete. Indeed, since the mid-nineteenth century, carbon intensive steel has been paired with concrete to form fundamental structural members that are high in both strength and durability [4] and that have become one of the most highly regarded construction materials currently utilised [5]. As steel usage in construction is projected to increase in coming years [6, 7], the construction industry needs to consider utilising more sustainable building materials to reduce its carbon footprint, with considerations of alternatives to steel for reinforcement in concrete representing an important component of this.

Steel reinforcement is an iron-carbon alloy hot rolled into bars, 6–50 mm in diameter [8]. Since its adoption in the mid-nineteenth century, carbon steel has become the most common material for reinforcing concrete structures. The high tensile strength and ductility, enhanced constructability, economic availability, and thermal expansion properties allow for its synergistic relationship with reinforced concrete, a composite construction material unbounded in its shape and size [9]. Despite its desirable properties and widespread use, steel reinforcement is a product of the iron and steel industry, the largest energy consuming and carbon dioxide emitting manufacturing sector in the world [10] and steel reinforcement production processes contribute a significant portion of the embodied carbon emissions of construction projects [11]. A potential benefit of steel reinforcement is that it is highly recyclable, a property which provides an opportunity to utilise more efficient production technologies. Unfortunately, the time-lag of scrap availability restricts the feasibility for greener (recycled) production methods being able to meet the world's growing demand for steel reinforcement and there remains a reliance on virgin (primary) sourced materials [12]. This means that the construction industry needs to go beyond steel recycling and needs to do more than wait for the development of net-zero carbon "green steel" [13] if it is going to address the immediate issues surrounding its current high carbon emissions.

Identification of the high levels of carbon dioxide (CO_2) being emitted by the steel manufacturing industry has led to investigations into potential sustainable concrete reinforcement alternatives [14]. Studies into these alternatives indicate they can be grouped into three categories: recycled plastic, recycled carbon fibres and natural materials [15–20]. Whilst these studies have revealed capacity for each of these alternatives to provide reinforcement to concrete, there are yet to be any investigations that compare the performance of these sustainable reinforcement alternatives to not just traditional steel, but between each other to determine the most appropriate technology for the construction industry to adopt. Thus, the aim of this study is to redress this limitation in the existing literature and to provide clarity for an industry that is wanting to shift to more sustainable practices.

Identifying sustainable alternatives for concrete reinforcement is not a new concept. Research into the development of recycled plastics for reinforcing concrete dates to the 1960s [21], while carbon fibre reinforced polymers (CFRPs), developed using recycled carbon fibres, began being utilised for concrete reinforcement in the early 1970s [22]. Even more recently, the availability of certain recycled materials has prompted further research into the possibility of natural reinforcements for concrete that include basalt, bamboo and other plant fibres. However, the literature indicates that the performances of these sustainable concrete reinforcement alternatives show considerable variation. For example, studies indicate that the use of plastics and carbon fibres in concrete reinforcement enhances the mechanical characteristics of plain concrete, such as compressive, tensile, flexural, and impact strengths, and provides increased resistance to cracking [23, 24]. Likewise, natural reinforcements such bamboo provide a tensile strength that is almost identical to steel [25]. However, there are limitations to most of the alternatives to using steel for concrete reinforcement that suggest they are suitable for use only as non-structural members [20]. Thus, although there are possible alternatives to steel for concrete reinforcement there is also a need to consider the extent to which they can legitimately replace the current product.

Several types of plastic have been used in the construction industry to reinforce concrete, with different applications reflecting their inherent properties. For example, Glass Fiber Reinforced Polymer (GFRP) rods or sheets are resistant to corrosion and are therefore often used in environments where traditional steel reinforcement would corrode (e.g., bridges or marine decks) [26]. Carbon Fiber Reinforced Polymer (CFRP), on the other hand, has a high strength-to-weight ratio and is used to wrap concrete columns and beams, increasing their load capacity and seismic resistance [27]. Polyvinyl Chloride (PVC) and High-Density Polyethylene (HDPE) provide lightweight alternatives to using a coarse aggregate in concrete mixtures, potentially reducing pressure on the sourcing of natural materials [28]. These plastic-based reinforcements enhance the overall performance and longevity of concrete structures while reducing maintenance costs and improving resistance to environmental stressors.

Using plastic as reinforcement in concrete offers several benefits beyond its capacity to reduce the need for emissions intensive steel. Recycled plastic can be cost-effective and adaptable [29] and its redeployment for concrete reinforcement decreases the volume of plastic material that is sent to landfill [30–32]. These benefits indicate that reprocessed plastics have several benefits that make them appealing and may result in them being implemented more frequently as reinforcement in concrete in the future [33].

Of the plastics mentioned above, GFRP, PVC and HDPE can all be readily sourced from recycled (post consumer and industrial) materials. CFRP, however, is less likely to incorporate recycled plastic because the thermosetting materials used in their production create permanent chemical bonds that are often not recoverable after use. Currently, only 2% of carbon fibre (CF) products are recycled and reused each year in the UK [34], with the remaining being sent to landfill or incineration factories. Providing a market for decommissioned CF waste products represents an environmentally friendly approach that is also cost effective and energy efficient [35] and these products have some potential to work as alternatives for steel reinforcement in concrete. Indeed, several studies have proven that the addition of CF and CFRPs can greatly enhance the mechanical properties of plain concrete [1, 19] and point to the potential for CFRPs to provide reinforcement for concrete. However, ongoing work is required to develop systems that will more effectively enable the plastic matrix in CFRPs to be recycled.

In addition to using recycled plastic and CF products for concrete reinforcement, several naturally occurring elements are also available. This approach of natural concrete reinforcement employs a variety of natural materials such as different woods, stone, and plants. Making use of these materials will advance the construction industry's sustainable development agenda while also lowering waste production, conserving resources, and fostering a greener society. However, more research is required to ensure these materials are able to survive the intense strain of the structures they need to be used in. There are several natural materials that have been used for concrete reinforcement. The most common is bamboo, which has strong fibres that exhibit an impressive compressive strength, which is approximately twice that of concrete. Furthermore, its tensile strength is almost identical to that of steel [25]. Basalt, on the other hand, is a naturally occurring volcanic rock that is has extraordinary chemical and thermal stabilities and its production generates no harmful gases or waste residue [36]. Finally, plant fibres obtained from various sources, including jute, hemp, flax, cotton, and sisal, are extremely cost effective while exhibiting a large range of mechanical, sustainable, thermal, and physical properties [37]. Thus, the use of natural products to reinforce concrete has immense potential to help the construction industry reduce its carbon emissions into the future.

This research presents an in-depth analysis of existing literature examining the possibilities of using alternative sustainable materials in concrete as a viable replacement for steel. The study will consider three dimensions for each concrete reinforcement material: (1) its structural performance, (2) its environmental performance, and (3) its cost. Providing information about these dimensions will enable more informed decision making for those in the construction industry who are seeking more sustainable alternatives to their building practices.

16.2 Methods

To achieve the research aims, this study utilizes a series of nested systematic literature reviews that explore current research into concrete reinforcement. In total four systematic literature reviews were conducted (Fig. 16.1). These investigated the existing literature examining (1) traditional steel, (2) recycled plastic, (3) recycled carbon fibre and (4) natural materials. For each reinforcement type, studies that investigate structural performance, environmental performance, cost and suitability for different applications were included. This provided a comprehensive analysis of existing knowledge and research into the use of various materials for concrete reinforcement, while enabling the identification of possible gaps in the literature that might warrant further consideration [38].

The four systematic literature reviews were undertaken using a series of keywords entered into the large, multidisciplinary database known as Scopus. As Table 16.1 indicates, these keywords began with a high-level search term ("reinforcement AND concrete") that identified nearly 61,000 documents. These were subsequently honed down using increasingly focused search terms that resulted in 85, 68, 75 and 63 papers that considered traditional steel, recycled plastic, recycled carbon fibres and natural materials, respectively. Each of these 291 papers was then examined to consider how the concrete reinforcement materials performed in terms of strength, cracking control, cost and carbon emissions.

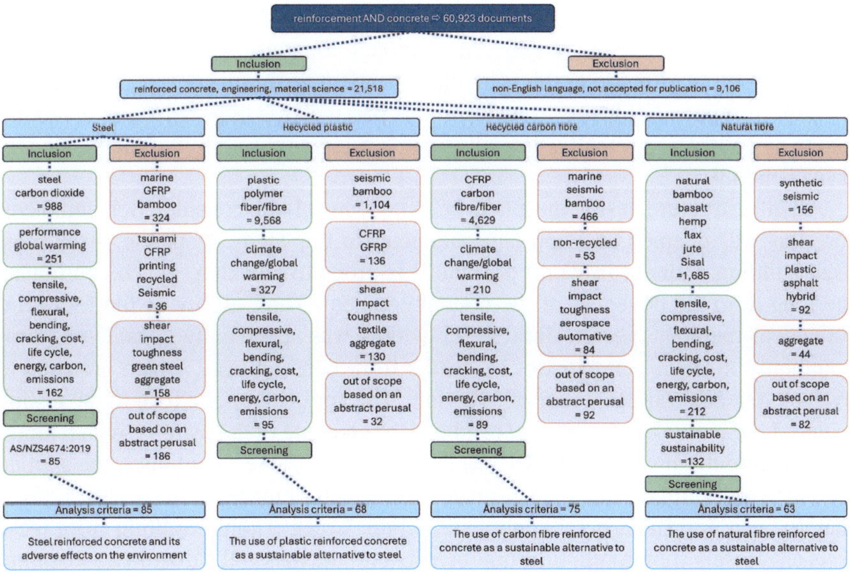

Fig. 16.1 Systematic literature review inclusions and exclusion criteria. Numbers indicate the document count for each criteria group

Table 16.1 Average carbon emissions for the various concrete reinforcement materials that were the focus of this review

Material type	Studies	Carbon emissions (Virgin product)		Carbon emissions (Recycled product)	
		Average	Range	Average	Range
	(n)	(kg CO_2/kg)	(kg CO_2/kg)	(kg CO_2/kg)	(kg CO_2/kg)
Steel	11	2.31	1.50–3.50	0.40	0.18–1.30
Carbon fibre	6	29.75	22.40–38.80	6.82	2.80–16.10
Plastic (combined)	8	4.71	2.40–8.02	2.23	1.12–3.50
PVC	5	5.41	3.64–8.02	2.59	1.54–3.50
HDPE	3	3.53	2.40–4.62	1.63	1.12–2.58
Bamboo fibre	7	0.30	0.13–0.58	N/D	N/D
Basalt fibre	5	0.33	0.08–0.49	N/D	N/D

Concrete, in general, is extremely weak in tension [39] and thus, reinforcement members are required to adequately withstand the forces under which it is regularly placed. There are several aspects to strength, all of which are important when evaluating the impacts of using concrete reinforcement. In this study, split tensile, compressive and flexural strengths were all considered. The split tensile strength represents the ability of a reinforcement member to resist splitting under vertical loading [40] whereas compressive strength represents the ability of the member to withstand being compressed when placed under pressure. A main consideration when investigating compressive strength is that although concrete is generally good at withstanding compression forces this is highly dependent on the mix design adopted [41]. To minimise potential inconsistencies in this parameter, a large selection of studies had to be considered. Finally, flexural strength refers to the ability of the member to resist bending.

All infrastructure design must take cracking control into consideration and therefore this was deemed an important parameter in this study. Concrete is the most used building material in world [42] but it is prone to cracking, which can impact the long-term effectiveness of the poured concrete. Thus, ability of the four reinforcement member types to limit crack widths was specifically considered, most commonly through comparisons of results from bending tests, which are used to evaluate the flexural behaviour of materials. The application of a load to the testing material (i.e., reinforced concrete) results in the cracking of the material which represents its initial failure point under bending stress. As the test continues, the applied load increases until reaching the ultimate load, which signifies the maximum load the material can withstand before complete failure. Crack widths resulting from these tests are closely monitored, offering insights into the material's ductility and ability to absorb stress. Although this is a widely reported testing method, it should be noted that conditions between studies can vary in terms of beam sizing, and the number and spacing of stirrups used in the test, which can impact on final performance. These discrepancies will be reported when and if they occur.

A primary consideration of any construction project is economic cost and this is a major driver for the selection of construction materials. Different reinforcement materials have the potential to improve the economic performance of reinforced concrete technology and, in doing so, reduce the total construction costs of projects. To understand the total economic costs of the alternative concrete reinforcements, raw material and reinforced concrete element costs were analysed separately.

Whereas strength, cracking control and cost dimensions of concrete reinforcement materials are of immediate interest to the construction industry, the control of carbon emissions is driven more by external pressures to reduce atmospheric pollution. Reporting about this parameter is now commonly expected and for this study estimates of the volume of carbon emitted by the different reinforcement materials are presented in terms of kg CO_2/kg.

16.3 Results

An analysis of existing literature was undertaken to determine the performance of sustainable materials as concrete reinforcement. The alternative materials included recycled plastics, carbon fibres and natural materials, with black carbon steel reinforcement as the control. In total, 180 articles were reviewed to assess the performance of the reinforcement materials across four key criteria, that included: (1) strength; (2) cracking control; (3) cost analysis; and (4) carbon emissions. Of these 180 articles, 65 analysed strength performance, 58 tested cracking control, 25 presented cost data and 32 analysed the carbon emissions of the alternative reinforcements.

16.3.1 Strength Performance of Alterantive Reinforcements

To guarantee adequate performance of concrete members, testing techniques have been developed to assess the tensile, compressive, and flexural strengths of the various materials [43]. Such testing mechanisms are pivotal in deriving suitable applications for a material and outlining any possible limitations or defects [44]. To successfully replace carbon intensive steel in concrete reinforcement, alternative reinforcements must exhibit similar strength characteristics.

16.3.1.1 Split Tensile Strength

The split tensile strength of steel and the alternative concrete reinforcement materials as determined by the systematic literature review are presented in Fig. 16.2. In the context of this study, split tensile strength indicates the ability of concrete to resist cracking and shrinking. Findings are returned for both plain concrete and

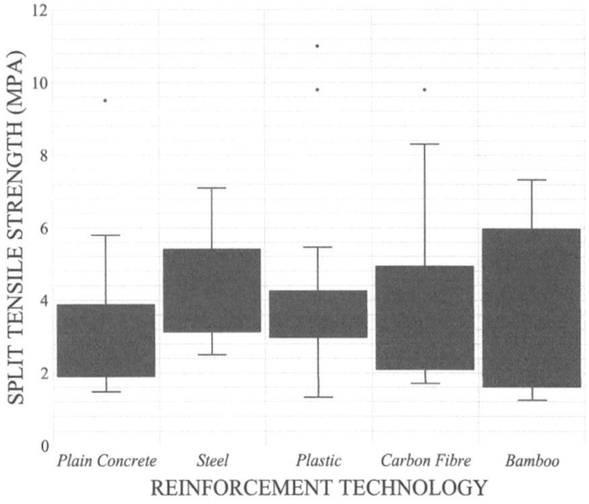

Fig. 16.2 Split tensile strength analysis of materials used to reinforce concrete

for concrete that has been reinforced with the various products considered in this study. The results show that natural reinforcement materials, specifically bamboo, exhibit the best median value for tensile strength. Indeed, under splitting tensile loads bamboo has a median tensile strength of 4.77 MPa, which is an improvement of 54% over the results for plain concrete. Other natural products also performed strongly in this category. For example, although there was some variation in responses for the highly ductile basalt fibre reinforced concrete, on average this product showed improvements of up to 30% over plain concrete. Plant fibres like hemp, sisal, jute, and flax were collectively shown to improve plain concrete splitting resistance by up to 25%. Traditional steel returned a median split tensile strength of 4.20 MPa followed by recycled plastic at 3.60 MPa and carbon fibres at 3.46 MPa. All of the reinforcing materials represent an improvement over plain concrete, which has a median result of 3.11 MPa. These results show that split tensile loading is a non-limiting factor for alternative materials as all alternatives improved the split tensile strength of concrete to some degree.

16.3.1.2 Compressive Strength

The compressive strength of concrete and its various reinforcement materials is presented in Fig. 16.3. Compressive strength indicates a material's ability to withstand being crushed or squashed when placed beneath a weight. This parameter is important for load bearing elements such as columns and walls. As was the case with the split tensile loading results, bamboo exhibited the largest increase in strength over the bare concrete, with compression improved by up to 52% (median value of 59.45 MPa). The next best performing natural materials were basalt fibres, improving compressive results by 10%, and natural plant fibres, which increased the compressive strength by 3–6%. Carbon fibres demonstrated a 46% increase in compressive

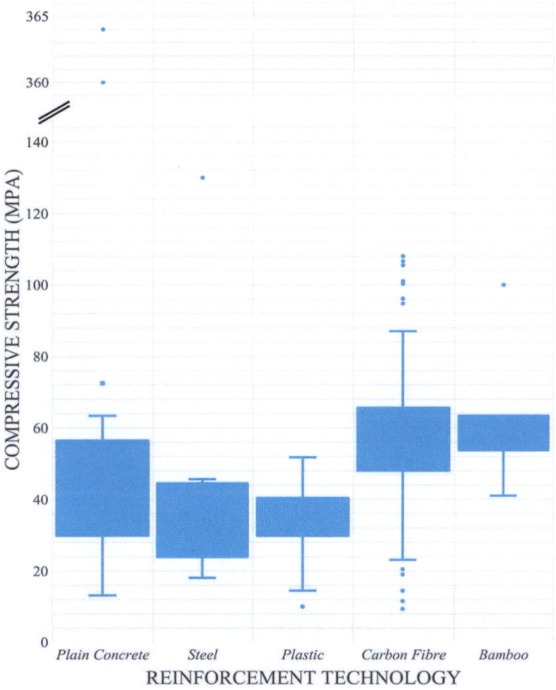

Fig. 16.3 Compressive strength analysis of materials used to reinforce concrete

behaviour with a median of 56.92 MPa. In contrast, steel only increased compressive strength by 2% (39.93 MPa), while the recycled plastic actually reduced compressive strength. Indeed, the recycled plastic had a median compressive strength value of 32.91 MPa, which represents a 16% reduction in strength compared to plain concrete.

16.3.1.3 Flexural Strength

The flexural strength of concrete and its reinforcement materials is presented in Fig. 16.4. Understanding flexural strength, which measures the ability of a material to withstand bending, is critical for the design of beams and slabs. The median flexural strength for steel reinforced beams was an impressive 39.90 MPa in comparison to plain concrete median results of 4.50 MPa in flexural capacity. All of the alternative materials improved the flexural loading of basic concrete, although to a much lower extent than steel. Of these, recycled plastic improved the flexural strength by up to 24% (5.60 MPa), carbon fibres the flexural strength by up 28% (5.80 MPa) and bamboo reinforcement improved flexural strength by 30% (5.90 MPa). Lastly, basalt fibres also increased the flexural strength of concrete with improvements of up to 30%, while natural fibres such as hemp, sisal, jute, and flax increased the flexural strength of the concrete by only 3–10%.

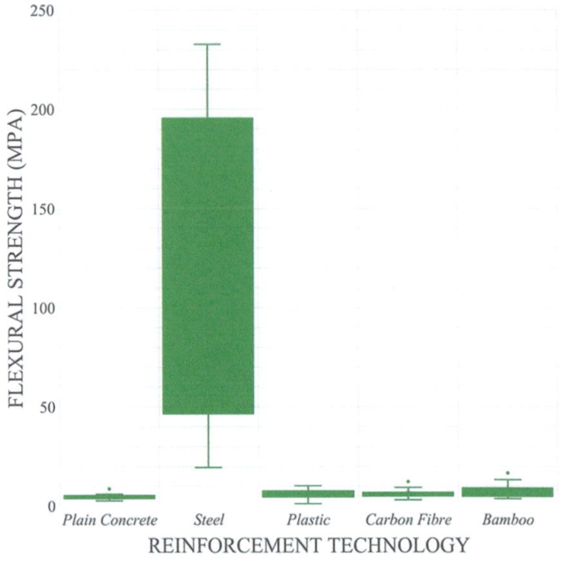

Fig. 16.4 Flexural strength analysis of materials used to reinforce concrete

16.3.2 Cracking Control

Cracking control results can be separated into first crack load (Fig. 16.5) and ultimate crack load (Fig. 16.6). Each bar in Fig. 16.5 and Fig. 16.6 represents a different study examined in the systematic literature review (i.e., 16 papers in total). The results reveal that steel has the best first crack load of any reinforcement type (average = 47.64 kN). Indeed, all of the alternative reinforcement materials averaged below 50% of the value obtained for steel, with bamboo averaging 22.13 kN, carbon fibre averaging 17.03 kN, and recycled plastic materials averaging 21.13 kN. A similar pattern is observed for ultimate load with steel performing best (average = 107.84 kN) followed by carbon fibre (average = 79.3 kN), plastic materials (average = 69.38 kN), and bamboo (average = 44.7 kN).

16.3.3 Cost Analysis

The data for the economic costs of the raw materials and the reinforced concrete elements are presented in Figs. 16.7 and 16.8, respectively. In both cases, costs are given relative to steel, which is set as 1.0 on each graph. For raw materials, the median costs relative to the cost of steel are 0.13, 1.84 and 0.35 for plastic, carbon fibre and natural material alternatives, respectively. Similar results can be seen for the reinforced concrete element costs, which relatively to steel are 0.60, 1.37 and 0.64 for plastic, carbon fibre and natural material alternatives, respectively.

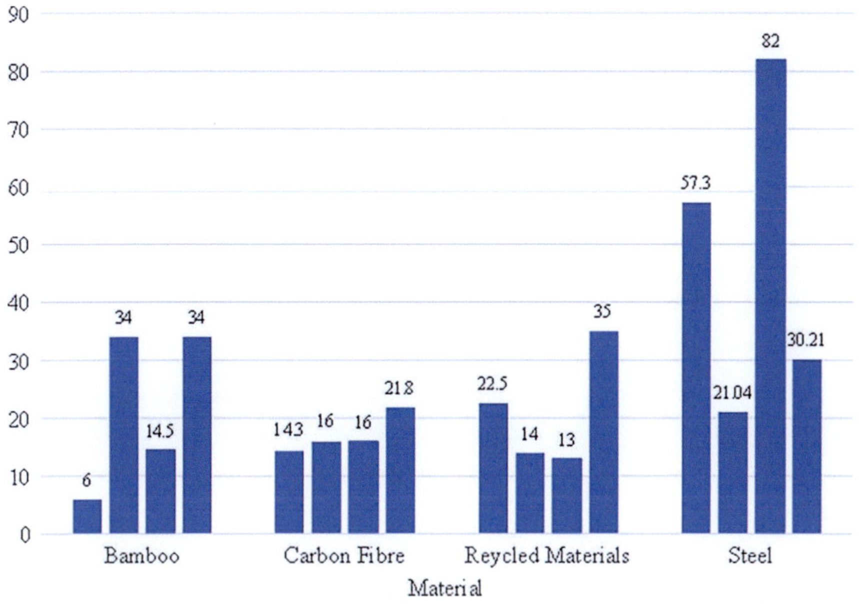

Fig. 16.5 Combined results for first crack load

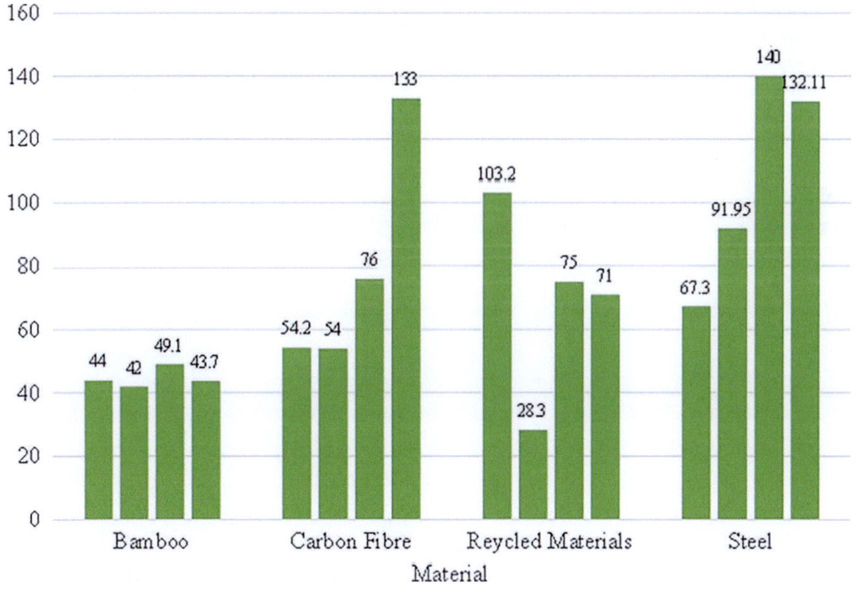

Fig. 16.6 Combined results for ultimate load

Fig. 16.7 Reinforcement material cost of each reinforcement alternative

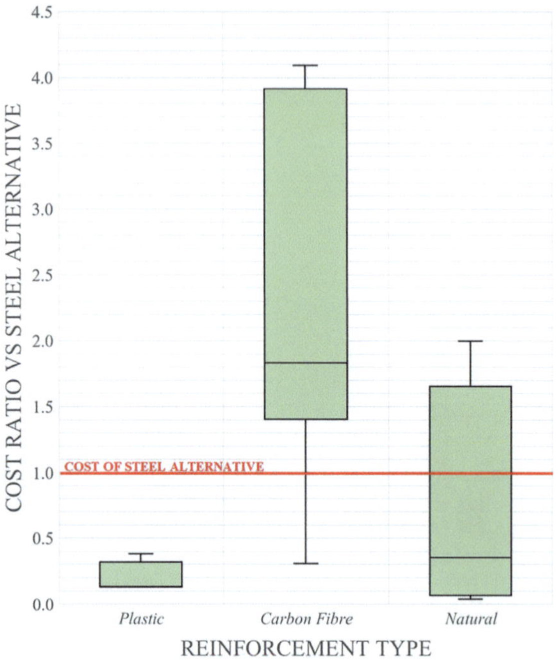

Fig. 16.8 RC element cost of each reinforcement alternative

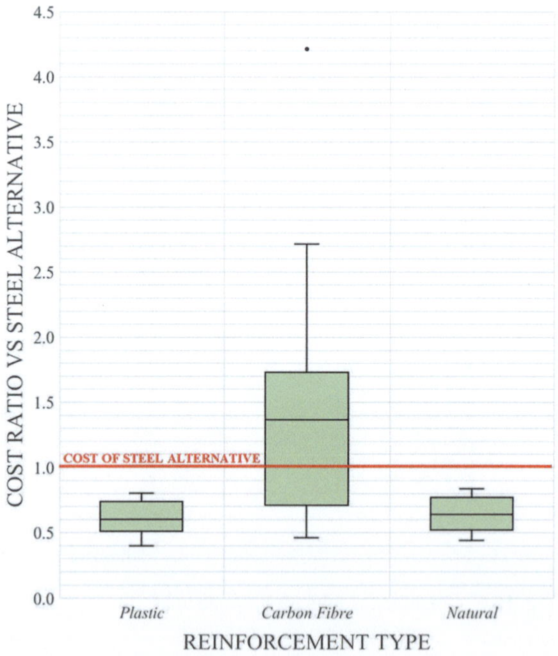

16.3.4 Carbon Emissions

Carbon emissions provide some evidence of variations in the environmental performances of the different materials used to reinforce concrete. Some papers within the systematic literature review provided information about this parameter and these results are presented in Table 16.1. The research indicates that the production of virgin (primary) and recycled (secondary) steel generates an average of 2.31 and 0.40 kg CO_2/kg, respectively. The carbon fibre and plastic alternatives to steel have considerably higher carbon emissions for both virgin and recycled products. Even considering that the research revealed a range of results, these products are consistently more carbon intensive than steel. In contrast, both the bamboo and basalt fibre alternatives recorded lower carbon emissions than either virgin or recycled steel. These findings also indicate that the production of recycled steel is a viable alternative for reducing carbon emissions, although the often limited availability of supply is a factor that reduces its widespread use.

16.4 Discussion

The ongoing production of steel for use within reinforced concrete structures is accelerating the detrimental impacts of global warming. As such, the adoption of alternative sustainable reinforcements presents an opportunity to promote sustainability within the construction industry and negate the impending consequences of climate change that are resulting from it. This study aims to determine the potential of recycled plastics, recycled carbon fibre and natural materials as sustainable and viable alternatives to using steel for reinforcing concrete. This study investigates this according to four key criteria, namely: (1) strength; (2) cracking control; (3) cost; and (4) carbon emissions. Comparisons are made between reinforcement material types and between the criteria to provide an interpretation of which material the construction industry should be considering.

16.4.1 Strength Analysis

Currently, no studies provide a succinct analysis of the ability of alternative reinforcements to improve the strength parameters of concrete. This investigation provides a summary of different tests performed across multiple studies using a variety of techniques and materials. The results of this study showed that all alternatives offer similar performance to steel in reinforcing concrete members when subject to splitting and compressive forces. However, the superiority of steel is evident in the flexural testing results with continuous steel bars able to withstand flexural loads up to eight time larger than the other reinforcement options.

The split tensile strength of a reinforcement member is heavily affected by the bond between the reinforcing materials and the concrete [45]. Natural materials, specifically bamboo, displayed impressive results in split tensile testing scenarios that were similar to the performances exhibited by steel. The unique anatomical structure of bamboo can help explain this result as staggered arrangements of vascular bundles effectively distribute the stresses imposed [40]. Furthermore, the fibre composition of bamboo is oriented in a longitudinal direction meaning applied loads in the same direction allow higher resistance to splitting forces. Also, the fibre density of bamboo allows great stiffness in small volumes of members [46]. Lastly, bamboo has great bonding capabilities to concrete which helps prevent members from splitting apart and results in the impressive splitting resistance [47]. Meanwhile, basalt and plant fibre reinforced concrete also represent versatile and sustainable reinforcement options when under splitting loads with both of these improving the properties of plain concrete. This suggests there is a potential target market for natural reinforcing materials in countries where steel is a difficult or expensive to procure. Possible regions for sustainable infrastructure that utilise local plant and rock fibres without hindering performance may be countries like China, India, and Bangladesh [48].

Similar to bamboo in split tensile loading results, recycled plastic provides an improvement over plain concrete strength with a median that is only slightly below that returned for steel. However, recycled plastic displays ductile properties that allow visual deformation prior to failure, which is an important serviceability factor [49]. However, plastic is a cheap material that can be sourced for landfills and existing waste streams. This could provide an opportunity for the construction industry to reduce pollution by diverting these materials into repurposed applications through the reinforcing of concrete to be used under certain circumstances. For example, plastic fibre reinforcement could lead to a reduction of steel reinforcement requirements in members subject to splitting forces such as bicycle lanes and footpaths [50]. Lastly, carbon fibre reinforcement materials returned the lowest split tensile median of all of the possible alternatives to steel. This can be associated with the material properties of carbon fibres which can be brittle in failure [51]. In addition to this, single strip carbon fibres generally have a weaker bond to concrete which prevents effective transfer mechanisms of stress through the member in tension [52].

The compressive strength values largely depend on the mix design of the concrete being used. For this study, testing results were taken at their 28-day strength value. Also, for compressive strength results it was assumed that the water to cement (w/c) ratio and type of aggregate were kept relatively consistent across studies to minimize discrepancies. Overall, the median value of plain concrete compressive strength was set to 40 MPa which is typically used in structural members like footings, industrial flooring and retaining walls [53]. Plastic fibres were shown not to increase the median value of concrete under compression for this strength value. However, there may be a limitation in this assumption that plastics have an adverse effect on compressive results due to the mix designs and fibre inclusion percentages explored in the literature. For example, [45] tested the strength of 30 MPa concrete with polypropylene (plastic) fibres and showed that the compressive strength was increased by about 9%. Similarly, [54] analyzed compressive results of the same strength grade (30 MPa)

Table 16.2 Effect of various fibre percentages on compressive strength results on concrete specimens

	% volume of member		
Compressive strength (MPa)	< 1%	1%	> 1%
Plastic fibres	36.79	39.70	36.08
Carbon fibres	41.80	46.46	41.66
Bamboo fibres	59.83	61.15	54.92

with polypropylene fibres and found that compressive strength increased by up to 23%. However, when the strength grade of the concrete was increased to 50 MPa, plastic fibres were found to decrease the compressive strength of the concrete.

Carbon fibres were shown to have a positive impact on the strength of compressed concrete due to its ductile nature under loading conditions. Similarly, the natural reinforcement options also displayed relative increases in compressive capacity with bamboo being the best performing natural material in concrete reinforcement. Carbon fibre, bamboo and basalt all had similar performance in terms of compressive strength improvement, meaning all offer similar benefits relative to plain concrete.

An issue that can arise with fibre additions to concrete is that the workability of concrete mixes lessen as the volume percentage of fibres is increased. Higher contents of fibre within cement may result in poorer distribution and decreased level of compaction leading to a lessened compressive strength [55]. Table 16.2 shows how the optimal percentage of included fibres by volume seems to be 1%, (which maximises compressive strength) with values above or below 1% resulting in reductions in compressive strength. Ultimately, to best determine whether plastic and other fibres improve or reduce the compressive strength of concrete, a wider variety of studies will need to be conducted, testing across more potential concrete strength and fibre inclusion percentages (Table 16.2).

Flexural strength is important for long spanning members such as concrete beams, columns, pipes, or culverts [56]. Based on the results from 70 studies, poor flexural strength outcomes for all concrete reinforcement alternatives is the main limitation for the replacement of steel as reinforcement for structural concrete. However, this does not mean that there is no place for such alternatives. There are instances where the flexural strength of a member is not critical, and in these circumstances recycled plastic, carbon fibres and natural materials could be used [57]. These could include secondary beams and/or footings although further research would need to validate the utility of steel replacements in these instances.

16.4.2 Cracking Control

The cracking control of the sustainable alternatives to steel for concrete reinforcement plays a key role in determining the application of each material. For bamboo, several important considerations arise when analysing its performance in concrete. For example, [58] found that bamboo culms in their natural state can withstand large loads. Meanwhile, [59] found that treated bamboo delivered even better results. These findings indicate that the treatment process significantly improves bamboo's load-bearing capacity. [60] and [61] further reinforced these findings, revealing that bamboo evidences consistent average first crack loads, although there were variations in deflection between the specimens, suggesting there are some potential differences in structural integrity. These results have important significance for the construction sector, suggesting that bamboo may be an effective replacement for steel in certain scenarios, such as the small-scale construction of slabs, particularly in areas where bamboo is prevalent. The findings of this study suggest that bamboo, especially when treated, holds promise as a sustainable reinforcement material for concrete.

Carbon fibre reinforced polymers (CFRP) tend to perform well in terms of cracking control, especially for ultimate load. These results were consistent across studies. The significance of these findings lies in the promising application of carbon fibre rebar in enhancing the structural integrity of concrete. With consistent and robust performances demonstrated across studies, the research highlights this material's potential to modernise construction practices, offering a lighter yet stronger alternative to traditional steel reinforcement. While there are areas for future exploration, the data indicates a potential future for using carbon fibre on its own, but also including it as a hybrid material with steel.

There are a wide variety of recycled materials that have been explored in the literature. For most of these, only limited observations have been made. For example, [62] were able to demonstrate the exceptional potential of synthetic waste wire, which exhibited a high first crack average. These cracking results suggest the feasibility of synthetic waste wire as a strong reinforcement material, challenging traditional opinions about using recycled materials. Meanwhile, [63] and [64] explored the utility of glass in different forms (rebar and fibre), with glass rebar outperforming glass fibres in terms of cracking control. These results suggest that glass, as a recycled material, might offer an environmentally friendly alternative to traditional steel reinforcement. For plastics, cracking control results are mixed, ranging from second best after steel, to amongst the lowest cracking control. This obviously depended on the nature of the product being considered.

The best performing material in terms of cracking control was steel, although the range of outcomes was again quite large between studies. These variations arise from differing test conditions, such as variations in the number or reinforcing bars or specimen size. The significance of these findings lies in reaffirming the critical role steel plays in construction. Because of these results, steel reinforcement continues to be a reliable choice for construction, offering relatively consistent and generally high-performance outcomes.

16.4.3 Cost Analysis

Existing literature provides limited discussion on the economic performance of concrete reinforcement alternatives. It is also very difficult to compare costs of reinforcement across studies given differences in material densities, quantity of reinforcement material, units of costing, and availability of material in varying locations. This objective of this study to incorporate numerous sources of data and compare the material and element costs of the various reinforcement materials provides industry professionals insight into the actual costings of recycled plastics, recycled carbon fibres and natural materials in comparison to the industry standard of steel reinforcement.

When considering economic costs, traditional black carbon steel reinforcement was used as the control in this study. The cost of black carbon steel reinforcement is relatively stable in the market given its use as the traditional material for inclusion in reinforced concrete. Material costs are heavily dependent on the grade of steel with prices ranging from USD $660 [65] to $1361 [66] per tonne. The most common steel reinforcement grade is the 500N (Type 60) steel costing on average USD $1065 per tonne. The results of this study show that steel reinforcement is generally more expensive than recycled plastic and natural alternatives but less costly than carbon fibre reinforcements. However, the use of black steel reinforcement engenders greater costs during the construction period relative to carbon fibre reinforcement due to its high density requiring higher labour costs and larger construction equipment to place the reinforcement [67]. The susceptibility of black steel reinforcement to corrosion also requires ongoing maintenance costs and greater life cycle costs in comparison to carbon fibre reinforcement [66]. This is considered one of the greatest limitations of steel reinforcement often requiring ongoing inspection and maintenance activities with a reduced design life resulting in more frequent demolition and reconstruction when compared to the use of alternative non-corrosive reinforcements such as carbon fibres and recycled plastics [68]. Black steel's susceptibility to corrosion also necessitates more stringent design requirements, often resulting in the inclusion of larger sections and superfluous reinforcement designs to meet the design life for the steel reinforced concrete structures [69, 70]. Despite these limitations, black steel reinforcement remains the most utilised concrete reinforcement in the world and is widely available across the construction industry.

In comparison to steel, recycled plastic reinforcement proved to be the least costly alternative of those analysed in this study, with an average price of USD $3273 per m^3. This compares to steel fibres which cost on average USD $9263 per m^3. The significant price difference makes recycled plastic an attractive alternative to steel and, not surprisingly, its use has grown significantly in recent years [71]. The density of plastic fibres is also beneficial to its mass per cost efficiency in comparison to dense steel fibres. Plastic fibres have densities ranging from 900 to 1800 kg/m^3, which is significantly less than steel fibres that have a density of 7860 kg/m^3. As such, the use of plastic fibres requires less mass of reinforcement material per volume of concrete and saves additional transportation costs in comparison to steel due to the reduced weight of the concrete mix [72]. However, the use of plastic fibres to reinforce concrete also requires increased costs associated with the use of superplasticizers, a concrete additive required to maintain the workability of the concrete and which can cost up to 10 times the price of a standard cement binder [72, 73].

Significantly, recycled plastic reinforcement is only appropriate for use as concrete reinforcement fibres and not in continuous reinforcement bars and, as such, all comparisons in this study are made to steel fibre reinforcements. Plastic fibres proved to be on average 39% cheaper than steel fibres when utilised for reinforced concrete pavements, such as for concrete footpaths and roads. In comparison to non-reinforced concrete pavements, pavements reinforced by plastic fibres were on average 5–15% cheaper due to the reduced thickness (10–20%) of the pavements. In contrast, the significant cost of steel fibres increased the cost of reinforced pavements by 8–20%, even considering a 32–35% average reduction in the pavement thickness due to its structural properties [72–74]. As such, the cost efficiency of recycled plastic in fibres to reinforce concrete pavements is clearly a cost-effective alternative in comparison to steel.

The cost of carbon fibre for concrete reinforcement varies widely depending on the type of fibre used, the manufacturing process utilised and the capacity of the manufacturing operation. The use of carbon fibre reinforced bars proved to be the most expensive reinforcement analysed in this study, costing a median of 223% above the price of steel reinforced bars. This high cost is due to the larger material costs associated with producing continuous carbon fibre reinforcement bars. The manufacturing of carbon fibre reinforcement is a complex process requiring specialised equipment and expertise. It is considered more labour-intensive and technologically advanced than steel production resulting in greater production costs. Given its use as a high-performance material with an impressive strength-to-weight ratio, carbon fibre reinforcement bars tend to be used in more specialist applications where cost considerations are secondary to performance. This lower economy of scale in comparison to steel reinforcement further accounts for the higher material costs of carbon fibres. Despite their high production costs, carbon fibre bars are lighter than steel reinforcement bars, allowing for greater prefabrication, reduced transport costs and reduced placement costs [69]. As such, the use of carbon fibre reinforcement generally entails reduced construction costs in comparison to steel reinforcement [65, 75]. Overall, carbon fibre reinforcement is not considered a cost-effective alternative to steel reinforcement in most applications. Its significant upfront costs generally make carbon fibre reinforcement too costly in comparison to other alternatives. However, given the large upfront costs, this material may be considered an economic option for use in large-scale reinforced concrete structures that have a design life in excess of 50 years.

In comparison to carbon fibre reinforcement, natural materials such as bamboo, basalt and various plant fibres are highly cost effective in comparison to steel reinforcement. The median price of these materials was 35% lower than the cost of steel. Similarly, the cost of reinforced concrete elements utilising natural reinforcements were an average of 36% lower than the cost of steel. Natural materials have applications in both reinforcement fibres and continuous reinforcing bars and bamboo and basalt are key natural materials utilised to reinforce continuous concrete bars. Bamboo particularly shows impressive economic savings with a cost ratio of 0.63

in comparison to steel. Applications such as its use in low-cost earthquake-resistant buildings has the potential to reduce total building costs by 5.5%, with the cost of reinforced concrete stiffening members, such as sloof and ring beams, reduced by up to 38% in comparison to a steel reinforced alternative [76]. As such, the use of bamboo has significant potential for employment in earthquake prone areas such as Indonesia where stiffening members are often ignored due to the high price of steel reinforcement. In contrast, basalt fibre reinforced polymer bars showed a 200% cost increase in comparison to the cost of steel. This is mainly due to its relatively recent adoption in concrete reinforcement (reflecting a low economy of scale) and limited availability [77].

Natural reinforcement fibres were the best performing natural material with a median cost ratio of 0.24 relative to steel. However, this result is based on a very limited literature. [78] suggest these materials are available in abundance at reduced prices, while acknowledging the price of natural fibre reinforcement varies greatly across geographic regions. Another factor that helps keep costs down for natural reinforcement materials is that some natural fibres, such as cellulose, do not impede the slump and workability of concrete in comparison to other fibre alternatives. This reduces the need for expensive superplasticizer additives [79].

16.4.4 Carbon Emissions

For a concrete reinforcement material to be considered sustainable, having low CO_2 emissions relative to traditional virgin steel is essential. In response to global warming and to reduce the carbon footprint of the construction industry, research has been conducted into the use of recycled, as opposed to virgin, steel for reinforcing concrete [80]. The unique ability of steel to retain its metallurgic properties over repeated recycling, establishes it as the world's most recyclable material and assures its consistent quality [81]. Some of the largest benefits of recycling steel can be seen in the reduction of energy consumption throughout the production process, which also directly corresponds to lower carbon emissions. A study by [82], showed that for every 1 kg of steel scrap recycled at end of life, 1.5 kg CO_2 equivalent emissions could be saved by voiding virgin production processes. These results were supported by other studies, including those by [83] and [84] who found that a saving of 1.6 kg CO_2/kg was observed in steel recycling processes when compared to virgin production.

Although recycled steel is preferred to virgin steel in terms of its environmental performance, certain recycled plastics can provide even lower emissions than recycled steel, while having the additional benefit of removing that plastic (at least temporarily) from the waste stream. The computation of average material carbon emissions from the literature reviewed in this study revealed that recycled plastic was comparable to virgin steel in terms of its associated carbon emissions. This suggests this is a potentially viable material to replace virgin steel and has the added benefit of removing plastic from the waste stream. Although recycled steel is preferred to virgin steel in terms of its environmental performance, certain recycled plastics can provide even lower emissions than recycled steel, while having the additional benefit of removing that plastic (at least temporarily) from the waste stream. The computation of average material carbon emissions from the literature reviewed in this study revealed that recycled plastic was comparable to virgin steel in terms of its associated carbon emissions. This suggests this is a potentially viable material to replace virgin steel and has the added benefit of removing plastic from the waste stream. However, unlike steel, recycled plastic can only be used once before becoming unfit for further use in the construction industry. This is because recycling plastic into fibres cannot be reversed [85] and indicates that plastic fibres are likely to end up in landfill at its end of life. This clearly limits the potential for plastic to be used as a meaningfully sustainable alternative to virgin steel.

In addition to recycled plastics, an investigation into the sustainable performance of recycled carbon fibre (CF) was also undertaken. Unlike steel, the concept of using recycled CF in the construction industry is relatively new, with the demand for this material increasing over the last decade [86]. Studies performed by [87] and [88] established that virgin production of CF created emissions of between 22 and 25 kg CO_2/kg produced. This figure was significantly reduced to between 3.73 and 4.65 kg CO_2/kg produced when utilising mechanical recycling [87, 88] and resulting in a reduction in carbon emissions of 82.2%. Despite these significant results, recycled CF emissions were still much higher than those for virgin steel, suggesting it is not a viable alternative to steel in terms of its impact on greenhouse gas emissions.

Products such as bamboo and basalt do not have existing recycling systems and, as a consequence, this study could not investigate emissions associated with recycling these materials. As such, this study only considers the virgin production of these materials. Bamboo only grows in certain climates and requires specific soil conditions, which means it is not an easily accessible material for all locations [89]. This means that bamboo fibres that have travelled long distances will have higher CO_2 emissions in their production process. Producing bamboo fibres can emit as little as 0.13 kg CO_2/kg, but this figure will substantially increase if transportation distances are high [90]. Despite this, bamboo fibres that have been locally sourced and produced seem to be one of the most environmentally friendly reinforcement alternatives. An additional reinforcement option that has been widely researched is basalt fibres. There aren't many locations where basalt cannot be found, making it a widely accessible reinforcement option when produced in fibrous form [91]. [92] and [93] examined the life cycle analysis of basalt fibres and found that the emissions associated with production varied between 0.43 and 0.49 kg CO_2/kg. Although the energy requirements to produce basalt fibres are considerably higher than that of bamboo fibres, the CO_2 emissions are comparable and choosing either of these natural reinforcements would result in significantly lower emissions than virgin steel and comparable or lower emissions than recycled steel [91].

16.4.5 Synthesis

The detrimental contributions of steel production to global warming are challenging the construction industry to consider the use of alternatives to it. The examination of three alternative concrete reinforcement materials, namely recycled plastics, carbon fibres and natural alternatives, presents an opportunity to produce a more sustainable built environment, although it also highlights limitations for each material in terms of its the potential to replace the consistently high performing traditional steel reinforcement. This section provides a synthesis of the findings for these proposed materials.

Recycled plastic fibres, both PVC and HDPE, show the lowest strength performance of the reinforcement materials analysed, proving to be considerably weak in some parameters in comparison to steel reinforcement. The flexural strength of recycled plastic fibre reinforcement was significantly less than that of steel, which is a major limitation on the application of plastic reinforcement for concrete structures. The materials performance through cracking control tests proved more promising, although it is still inferior in comparison to structural steel reinforcement. Recycled plastic did prove to be the most cost-effective reinforcement material, being significantly cheaper than steel. Finally, in terms of its sustainability credentials, recycled plastic reinforcements outperformed steel with regards to CO_2 emissions during their production, while simultaneously having the benefit of removing plastic from the waste stream. The production of virgin plastics, however, is not a sustainable alternative to steel reinforcement as its emissions are much higher than that of virgin steel. Overall, recycled plastic reinforcement should be considered as a sustainable and cost-effective alternative to steel reinforcement in lighter loading applications such as footpath pavements and bicycle lanes, with additional strength and serviceability tests required prior to each application.

In comparison to plastics, recycled carbon fibres showed little promise as a truly sustainable reinforcement alternative. This was due to the carbon emissions of their production, more than doubling the carbon emissions of virgin steel reinforcement production. Carbon fibres did show strong strength characteristics for split tensile and compressive strengths, but similarly to plastics, lacked the flexural performance of steel reinforcement. The brittle nature of carbon fibres also resulted in the greatest cracking of the materials tested, a well-documented limitation of the material given its poor ductility. Carbon fibre also proved to be the costliest material analysed, a significant setback for its application in a cost centric construction industry.

Similarly, to recycled plastics, natural materials were also found to be effective as sustainable materials in concrete reinforcement. The use of bamboo outperformed the characteristics of steel in both split tensile and compressive strength, with basalt also producing competitive results. Again, steel proved to be superior with regards to the flexural strength performance, clearly outperforming bamboo and basalt, both of which showed greater flexural strengths in comparison to recycled plastics and carbon fibre reinforcements. Natural fibres however showed the least flexural strength of all the three alternative materials, perhaps the greatest limitation in their application in reinforced concrete members. The cracking resistance of bamboo was again inferior to steel, raising concerns on its ability to satisfy the serviceability requirements of reinforced concrete members. Despite this, the cost performance of bamboo, basalt and other natural fibres was significantly more attractive than steel reinforcement, a substantial benefit to increase its supply chains and application within the industry. Natural materials were proven to be the most sustainable alternative material studied, showing the least embodied energy and carbon emissions when utilised as concrete reinforcement, with the majority of their energy inputs and carbon emissions associated with their supply chains. Overall, natural materials have exceptional potential as an alternative to steel reinforcement, with bamboo showing the most promising results with regards to its strength contributions, cracking durability, economic performance and reduced embodied energy and carbon emissions.

16.4.6 Recommendations

Despite some of the limitations of the materials discussed throughout this study the results show a clear benefit when adopting alternative reinforcements such as recycled plastics and natural materials. The results are particularly encouraging across the economic and sustainability parameters aligning with the interests of construction professionals and the aims and objectives of this study. However, this study indicates the need for more work on this topic, particularly surrounding the strength and cracking control performance of the sustainable reinforcement materials to increase confidence in their application. Recycled plastics and natural materials were proven to show inferior strength and cracking characteristics in comparison to carbon steel reinforcement. As such, it is not feasible to suggest they are superior alternatives to safely provide the required reinforcement in all concrete structures and further investigation is required to determine the limits of their possible application. Existing literature shows strong potential in reinforced pavement applications however, further research should be undertaken to broaden their potential use with strong reference to the economic and sustainable benefits demonstrated through this study.

Further research should also be focused on developing new recycling methods for materials such as plastics, steel, and carbon fibres to improve the embodied energy and carbon emissions of their production. The construction industry needs to continuously find ways to improve the sustainability of its supply chain and as such, should seek for the continuous improvement of recycling processes to reduce overall energy inputs and carbon emissions when utilising recycled materials as concrete reinforcement.

Finally, the use of hybrid reinforced structures should be investigated for its potential to provide immediate sustainability improvements in the construction industry. The ability to combine the structural and durability characteristics of steel, with the sustainable and economic benefits of recycled plastics and natural materials presents potentially viable options that could help to negate the exponential growth of steel reinforcements. As such, for the successful application of sustainable reinforcement it is necessary for future studies to build upon the findings of this research and fully understand the capabilities of sustainable materials for concrete reinforcement.

16.4.7 Research Limitations

The capacity to identify which materials represent the preferable alternative for reinforcing concrete was limited by the lack of uniformity between the studies that were investigated. This is especially evident in the wide ranges of concrete mixes tested, which often resulted in vast differences in results between studies, particularly in terms of strength. To improve the capacity of companies to compare data about different forms of concrete reinforcement, standard measurements of these materials are required.

In addition, the measurement of split tensile strength in natural elements like bamboo must consider the variability of the material, with a requirement that the temperature at the time of testing be controlled [94]. The moisture content must also be consistent amongst samples and the species and age of the plant must be controlled to obtain reliable data for comparisons to typical homogenous steel [95]. Unfortunately, these factors are rarely controlled for or reported on in the literature, once again limiting the capacity to undertake a systematic analysis of the behaviour of bamboo. Similarly, the measurement of the strength of plastic and carbon fibres must also be considered and the material quality needs to be reported as the purity of recycled materials can lessen strength capacities [49]. Lastly, due to a lack of specific standardization across the industry no single reinforcement configuration was able to be studied. Essentially allowing for a range of variable factors through irregularity in reinforcement sizing, orientation, quality, volume percentage and composition.

Despite the promising results, limitations exist within the study of cracking. Testing conditions often varied between studies, such as whether stirrups were used or the number and spacing of the stirrups, can influence results and give values outside the expected ranges. This introduces variability, affecting the comparability of results. Furthermore, the number of reinforcing bars also differs between tests. Although similar results were found using mostly two bottom reinforcing bars, other tests utilised one or three. Ultimately, this study has highlighted a lack of consistency across all concrete reinforcement types in terms of how the materials were treated, the concrete mixes used and even the testing setups. Until research addresses these issues, it remains unclear exactly how these alternatives will perform under standardised conditions.

Finally, a lack of data on materials like carbon fibre, bamboo and basalt make it difficult to estimate accurately estimate CO_2 emissions associated with their production processes. For emissions, transportation was a major contributor to the varying results seen in this study and this is challenging to fully account for. Longer distances travelled by materials will results in higher CO_2 emissions, but it is difficult to directly compare these results because they are relative to the site of application of the material [96]. Finally, the production or recycling method utilised was the biggest factor in energy requirements observed during the production process. Some recycling methods, such as pyrolysis, will use up to 30 times more energy [97] than simpler recycling methods, such as mechanical recycling, leading to much higher estimated CO_2 emissions. This complicates the comparison of emissions benefits derived from using alternative materials for concrete reinforcement.

16.5 Conclusion

The ongoing production of carbon intensive steel reinforcement for use in reinforced concrete technology continues to exacerbate the detrimental impacts of global warming. There is a growing need for the application of alternative materials to provide the necessary structural reinforcement in concrete structures with potential to reduce the consequential embodied energy inputs and carbon dioxide emissions associated with the production of steel reinforcement. The use of recycled plastic, recycled carbon fibre and natural alternatives may provide an opportunity to negate the increasing CO_2 emissions resulting from the increasing use of reinforced concrete by the construction industry. However, to date a direct comparison of the potential of these different materials has not been undertaken.

This study aimed to determine the suitability of recycled plastics, recycled carbon fibre, and natural materials as sustainable reinforcements to provide viable alternatives to the use of steel in reinforced concrete structures. A systematic literature review was undertaken on these alternative reinforcement materials with respect to their: (1) strength; (2) cracking control; (3) cost; and (4) carbon emissions. For an alternative reinforcement material to be deemed a viable alternative to steel, they must meet stringent strength requirements to ensure the continued safety of reinforced concrete technology. This study analysed the strength of reinforcement through a review of the split tensile, compressive, and flexural strengths of the materials, all of which represent integral performance criteria in the selection of reinforcement materials. The flexural strength of steel reinforcement was shown to be unmatched by the alternative materials, immediately placing limitations on the use of the other materials within structural applications, particularly for concrete members subject to large flexural loads. Alternative materials did however outperform steel across split tensile and compressive strength tests, validating their application as a viable reinforcement alternative where flexural loads are not limiting. Concrete reinforcement is also crucial to limit the propagation of cracks throughout concrete structures. While it is reasonable to expect that cracking will occur in concrete, effective reinforcement designs limit the onset and widths of cracking, ensuring reinforced concrete members meet both strength and serviceability requirements. The study found steel to be the dominant material with the largest first cracking load. Although recycled plastics, recycled carbon fibre and natural materials showed inferior performance in comparison to steel reinforcement, they performed well enough to suggest that they could still be suitable for certain applications.

The alternative materials showed greater promise in their economic performances, a major consideration in the selection of construction materials. Recycled plastics and natural materials significantly outperformed steel across material costing and reinforced concrete element costing. In contrast, recycled carbon fibres exhibited increased economic costs across both categories. To ensure each material met the aims and objectives of this study a sustainability analysis was also undertaken to determine carbon emissions associated with the production of each material. Recycled plastic fibres and natural reinforcements indicated significant improvements over the sustainability of steel, whereas recycled carbon fibres showed an increase in carbon emissions in comparison to steel. As such, recycled carbon fibres are not considered a sustainable alternative given they do not meet the aims of this study.

This research investigated multiple concrete reinforcement alternatives across key performance criteria, accentuating the need to consider sustainable concrete reinforcement and its application within a greener construction industry. The study reveals there is a place for recycled plastics and natural material in concrete reinforcement and encourages industry standards and engineers to pursue the specific application of these materials to ensure a greener future. However, much research is still needed, especially with respect to standardised testing conditions across materials, to determine accurate behaviours for all reinforcement materials under similar conditions and to identify the best applications for alternative reinforcement materials.

References

1. J. Min, G. Yan, M.A. Abed, S. Elattar, M.A. Khadimallah, A. Jan, H.E. Ali, The effect of carbon dioxide emissions on the building energy efficiency. Fuel (Guildford) **326**, 124 (2022)
2. C. Xiong, Q. Li, T. Lan, H. Li, W. Long, F. Xing, Sustainable use of recycled carbon fibre reinforced polymer and crumb rubber in concrete. Mech. Prop. Ecol. Eval. **279**, 1–3 (2021)
3. M. Bararzadeh Ledari, H. Khajehpour, H. Akbarnavasi, S. Edalati, Greening steel industry by hydrogen: lessons learned for the developing world. Int. J. Hydrogen Energy **48**(94), 36623–36649 (2023)
4. Van Mead, N., *A brief history of concrete: from 10,000BC to 3D printed houses* (The Guardian, London, UK, 2019)
5. A. Chahar, A. Pal, Study on various properties of reinforced concrete. Mater. Today Proc. **65**, 597–602 (2022)
6. P. Wang, M. Ryberg, Y. Yang, K. Feng, S. Kara, M. Hauschild, W.-Q. Chen, Efficiency stagnation in global steel production urges joint supply-and demand-side mitigation efforts. Nat Commun. **12**(1), 2066 (2021)
7. M.J. de Villafranca Casas, S. Smit, A. Nilsson, T. Kuramochi, Climate targets by major steel companies: an assessment of collective ambition and planned emission reduction measures. Energy Clim. Chang. **5**, 100120 (2024)
8. Infrabuild, *Environmental Product Declaration* (InfraBuild, Melbourne, Australia, 2022)
9. T. Achamyeleh, Y. Şahin, Investigation of mechanical properties of ribbed reinforcement steel bars: a case study on Ethiopian construction industry. Int. J. Steel Struct. **19**, 1682–1693 (2019)

10. M.A. Quader, S. Ahmed, S.Z. Dawal, Y. Nukman, Present needs, recent progress and future trends of energy-efficient ultra-low carbon dioxide (CO_2) steelmaking (ULCOS) program. Renew. Sustain. Energy Rev. **55**, 537–549 (2016)
11. J. Goggins, T. Keane, A. Kelly, The assessment of embodied energy in typical reinforced concrete building structures in Ireland. Energy Build. **42**, 735–744 (2010)
12. Z. Li, F. Andersson, L. Nilsson, M. Ahman, Steel decarbonization in China—a top-down optimization model for exploring the first steps. J. Clean. Prod. **384**, 135550 (2023)
13. A. Jenkin, in *Steeling Ourselves: How Australia Can Support the Transition to Net-Zero Steel. Issue 26. Implementation Manual* (CSIRO 2022, Victoria, Australia)
14. The world counts, in *Environmental Impact of Steel Production.* Viewed 23 Apr 2023
15. K. Ogi, T. Shinoda, M. Mizui, Strength in concrete reinforced with recycled CFRP pieces. Compos. A Appl. Sci. Manuf. **36**, 893–902 (2005)
16. A.B. Kizilkanat, N. Kabay, V. Akyuncu, S. Chowdhury, A.H. Akca, Mechanical properties and fracture behaviour of basalt and glass fibre reinforced concrete: an experimental study. Constr. Build. Mater. **100**, 218–224 (2015)
17. M. Inman, E.R. Thorhallsson, K. Azrague, A mechanical and environmental assessment and comparison of basalt fibre reinforced polymer (BFRP) rebar and steel rebar in concrete beams. Energy Procedia **111**, 31–40 (2017)
18. S. Yin, *Development of Recycled Polypropylene Plastic Fibres to Reinforce Concrete* (Springer, 2016)
19. M. Mastali, A. Dalvand, A. Sattarifard, The impact resistance and mechanical properties of the reinforced self-compacting concrete incorporating recycled CFRP fibre with different lengths and dosages. Compos. B Eng. **112**, 74–92 (2017)
20. A.H. Ali, A. Gouda, H.M. Mohamed, H.M. Esmael, Experimental and numerical analysis of steel and fibre-reinforced polymer concrete beams under transverse load. ACI Struct. J. **199**, 109–121 (2022)
21. A.E. Naaman, Fiber reinforced concrete: five decades of progress, in *Proceedings of the 4th Brazilian Conference on Composite Materials* (Rio de Janeiro, Brazil) pp. 22–25
22. S. Ghanem, J. Bowling, Mechanical properties of carbon fibre reinforced concrete. Morehead State Univ. **1**, 1–44 (2019)
23. S. Yin, R. Tuladhar, T. Collister, M. Combe, N. Sivakugan, Mechanical properties and post-crack behaviours of recycled PP fibre reinforced concrete, in *Proceedings of the, 27th Biennial National Conference of the Concrete Institute of Australia* (Construction Innovations, Melbourne, 2015)
24. M. Mastali, A. Dalvand, The impact resistance and mechanical properties of self-compacting concrete reinforced with recycled CFRP pieces. Compos. B Eng. **92**, 360–376 (2016)
25. M. Yadav, A. Mathur, Bamboo as a sustainable material in the construction industry: an overview. Mater. Today Proc. **43**, 2872–2876 (2021)
26. J. Zhao, X. Luo, Z. Wang, S. Feng, X. Gong, E.D. Shumuye, Experimental study on bond performance of carbon-and glass-fiber reinforced polymer (CFRP/GFRP) bars and steel strands to concrete. Materials **14**, 1268 (2021)
27. A. Jahami, C.A. Issa, An updated review on the effect of CFRP on flexural performance of reinforced concrete beams. Int. J. Concr. Struct. Mater. **18**, 14 (2023)
28. M. Manjunatha, S. Dinesh, K.V.G.D. Balaji, C. Srilakshmi, Influence of PVC waste powder and silica fume on strength and microstructure properties of concrete: an experimental study. Case Stud. Constr. Mater. **15**, e00610 (2021)
29. U. Persson, *Construction for a Regenerative Future* (Routledge, Abingdon, UK, 2022)
30. Q. Yue, X. Chai, Y. Zhang, Q. Wang, H. Wang, F. Zhao, W. Ji, Y. Lu, Analysis of iron and steel production paths on the energy demand and carbon emission in China's iron and steel industry. Environ. Dev. Sustain. **25**, 4065–4085 (2022)
31. H. Ghanem, S. Chahal, J. Khatib, A. Elkordi, Flexural behavior of concrete beams reinforced with recycled plastic mesh. Buildings **12**, 1–21 (2022)
32. R. Geyer, J.R. Jambeck, K.L. Law, Production, use, and fate of all plastics ever made. Sci. Adv. **3**, 340–345 (2017)

33. T. Imjai, R. Garcia, M. Guadagnini, K. Pilakoutas, Strength degradation in curved fibre-reinforced polymer (FRP) bars used as concrete reinforcement. Polymers **12**, 16–53 (2020)
34. S. Gopalraj, T. Karki, A review on the recycling of waste carbon fibre/glass fibre-reinforced composites: fibre recovery, properties and life cycle analysis. SN Appl. Sci. **2**, 433 (2020)
35. Hazell, J., 2021. *Getting it right from the start: Developing a circular economy for novel materials.* London, UK: Green Alliance.
36. H. Zhou, B. Jia, H. Huang, Y. Mou, Experimental study on basic mechanical properties of basalt fibre reinforced concrete. Materials **13**, 1362 (2020)
37. H.M. Hamada, J. Shi, M.S. Jawahery, A. Majdi, S.T. Yousif, G. Kaplan, Application of natural fibres in cement concrete: a critical review. Mater. Today Commun. **35**, 55–72 (2023)
38. H. Snyder, Literature review as a research methodology: an overview and guidelines. J. Bus. Res. **104**, 333–339 (2019)
39. Z. Marcalikova, M. Racek, P. Mateckova, R. Cajka, Comparison of tensile strength fiber reinforced concrete with different types of fibers. Procedia Struct. Integr. **28**, 950–956 (2020)
40. A. Yusra, T. Triwulan, M. Safriani, M. Ikhsan, Use of bamboo fiber on the relationship between compressive strength and split tensile strength of high strength concrete. Mater. Sci. Eng. **933**, 012010 (2020)
41. P.S. Song, S. Hwang, B.C. Sheu, Strength properties of nylon- and polypropylene-fiber-reinforced concretes. Cem. Concr. Res. **35**, 1546–1550 (2005)
42. C. Meyer, Concrete and sustainable development. Am. Concr. Inst. Spec. Publ. **206**, 501–512 (2002)
43. Y. Han, T. Zhou, Performance analysis of high-performance concrete materials in civil construction. Materials **16**, 5711 (2023)
44. L. Mello, J.-L. Le, R. Ballarini, Effect of time-dependent bond slip on delayed failure of reinforced concrete frames. J. Eng. Mech. **148**(9), 04022047 (2022)
45. R. Abdulwahab, T.M. Akinleye, H.S. Taiye, Effects of polypropylene fibre on the compressive and splitting tensile strength of concrete. J. Mater. Eng. Struct. **5**, 15–22 (2017)
46. M.K. Chahrour, A. Hosen, Y. Goh, T.Y. Tong, S.P. Yap, M.A. Khadimallah, Failure mechanisms of structural bamboo using microstructural analyses. Adv. Civ. Eng. **2021**, 1–10 (2021)
47. O.D. Atoyebi, S.O. Odeyemi, J.A. Orama, Experimental data on the splitting tensile strength of bamboo reinforced lateritic concrete using different culm sizes. Data Brief **20**, 1960–1964 (2018)
48. A. Harison, A. Agrawal, A. Imam, Bamboo as an alternative to steel for green construction towards low cost housing. J. Environ. Nanotechnol. **6**, 100–104 (2017)
49. S. Yin, R. Tuladhar, M. Combe, T. Collister, M. Jacob, A.R. Shanks, Mechanical properties of recycled plastic fibres for reinforcing concrete. Fibre Concr. **2013**(1), 1–10 (2013)
50. A. Behnood, K.P. Verian, M.M. Gharehveran, Evaluation of the splitting tensile strength in plain and steel fiber-reinforced concrete based on the compressive strength. Constr. Build. Mater. **98**, 519–529 (2015)
51. T.V. Muthukumarana, M.A.V.H.M. Aranchchi, H.M.C.C. Somarathna, S.N. Raman, A review on the variation of mechanical properties of carbon fibre-reinforced concrete. Constr. Build. Mater. **366**, 130173 (2023)
52. J. Zhang, V.S. Chevali, H.M.C.C. Wang, C.H. Wang, Current status of carbon fibre and carbon fibre composites recycling. Compos. B **193**, 108053 (2020)
53. A. Raza, B. Ali, M.A. Nawaz, I. Ahmed, Structural performance of FRP-RC compression members wrapped with FRP composites. Structures **27**, 1693–1709 (2020)
54. H.A.A. Al-Katib, H.H. Alkhudery, H.A. Al Tameemi, Behavior of polypropylene fibers reinforced concrete modified with high performance cement. Int. J. Civ. Eng. Technol. (IJCIET) **9**, 1066–1074 (2018)
55. T.P. Huynh, T. Ho Minh Le, N. Vo Chau Ngan, An experimental evaluation of the performance of concrete reinforced with recycled fibers made from waste plastic bottles. Results Eng. **18**, 101205 (2023)
56. M.A. Mashrei, A. Sultan, A.M. Mahdi, Effects of polypropylene fibers on compressive and flexural strength of concrete material. Int. J. Civ. Eng. Technol. (IJCIET) **9**, 2208–2217 (2018)

57. K. Kumarasamy, G. Shyamala, H. Gebreyowhanse, Strength properties of bamboo fiber reinforced concrete. Mater. Sci. Eng. **981**, 032063 (2020)
58. A. Mark, A. Russell, A comparative study of bamboo reinforced concrete beams using different stirrup materials for rural construction. Int. J. Civ. Struct. Eng. **2** (2011)
59. U. Sayed, A. Dauletbek, X. Xin, R. Lorenzo, H. Li, A Review on the mechanical behaviour of bamboo reinforced concrete beams. J. Renew. Mater. **10**(12), 3629–3657 (2022)
60. A. Javadian, I. Smith, D. Hebel, Application of sustainable bamboo-based composite reinforcement in structural-concrete beams: design and evaluation. Materials **13**, 696 (2020)
61. P.R. Mali, D. Datta, Experimental evaluation of bamboo reinforced concrete beams. J. Build. Eng. **28**, 45–53 (2020)
62. S. Khalid, J.M. Irwan, M.H. Wan Ibrahim, N. Othman, S. Shahidan, Performance of plastic wastes in fiber-reinforced concrete beams. Constr. Build. Mater. **183**, 451–464 (2018)
63. C. Barris, L. Torres, I. Vilanova, C. Miàs, M. Llorens, Experimental study on crack width and crack spacing for Glass-FRP reinforced concrete beams. Eng. Struct. **131**, 231–242 (2017)
64. J. Zhao, G. Li, Z. Wang, X. Zhao, Fatigue behavior of concrete beams reinforced with glass- and carbon-fiber reinforced polymer (GFRP/CFRP) bars after exposure to elevated temperatures. Compos. Struct. **229** (2019)
65. A. Younis, U. Ebead, S. Judd, Life cycle cost analysis of structural concrete using seawater, recycled concrete aggregate, and GFRP reinforcement. Constr. Build. Mater. **175**, 152–160 (2018)
66. T. Cadenazzi, G. Dotelli, M. Rossini, S. Nolan, A. Nanni, Cost and environmental analyses of reinforcement alternatives for a concrete bridge. Struct. Infrastruct. Eng. **16**, 787–802 (2020)
67. A. Berg, L. Bank, M. Oliva, J. Russell, Construction and cost analysis of an FRP reinforced concrete bridge deck. Constr. Build. Mater. **20**, 515–526 (2006)
68. A. Younis, U. Ebead, P. Suraneni, A. Nanni, Cost effectiveness of reinforcement alternatives for a concrete water chlorination tank. J. Build. Eng. **27**, 38–50 (2020)
69. R. Bohm, M. Thieme, D. Wohlfahrt, D.S. Wolz, B. Richter, Reinforcement systems for carbon composites based on low-cost carbon fibers. Fibers **6**, 402–424 (2018)
70. L. Shen, M. Soliman, S. Ahmed, A probabilistic framework for life-cycle cost analysis of bridge decks constructed with different reinforcement alternatives. Eng. Struct. **245**, 112879 (2021)
71. S. Yin, R. Tuladhar, T. Collister, M. Combe, N. Sivakugan, Use of macro plastic fibres in concrete: a review. Constr. Build. Mater. **93**, 180–184 (2015)
72. B. Ali, H. Ahmed, H. Hafez, A. Brahmia, M. Ouni, A. Raza, Life cycle impact assessment (cradle-to-gate) of fibre reinforced concrete application for pavement use: a case study of Islamabad city. Int. J. Pavement Res. Technol. **16**, 247–263 (2022)
73. I. Hussain, B. Ali, T. Akhtar, M. Jameel, S. Raza, Comparison of mechanical properties of concrete and design thickness of pavement with different types of fiber-reinforcements (steel, glass, and polypropylene). Case Stud. Constr. Mater. **13**, 48–69 (2020)
74. S. Yin, R. Tuladhar, M. Sheehan, M. Combe, T. Collister, A life cycle assessment of recycled polypropylene fibre in concrete footpaths. J. Clean. Prod. **112**, 2231–2242 (2015)
75. J. Brown, Glass fibre reinforced polymer bars in concrete compression members (2015). Retrieved at: https://core.ac.uk/download/pdf/83963077.pdf
76. N. Hidayat, A. Aldo, Bamboo reinforcement concrete beam as innovation for low-cost earthquake resistant house. J. Phys. Conf. Ser. **1625**, 12–20 (2020)
77. A. Hammad, C. Gyan, H. Kasun, S. Rehan, A comprehensive review on construction applications and life cycle sustainability of natural fibre biocomposites. Sustainability **14**, 159–170 (2022)
78. J. Ahmad, Z. Zhou, Mechanical properties of natural as well as synthetic fibre reinforced concrete: a review. Constr. Build. Mater. **333**, 43–47 (2022)
79. E. Mello, C. Ribellato, E. Mohamedelhassan, Improving concrete properties with fibres addition. Int. J. Civ. Environ. Eng. **8**, 249–254 (2014)
80. P. Zhang, C. Wang, C. Wu, Y. Guo, Y. Li, A review on the properties of concrete reinforced with recycled steel fiber from waste tires. Rev. Adv. Mater. Sci. **29**, 4–7 (2022)

81. B. Bjorkman, C. Samuelsson, Recycling of Steel, in *Handbook of Recycling* (2014), pp.65–83
82. B. Lee, M. Trcka, J. Hensen, Embodied energy of building materials and green building rating. A Case Study Ind. Halls **2**, 67–71 (2011)
83. J. Pandit, M. Watson, A. Qader, Reduction of greenhouse gas emissions in steel production. CO2CRC **3**, 3–110 (2020)
84. P. Woodall, What is the carbon footprint of steel? A Res. Pap. **2**, 11–23 (2023)
85. R.P. Borg, O. Baldacchino, L. Ferrara, Early age performance and mechanical characteristics of recycled PET fibre reinforced concrete. Constr. Build. Mater. **108**, 29–47 (2016)
86. A. Isa, N. Nosbi, M. Ismail, H. Akil, Methods to reinforce and expect fibre composite degradations. A Rev. Recycl. Carbon Fibres **14**, 15–18 (2022)
87. K. Kawajiri, K. Sakamoto, Environmental impact of carbon fibers fabricated by an innovative manufacturing process on life cycle greenhouse gas emissions. An Investig. Sustain. Mater. Technol. **31**, 286–365 (2022)
88. M. Kanemasu, W. Nishimaru, A. Yoshikawa, T. Yamamori, Reduction of environmental impact by recycling waste composite material for aircraft. Mitsubishi Heavy Ind. Tech. Rev. **55**, 5 (2018)
89. E. Ferreira, R. Kalliola, K. Roukolainen, Bamboo, climate change and forest use: a critical combination for southwestern Amazonian forests? Ambio **49**, 1353–1363 (2019)
90. Plymouth University, *Natural Fibres—Environmental, Technical, and Economic Issues* (Plymouth University, Plymouth, England, 2023)
91. M. Mahmut, Extrusive igneous rocks: basalt. An Investig. Igneous Rocks **6**, 32–44 (2023)
92. S. Grzesiak, M. Pahn, A. Klinger, E. Akpan, Mechanical and thermal properties of basalt fibre reinforced polymer lamellas for renovation of concrete structures. Polymers **14**, 790 (2022)
93. A. Pavlovic, T. Donchev, D. Petkova, N. Staletovic, Sustainability of alternative reinforcement for concrete structures: life cycle assessment of basalt FRP bars. A Rev. Previous Res. **334**, 127–424 (2022)
94. A. Bala, S. Gupta, Engineered bamboo and bamboo-reinforced concrete elements as sustainable building materials: a review. Constr. Build. Mater. **394**, 132116 (2023)
95. N.S. Kathiravan, R. Manojkumar, P. Jayakumar, J. Kumaraguru, V. Jayanthi, State of art of review on bamboo reinforced concrete. Mater. Today: Proc. **45**, 1063–1066 (2021)
96. Vengala, Comparison of embodied energy in different bamboo-based houses, in *Advances in Sustainable Construction Materials: Select Proceedings of ASCM 2019*, vol. 68 (2020), pp. 197–208
97. H. Cheng, L. Guo, L. Zheng, Z. Qian, S. Su, A closed-loop recycling process for carbon fiber-reinforced polymer waste using thermally activated oxide semiconductors: carbon fiber recycling, characterization and life cycle assessment. Waste Manag. **153**, 283–292 (2022)

Chapter 17
Lightweight Alkali-Activated Materials: Sustainable Route to Passively Decarbonize the Built Environment

Rui M. Novais and João A. Labrincha

Abstract Over recent years, alkali-activated materials (AAMs) have been regarded as potentially lower carbon footprint alternatives to Portland cement binders, with a role to play towards the decarbonization of the built environment. The use of low density AAMs seem particularly promising extending the application range of these novel binders to other high added value applications including their use as thermal and acoustic barriers, moisture regulators, and thermal energy storage materials. These additional functionalities have the potential to further contribute to the reduction of energy consumption and the improvement of living conditions. Herein, the most relevant and recent research focusing the use of lightweight AAMs in the built environment is summarised and critically discussed.

17.1 Introduction

The building sector is one of the main contributors to global final energy consumption (30%) and energy derived emissions (26%) [1], this being particularly relevant under the EU climate-neutrality strategy by 2050. The sector is currently underperforming, and the operational emission must fall by ca. 50% to get on track with the net zero emission scenario by 2030. Urgent actions to decarbonize the building stock and realign with the forecasted emissions path are imperative. The adoption of energy efficiency measures can play a significant role in the decarbonization scenario. Amongst the various approaches the use of thermal insulating materials is particularly attractive as they mitigate the energy losses inside buildings, therefore reducing the need for heating, ventilation, and air conditioning (HVAC) systems [2, 3]. This is critical as heating and colling needs account for a significant share of the energy used in buildings [4]. Conventional thermal insulating materials encompass inorganic (e.g., rock and glass wool) and organic materials (e.g., expanded polystyrene

R. M. Novais (✉) · J. A. Labrincha
Department of Materials and Ceramic Engineering/CICECO-Aveiro Institute of Materials, University of Aveiro, Campus Universitário de Santiago, Aveiro, Portugal
e-mail: ruimnovais@ua.pt

D. A. H. Hanaor (ed.), *Sustainable Materials for the Built Environment*, Springer Series in Materials Science 352, https://doi.org/10.1007/978-3-031-97818-0_17

and polyurethane foams). The former are gradually being replaced due to their high embodied energy, while the latter are known to present poor thermal stability and potential safety hazards [5]. In addition, in the era of unprecedented use of resources [6, 7], with still not fully perceived consequences on the environment and human health [8], other parameters including the embodied energy, the carbon footprint and the recyclability of the materials are becoming hugely important [9, 10]. Lightweight alkali activated materials (AAMs) emerge as an excellent alternative considering both technical (e.g., high temperature stability, low thermal conductivity, non-flammable) and environmental reasons (e.g., green and low cost technology enabling the use of a wide range of industrial wastes as precursors) [11–13]. Their dense counterparts have been mainly considered as a possible eco-friendlier alternative to Portland cement [14, 15] provided that the mixture design is property considered [16]. Nevertheless, lightweight AAMs present additional features suggesting their use in less explored, but promising, applications including non-structural building materials (e.g., thermal [17–19] and acoustic panels [20–22], moisture regulators [23–25]; fire resistant materials [26–28], EMI shielding [29, 30] and thermal energy storage panels [31–33]), pH regulators [34–36], adsorbents envisioned for wastewater treatment [37, 38] and catalysts or catalytic supports [39–41]. As a result, the interest in such materials has increased dramatically in recent years as demonstrated by the data shown in Fig. 17.1. These novel applications might be a key driver to foster the use of this technology (alkali activation). It should be noted that the survey was performed by using the keywork "geopolymer" and not "alkali activated material" due to the much higher number of articles using this terminology. The differences between both systems are beyond the scope of the present chapter, the distinction can be found in relevant literature [42–44]. Herein, the term AAM will be employed, geopolymers being considered subsets of AAMs.

The connection between the identified 383 articles focusing "lightweight geopolymers" and the dominant keywords was then evaluated using the *VOS viewer* software [45]. A co-occurrence map is presented in Fig. 17.2. The map was generated by defining the minimum number of occurrences as ten. From the initial 2630 keywords, only 91 met the defined criterion. Disregarding the expected keywords associated with the nature of the binders (e.g., "inorganic polymers", "geopolymers", "geopolymer"), the most frequently used keyword is "compressive strength", followed by "fly ash", "thermal conductivity", "light weight concrete" and "lightweight geopolymer". Interestingly, the ninth most frequently used term is "aggregates", and this is associated with one of the strategies used to produce lightweight AAMs (as detailed in Sect. 17.2).

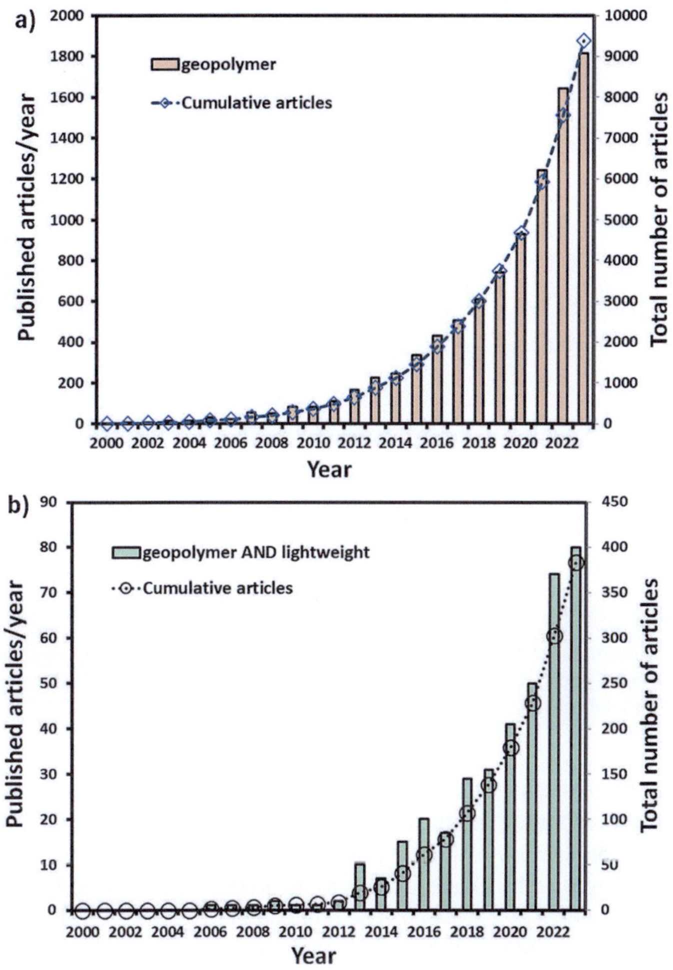

Fig. 17.1 Number of publications on **a** geopolymers and **b** lightweight geopolymers since the year 2000. Information retrieved from Scopus database in July 2024. The keyword "geopolymer" was applied to generate (**a**), while in (**b**) two keywords were combined ("geopolymer" AND "lightweight"). A total of 9384 papers, including articles and reviews, and excluding conference papers, book chapters and reprints, were identified

17.2 Different Strategies to Produce Lightweight Alkali-Activated Materials

One of the first investigations regarding the production of lightweight AAMs was reported by Le-ping et al. in 2010 using aluminium powder as a pore-forming agent [46]. Bodies with porosity ranging from 40 to 83% and compressive strength comprehended between 6.0 and 13.7 MPa were produced. A similar approach was then

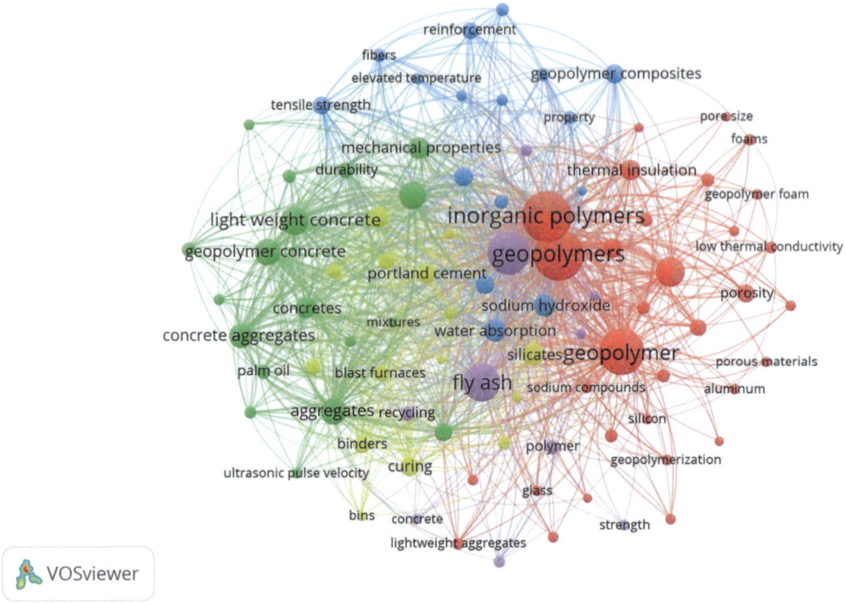

Fig. 17.2 Co-occurrence map of the keywords reported in literature for studies dealing with lightweight geopolymers. The size of the label is related to the keyword frequency, while the connection lines indicate the link strength with other keywords

reported by Prud'homme et al. [47] but in this case in situ foaming was achieved through the incorporation of silica fume in the slurries, generating dihydrogen and leading to the production of highly porous bodies. In the same year a pioneering approach was reported by Okada et al. [48] involving the production of composites containing PLA fibres by extrusion. The production of porous AAMs was achieved after extracting the polymer fibres using alkali treatment and/or by thermal decomposition. Since then, a large number of articles has been published and various techniques employed. In this section, the most common routes to produce lightweight AAMs will be briefly described, the pros and cons of each technique will be highlighted. It should be noted that the emphasis is attributed to protocols envisioning the production of materials for the built environment. Detailed descriptions of the existing protocols can be found in recent review papers [49, 50].

The two most commons, and mature, methods that have been employed in literature to produce lightweight AAMs are by physical and chemical foaming or by incorporating lightweight and/or porous aggregates. In the physical foaming (also known as mechanical foaming) gas bubbles are introduced in the slurry and then vigorously mixed to produce bodies containing a large number of voids [21, 51], the mixing speed being a key parameter to control the final porosity of the specimens [50]. The chemical route involves the use of small amounts of pore forming agents that trigger chemical reactions and generate gas bubbles. Common agents include fine metallic powders (e.g., Si, Al, Zn) and hydrogen peroxide [11]. Figure 17.3

presents a schematic representation of the chemical and mechanical foaming [52]. The figure also illustrates the role a surfactant, typically employed to stabilize the liquid–gas interface (e.g., reduce the surface tension) and ensure a more homogeneous pore size distribution. Recent studies have shown that the amount and nature of the surfactant can play a significant role on the type and pore size distribution of the produced pores [53–55].

The physical and chemical foaming approaches enable the production of highly porous specimens with properties (e.g., mechanical, thermal and acoustic) that can be tailored according to the envisioned application. However, the production of bodies with homogeneous pore size distribution by these methods is challenging as porosity is dependent on several parameters including the nature of the precursors (e.g., reactivity/dissolution rate that affect the viscosity and the setting time of the pastes) [56], the mixing speed [57], the type and amount of foaming agent [57–59], but also on the geometry of the mould (e.g., the height and volume affect the expansion of the slurry in the chemical route) [60–62]. This said, a precise control of the synthesis

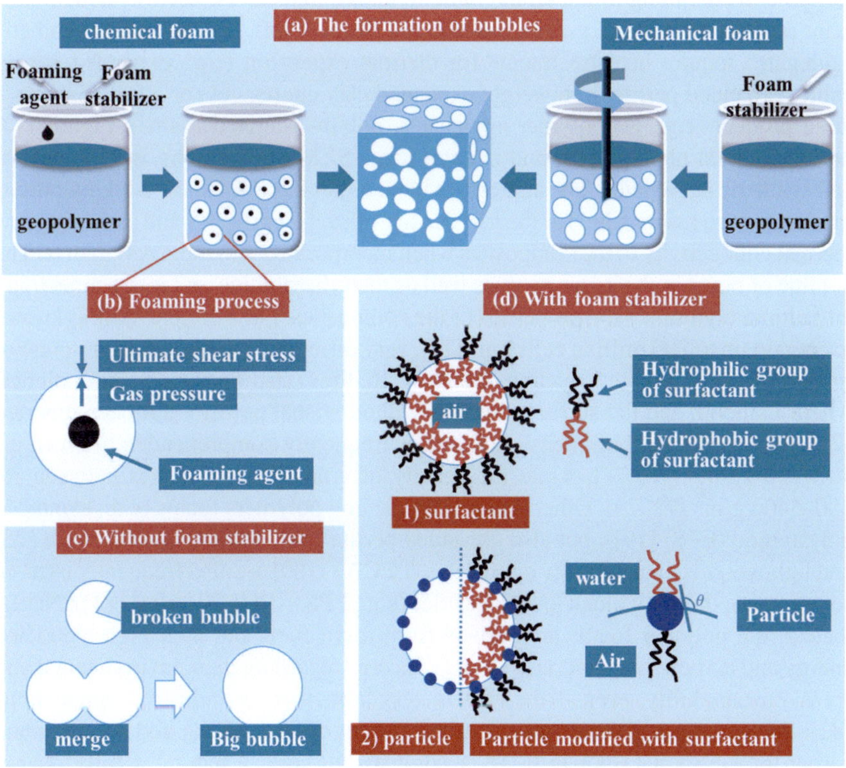

Fig. 17.3 Schematic representation of the chemical and mechanical foaming methods with or without using surfactants. Reprinted from [52] with permission from Elsevier (license: 5835860522138)

conditions is feasible and bodies with fairly homogenous pore size distribution can be produced [63, 64].

An alternative strategy to decrease the density of AAMs is by the incorporation of porous and/or lightweight aggregates. This approach has long been established for Portland cement-based composites [65–68], and their AAM counterparts are produced by a similar manner. Typically, the solid precursors and the activators are initially mixed for a specific time until a homogenous slurry is produced. Afterwards, the aggregates are added to the paste, mixed, and then the paste is poured into moulds [69, 70]. Other studies follow a different protocol, in which the aggregates are initially mixed with the solid precursors and only then the water and the alkali activator is added to the solid materials (e.g., aggregates and raw materials) [69, 71]. This methodology is straightforward compared to the previous avenue (e.g., physical or chemical foaming), allowing a precise control of the properties of the composites simply by controlling the recipe of the binder (e.g., liquid-to-solid ratio) and the nature and volume of the aggregates [20, 72, 73]. The aggregates can be divided according to their origin in two classes: (i) natural or (ii) synthetic. The former corresponding to materials found in nature that can be used without significant processing steps (e.g., cork, rice husk, pumice, tuff, scoria), while synthetic aggregates require heat treatments for thermal expansion (e.g., expanded vermiculite, expanded perlite, lightweight expanded clay aggregates) or a fully manufactured product (e.g., geopolymer aggregates, polymer-derived foams). Figure 17.4 presents digital photographs, and in some cases SEM micrographs, of both natural and synthetic aggregates. The aggregates included in Fig. 17.4 show a highly porous microstructure responsible for the low density of such aggregates and endowing low thermal conductivity to the composites when incorporated into the AAM matrix. The volume of pores in the aggregates, as well as their distribution and interconnectivity will ultimately impact the properties of the composites. For example, cork is known to contain up to 200 million cells/cm^3 [74], very different from the microstructure of rubber derived from the recycling of end-of-life tires (also known as crumb rubber) where a smooth and flat surface structure without visual porosity has been reported [75]. As a result, the bulk density of cork is frequently comprehended between ca. 100–130 kg/m^3 [76, 77], while the density of crumb rubber ranges between ca. 400–540 kg/m^3 [78, 79]. Other materials including polymers foams (e.g., expanded polystyrene (EPS) [19]), but also expanded perlite [28], expanded vermiculite [73] and lightweight expanded clay aggregates (LECA®) [80] are also frequently used, in some cases exhibiting much lower densities (e.g., EPS—10 to 20 kg/m^3 [81]). Nevertheless, the polymer foams are derived from fossil fuels and their high embodied energy and carbon footprint raise strong concerns regarding their environmental and social sustainability, even if the cost is very attractive for industrial applications [82]. For example, EPS has an embodied energy of 88.6 MJ/kg and an embodied carbon of 2.55 kgCO_2/kg [83]. As a comparison, cork, a natural a fully renewable resource, shows ca. 22 times lower embodied energy (4.0 MJ/kg) and 13 times lower embodied carbon (0.19 kgCO_2/kg) [83]. As for the mentioned synthetic aggregates their expansion requires very high temperatures (e.g., expanded vermiculite (800–1000 °C) [84], expanded perlite (700–1200 °C) [85], expanded glass aggregate

(900–1300 °C) [86]) which again result in high embodied energy and carbon (e.g., expanded perlite—10.0 MJ/kg and 2.55 $kgCO_2/kg$) [83]. As a result, there is a need to develop new synthetic aggregates that show lower embodied energies and reduced carbon footprint. Alkali activation can also be employed to synthesize lightweight aggregates [87–89]. Conventional methods include sintering [88] and cold bonding [89], the latter being desirable from an environmental perspective. In a recent work, a simple and scalable procedure (e.g., suspension-solidification) was used to fabricate highly porous mm-size spheres by the alkali activation of red mud (challenging industrial waste derived from the Bayer process [90]) [28]. The spheres showed a geometric density of 570 kg/m^3, enabling their use as lightweight aggregate. A digital photograph and a representative SEM micrograph is shown in Fig. 4d.

The use of lightweight aggregates also raises technical challenges including the higher water-to-binder ratio required for synthesising the AAM composites, which subsequently reduce their mechanical performance [79]. The expected increase in the viscosity of the slurries when rising the volume of aggregate in the formulations might lead to the partial destruction of the aggregates, this being particularly relevant when dealing with fragile aggregates such as expanded perlite [28]. Another critical parameter that can strongly impact the properties of the composites is the possible formation of a interfacial transition zone (ITZ) [91], even if this feature seems much more problematic in Portland cement composites [92]. Other techniques including the use of sacrificial fillers [48, 93], freeze casting [94, 95], additive manufacturing [96], microwave irradiation [97, 98], can also be used to produce lightweight AAMs. Nevertheless, and despite their merits they will not be considered in this chapter as their synthesis is more complex (e.g. sacrificial fillers, freeze casting) or the technology is at an embryonic stage (e.g., additive manufacturing, microwave irradiation), factors that hinder their mass production (at least in the current scenario) at a competitive cost compared to the chemical foaming or the use of lightweight aggregates.

17.3 Thermal and Acoustic Barriers

The prospects of using AAMs in the built environment as low thermal conductivity materials alternative to the conventional fossil fuel-derived foams is explained by a multitude of potential benefits including high thermal stability, non-flammability and green synthesis. Since the original works in the field [46–48], the impact of key influential parameters such as nature and amount of the foaming agent [64, 99] or the aggregate [20, 25, 75] has been deeply studied. In most cases, an increase on the amount of foaming agent or aggregate induces a reduction on the density of the AAMs. Nevertheless, in the former there is an upper limit after which the foam is unstable and collapses [100], while in the latter the maximum amount of the aggregate is controlled by the drop on the workability of the samples [78] but might be surpassed by modifying the recipe of the binder (e.g., higher liquid phase) or by adding superplasticizers [52]. Table 17.1 summarises literature results on lightweight

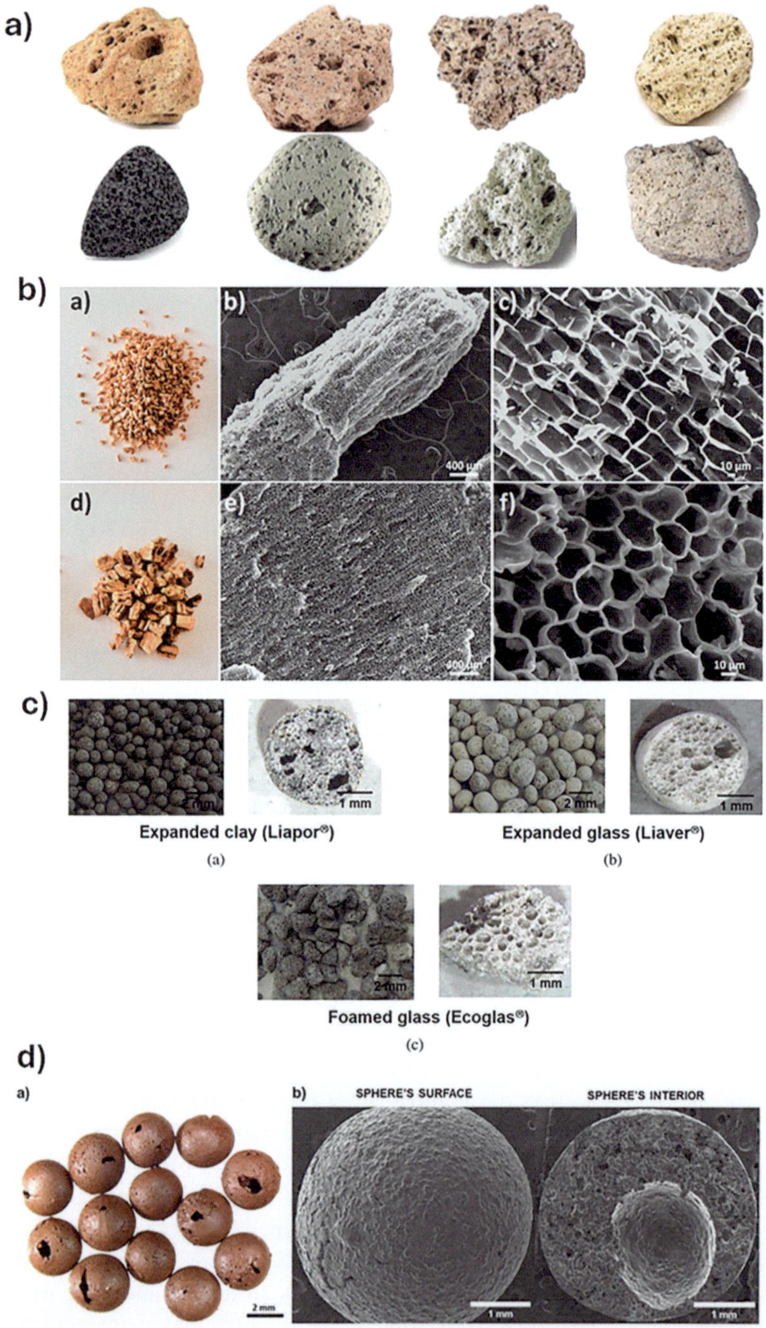
a)
b)
a)
b)
c)
400 µm
10 µm
d)
e)
f)
400 µm
10 µm
c)
2 mm
1 mm
Expanded clay (Liapor®)
(a)
2 mm
1 mm
Expanded glass (Liaver®)
(b)
2 mm
1 mm
Foamed glass (Ecoglas®)
(c)
d)
a)
b)
SPHERE'S SURFACE
SPHERE'S INTERIOR
2 mm
1 mm
1 mm

◂**Fig. 17.4** Digital images (and in some cases SEM micrographs) of natural and synthetic lightweight aggregates: **a** pumice stone [155], **b** cork granules [25], **c** expanded clay, expanded glass and foamed glass [156] and d) red mud-containing geopolymer spheres (reprinted from [28] under a Creative Commons license (CC BY-NC-ND 4.0)). a–c were reprinted with permission from Elsevier (IDs: 5836411480758, 835980600524, 5836410464598)

AAMs envisioned for the building sector. Besides the thermal conductivity, the table compiles the bulk density, total porosity, compressive strength and the production method. It should be noted that the studies were sorted in ascending order of thermal conductivity.

To provide additional insights, the thermal conductivity and the bulk density of the most relevant studies (selecting criterion: thermal conductivity below 300 mW/m K) were plotted in Fig. 17.5. It should be highlighted that for clarity only the lowest thermal conductivity value reported in each study was included.

As can be seen both strategies enable the production of ultra lightweight materials. Interestingly, the lowest value reported in the table corresponds to the work of Duan et al. [101] where hydrogen peroxide (H_2O_2) was used to produce a foamed paste and then polystyrene beads were used as lightweight aggregate. The combination of chemical foaming with the use of a lightweight aggregate lead to the production of specimens with ultra-low density (100 kg/m^3). The idea behind the study is particularly attractive even if in this case a polymer foam has been used in very high amounts, this strategy deviating from the environmental sustainability paradigm envisioned for future building materials. A similar approach was reported in [125] but using cork as a lightweight aggregate and aluminium powder as foaming agent. Surprisingly, the density of the cork foams was slightly higher than those seen in the foams prepared using the same amount of foaming agent but without adding cork, and this shows that the presence of the aggregate disrupts the formation of gas bubbles and hinders their entrapping in the mixture. As mentioned previously, the chemical foaming method is challenging and the combination with lightweight aggregates raises complexity to the system. Nevertheless, this might be an interesting avenue to produce bodies with very low thermal conductivity. The results included in Table 17.1 also show the feasibility of producing AAMs combining suitable mechanical performance with low thermal conductivity values. For example in [105] bodies showing compressive strength of 10 MPa and thermal conductivity of 110 mW/m K were produced using expanded glass combined with an air entraining agent.

An additional advantage of the AAMs compared to polymer foams is their much higher thermal stability. Alves et al. evaluated the thermal stability of AAM composites prepared using various aggregates, including EPS, cork, crumb rubber, expanded vermiculite and LECA® [75]. The composites containing expanded vermiculite and LECA® showed remarkable thermal stability up to 600 °C (for 2 h), while the other composites showed lower thermal stability due to the organic nature of the aggregates. Recently, the fire resistance of composites containing alkali activated red mud spheres was evaluated according to the ISO-834 test [28]. The spheres-containing composite maintained their structural integrity, showing no visible cracks and remaining unbent during the fire test (maximum temperature 1050 °C).

Table 17.1 Summary of bulk density, total porosity, compressive strength and thermal conductivity of various AAMs reported in literature and envisioned for the built environment

Production method	Foaming agent or aggregate	Bulk density (g/cm^3)	Total porosity (%)	Compressive strength (MPa)	Thermal conductivity (mW/m K)	References
LA	EPS + H_2O_2	0.10–0.40	–	3–4	34–83	[101]
CF	Al	0.64	83	1.3	45	[102]
CF	H_2O_2	0.15–0.30	72–85	0.7–2.2	62–85	[103]
LA	EPS	0.17–0.29	–	0.24–0.57	63–82	[75]
MF	Surfactant	0.35–0.46	–	0.09–0.13	64–72	[104]
LA	Cork	0.30–0.69	–	0.33–2.58	68–161	[75]
LA	Expanded glass + air entraining agent	0.65–0.90	–	6–12	70–174	[105]
LA	Cork	0.26–0.83	–	0.23–2.86	72–156	[70]
CF	H_2O_2	0.24–0.37	75–81	0.4–1.4	74–90	[106]
CF	Al + surfactant	0.43–0.85	74–87	0.6–4.3	79–170	[64]
CF	H_2O_2	0.44–1.10	48–81	0.26–21.0	82–227	[99]
CF	H_2O_2 + surfactant	0.26–0.48		0.5–3.3	84–140	[107]
CF	H_2O_2 + surfactant	0.31–0.36	70–82	1.45–1.65	95–139	[108]
CF	H_2O_2	0.56–1.20	42–73	1.2–8.9	107–432	[18]
CF	H_2O_2	0.33–0.85	–	1.8–5.2	109–170	[109]
CF	H_2O_2 + surfactant	0.37–0.74	66–83	0.3–11.6	110–170	[110]
LA	EPS	0.50–0.80	–	1.8–6.0	121–207	[19]
CF	Aluminium scrap recycling waste	0.38–0.47	83–86	1.1–2.0	140–150	[111]
LA	Expanded perlite	0.92–1.74	–	0.21–14.9	140–318	[28]
LA	Expanded vermiculite	0.55–0.91	–	0.58–2.74	150–254	[75]
CF	Al	–	–	–	150–600	[112]
MF	Foam concentrate	0.60–1.50	–	3–50	150–500	[17]
CF	Al	–	30–70	–	150–600	[113]
LA	Crumb rubber	0.74–0.99	–	0.97–1.93	159–234	[75]
CF	Silica fume	0.5	80	–	160	[61]
LA	Expanded vermiculite	0.87–1.45	–	0.55–11.7	171–321	[28]

(continued)

Table 17.1 (continued)

Production method	Foaming agent or aggregate	Bulk density (g/cm^3)	Total porosity (%)	Compressive strength (MPa)	Thermal conductivity (mW/m K)	References
LA	Alkali-activated spheres	0.84–1.19	–	0.98–2.36	175–234	[28]
CF	H_2O_2 + surfactant	0.14–1.02	32–63	3.2–44.8	183–646	[114]
CF	H_2O_2	1.0–1.7	44–56	3.6–11.9	190–440	[24]
MF	Surfactant	-	–	1.7–2.4	200–260	[115]
LA	LECA®	0.70–0.86	–	2.89–7.63	218–261	[75]
CF	Silica fume	0.53	–	–	220–240	[47]
LA	Crumb rubber	1.07–1.28	29.7–36.3	2.07–3.29	237–298	[116]
CF	H_2O_2	-	–	1.55–1.90	250–270	[115]
CF	H_2O_2	0.86–1.09	54.8–63.1	1.57–2.97	267–334	[117]
LA	Cenospheres	0.98	–	17.5	280	[118]
LA	Polyurethane	0.89–1.15	10.3–12.7	–	300–360	[119]
CF	SiC	–	32–52	–	420–670	[120]
LA	Oil palm shell	1.29–1.29	–	8.3–30.1	470–580	[121]
CF	Recycled Al foil powders	0.63–0.79	60–75	3.5	581–700	[122]
CF	H_2O_2 + surfactant	0.23–1.0	–	0.26–5.9	–	[123]
CF	Silica fume	0.97–1.17	–	4–12	–	[124]
MF	Surfactant + H_2O_2	0.40–0.43	78–83	0.64–0.81	–	[51]

Note only studies dealing with the chemical (CF) or mechanical foaming (MF) and the incorporation of lightweight aggregates (LA) were included

Besides the low thermal conductivity and high thermal stability that can be observed in lightweight AAMs, these materials can also present other key features including sound absorption properties. This is particularly relevant as noise pollution is now a key global public health concern [126, 127]. However, the number of studies addressing the use of AAMs as acoustic barriers in building applications is much lower than those considering their use as low thermal conductivity materials as highlighted in our recent review [11]. One of the first investigation on the topic was reported in 2014 by Hung et al. [128]. The mechanical foaming route (e.g., incorporation of air bubbles produced by a foam generating tank) was employed to fabricate porous AAMs. The impact of the density of the foams on the sound absorption coefficient (α) was evaluated, results showing that lower densities significantly increase the α across the studied frequency range (100–4000 Hz). The lightest specimen (0.4 g/cm^3) shows a α ranging from 0.5 (100 Hz) to 0.9 (ca. 4000 Hz). Other studies follow, and in 2015 Zhang et al. [17] used a similar protocol (e.g., mechanical foaming)

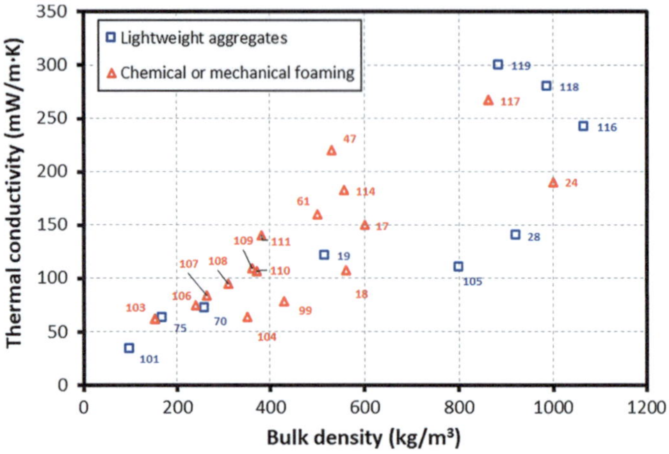

Fig. 17.5 Thermal conductivity and bulk density of various lightweight AAMs reported in literature

however a very different behaviour was observed. The sound absorption coefficient of the specimens has very high in the low frequency range (below 150 Hz), being low ($\alpha < 0.3$) beyond this frequency range and up to 1600 Hz. Chemical foaming was employed by Papa et al. [61] to produce silica-fume based foams. The samples showed two absorption peaks: one between 500 to 2500 Hz, and the second between 4000 to 5000 Hz. Interestingly, the studied samples, comprising the use of silica fume and/or metakaolin and different activating solutions (sodium or potassium based), showed very different behaviour suggesting that the properties of the AAMs can be tuned by modifying the formulations possibly due to the modification the porosity and tortuosity of the foams. The sample produced using metakaolin and potassium silicate solution reached an impressive $\alpha = 0.9$ (1210 Hz).

AAMs prepared by adding light aggregates can also exhibit very interesting sound absorption coefficient [75, 80]. As an example, cork-containing AAM composites prepared with 85 vol% of black expanded cork show $\alpha = 1.0$ (ca. 2100 Hz) [20]. The same group also showed the viability of tuning the sound absorption coefficient of the AAM composites simply by adjusting the volume and size of the cork (e.g., expanded cork granules) as demonstrated in Fig. 17.6, showing the feasibility to tune the samples according to the frequency of the noise pollution source [25].

Other recent approaches include the synthesis of composites containing silica aerogel, material know to present extremely low thermal conductivity (12–18 mW/m K) [129]. The composites were prepared using H_2O_2 coupled with two surfactants (e.g., sodium dodecyl sulphate and triton X 100) and with 10 or 20 vol% of silica aerogel particles (four different particle sizes were used: 2–40 μm, 100–700 μm, 100–120 μm and 700–4000 μm). Figure 17.7 presents the acoustic absorption of the studied samples. Most of the samples enriched with aerogel particles showed enhanced sound absorption coefficient compared with the reference composition (foamed AAM without silica aerogel) and this was attributed to the higher open

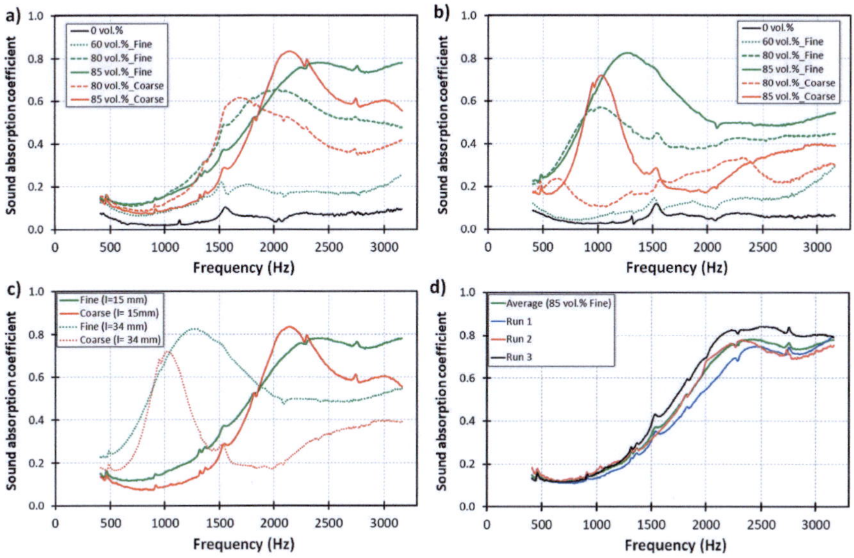

Fig. 17.6 Sound absorption coefficient for cork-containing composites measured in samples having **a** 15 mm and **b** 34 mm thickness (reprinted from [25] with permission from Elsevier (ID: 5838730101350)). **c** Illustrates the influence of the specimens' thickness in the sound absorption coefficient for composites prepared with 85 vol% cork, while **d** shows different runs and the average curve for the fine cork granules. *Note* Fine granules (0.5 < d < 2.0 mm) and coarse granules (3.5 < d < 6.0 mm)

porosity and the inherent acoustic absorption of the silica aerogel. The sample coded as "AG4-10" (containing 10 vol.% of silica aerogel particles with 700–4000 μm) showed α = 0.88 (1307 Hz) when using 20 mm thickness specimens, while in thicker samples (50 mm) the maximum sound absorption coefficient dropped to 0.79 and shifted to lower frequencies (505 Hz). These are interesting results that provide additional insights regarding the feasibility of tuning the AAMs sound absorption properties. Nonetheless, the economic and environmental sustainability of this approach needs to be carefully considered, at least when considering a global implementation. It should be noted that the economic argument is also valid when using natural, but finite, resources such as cork.

These studies provide valuable information regarding the feasibility of lightweight AAMs to act as acoustic barriers. Nevertheless, additional investigations are required not only to validate previous findings but to provide additional clues regarding the impact of the samples pore size distribution and tortuosity, or the nature and amount of the aggregate on the sound absorption coefficient.

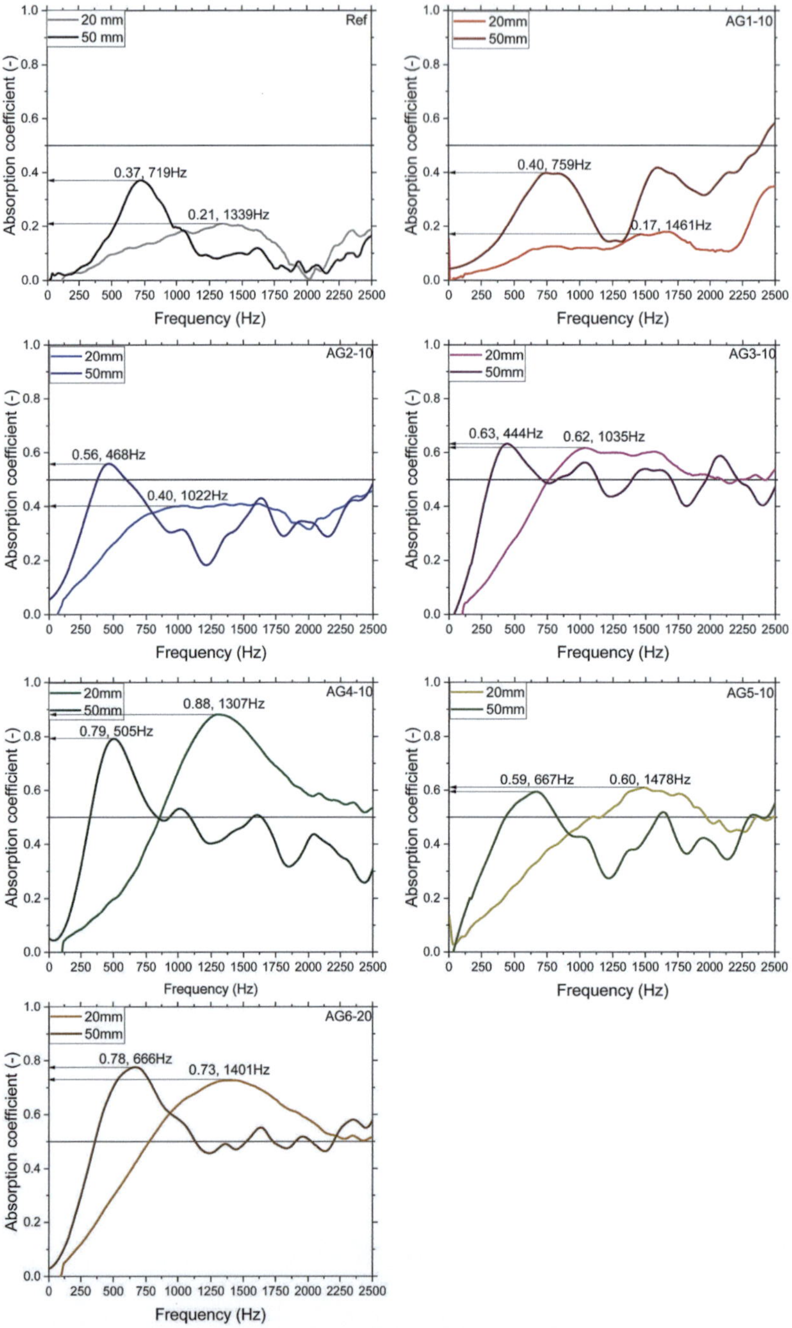

Fig. 17.7 Sound absorption coefficient for the foamed AAM and the silica aerogel containing AAMs (reprinted from [129] under a Creative Commons license (CC BY 4.0))

17.4 Moisture Regulators

Besides the thermal and acoustic properties there are other important characteristics that should be considered when designing future building materials. A key feature to deploy novel materials is there multifunctionally, i.e. a material that can fulfil several tasks simultaneously. As mentioned in the introduction, one strategy that can mitigate the energy consumption of buildings is using thermal insulators, however, a downside of this option is that highly insulated walls pose risk of condensation and moisture related issues [130] known to affect human health [131]. The air quality inside buildings is currently a major source of concern, particularly in the post-pandemic scenario (Covid-19) [132] as literature suggest links between the humidity level and the spread of the virus [130]. In this context, a very desirable property of building materials is their ability to passible adjust the humidity levels [11]. The moisture buffer value (MBV) is a key performance indicator quantifying the capacity of the given material to absorb and release moisture upon daily fluctuations of humidity [133]. Common hygroscopic materials include wood and cellulose insulation [134], but modified mortars can also be used to adjust the humidity level inside buildings particularly when doped with superabsorbent polymers [135] or porous aggregates (e.g., perlite, zeolite) [136]. Studies evaluating the feasibility of lightweight AAMs to act as moisture regulators are limited, the few exceptions include the pioneer work of Rossi et al. [24] where AAM mortars were produced using construction and demolition wastes as aggregate and hydrogen peroxide as foaming agent. The MBV of the non-foamed mortar (0.8 g/m^2 Δ%RH) was substantially increased to 5.6 g/m^2 Δ%RH in the sample prepared with higher amount of foaming agent. Despite the promising result, the mass evolution of the mortars upon cycling followed an increasing trend which might suggest a moderate saturation tendency. A much lower MBV (1.64 g/m^2 Δ%RH) was reported for black expanded cork-containing AAMs [20]. Nevertheless, the composition containing 87.5 vol% of cork showed a very stable moisture sorption and desorption behaviour throughout the studied cycles, as can be seen in Fig. 17.8.

A follow-up study by the same authors evaluated the impact of the cork granules size (expanded cork was used) on the composites MBV [25]. The highest MBV reached 2.22 g/m^2 Δ%RH for the composition containing 80 vol.% of cork granules with particle sizes comprehended between 0.5 and 2.5 mm, and this ranks this composite in the top classification ascribed in the Nordtest protocol (MBV > 2.0 g/m^2 Δ%RH, "Excellent") [137]. It should be noted that besides the excellent ability to regulate the moisture level, this composite also exhibited low thermal conductivity (79 mW/m K), and high sound absorption ability ($\alpha = 0.65$ at ca. 2000 Hz), demonstrating the potential of this route (alkali activation) for producing multifunctional building materials.

Another study evaluated the impact of the sodium hydroxide/sodium silicate ratio on the hygric properties of porous AAMs [138], reported MBV values comprehended between 4.58 and 5.20 g/m^2 Δ%RH. The MBV initially increased when rising the sodium silicate content up to a specific threshold and then decreased. This

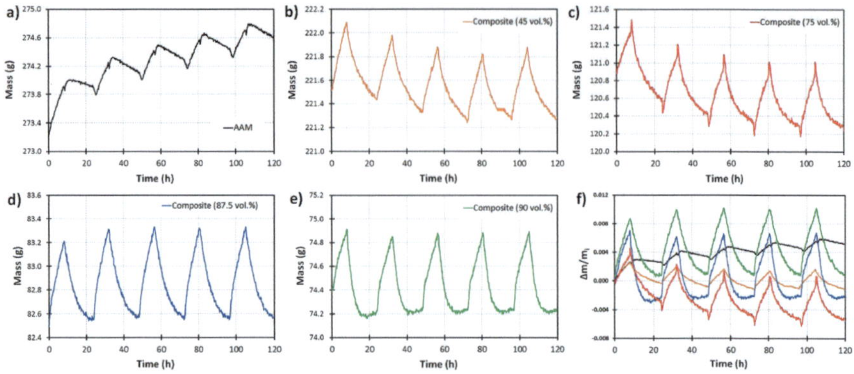

Fig. 17.8 Mass fluctuation of the alkali-activated fly ash (**a**) and the cork –alkali-activated composites (**b**–**e**) upon cyclic variation of the humidity, as prescribed by the Nordtest protocol [137]. **f** Represents the ratio between the mass gradient and the specimens' initial mass. Figure reprinted from [20] with permission from Elsevier (ID: 5838831064440)

phenomenon was attributed to changes in the pore network (size and connectivity) of the produced AAMs. Recently, the same group intended to correlate the paste chemistry (initial water content and alkalinity) with the hygrothermal performance of AAM monoliths [23]. Results show that the water content does not play a significant role in the samples ability to adsorb and release humidity, at least in the conditions reported in the study, MBV values ranging from 4.93 to 5.14 g/m^2 Δ%RH. On the contrary, the molarity of the activating solution does play a role in the sample's moisture buffering ability, even if no clear tendency has been observed: 10 M (4.65 g/m^2 Δ%RH), 12 M (5.65 g/m^2 Δ%RH) and 14 M (4.79 g/m^2 Δ%RH).

A different approach to measure the hygroscopic properties of porous AAMs has been reported in [139]. The authors stated that the Nordtest protocol, and others including the ISO 2425, merely quantify the effects of cyclic humidity fluctuations on the material mass. To surpass this limitation the authors have collected data for one year to define the principal climatic periods. Two main periods of relative humidity were identified: low relative humidity (35–45%) and high relative humidity (75–85%). The hygroscopic performance of the AAMs was then evaluated as follows: (i) the samples were exposed to the laboratory environment after saturation in moisture (e.g., water immersion for 24 h); and (ii) dried at 100 °C for 24 h, followed by exposition to laboratory environment. In each cycle (saturation and drying) the mass of the samples was monitored every hour for the first 12 h, and then every 24 h until reaching 200 h. The role of nano and mesoporosity in the samples (promoted by tuning the SiO_2/Al_2O_3 and the amount of hydrogen peroxide) on the moisture buffering capacity was also studied. Not surprisingly, the samples showed higher moisture saturation in the high relative humidity period reaching 27.5% for the samples foamed with 0.9 vol.% of H_2O_2 and with $SiO_2/Al_2O_3 = 2.45$. At this regime (RH = 75–85%) the SiO_2/Al_2O_3 ratio in the samples also plays a slight role on the moisture saturation of the materials, increasing from 22.33 to 24.10% as the SiO_2/Al_2O_3 ratio rises from 1.93 to

2.71 (for samples foamed with 0.6 vol.% of H_2O_2). After the drying step the moisture accumulation was evaluated, results showing low accumulation in the low humidity regime (ranging from 5 to 7%). In the high humidity regime, the samples show a higher ability to accumulate moisture (values ranging from 8 to 13%). Interestingly, the accumulation is dependent on the volume of millipores. The proper combination of gel pores (nano and meso) with millipores (induced by the foaming agent) allows a very efficient sorption and desorption behaviour. According to the authors the millipores play an important role in aeration by limiting the moisture condensation potential, while facilitating desorption [139]. These are interesting findings that suggest the feasibility of tuning the AAMs hygroscopic capacity by adjusting the formulation and design materials with hierarchical porosity. Nevertheless, additional studies are suggested as the conditions imposed to the samples (e.g., water soaking and drying at 100 °C) deviate from real life conditions.

17.5 Thermal Energy Storage Materials

The use of thermal energy storage (TES) materials in the built environment might be a key strategy to tackle the energy consumption of the sector. TES systems can be divided in three classes considering the type of TES material: sensible, latent and chemical heat storage systems, the pros and cons of each method are detailed in [140]. The implementation of such systems in buildings might be performed using passive and active solutions [141]. Considering the built environment one of the most promising methods is to store latent heat in phase change materials (PCMs) [142]. PCMs can store and release thermal energy upon indoor temperature swings reducing the peak temperatures and improving the thermal comfort for inhabitants [141].

The incorporation of PCMs in Portland cement matrix has been widely reported as a strategy to enhance the concrete heat storage capacity [143, 144], while this avenue is much less explored in AAMs [145, 146]. Moreover, the vast majority of studies has explored PCM incorporation in dense AAMs envisioned for structural applications [31, 32], possibly due to the high thermal conductivity requirement for efficient heat exchange. A recent example reports the incorporation of PCMs in 3D printed AAMs [147]. Nevertheless, herein the emphasis is given to lightweight AAMs as a solution to decarbonize the built environment, their dense counterparts being beyond the scope of this chapter. The number of studies addressing the production of lightweight PCM-containing AAMs is scarce, virtually all existing literature covering the impregnation of porous aggregates with PCMs (e.g., waste aerated concrete block [148], lightweight expanded clay aggregate [149, 150]) prior to their addition to the AAM slurry. The encapsulation of the PCM in the host aggregate must be ensured and this usually requires the use of a coating material (e.g., polyester resin [149], alkali activated paste [150], expanded graphite [151], hydrophobic expanded perlite [152] and polyurethane foam [153]) adding complexity to the system. While this is an interesting approach, some studies address only the efficiency of the impregnation process [150], while others evaluate the impact of the PCM content on the physical properties of the AAMs (e.g., compressive strength and thermal conductivity) [149], neglecting the actual TES capacity of the produced materials. In [153], differential scanning calorimetry (DSC) was used to measure the TES properties of the PCM (encapsulated in a polyurethane foam) containing AAM composite. Not surprisingly, the sample containing the highest amount of PCM (20 wt%) showed the highest latent heat of fusion and crystallization, 6.4 and 6.2 J/g, respectively. The thermal stability of this composite over time was evaluated by performing twenty DSC heating/cooling measurements, results showing no impact on the thermal properties over time. While DSC measurements can provide relevant information (e.g., specific heat, melting temperature and enthalpy) more robust tests to characterise the TES performance of building materials are advisable. Hassan et al. [154] built small size test cells $40 \times 40 \times 40$ cm. An 80 W silicon heating mat was used to supply heat to the setup. The samples were placed at a predefined distance from the heat source (220 mm) and then the temperature in the front and the back surface of the samples was measured during one heating cycle. The PCM-containing composites showed lower maximum temperature in both the front and back surfaces. A similar protocol was used in [148]. While such tests provide information regarding the TES capacity of the materials it deviates from the real-life scenario where daily fluctuations of temperature are to be expected. Ramakrishnan et al. [152] constructed cubic test cells ($30 \times 30 \times 30$ cm) intended to ensure one-dimensional heat transfer process through the samples under study. Five surfaces were isolated with EPS, the remaining surface corresponding to the test panel. The test cells operated in a cyclic temperature variation to represent a summer diurnal temperature profile, two heating and cooling cycles were imposed. Results showed a reduction of 1.85 °C and 3.76 °C in the maximum temperature of the test cells in the 15 and 30 wt% PCM composites compared to the reference composition, respectively. Nevertheless, very similar values were observed in the minimum temperature of the cells but no explanation for this feature was given.

An innovative approach to produce multifunctional lightweight AAMs with TES capacity has been reported by Gonçalves et al. [33], the production of bodies combining a highly porous top-layer with a dense bottom-layer containing PCMs as illustrated by the digital images included in Fig. 17.9.

The idea behind the approach is that the top porous layer provides thermal insulation and moisture regulation ability, while the PCM-containing layer ensures TES. Indeed, the composite containing 30 wt% of PCM in the bottom-layer had a thermal conductivity of 172 mW/m K and a MBV of 2.71 g/m^2 Δ%RH. The TES capacity was evaluated by performing thermal cycles under realistic temperature and humidity conditions by placing small size tests cells ($30 \times 30 \times 30$ cm) in a climatic chamber. Ten heating/cooling cycles were then imposed at a constant relative humidity (60%). Figure 17.10 presents the temperature evolution inside the test cells within the first five cycles for the reference composition (0 wt% PCM) and the PCM-containing specimens. A decrease in the temperature swing was observed in the PCM-coated cells, and an increase in the PCM content in the samples decreased the temperature fluctuation. The lowest temperature swing (13 °C) being observed for the 30 wt% PCM, this being 5 °C lower than the reference (18 °C). The impact of the heating/

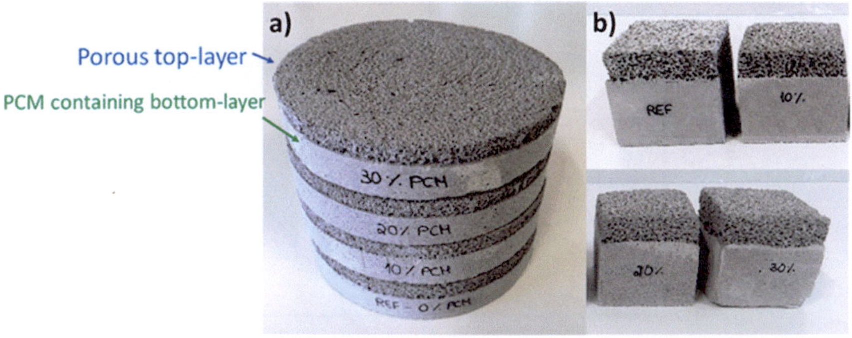

Fig. 17.9 Digital pictures of the bi-layered AAM specimens illustrating the porous top-layer and dense bottom layer: **a** cylindrical specimens (12 cm diameter and 2 cm thickness) and **b** cubic specimens (4 cm). Figure reprinted from [33] with permission from Elsevier (ID: 5840130777397)

cooling rate on the TES was also explored, results showing that under the studied conditions a minor impact was observed as in all cases fusion and solidification of the PCM occurred.

Despite the promising results reported in existing literature on the topic, there is a knowledge gap regarding the long-term TES performance of PCM-containing AAMs under realistic conditions. Additional and deeper investigations are required to validate the use of these materials in the built environment.

17.6 Conclusions and Future Trends

The use of lightweight AAMs in the built environment emerges as a very promising avenue to decarbonize the sector. Results reported to date demonstrate the possibility of producing materials with multifunctional behaviour including thermal and acoustic insulation, moisture regulation ability, and in some cases fire resistance. The multifunctionally exhibited by these materials combined with their green synthesis protocol sets them apart from conventional thermal insulating materials. Despite the unique combination of properties of these novel materials, their industrial deployment is hindered by the lack of standards regarding synthesis protocols and formulations, the latter being particularly sensitive as the raw materials will exert a major role on the physical properties of such materials. Future research should also address the use of alternative waste-based activators as these reagents are currently the main bottleneck when considering the embodied energy and carbon of AAMs. Regardless of the synthesis protocol and the composition of the mixture, the environmental, social and economic sustainability of any of the proposed solutions requires in-deep life cycle assessments (LCA) and life cycle cost analysis (LCCA). This is expected to be a hot topic in the near future.

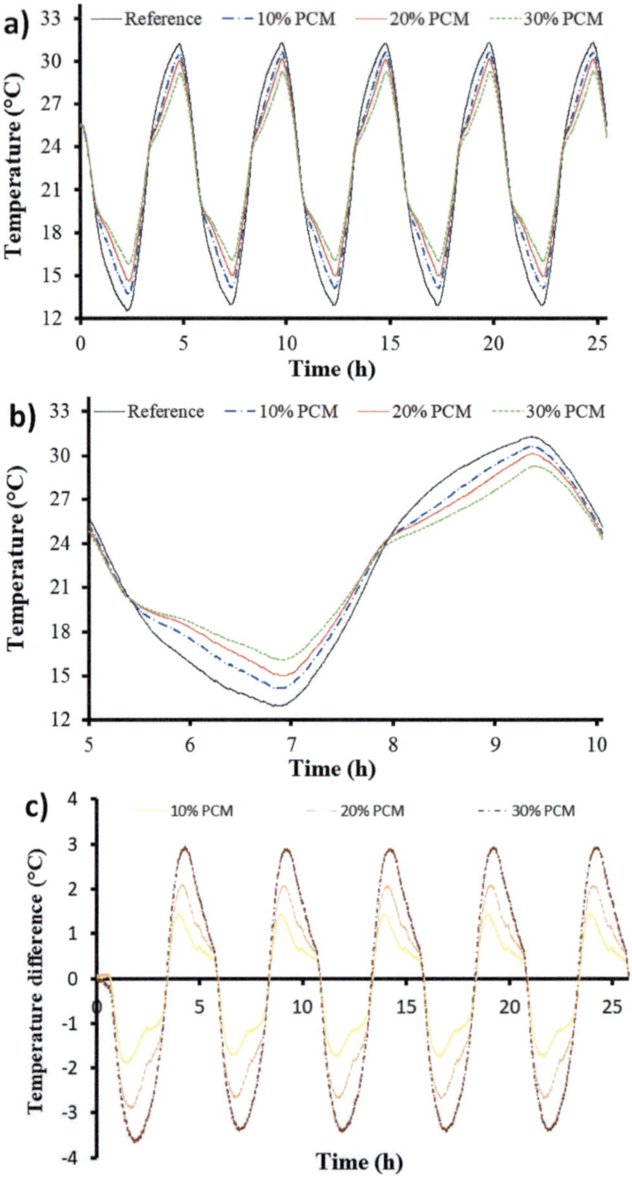

Fig. 17.10 Thermal tests results: **a** temperature evolution inside the test cells, **b** detail of a heating/cooling cycle, and **c** temperature difference in the various test cells compared to the reference cell. Figure reprinted from [33] with permission from Elsevier (ID: 5840130777397)

The potential of lightweight AAMs to act as moisture regulators has yet to be fully explored, despite the relevance of this property in indoor air quality and inhabitants' health conditions. It should also be noted that existing protocols to measure the moisture buffering capacity of building materials present significant shortcomings, robust approaches must include deeper characterizations under various, but realistic, scenarios (e.g., impact of humidity, temperature) and study the long-term performance of the materials. Investigations focusing the optimization of the AAM pore size distribution to maximize the performance of these materials are suggested.

The TES capacity of lightweight AAMs remains virtually unexplored as a significant effort has been devoted to the physical characterization (e.g., compressive strength, density, thermal conductivity) of the PCM-containing AAMs, rather than on their TES capacity. A major bottleneck of existing literature is that the TES evaluation is performed in unrealistic conditions, significant steps must be taken to evaluate the TES capacity under realistic temperature fluctuations mimicking their real-life operation. In a nutshell, additional studies to demonstrate the viability of these materials in energy storage applications are of the utmost importance.

Acknowledgements This work was developed within the scope of the project CICECO-Aveiro Institute of Materials, UIDB/50011/2020 (https://doi.org/10.54499/UIDB/50011/2020), UIDP/50011/2020 (https://doi.org/10.54499/UIDP/50011/2020) & LA/P/0006/2020 (https://doi.org/10.54499/LA/P/0006/2020), financed by national funds through the FCT/MCTES (PIDDAC).

Declarations of Interest None.

References

1. International Energy Agency (IEA), *Buildings* (2024). https://www.iea.org/energy-system/buildings#tracking. Accessed 23 July 2024
2. B. Abu-Jdayil, A.H. Mourad, W. Hittini, M. Hassan, S. Hameedi, Traditional, state-of-the-art and renewable thermal building insulation materials: an overview. Constr. Build. Mater. **214**, 709–735 (2019). https://doi.org/10.1016/j.conbuildmat.2019.04.102
3. Y. Ding, C.H. Dreimol, R. Zboray, K. Tu, S. Stucki, T. Keplinger, G. Panzarasa, I. Burgert, Passive climate regulation with transpiring wood for buildings with increased energy efficiency. Mater. Horizons. **10**, 257–267 (2022). https://doi.org/10.1039/d2mh01016j
4. European Environment Agency. Decarbonising heating and cooling—a climate imperative, (2023). https://www.eea.europa.eu/publications/decarbonisation-heating-and-cooling. Accessed 23 July 2024
5. Y. Song, C. Xue, W. Guo, Y. Bai, Y. Shi, Q. Zhao, Foamed geopolymer insulation materials: Research progress on insulation performance and durability. J. Clean. Prod. **444**, 140991 (2024). https://doi.org/10.1016/j.jclepro.2024.140991
6. *Global Material Resources Outlook to 2060* (2019). https://doi.org/10.1787/97892643074522-en
7. J. Wang, W. Azam, Natural resource scarcity, fossil fuel energy consumption, and total greenhouse gas emissions in top emitting countries. Geosci. Front. **15**, 101757 (2024). https://doi.org/10.1016/j.gsf.2023.101757
8. J. Rising, M. Tedesco, F. Piontek, D.A. Stainforth, The missing risks of climate change. Nature **610**, 643–651 (2022). https://doi.org/10.1038/s41586-022-05243-6

9. Z. Peng, W. Lu, C.J. Webster, Quantifying the embodied carbon saving potential of recycling construction and demolition waste in the Greater Bay Area, China: status quo and future scenarios. Sci. Total. Environ. **792**, 148427 (2021). https://doi.org/10.1016/j.scitotenv.2021.148427
10. C. Economy, The circularity report 2022. Circ. Econ. 1–49 (2022). https://www.circularity-gap.world/2022#Download-the-report
11. R.M. Novais, R.C. Pullar, J.A. Labrincha, Geopolymer foams: an overview of recent advancements. Prog. Mater. Sci. **109**, 100621 (2020). https://doi.org/10.1016/j.pmatsci.2019.100621
12. X. Zhang, C. Bai, Y. Qiao, X. Wang, D. Jia, H. Li, P. Colombo, Porous geopolymer composites: a review. Compos. Part A Appl. Sci. Manuf. **150**, 106629 (2021). https://doi.org/10.1016/j.compositesa.2021.106629
13. Z. Zhang, J.L. Provis, A. Reid, H. Wang, Geopolymer foam concrete: an emerging material for sustainable construction. Constr. Build. Mater. **56**, 113–127 (2014). https://doi.org/10.1016/j.conbuildmat.2014.01.081
14. P. Zhang, Z. Gao, J. Wang, J. Guo, S. Hu, Y. Ling, Properties of fresh and hardened fly ash/slag based geopolymer concrete: a review. J. Clean. Prod. **270**, 122389 (2020). https://doi.org/10.1016/j.jclepro.2020.122389
15. A. Hassan, M. Arif, M. Shariq, Use of geopolymer concrete for a cleaner and sustainable environment – A review of mechanical properties and microstructure. J. Clean. Prod. **223**, 704–728 (2019). https://doi.org/10.1016/j.jclepro.2019.03.051
16. W.M. Kriven, C. Leonelli, J.L. Provis, A.R. Boccaccini, C. Attwell, V.S. Ducman, C. Ferone, S. Rossignol, T. Luukkonen, J.S.J. van Deventer, J.V. Emiliano, J.E. Lombardi, Why geopolymers and alkali-activated materials are key components of a sustainable world: a perspective contribution. J. Am. Ceram. Soc. **107**, 5159–5177 (2024). https://doi.org/10.1111/jace.19828
17. Z. Zhang, J.L. Provis, A. Reid, H. Wang, Mechanical, thermal insulation, thermal resistance and acoustic absorption properties of geopolymer foam concrete. Cem. Concr. Compos. **62**, 97–105 (2015). https://doi.org/10.1016/j.cemconcomp.2015.03.013
18. R.M. Novais, L.H. Buruberri, G. Ascensão, M.P. Seabra, J.A. Labrincha, Porous biomass fly ash-based geopolymers with tailored thermal conductivity. J. Clean. Prod. **119**, 99–107 (2016). https://doi.org/10.1016/j.jclepro.2016.01.083
19. F. Colangelo, G. Roviello, L. Ricciotti, V. Ferrándiz-Mas, F. Messina, C. Ferone, O. Tarallo, R. Cioffi, C.R. Cheeseman, Mechanical and thermal properties of lightweight geopolymer composites. Cem. Concr. Compos. **86**, 266–272 (2018). https://doi.org/10.1016/j.cemconcomp.2017.11.016
20. R.M. Novais, J. Carvalheiras, L. Senff, A.M. Lacasta, I.R. Cantalapiedra, J. Giro-Paloma, M.P. Seabra, J.A. Labrincha, Multifunctional cork—alkali-activated fly ash composites: a sustainable material to enhance buildings' energy and acoustic performance. Energy Build. **210**, 109739 (2020). https://doi.org/10.1016/j.enbuild.2019.109739
21. J. Stolz, Y. Boluk, V. Bindiganavile, Mechanical, thermal and acoustic properties of cellular alkali activated fly ash concrete. Cem. Concr. Compos. **94**, 24–32 (2018). https://doi.org/10.1016/j.cemconcomp.2018.08.004
22. H. Gao, H. Liu, L. Liao, L. Mei, F. Zhang, L. Zhang, S. Li, G. Lv, A bifunctional hierarchical porous kaolinite geopolymer with good performance in thermal and sound insulation. Constr. Build. Mater. **251**, 118888 (2020). https://doi.org/10.1016/j.conbuildmat.2020.118888
23. D.M. Degefu, Z. Liao, U. Berardi, G. Labbé, I. Akhmetova, Geopolymer concrete for net-zero buildings: Correlating paste chemistry with monolith hygrothermal performance. Resour. Conserv. Recycl. **189** (2023). https://doi.org/10.1016/j.resconrec.2022.106743.
24. A. De Rossi, J. Carvalheiras, R.M. Novais, M.J. Ribeiro, J.A. Labrincha, D. Hotza, R.F.P.M. Moreira, Waste-based geopolymeric mortars with very high moisture buffering capacity. Constr. Build. Mater. **191**, 39–46 (2018). https://doi.org/10.1016/j.conbuildmat.2018.09.201
25. R.M. Novais, L. Senff, J. Carvalheiras, A.M. Lacasta, I.R. Cantalapiedra, J.A. Labrincha, Simple and effective route to tailor the thermal, acoustic and hygrothermal properties of cork-containing waste derived inorganic polymer composites. J. Build. Eng. **42** (2021). https://doi.org/10.1016/j.jobe.2021.102501

26. P. Nuaklong, P. Jongvivatsakul, T. Pothisiri, V. Sata, P. Chindaprasirt, Influence of rice husk ash on mechanical properties and fire resistance of recycled aggregate high-calcium fly ash geopolymer concrete. J. Clean. Prod. **252**, 119797 (2020). https://doi.org/10.1016/j.jclepro.2019.119797
27. T.W. Cheng, J.P. Chiu, Fire-resistant geopolymer produce by granulated blast furnace slag. Miner. Eng. **16**, 205–210 (2003). https://doi.org/10.1016/S0892-6875(03)00008-6
28. Z. Alves, L. Senff, K. Sakkas, I. Yakoumis, J.A. Labrincha, R.M. Novais, Synthesis of geopolymer composites using bauxite residue-based spheres as aggregate: novel and eco-friendly strategy to produce lightweight building materials. Cem. Concr. Compos. **148**, 105478 (2024). https://doi.org/10.1016/j.cemconcomp.2024.105478
29. R.M. Novais, M. Saeli, A.P.F. Caetano, M.P. Seabra, J.A. Labrincha, K.P. Surendran, R.C. Pullar, Pyrolysed cork-geopolymer composites: a novel and sustainable EMI shielding building material. Constr. Build. Mater. **229**, 116930 (2019). https://doi.org/10.1016/j.conbuildmat.2019.116930
30. Y. Zhu, B. Bai, E. Ding, S. Bi, W. Liu, L. Zhang, Enhanced electromagnetic interference shielding performance of geopolymer nanocomposites by incorporating carbon nanotubes with controllable silica shell. Ceram. Int. **48**, 11103–11110 (2022). https://doi.org/10.1016/j.ceramint.2021.12.330
31. V.D. Cao, S. Pilehvar, C. Salas-Bringas, A.M. Szczotok, L. Valentini, M. Carmona, J.F. Rodriguez, A.L. Kjøniksen, Influence of microcapsule size and shell polarity on thermal and mechanical properties of thermoregulating geopolymer concrete for passive building applications. Energy Convers. Manag. **164**, 198–209 (2018). https://doi.org/10.1016/j.enconman.2018.02.076
32. S. Pilehvar, V.D. Cao, A.M. Szczotok, L. Valentini, D. Salvioni, M. Magistri, R. Pamies, A.L. Kjøniksen, Mechanical properties and microscale changes of geopolymer concrete and Portland cement concrete containing micro-encapsulated phase change materials. Cem. Concr. Res. **100**, 341–349 (2017). https://doi.org/10.1016/j.cemconres.2017.07.012
33. M. Gonçalves, R.M. Novais, L. Senff, J. Carvalheiras, J.A. Labrincha, PCM-containing bi-layered alkali-activated materials: A novel and sustainable route to regulate the temperature and humidity fluctuations inside buildings. Build. Environ. **205** (2021). https://doi.org/10.1016/j.buildenv.2021.108281
34. R.M. Novais, L.H. Buruberri, M.P. Seabra, D. Bajare, J.A. Labrincha, Novel porous fly ash-containing geopolymers for pH buffering applications. J. Clean. Prod. **124**, 395–404 (2016). https://doi.org/10.1016/j.jclepro.2016.02.114
35. T. Gameiro, R.M. Novais, C.L. Correia, J. Carvalheiras, M.P. Seabra, J.A. Labrincha, A.C. Duarte, I. Capela, Red mud-based inorganic polymer spheres: Innovative and environmentally friendly anaerobic digestion enhancers. Bioresour. Technol. **316**, 123904 (2020). https://doi.org/10.1016/j.biortech.2020.123904
36. X. Li, J. Zheng, K. Zheng, F. Su, Z. Zhao, C. Bai, T. Zheng, X. Wang, Y. Qiao, P. Colombo, Rapid fabrication of coal gangue-based alkali activated foams and application as pH regulators. Mater. Lett. **338**, 134020 (2023). https://doi.org/10.1016/j.matlet.2023.134020
37. R.M. Novais, J.A. Labrincha, Alkali-activated materials as adsorbents for water and wastewater treatment, in *Alkali-Activated Materials in Environmental Technology Applications*, ed. by T. Luukkonen (Woodhead Publishing, Kidlington, United Kingdom, 2022), pp. 143–166. ISBN: 978-0-323-88438-9
38. N.P.F. Gonçalves, E.F. da Silva, L.A.C. Tarelho, J.A. Labrincha, R.M. Novais, Simultaneous removal of multiple metal(loid)s and neutralization of acid mine drainage using 3D-printed bauxite-containing geopolymers. J. Hazard. Mater. **462** (2024). https://doi.org/10.1016/j.jhazmat.2023.132718
39. Y. Zhang, L. Liu, Fly ash-based geopolymer as a novel photocatalyst for degradation of dye from wastewater. Particuology. **11**, 353–358 (2013). https://doi.org/10.1016/j.partic.2012.10.007
40. S. Sharma, D. Medpelli, S. Chen, D.K. Seo, Calcium-modified hierarchically porous aluminosilicate geopolymer as a highly efficient regenerable catalyst for biodiesel production. RSC Adv. **5**, 65454–65461 (2015). https://doi.org/10.1039/c5ra01823d

41. M.I.M. Alzeer, K.J.D. MacKenzie, R.A. Keyzers, Porous aluminosilicate inorganic polymers (geopolymers): a new class of environmentally benign heterogeneous solid acid catalysts. Appl. Catal. A Gen. **524**, 173–181 (2016). https://doi.org/10.1016/j.apcata.2016.06.024
42. C. Shi, A.F. Jiménez, A. Palomo, New cements for the 21st century: the pursuit of an alternative to Portland cement. Cem. Concr. Res. **41**, 750–763 (2011). https://doi.org/10.1016/j.cemconres.2011.03.016
43. J.L. Provis, A. Palomo, C. Shi, Advances in understanding alkali-activated materials. Cem. Concr. Res. **78**, 110–125 (2015). https://doi.org/10.1016/j.cemconres.2015.04.013
44. P. Duxson, J.L. Provis, G.C. Lukey, S.W. Mallicoat, W.M. Kriven, J.S.J. Van Deventer, Understanding the relationship between geopolymer composition, microstructure and mechanical properties, colloids surfaces a physicochem. Eng. Asp. **269**, 47–58 (2005). https://doi.org/10.1016/j.colsurfa.2005.06.060
45. L.W. Nees Jan van Eck, Software survey : VOSviewer, a computer program for bibliometric mapping. 523–538 (2010). https://doi.org/10.1007/s11192-009-0146-3.
46. L. Le-ping, C. Xue-min, Q. Shu-heng, Y. Jun-li, Z. Lin, Preparation of phosphoric acid-based porous geopolymers. Appl. Clay Sci. **50**, 600–603 (2010). https://doi.org/10.1016/j.clay.2010.10.004
47. E. Prud'homme, P. Michaud, E. Joussein, C. Peyratout, A. Smith, S. Arrii-Clacens, J.M. Clacens, S. Rossignol, Silica fume as porogent agent in geo-materials at low temperature. J. Eur. Ceram. Soc. **30**, 1641–1648 (2010). https://doi.org/10.1016/j.jeurceramsoc.2010.01.014.
48. K. Okada, A. Imase, T. Isobe, A. Nakajima, Capillary rise properties of porous geopolymers prepared by an extrusion method using polylactic acid (PLA) fibers as the pore formers. J. Eur. Ceram. Soc. **31**, 461–467 (2011). https://doi.org/10.1016/j.jeurceramsoc.2010.10.035
49. C. Bai, P. Colombo, Processing, properties and applications of highly porous geopolymers: a review. Ceram. Int. **44**, 16103–16118 (2018). https://doi.org/10.1016/j.ceramint.2018.05.219
50. K. Dhasindrakrishna, K. Pasupathy, S. Ramakrishnan, J. Sanjayan, Progress, current thinking and challenges in geopolymer foam concrete technology. Cem. Concr. Compos. **116**, 103886 (2021). https://doi.org/10.1016/j.cemconcomp.2020.103886
51. C. Bai, A. Conte, P. Colombo, Open-cell phosphate-based geopolymer foams by frothing. Mater. Lett. **188**, 379–382 (2017). https://doi.org/10.1016/j.matlet.2016.11.103
52. J. Zhao, L. Tong, B. Li, T. Chen, C. Wang, G. Yang, Y. Zheng, Eco-friendly geopolymer materials: A review of performance improvement, potential application and sustainability assessment. J. Clean. Prod. **307**, 127085 (2021). https://doi.org/10.1016/j.jclepro.2021.127085
53. S. Bilici, J. Carvalheiras, J.A. Labrincha, R.M. Novais, Evaluation of the nature and concentration of the surfactant on the properties of red mud/metakaolin porous geopolymers foamed with aluminium. Materials (Basel). **15**, 1–23 (2022). https://doi.org/10.3390/ma15217486
54. Y. Qiao, X. Li, C. Bai, H. Li, J. Yan, Y. Wang, X. Wang, X. Zhang, T. Zheng, P. Colombo, Effects of surfactants/stabilizing agents on the microstructure and properties of porous geopolymers by direct foaming. J. Asian Ceram. Soc. **9**, 412–423 (2021). https://doi.org/10.1080/21870764.2021.1873482
55. K.M. Klima, C.H. Koh, H.J.H. Brouwers, Q. Yu, Synergistic effect of surfactants in porous geopolymer: Tailoring pore size and pore connectivity. Cem. Concr. Compos. **134**, 104774 (2022). https://doi.org/10.1016/j.cemconcomp.2022.104774
56. R.M. Novais, G. Ascensão, D.M. Tobaldi, M.P. Seabra, J.A. Labrincha, Biomass fly ash geopolymer monoliths for effective methylene blue removal from wastewaters. J. Clean. Prod. **171**, 783–794 (2018). https://doi.org/10.1016/j.jclepro.2017.10.078
57. M. Strozi Cilla, P. Colombo, M. Raymundo Morelli, Geopolymer foams by gelcasting, Ceram. Int. **40**, 5723–5730 (2014). https://doi.org/10.1016/j.ceramint.2013.11.011
58. X. Zhang, X. Zhang, X. Li, Y. Liu, H. Yu, M. Ma, Porous geopolymer with controllable interconnected pores—a viable permeable reactive barrier filler for lead pollutant removal. Chemosphere **307**, 136128 (2022). https://doi.org/10.1016/j.chemosphere.2022.136128
59. G. Masi, W.D.A. Rickard, L. Vickers, M.C. Bignozzi, A. Van Riessen, A comparison between different foaming methods for the synthesis of light weight geopolymers. Ceram. Int. **40**, 13891–13902 (2014). https://doi.org/10.1016/j.ceramint.2014.05.108

60. S.E. Abo Sawan, M.F. Zawrah, R.M. Khattab, A.A. Abdel-Shafi, In-situ formation of geopolymer foams through addition of silica fume: preparation and sinterability. Mater. Chem. Phys. **239**, 121998 (2020). https://doi.org/10.1016/j.matchemphys.2019.121998
61. E. Papa, V. Medri, D. Kpogbemabou, V. Morinière, J. Laumonier, A. Vaccari, S. Rossignol, Porosity and insulating properties of silica-fume based foams. Energy Build. **131**, 223–232 (2016). https://doi.org/10.1016/j.enbuild.2016.09.031
62. A.P.F. Caetano, J. Carvalheiras, L. Senff, M.P. Seabra, R.C. Pullar, J.A. Labrincha, R.M. Novais, Unravelling the affinity of alkali-activated fly ash cubic foams towards heavy metals sorption. Materials (Basel). **15** (2022). https://doi.org/10.3390/ma15041453
63. N. Cristelo, J. Maia, N.M.M. Ramos, J. Rivera, J. Ventura, R.C. Veloso, A. Fernández-Jiménez, Development of highly porous alkaline cements from industrial waste for thermal insulation of building envelops. Constr. Build. Mater. **409** (2023). https://doi.org/10.1016/j.conbuildmat.2023.134068.
64. R.M. Novais, G. Ascensão, N. Ferreira, M.P. Seabra, J.A. Labrincha, Influence of water and aluminium powder content on the properties of waste-containing geopolymer foams. Ceram. Int. **44**, 6242–6249 (2018). https://doi.org/10.1016/j.ceramint.2018.01.009
65. R. Demirboğa, I. Örüng, R. Gül, Effects of expanded perlite aggregate and mineral admixtures on the compressive strength of low-density concretes. Cem. Concr. Res. **31**, 1627–1632 (2001). https://doi.org/10.1016/S0008-8846(01)00615-9
66. T.Y. Lo, H.Z. Cui, Effect of porous lightweight aggregate on strength of concrete. Mater. Lett. **58**, 916–919 (2004). https://doi.org/10.1016/j.matlet.2003.07.036
67. Y.W. Choi, D.J. Moon, J.S. Chung, S.K. Cho, Effects of waste PET bottles aggregate on the properties of concrete. Cem. Concr. Res. **35**, 776–781 (2005). https://doi.org/10.1016/j.cemconres.2004.05.014
68. R. Demirboğa, R. Gül, The effects of expanded perlite aggregate, silica fume and fly ash on the thermal conductivity of lightweight concrete. Cem. Concr. Res. **33**, 723–727 (2003). https://doi.org/10.1016/S0008-8846(02)01032-3
69. P. Posi, C. Teerachanwit, C. Tanutong, S. Limkamoltip, S. Lertnimoolchai, V. Sata, P. Chindaprasirt, Lightweight geopolymer concrete containing aggregate from recycle lightweight block. Mater. Des. **52**, 580–586 (2013). https://doi.org/10.1016/j.matdes.2013.06.001
70. R.M. Novais, L. Senff, J. Carvalheiras, M.P. Seabra, R.C. Pullar, J.A. Labrincha, Sustainable and efficient cork - inorganic polymer composites: An innovative and eco-friendly approach to produce ultra-lightweight and low thermal conductivity materials. Cem. Concr. Compos. **97**, 107–117 (2018). https://doi.org/10.1016/J.CEMCONCOMP.2018.12.024
71. Z. Li, W. Chen, H. Hao, N.S. Ha, T.M. Pham, Static and dynamic properties of novel ambient-cured lightweight geopolymer composites with fibre reinforced epoxy coated EPS aggregates. Compos. Part B Eng. **250**, 110439 (2023). https://doi.org/10.1016/j.compositesb.2022.110439
72. V. Vaou, D. Panias, Thermal insulating foamy geopolymers from perlite. Miner. Eng. **23**, 1146–1151 (2010). https://doi.org/10.1016/j.mineng.2010.07.015
73. V. Medri, E. Papa, M. Mazzocchi, L. Laghi, M. Morganti, J. Francisconi, E. Landi, Production and characterization of lightweight vermiculite/geopolymer-based panels. Mater. Des. **85**, 266–274 (2015). https://doi.org/10.1016/j.matdes.2015.06.145
74. O. Anjos, H. Pereira, M.E. Rosa, Tensile properties of cork in the tangential direction: Variation with quality, porosity, density and radial position in the cork plank. Mater. Des. **31**, 2085–2090 (2010). https://doi.org/10.1016/j.matdes.2009.10.048
75. Z. Alves, J. Carvalheiras, L. Senff, A.M. Lacasta, I.R. Cantalapiedra, J.A. Labrincha, R.M. Novais, A comparison between the use of cork and synthetic aggregates in the production of geopolymer composites. Constr. Build. Mater. **438**, 1–10 (2024). https://doi.org/10.1016/j.conbuildmat.2024.137147
76. M.C. Pacheco Menor, P. Serna Ros, A. Macías García, M.J. Arévalo Caballero, Granulated cork with bark characterised as environment-friendly lightweight aggregate for cement based materials. J. Clean. Prod. **229**, 358–373 (2019). https://doi.org/10.1016/j.jclepro.2019.04.154
77. A. Moreira, J. António, A. Tadeu, Lightweight screed containing cork granules: mechanical and hygrothermal characterization. Cem. Concr. Compos. **49**, 1–8 (2014). https://doi.org/10.1016/j.cemconcomp.2014.01.012

78. S.M.A. Qaidi, A.S. Mohammed, H.U. Ahmed, R.H. Faraj, W. Emad, B.A. Tayeh, F. Althoey, O. Zaid, N.H. Sor, Rubberized geopolymer composites: a comprehensive review. Ceram. Int. **48**, 24234–24259 (2022). https://doi.org/10.1016/j.ceramint.2022.06.123
79. M.S. Tale Masoule, N. Bahrami, M. Karimzadeh, B. Mohasanati, P. Shoaei, F. Ameri, T. Ozbakkaloglu, Lightweight geopolymer concrete: a critical review on the feasibility, mixture design, durability properties, and microstructure. Ceram. Int. **48**, 10347–10371 (2022). https://doi.org/10.1016/j.ceramint.2022.01.298
80. H.A. Mahmoud, T.A. Tawfik, M.M. Abd El-razik, A.S. Faried, Mechanical and acoustic absorption properties of lightweight fly ash/slag-based geopolymer concrete with various aggregates Ceram. Int. **49**, 21142–21154 (2023). https://doi.org/10.1016/j.ceramint.2023.03.244
81. A.A. Sayadi, J.V. Tapia, T.R. Neitzert, G.C. Clifton, Effects of expanded polystyrene (EPS) particles on fire resistance, thermal conductivity and compressive strength of foamed concrete. Constr. Build. Mater. **112**, 716–724 (2016). https://doi.org/10.1016/j.conbuildmat.2016.02.218
82. W. Villasmil, L.J. Fischer, J. Worlitschek, A review and evaluation of thermal insulation materials and methods for thermal energy storage systems. Renew. Sustain. Energy Rev. **103**, 71–84 (2019). https://doi.org/10.1016/j.rser.2018.12.040
83. G. Hommond, C. Jones, Embodied Carbon: the inventory of carbon and energy (ICE), BSRIA Guid (2011)
84. P.C. de Assis Neto, L.P.B. Sales, P.K.S. Oliveira, I.C. da Silva, I.M. da S. Barros, A.F. da Nóbrega, A.M.P. Carneiro, Expanded vermiculite: a short review about its production, characteristics, and effects on the properties of lightweight mortars. Buildings **13** (2023). https://doi.org/10.3390/buildings13030823.
85. E. Mercan, S. Yilmazer, Study of physical and mechanical properties of aerogel-modified expanded perlite aggregate and clay (AEP/C) board. Constr. Build. Mater. **361**, 129602 (2022). https://doi.org/10.1016/j.conbuildmat.2022.129602
86. S.K. Adhikary, D.K. Ashish, Ž. Rudžionis, Expanded glass as light-weight aggregate in concrete—a review. J. Clean. Prod. **313** (2021). https://doi.org/10.1016/j.jclepro.2021.127848
87. L.P. Qian, L.Y. Xu, B.T. Huang, J.G. Dai, Pelletization and properties of artificial lightweight geopolymer aggregates (GPA): One-part vs. two-part geopolymer techniques. J. Clean. Prod. **374**, 133933 (2022). https://doi.org/10.1016/j.jclepro.2022.133933
88. M. Balapour, W. Zhao, E.J. Garboczi, N.Y. Oo, S. Spatari, Y.G. Hsuan, P. Billen, Y. Farnam, Potential use of lightweight aggregate (LWA) produced from bottom coal ash for internal curing of concrete systems. Cem. Concr. Compos. **105**, 103428 (2020). https://doi.org/10.1016/j.cemconcomp.2019.103428
89. C. Narattha, A. Chaipanich, Phase characterizations, physical properties and strength of environment-friendly cold-bonded fly ash lightweight aggregates. J. Clean. Prod. **171**, 1094–1100 (2018). https://doi.org/10.1016/j.jclepro.2017.09.259
90. R.M. Novais, J. Carvalheiras, M.P. Seabra, R.C. Pullar, J.A. Labrincha, Red mud-based inorganic polymer spheres bulk-type adsorbents and pH regulators. Mater. Today **23**, 2017–2018 (2019). https://doi.org/10.1016/j.mattod.2019.01.014
91. M. Almadani, R.A. Razak, M.M.A.B. Abdullah, R. Mohamed, Geopolymer-based artificial aggregates: a review on methods of producing, properties, and improving techniques. Materials (Basel). **15** (2022). https://doi.org/10.3390/ma15165516
92. Z. Luo, W. Li, K. Wang, A. Castel, S.P. Shah, Comparison on the properties of ITZs in fly ash-based geopolymer and Portland cement concretes with equivalent flowability. Cem. Concr. Res. **143**, 106392 (2021). https://doi.org/10.1016/j.cemconres.2021.106392
93. H.R. Rasouli, F. Golestani-Fard, A.R. Mirhabibi, G.M. Nasab, K.J.D. MacKenzie, M.H. Shahraki, Fabrication and properties of microporous metakaolin-based geopolymer bodies with polylactic acid (PLA) fibers as pore generators. Ceram. Int. **41**, 7872–7880 (2015). https://doi.org/10.1016/j.ceramint.2015.02.125

94. E. Papa, V. Medri, P. Benito, A. Vaccari, S. Bugani, J. Jaroszewicz, W. Swieszkowski, E. Landi, Synthesis of porous hierarchical geopolymer monoliths by ice-templating. Microporous Mesoporous Mater. **215**, 206–214 (2015). https://doi.org/10.1016/j.micromeso.2015.05.043
95. E. Papa, V. Medri, P. Benito, A. Vaccari, S. Bugani, J. Jaroszewicz, E. Landi, Insights into the macroporosity of freeze-cast hierarchical geopolymers. RSC Adv. **6**, 24635–24644 (2016). https://doi.org/10.1039/c6ra02232d
96. H. Alghamdi, N. Neithalath, Synthesis and characterization of 3D-printable geopolymeric foams for thermally efficient building envelope materials. Cem. Concr. Compos. **104**, 103377 (2019). https://doi.org/10.1016/j.cemconcomp.2019.103377
97. S. Onutai, S. Jiemsirilers, P. Thavorniti, T. Kobayashi, Fast microwave syntheses of fly ash based porous geopolymers in the presence of high alkali concentration. Ceram. Int. **42**, 9866–9874 (2016). https://doi.org/10.1016/j.ceramint.2016.03.086
98. J. Zheng, X. Li, C. Bai, K. Zheng, X. Wang, G. Sun, T. Zheng, X. Zhang, P. Colombo, Rapid fabrication of porous metakaolin-based geopolymer via microwave foaming. Appl. Clay Sci. **249**, 107238 (2024). https://doi.org/10.1016/j.clay.2023.107238
99. R.M. Novais, G. Ascensão, L.H. Buruberri, L. Senff, J.A. Labrincha, Influence of blowing agent on the fresh- and hardened-state properties of lightweight geopolymers. Mater. Des. **108**, 551–559 (2016). https://doi.org/10.1016/j.matdes.2016.07.039
100. G. Roviello, C. Menna, O. Tarallo, L. Ricciotti, F. Messina, C. Ferone, D. Asprone, R. Cioffi, Lightweight geopolymer-based hybrid materials. Compos. Part B Eng. **128**, 225–237 (2017). https://doi.org/10.1016/j.compositesb.2017.07.020
101. P. Duan, L. Song, C. Yan, D. Ren, Z. Li, Novel thermal insulating and lightweight composites from metakaolin geopolymer and polystyrene particles. Ceram. Int. **43**, 5115–5120 (2017). https://doi.org/10.1016/j.ceramint.2017.01.025
102. H.S. Hassan, H.A. Abdel-Gawwad, S.R.V. García, I. Israde-Alcántara, Fabrication and characterization of thermally-insulating coconut ash-based geopolymer foam. Waste Manag. **80**, 235–240 (2018). https://doi.org/10.1016/j.wasman.2018.09.022
103. J. Wu, Z. Zhang, Y. Zhang, D. Li, Preparation and characterization of ultra-lightweight foamed geopolymer (UFG) based on fly ash-metakaolin blends. Constr. Build. Mater. **168**, 771–779 (2018). https://doi.org/10.1016/j.conbuildmat.2018.02.097
104. J. Jouin, J.N. Nouping Fekoua, L. Ouamara, E. Piolet, A. Gharzouni, S. Rossignol, Insulating phosphoric acid-based geopolymer foams with water and high temperature resistance. Constr. Build. Mater. **398**, 132406 (2023). https://doi.org/10.1016/j.conbuildmat.2023.132406
105. D.M.A. Huiskes, A. Keulen, Q.L. Yu, H.J.H. Brouwers, Design and performance evaluation of ultra-lightweight geopolymer concrete. Mater. Des. **89**, 516–526 (2016). https://doi.org/10.1016/j.matdes.2015.09.167
106. J. Feng, R. Zhang, L. Gong, Y. Li, W. Cao, X. Cheng, Development of porous fly ash-based geopolymer with low thermal conductivity. Mater. Des. **65**, 529–533 (2015). https://doi.org/10.1016/j.matdes.2014.09.024
107. G. Samson, M. Cyr, X.X. Gao, Thermomechanical performance of blended metakaolin-GGBS alkali-activated foam concrete. Constr. Build. Mater. **157**, 982–993 (2017). https://doi.org/10.1016/j.conbuildmat.2017.09.146
108. Y. Cui, D. Wang, J. Zhao, D. Li, S. Ng, Y. Rui, Effect of calcium stearate based foam stabilizer on pore characteristics and thermal conductivity of geopolymer foam material. J. Build. Eng. **20**, 21–29 (2018). https://doi.org/10.1016/j.jobe.2018.06.002
109. P. Palmero, A. Formia, P. Antonaci, S. Brini, J.M. Tulliani, Geopolymer technology for application-oriented dense and lightened materials. Elaboration and characterization. Ceram. Int. **41**, 12967–12979 (2015). https://doi.org/10.1016/j.ceramint.2015.06.140.
110. C. Bai, T. Ni, Q. Wang, H. Li, P. Colombo, Porosity, mechanical and insulating properties of geopolymer foams using vegetable oil as the stabilizing agent. J. Eur. Ceram. Soc. **38**, 799–805 (2018). https://doi.org/10.1016/j.jeurceramsoc.2017.09.021
111. L. Dembovska, D. Bajare, V. Ducman, L. Korat, G. Bumanis, The use of different by-products in the production of lightweight alkali activated building materials. Constr. Build. Mater. **135**, 315–322 (2017). https://doi.org/10.1016/j.conbuildmat.2017.01.005

112. Z.N.M. Ngouloure, B. Nait-Ali, S. Zekeng, E. Kamseu, U.C. Melo, D. Smith, C. Leonelli, Recycled natural wastes in metakaolin based porous geopolymers for insulating applications. J. Build. Eng. **3**, 58–69 (2015). https://doi.org/10.1016/j.jobe.2015.06.006
113. E. Kamseu, B. Nait-Ali, M.C. Bignozzi, C. Leonelli, S. Rossignol, D.S. Smith, Bulk composition and microstructure dependence of effective thermal conductivity of porous inorganic polymer cements. J. Eur. Ceram. Soc. **32**, 1593–1603 (2012). https://doi.org/10.1016/j.jeurceramsoc.2011.12.030
114. F. Xu, G. Gu, W. Zhang, H. Wang, X. Huang, J. Zhu, Pore structure analysis and properties evaluations of fly ash-based geopolymer foams by chemical foaming method. Ceram. Int. **44**, 19989–19997 (2018). https://doi.org/10.1016/j.ceramint.2018.07.267
115. A. Hajimohammadi, T. Ngo, P. Mendis, Enhancing the strength of pre-made foams for foam concrete applications. Cem. Concr. Compos. **87**, 164–171 (2018). https://doi.org/10.1016/j.cemconcomp.2017.12.014
116. A. Wongsa, V. Sata, B. Nematollahi, J. Sanjayan, P. Chindaprasirt, Mechanical and thermal properties of lightweight geopolymer mortar incorporating crumb rubber. J. Clean. Prod. **195**, 1069–1080 (2018). https://doi.org/10.1016/j.jclepro.2018.06.003
117. S.F.A. Zaidi, E. Ul Haq, K. Nur, N. Ejaz, M. Anis-ur-Rehman, M. Zubair, M. Naveed, Synthesis and characterization of natural soil based inorganic polymer foam for thermal insulations. Constr. Build. Mater. **157**, 994–1000 (2017). https://doi.org/10.1016/j.conbuildmat.2017.09.112.
118. A. Hajimohammadi, T. Ngo, J.L. Provis, T. Kim, J. Vongsvivut, High strength/density ratio in a syntactic foam made from one-part mix geopolymer and cenospheres. Compos. Part B Eng. **173**, 106908 (2019). https://doi.org/10.1016/j.compositesb.2019.106908
119. L. Bergamonti, R. Taurino, L. Cattani, D. Ferretti, F. Bondioli, Lightweight hybrid organic-inorganic geopolymers obtained using polyurethane waste. Constr. Build. Mater. **185**, 285–292 (2018). https://doi.org/10.1016/j.conbuildmat.2018.07.006
120. M. Lassinantti Gualtieri, A. Cavallini, M. Romagnoli, Interactive powder mixture concept for the preparation of geopolymers with fine porosity. J. Eur. Ceram. Soc. **36**, 2641–2646 (2016). https://doi.org/10.1016/j.jeurceramsoc.2016.03.030
121. M.Y.J. Liu, U.J. Alengaram, M.Z. Jumaat, K.H. Mo, Evaluation of thermal conductivity, mechanical and transport properties of lightweight aggregate foamed geopolymer concrete. Energy Build. **72**, 238–245 (2014). https://doi.org/10.1016/j.enbuild.2013.12.029
122. A. Font, M.V. Borrachero, L. Soriano, J. Monzó, J. Payá, Geopolymer eco-cellular concrete (GECC) based on fluid catalytic cracking catalyst residue (FCC) with addition of recycled aluminium foil powder. J. Clean. Prod. **168**, 1120–1131 (2017). https://doi.org/10.1016/j.jclepro.2017.09.110
123. S. Petlitckaia, A. Poulesquen, Design of lightweight metakaolin based geopolymer foamed with hydrogen peroxide. Ceram. Int. **45**, 1322–1330 (2019). https://doi.org/10.1016/j.ceramint.2018.10.021
124. Y. Luna-Galiano, C. Leiva, C. Arenas, C. Fernández-Pereira, Fly ash based geopolymeric foams using silica fume as pore generation agent. Physical, mechanical and acoustic properties, J. Non. Cryst. Solids. **500**, 196–204 (2018). https://doi.org/10.1016/j.jnoncrysol.2018.07.069
125. R.M. Novais, L. Senff, J. Carvalheiras, J.A. Labrincha, Bi-layered porous/cork-containing waste-based inorganic polymer composites : innovative material towards green buildings. Appl. Sci. **10**, 2995 (2020).
126. M. Yuan, C. Yin, Y. Sun, W. Chen, Examining the associations between urban built environment and noise pollution in high-density high-rise urban areas: a case study in Wuhan, China. Sustain. Cities Soc. **50**, 101678 (2019). https://doi.org/10.1016/j.scs.2019.101678
127. R. Thompson, R.B. Smith, Y. Bou Karim, C. Shen, K. Drummond, C. Teng, M.B. Toledano, Noise pollution and human cognition: an updated systematic review and meta-analysis of recent evidence, Environ. Int. **158**, 106905 (2022). https://doi.org/10.1016/j.envint.2021.106905

128. T.C. Hung, J.S. Huang, Y.W. Wang, K.Y. Lin, Inorganic polymeric foam as a sound absorbing and insulating material. Constr. Build. Mater. **50**, 328–334 (2014). https://doi.org/10.1016/j.conbuildmat.2013.09.042
129. Y.X. Chen, K.M. Klima, H.J.H. Brouwers, Q. Yu, Effect of silica aerogel on thermal insulation and acoustic absorption of geopolymer foam composites: the role of aerogel particle size. Compos. Part B Eng. **242**, 110048 (2022). https://doi.org/10.1016/j.compositesb.2022.110048
130. M. Bayat Pour, J. Niklewski, A. Naghibi, E. Frühwald Hansson, A literature review of probabilistic hygrothermal assessment for building envelopes. Build. Environ. **261**, 111756 (2024). https://doi.org/10.1016/j.buildenv.2024.111756
131. P. Wolkoff, Indoor air humidity, air quality, and health—an overview. Int. J. Hyg. Environ. Health **221**, 376–390 (2018). https://doi.org/10.1016/j.ijheh.2018.01.015
132. N.A. Megahed, E.M. Ghoneim, Indoor Air Quality: rethinking rules of building design strategies in post-pandemic architecture. Environ. Res. **193**, 110471 (2021). https://doi.org/10.1016/j.envres.2020.110471
133. V. Gentile, J.D. Vargas Velasquez, S. Fantucci, G. Autretto, R. Gabrieli, P.K. Gianchandani, M. Armandi, F. Baino, 3D-printed clay components with high surface area for passive indoor moisture buffering, J. Build. Eng. **91**, 109631 (2024). https://doi.org/10.1016/j.jobe.2024.109631
134. O.F. Osanyintola, C.J. Simonson, Moisture buffering capacity of hygroscopic building materials: experimental facilities and energy impact. Energy Build. **38**, 1270–1282 (2006). https://doi.org/10.1016/j.enbuild.2006.03.026
135. L. Senff, G. Ascensão, D. Hotza, V.M. Ferreira, J.A. Labrincha, Assessment of the single and combined effect of superabsorbent particles and porogenic agents in nanotitania-containing mortars. Energy Build. **127**, 980–990 (2016). https://doi.org/10.1016/j.enbuild.2016.06.048
136. F. Tittarelli, C. Giosuè, A. Mobili, M.L. Ruello, K.M. E, C. Adsorption, D. Aggregate, C.B. Vocs, Influence of binders and aggregates on VOCs adsorption and moisture buffering activity of mortars for indoor applications, Cem. Concr. Compos. **57**, 75–83 (2015). https://doi.org/10.1016/j.cemconcomp.2014.11.013
137. C. Rode, *Moisture Buffering of Building Materials* (Technical University of Denmark, 2006)
138. D.M. Degefu, Z. Liao, U. Berardi, G. Labbé, The effect of activator ratio on the thermal and hygric properties of aerated geopolymers, J. Build. Eng. **45** (2022). https://doi.org/10.1016/j.jobe.2021.103414
139. Z.N.M. Ngouloure, E. Kamseu, L.M.B.À. Moungam, H.K. Tchakoute, L. Valentini, C. Leonelli, Design of porous Geopolymers for hygrothermal applications: role of nano and meso porosity. SILICON **14**, 10045–10059 (2022). https://doi.org/10.1007/s12633-022-01741-4
140. G. Alva, Y. Lin, G. Fang, An overview of thermal energy storage systems. Energy **144**, 341–378 (2018). https://doi.org/10.1016/j.energy.2017.12.037
141. J. Lizana, R. Chacartegui, A. Barrios-Padura, J.M. Valverde, Advances in thermal energy storage materials and their applications towards zero energy buildings: a critical review. Appl. Energy **203**, 219–239 (2017). https://doi.org/10.1016/j.apenergy.2017.06.008
142. C. Fabiani, A.L. Pisello, M. Barbanera, L.F. Cabeza, Palm oil-based bio-PCM for energy efficient building applications: Multipurpose thermal investigation and life cycle assessment. J. Energy Storage. **28** (2020). https://doi.org/10.1016/j.est.2019.101129
143. T.C. Ling, C.S. Poon, Use of phase change materials for thermal energy storage in concrete: an overview. Constr. Build. Mater. **46**, 55–62 (2013). https://doi.org/10.1016/j.conbuildmat.2013.04.031
144. U. Berardi, A.A. Gallardo, Properties of concretes enhanced with phase change materials for building applications. Energy Build. **199**, 402–414 (2019). https://doi.org/10.1016/j.enbuild.2019.07.014
145. M. Muraleedharan, Y. Nadir, Geopolymer mortar integrated with phase change materials for improvement of thermal efficiency in buildings: a review. Mater. Today Proc. **44**, 878–885 (2021). https://doi.org/10.1016/j.matpr.2020.10.791

146. I. Asadi, M.H. Baghban, M. Hashemi, N. Izadyar, B. Sajadi, Phase change materials incorporated into geopolymer concrete for enhancing energy efficiency and sustainability of buildings: a review. Case Stud. Constr. Mater. **17**, e01162 (2022). https://doi.org/10.1016/j.cscm.2022.e01162
147. S. Rahemipoor, M. Hasany, M. Mehrali, K. Almdal, N. Ranjbar, M. Mehrali, Phase change materials incorporation into 3D printed geopolymer cement: a sustainable approach to enhance the comfort and energy efficiency of buildings. J. Clean. Prod. **417**, 138005 (2023). https://doi.org/10.1016/j.jclepro.2023.138005
148. P. Sukontasukkul, N. Nontiyutsirikul, S. Songpiriyakij, K. Sakai, P. Chindaprasirt, Use of phase change material to improve thermal properties of lightweight geopolymer panel. Mater. Struct. Constr. **49**, 4637–4645 (2016). https://doi.org/10.1617/s11527-016-0812-x
149. G. Kastiukas, X. Zhou, J. Castro-Gomes, Development and optimisation of phase change material-impregnated lightweight aggregates for geopolymer composites made from aluminosilicate rich mud and milled glass powder. Constr. Build. Mater. **110**, 201–210 (2016). https://doi.org/10.1016/j.conbuildmat.2016.02.029
150. A. Hassan, N. Ismail, A.H.I. Mourad, Y. Rashid, M.S. Laghari, Preparation and characterization of expanded clay-paraffin wax-geo-polymer composite material. Materials (Basel). **11** (2018). https://doi.org/10.3390/ma11112191
151. L.O. Afolabi, Z.M. Ariff, P.S.M. Megat-Yusoff, H.H. Al-Kayiem, A.I. Arogundade, O.T. Afolabi-Owolabi, Red-mud geopolymer composite encapsulated phase change material for thermal comfort in built-sector. Sol. Energy **181**, 464–474 (2019). https://doi.org/10.1016/j.solener.2019.02.029
152. S. Ramakrishnan, K. Pasupathy, J. Sanjayan, Synthesis and properties of thermally enhanced aerated geopolymer concrete using form-stable phase change composite. J. Build. Eng. **40**, 102756 (2021). https://doi.org/10.1016/j.jobe.2021.102756
153. E. Haily, H. Ait Ousaleh, N. Zari, A. Faik, R. Bouhfid, A. Qaiss, Use of a form-stable phase change material to improve thermal properties of phosphate sludge-based geopolymer mortar. Constr. Build. Mater. **386**, 131570 (2023). https://doi.org/10.1016/j.conbuildmat.2023.131570
154. A. Hassan, Y. Rashid, A.H.I. Mourad, N. Ismail, M.S. Laghari, Thermal and structural characterization of geopolymer-coated polyurethane foam-phase change material capsules/geopolymer concrete composites. Materials (Basel). **12**, 1–16 (2019). https://doi.org/10.3390/MA12050796
155. A.M. Rashad, A short manual on natural pumice as a lightweight aggregate. J. Build. Eng. **25**, 100802 (2019). https://doi.org/10.1016/j.jobe.2019.100802
156. S.Y. Chung, P. Sikora, D.J. Kim, M.E. El Madawy, M. Abd Elrahman, Effect of different expanded aggregates on durability-related characteristics of lightweight aggregate concrete. Mater. Charact. **173** (2021). https://doi.org/10.1016/j.matchar.2021.110907

Chapter 18
Incorporating Residues for Sustainable Road Pavement Construction: An Analysis of Asphalt Mixtures with Dolomite, Limestone, Fly Ash, and Red Mud

Mayara S. Siverio Lima, Liseane P. Thives, Arturs Riekstins, Remo Merijs-Meri, and Viktors Haritonovs

Abstract The pursuit of sustainability within the construction industry has led to innovative approaches in materials engineering, particularly in road pavement construction. This paper explores the utilization of industrial by-products and mineral fillers—specifically red mud (RM), limestone (LS), dolomite (DL), and fly ash (FA)—as substitutes for conventional fillers in asphalt mixtures. Various proportions of these fillers were incorporated into asphalt, following the Superpave mix design guidelines. The selection of filler percentages was strategically aligned with both the theoretical framework of optimal filler-to-asphalt ratios and regulatory limits on filler content. This approach takes into account the potential for large-scale application in countries like Brazil, a major producer of Red Mud. A comprehensive materials characterization has been carried out, including physical, chemical, and performance properties. The asphalt mixtures mechanical performance was evaluated

M. S. Siverio Lima (✉)
Unit of Construction Management and Tunneling, University of Innsbruck, Innsbruck, Austria
e-mail: mayara.siverio-lima@uibk.ac.at

L. P. Thives
Department of Civil Engineering, Federal University of Santa Catarina, Florianópolis, Brazil
e-mail: liseane.thives@ufsc.br

A. Riekstins · V. Haritonovs
Institute of Civil Engineering, Faculty of Civil and Mechanical Engineering, Riga Technical University, Riga, Latvia
e-mail: riekstins.arturs@gmail.com

V. Haritonovs
e-mail: viktors.haritonovs@rtu.lv

R. Merijs-Meri
Institute of Chemistry and Chemical Technology, Faculty of Natural Sciences and Technology, Riga Technical University, Riga, Latvia
e-mail: remo.merijs-meri@rtu.lv

D. A. H. Hanaor (ed.), *Sustainable Materials for the Built Environment*, Springer Series in Materials Science 352, https://doi.org/10.1007/978-3-031-97818-0_18

through a series of tests designed to assess their suitability for road pavement applications. These tests included the Semi-Circular Bending test (SCB), Thermal Stress Restrained Specimen Test (TSRST), and Wheel Tracking test (WTT), providing insights into the mixtures' crack resistance, thermal cracking resistance, and rutting resistance, respectively. The findings reveal that the incorporation of Red Mud and Fly Ash enhances the adhesion properties between bitumen and aggregate, with notable improvements in bitumen coverage. The mixture with Fly Ash demanded higher amounts of bitumen, which has led to higher permanent deformation results. On the other hand, mixtures utilizing Dolomite and Limestone powders exhibited superior performance in the WTT, suggesting their potential applicability in high-traffic roads. The TSRST results underscored Red Mud's satisfactory performance, attributed to its granulometry and the stiffness it given to the mixture.

18.1 Introduction

Flexible pavement is extensively used in road infrastructure globally, and is selected to provide rolling comfort. In the UK, 95% [1] of paved roads are flexible pavements; in the US, 94% [2, 3]; and in Brazil, 98% [4], underscoring its significance. The pavement structure comprises layers like surface, base, and sub-base over the subgrade to support traffic loads and climate [5]. Hot mix asphalt (HMA) is the predominant surface layer material, typically made of fine and coarse aggregates, filler, and asphalt mixed at temperatures above 150 °C. Asphalt, the primary component of HMA, is mostly derived from petroleum crude oil refining, though it occurs naturally in some places and is graded for specific engineering properties [6].

Asphalt's waterproof and inert nature makes it valuable for various applications involving waste reuse or recycling [7]. Asphalt mixtures are among the most recycled construction materials. Its reused in new pavements, reduces the need for virgin materials, contributing to a circular economy [8, 9]. Reducing the carbon footprint of asphalt pavements requires understanding the impacts generated over the entire pavement life cycle, employing low-carbon materials, and extending its lifetime [10].

The pavement material enhance its capacity in bearing the structure's load. Additives like Portland cement, lime, and fly ash are used to improve strength and performance. The filler, consisting of fine particles, is crucial for HMA's behavior and mechanical performance, and are classified into the following categories: (i) mineral powders obtained through screening and crushing of rocks, like limestone, granite and dolomite; (ii) industrial products or by-products, for instance, hydrated lime, cement, and fly ash, and (iii) ashes or fine particles derived from waste materials, as steel slag powder, as rice husk ash and glass powder [11].

These materials enhance the stability of the mixture and improve the adhesion between the asphalt and aggregate interface [12]. Previous studies have confirmed that the inclusion of fillers, both mineral and by-product materials, can significantly improve the moisture resistance, mechanical performance, and durability of HMA and mastics [11, 13–17].

Al-Bayati et al. [18] observed that the addition of fillers in asphalt mixtures impacts two primary aspects: mastic properties and consequently asphalt mixture's mechanical performance. Studies have shown that optimal filler concentrations result in stronger mixtures due to their ability to promote better aggregate-asphalt cohesiveness, minimize oxidative aging, improve moisture damage resistance, increase modulus stiffness, and enhance mechanical properties such as fatigue resistance and rutting performance [19–22].

Regarding the effects on asphalt and mastic, filler addition significantly stiffens the mixture [22, 23]. Diab and Enieb [24] investigated the mechanisms of mineral fillers in asphalt at both the mixture and mastic scales. They concluded that the type and amount of filler strongly influence the mechanical properties of asphalt mixtures, and the physical properties of mastics are highly dependent on the filler type and concentration. The filler-asphalt interaction significantly affects asphalt mastics, increasing their rigidity even under varying temperatures and load frequencies [25].

Researchers have examined the suitability of industrial waste and recycled materials in pavement construction worldwide during the last decades. One prevalent method for employing circular economy practices in pavement production is to use waste materials as a substitute for non-renewable resources or raw materials [26]. Yaro et al. [27] asserted that using waste materials and inserting new recycling techniques in asphalt pavement is critical. Materials derived from industrial activities, such as fly ash from coal power plants, slag from steel manufacture, or red mud from aluminium production, can likewise be utilized as alternative materials that contribute to the sustainability of the road area.

Although the effects of mineral fillers on the properties of asphalt mixtures are continuously investigated, the potential of reusing industrial waste materials as fillers, such as red mud and fly ash, remains underexplored. This is particularly significant considering the potential benefits associated with their reuse. As industrial waste generation increases worldwide, these materials present a promising alternative for use in various infrastructure projects [28, 29].

Wong et al. [30] emphasized that substituting solid waste products for conventional fillers can reduce the demand for raw materials. However, it is crucial to assess the potential impacts, such as the leaching of chemical compounds, to minimize negative environmental outcomes. Evaluating these factors is essential to ensure the safe and sustainable use of industrial waste materials in construction.

According to the United States Geological Survey [31], bauxite ore is a crucial raw material for many industries, including chemical, metallurgical, and construction sectors. Brazil ranks as the fourth-largest global producer of bauxite, with an annual output of approximately 33 million metric tons. The primary bauxite deposits are located in the northern region of the country, with the State of Pará being the leading producer. Brazilian mining companies face significant environmental and sustainability challenges in managing and reusing red mud.

Red mud is a by-product of refining bauxite ore to produce alumina for the aluminium industry, primarily through the Bayer process. This process involves reacting bauxite with sodium hydroxide under high temperature and pressure conditions. For every ton of alumina produced, around 1.5–1.6 tonnes of red mud are generated [32]. According to the International Aluminium Institute, the amount of red mud produced per ton of alumina can range from 0.55 to 2.21 tonnes [33].

Due to its variable composition, red mud is often considered a hazardous material. However, its potential as a valuable resource merits further exploration [34]. Numerous studies have investigated the use of red mud as a filler in asphalt mixtures [34–37]. Nevertheless, additional analysis is necessary to fully address and satisfy environmental concerns.

In countries with extensive construction activities, such as China, the demand for raw construction materials, like fillers to produce asphalt mixtures, is exceptionally high. Utilizing by-products like red mud not only addresses this demand but also offers significant cost savings in construction. Moreover, it contributes to sustainability by repurposing industrial waste and reducing the reliance on traditional raw materials [35].

Choudhary et al. [38] compared the performance of asphalt mixtures using stone dust and red mud as fillers. Their findings indicated that the red mud mixture exhibited higher Marshall stability, as well as improved resistance to rutting and cracking. This enhanced performance was attributed to the characteristics of red mud, such as its fineness and porosity, which facilitate uniform filler distribution and greater asphalt absorption.

Wu et al. [37] incorporated red mud into asphalt and compared it with the standard asphalt base. They found that the penetration of the red mud-asphalt blend was lower, the softening point was significantly improved, and ductility was reduced. Additionally, Ou et al. [39] demonstrated that using red mud instead of limestone as a filler enhanced the fatigue resistance of asphalt mastics.

Another valuable waste material is fly ash, a by-product of burning pulverized coal in thermal power plants. Over the years, researchers have explored numerous ways to reuse fly ash across various sectors, including as a filler material, to reduce environmental burdens and enhance economic benefits. Globally, coal combustion for power generation produces more than 500 million tons of fly ash annually, yet only 25–30% is reused and reintegrated into the production chain [40]. Fly ash primarily consists of silica, aluminium, and iron oxide. Studies have shown that using fly ash as a filler in asphalt mixtures improves their mechanical properties [41]. Giri and Priyadarshini [42] concluded that fly ash not only enhances asphalt mixture performance but also has the potential to replace conventional filler materials in asphalt pavement construction.

To comprehensively evaluate the performance of asphalt mixtures containing various by-products, this study conducted a detailed laboratory assessment. The focus was on analysing the characteristics and mechanical properties of asphalt mixtures using different fillers, including red mud, dolomite, limestone, and fly ash. The tests were conducted at the Latvian State Road Laboratory in Riga. The primary objective was to assess the effectiveness of red mud as a filler and to evaluate its impact when used together with other fillers.

18.2 Materials and Methods

18.2.1 Aggregates

For the production of asphalt mixtures, granitic aggregates with nominal maximum sizes of ¾″ (19.0 mm) and 3/8″ (9.5 mm) were sourced from a quarry in Latvia. The characterization of these aggregates was conducted using the following tests: Los Angeles abrasion (EN 1097-2) [43], density (EN 1097-6) [43], flakiness index (EN 933-3) [44], and granulometry (EN 933-1) [45].

The Los Angeles abrasion test results were 10.4% for the ¾″ aggregates and 11.36% for the 3/8″ aggregates. These values are significantly lower than the threshold of 25% specified by EN 1097-2 [43], indicating that the aggregates are well-suited for use under high and heavy traffic conditions.

The measured density for both the ¾″ and 3/8″ aggregates was 2.78 g/cm^3. The flakiness index, determined according to the EN 933-3 standard [45], was 14, which complies with Latvia's requirements (<20). The granulometry results are detailed in Sect. 18.2.5, Fig. 18.3.

18.2.2 Asphalt

The asphalt used was a PEN 50/70, classified by penetration, and supplied by an asphalt distributor from Poland. The characterization test results are presented in Table 18.1:

Table 18.1 PEN 50/70 characterization test results

Test	Result	Standard
Penetration (0,1 mm)	59	EN 1426 [46]
Softening point (°C)	49	EN 1427 [47]
Fraass breaking point (°C)	– 15	EN 12593 [48]
Kinematic viscosity 135 °C (mm^2/s)	483.4	EN 12595 [49]
Dynamic viscosity 60 °C (Pa.s)	60	EN 12596 [50]
Elastic recovery (%)	11.8%	EN 13398 [51]
RTFOT		EN 12607 [52]
Penetration (0.1 mm)	36	EN 1426 [46]
Softening point (°C)	55	EN 1427 [47]

18.2.3 Fillers

Four different filler materials were used in this study: red mud (RM), limestone (LS), dolomite (DL), and fly ash (FA). The RM was sourced from northern Brazil, while the LS, DL, and FA were obtained from Latvia. An extensive and comprehensive analysis was conducted to characterize these fillers through various laboratory tests. Prior to testing, the filler samples were dried in a laboratory oven at 110 °C.

The particle size distributions of the fillers were determined using Laser Diffraction (LD) analysis with the Analysette 22 NanoTec device, which has a detection range of 0.1–1000 mm. The granulometric curves are presented in Fig. 18.1.

The ASTM D242 standard [53] specifies that filler materials should pass through sieves No. 16 (1.18 mm), No. 30 (600 μm), No. 50 (300 μm), and No. 200 (75 μm) with the requirements of 100%, 97%–100%, 95%–100%, and 70%–100% passing, respectively. Figure 18.1 illustrates that all fillers met the specification requirements for granulometry except for fly ash (FA), with only 54% passing through the No. 200 sieve.

The filler densities were measured following the EN 1097-7 standard [43], with results presented in Table X. The specific surface areas of the fillers were determined using the BET (Brunauer–Emmett–Teller) method. The test results (Table 18.2) indicated that red mud from Brazil exhibited the highest specific surface area (8.3 m^2/g), while limestone showed the lowest (2.5 m^2/g). According to Géber and Gömze [54], the presence of clay minerals accounts for the high specific surface area observed in red mud. Similarly, Zhang et al. [29] found that red mud had a significantly higher specific surface area compared to other mineral fillers. The authors concluded that this

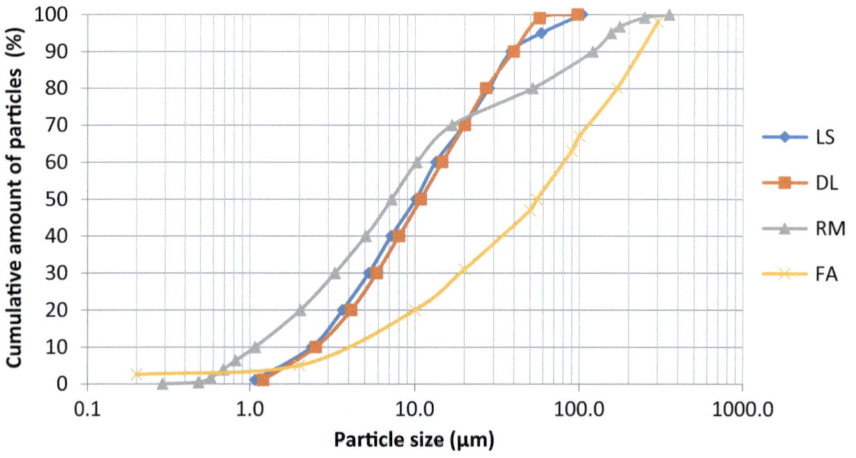

Fig. 18.1 Fillers granulometric curves

Table 18.2 Fillers characterization test results

Filler	Density (g/cm^3)	BET (m^2/g)	RV (g/cm^3)
RM	2.833	8.30	44.80
LS	2.709	2.50	32.60
DL	2.682	4.10	30.90
FA	2.917	2.90	42.50

difference in specific surface area enhances red mud's adsorption to asphalt components, positively influencing the mechanical properties of asphalt mastic, particularly in terms of improved resistance to permanent deformation.

The void characteristics of fillers influence asphalt absorption, which was evaluated in this study using the Rigden Voids (RV) method according to the EN 1097-4 standard. Harris and Stuart (1995) stated that the fractional voids in dry compaction correlate well with the stiffening effect of filler on asphalt, and the RV result is a tool for selecting a filler. The results (Table 18.2) showed that RM (44.8%) and FA (42.4%) had higher RV values compared to LS (32.6%) and DL (30.9%), which were similar. Romeo et al. [55] affirmed that RV is a valuable indicator of the stiffening effect of fillers on asphalt. According to Zhang et al. [56], fillers with elevated RV can absorb and fix more free asphalt, thereby enhancing the friction effect among filler particles.

The chemical composition of the fillers was evaluated using X-ray fluorescence (XRF) through the fusion method, an accurate technique for analysing powder samples. This method mitigates heterogeneity caused by grain size and mineralogical effects, ensuring consistent results. The fusion process homogenizes the material properties by vitrification, allowing for an expanded calibration range using either synthetic calibration curves with reagents or calibration to varied materials Yamada [57].

The tests were performed using a benchtop wavelength-dispersive X-ray fluorescence (WDXRF) spectrometer, specifically the Supermini model. The procedure, as described by Yamada [57], involves mixing more than two selected reagents to create a fusion bead, followed by material calibration. The filler samples were fused using a benchtop high-frequency fusion machine (Bead Sampler) at 1200 °C. The spectrometer, equipped with an end-window Pd-target 200 W X-ray tube, operated at 50 kV–4.0 mA, with a measurement area of 30 mm in diameter. Table 18.3 presents the chemical composition of each filler analysed in the study.

Table 18.3 reveals the primary chemical compounds of the fillers analysed. Limestone (LS) predominantly consists of calcium oxide (CaO), while fly ash (FA) has elevated percentages of silicon dioxide (SiO_2) and CaO. Red mud (RM) is mainly composed of Fe_2O3, Al_2O_3, and SiO_2, which together account for over 80% of its total weight. Additionally, RM contains a significant amount of Na_2O (11.31%), contributing to its alkaline character and enhancing asphalt adhesion. Studies have confirmed that higher percentages of Na_2O in red mud result in higher pH values compared to other mineral fillers [28, 29].

Table 18.3 X-ray fluorescence test results

Chemical compounds	Filler			
	RM	LS	DL	FA
Fe_2O_3 (Iron(III) oxide) (%)	29.86	–	0.34	1.15
Al_2O_3 (Aluminium oxide) (%)	21.96	0.54	0.64	5.03
SiO_2 (Silicon dioxide) (%)	18.31	3.0	2.5	46.6
Na_2O (Sodium oxide) (%)	11.33	–	0.82	1.22
TiO_2 (Titanium dioxide) (%)	5.50	–	–	–
CaO (Calcium oxide) (%)	1.53	51.81	31	24.68
MgO (Magnesium oxide) (%)	0.23	0.5	17	2.11
MnO (Manganese(II) oxide) (%)	0.17	–	–	–
P_2O_5 (Phosphorus pentoxide) (%)	0.10	–	–	–
SO_3 (Sulfur trioxide) (%)	0.09	0.11	–	8.26
K_2O (Potassium oxide) (%)	0.07	–	0.76	4.35
Cr_2O_3 (Chromic oxide) (%)	0.04	–	–	–

The presence of silicon dioxide (SiO_2), particularly in FA and RM, can negatively impact adhesion with asphalt, leading to lower moisture resistance in mixtures [58]. However, compounds such as SiO_2, Al_2O_3, and Fe_2O_3, which are abundant in RM, have been shown to significantly affect the stiffness of asphalt mastics, thereby improving rutting resistance [59].

18.2.4 Mastics

Seven types of mastics were produced, each containing 40% filler by volume mixed with bitumen. To evaluate the influence of RM, different proportions and combinations were created. These included: 40% red mud (40RM), 40% limestone (40LS), 40% dolomite (40DL), 40% fly ash (40FA), a mixture of 20% red mud and 20% limestone (20RM20LS), a mixture of 20% red mud and 20% dolomite (20RM20DL), and a mixture of 20% red mud and 20% fly ash (20RM20FA).

The mastics were prepared by incorporating a specific amount of each filler or filler combination into the bitumen. Initially, the bitumen PEN 50/70 and the fillers were heated to 155 °C and 160 °C, respectively, following the EN 12697-35 standard [60]. For the bitumen PEN 50/70, the compaction temperature was set at 150 °C, while the mixing temperature was maintained within 20 °C above the compaction temperature, with aggregates and fillers heated to the mixing temperature ± 5°C.

The fillers were then gradually added to the bitumen and manually mixed using a glass rod for approximately 3–4 min to ensure homogeneous mastics. The prepared mastics were stored at room temperature for subsequent laboratory tests.

Laboratory tests were conducted to evaluate the characteristics of the mastics. As shown in Table 18.4, adding filler to the bitumen increased the softening point while decreasing the penetration values of the mastics. FA was the filler that resulted in the highest softening point value. Conversely, the inclusion of RM with other fillers also elevated the softening point of the mastics. The Fraass breaking point was reduced for all the mastics. As expected, the mastics became stiffer after aging, and the softening point increased.

The elastic recovery test was performed according to the standard EN 13398 [51], and a storage stability test was conducted following EN 13399 [61].

The adhesion of mastic/aggregate was evaluated using the EN 12697-11 standard [60]. In this study, the boiling water stripping method was adapted without using chemicals. The preparation of materials for the test involved the following steps: (i) the granitic aggregates, size fraction 8–11 mm, were dried in an oven at 110 °C for 3 h; (ii) after drying, the aggregates were heated at 160 °C for 3 h; (iii) the mastics were heated at 155 °C for 3 h. The bowl used for mixing was also heated with the aggregates to maintain the temperature during the mixing process.

Table 18.4 Mastics characterization test results

Tests	Mastics						
	40RM	40LS	40DL	40FA	20RM + 20LS	20RM + 20DL	20RM + 20FA
Penetration (0,1 mm)	32.54	47.00	38.63	27.75	30.83	29.65	25.48
Softening point (°C)	59.7	56.4	56.3	62.3	58.5	57.8	61.0
Fraass breaking point (°C)	– 12.0	– 13.0	– 12.5	– 12.0	– 13.0	– 11.0	– 11.5
Elastic recovery (%)	0	15	16	0	0	0	15
RTFOT							
Penetration (0.1 mm)	23.2	18.9	23.3	24.7	29.5	26.8	23.8
Softening point (°C)	69.9	61.9	61.8	66.5	61.9	61.7	63.2
Storage							
Penetration (0.1 mm)—top	52.7	59.3	57.6	54.3	64.3	59.3	54.5
Penetration (0.1 mm)—bottom	14.3	27.7	31.5	15.7	35.0	34.0	17.5
Softening point (°C)—top	51.2	50.1	50.2	50.9	50.8	50.8	52.6
Softening point (°C)—bottom	80.0	58.7	52.9	69.2	59.5	61.8	66.3

Table 18.5 Affinity between aggregate and asphalt

Test	PEN 50/70	Mastics						
		40RM	40LS	40DL	40FA	20RM + 20LS	20RM + 20DL	20RM + 20FA
Adhesivity (%)	X	90	64	47	99	78	73	91

Expressed as the degree of bitumen coverage in percentage

To perform the tests, 510 g of aggregates were used per mix. The amount of mastics added depended on the filler type, ranging from 16 to 20 g. Aggregates and mastics were mixed by hand until the aggregates were fully coated. After mixing, the coated aggregates were spread on aluminium foil to cool down for 12 h at 20 ± 5 °C. Thirty aggregate particles were meticulously chosen, ensuring complete coverage, to enhance the accuracy of the process. These particles were placed in supporting netting with hooks.

A beaker containing 700 mL of water was placed on an electric stove, regulated to maintain the water at 100 °C. When the water reached boiling point, the supporting netting with aggregates was inserted for 30 min. Subsequently, the particles were visually evaluated by two different individuals separately; each person determined the coating percentage for each aggregate particle based on the guidance established in the EN 12697-11 standard [60]. An average of the individual evaluations was calculated to express the results (Table 18.5).

The results (Table 18.5) showed that LS and DL mastics did not promote adequate aggregate adhesion. Adding RM to the limestone and dolomite improved the adhesion, while mastics composed only of RM and FA were enough to promote adhesion to the aggregates. In this way, it was necessary to incorporate an anti-stripping additive for mixtures composed with the following mastics: 40LS, 40DL, 20RM20LS and 20RM20DL.

Figure 18.2 illustrates the Thermal Susceptibility Index (TSI) of mastics, a parameter derived from penetration and softening point tests. TSI quantifies asphalt's consistency sensitivity to temperature variations, typically ranging from − 1.5 to + 0.7 for Brazilian bitumen standards [53, 62]. A higher TSI value indicates lower thermal susceptibility, meaning the asphalt is less sensitive to temperature changes.

18.2.5 Asphalt Mixtures

The percentage of filler added to the mixture was chosen to meet SUPERPAVE specifications [63], particularly the filler/asphalt ratio, which must fall within the range of 0.8–1.6. Furthermore, in accordance with Brazilian standards DNIT [62, 64], which set a maximum filler content limit of 12%, the formulation adhered to this constraint, considering Brazil's significant global production of red mud (RM).

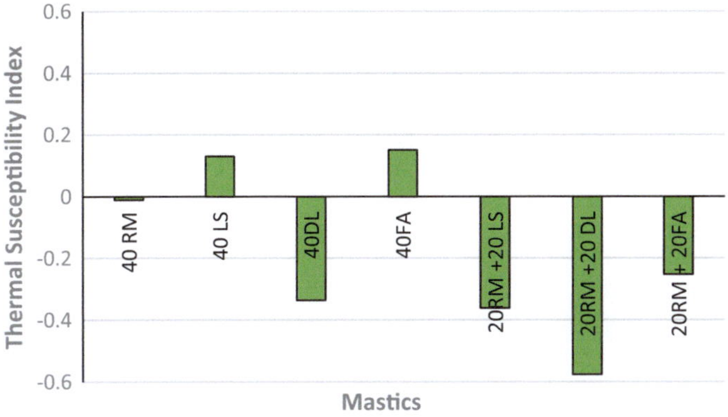

Fig. 18.2 TSI of mastics

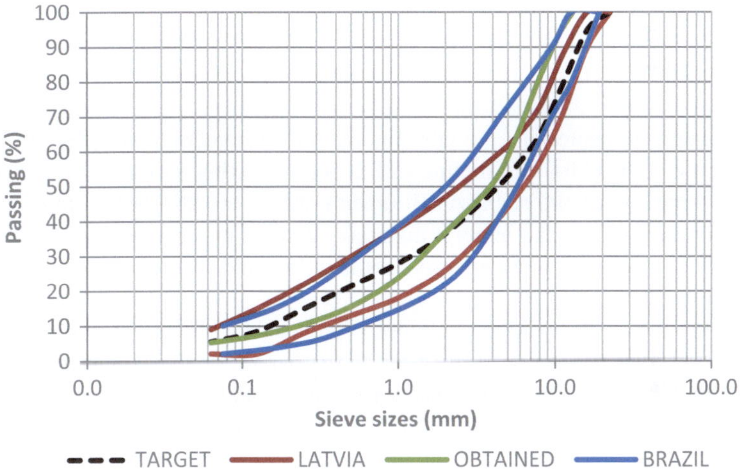

Fig. 18.3 Mixture granulometric curve

Figure 18.3 depicts the granulometric curve of the aggregates utilized in the asphalt mixture.

The asphalt mixture was dosed following the Marshall method (EN 12697-30) [60]. Table 18.6 details the quantities of aggregates, bitumen, and the concentration of the anti-stripping (AS) additive utilized in fabricating the asphalt mixtures. The AS content was determined relative to the mass of bitumen employed. The filler/bitumen ratio (F/B) was carefully adjusted to achieve optimal performance and adherence to applicable standards.

Table 18.6 Asphalt mixtures dosage results

Mastics	Aggregates (%)	Mastic content (%)	AS (%)	F/B (%)
40RM	95.3	4.7	–	1.11
40LS	95.5	4.5	0.3	1.06
40DL	95.3	4.7	0.5	1.05
40FA	95	5.0	–	1.14
20RM + 20LS	95	5.0	0.3	1.08
20RM + 20DL	95	5.0	0.3	1.08
20RM + 20FA	95	5.0	–	1.12

18.2.6 Mixing

Initially, the mineral material, filler, and bitumen are divided into appropriate portions based on the specific mix recipe. The size of these portions is determined by the mix design requirements. The mixing temperature is dependent on the bitumen class; for road bitumen B 50/70, it is set at 150 °C, with a maximum allowable temperature of 170 °C.

To prepare the materials, the portions of mineral material and filler are placed in an oven at 155 °C for a minimum of 8 h before mixing, ensuring they reach the required temperature and stabilize. Bitumen portions are also heated in an oven at 155 °C for 3–5 h. Both the mineral material/filler and bitumen are heated 5° above the mixing temperature to compensate for heat loss and maintain optimal viscosity during mixing.

Prior to mixing, the mixing container is preheated to 150 °C. Once the mineral material and filler have sufficiently aged in the ovens, they are transferred into the mixing container. The bitumen (mastic) is immediately poured over the mineral material, and stirring commences promptly. The mixing duration is set at 4 min.

After mixing, the asphalt concrete mixture is portioned as needed for subsequent tests or applications. This process ensures uniformity and proper preparation of the asphalt mixture according to established standards.

18.2.7 Performance Tests

18.2.7.1 Semi Circular Bending Test (SCB)

The SCB test was employed to evaluate the fracture toughness of asphalt mixtures, providing crucial insights into crack propagation under load. This test is instrumental in designing mixtures to enhance crack resistance, particularly against thermal and fatigue stresses. The SCB test adhered to the standard EN 12697-44.

For each asphalt mixture, a single slab was compacted, and a core with dimensions of 150 ± 1 × 80 mm was extracted. Two specimens were prepared from each core. These specimens were then shaped to appropriate dimensions (diameter 150 ± 1 mm, height 74 ± 1 mm, thickness 50 ± 3 mm), with a notch (width 0.40 ± 0.2 mm, depth 10.0 ± 1.0 mm) cut into the middle of the flat edge.

Prior to testing, specimens were conditioned in a climatic chamber at 0°C for 4 h. Subsequently, they were subjected to three-point bending at a constant deformation speed of 5.0 ± 0.2 mm/min. During testing, force (kN) and displacement (mm) values were continuously recorded.

The fracture toughness (N/mm^3/2) of the specimens was calculated using Eqs. (18.1–18.2), where factors such as specimen height and notch depth influence the computed values of fracture toughness.

$$K_{IC,i} = \sigma_{max,i} Y_1 \sqrt{\pi a_1} \tag{18.1}$$

$$Y_1 = 4.782 - 1.219(a_1/r_1) + 0.063 \exp(7.045\, a_1/r_1) \tag{18.2}$$

where

$K_{IC,i}$ Fracture toughness of specimen ($N/mm^{1.5}$)
$\sigma_{max,i}$ Maximum stress at failure of specimen (N/mm^2)
Y_1 Normalised mode I stress intensity factor
a_1 Notch depth of specimen (mm)

18.2.7.2 Thermal Stress Restricted Specimen Test (TSRST)

The TSRST was employed to assess the thermal cracking resistance of asphalt mixtures, focusing on determining the minimum temperature at which asphalt can resist cracking. The test methodology adhered to the standard EN 12697-33 [60].

Initially, asphalt slabs were compacted using a roller compactor. From each mixture, two prismatic samples measuring 160 × 40 × 40 mm were cut longitudinally from the slab, following the guidelines specified in EN 12697-46 [60] where the longitudinal axis of the specimens must be orthogonal.

Subsequently, each sample was securely glued to two aluminium plates and left to dry completely. The specimens were then positioned in a fixed frame within a thermostatic chamber. According to the standard protocol, the test commenced at 20 °C, and the temperature was decreased at a rate of 10 °C per hour, precisely following these parameters throughout the test duration.

At the conclusion of the test, failure stress and failure temperature were determined, providing essential data on the asphalt mixture's ability to withstand thermal stresses without cracking. This approach ensures standardized testing conditions and reliable evaluation of thermal cracking resistance in asphalt materials.

18.2.7.3 Wheel Tracking Test

The wheel tracking test was employed to assess the rutting resistance of asphalt mixtures in accordance with standard EN 12697-22 [60]. Each mixture underwent testing using a single slab compacted by a roller compactor to dimensions of 320 × 260 × 47 mm, as specified in EN 12697-33 [60].

The test procedure followed method B, where a small device tracks a wheel across the sample continuously for either 20,000 crossings or until a rut depth of 20 mm is attained. Testing was conducted in ambient air conditions at a temperature of 60 °C, replicating conditions comparable to those experienced by truck tires under real-world conditions.

During the test, the load applied to the sample by the wheel was designed to simulate the stress exerted by a truck tire. The depth of the wheel track was measured using LVDT (Linear Variable Differential Transformer) displacement indicators, with measurements recorded every 5 cycles throughout the test duration.

At the conclusion of the test, the tendency of rut depth accumulation provided critical data on the asphalt mixture's ability to resist permanent deformation caused by repeated loading cycles, thereby informing decisions related to material performance and suitability for road construction applications.

18.3 Results and Discussion

18.3.1 Semi Circular Bending Results

The SCB results (Fig. 18.4) provide insights into the fracture toughness and resistance to cracking of asphalt mixtures, revealing distinct performances influenced by the type of filler used. The DL exhibited notable strength, achieving a resistance value of 43.0 N/mm$^{3/2}$, indicating its suitability for demanding applications. The FA showed slightly lower resistance at 38.4 N/mm$^{3/2}$, yet still contributed significantly to fracture resistance.

The LS demonstrated competitive performance with a resistance value of 40.4 N/mm$^{3/2}$, comparable to conventional fillers. Red mud (RM) exhibited a resistance of 39.3 N/mm$^{3/2}$, highlighting its potential as a sustainable filler option. Mixtures combining RM-DL and RM-FA showed resistance values of 39.0 N/mm$^{3/2}$ and 38.8 N/mm$^{3/2}$, respectively, suggesting viable alternatives with substantial cracking resistance.

The combination of RM with LS yielded a resistance of 39.4 N/mm $^{3/2}$, surpassing that of RM alone, indicating a promising balance between mechanical performance and sustainability in road construction.

It is important to note that the observed differences among these mixtures are statistically marginal, requiring further rigorous statistical analysis to conclusively

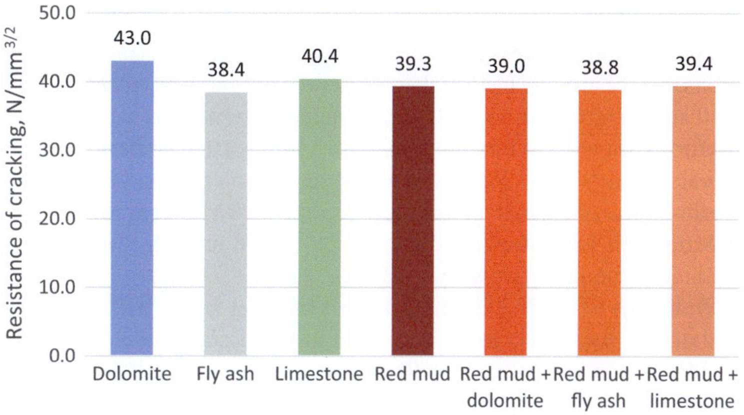

Fig. 18.4 Resistance to cracking of asphalt mixtures given by SCB test results

discern significant distinctions in cracking resistance performance between the various filler types.

18.3.2 Thermal Stress Restricted Specimen Test (TSRST)

The TSRST results provide insights into the thermally-induced stress resistance of various asphalt mixtures, crucial for assessing their performance in colder climates where susceptibility to thermal cracking is a significant concern. Figure 18.5 illustrates stress values measured in megapascals (MPa) alongside corresponding temperatures in degrees Celsius (°C).

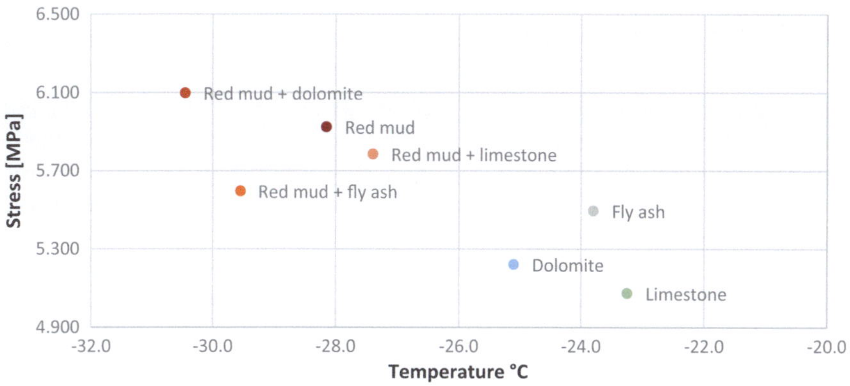

Fig. 18.5 TSRST results

- **Red Mud with Dolomite**: This mixture shows the highest resistance to thermal stress, withstanding a stress of over 6.1 MPa at a temperature nearing – 31 °C. This suggests that the mixture has a high tolerance to cold-induced tension, which could make it suitable for environments experiencing extreme cold weather conditions.
- **Red Mud**: The pure RM mixture resisted to a stress of approximately 5.9 MPa at a temperature just below – 28 °C, eventually suggesting that the DL could optimize the performance of asphalt mixtures produced with RM at very low temperatures.
- **Red Mud with Limestone**: A combination of RM and LS resisted around 5.8 MPa at temperatures approaching – 28 °C.
- **Red Mud with Fly Ash**: Exhibiting a stress of about 5.6 MPa at roughly – 26 °C, this mixture ranks slightly lower in thermal stress resistance compared to the other Red Mud mixtures. However, it still maintains a considerable level of resistance, underscoring the viability of utilizing a combination of industrial by-products in asphalt mixtures.
- **Fly Ash**: The mixture with Fly Ash as the sole filler shows resilience to thermal stress at a higher temperature, close to – 22 °C, with a stress resistance near 5.4 MPa. Although its performance is commendable, it suggests that FA may be less suited to extremely cold climates compared to the other fillers tested.
- **Dolomite and Limestone**: Both mixtures appear to have the lowest resistance to thermal stress among the evaluated materials, with DL exhibiting slightly better performance than LS. They withstood stress levels in the vicinity of 5.3 MPa at temperatures around – 22 °C and – 24 °C, respectively.

Hence, the results suggest that the RM promotes thermal cracking resistance at lower temperatures, which could be beneficial for paving in regions with cold climates. Conversely, conventional fillers like DL and LS alone, while still resisting thermal stress, may be more suitable for environments where the temperatures do not fall as low.

18.3.3 Wheel Tracking Test

In evaluating the rutting resistance of asphalt mixtures, the WTT serves as a critical assessment, closely representing the effect of continuous vehicular traffic on pavement structures. The empirical data from laboratory testing presents an instructive comparison among various asphalt mixtures enhanced with different fillers. Figure 18.6 shows the rutting resistance of the diverse asphalt mixtures tested:

In the WTT, the best results were shown by the mixtures that used classic fillers—DL and LS powders. Both mixtures presented very similar results, and according to the Latvian road specifications could even be applied to main roads with the high traffic loads, where the WTS (Wheel Track Formation Speed) requirement is 0.1 mm/1000 cycles.

Either the reference mixture with LS as the one produced with DL show good resistance with minimal rut depth after extensive wheel crossings, suggesting longer

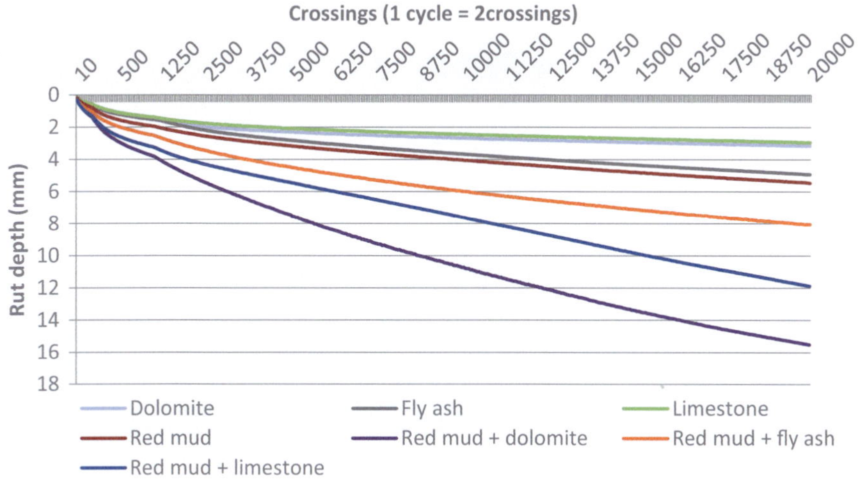

Fig. 18.6 WTT results

pavement's lifecycle and potentially less frequent maintenance required. Mixtures produced with FA and RM separately come after with slightly less performance than DL, with approximately 5 mm rutting deformation after 20,000 crossings.

Mixtures that used by-products in their formulations also showed similar results to each other. The results of these mixtures were around 0.3 mm/1000 cycles. When mixed with FA, LS and DL, the RM asphalt mixture's rutting resistance substantially decreases showing potential low durability over a pavement's lifetime. This doesn't necessarily mean that they can not be used at all, but eventually the adoption of such mixtures may be particularly strategic in lower-traffic settings, for example.

As a reference, the Figs. 18.7, 18.8, 18.9 and 18.10 show the cross-sectional outcomes of the WTT across various Asphalt Binder Mixtures (ABM), each incorporating a unique type of filler.

Fig. 18.7 ABM with RM

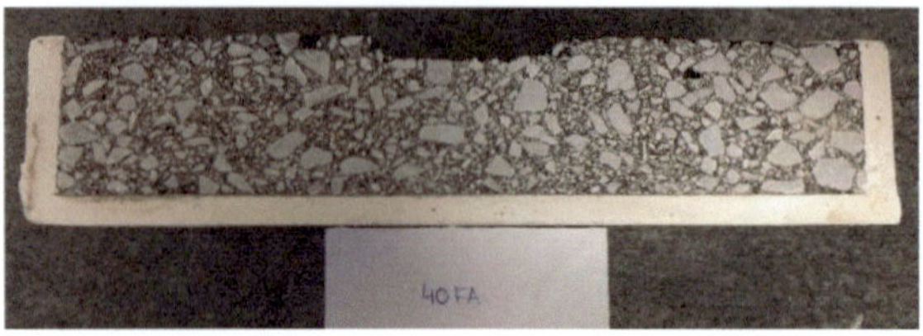

Fig. 18.8 ABM with FA

Fig. 18.9 ABM with DL

Fig. 18.10 ABM with LS

18.4 Hypothesis

These findings suggest that while the performance of mixtures with solely industrial by-products as fillers might be slightly inferior to conventional fillers, they present a potential alternative with acceptable mechanical properties for road pavement construction.

(a) Wheel Tracking Test:

- Dosage: The results are primarily influenced by the amount of asphalt used in each mixture. Mixtures with higher percentages (5% vol.) of asphalt, such as FA, RM + FA, RM + DL and RM + LS have automatically higher susceptibility to temperature and, therefore, are more prone to deformation. The least amount of bitumen presented at the mixture with 40% LS is potentially the main reason for its best mechanical performance, in comparison with other mixtures.
- Granulometry Influence: The granulometry of the fillers also impacts their resistance to permanent deformation. Mixtures with dolomite and limestone showed the best performance, likely due to their relatively continuous, fine, and stable granulometry (with minimal variations). This allows the mastic to be better mixed and ensures that the binder coats all grains more effectively. The mixture with red mud showed the worst WTT result—0.28 mm/1000

cycle. This is probably explained by the granulometry of the red mud. Unlike classic fillers, this powder contains more particles that are relatively coarser. In this way, potentially contributing to the overall adhesion of the mixture.

(b) Thermal Stress Restrained Specimen Test (TSRST):

- Anti-aging Agents: Some studies [65–67] show that certain elements, such as Fe_2O_3, Al_2O_3, and SiO_2, are directly associated with anti-aging properties due to their ability to enhance the mechanical, thermal, and oxidation resistance of asphalt. Therefore, the presence of these compounds in RM and FA can explain the positive results observed in aging resistance and adhesion properties. Studies suggest [36, 68–70] that the red mud's alkalinity and hydrophobicity promote a strong bond with asphalt due to its acidic nature, offering a satisfactory resistance against moisture damage and consequently, benefiting the low temperature performance of mastics with red mud in its composition.
- Contribution of Other Fillers: Meanwhile, the presence of CaO and MgO in dolomite (DL) and limestone (LS) contributes to the rigidity and mechanical stability of the asphalt. The MgO presented in dolomite's composition, has an affinity to water, and promotes a volumetric expansion reaction, which allows more water to come inside the mastic and remove the binder film from the aggregate. Thus, the adhesive and consequently low temperature properties of the dolomite are strongly dependent on its hydrophilic characteristics. These results are supported by adhesion and anti-aging analyses conducted in a previously published article [36].

(c) Semi-Circular Bending (SCB) Test:

- Sample Size: Due to the limited number of samples tested for each mixture, it is not possible to statistically confirm significant differences in the results among the various asphalt mixtures.

18.5 Conclusions

This study demonstrates that incorporating industrial by-products, such as red mud and fly ash, into asphalt concrete mixtures can achieve performance levels comparable to traditional fillers like dolomite and limestone. Even though it is not possible to statistically confirm significant differences between results due to the number of samples, the Semi-Circular Bending (SCB) test results revealed that dolomite exhibited the highest resistance to cracking, followed by limestone and mixtures containing red mud, suggesting their potential for high-performance applications.

The Thermal Stress Restrained Specimen Test (TSRST) findings highlighted the superior thermal stress resistance of red mud mixtures, particularly when combined with dolomite, making them potentially suitable for cold climates. The presence of anti-aging agents compounds in red mud and fly ash, such as Fe_2O_3, Al_2O_3, and SiO_2, potentially contributed to their thermal stability.

The Wheel Tracking Test results showed that mixtures with dolomite and limestone have a good rutting resistance, meeting high-traffic road requirements. On the other hand, red mud and fly ash mixtures showed slightly lower rutting resistance, which would be probably viable for lower-traffic applications. These findings underscore the feasibility of using industrial by-products in asphalt mixtures, promoting sustainability in road construction.

To confirm the potential use of alternative materials, such as red mud and fly ash, in road construction, it is essential to conduct comprehensive environmental tests. These tests should ensure that any hazardous components are effectively encapsulated by the bitumen film, thereby mitigating environmental risks. This would provide a safe and durable alternative for road construction.

References

1. Asphalt Industry Alliance. *Key Facts* (2024). https://www.asphaltuk.org/key-facts/
2. Y. Qiao, A.R. Dawson, T. Parry, G. Flintsch, W. Wang, Flexible pavements and climate change: a comprehensive review and implications. Sustainability (2020). https://doi.org/10.3390/su12031057
3. Federal Highway Administration. *Highway Statistics 2019, Table HM-12* (2020). https://www.fhwa.dot.gov/policyinformation/statistics/2019/hm12.cfm
4. Confederação Nacional de Transportes (In Portuguese). *Pequisa CNT de rodovias 2023* (2023)
5. Federal Highway Administration. *Geotechnical Aspects of Pavements.* Publication No. FHWA NHI-05-037 (2006). https://highways.dot.gov/sites/fhwa.dot.gov/files/FHWA-NHI-05-037.pdf.
6. Virginia Asphalt Association. *Asphalt binder/cement* (2024). https://vaasphalt.org/liquid-asphalt/
7. European Asphalt Pavement Industry. *The European Asphalt Pavement Industry's Position Statement on the Use of Secondary Materials, By-Products and Waste in Asphalt Mixtures* (2017). https://eapa.org/advantages-of-asphalt/
8. Environmental Protection Agency. *Construction and Demolition Debris Management in the United States* (2020). https://www.epa.gov/sites/default/files/2020-03/documents/final_cd-eol-management_2015_508.pdf
9. National Asphalt Pavement Association. *The Role of Asphalt Pavements in a Low-Carbon Transportation Network* (2023). https://www.asphaltpavement.org/uploads/documents/GovAffairs/NAPA%20Carbon%20Footprint%20Priorities.pdf
10. National Asphalt Pavement Association. *The Carbon Footprint of Asphalt Pavements: A Reference Document for Decarbonization* (2024). https://www.asphaltpavement.org/uploads/documents/Climate/NAPA-SIP109-TheCarbonFootprintOfAsphaltPavements-March2024.pdf
11. Y. Chen, S. Xu, G. Tebaldi, E. Romeo, Role of mineral filler in asphalt mixture. Road Mater. Pavement Des. **23**, 247–286 (2022). https://doi.org/10.1080/14680629.2020.1826351
12. Y. Liang, T. Bai, X. Zhou, F. Wu, C. Chenxin, C. Peng et al., Assessing the effects of different fillers and moisture on asphalt mixtures' mechanical properties and performance. Coatings (2023). https://doi.org/10.3390/coatings13020288
13. E. Remisova, *Study of Mineral Filler Effect on Asphalt Mixtures Properties* (2015)
14. A. Mohammed, A. Fadhil, Effect of filler type on the durability of asphalt concrete mixes. Int. J. Curr. Eng. Technol. (2018). https://doi.org/10.14741/ijcet/v.8.3.36

15. A. Ali, Effect of type and percentage of unconventional mineral fillers on the performance of hot mixed asphalt. Int. J. Eng. Adv. Technol. **8**, 34–38 (2018)
16. Z. Liu, X. Yang, Y. Wang, S. Luo, Engineering properties and microwave heating induced ice-melting performance of asphalt mixture with activated carbon powder filler. Constr. Build. Mater. **197**, 50–62 (2019). https://doi.org/10.1016/j.conbuildmat.2018.11.094
17. M.S. Lima, L.P. Thives, Evaluation of red mud as filler in Brazilian dense graded asphalt mixtures. Constr. Build. Mater. **260**, 119894 (2020). https://doi.org/10.1016/j.conbuildmat.2020.119894
18. H.K.A. Al-Bayati, A.G. Oyeyi, S.L. Tighe, Experimental assessment of mineral filler on the volumetric properties and mechanical performance of HMA mixtures. Civ. Eng. J. **6**, 2312–2331 (2020). https://doi.org/10.28991/cej-2020-03091619
19. A. Zulkati, D.W. Yiik, D.D. Sun, Effects of fillers on properties of asphalt-concrete mixture. J. Transp. Eng. **138**, 902–910 (2012). https://doi.org/10.1061/(ASCE)TE.1943-5436.0000395
20. G.H. Hamedi, Comparing the effect of nanomaterial and traditional fillers on the asphalt mixture properties. Civ Eng J. **5**, 320 (2019). https://doi.org/10.28991/cej-2019-03091247
21. Y.M. Jebur, M.W. Falah, M.A. Kadhim, M.K. Imran, T.M. Hashim, Z.S. Obeid et al., Effect of filler type and content on the mechanical properties of asphalt mixtures, in *2nd International Conference for Engineering Sciences and Information technology (ESIT 2022): ESIT2022 Conference Proceedings* 17–18 Aug 2022 (AIP Publishing, Al Anbar, Iraq, 2024). p. 30044. https://doi.org/10.1063/5.0190471
22. D.S. Kathem Taeh Alnealy, Effect of using waste material as filler in bituminous mix design. AJCE **3**, 88 (2015).https://doi.org/10.11648/j.ajce.20150303.16
23. R. Muniandy, E. Aburkaba, L. Mahdi, Effect of mineral filler type and particle size on asphalt-filler mastic and stone mastic asphalt laboratory measured properties. Aust. J. Basic Appl. Sci. **7**, 475–487 (2013)
24. A. Diab, M. Enieb, Investigating influence of mineral filler at asphalt mixture and mastic scales. Int. J. Pavement Res. Technol. **11**, 213–224 (2018). https://doi.org/10.1016/j.ijprt.2017.10.008
25. M. Lagos-Varas, D. Movilla-Quesada, A.C. Raposeiras, J.P. Arenas, M.A. Calzada-Pérez, A. Vega-Zamanillo, P. Lastra-González, Influence of limestone filler on the rheological properties of bituminous mastics through susceptibility master curves. Constr. Build. Mater. **231**, 117126 (2020). https://doi.org/10.1016/j.conbuildmat.2019.117126
26. A. Usman, M.H. Sutanto, M. Napiah, S.E. Zoorob, N.S. Yaro, M.I. Khan, Comparison of performance properties and prediction of regular and gamma-irradiated granular waste polyethylene terephthalate modified asphalt mixtures. Polymers (2021). https://doi.org/10.3390/polym13162610
27. N.S. Yaro, M.H. Sutanto, L. Baloo, N.Z. Habib, A. Usman, A.K. Yousafzai et al., A comprehensive overview of the utilization of recycled waste materials and technologies in asphalt pavements: towards environmental and sustainable low-carbon roads. Processes (2023). https://doi.org/10.3390/pr11072095
28. J. Zhang, S. Liu, Z. Yao, S. Wu, H. Jiang, M. Liang, Y. Qiao, Environmental aspects and pavement properties of red mud waste as the replacement of mineral filler in asphalt mixture. Constr. Build. Mater. **180**, 605–613 (2018). https://doi.org/10.1016/j.conbuildmat.2018.05.268
29. H. Zhang, H. Li, Y. Zhang, D. Wang, J. Harvey, H. Wang, Performance enhancement of porous asphalt pavement using red mud as alternative filler. Constr. Build. Mater. **160**, 707–713 (2018). https://doi.org/10.1016/j.conbuildmat.2017.11.105
30. T.L.X. Wong, M.R. Mohd Hasan, L.C. Peng, Recent development, utilization, treatment and performance of solid wastes additives in asphaltic concrete worldwide: a review. J. Traffic Transp. Eng. (Engl. Ed.) **9**, 693–724 (2022). https://doi.org/10.1016/j.jtte.2022.06.003
31. USGS. Bauxite production by country. United States Geological Survey (2024)
32. C. Brunori, C. Cremisini, P. Massanisso, V. Pinto, L. Torricelli, Reuse of a treated red mud bauxite waste: studies on environmental compatibility. J. Hazard. Mater. **117**, 55–63 (2005). https://doi.org/10.1016/j.jhazmat.2004.09.010
33. IAI. BRManagementGuidance, International Aluminium Institute (2022)

34. Y. Qin, K. Xie, Y. Meng, T. Fu, G. Fang, X. Luo, Q. Wang, Feasibility and environmental assessment of reusing aluminum tailing slurry in Asphalt. Constr. Build. Mater. **411**, 134737 (2024). https://doi.org/10.1016/j.conbuildmat.2023.134737
35. L. Yao, W. Gao, X. Ma, H. Fu, Properties analysis of asphalt binders containing bayer red mud. Materials (2020). https://doi.org/10.3390/ma13051122
36. M.S.S. Lima, L.P. Thives, V. Haritonovs, F. Gschösser, The influence of alternative fillers on the adhesive properties of mastics fabricated with red mud. Materials (2020). https://doi.org/10.3390/ma13020484
37. D. Wu, Y. Yin, T. Fu, F. Liu, Simulation study on basic road performance and modification mechanism of red mud modified asphalt mixture. Sci. Eng. Compos. Mater. (2024). https://doi.org/10.1515/secm-2024-0003
38. J. Choudhary, B. Kumar, A. Gupta, Performance evaluation of bauxite residue modified asphalt concrete mixes. Eur. J. Environ. Civ. Eng. **26**, 978–994 (2022). https://doi.org/10.1080/19648189.2019.1691662
39. L. Ou, H. Zhu, R. Chen, C. Su, X. Yang, Effect of industrial solid waste as fillers on the rheology and surface free energy of asphalt mastic. Materials (2024). https://doi.org/10.3390/ma17051125
40. M. Mathapati, K. Amate, C. Durga Prasad, M.L. Jayavardhana, R.T. Hemanth, A review on fly ash utilization. Mater. Today: Proc. **50**, 1535–1540 (2022). https://doi.org/10.1016/j.matpr.2021.09.106
41. A. Mondal, G.D. Ransinchung R.N., S. Marik, Investigation on ageing susceptibility of asphalt mix containing industrial by-products and reclaimed asphalt pavement. Iran. J. Sci. Technol., Trans. Civ. Eng. **48**, 2741–2758 (2024). https://doi.org/10.1007/s40996-023-01262-z
42. J.P. Giri, M. Priyadarshini, Utilization of building and agriculture wastes in bituminous pavement: a sustainable solution. J. Build. Pathol. Rehabil. **9**, 2 (2023). https://doi.org/10.1007/s41024-023-00351-w
43. EN 1097. EN 1097, *Tests for Mechanical and Physical Properties of Aggregates; EN 1097* (British Standards Institution, London)
44. EN. EN 933-3, *Tests for Geometrical Properties of Aggregates* (BSI British Standards, London). https://doi.org/10.3403/BSEN933
45. EN. DIN EN 933-1:2012-03, *Prüfverfahren für geometrische Eigenschaften von Gesteinskörnungen_- Teil_1: Bestimmung der Korngrößenverteilung_- Siebverfahren; Deutsche Fassung EN_933-1* (Beuth Verlag GmbH, Berlin, 2012). https://doi.org/10.31030/1863141.
46. EN. DIN EN 1426:2015-09, *Bitumen und bitumenhaltige Bindemittel_- Bestimmung der Nadelpenetration; Deutsche Fassung EN_1426* (Beuth Verlag GmbH, Berlin, 2015). https://doi.org/10.31030/2292039
47. EN. DIN EN 1427:2015-09, *Bitumen und bitumenhaltige Bindemittel_- Bestimmung des Erweichungspunktes_- Ring- und Kugel-Verfahren; Deutsche Fassung EN_1427* (Beuth Verlag GmbH, Berlin, 2015). https://doi.org/10.31030/2292069
48. EN. EN 12593, *Bitumen and Bituminous Binders—Determination of the Fraass Breaking Point* (European Committee for Standardization, Brussels, 2015)
49. EN. EN 12595, *Bitumen and Bituminous Binders. Determination of Kinematic Viscosity* (British Standards Institution, London, 2023)
50. EN. DIN EN 12596:2015-01, *Bitumen und Bitumenhaltige Bindemittel_-Bestimmung der Dynamischen Viskosität mit Vakuum-Kapillaren; Deutsche Fassung EN_12596* (Beuth Verlag GmbH, Berlin, 2014). https://doi.org/10.31030/2143331
51. EN. EN 13398, *Bitumen and Bituminous Binders—Determination of the Elastic Recovery of Modified Bitumen 2017* (European Committee for Standardization, Brussels, 2017)
52. EN. EN 12607, *Bitumen and Bituminous Binders—Determination of the Resistance to Hardening Under the Influence of Heat and Air 2014* (European Committee for Standardization, Brussels, 2014)
53. D04 Committee. *Specification for Mineral Filler For Bituminous Paving Mixtures* (ASTM International, West Conshohocken, PA). https://doi.org/10.1520/D0242_D0242M-19

54. L. Gömze, R. Géber, D. Charles, Investigation of hydrophilic and hydrophobic properties of different mineral fillers for asphalt mixtures, in *11th International Conference and Exhibition of the European Ceramic Society* 2009, vol. 2 (2009)
55. E. Romeo, V. Ghizzardi, S. Rastelli, A. Montepara, Influence of mineral fillers and their fractional voids on mastic rheological and mechanical properties. RILEM Bookseries. **11**, 681–692 (2016). https://doi.org/10.1007/978-94-017-7342-3_55
56. J. Zhang, C. Sun, P. Li, M. Liang, H. Jiang, Z. Yao, Experimental study on rheological properties and moisture susceptibility of asphalt mastic containing red mud waste as a filler substitute. Constr. Build. Mater. **211**, 159–166 (2019). https://doi.org/10.1016/j.conbuildmat.2019.03.252
57. Y. Yamada, XRF analysis by the fusion method for oxide powder on a benchtop WDXRF spectrometer Supermini. Rigaku J. **28**(2), 15–21 (2012)
58. A. Kuity, A. Das, Homogeneity of filler distribution within asphalt mix—a microscopic study. Constr. Build. Mater. **95**, 497–505 (2015). https://doi.org/10.1016/j.conbuildmat.2015.07.043
59. E.G. Bautista, J. Flickinger, R. Saha, I. Flores-Vivian, A.F. Faheem, K. Sobolev, Effect of coal combustion products on high temperature performance of asphalt mastics. Constr. Build. Mater. **94**, 572–578 (2015). https://doi.org/10.1016/j.conbuildmat.2015.07.022
60. EN. EN 12697, *Bituminous Mixtures—Test Methods for Hot Mix Asphalt* (BSI British Standards, London). https://doi.org/10.3403/BSEN12697
61. EN. EN 13399, *Bitumen and Bituminous Binders—Specification Framework for Cationic Bituminous Emulsions 2017* (European Committee for Standardization, Brussels, 2017)
62. DNIT 095/2006. Cimentos asfálticos de petróleo - Especificação de material. (In Portuguese)
63. AASHTO, *AASHTO MP1: Standard Specification for Performance-Graded Asphalt Binder 2010*. (American Association of State Highway and Transportation Officials (AASHTO, Washington, 2010)
64. DNIT 031/2006. Pavimentos flexíveis - Concreto asfáltico -Especificação de serviço. (In Portuguese)
65. J. Xu, Z. Fan, J. Lin, P. Liu, D. Wang, M. Oeser, Study on the effects of reversible aging on the low temperature performance of asphalt binders. Constr. Build. Mater. **295**, 123604 (2021). https://doi.org/10.1016/j.conbuildmat.2021.123604
66. M. Guo, M. Guan, Y. Tan, X. Du, Study on the change rule of low-temperature properties of asphalt binder with aging time. Int. J. Pavement Eng. **25**, 2324886 (2024). https://doi.org/10.1080/10298436.2024.2324886
67. G. Shafabakhsh, M. Sadeghnejad, S.K. Hejazi, A. Shirazi, Evaluation the rheological behavior of asphalt binder, fracture resistance and moisture susceptibility of asphalt mixtures: before and after adding nano Fe_2O_3. Int. J. Pavement Res. Technol. (2024). https://doi.org/10.1007/s42947-024-00437-9
68. A. Woszuk, L. Bandura, W. Franus, Fly ash as low cost and environmentally friendly filler and its effect on the properties of mix asphalt. J. Clean. Prod. **235**, 493–502 (2019). https://doi.org/10.1016/j.jclepro.2019.06.353
69. J. Li, F. Tang, Effects of two metal nanoparticles on performance properties of asphalt binder and stone matrix asphalt mixtures containing waste high density polyethelene. Constr. Build. Mater. **401**, 132787 (2023). https://doi.org/10.1016/j.conbuildmat.2023.132787
70. Y. He, Q. Zeng, Y. Liu, P. Liu, Y. Zeng, Z. Xu, Q. Liu, Evaluation of the composite mechanism of nano-Fe_2O_3/asphalt based on molecular simulation and experiments. Materials (2021). https://doi.org/10.3390/ma14123425

Chapter 19
Sugarcane Enzymes in Construction: Sustainable Alternatives for Earth Material Stabilization

Ana Cervilla-Maldonado, Francisco Martín-Villegas, Ignacio Valverde-Palacios, and Raquel Maria Fuentes-García

Abstract Most contemporary construction systems rely heavily on the use of steel and concrete as their primary building material, with natural materials being less frequently employed. Earth or soil is a key natural material that is readily accessible, and which was once of great importance as a structural material. This material is well-known and is employed in the construction of simpler architectural structures, primarily in developing countries. However, this material is currently in disuse as a structural material in Europe. The use of earthen constructions does not necessarily imply a certain social level, as this material has been employed successfully in the construction of simple houses as well as palaces, walls and churches. Earthen construction is often used with stabilizing binders including lime and cement. The use of such additives enhances the structural integrity of so-called stabilized-earths or stabilized soils. The objective of this research is to demonstrate how the optimized use of binding additives, including ecological enzymes, lime, clay and cement can improve the mechanical performance of stabilized soils. The physical–mechanical properties of stabilized soils examined here, including uniaxial compression and longitudinal strain, are evaluated in accordance with the European Standard EN ISO 17892-7:2019. The average compressive strength values obtained are between 1.0 MPa and 2.3 MPa for soils stabilized lime and organic enzymes with clay additives, respectively. These values compare favorably with compressed earth blocks and adobe blocks [1]. On the other hand, the values of the Modulus of Elasticity,

A. Cervilla-Maldonado · F. Martín-Villegas · I. Valverde-Palacios (✉) · R. M. Fuentes-García
Departamento de Construcciones Arquitectónicas, Escuela Técnica Superior de Arquitectura, Universidad de Granda, Campo del Principe, 18071 Granada, Spain
e-mail: nachoval@ugr.es

A. Cervilla-Maldonado
e-mail: anaecervilla@correo.ugr.es

F. Martín-Villegas
e-mail: fracimartin@correo.ugr.es

R. M. Fuentes-García
e-mail: rfuentes@ugr.es

D. A. H. Hanaor (ed.), *Sustainable Materials for the Built Environment*, Springer Series in Materials Science 352, https://doi.org/10.1007/978-3-031-97818-0_19

between 574 and 1852 MPa, are higher than those of the reference samples without additives.

19.1 Introduction

Construction materials are considered environmental resources that have been obtained to be processed to shape the buildings around us. These materials undergo, to a greater or lesser extent, a series of natural processes that help us to better understand how they will be used, as well as their behaviour throughout their useful life. Over time, the methods of extraction and treatment of these materials have evolved, enhancing their inherent qualities and facilitating their adaptation to the diverse requirements to which they are subjected [2].

The Industrial Revolution marked a turning point in history, during which new materials such as concrete and steel emerged. These new materials have gradually supplanted soil as a historical construction material. Nevertheless, despite the significant advancements these materials have brought to the construction sector, they remain among the most polluting materials globally. Consequently, it is imperative to take all possible measures to minimise this damage [3–8]. One such measure is the return of earth to the construction scenario, as it can contribute significantly to sustainable development, and to the fulfilment of the Sustainable Development Goals (SDGs) of the UN.

Furthermore, in order to provide the soil with resistance at an early stage, white cements of low/medium strength are often used at low levels to provide improved robustness while maintaining a certain permeability to gases [9]. Historically, numerous binders, such as lime, have been employed with considerable success in conjunction with soil. This can be observed in numerous rammed earth constructions that were built in Granada, encompassing both residential and palatial edifices, as well as defensive structures such as walls and ramparts [9].

In recent years, interest has grown in the development of more sustainable alternatives to Portland cement based binders, with a particular focus on those with lower energy consumption [10]. In the field of road construction, enzymatic stabilisers have gained considerable popularity due to their simplicity of application and effectiveness in stabilization [11–14]. These stabilisers are environmentally benign and free from toxic substances, which has led to their being regarded as a potentially valuable solution for soil stabilisation by numerous studies.

The findings of this study have the potential to inform and enhance the utilisation of enzyme-enhanced soil as a construction material. Consequently, the principal objectives of this research are to characterize and compare the mechanical properties of soil with varying dosages of ecological enzymes derived from sugar cane, clay, lime and cement. Furthermore, the use of strain gauges enables the determination of the Poisson Ratio and the Dynamic Modulus of Elasticity, among other parameters.

19.2 Materials and Methods

19.2.1 Materials

19.2.1.1 Alhambra Formation Soil

The soil layers or horizons defined by Perez-Acosta [15] are indicated in Fig. 19.1. These comprise the A horizon (the shallowest layer) to the D horizon (the deepest layer). The colour of Horizon A is darker due to the presence of decomposed organic matter near the surface. The accumulation layer, or Horizon B, is the layer in which minerals transported by water accumulate. Consequently, this layer contains a significant quantity of clay. Horizon C is formed on top of the parent rock and is composed of fragments of the parent rock. The deepest layer, Horizon D, is known as the source rock.

The base material for this investigation is the Alhambra Formation soil (Fig. 19.2a), which was extracted in the north-eastern part of Granada, southern Spain. For the purposes of this investigation, the B horizon of the edaphic soil, mixed with the top of the C horizon, was used.

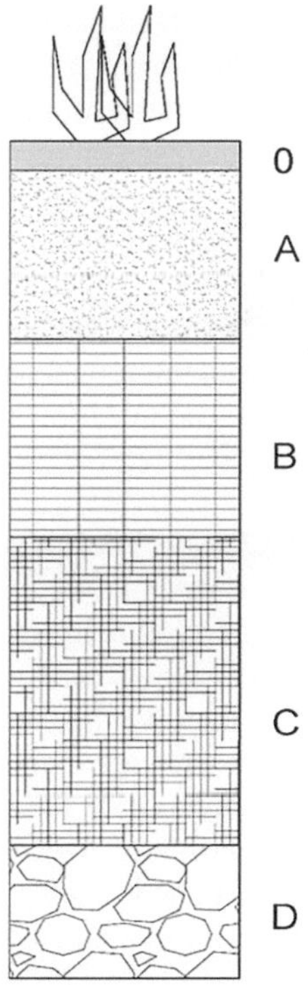

Fig. 19.1 Detail of Alhambra Formation horizons

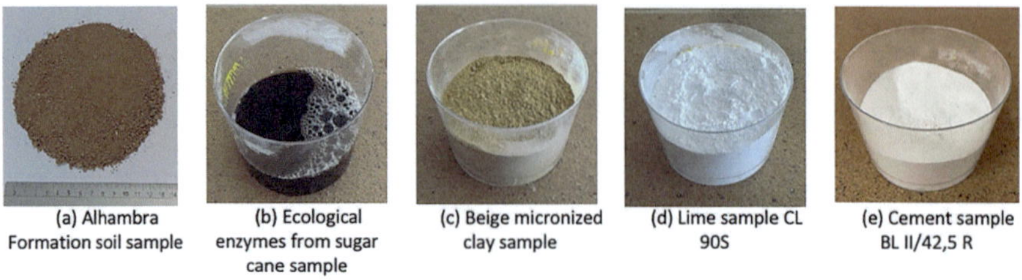

Fig. 19.2 Detail of the materials used for the research

In order to avoid contamination of the B horizon with organic remains, the A layer or horizon was first removed for the extraction of the soil samples. The samples were stored in 20 kg bags and transported to the laboratory, where they were subjected to a preliminary cleaning and drying treatment, prior to carrying out the tests.

19.2.1.2 Organic Sugar Cane Enzymes

The organic enzymes utilized, designated as COMPAT-TO® (Fig. 19.2b), are manufactured by the company Nanosystems. The mixture of organic enzymes acts as a catalyst, facilitating the binding action on the plastic particles of cohesive soils. These enzymes have been employed in previous research with the objective of identifying a solution that would facilitate the compaction, conservation and protection of roads or soils against wear and deformation. The properties and applications have been interpreted from the technical data sheets provided by the manufacturer.

Given the high clay content of the selected horizon of the Alhambra Formation soil, which is indicative of a cohesive soil, it may be reasonably anticipated that the mechanical behaviour of the material will be improved. It is therefore postulated that the soil reinforced with the enzymatic additives derived from sugar cane will exhibit enhanced mechanical resistance values and a reduction in deformation.

19.2.1.3 Micronized Beige Clay

Micronized beige clay (Fig. 19.2c) was employed in conjunction with the ecological enzymes derived from sugar cane to introduce additional fines to the mixture and improve the mechanical performance of the material. The dosage of clay used varies between 4 and 8%, in order to define the optimal amount that combines most effectively with the organic enzymes and the Alhambra Formation Soil.

The mineralogical composition of the micronized beige clay is principally constituted of quartz (43%), kaolinite (9%), calcite (7%) and plagioclase (5%), in addition to lesser quantities of other components, including k-feldspar, hematite and montmorillonite, which account for less than 5% of the total analysis. The sample was provided by the company Argiles Colades.

19.2.1.4 Hydrated Lime CL 90 S

In this study, the type of hydrated lime CL 90 S (Fig. 19.2d), produced by the company Ancasa, has been employed. Hydrated calcium lime 90 powder is an aerial lime that is primarily composed of CaO (quicklime) and/or $Ca(OH)^2$ (calcium hydroxide), with no hydraulic or pozzolanic additives [16].

19.2.1.5 Cement

The cement employed was of the white cement BL II/A-L 42.5 R [17] (Fig. 19.2e). No mechanical tests were conducted, as the requisite information was available in the form of a quality certificate from the supplier. Moreover, the quantity of cement used in this research in relation to the weight of the soil is notably low. This is because its role as a binder is only temporary until the main binder, lime, begins to carbonate, providing strength.

19.2.2 Methods

19.2.2.1 Conformation of Compressed Earth Cylinders (CEC)

Once the previous cleaning treatment of the extracted and oven-dried soil has been carried out, the manufacture of test specimens is carried out by compacting them to normal proctor density and with an optimum moisture content of 8.5% established in previous research [9] for soil of the same nature and granulometry.

Cylindrical samples of compacted soil with dimensions 100 × 120 mm were made up of Alhambra Formation soil (AF soil) without additives, and other samples mixed with specific additives. It should be noted that all the samples were composed of a base of 2500 g of AF soil, to which different dosages of ecological enzymes and clay were added (7.5 ml of enzymes and 4%, 6% and 8% of clay for each sample), as defined in Table 19.1. In the case of the samples that were combined with lime, the lime contents used were 2, 4, 6, 8 and 10% of the weight of the soil. Finally, for the samples added with lime and cement, the dosage percentages were 8% lime and 1 and 2% cement, calculated with respect to the weight of the soil.

The CECs were manufactured in accordance with the specifications set forth in Standard UNE 103500:1994 [18] and allowed to reach a constant weight through a drying process. The required drying time is defined as a period of between 30 and 45 days, with a temperature of 20–24 °C and humidity of 30–40%, contingent upon the specific conditions of the laboratory environment.

Table 19.1 Representative dosage of CEC added with organic enzymes derived from sugar cane and clay

	AF Soil + 7.5mlE + 4%Clay	AF Soil + 7.5mlE + 6%Clay	AF Soil + 7.5mlE + 8%Clay
Soil (G)	2500	2500	2500
Clay (%)	4	6	8
Clay (G)	100	150	200
Enzymes (mL)	7.5	7.5	7.5
Optimum moisture (%)	8.5	8.5	8.5
Water (G)	214	214	214

19.2.2.2 Measurement of Ultrasonic Pulse Velocity

Once the drying process of the samples has been completed, the propagation speed of the longitudinal wave impulses is determined in accordance with the requirements set out in the UNE-EN 12504-4 Standard. In this way, the propagation time of the wave through the test specimen is obtained. The wave propagation speed through the sample is calculated using Eq. 19.1, where the distance between sensors (corresponding to the length of the test specimen, 120 mm) and the time t of transmission of the wave between sensors (s) are used to obtain the longitudinal wave propagation speed (m/s).

$$v = \frac{d}{t} \tag{19.1}$$

19.2.2.3 Compression Test, Modulus of Elasticity and Poisson Ratio

Subsequently, in accordance with the provisions of the UNE-EN ISO 17892-7:2019 standard, the compressive strength of the samples is carried out. The test entails applying a constant force of 1 mm per minute until the specimen breaks. The force is applied through the plates of the 10-tonne capacity multi-test press manufactured by Controls S.A. Furthermore, some specimens have been instrumented with strain gauges connected to the HBM QuantumX data acquisition equipment, with directions parallel and perpendicular to the axial load, to obtain the Poisson ratio (ν) and the Modulus of Elasticity (E, 0).

The Modulus of Elasticity is then calculated using Eq. 19.2, based on the recorded values from the gauges. In this equation, the parameters x and y represent the intervals within which both the stress (σx, σy) and the longitudinal or transverse deformation (ε) are calculated prior to reaching the maximum value of rupture. Stress (σ) is defined as the internal resistance that a material presents to an applied external force. Deformation (ε) has a non-dimensional value that represents the change in the length of a material with respect to its initial length. Changes in the length of the samples are usually minimal and strain gauges with a gauge factor of 2.11 ± 1.0% were employed to measure them. The gauge type used was 50/120 CLY41-3L-0.5M from HBM.

$$E = \frac{\sigma x\% - \sigma y\%}{\varepsilon x\% - \varepsilon y\%} \tag{19.2}$$

In order to calculate the Poisson ratio (v) according to Eq. 19.3, it is necessary to determine the ratio between the longitudinal and transverse deformation.

Poisson's coefficient (v) is the measure of the relationship between transverse and longitudinal strains, defined as the elongation that occurs when a force is applied in the longitudinal direction as opposed to the strain in the transverse direction. This coefficient is calculated using the same interval values as those employed to calculate the Modulus of Elasticity.

$$\nu = \frac{\varepsilon Tx\% - \varepsilon Ty\%}{-\varepsilon Lx\% - \varepsilon Ly\%} \tag{19.3}$$

19.2.2.4 Analysis with the Inverted Plate Microscope

A detailed microscopic analysis was conducted to compare the samples of CECs that achieved the highest compressive strength results and exhibited the best deformation behaviour with a reference sample of CEC without additives. For this purpose, an inverted plate microscope (Nikon Epiphot 200) with polarised light was employed.

Following the removal of the breakage cones from the selected samples, they were subsequently embedded in resin. A vacuum machine (Struers CitoVac) was employed to distribute the Epofix resin evenly throughout the sample and to fill all the pores. After a 24 h period of resin curing, the samples were cut and polished. The samples were polished with diamond discs (MD-Piano 220, MD-Allegro and MD-Dac). The duration of polishing was determined individually until the samples were correctly visualised in the microscope, with the objective of eliminating the marks of the cutting process.

19.3 Experimental Results

19.3.1 Relationships of Compression Strength Versus Time and Longitudinal Strain

The load capacity of each specimen is studied until reaching its breaking point. In addition, the relationship between compressive strength and longitudinal strain is studied, obtained using a strain gauge placed in the longitudinal direction of the applied load. This allows the behaviour and deformation of the specimen to be analysed over time.

The mean compressive strength of the reference samples with FA soil (without additives) is 1.9 MPa. A linear behaviour is also observed from 0.5 MPa, up to the points of maximum compressive strength of each sample, at which point the sample passes into a plastic state and the load begins to decrease caused by the collapse of the material. It can be observed that the deformation of the samples is linear as the compressive force is applied and increases (Fig. 19.3).

Figures 19.4 and 19.5 present the results of tests conducted on samples composed of 7.5 ml of ecological enzymes derived from sugar cane and micronized clay, with a percentage of 4% and 6%, respectively. The objective of this study was to analyse the behaviour of the material, AF soil, when lime was replaced by a more ecological binder such as the enzymes used. The results were positive, with the highest compressive strength achieved by the sample composed of 7.5 ml of enzymes and 4% of micronized clay, which exhibited an average compressive strength value of 2.3 MPa.

The samples containing 8% hydrated lime by weight of soil (Fig. 19.6) showed higher compressive strength values compared to the rest of the soil and hydrated lime dosages. Conversely, the deformations in tension demonstrate a more linear behaviour. This sample reaches the highest tension and exhibits the best linearity in the elastic state until breakage.

The samples containing 8% hydrated lime and 1% cement in relation to the weight of the soil (Fig. 19.7) exhibited higher compressive strength values than the other dosages studied, which contained 2% cement.

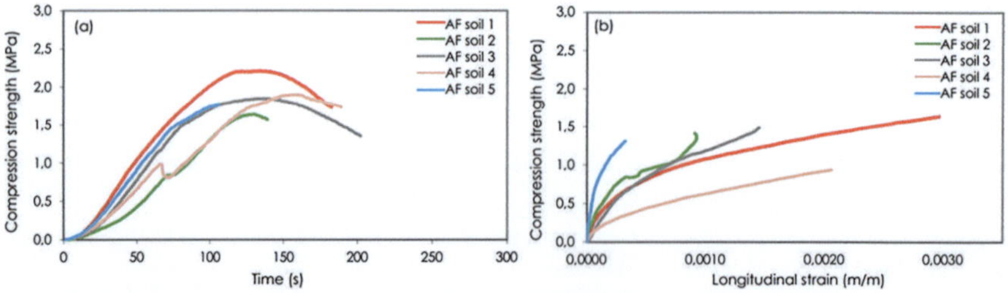

Fig. 19.3 **a** Compression stress-time curve of the reference CECs without additives. **b** Stress–strain curve of the reference CECs without additives

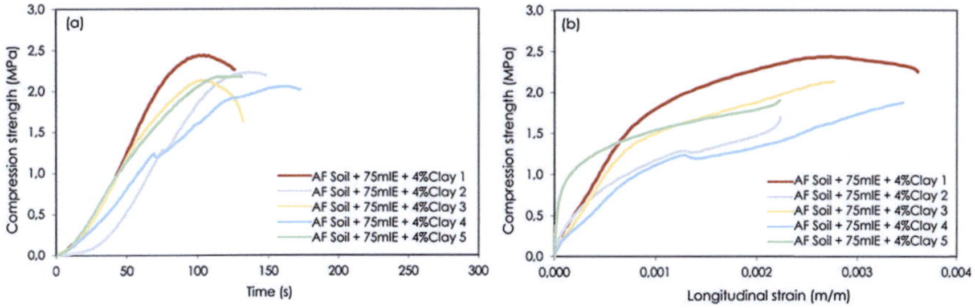

Fig. 19.4 **a** Compression stress-time curve of CECs added with 75 ml of enzymes and 4% of clay. **b** Stress–strain curve of CECs added with 75 ml of enzymes and 4% of clay

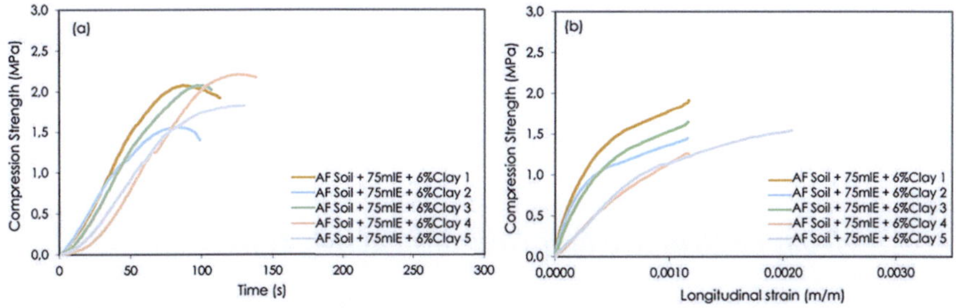

Fig. 19.5 **a** Compression stress-time curve of CECs added with 75 ml of enzymes and 6% of clay. **b** Stress–strain curve of CECs added with 75 ml of enzymes and 6% of clay

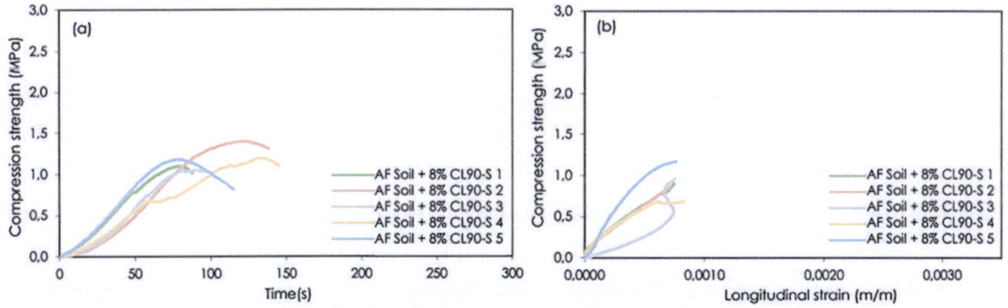

Fig. 19.6 **a** Compression stress-time curve of CECs added with 8% hydrated lime CL 90 S. **b** Stress–strain curve of CECs added with 8% hydrated lime CL 90 S

As illustrated in Figs. 19.4 and 19.5, the soil sample containing 75 ml of ecological enzymes and 4% of clay shows improved strength, not only within the category of these samples but also in the entire group. The average compressive strength achieved exceeds the maximum values of those without additives, as well as those of AF soil, lime and cement. Regarding the analysis of their longitudinal deformations as a function of time, these show greater deformations before failure. Initially they are irregular, but close to the moment of failure they present a more linear behavior. In

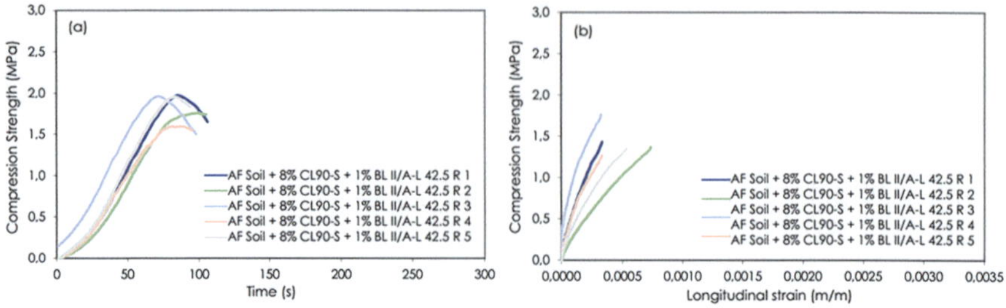

Fig. 19.7 **a** Compression stress-time curve of CECs added with 8% hydrated lime CL 90 S and 1% cement BL II/42,5 R. **b** Stress–strain curve of CECs added with 8% hydrated lime CL 90 S and 1% cement BL II/42,5 R

addition, the AF soil samples with organic enzymes and 4% clay show deformations similar to the reference samples without additives.

19.3.2 Physical–Mechanical Parameters Derived from Destructive and Non-destructive Tests

This section presents the physical–mechanical parameters of the samples following the ultrasonic pulse propagation velocity tests and the simple compression rupture tests. Table 19.2 shows the mean value of the results obtained for each typology.

As can be seen in Table 19.2, the mean values of compressive strength, transverse and longitudinal elastic modulus, Poisson ratio, density and dynamic modulus of elasticity calculated from the ultrasonic pulse propagation time are shown.

The CECs exhibiting the highest compressive strength are those reinforced with 7.5 ml of ecological enzymes derived from sugar cane and 4% of micronized clay. Additionally, these specimens demonstrate the highest densities in comparison to those with additives. Furthermore, a correlation is observed between the average values of compressive strength and the densities obtained in this typology. Conversely, the propagation velocity of the ultrasonic wave is inversely proportional to the proportion of clay added to the mixture. About the analysis of the elastic moduli and Poisson's ratio, the average values are generally improved compared to the specimens formed with soil only (Table 19.2).

Regarding the analysis of the dynamic elasticity modulus, it can be observed that the values are lower in the lime and cement samples, since they require a longer time for the wave to pass through them due to the low percentage of binder. On the other hand, the samples containing ecological enzymes and clay present a higher value of elasticity modulus, since the propagation of the ultrasonic wave through the material is faster, showing that a more homogeneous and more compact mixture is achieved.

Table 19.2 Results obtained for the different dosages: compressive strength, longitudinal and transverse elastic modulus, Poisson ratio, density, ultrasonic pulse propagation velocity and modulus of dynamic elasticity

	Compressive strength (MPa)	Longitudinal elastic modulus (MPa)	Transverse elastic modulus (MPa)	Poisson ratio	Density (kg/m^3)	Ultrasonic pulse propagation velocity ($m/s \cdot 10^{-9}$)	Modulus of dynamic elasticity (MPa)
AF Soil	1.9	572	2575	0.1	1865.8	1.62	4903
AF Soil + 75mlE + 4%Clay	2.3	1654	8645	0.2	1869.1	1.64	5086
AF Soil + 75mlE + 6% Clay	1.9	1404	3855	0.1	1856.4	1.67	5169
AF Soil + 75mlE + 8% Clay	1.9	1349	19,339	0.1	1825.1	1.57	4491
AF Soil + 2% CL90-S	1.2	574	799	0.2	1807.7	1.30	3056
AF Soil + 4% CL90 S	1.0	881	4521	0.2	1767.0	1.32	3089
AF Soil + 6% CL90-S	1.2	999	2439	0.2	1740.0	1.24	2695
AF Soil + 8% CL90-S	1.3	987	1758	0.2	1710.8	1.17	2354
AF Soil + 10% CL90-S	1.2	1145	2481	0.2	1688.8	1.10	2032

(continued)

Table 19.2 (continued)

	Compressive strength (MPa)	Longitudinal elastic modulus (MPa)	Transverse elastic modulus (MPa)	Poisson ratio	Density (kg/m^3)	Ultrasonic pulse propagation velocity (m/s $\cdot 10^{-9}$)	Modulus of dynamic elasticity (MPa)
AF Soil + 8% CL90-S + 1% BL II/A-L 42.5 R	1.9	1430	3340	0.3	1721.1	1.33	3050
AF Soil + 8% CL90-S + 2% BL II/A-L 42.5 R	1.7	1852	2955	0.3	1721.3	1.27	2788

19.3.3 Microstructural Analysis

Finally, the porosity, the bond between the matrix and the natural aggregates, as well as the effect of the additives on the microscopic level in the mixture were analysed. In order to conduct this analysis, the sample containing 7.5 ml of ecological enzymes and 4% micronized clay was compared with the sample that had been added with 8% hydrated lime CL 90 S. These two samples were then compared with the sample without additives. The enzyme-reinforced sample was selected for analysis because it had the highest compressive strength of all the samples. In addition, it is essential to compare it with the sample reinforced with 8% hydrated lime, since it has the highest compressive strength of its type and the one that presents the most favourable deformations. In addition, the microscopic effects of replacing lime with ecological enzymes derived from sugar cane and micronized clay will be investigated. Figure 19.8 shows the details of the resin-embedded breakage cone of the analysed samples. At the macroscopic level, the aggregate particles, which are predominantly elongated and oval, are dispersed within the soil matrix. Furthermore, the heterogeneity of the aggregates is observed, with different tonality depending on their nature and composition. For instance, in the FA soil sample (Fig. 19.8a), the whitish natural stones are quartz, dolomite, or calcite, as the dominant mineralogical elements. In contrast, the grey or brown natural stones are phyllosilicates, as indicated by reference studies (Valverde-Palacios et al., 2022). At the sample added with enzymes and clay (Fig. 19.8b), areas with more golden or yellow staining are additionally observed. In contrast, in the lime-reinforced sample (Fig. 19.8c), white crystallisations of heterogeneous sizes are randomly distributed in the sample.

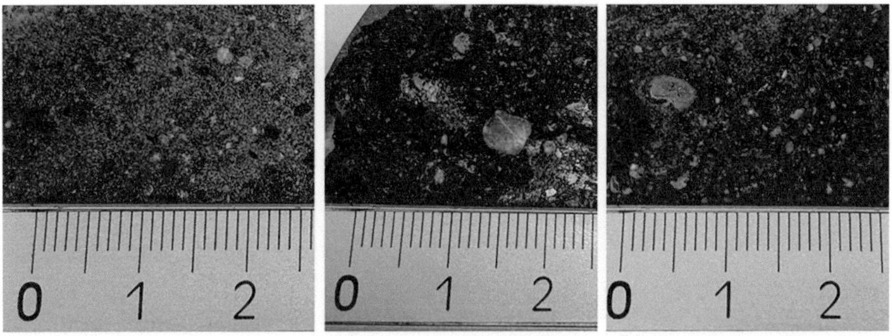

(a) Subsample of CEC without additives (only FA soil). (b) Subsample of CEC added with 75ml enzymes and 4% micronized clay. (c) Subsample of CEC added with 8% hydrated lime.

Fig. 19.8 Detail of the subsamples. All the samples were bedded in epoxy resin and polished, on a scale of 1:100

Microstructure of the CEC added with 7.5 ml of organic enzymes from sugar cane and 4% micronized clay

The microstructure of the interfacial zone is contingent upon a few factors, including the type of soil, the size and composition of the aggregates, as well as the moisture content during the dosing process. In this case, the sample exhibits areas with heightened golden hues, which accentuate the brownish-pink hue characteristic of FA soil (Fig. 19.11).

Low porosity is evident in the sample added with enzymes and clay and is quantified at 0.58% (Fig. 19.12). The discontinuity between the natural stones, matrix, and additives is minimal and is mainly marked by irregular shapes (42%). The porosity of the sample is isolated and randomly distributed in the matrix, corresponding to occluded air (Fig. 19.9a, c). Furthermore, the formation of air pores during the mixing and compaction process, which become trapped beneath the natural aggregates with an irregular surface area exceeding 1 mm2, has also been observed in previous studies, in that case, in concrete samples with recycled aggregates [19].

The accumulation of water at the perimeter of the surface of the aggregates continuing in the matrix indicates a point of weakness between the interfacial matrix-aggregate bonding. This zone is defined as a porous region surrounding the aggregate with a maximum thickness of 23 μm (Fig. 19.9b).

Microstructure of the CEC added with 8% hydrated lime CL 90 S

The most representative feature of the sample is the presence of bright white crystals, which correspond to the crystallisation of the hydrated lime was added to the sample. These crystallisations are distributed throughout the sample and present diameters below 20 μm in the case of the smaller crystals and reaching 155 μm in the case of the larger crystals (Figs. 19.10 and 19.11).

As shown in Fig. 19.12, this sample exhibits a higher porosity (1.51%) than the samples reinforced with enzymes and clay, as well as the sample without additives.

(a) Set of approximately 9 images (about 56 mm2)

(b) Crack discontinuity between natural aggregate and matrix

(c) Air bubbles trapped under natural aggregates and in the matrix

Fig. 19.9 Analysis of images of the sample of CEC modified with 7.5 ml of organic enzymes from sugar cane and 4% micronized clay

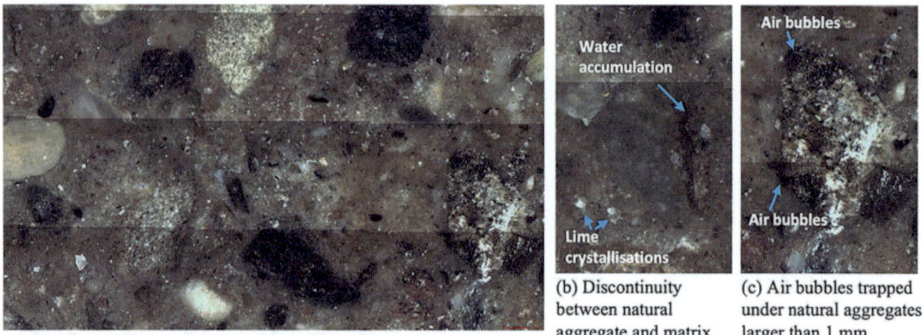

(a) Set of approximately 9 images (about 56 mm2)

(b) Discontinuity between natural aggregate and matrix

(c) Air bubbles trapped under natural aggregates larger than 1 mm

Fig. 19.10 Analysis of images of the sample of CEC modified with 8% hydrated lime CL 90 S

Nevertheless, the porosity remains relatively low, resulting in a minimal discontinuity between natural stones and the matrix (Fig. 19.10). This discontinuity is predominantly marked by elliptical (69%) and circular (17%) shapes, which correspond to occluded air produced during the mixing process (Fig. 19.13).

On the other hand, the proportion of elongated pores (3%), which correspond to cracks, is notably lower than that observed in the samples reinforced with enzymes and clay, as well as the reference sample without additives, which exhibited values between 15 and 21% (Fig. 19.13).

The accumulation of air bubbles at the periphery of the large aggregates is evident in the sample (Fig. 19.10c), as are smaller air bubbles distributed randomly throughout the mixture. This observation aligns with previous studies [19]. In this instance, point water accumulations with a thickness of 32 μm were also identified at the periphery of the aggregates (Fig. 19.10b), although to a much lesser extent than in the case of the other samples analysed.

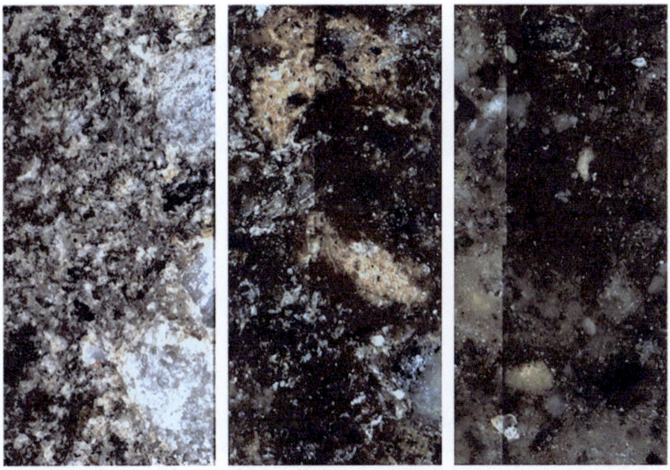

(a) CEC without additives.

(b) CEC added with 7.5ml of organic enzymes from sugar cane and 4% micronized clay.

(c) CEC added with 8% hydrated lime CL 90 S.

Fig. 19.11 Microscopic comparison of the samples

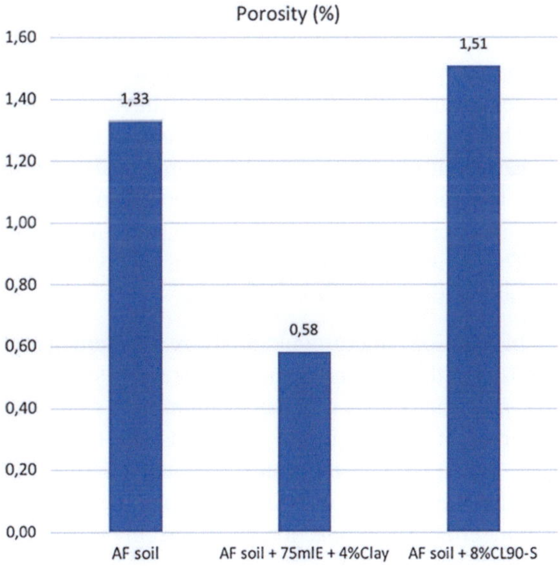

Fig. 19.12 Porosity of the subsamples (%)

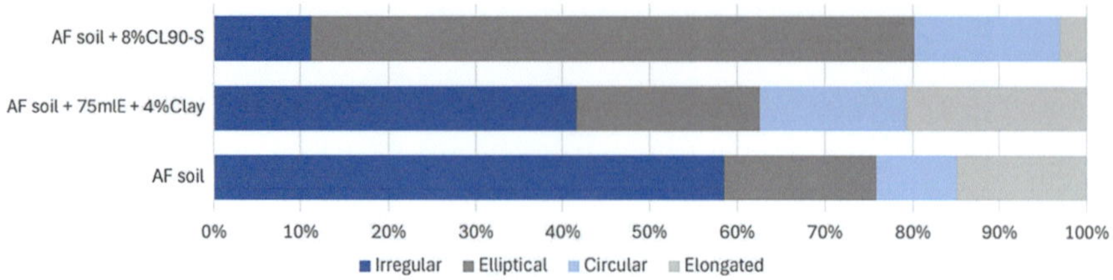

Fig. 19.13 Study of pore size (%)

19.4 Conclusions

This research focuses on analysing the results obtained from each dosage of test specimens studied, with special emphasis on the use of ecological enzymes from sugar cane and clays as a substitute for hydrated lime. In order to ascertain the material with the optimal performance, a series of tests were conducted, including ultrasonic pulse propagation velocity and compression strength tests. Considering these tests, the findings of this study can be summarised as follows:

- The CECs added with 75 ml of enzymes and 4% of micronized clay exhibited the highest compressive strength (2.3 MPa) when compared to the remaining types. It is important to note that these specimens also exhibit greater deformations, as they lack the addition of lime or cement.
- The addition of CECs to a hydrated lime mixture (CL 90 S) with a percentage of 8% results in the achievement of the requisite properties for the material. Among the samples containing lime, those with 8% exhibit the highest resistance and demonstrate the most favourable behaviour against deformations.
- The favourable deformation behaviour of CECs combined with 8% hydrated lime can be attributed to the microscopic analysis of this specimen, which revealed a low ratio of elongated cracks or microcracks. In this specimen, 3% of elongated pores were observed, in comparison to 15% in the CEC without additives. This is attributed to the stability of this typology, which is a consequence of the carbonation processes of the lime.

In addition, microscopic analysis has demonstrated a correlation between the porosity of the specimens, their compressive strength and deformation. This indicates that specimens with a higher compressive strength exhibited a lower porosity, as observed in the case of the specimens that reached a higher compressive strength: CEC added with 75 ml of enzymes and 4% of micronized clay. This is due to the fact that the presence of fines (clays) and the binding effect of the enzymes result in a more effective compaction of the specimens.

This research demonstrates the impact of additives, including organic enzymes derived from sugar cane, clays, lime, and cement, on the physical and mechanical properties of CECs. It is also important to note that the soil, with or without additives, exhibits very satisfactory physic mechanical properties.

On the other hand, this study contributes to the advancement of the mechanical properties of rammed earth, with the objective of providing a superior, cost-effective and environmentally sustainable building material.

References

1. A. E. Aenor, *UNE 41410 Bloques de Tierra Comprimida Para Muros y Tabiques*, Definiciones, Especificaciones y Métodos de Ensayo, vol. 26 (2008)
2. I. Valverde-Palacios, R. Fuentes-García, A. Cervilla-Maldonado, I. Valverde-Espinosa, Earth as a sustainable construction material. Characterization of different mixtures and implementation using the projected earth system, in *New Technologies in Building and Construction: Towards Sustainable Development* (Springer, 2022), pp. 299–320.
3. C.A.D. Chazelles, Savoir-Faire Indigènes et Influences Coloniales Dans l'architecture de Terre Antique de l'extrême-Occident (Afrique Du Nord, Espagne, France Méridionale), in *Conferência Internacional Sobre o Estudo e Conservaçao Da Arquitectura de Terra* (1993)
4. C. Costa, Â. Cerqueira, F. Rocha, A. Velosa, The sustainability of adobe construction: past to future. Int. J. Archit. Herit. **13**, 639 (2019)
5. L.F. Guerrero Baca, *Arquitectura en tierra: Hacia la recuperación de una cultura constructiva*, Apuntes: Revista de Estudios sobre Patrimonio Cultural. J. Cult. Herit. Stud. **20**, 182 (2007)
6. H. Houben, H. Guillaud, *Earth Construction. A Comprehensive Guide* (London, 1994)
7. L. Keefe, *Earth Building: Methods and Materials, Repair and Conservation* (2012)
8. G. Minke, *Manual de Construcción Para Viviendas Antisísmicas de Tierra* (2001)
9. R. Fuentes-García, Construcciones de Tierra. El Tapial. Nuevo Sistema Para Construcción y Restauración Mediante La Técnica de "Tierra Proyectada", PhD Thesis, Universidad de Granada (2010)
10. P. Bajpai, Non-conventional soil stabilization techniques the way forward to an aggregate free pavement and a cost effective method of road construction (unpublished)
11. R. Brazetti, Considerations about the influence of different additives in organic micromorphological, mineralogical, physical, mechanical and hydraulics characteristics of a lateritic soil (unpublished)
12. W.S. Guthrie, D.O. Simmons, D.L. Eggett, Enzyme stabilization of low-volume gravel roads. Transp. Res. Rec. **2511**, 112 (2015)
13. D. Kumar Das, M. Maharana, Terrazyme stabilized soil-fly ash mix as a road construction material (2019)
14. V. Ortega-Ruiz, Caracterización de La Compactación de Suelos Edáficos de La Formación Alhambra Utilizando Un Estabilizador de Enzimas Ecológicas En Sustitución de La Cal (2017)
15. R. Perez-Acosta, Caracterización de Los Suelos Coluviales o Vertidos Antrópicos de La Formación Alhambra Para Su Uso En Conservación y Mantenimiento Del Patrimonio y Su Identidad Cultural (2016)
16. UNE-EN 459-1:2016 Cales Para La Construcción. Parte 1: Definic..., https://www.une.org/encuentra-tu-norma/busca-tu-norma/norma/?c=N0056852.
17. UNE-EN 197-1:2011 Cemento. Parte 1: Composición, Especificacio..., https://www.une.org/encuentra-tu-norma/busca-tu-norma/norma/?c=N0048623
18. UNE 103500:1994 Geotechnic. Compactation Test. Standard Proctor, https://en.tienda.aenor.com/norma-une-103500-1994-n0007850
19. Z. Sánchez Roldán, I. Valverde Palacios, I. Valverde Espinosa, M. Martín-Morales, Microstructural analysis of concretes manufactured with recycled coarse aggregates pre-soaked using different methods (2020)

Chapter 20
Photocatalytic, Antimicrobial and Conductive Functional Nanomaterial-Doped Cement Composites for Sustainable Construction

Agnieszka Ślosarczyk and Łukasz Klapiszewski

Abstract Due to the high carbon footprint of building materials, including those based on cementitious binders, high durability and very often new functionalities are expected. This chapter provides an overview of recent research on the role of nanomaterials in shaping new properties of cementitious composites, including self-cleaning, antimicrobial and self-sensing properties. Important parameters in the design of these solutions are the nature and physicochemical properties of the nanoadditives, their proper dispersion in cement composites and the shaping of new material properties in terms of their durability.

20.1 Introduction

It will soon be 25 years since the Roadmap for Nanotechnology in Construction (RoNaC) was drawn up. It envisaged certain developments in nanotechnology in the construction industry, as shown schematically in Fig. 20.1. An important aspect of this issue has been the understanding of the mechanisms of nanoscience, i.e. the modelling of the structure of materials at the nanoscale using bottom-up or top-down methods, and the application of the resulting nanomaterials in construction. These directions have enabled the gradual development of materials nanotechnologies, including the nanocoating of building materials with special functions, the modification of structural materials such as steel, wood, concrete and glass with nanoparticles, or the design of materials with super-insulating properties such as silica aerogel. However,

A. Ślosarczyk
Institute of Building Engineering, Faculty of Civil and Transport Engineering, Poznan University of Technology, PL-60965 Poznan, Poland
e-mail: agnieszka.slosarczyk@put.poznan.pl

Ł. Klapiszewski (✉)
Institute of Chemical Technology and Engineering, Faculty of Chemical Technology, Poznan University of Technology, PL-60965 Poznan, Poland
e-mail: lukasz.klapiszewski@put.poznan.pl

D. A. H. Hanaor (ed.), *Sustainable Materials for the Built Environment*, Springer Series in Materials Science 352, https://doi.org/10.1007/978-3-031-97818-0_20

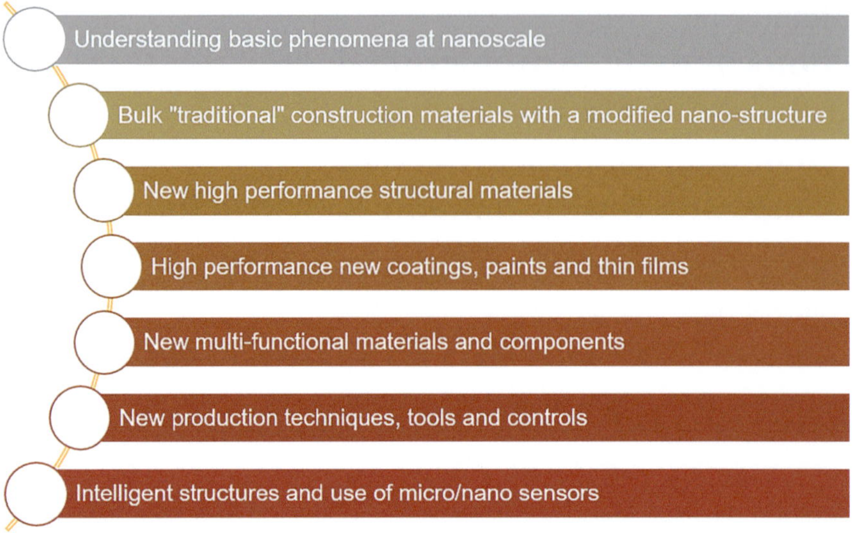

Fig. 20.1 Aims and destinations, which are expected to be of significance for construction in the medium-long term, based on [1]

for the first decade or so, mainly due to the size of the construction sector and socio-economic factors, these developments were not as spectacular as in other industries such as medicine, agriculture or electronics [1–3].

This trend has been changing over the last few years and has been particularly noticeable in the last three to four years. The move towards nanotechnology-based solutions is being driven by a number of factors, the most important of which is the drive to reduce carbon dioxide emissions and move towards a near zero carbon economy. This trend in all areas of business and life has also led the construction industry to look for low-carbon material solutions with improved performance and durability to ensure a long service life of building materials. The increased performance and durability of materials also contributes to lower maintenance costs for building infrastructure and lower deferred repair costs. This also has the effect of postponing the problem of recycling and reusing these materials at the end of their service life [4, 5]. Figure 20.2, based on publications from the last three years, provides a summary of the most frequently used keywords reflecting the topics of articles on nanomaterial modifications of cementitious composites. Three groups of relationships can be observed between nanoparticles and self-cleaning and antibacterial properties. The same relationships can be seen regarding biocompatibility. These areas are mainly related to metallic nanoparticles, of which silver nanoparticles and zinc oxide have been the most studied in the last few years.

Slightly less pronounced is the third record, sensing, which is mainly associated with carbon nanostructures. This is an issue that has emerged in the last few years and is gaining in popularity with smart concretes, which make it possible to monitor the gradual degradation at the nano and micro levels. Nevertheless, it is worth noting

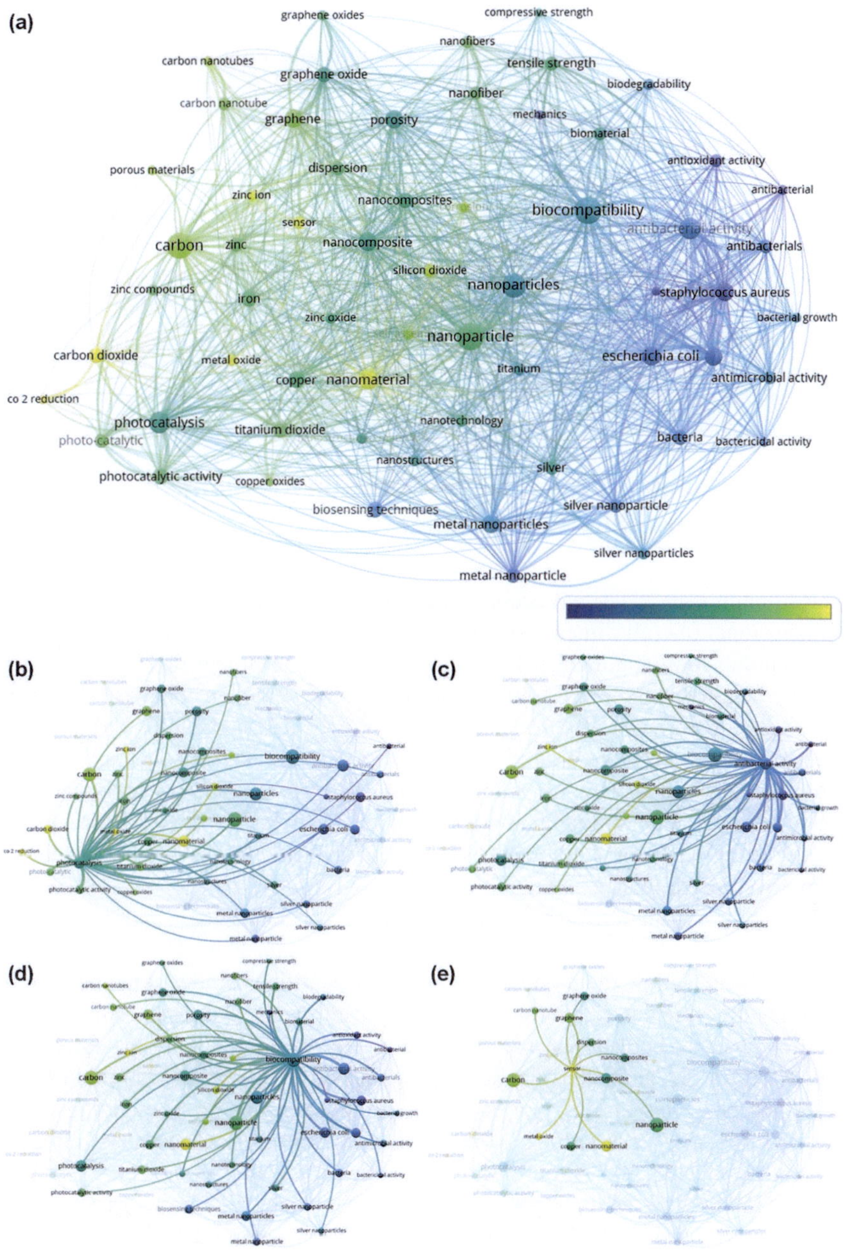

Fig. 20.2 Bibliometric relationships of selected keywords with the highest repetition rate in scientific articles published in 2023–2025, based on approximately 12,400 records (**a**), highlighting the relationships of photocatalytic (**b**), antibacterial (**c**), bio-compatibility (**d**), and sensing (**e**) to other keywords, based on the Scopus database

the correlation of all these keywords with the concept of carbon dioxide emissions and the durability of the cement matrix. The intensification of the above records in recent years indicates a widespread interest in nanomodification of cementitious composites to obtain specific new functionalities, which could lead to their increasing importance in the construction sector in the near future [6].

The main objective of this chapter is to present the latest trends in nanomodified cement composites. The first section presents the influence of nanomaterials on the physical, mechanical and durability properties of cement composites, while the following sections focus on the new functionalities that cement composites acquire as a result of nanomodification. The mechanisms of action and the most commonly used nanostructures for self-cleaning, antibacterial and self-sensing properties of engineered cementitious composites are presented.

20.2 Nanomaterials and Its Influence on Mechanical and Durability Performance

Nanostructures are classified according to the number of dimensions in the nano: 0-dimensional 0D (all dimensions in the nano, e.g. nanoparticles of silver, copper, silica, alumina or titanium dioxide), 1-dimensional 1D (1 dimension in the nano, e.g. metallic nanorods, carbon nanotubes and nanofibers), 2-dimensional 2D (2 dimensions in the nano, e.g. metallic nanoshells or graphene, graphene oxide) and bulk nanostructures, which are bulk materials with nanoscale features such as grains or pores, an example of which is silica aerogel. All of these types of nanostructures can be used in cementitious composites to influence various material properties, ranging from mechanical, photocatalytic, antimicrobial and self-sensing properties to improved performance and durability [7].

The improvement of mechanical performance has been the main and initial direction for the use of nanoadditives in cementitious composites. The use of nanostructures, which allow the modification of the cementitious binder at the nanoscale, instead of the widely used microadditives, has given spectacular results in improving the strength of the cement matrix and, consequently, the durability of the composite. The most widely studied nanoadditive is nanosilica. Compared to microsilica, nanosilica has a larger BET surface area and smaller size, corresponding to the size of the C–S–H crystallites. Nanosilica is therefore ten times more effective than microsilica [8, 9]. The effect of nanosilica in cement composites is multifaceted. Firstly, due to its nanometric dimensions, nanosilica provides additional active centers that act as growth sites for the silicate phases of the cement binder and accelerate the hydration of the cement. This is the so-called filling or nucleation effect, which results in a thickening of the microstructure of the cement composite, a refinement of the porous structure, an increase in adhesion to the aggregate and an improvement in both the mechanical and durability parameters of the cement composite. In addition, due to its pozzolanic properties, nanosilica can react with calcium hydroxide to form additional C–S–H phases, resulting in significant increases, particularly in compressive strength. The effect of nanosilica can be seen in the early stages of strength development. The positive effect of nanosilica can be seen at low levels, up to 3% by weight of the cement; typically nanosilica in powder form with a size of 5–25 nm has been used [8, 10]. Similar effects of accelerating the hydration of the cementitious binder, thickening the microstructure and improving the mechanical performance of the cement composite can be achieved by using other nanoadditives. However, these are often inert additives, i.e. they do not participate in pozzolanic reactions like nano- or microsilica. Their action is mainly physical, according to the filling mechanism or the so-called seeding effect. In studies into nanoadditives in concrete, metallic nanoxides, a portmanteau of nano-oxide referring to nanoparticles of metal oxides (Fe_2O_3, NiO, TiO_2), nanoscale calcium carbonates, or carbon nanostructures (carbon nanotubes and nanofibers, graphene, graphene oxide) have been most commonly used [7, 11–15]. The effects of nanosilica with a diameter of 15 and 50 nm, iron nanoxide with a diameter of 50 nm and nickel nanoxide with a particle diameter of 15 nm at a level of 1 wt% on the properties of cement mortars have been studied by Zhang et al. [16]. They showed that the nanoadditives accelerate the hydration of the cement binder due to their higher BET surface area and accelerate the initial and final setting of the cement binder. They also show an effect on early compressive strength values after 3 days of setting, increasing them by an average of about 20% compared to mortars without nanomodifiers. Final compressive strengths after 28 days of setting were also higher, by an average of 10%. The small diameter nanosilica was the most effective, followed by the 50 nm diameter nanosilica, and the nickel and iron nanoxides were slightly less effective [16]. In addition to improving compressive strength, the metallic nanoparticles also had a positive effect on tensile and flexural strength [9, 17, 18]. Recently, metallic nanoxides such as ZnO or CuO have been used in cementitious composites, which can delay the setting process of the cement binder during the first days of curing. In the long term, however, the presence of these nanoparticles contributes to the improvement of both flexural and compressive strength, through the nucleation effect. The increase in the C–S–H phase on these nanostructures leads to a reduction in the porosity of the cement matrix and a reduction in the size of the calcium hydroxide crystallites, which contributes to better adhesion of the cement matrix to the aggregate and consequently to greater durability of the whole system [19–21].

Spectacular improvements in the mechanical performance of cementitious composites can be achieved by introducing 1D and 2D carbon-based nanomaterials in the form of single or multi-walled carbon nanotubes (single-walled carbon nanotubes SWCNTs and multi-walled carbon nanotubes MWCNTs, respectively) and graphene and graphene oxide nanostructures. Carbon nanotubes are characterized by very good mechanical performance, with tensile strength ranging from 50 to 500 GPa for SWCNTs and 10–60 GPa for MWCNTs. Young's modulus values are 1 TPa and 0.3–1.0 TPa for SWCNTs and MWCNTs respectively. In addition, their high aspect ratio of over 1000 means that the carbon nanotubes not only accelerate the hydration process through the nucleation effect, but also act as nano-reinforcement, bridging the cracks that form in the cement matrix. In this way, the introduction of even small amounts of nanotubes, at a level of 0.1% by weight, leads to an improvement in compressive strength of up to 30% and in tensile and flexural strength of up to 50% [22–24]. Graphene, which has a modulus of elasticity of 1000 GPa, a tensile strength of around 130 GPa and a high aspect ratio of over 6000, also has a crack-inhibiting effect at the nanometer level. However, it has a very high BET surface area of around 2600 m^2/g, which, like carbon nanotubes, makes it very difficult to disperse in a cement matrix. More easily dispersed is graphene oxide, which, due to the way it is obtained (oxidation, purification and exfoliation of graphite), has hydrophilic oxygen-rich functional groups built into the structure that allow electrostatic repulsion. These groups also provide additional active centers on which the C–S–H phase begins to grow and a strong bond is formed. Due to oxidation, graphene oxide has lower parameters than graphene; these are: Young's modulus 23–42 GPa, tensile strength around 0.13 GPa, aspect ratio over 1500 and BET surface area 700–1500 m^2/g. Nevertheless, even at low levels of 0.02–0.05 wt%, it is able to bridge repeated cracks and accelerate the hydration of the cementitious binder, reducing the overall porosity of the system by about 13% and improving the compressive strength by 15–33% and the flexural strength by 41–59% of the cementitious composite [25, 26]. The increase in material strength and, in particular, the thickening of the microstructure and reduction in the porosity of the cement matrix in nanomodified cement composites results in increased resistance to external factors. Nanomodification of materials has been shown to significantly reduce water sorption, carbonate and chloride penetration, sulfate attack and improve corrosion of steel reinforcement [27, 28].

The main disadvantage of all nanomaterials, from nanosilica to carbon nanostructures, is their large BET surface area, which gives them a high affinity for water, leading to significant limitations in the workability of the cement matrix and the tendency of nanoadditives to aggregate and agglomerate. The uneven distribution of additives very often contributes to the deterioration of mechanical properties and special properties, including photocatalytic and antibacterial properties and, in the case of conductive materials, an increase in the percolation threshold and the need for increased use of carbon nanostructures. Therefore, an important element of nanoengineered cementitious composites is the uniform dispersion of nanoadditives, which can be achieved by different techniques: mechanical or ultrasonic mixing, with or without the addition of water-reduced admixtures, and, in the case of carbon-based materials, it is possible to use surface modification by functionalization with COOH and OH groups [29–32]. Another more sophisticated method is the use of core–shell structures, in which a hydrophobic nanoxide is coated with a hydrophilic one to improve its dispersion in the cement matrix. Examples of such solutions are $Fe_2O_3@SiO_2$ or $Fe_3O_4@SiO_2$ systems [33, 34]. Good dispersion can also be achieved with hybrid systems. In their work, Klapiszewska et al. used lignin as a nanoparticle carrier to achieve a satisfactory dispersion of zinc nanoxide and $ZnO\text{-}SiO_2$ bicomponent systems [35]. It is worth noting that the efficient dispersion of nanostructures in cementitious composites results in the possibility of using smaller amounts of nano additives, which in turn leads to lower composite manufacturing costs, increased durability and more efficient use of special functionalities of these materials, such as antibacterial, self-cleaning or self-sensing properties.

20.3 Self-cleaning Properties of Cement Composites

The most commonly used metal nanoxide in engineered cement composites with self-cleaning properties is titanium(IV) oxide. This is mainly due to the photocatalytic properties of the oxide in the presence of solar radiation; the band gap energy for titania in the most effective crystalline form of anatase is 3.2 eV, while for the rutile phase equals 3.0 eV. Other semiconducting materials that can be used in these solutions include zinc oxide, which has similar photocatalytic properties to titania, and ferric oxide, which has slightly lower photocatalytic properties (band gap energy is 2.2 eV). In addition, titanium dioxide exhibits superhydrophilic properties and is known to be inexpensive, non-toxic and chemically stable, making it compatible with most building materials, including cement composites. These properties enable titanium dioxide to be used in building materials, particularly in self-cleaning surfaces and in the reduction of organic and inorganic pollutants in the air, especially nitrogen and sulfur oxides and volatile organic compounds (VOCs) [36–38]. In addition, its optical properties mean that titanium dioxide is used in roofing and road systems in the form of passive cooling coatings that can reduce the surface temperature of materials by several degrees. The future use of such solutions in large cities could contribute to reducing the negative effect of urban heat islands [39, 40]. In turn, the antibacterial properties resulting from the photocatalytic properties of titanium dioxide, combined with its self-cleaning properties, make titanium dioxide suitable for use in organic and mineral coatings on heritage buildings or in public facilities (hospitals, schools). The use of such coatings counteracts biological corrosion of building materials and prolongs the life of buildings [41–43].

Figure 20.3a shows potential directions of application for materials with self-cleaning properties, while Fig. 20.3b shows a mechanism for the self-cleaning action of semiconducting materials.

Semiconducting oxides, including titanium dioxide, are characterized by a specific electronic structure that includes an electron-filled valence band and an empty conduction band. As a result of the absorption of a quantum of radiation corresponding to the energy difference between the bands, an electron is excited and moves from the valence band to the conduction band, leaving a positively charged empty hole in the valence layer. The resulting electrons and holes can migrate to the surface of the photocatalyst and react with the donors and acceptors adsorbed on the oxide surface. In this way, positively charged holes can oxidize water or hydroxyl groups adsorbed on the surface to form reactive hydroxyl radicals. At the same time, the electrons excited to the conduction band react with the oxygen adsorbed on the TiO_2 surface and reduction to the superoxide radical anion takes place. This anion can then react with the hydronium cation to form a hydroperoxyl radical and cause further

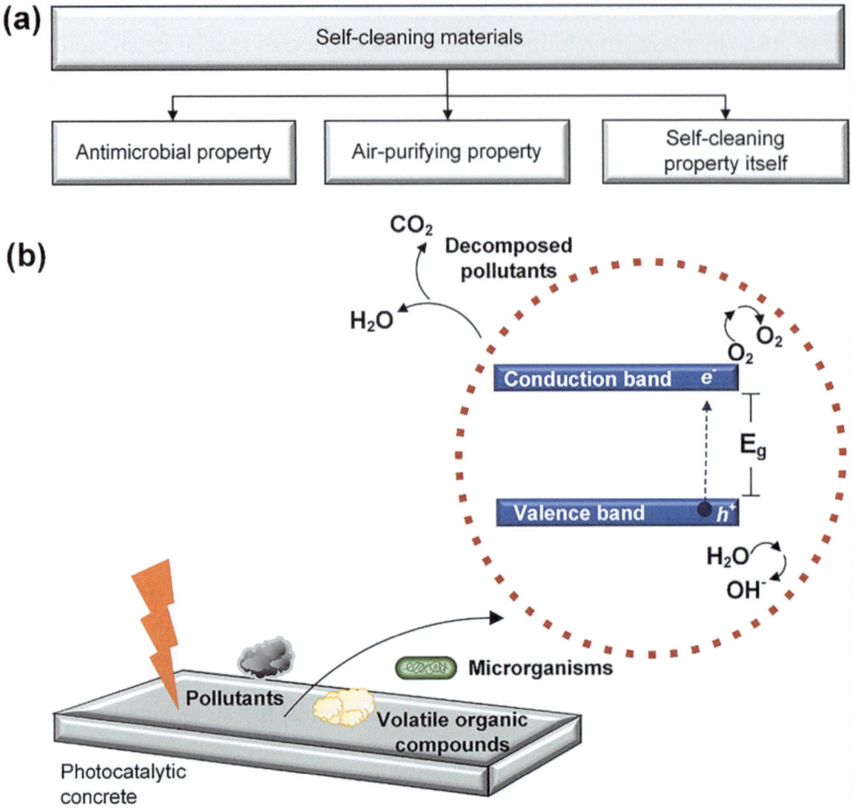

Fig. 20.3 Potential directions of application for materials with self-cleaning properties (**a**) and a mechanism for the cleaning action of semiconducting materials (**b**), based on [44]

reduction to H_2O_2. The resulting reactive oxygen species (ROS) are involved in the oxidation of organic compounds to fully mineralized forms. The greatest contribution to the reduction of nitrogen and sulfur oxides is made by reactive hydroxyl and superoxide anion radicals, which can oxidize the oxides to form nitric acid (V) and sulfuric acid (VI), which are easily removed from the surface of the material [44, 45].

In self-cleaning applications, commercial P25 is most commonly used as a mixture of rutile and anatase, suggesting that its properties are sufficient for the cement matrix to achieve photocatalytic properties. However, studies indicate that there are several factors that influence the photocatalytic performance of titania nanoparticles. Of the many factors, the most important are the proper dispersion of the nanoxide, with the anatase form being more effective than the rutile form, and the size of the nanoparticles and their BET surface area. Smaller particle sizes and larger BET surface areas are more effective and mean that high photocatalytic efficiencies in cement composites can be achieved with lower levels of nanoadditive, below 1.5 wt% [46, 47].

Titanium nanoxide can be incorporated into the cement matrix in two ways, either as a surface impregnation layer or as an admixture to mortar or concrete. In the first case, the nanoxide is usually mixed with organic substances such as silanes and mechanically applied to the surface of the cement composite and/or other building materials, most commonly stone, ceramics or wood. In the second case, nanoxide is added to mortars and concrete mixtures during the mixing process, usually in quantities of less than 5 wt%. Both methods have their drawbacks. In the case of silane coatings, it is their short durability, while in the case of TiO_2 admixture to cementitious composites, it is the loss of photocatalytic properties due to adsorption of the C–S–H phase onto the oxide grains and impaired access of UV radiation to the oxide surface, which often blocks the photocatalytic reaction [48–50]. Another drawback is the small range of visible light in which titanium dioxide is activated. Therefore, work has been done in recent years to increase the photocatalytic efficiency of titanium dioxide in the visible light range by enhancing its action with other semiconducting oxides such as ZnO, Fe_2O_3, SiO_2 or by modifying noble metals with Ag, Au, Cu or elements such as C, S or N [51–55].

20.4 Antimicrobial Properties of Cement Composites

Cement composites are exposed to many external factors, the most common of which are physical hazards (temperature effects, cyclic mechanical loading, erosion) or chemical hazards (carbonization, salt and soft water, nitrogen and sulfur oxides in the urban atmosphere). All these external factors contribute to the progressive disintegration of the cement binder, an increase in its porosity and a consequent loss of durability and performance [56]. An additional factor that can significantly accelerate these processes is the presence of micro-organisms (algae, mosses, fungi, bacteria), which affect the durability of surfaces based on cement binders, particularly in structures exposed to conditions of variable humidity. The growth of microorganisms on the surface of cementitious composites, initially forming a thin film and over time numerous colonies, affects their visual assessment, while the presence of toxic substances secreted by the microorganisms causes acidification of the environment. The gradual leaching of cementitious binder components leads to a reduction in pH and degradation of the cementitious composite [57–60].

One way to counteract this phenomenon is to modify cement composites with nanoadditives with antimicrobial properties. Most commonly, silver and copper nanoparticles or, increasingly, titanium, zinc, calcium, aluminum or magnesium oxides are used in solutions for the construction industry [61–67]. These nanoxides are incorporated into cementitious composites in two ways: either by using silane dispersions of the nanoadditives on the surface, or by mixing them into the mortar or concrete mix before setting in small quantities of less than 5 wt% [68–70]. These solutions are most commonly used in heritage buildings, but due to their potential fungicidal and bactericidal properties, they can also be used in public buildings, hospitals and food storage areas [71–74].

The mechanism of microbial cell destruction and the effects of microbial inactivation by selected nanostructures are shown in Fig. 20.4. The types of nanostructures mentioned above are divided into three types that determine their interaction with microorganisms, namely pure nanoparticles (e.g. Ag, Cu), nanoxides with photocatalytic properties (e.g. ZnO, TiO_2) and other nanoxides (e.g. Al_2O_3, MgO).

The action of the nanoparticles in the first group and the nanoxides in the third group is mainly based on their mechanical interaction with the cell wall. Thanks to their nanometric size, Ag and Cu, as well as the Al^{3+}, Mg^{2+}, Zn^{2+} cations released from the oxides, are able to disrupt the cell membrane of the micro-organism and, thus introduced into the structure, are able to react with the cell's DNA and enzymes, leading to their deactivation. Nanoparticles also have the ability to accumulate in the structure of the cell membrane, thus preventing the replication of the cell's DNA code. In the case of nanoxides with photocatalytic properties, the disruption of the cell membrane occurs through the action of reactive oxygen species released by the photocatalytic reaction. ROS have a tendency to disrupt the cell membrane and penetrate the cell structure, leading to the deactivation of DNA, proteins and enzymes, resulting in a decrease in cellular metabolic stability and reduced ATP levels [75, 76].

Table 20.1 shows nanoparticles and nanoxides and their inhibitory effects on selected micro-organisms in cementitious composites. Nanostructures have been shown to have a beneficial effect on both gram-positive and gram-negative bacteria, fungi and algae, although the efficacy depends on several factors, including the type and size of nanoparticles, morphology and dispersion in the cement matrix. Nanoparticles are the most effective in inhibiting bacteria. Fungi and algae are more difficult to deactivate [30, 77–79]. In this regard, it is worth noting that in the future, when designing composites with antimicrobial properties, attention should be paid to the selective selection of the right nanoadditives in order to use the most effective nanostructures for deactivating a given microorganism. This will result in slower microbial immunization against the selected nanoadditives and better microbial resistance of the cementitious composite over a longer service life.

20.5 Self-sensing Properties of Cement

The durability of cementitious composites, their longevity and, therefore, their overall life cycle, depends to a large extent on the maintenance and repair phase of the material during its service life. The more durable and resistant the material, the longer it can be used and the repair phase can be postponed. As a result, recycling and reusing into new, fully functional products, which is the basis of the circular economy, is delayed. This is why the maintenance and condition monitoring phase of building components and structures is so important [88, 89]. In recent years, the idea of smart concrete has been developed to monitor the condition of structures. Smart concrete is an engineered material based on high performance, low emissivity cementitious composites that incorporate nanostructures with conductive properties into their structure [90]. Such structures can also include nanomaterials that tend to self-heal or self-adjust

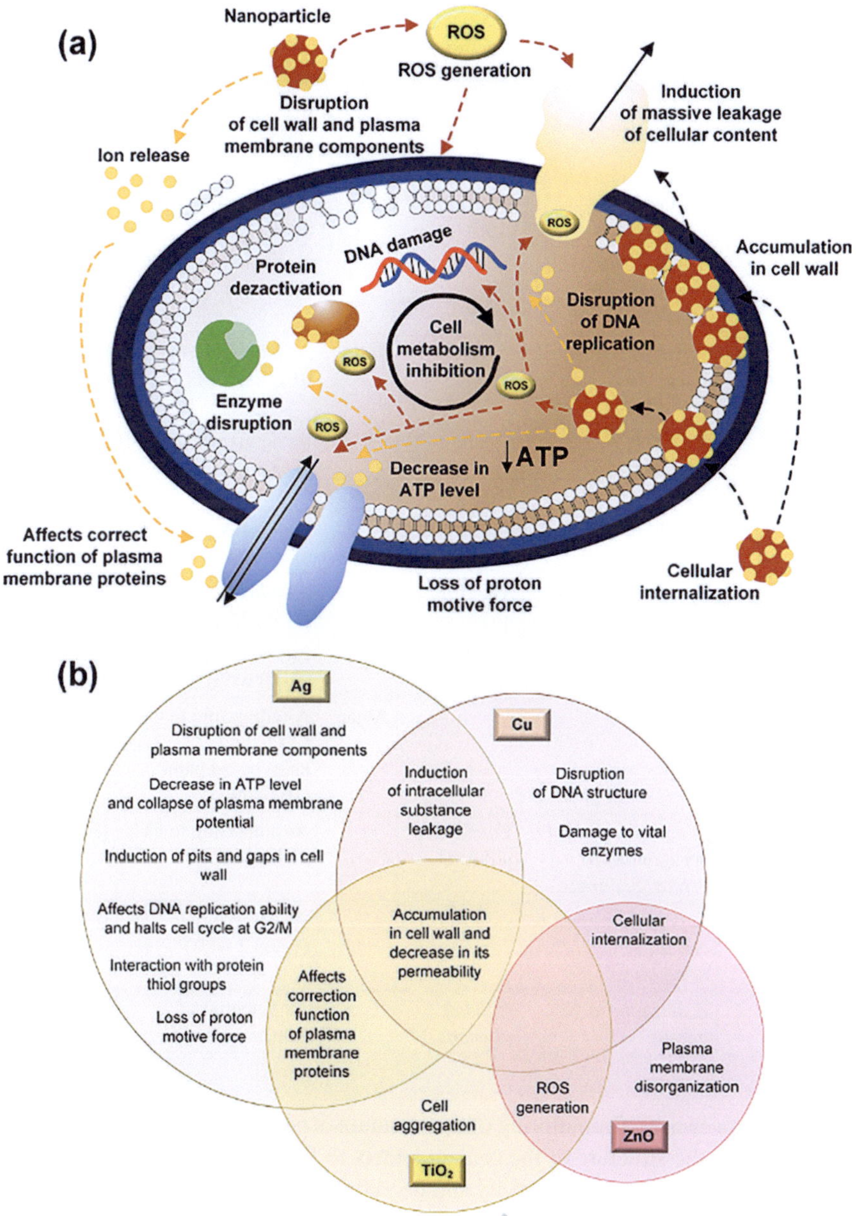

Fig. 20.4 Mechanism of microbial cell destruction (**a**) and effects of microbial inactivation by selected nanostructures (**b**), based on [75]

Table 20.1 Overview of selected nanoparticles and nanoxides and their inhibitory effects on selected microorganisms in cementitious composites

Nanoparticle or nanoxide (amount)	Bacteria	Fungi	Algae	Way of introducing to cement composites CC and amount	References
Ag	*S. aureus, P. aeruginosa, E. coli*	*Aspergillus niger*	–	As admixture to grout	[80]
CuO	*E. coli, B. aureus, B. cereus, P. aeruginosa*	–	–	As admixture to CC (0.5 wt%)	[81]
ZnO	*E. coli, B. aureus, B. cereus, P. aeruginosa*	–	–	As admixture to CC (0.1, 0.3 wt%)	[81]
	E. coli, P. aeruginosa, S. aureus	*Candida albicans*	–	As admixture to CC (1, 3 wt%)	[82]
	E. coli, B. cereus	–	–	Dip-coated solution of ZnO (1, 3 wt%)	[83]
	E. coli, B. subtilis	*Aspergillus niger*	–	As admixture to CC (5, 10, 15 wt%)	[20]
	S. aureus, P. aeruginosa	*Candida albicans*	–	As admixture to CC (5–20 wt%)	[84]
	–	–	Algae	As admixture to CC (20%) and latex-based paint (16.67 g/L)	[85]
	E. coli, P. putida, P. aeruginosa, B. cereus, S. aureus	*Candida albicans*		As admixture to CC (0.1 wt%)	[86]
TiO_2	*E. coli, B. aureus, B. cereus, P. aeruginosa*	–	–	As admixture to CC (1 and 2 wt%)	[81]
Al_2O_3	*P. aeruginosa, S. aureus*	*Candida albicans*	–	As potential admixture to CC	[87]

[91–93]. However, self-monitoring of the structure is possible by introducing carbon materials into the structure of the cement matrix to form a conductive network that responds to changes in the resistivity and degradation of the concrete. This makes it possible to predict the condition of the concrete structure and apply appropriate repair in time to prevent permanent, irreversible material degradation [90]. Figure 20.5 illustrates the idea of smart concrete, with an indication of the development directions of the material solution and the positive pro-environmental effects that can be expected in the future due to the introduction of carbon nanostructures in the cement matrix. The design of high performance cementitious composites with self-sensing properties, through the use of low carbon binders and a more durable structure, will enable a reduction in CO_2 emissions to the atmosphere and a reduction in the consumption of natural raw materials, as well as timely repairs due to their self-monitoring ability. This in turn will result in a longer service life of the material and lower repair and maintenance costs for the structure.

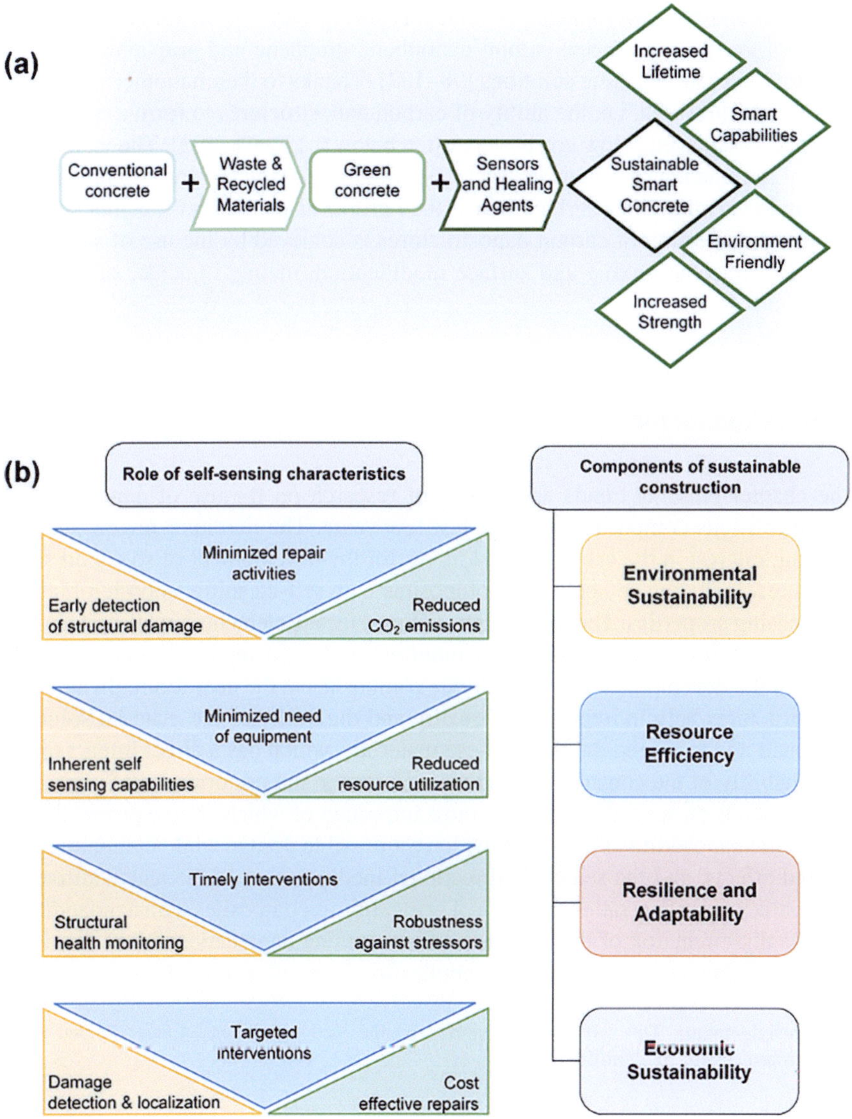

Fig. 20.5 Components of smart concrete (**a**) and role of smart concrete in construction (**b**), based on [90]

The possibility of using carbon materials to monitor the condition and degradation of the cement matrix by studying the conductivity of the cement matrix (and indeed changes in its resistivity) was introduced in the early 1990s [94]. Initially, the main conductive materials used were short carbon fibers, steel fibers or carbon black, added to the cement matrix in quantities of less than 0.5% by weight [95–97]. Today, as a result of the intensive development of different allotropic varieties of carbon and the development of relatively cheap ways to obtain carbon nanostructures, carbon nanotubes, carbon nanofibers, graphene and graphene oxide are proposed in smart concrete solutions [98–102]. Thanks to their nanometric size, the percolation threshold, i.e. the ability of carbon nanostructures to form a conductive network, is achieved at low levels, very often below 0.15 wt% [103]. The main drawback of introducing carbon nanostructures, which are characterized by a high aspect ratio and a very high BET surface area, is their proper dispersion in the cement matrix. Improved dispersion of carbon nanostructures is achieved by the use of superplasticisers, ultrasonic mixing and surface modification, mainly in acids, sulfuric and nitric acids [104].

20.6 Conclusion

The chapter presents trends and results of research on the use of nanostructures in cementitious composites over the last few years. The literature review shows a growing interest in the use of nanoadditives for the development of so-called smart structures, including cementitious composites with self-cleaning, antimicrobial and self-sensing properties. The use of such coatings, façade elements, small-scale infrastructure and structures capable of self-monitoring and self-repair are potential directions for the development of cementitious composites in the near future. In addition, nanostructures help to improve the quality and durability of the material solutions presented and to extend the life of these materials, which has a direct impact on the sustainability of the construction industry. However, the performance of such solutions depends on several factors, the most important of which are the proper dispersion of the nanoadditives to ensure the lowest possible dosage while maintaining the desired effects, and the selective, directional incorporation of the nanoadditive for the desired effect. The rational design of cementitious composites with nanoadditives and the dissemination of their positive effects among engineers and architects will certainly influence a wider and safer application in practice in the future.

Acknowledgements This work was supported by the National Science Center Poland under research project no. 2022/45/B/ST8/02288.

References

1. P.J.M. Bartos, Nanotechnology in construction: a roadmap for development, in *Nanotechnology in Construction 3*. ed. by Z. Bittnar, P.J.M. Bartos, J. Němeček, V. Šmilauer, J. Zeman (Springer, Berlin, Heidelberg, 2009), pp.15–26
2. F. Pacheco-Torgal, S. Jalali, Nanotechnology: Advantages and drawbacks in the field pf construction and building materials. Constr. Build. Mater. **25**, 582–590 (2011)
3. R. Sharma, V. Kumar, Nano enabled agriculture for sustainable soil. Waste Manag. Bull. **2**, 152–161 (2024)
4. S.P. Shah, P. Hou, M.S. Konsta-Gdoutos, Nano-modification of cementitious material: toward a stronger and durable concrete. J. Sustain. Cem-Based Mater. **5**, 1–22 (2016)
5. M. Thanmanaselvi, V. Ramasamy, A study on durability characteristics of nano-concrete. Mater. Today Proc. **80**, 2360–2365 (2023)
6. Y. Hao, C. Shi, Z. Bi et al., Recent advances in properties and applications of carbon fiber-reinforced smart cement-based composites. Materials **16**, 2552 (2023)
7. S.K. Sood, G.R. Vesmawala, An overview of recent advances in fracture performance of nano engineered cement composites. Constr. Build. Mater. **431**, 136489 (2024)
8. P.P. Abhilash, D.K. Nayak, B. Sangoju et al., Effect of nano-silica in concrete: a review. Constr. Build. Mater. **278**, 122347 (2021)
9. M. Gesoglu, E. Güneyisi, D.S. Asaad, Properties of low binder ultra-high performance cementitious composites: comparison of nanosilica and microsilica. Constr. Build. Mater. **102**, 706–713 (2016)
10. H. AlTawaiha, F. Alhomaidat, T. Eljufout, A review of the effect of nano-silica on the mechanical and durability properties of cementitious composites. Infrastructures **8**, 132 (2023)
11. A.E. Douba, P. Hou, S. Kawashima, Hydration and mechanical properties of high content nano-coated cements with nano-silica, clay and calcium carbonate. Cem. Concr. Res. **168**, 107132 (2023)
12. A. Dinesh, A. Yuvaraj, S. Abinaya et al., Nanopowders as an additive for strength and durability enhancement of cement composite: review and prospects. Mater. Today Proc. (2023). https://doi.org/10.1016/j.matpr.2023.03.651
13. J.A. Abdalla, B.S. Thomas, R.A. Hawileh, Influence of nano-TiO_2, nano-Fe_2O_3, nanoclay and nano-$CaCO_3$ on the properties of cement/geopolymer concrete. Clean. Mater. **4**, 100061 (2022)
14. Y. Lin, H. Du, Graphene reinforced cement composites: a review. Constr. Build. Mater. **265**, 120312 (2020)
15. H. Saleem, S.J. Zaidi, N.A. Alnuaimi, Recent advancements in the nanomaterial application in concrete and its ecological impact. Materials **14**, 6387 (2021)
16. A. Zhang, Y. Ge, W. Yang et al., Comparative study on the effects of nano-SiO_2, nano-Fe_2O_3 and nano-NiO on hydration and microscopic properties of white cement. Constr. Build. Mater. **228**, 116767 (2019)
17. E. Horszczaruk, Properties of cement mortars modified with commercial nanosilica. Struct. Environ. **11**, 247–255 (2019)
18. V.P. Dang, D.J. Kim, Effects of nanoparticles on the tensile behavior of ultra-high-performance fiber-reinforced concrete at high strain rates. J. Build. Eng. **63**, 105513 (2023)
19. M. Kumar, M. Bansal, R. Garg, An overview of beneficiary aspects of zinc oxide nanoparticles on performance of cement composites. Mater. Today Proc. **43**, 892–898 (2021)
20. V.P. Singh, K. Sandeep, H.S. Kushwaha, Photocatalytic, hydrophobic and antimicrobial characteristics of ZnO nano needle embedded cement composites. Constr. Build. Mater. **158**, 285–294 (2018)
21. A. Ślosarczyk, I. Klapiszewska, A. Parus et al., Antimicrobial action and chemical and physical properties of CuO-doped engineered cementitious composites. Sci. Rep. **13**, 10404 (2023)
22. L. Silvestro, P.J.P. Gleize, Effect of carbon nanotubes on compressive, flexural and tensile strengths of Portland cement-based materials: a systematic literature review. Constr. Build. Mater. **264**, 120237 (2020)

23. V.V. Rocha, P. Ludvig, A.C.C. Trindade et al., The influence of carbon nanotubes on the fracture energy, flexural and tensile behavior of cement based composites. Constr. Build. Mater. **209**, 1–8 (2019)
24. G. Sun, R. Liang, Z. Lu et al., Mechanism of cement/carbon nanotube composites with enhanced mechanical properties achieved by interfacial strengthening. Constr. Build. Mater. **115**, 87–92 (2016)
25. M. Tiong, Q. Liu, Y. Xie et al., A review on cementitious composite incorporating carbon nanomaterials: the self-sensing functionality as an implication for geological CO_2 storage monitoring. Geoenergy Sci Eng **240**, 212997 (2024)
26. S. Chuah, Z. Pan, J.G. Sanjayan, Nano reinforced cement and concrete composites and new perspective from graphene oxide. Constr. Build. Mater. **73**, 113–124 (2014)
27. F.U.A. Shaikh, S.W.M. Supit, Chloride induced corrosion durability of high volume fly ash concretes containing nano particles. Constr. Build. Mater. **99**, 208–225 (2015)
28. L.P. Singh, D. Ali, I. Tyagi et al., Durability studies of nano-engineered fly ash concrete. Constr. Build. Mater. **194**, 205–215 (2019)
29. Y. Reches, K. Thomson, M. Helbing et al., Agglomeration and reactivity of nanoparticles of SiO_2, TiO_2, Al_2O_3, Fe_2O_3, and clays in cement pastes and effects on compressive strength at ambient and elevated temperatures. Constr. Build. Mater. **167**, 860–873 (2018)
30. I. Klapiszewska, Ł Ławniczak, S. Balicki et al., Influence of zinc oxide particles dispersion on the functional and antimicrobial properties of cementitious composites. J. Mater. Res. Technol. **24**, 2239–2264 (2023)
31. X. Li, A.H. Korayem, C. Li et al., Incorporation of graphene oxide and silica fume into cement paste: a study of dispersion and compressive strength. Constr. Build. Mater. **123**, 327–335 (2016)
32. F.T. Isfahani, W. Li, E. Redaelli, Dispersion of multi-walled carbon nanotubes and its effects on the properties of cement composites. Cem. Concr. Compos. **74**, 154–163 (2016)
33. F.K. Tazune, H.K. Tchakouté, C.H. Rüscher et al., Effects of Fe_2O_3/SiO_2 molar ratios in the Fe-silica on the compressive strengths and microstructural properties of geopolymer materials derived from waste fired clay brick and metakaolin. J. Inorg. Organomet. Polym. Mater. **34**, 1725–1737 (2024)
34. E. Horszczaruk, M. Aleksandrzak, K. Cendrowski et al., Mechanical properties cement based composites modified with nano-Fe_3O_4/SiO_2. Constr. Build. Mater. **251**, 118945 (2020)
35. I. Klapiszewska, A. Parus, Ł Ławniczak et al., Production of antibacterial cement composites containing ZnO/lignin and ZnO–SiO_2/lignin hybrid admixtures. Cem. Concr. Compos. **124**, 104250 (2021)
36. J. Chen, C. Poon, Photocatalytic construction and building materials: from fundamentals to applications. Build. Environ. **44**, 1899–1906 (2009)
37. A. Folli, C. Pade, T.B. Hansen et al., TiO_2 photocatalysis in cementitious systems: Insights into self-cleaning and depollution chemistry. Cem. Concr. Res. **42**, 539–548 (2012)
38. J. Wu, Y. Luo, Z. Qin, Composite-modified nano-TiO_2 for the degradation of automobile exhaust in tunnels. Constr. Build. Mater. **408**, 133805 (2023)
39. Z. Gholami, S. Jalilisadrabad, R. Amrollahi, Investigating the impact of using modified cool materials by titanium dioxide (TiO_2)-based photocatalytic self-cleaning nanoparticles in urban facades on urban microclimate parameters. Case Stud Constr Mater **19**, e02268 (2023)
40. Z. Wang, Y. Xie, M. Mu et al., Materials to mitigate the urban heat island effect for cool pavement: a brief review. Buildings **12**, 1221 (2022)
41. S.J. Dancer, Hospital cleaning in the 21st century. Eur. J. Clin. Microbiol. Infect. Dis. **30**, 1473–1481 (2011)
42. P. Sikora, A. Augustyniak, K. Cendrowski et al., Characterization of mechanical and bactericidal properties of cement mortars containing waste glass aggregate and nanomaterials. Materials **9**, 701 (2016)
43. S.M. Zacarías, S. Marchetti, O.M. Alfano et al., Photocatalytic paint for fungi growth control under different environmental conditions and irradiation sources. J. Photochem. Photobiol. A Chem. **364**, 76–87 (2018)

44. I.N.R. Prudente, H.C. dos Santos, J.L. Fonseca et al., Advancements in self-cleaning building materials: photocatalysts, superhydrophobic surfaces, and biocides approaches. Constr. Build. Mater. **434**, 136700 (2024)
45. E.I. Cedillo-González, J.M. Hernández-López, J.J. Ruiz-Valdés et al., Self-cleaning TiO_2 coatings for building materials: the influence of morphology and humidity in the stain removal performance. Constr. Build. Mater. **237**, 117692 (2020)
46. P. Jędrzejczak, A. Parus, S. Balicki et al., The influence of various forms of titanium dioxide on the performance of resultant cement composites with photocatalytic and antibacterial functions. Mater. Res. Bull. **160**, 112139 (2023)
47. E. Franzoni, M.C. Bignozzi, E. Rambaldi, TiO_2 in the building sector, in *Titanium Dioxide (TiO_2) and Its Applications*, ed. by F. Parrino, L. Palmisano, 1st ed (Elsevier, 2021), pp. 449–479
48. A.M. Castro-Hoyos, M.A.R. Manzano, A. Maury-Ramírez, Challenges and opportunities of using titanium dioxide photocatalysis on cement-based materials. Coatings **12**, 968 (2022)
49. D. Enea, M. Bellardita, P. Scalis et al., Effects of weathering on the performance of self-cleaning photocatalytic paints. Cem. Concr. Compos. **96**, 77–86 (2019)
50. M. Pérez-Nicolás, J. Plank, D. Ruiz-Izuriaga et al., Photocatalytically active coatings for cement and air lime mortars: enhancement of the activity by incorporation of superplasticizers. Constr. Build. Mater. **162**, 628–648 (2018)
51. J.D. Bersch, A.B. Masuero, D.C. Coitinho Dal Molin, Photocatalytic coloured rendering mortars: effect of TiO_2 and iron oxide pigments on the physical, mechanical, hygric, and photoactive behaviour. Mater. Struct. **56**, 146 (2023)
52. I.P. Parkin, R.G. Palgrave, Self-cleaning coatings. J. Mater. Chem. **15**, 1689–1695 (2005)
53. P. Sikora, K. Cendrowski, A. Markowska-Szczupak et al., The effects of silica/titania nanocomposite on the mechanical and bactericidal properties of cement mortars. Constr. Build. Mater. **150**, 738–746 (2017)
54. P. Jędrzejczak, M. Janczarek, A. Parus et al., Carbon-modified TiO_2 as a promising and efficient admixture for cementitious composites: a comprehensive study of photocatalytic, mechanical and structural properties. J Build Eng **78**, 107747 (2023)
55. Z. Gholami, S. Jalilisadrabad, R. Amrollahi, The effect of using nanomaterial at the ground level of urban space on the thermal comfort. Int. J. Environ. Sci. Technol. **21**, 2475–2488 (2024)
56. L. Qiu, S. Dong, A. Ashour et al., Antimicrobial concrete for smart and durable infrastructures: a review. Constr. Build. Mater. **260**, 120456 (2020)
57. T. Noeiaghaei, A. Mukherjee, N. Dhami et al., Biogenic deterioration of concrete and its mitigation technologies. Constr. Build. Mater. **149**, 575–586 (2017)
58. N. Kip, J. van Veen, The dual role of microbes in corrosion. ISME J. **9**, 542–551 (2015)
59. S. Wei, Z. Jiang, H. Liu et al., Microbiologically induced deterioration of concrete—a review. Braz. J. Microbiol. **44**, 1001–1007 (2013)
60. J.D. Gu, T.E. Ford, N.S. Berke et al., Biodeterioration of concrete by the fungus *Fusarium*. Int. Biodeterior. Biodegrad. **41**, 101–109 (1998)
61. T. Haile, G. Nakhla, E. Allouche, Evaluation of the resistance of mortars coated with silver bearing zeolite to bacterial-induced corrosion. Corrosion. Sci. **50**, 713–720 (2008)
62. K. Sharafutdinov, K. Saraykina, G. Kashevarova et al., The use of copper nanomodified calcium carbonate as a bactericidal additive for concrete. Int. J. Comput. Civ. Struct. Eng. **18**, 143–155 (2022)
63. V.P. Klienchen de Maria, F.F. Guedes de Paiva, J.R. Tamashiro et al., Advances in ZnO nanoparticles in building material: Antimicrobial and photocatalytic applications—systematic literature review. Constr. Build. Mater. **417**, 135337 (2024)
64. B.A. Dehkordi, M.R. Nilforoushan, N. Talebian et al., A comparative study on the self-cleaning behavior and antibacterial activity of Portland cement by addition of TiO_2 and ZnO nanoparticles. Mater. Res. Express **8**, 035403 (2021)
65. K. Loh, C.C. Gaylarde, M.A. Shirakawa, Photocatalytic activity of ZnO and TiO_2 'nanoparticles' for use in cement mixes. Constr. Build. Mater. **167**, 853–859 (2018)

66. S.V. Gudkov, D.E. Burmistrov, V.V. Smirnova et al., A mini review of antibacterial properties of Al_2O_3 nanoparticles. Nanomaterials **12**, 2635 (2022)
67. M.A. Gatou, E. Skylla, P. Dourou et al., Magnesium oxide (MgO) nanoparticles: synthetic strategies and biomedical applications. Crystals **14**, 215 (2024)
68. F. Hamidi, F. Aslani, TiO_2-based photocatalytic cementitious composites: Materials, properties, influential parameters, and assessment techniques. Nanomaterials **9**, 1444 (2019)
69. M. Janczarek, Ł Klapiszewski, P. Jędrzejczak et al., Progress of functionalized TiO_2-based nanomaterials in the construction industry: a comprehensive review. Chem. Eng. J. **430**, 132062 (2022)
70. J. Chen, S. Kou, C. Poon, Hydration and properties of nano-TiO_2 blended cement composites. Cem. Concr. Compos. **34**, 642–649 (2012)
71. I. Ielo, F. Giacobello, A. Castellano et al., Development of antibacterial and antifouling innovative and eco-sustainable sol–gel based materials: from marine areas protection to healthcare applications. Gels **8**, 26 (2021)
72. X. Bai, S. Yang, C. Tan et al., Synthesis of TiO_2 based superhydrophobic coatings for efficient anti-corrosion and self-cleaning on stone building surface. J. Clean. Prod. **380**, 134975 (2022)
73. A. Rabajczyk, M. Zielecka, W. Klapsa et al., Self-cleaning coatings and surfaces of modern building materials for the removal of some air pollutants. Materials **14**, 2161 (2021)
74. L. Bergamonti, G. Predieri, Y. Pa et al., Enhanced self-cleaning properties of N-doped TiO_2 coating for cultural heritage. Microchem. J. **133**, 1–12 (2017)
75. G. Wyszogrodzka, B. Marszałek, B. Gil et al., Metal-organic frameworks: mechanisms of antibacterial action and potential applications. Drug Discov. Today **21**, 1009–1018 (2016)
76. A. Ślosarczyk, I. Klapiszewska, D. Skowrońska et al., A comprehensive review of building materials modified with metal and metal oxide nanoparticles against microbial multiplication and growth. Chem. Eng. J. **466**, 143276 (2023)
77. L.K. Babu, E. Sarala, O. Audiseshaiah et al., Synthesis, characterization of nanocrystalline ZnO via two different chemical methods and its antibacterial activity. Surf. Interface **16**, 93–100 (2019)
78. I. Klapiszewska, A. Kubiak, A. Parus et al., The in situ hydrothermal and microwave syntheses of zinc oxides for functional cement composites. Materials **15**, 1069 (2022)
79. N.K. Dhiman, N. Sidhu, S. Agnihotri, A. Mukherjee, M.S. Reddy, Role of nanomaterials in protecting building materials from degradation and deterioration, in *Biodegradation and Biodeterioration At the Nanoscale. A volume in Micro and Nano Technologies*, ed. by H.M.N. Iqbal, M. Bilal, T.A. Nguyen, G. Yasin (Elsevier, 2022), pp. 405–475
80. G.D. da Silva, E.J. Guidelli, G.M. de Queiroz-Fernandes et al., Silver nanoparticles in building materials for environment protection against microorganisms. Int. J. Environ. Sci. Technol. **16**, 1239–1248 (2019)
81. P. Jędrzejczak, Ł Ławniczak, A. Ślosarczyk et al., Physicomechanical and antimicrobial characteristics of cement composites with selected nano-sized oxides and binary oxide systems. Materials **15**, 661 (2022)
82. A. Augustyniak, J. Jablonska, K. Cendrowski et al., Investigating the release of ZnO nanoparticles from cement mortars on microbiological models. Appl. Nanosci. **12**, 489–502 (2022)
83. T. Noeiaghaei, N. Dhami, A. Mukherjee, Nanoparticles surface treatment on cemented materials for inhibition of bacterial growth. Constr. Build. Mater. **150**, 880–891 (2017)
84. G. Voicu, G.A. Tiuca, A.I. Badanoiu et al., Nano and mesoscopic SiO_2 and ZnO powders to modulate hydration, hardening and antibacterial properties of Portland cements. J. Build. Eng. **57**, 104862 (2022)
85. A. Alum, A. Rashid, B. Mobasher et al., Cement-based biocide coatings for controlling algal growth in water distribution canals. Cem. Concr. Compos. **30**, 839–847 (2008)
86. I. Klapiszewska, Ł Ławniczak, A. Parus et al., Zinc oxide as a functional admixture to cement composites. Physicochem. Probl. Min. Process. **58**, 145565 (2022)
87. P. Sikora, A. Augustyniak, K. Cendrowski et al., Antimicrobial activity of Al_2O_3, CuO, Fe_3O_4, and ZnO nanoparticles in scope of their further application in cement-based building materials. Nanomaterials **8**, 212 (2018)

88. M. Stanitsas, K. Kirytopoulos, Investigating the significance of sustainability indicators for promoting sustainable construction project management. Int. J. Constr. Manag. **23**, 434–448 (2023)
89. N. Makul, Advanced smart concrete—a review of current progress, benefits and challenges. J. Clean. Prod. **274**, 122899 (2020)
90. Q. Tian, J. Zhou, J. Hou et al., Building the future: smart concrete as a key element in next-generation construction. Constr. Build. Mater. **429**, 136364 (2024)
91. M. Seifan, A.K. Sarmah, A.K. Samani et al., Mechanical properties of bio self-healing concrete containing immobilized bacteria with iron oxide nanoparticles. Appl. Microbiol. Biotechnol. **102**, 4489–4498 (2018)
92. S. Rajadesingu, K.C. Mendonce, N. Palani et al., Exploring the potential of bacterial concrete: a sustainable solution for remediation of crack and durability enhancement—a critical review. Constr. Build. Mater. **439**, 137238 (2024)
93. J.H. Yeon, Thermal behavior of cement mortar embedded with low-phase transition temperature PCM. Constr. Build. Mater. **252**, 119168 (2020)
94. P.W. Chen, D.D.L. Chung, Carbon fiber reinforced concrete for smart structures capable of non-destructive flaw detection. Smart Mater. Struct. **2**, 22 (1993)
95. D.D.L. Chung, A critical review of electrical-resistance-based self-sensing in conductive cement-based materials. Carbon **203**, 311–325 (2023)
96. A.K. Cholker, P. Kulkarni, V. Ravekar, Experimental behaviour of smart concrete embedded with micro carbon fibres as a sensing material. J. Phys. Conf. Ser. **1921**, 012095 (2021)
97. Y. Huang, H. Li, S. Qian, Self-sensing properties of engineered cementitious composites. Constr. Build. Mater. **174**, 253–262 (2018)
98. G. Wang, Y. Wang, P. Zhang et al., Structure dependent properties of carbon nanomaterials enabled fiber sensors for *in situ* monitoring of composites. Compos. Struct. **195**, 36–44 (2018)
99. D.Y. Yoo, I. You, S.J. Lee, Electrical properties of cement-based composites with carbon nanotubes, graphene, and graphite nanofibers. Sensors **17**, 1064 (2017)
100. D.Y. Yoo, I. You, G. Zi et al., Effects of carbon nanomaterial type and amount on self-sensing capacity of cement paste. Measurement **134**, 750–761 (2019)
101. W. Li, F. Qu, W. Dong et al., A comprehensive review on self-sensing graphene/cementitious composites: a pathway toward next-generation smart concrete. Constr. Build. Mater. **331**, 127284 (2022)
102. Y. Wang, L. Zhang, B. Han et al., Advances in self-sensing cement-based composites containing nano materials for smart civil infrastructures. Measurement **230**, 114514 (2024)
103. P. Xie, P. Gu, J.J. Beaudoin, Electrical percolation phenomena in cement composites containing conductive fibres. J. Mater. Sci. **31**, 4093–4097 (1996)
104. M. Yaghobian, G. Whittleston, A critical review of carbon nanomaterials applied in cementitious composites—a focus on mechanical properties and dispersion techniques. Alexandria Eng. J. **61**, 3417–3433 (2022)

Part III
Lifecycles and Sustainability of Materials in the Built Environment

The sustainability of materials production and their implementation in the built environment is a complex topic to grapple with and there is no single right way to conduct an LCA for a system or material. Implementing more sustainable approaches towards materials cannot be done on gut-feeling alone and requires meaningful insights into the real-world consequences of procuring and implementing the alternative materials discussed in previous chapters of this book. Design for sustainability in the built environment requires depth and breadth of knowledge. Understanding the thermal, mechanical, and longevity performance of earth- and bio-based materials and gaining insights into the complexity of processing and applying these materials at the large scales encountered in the construction industry is instrumental towards bringing about positive change in this conservative sector.

The chapters in this part of the book *Sustainable Materials for the Built Environment* confront the complexity of material lifecycles, interrogating the metrics, and methodologies that define "sustainable" practices in construction. From extraction to demolition, each link in the construction value chain including processing, transportation, application, maintenance, and end-of-life, brings about cascading consequences for carbon budgets, biodiversity, and human well-being. Sustainability is regarded here not as a static destination but as a dynamic interface between society and nature. Designing more sustainable practices in the built environment requires methodical data-driven analysis, transparent accountability, and a willingness to dig deeper under the surface, reveal new information, and confront trade-offs. Scrutinies of materials performance across multiple indicators are presented here alongside approaches to conducting LCAs and an examination of frameworks required to foment sustainable practices, revealing new insights into the design and selection of sustainable materials.

Through case studies, materials reviews, and the examination of LCA frameworks, this part of the book reveals how innovators are redefining lifecycle thinking. Importantly, these studies demonstrate that using materials to create more sustainable built environments requires cross-disciplinary collaboration. Architects may seek to partner with ecologists or agriculturists to design buildings using sustainably sourced bio-based materials in a way that sequesters carbon and supports biodiversity. Engineers may work alongside sociologists or geologiests to ensure low-carbon materials align with cultural preferences, needs, and local availability of materials. Professionals in policy, data-science, and systems engineering meanwhile promote standardized LCA frameworks that help create transparency and catalyse effective decision making processes throughout the value chain.

The chapters in Part III showcase the multi-faceted and interconnected nature of materials and their use in construction, inspiring new ways of designing for sustainability. These chapters show that sustainability is not a constraint but a catalyst for creativity. By embracing complexity, confronting trade-offs, and prioritizing systems over silos, the built environment can evolve from a driver of planetary degradation to a pillar of regeneration. The path forward is neither simple nor singular, but as the following pages demonstrate, it is rich with possibility.

Dorian A. H. Hanaor

Chapter 21
Engineered Wood Products (EWP)—Resource Efficiency and Environmental Performance

Maximilian Pramreiter, Johannes Konnerth, and Paul Mayencourt

Abstract Today, the construction sector has access to a vast array of resources, ranging from non-renewable materials like steel and concrete to natural materials like stone and timber. These materials are either used alone or in combination as part of composites. Driven by the need to reduce the environmental impact of the built environment, Engineered Wood Products (EWPs) like Glued Laminated Timber (GLT) or Cross Laminated Timber (CLT) have gained increasing interest in the construction sector in recent years. This chapter explores the resource efficiency and environmental performance of these EWPs, by examining established literature on production processes, material yields, and life cycle assessments. The chapter highlights the role of different wood species, processing methods, and mechanical as well as environmental performances. It also discusses circularity and potential areas for future research to enhance the overall sustainability of the wood value chain. This comparative approach aims to support material selection decisions and foster more responsible use of forest resources in the built environment.

M. Pramreiter (✉) · J. Konnerth
Department of Natural Sciences and Sustainable Resources, Institute of Wood Technology and Renewable Materials, University of Natural Resources and Life Sciences Vienna, Austria (BOKU University), A-3430 Tulln a.d. Donau, Austria
e-mail: maxmilian.pramreiter@boku.ac.at

M. Pramreiter · P. Mayencourt
Department of Architecture, College of Environmental Design, University of California Berkeley, Berkeley, CA, USA

Department of Environmental Science, Policy, and Management, Rausser College of Natural Resources, University of California Berkeley, Berkeley, CA, USA

D. A. H. Hanaor (ed.), *Sustainable Materials for the Built Environment*, Springer Series in Materials Science 352, https://doi.org/10.1007/978-3-031-97818-0_21

21.1 Introduction

21.1.1 Is There Enough Wood for Construction?

The built environment fulfils a multitude of functions in our modern society. While in the Stone Age it primarily served as a shelter and mankind spent most of its time in nature, today we spend a large part of our day in or around man-made structures. Similarly, the materials we use to create these environments have significantly changed and expanded throughout the course of history. Today, the construction sector has access to a vast array of resources, ranging from non-renewable materials like steel and concrete to natural materials like stones and timber. These materials are either used alone or in combination in the form of composites and are either load-bearing or non-load bearing. One of these natural materials, wood, is currently experiencing a renaissance in construction.

This renaissance is fuelled by the need to create a sustainable living space for our society while at the same time tackling climate change [1, 2]. Wood as building material offers not only the capability to build high rise and large volume structures but it also sequesters carbon while being used as material. It is therefore no surprise that a race for the tallest wooden skyscraper has ensued in recent years, with height records sometimes being broken only by one metre to the previous champion (e.g. the Ascent Tower in the USA at 87 m, followed by the Mjøstårnet Building in Norway at 85 m, the HoHo Tower in Austria at 84 m as seen in Fig. 21.1b and the Sara Kulturhus Centre in Sweden at 75 m). It will not stop here, future projects like the "Rocket & Tigerli" in Switzerland (100 m, targeted completion in 2027) or the "C6" in Australia (planned with 183 m) in Australia are already in the pipeline. Considering humans' past, it is not the first time that wood is considered to be an excellent building material. Recent discoveries by Barham et al. [3] assume that wood was used as structural material already 476,000 years ago (see Fig. 21.1a).

However, compared to this distant past, the current material need for construction is on a totally different level [4] and when society talks about (partially) substituting steel and concrete with wood, the following question arises: "Is there enough wood for construction and how should we build with wood to preserve our forests?"

21.1.2 The Global Wood Value Chain

In an attempt to answer this question, it is necessary to take a look at the globally available woody biomass from forests, the current areas of use and the potential demand for building products in the future.

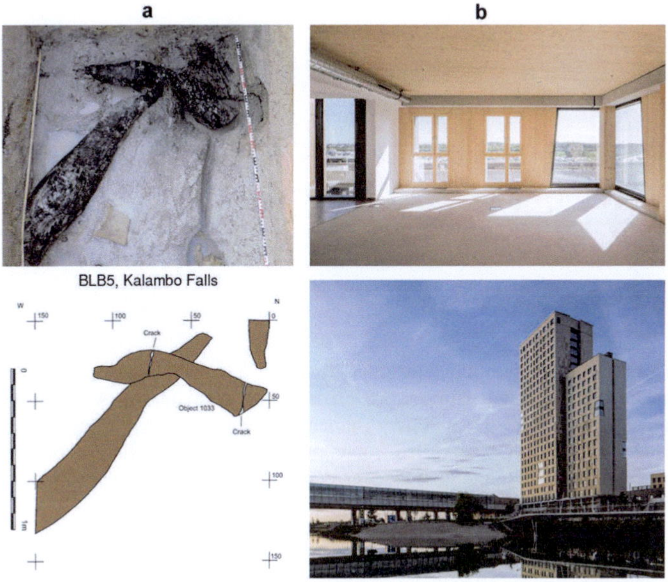

Fig. 21.1 Wood as oldest building material and current shooting star in architecture and design. **a** Earliest evidence of structural use of wood in archaeological records, dating back at least 476,000 years (adopted with permission from [3], copyright Prof. Larry Barham)). **b** HoHo in Vienna, high-rise flagship project using mass timber, completed in 2021 (with permission from KiTO photography https://www.kito.at/, copyright by Michael Baumgartner/KiTO)

21.1.2.1 Current Material Flows and Areas of Use

According to Bais et al. [5] roughly 158 megatons of carbon [MtC] of the annually extracted woody biomass (1936 MtC) end up in wood-based products (part of HWP in Fig. 21.2). This represents roughly 13% of the annually socioeconomically used biomass of the world's forests (1219 MtC, "Used extraction" plus "Recovered paper"), with the remaining 87% divided between wood energy use (48%), paper and paperboards (22%), losses (8%), wood waste (6%) and poles as well as posts among other niche products (3%).

Based on these current wood carbon flows (see Fig. 21.2), a significant utilisation potential for wood in construction can be identified, as almost 56% of the annually used forest biomass (1050 MtC) is used as fuel wood. A portion of these resources could be used for construction products with higher added value and longer carbon sequestration periods, while remaining available for further material use after re-use or recycling [8]. The material can then still be used as fuel at the end of its use cycles. However, it should be noted that there are considerable regional differences. For example, looking at the findings of Mantau [9] for wood flows in Europe (EU 27), only 36% of the annually used forest biomass is directly put into fuel wood.

In addition, there are still available reserves in our forests that could be sustainably harvested for the production of building materials. According to Bais et al. [5]

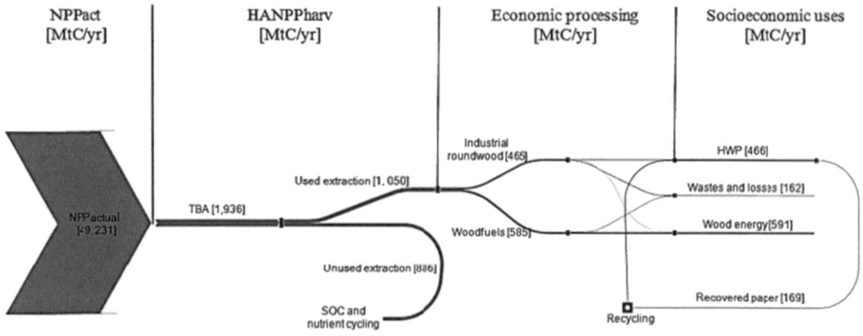

Fig. 21.2 Global wood carbon flows [MtC/yr] from net primary production (NPP_{actual}) to final consumption in 2010 (adopted with permission from [5], copyright by Anna Liza Bais). TBA (total biomass appropriation), HANPPharv (human appropriated net primary production), SOC (soil organic carbon) and HWP (harvested wood products). To estimate [m^3/yr] divide numbers by 0.5 (carbon to wood ratio [6]) and then by density of assumed wood species mix (e.g. 0.650 t/m^3 average density according to [7])

roughly 30% of the annual net primary production (currently 4% of 49,200 MtC/yr, globally) could be harvested without impairing the health of the forests. However, regional differences must also be considered here, as parts of the world have already exceeded this threshold [5]. However, in the European Union (EU 27) for example, only 75.7% of the annually available forest biomass is utilised [9]. A similar potential was recently reported by the European Commission [10]. In addition, the availability and accessibility of forest area for wood supply is also essential to mobilize these potential reserves in the future [11, 12]. Considering accessibility, it is worth mentioning that countries with large cities not necessarily also have access to vast amounts of forest resources (see Fig. 21.3). Therefore, the utilisation of wood might not be the best solution everywhere on the planet yet, due to increasing transport distances and potential exploitation of foreign forest resources without proper regulations and adequate forest management practices.

To sum up the supply side for wood in construction it can be said, that there is still potential for raw materials from the world's forests and not mentioned so far, forest plantations. However, sustainable forest management practices and accurate resource monitoring will be an absolute necessity in order to prevent overuse in the future [15].

21.1.2.2 Potential Demand for Construction

Taking a look on the demand side, projecting future growth for wood in construction is even more complex than the available forest resources. The various methods of collecting data on the current consumption and flows of building materials are already a much-researched topic [16]. However, future shares of different materials and their demand are often not even considered at this stage. Furthermore, it is also difficult

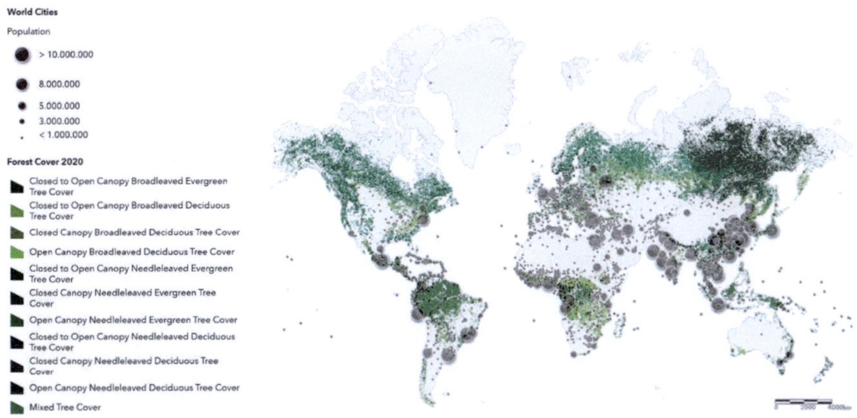

Fig. 21.3 The world's largest cities and global forest cover in 2020. (Created with ArcGIS online, copyright by ESRI [13, 14])

to provide a global perspective, as regional customs, policies and regulations play an important role when it comes to the demand side of the construction sector [17–21]. Nevertheless, a possible approach estimating future demand for wood in construction was recently provided by Churkina et al. [1]. They investigated different transition scenarios ("business as usual", "10% timber", "50% timber" and "90% timber") to provide a sustainable living space for the additional 2.3 billion urban residences until 2050 by incrementally substituting a steel- and concrete-based construction method with a timber-based building system. According to their findings, the additional wood demand could range from 20 MtC/yr in the "10% timber" scenario up to 1360 MtC/yr in the "90% timber" scenario, which would outstrip the current total roundwood supply reported by the FAO [22, 23]. This wide range of potential demand shows the high uncertainties linked to the different visions for the future of the construction sector and underlines the unpredictability that the forest-based industries have to deal with. Especially considering the wide range of EWPs and possible building systems they are used for. However, besides a singular focus on mass timber construction as assumed by Churkina et al. [1], a wide range of possibly more resource-efficient systems or combinations of them could be employed to create the future living space. For example, timber frame construction stands out in particular as a possible complementary construction method with potentially lower raw material consumption compared to systems based only on mass timber.

Paired with the push to also transition the energy sector to a more sustainable mix of energy sources it becomes clear that in order to prepare for this future market conditions the resource efficiency and environmental performance of processes and products needs to be part of the discussion. This will help provide the necessary baseline in order to support policymakers with the sustainably distribution of the available forest-based raw materials.

21.1.3 Scope of This Chapter

This chapter aims to present currently available Engineered Wood Products (EWPs) for structural applications (see Fig. 21.4) and summarizes the current state of the art in regard to their resource efficiencies as well as production processes (Sect. 21.2). The overview is complemented by a concise discussion of their mechanical (Sect. 21.3) and environmental (Sect. 21.4) performance.

These aspects of EWPs have recently been reviewed in depth by Pramreiter et al. [24]. This publication serves as the foundation for this chapter and is supplemented by a more detailed look into the by-products, sometimes referred to as residues or scrap, during the production processes. Furthermore, six additional EWPs are added to the comparison. It should be mentioned that a detailed description of the different products and production steps has been published multiple times up to this day and can be found most recently in Niemz et al. [25] and also in Dunky et al. [26], Ansell [27], Walker [28], Wagenführ [29], Williamson [30] and the Forest Products Laboratory [31]. These references serve as a baseline for the depicted production steps in Sect. 21.2.

It should be noted here that the chapter aims to present the products without the goal of promoting one over the other. The comparisons should rather give insight into the variety of available products and provide base-values for material selection. It is therefore not the aim of the authors to rank the available products.

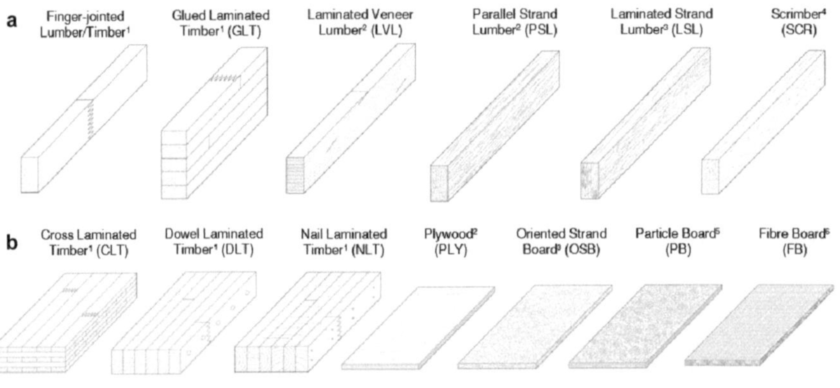

Fig. 21.4 Overview of established engineered wood products (EWP) for structural applications based on the used intermediary ([1]Lumber or Sawn timber, [2]Peeled veneers, [3]Strands, [4]Scrim-mats, [5]Particles, [6]Fibres). **a** Typically beam-shaped. **b** Typically panel-shaped. (Not to scale, drawings created by Frederic Lam)

21.2 Resource Efficiency

The first section of this chapter gives insight into the utilized raw materials and the resource efficiency from the first breakdown (the first disintegration of roundwood into intermediaries) to the final material yield of the finished EWP. Furthermore, it presents the most common occurrences of by-products during manufacturing. The statements are taken from the literature (see Sect. 21.1.3) and expert interviews in the field.

21.2.1 Definition of Resource Efficiency

Resource (process) efficiency, also referred to as material or product yield can be calculated based on the following equation:

$$\text{Resource efficiency or material yield} = \frac{\text{Intermediaries or material output}}{\text{Roundwoodor intermediary input}} \times 100[\%] \quad (21.1)$$

It expresses the ratio between the output of intermediaries or final material and the input of roundwood or intermediaries. The remaining % to 100% is typically by-product which still can be used in other wood processing streams and end up in different materials. This is an important merit of the wood processing industry. Basically, nothing of the input raw material is wasted in a classical sense but rather transferred to other process streams of typically lower added value compared to structural building products [32]. Furthermore, there are few additional factors that need to be considered when using this metric to compare products and their efficiency in utilizing raw material.

In regards to first roundwood breakdown into different intermediaries, the measuring principle (volume or weight), the measuring standard (different log scales) as well as the consideration of the bark content dictates how one unit of roundwood input is actually calculated. This can lead to significant differences in reported vs. actual roundwood input and significantly influences the reported process efficiencies [33, 34]. According to a recent study by the Food and Agriculture Organization of the United Nations (FAO) [35] different national log scales can cause a deviation from true log volume as high as 32.3% (8.4% on average). Furthermore, when converting 1000 Board Feet (bd. ft.), the traditional unit for sawmill processes in North America, the resulting volume in m^3 can range from 3.87 m^3 up to 17.23 m^3 depending on the log scale and small end diameter [35].

Additionally, the subsequent processing of intermediaries into finished EWPs, the unit (weight or volume) in which the throughput is monitored needs to be considered as well. While lumber-(which can also be referred to as sawn timber) and veneer-based processes are usually controlled only using volume (e.g. m^3 or bd. ft.), the strand-, particle- and fibre-based processes are monitored using mostly weight (e.g. bone-dry tons also referred to as ATRO) and to some extend also volume. This difference is especially critical when considering the different drying steps to achieve a target moisture content and the compaction (i.e. densification) during the pressing steps. The change in moisture content leads to a change in dimension (volume shrinkage, values for volume shrinkage of selected wood species can be found in literature [31, 36]) and weight (water extraction), while the compaction further influences the dimension (reduced cross-section). Both steps influence the actual density [kg/m^3] of the material, which is the driving parameter when converting output material into different units. The accurate determination of these factors is therefore key when comparing these materials.

21.2.2 Raw Material

The predominantly used raw material for the production of EWPs consists of different assortments of roundwood (typically without bark) from forests and, in the case of Particle Boards (PB) and to some extent Fibre Boards (FB), industry by-products (e.g. chips and sawdust) as well as recovered wood from consumers [25]. Depending on the required properties of the final product and the technological processing conditions, relevant quality factors of the utilised roundwood include:

- Wood species (coniferous or deciduous, also referred to as softwood or hardwood, see also Fig. 21.5a)
- Dimension (mainly diameter, length and volume)

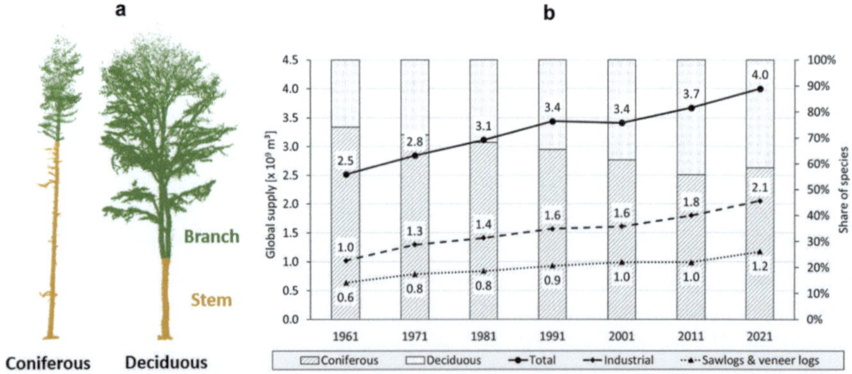

Fig. 21.5 Overview of global roundwood used for the production of engineered wood products (EWP). **a** Typical tree morphology of coniferous and deciduous trees (adopted with permission, copyright by Andreas Tockner, Institute of Forest Growth, University of Natural Resources and Life Sciences, Vienna). **b** Development of global total roundwood supply from 1961 to 2021 as reported by the FAO [22] and amount of industrially used roundwood (incl. the volume of sawlogs & veneer logs) for the production of EWPs

- Growth characteristics (the shape and the quality of the roundwood)
- Biological condition of the tree (is the wood in a healthy condition, e.g. fungal decay or other forms of biological attack).

For the sake of simplicity, the definitions laid out by the FAO [37] for different raw materials of the forest-based value chain will be used in the subsequent comparison of EWPs. More specifically:

- "Roundwood" (roundwood that is removed from the forest),
- "Wood fuel" (roundwood that will be used for the purposes of power production).
- "Industrial roundwood" (all roundwood removed for utilisation, except wood fuel).
- "Sawlogs and veneer logs" (roundwood that will be sawn lengthways or used for the production of peeled or sliced veneers). The terms "sawlogs" and "veneer logs" will also be used on their own.

Furthermore, general terms and definitions for roundwood can be found in ISO 24294 [38] and EN 844 [39], while the measurement of dimensions and volume calculation is standardised in ISO 13059 [40] and further described for Europe in EN 1309-2 [41]. Definitions in regards to growth characteristics (quality) can be found in ISO 24294 as well as in EN 1927 [42–44] for coniferous and in EN 1316 [45, 46] for deciduous species. Definitions concerning the biological condition of the tree are standardised in EN 1309-3 [47]. The previously mentioned quality aspects are to a certain extend already assessed in the forest, before the tree is harvested in order to decide on the target process and potential part of the industry sector the roundwood will be sold to. In that regard it needs to be mentioned that in some instances the market condition may favour certain process streams over others even if the roundwood would fulfil higher quality requirements. In some cases, this can go so far that individual production sites procure all the available roundwood in a region without pre-assessing the quality of the acquired material.

According to the FAO [22, 37], the global roundwood supply has almost constantly increased since 1961 (the earliest recording) from 2.5×10^9 m^3 to 4.0×10^9 m^3 in 2021 (see Fig. 21.5b). Similarly, the volume of industrial roundwood has increased from 1.0×10^9 m^3 in 1961 up to 2.1×10^9 m^3 in 2021. About 1.2×10^9 m^3 of "sawlogs & veneer logs" were harvested in 2021, which accounts for 57% of the total industrial roundwood supply and represents a slight decrease compared to the share of 63% (0.6×10^9 m^3) in 1961. This decrease in relative importance is partly caused by the development of particle-based EWPs like Laminated Strand Lumber (LSL) and Oriented Strand Board (OSB) in the late 1960s and the subsequent installation of production capacities in Europe and North America from the 1980s onwards, which led to the substitution of PLY and other EWPs in some regions [28]. However, from a structural perspective, sawn timber-based EWPs are still dominating the wood-based construction sector. This starts with strength-graded and sometimes finger-jointed timber for timber-frame construction at the lower end of added value, all the way up to prefabricated Cross-Laminated Timber (CLT) wall- and floor-elements, currently offering the highest added value for the wood industry. CLT has seen significant increases in market share in Europe since 2010 [48, 49] and is currently fuelling a hype of so-called mass timber construction in North America which started in 2018 [50]. The roundwood used for these products predominantly comprises of coniferous wood species like Spruce, Pine or Fir and usually requires a uniform/straight trunk shape with little taper and only tolerates small knots (see Fig. 21.5a). However, according to recent studies by Hanewinkel et al. [51] and Thuiller et al. [52] climate change will also impact the species mix of forests in the future, with an expected increasing share of climate-resilient deciduous trees at the cost of currently dominating conifers. This trend is already visible in the supply of industrial roundwood (see Fig. 21.5b), as the share of coniferous wood has decreased from 74% in 1961 to 58% in 2021, while the share of deciduous wood has increased from 26 to 42%. However, this past shift is probably caused by an increase in industrial roundwood used in the pulp and paper industry as well as other industrial roundwood used for poles, posts, fencing, pitprops, shingles and shakes, wood wool, tanning, distillation, among others. In these areas the share of deciduous species has continuously increased while the supply of conifers has decreased [22]. Nevertheless, this shifting species mix will lead to an increase of, currently considered low-quality, non-sawable assortments. This is due to the fact, that deciduous trees grow more heterogenous compared to conifers, with a higher share of branches and lower share of high-quality stem wood (see Fig. 21.5a) [11, 53]. This shift will require a more diverse combination of processes in order to efficiently cover the potential demand for structural EWPs [2]. It also influences other technological factors like adhesive technology, mechanical performance and dimensional stability.

21.2.3 First Breakdown and Intermediaries

As mentioned in the previous section, the predominant raw material source for EWPs is roundwood (without bark) from coniferous or deciduous tree stems (see also Fig. 21.5). An exception to this is recovered wood in the PB production. Similarly, by-products from other first breakdown processes like wood chips, sawdust or fines can be seen as raw material for the PB and in the case of wood chips also for the FB production.

Figure 21.6 summarizes the breakdown steps leading to different EWPs with their intermediaries in a simplified way. The raw material is put through first breakdown by either sawmilling, veneer peeling, stranding, scrimming or chipping. In some cases, a secondary breakdown will occur if the required size of the intermediary needs to be further reduced. This can include for example clipping for PSL, flaking for PB or refining for FB production. Steps like planing, finger-jointing or sanding should not be considered as part of a secondary breakdown, as it does not change the size of the material in a significant way. A concise discussion of the typical sizes of the different intermediaries and the dimensional freedom of the different EWPs can be found in Sect. 21.2.4.

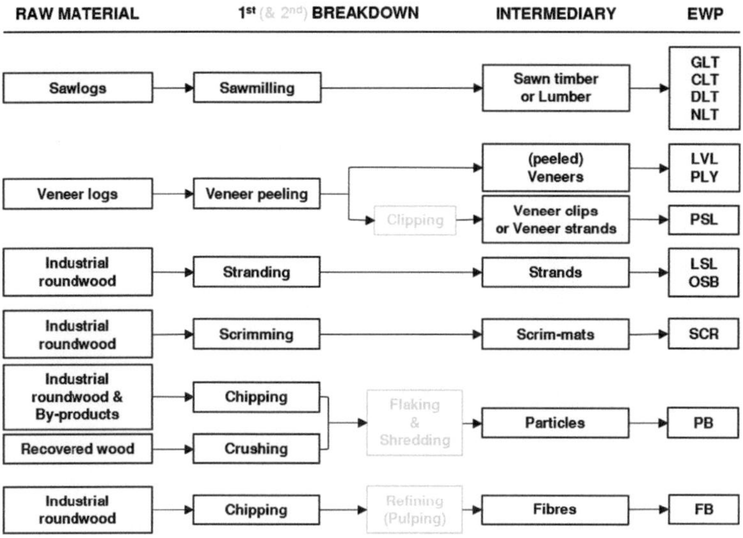

Fig. 21.6 Simplified summary of utilized raw material, first and secondary breakdown, intermediaries and finished EWPs (based on [25, 37])

This first roundwood breakdown is one of the most dominating factors when it comes to overall resource efficiency. The first breakdown gives the basic dimension of the intermediary and dictates a certain range of resource efficiency due to the underlying technology. Currently, only the sawmilling and veneer peeling (also slicing, for non-structural applications) processes create an intermediary that is also sold as its own product. The production of the strand-, particle- and fibre-based EWPs is normally a fully finished process and the intermediaries are usually not marketed. This is due to the fact that the sawmill and veneer industry has a very long tradition and has historically become more and more integrated by implementing additional value-adding processes over the years. In comparison, the other industries have mainly developed from the need to either utilize a certain raw material or substitute an existing product (e.g. OSB as substitute for PLY). For this reason, there is a lot of literature on the resource efficiency of lumber and veneer production and relatively little on the first breakdown of the other EWPs.

Figure 21.7 summarizes the values found in literature on the process efficiency of sawmilling and veneer peeling and the subsequent process steps to arrive at either dried and planed lumber or dried and clipped veneer sheets. A detailed list of literature dealing with the process efficiency of first-round wood breakdown can be found in the appendix of Pramreiter et al. [24].

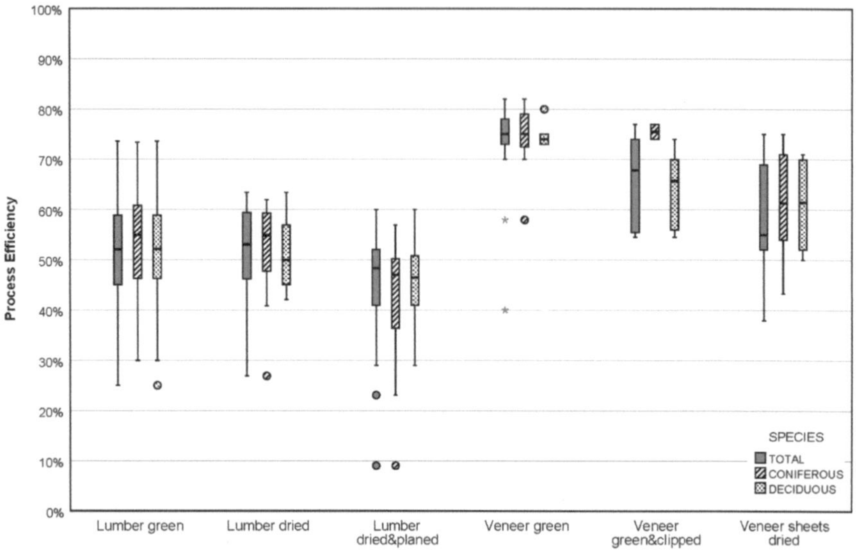

Fig. 21.7 Process efficiency [%] of the first roundwood breakdown and subsequent by-products until the final intermediary material. The base input material is roundwood (without bark), with the input unit being either cubic volume or weight. (Based on [24])

21.2.3.1 Sawmilling

The average total green lumber yield related to sawlogs under bark ranges from 25% up to 74% with an average of 52%. In general, sawmilling efficiency increased due to improved process technologies over the past decade's [54–57]. Nevertheless, various factors influencing sawmilling efficiency are summarized in Table 21.1.

Market characteristics (e.g. price of lumber vs. price of by-products) influence sawmilling efficiency by conscious decisions against or in favour of an improved lumber yield [58, 59]. In contrast, mill, roundwood, process and lumber characteristics are predominantly discussed to improve lumber yield. Roundwood characteristics, diameter, length, taper, curvature and local decay are the most important factors influencing process efficiency. Based on findings in the literature [33, 58, 63–65] lumber yield increases with increasing roundwood diameter, while yield generally decreases with increasing length, taper and curvature. To compensate for differing roundwood shapes and sizes, process characteristics like sawing patterns (opening faces) [33, 58, 68, 73], as well as rotation and positioning [74, 75] of the roundwood are usually a direct answer to raw material characteristics to maximize yield on a log-to-log basis. Additionally, general equipment properties like machine set-up [61, 63] and sawing kerf width [33, 58, 63] along with maintenance schedule [58] also influence the process efficiency. However, the ongoing improvements in 3D X-ray scanner technology have enabled the industry to obtain information about the internal

Table 21.1 Factors influencing sawmilling efficiency

Factor	Factors	References
Market	Product mix, prices, demand and price for by-products, round wood supply	[33, 56, 58, 59]
Mill	Production capacity, equipment type, age of the machinery, decisions making	[54–56, 58, 60, 61]
Roundwood	Species, origin (native or plantation forest), diameter, length, taper, curvature, knottiness, position of crosscut	[33, 54–56, 58, 62–71]
Process	Cutting pattern, accuracy, saw kerf, rotation and positioning, curve sawing capabilities, condition and maintenance of equipment	[56, 58, 59, 63, 68, 72–75]
Lumber	Species, target size (nominal vs. actual), moisture content (drying), visual characteristics (twist, bow, knots, ring width, checks, wane), applied sorting criteria	[33, 58, 59, 67, 70, 76–83]

structure of the raw material and have led to a shift towards value-focused processing rather than yield-based [84].

In order the discuss the subsequent processing steps of green lumber into dried & planed lumber, Fig. 21.8 gives an overview of typical sawmilling operations and indicates where by-products occur.

Literature reports an average total dried & planed lumber yield from 9% up to 60% with an average of 46%. The lowest reported dried & planned lumber yield of 9% [85] is based on an assumed distribution of process efficiencies and very likely reflects only a theoretical worst-case scenario. Similar to round wood volume,

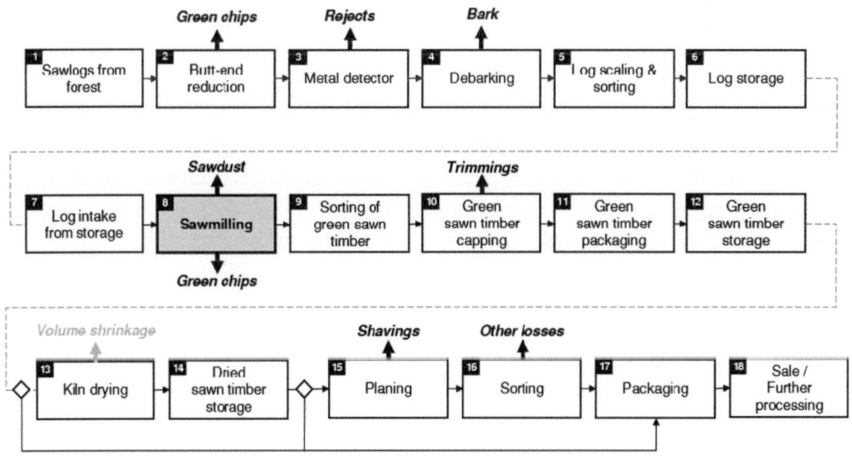

Fig. 21.8 Typical operations and by-products during sawmilling. Sawn timber can also be replaced with lumber

the methods of measuring lumber volume influence the reported process efficiency. Most commonly, actual dimensions of lumber are used to determine the cubic volume (length × width × height). One exception can be found in North America, where nominal dimensions are used to calculate board feet (bd. ft.) of lumber [33, 35, 54]. One bd. ft. is defined as a 1-inch (2.54 cm) thick and 12 by 12 in. (30.48 cm) long board, which would lead to 1 m^3 being 424 bd. ft. [33]. However, as these nominal sizes sometimes ignore dimensional changes due to drying and planing of the lumber actual volume could be as low as 0.66 m^3 [35]. Besides these variabilities, drying (6% average drying loss [35, 59, 64, 78, 85–90]), planing (13% average planning loss [35, 59, 64, 78, 85–93] and sorting [59, 78–83] significantly decreases the final intermediary yield of the sawmilling process (also see Fig. 21.8).

21.2.3.2 Veneer Peeling

The total green veneer yield ranges from 40% up to 82% with an average of 72%, showing lower variability compared to sawmilling (see Fig. 21.7). Similarly, technological advancements increased veneer recovery over the last decade significantly. The chipless peeling of veneers leads to a generally higher material yield compared to lumber from sawmilling and similarly veneer yield increases with increasing small end diameter. However, this requires uniform raw material of preferably high diameter and high quality. In addition, the machine configuration also plays an important role. Currently, spindleless lathe technology is the state of the art, enabling the log to be peeled down to a residual diameter of about 25 mm. In comparison, older literature discusses suitable core diameters to be around 80 mm and above, also depending on the potential utilisation of the peeler cores for other products (e.g. poles for fencing) [33]. After peeling, green veneers are usually clipped and dried to target moisture content as depicted in Fig. 21.9.

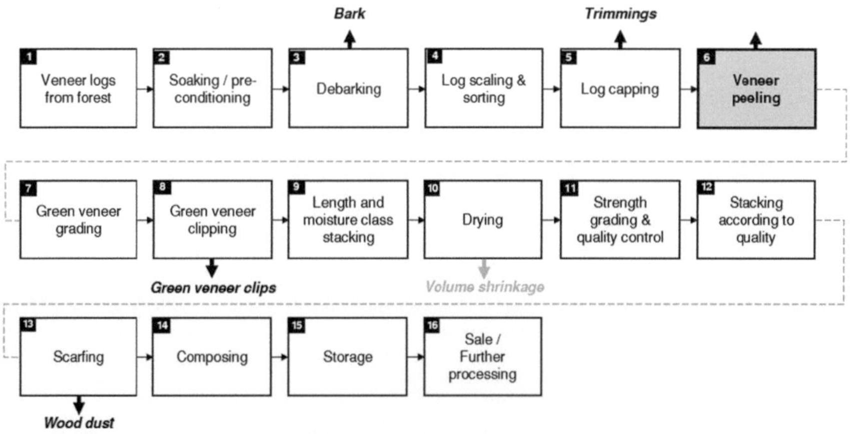

Fig. 21.9 Typical operations and by-products during veneer peeling

As depicted in Fig. 21.7 total yield of veneer sheets dried ranges from 38% up to 77% with an average of 57% showing higher variability (Fig. 21.3) compared to green veneers. As with the sawing process, there are also different ways in which the veneer volume is determined. The cubic volume is either measured based on actual dimensions (length × width × thickness) or based on the surface area on an equivalent thickness basis [35]. For example, a veneer with 1.25 m^2 and a 1 mm thickness would equal 0.00125 m^3, including a small plus or minus thickness tolerance [33]. As both methods are based on actual dimensions the variability in reported veneer yield should be negligible. More importantly, the loss due to clipping is reported [33, 64, 94, 95] to be the most severe factor in decreasing veneer yield. According to findings by Woodfin [94] as well as Fahey and Woodfin [64] roughly 20% of material is turned into by-products, whereby the clipped volume (mainly cracked and defective areas) is significantly influenced by the applied clipping strategy [96]. Final veneer drying further reduces the overall yield, with the target moisture content ranging from as low as 2% up to 15% [28]. The lower target moisture content is usually dictated by the target size [33] and possible bonding conditions for subsequent Laminated Veneer Lumber (LVL) and Plywood (PLY) production [26].

21.2.3.3 Influence of Wood Species on the First Breakdown

Based on Fig. 21.7, the influence of wood type (coniferous vs. deciduous) is negligible for the first roundwood breakdown. This underlines the greater influence of the principal technology on the level of achievable resource efficiency as compared to the processed wood species. Considering sawmilling and veneer peeling, a key factor in process efficiency [58, 97] is the optimization of mills to process either coniferous or deciduous roundwood. This emphasizes the importance of a plannable and steady supply of comparable roundwood assortments for these industries. Currently, the smaller share of deciduous wood species processed by the industry (see Fig. 21.5) is either subjected to a stricter degree of pre-sorting compared to softwoods, with the non-sawable and non-peelable assortments being either used for wood fuel or pulp and paper, or is directly felled for these low-added value sectors [22]. Therefore, the attainable roundwood quality of deciduous trees will be a key issue in the future [98, 99] and could significantly reduce the process efficiency of sawmilling [71] as well as veneer peeling [97]. Furthermore, a closer look at other process-related factors like productivity, energy demand, necessary pre-treatment and resulting intermediary quality is needed to get a full view of the effect of the changing raw material situation.

21.2.4 Typical Sizes of Intermediaries and Finished EWP

At this point, the typical sizes of the intermediaries and finished EWPs will be presented briefly in order to provide an overview of the achievable dimensions. To utilise different EWPs in construction, a certain dimensional freedom is required to support an array of different building types and designs, while also ensuring the necessary structural safety. Furthermore, the high degree of prefabrication in the wood-based construction industry also requires a certain flexibility when it comes to logistics. Based on currently available literature and building projects, Table 21.2 summarizes the typical sizes of the used intermediaries (sawn timber, peeled veneers, strands, scrim-mats, particles and fibres) and indicates the typical dimensional ranges for the finished EWPs.

Looking at the typically beam-shaped EWPs, Glued Laminated Timber (GLT) currently offers the highest flexibility in regards to the size of the final members. It ranges from standardized dimensions up to singles beams for timber engineering, not only covering great spans but also 3D-shaped constructions [121, 122]. Today, the construction industry can access a wide knowledge-base in regards to design and manufacturing but also in the field of connectors and logistics (on and off the construction site) when it comes to GLT. LVL, Parallel Strand Lumber (PSL) and Laminated Strand Lumber (LSL) on the other hand are predominantly used in standardised lengths. However, LVL recently gained interest for timber engineering due to the higher mechanical performance at smaller dimensions [123], while LSL and PSL are mainly used in North America with standardised dimensions. SCR is currently not produced on an industrial scale and it has been largely ignored since the last push to bring a product onto the market in 2003 [118]. However, the scrimber process has received renewed attention in recent years due to its ability to process small-diameter and low-quality roundwood [124].

Considering the panel-shaped EWPs, CLT is currently generating the most attention in the industry as well as research, which has significantly increased its dimensional flexibility while also offering multiple degrees of pre-fabrication. PLY, OSB, PB and FB are still mainly used with standardized dimensions and rather as one component of a construction element. However, companies started producing PLY [111] and OSB [116] as single load bearing element with bigger dimensions and design flexibility in the recent past. DLT and NLT are currently only manufactured to a limited extent with standardised dimensions [104, 105, 125]. However, they have a long history as structural EWP and offer an adhesive-free alternative to the established EWPs and are comparatively easy to manufacture and recycle.

21.2.5 Further Processing and Finished EWPs

After the first and secondary roundwood breakdown the intermediaries can be either stored for later use or directly processed into the finished EWPs. Figure 21.10 summarizes the values found in literature on the material yield of different EWPs.

Table 21.2 Typical sizes of intermediary material and engineered wood products (EWP)

Intermediary	Length (mm)	Width (mm)	Thickness (mm)	EWP	Length (m)	Width (m)	Thickness (m)	References
Sawn timber/Lumber	1000–6000	40–300	6–45	GLT	≤ 150	0.08–0.24	≤ 2.0	[25, 28, 100, 101]
			6–47	CLT	≤ 18–30	≤ 3.0–4.8	≤ 0.3–0.5	[100, 102, 103]
		≥ 80	≥ 20	DLT	≤ 20	0.305–4.3	0.09–0.4	[104–108]
		89–286	≤ 38–89	NLT	≤ 7.9	≤ 2.4	0.09–0.29	[109, 110]
(peeled) Veneers	800–2500	800–2500	≤ 7	LVL	≤ 25	≤ 2.5	0.02–0.62	[27, 28, 111, 112]
				PLY	1.2–14.6	1.2–3.7	0.01–0.31	[27, 28, 31, 111–113]
	≥ 600	≤ 19		PSL	≤ 20	≤ 0.28	≤ 0.48	[28, 31, 100, 112]
Strands	≤ 300	≤ 25	0.3–0.8 (≤ 2)	LSL	≤ 19	≤ 2.4	≤ 0.14	[26, 27, 114]
	50–150	4–15		OSB	≤ 20	≤ 2.8	0.006–0.25	[25, 26, 115, 116]
Scrim-mats	≥ 380	≤ 19	≤ 19	SCR	5.5	0.05	0.3	[117–120]
Particles	1–15	0.1–5	0.1–1	PB	≤ 5	≤ 2.5	0.006–0.04	[25, 26, 100, 115]
Fibres	1 -25	0.03–0.5		FB			0.001–0.03	

Glued laminated timber (GLT), cross laminated timber (CLT), dowel laminated timber (DLT), nail laminated timber (NLT), laminated veneer lumber (LVL), Plywood (PLY), parallel strand lumber (PSL), Scrimber (SCR), laminated strand lumber (LSL), oriented strand board (OSB), particle board (PB) and fibre board (FB)

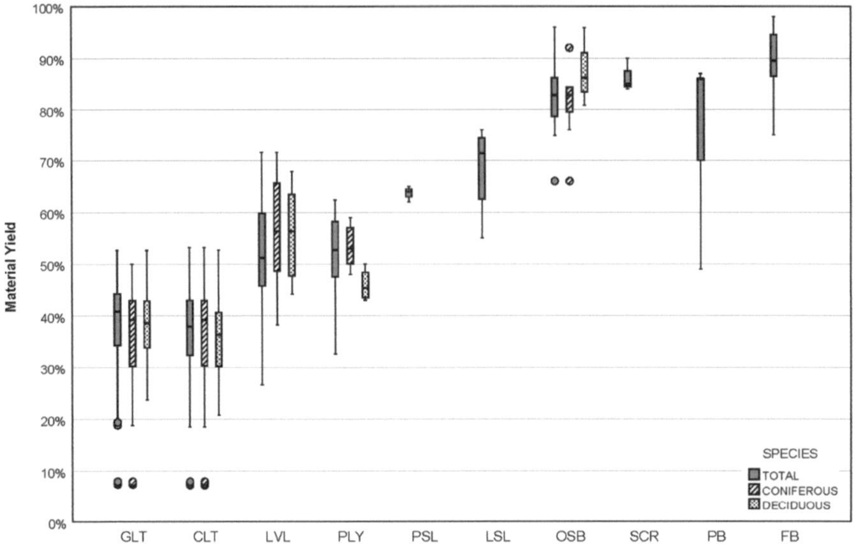

Fig. 21.10 Material yield [%] of established EWPs based on roundwood under bark (excluding bark). Yield for GLT, CLT is based on the minimum, maximum, and average process efficiency of lumber dried & planed (see Fig. 21.7) multiplied by the average further processing loss reported in the literature. Yield for LVL is based on literature and the minimum, maximum and average process efficiency of veneer dried & planed (see Fig. 21.7) multiplied by the average further processing loss reported in the literature. It can be assumed that DLT and NLT either lay in the same range or slightly above GLT and CLT (data taken from [24] and extended with [126–135])

21.2.5.1 Influence of Wood Species on Material Yield of EWPs

Depending on the EWP (see Fig. 21.10) no or too little literature is available to provide a general statement regarding the influence of wood species on the final material yield. Taking a look at the lumber-based products GLT and CLT, there was no literature on the processing losses of GLT made from deciduous species and the literature for CLT is primarily based on the same numerical simulation of low-grade yellow-poplar roundwood [76, 136] reporting a lower yield compared to conifers. In addition, the losses through pre-sorting and strength grading of the input lumber are reported to be significantly higher for deciduous [59, 78] than for coniferous species [79, 80]. In regards to the veneer-based EWPs, a similar statement can be made, with no literature available on deciduous LVL and little literature on PLY [35, 127], showing values significantly lower compared to conifers. In addition, total LVL and PLY includes literature [33, 55, 64, 137, 138] reporting a mixture of both coniferous as well as deciduous species, exhibiting a higher variability in yield compared to mere conifers. This indicates that the higher inhomogeneity of deciduous trees also influences the range of achievable material yields of LVL and PLY. In contrast to the lumber and veneer-based EWPs, Fig. 21.10 shows a different picture for OSB, with deciduous OSB having a higher material yield compared to OSB made from conifers.

In that regard, results from Feng and Knudson [139] show a significantly higher share of fines from green pine (15.4%) compared to green aspen (4.1%). This indicates, that the production of deciduous OSB could result in a generally higher material yield. Nevertheless, aspen can be considered a low-density deciduous species [140] with technological properties similar to conifers [141]. Therefore, this statement should not be transferred to other deciduous species like oak or beech.

In general, most of the reported values either stem from a process that was optimized for the utilized tree species or are the result of an optimization to utilise a new species. This wide array of investigations underlines the necessity to approach potential changes in the raw material basket with an open eye in regards to the potential processing technologies.

21.2.5.2 Glued Laminated Timber (GLT) and Cross Laminated Timber (CLT)

Taking a look at GLT and CLT, comparable efficiencies can be observed, with total GLT yield (GLT dimensional or custom in Fig. 21.11) ranging from 7% up to 53% and an average of 39% and CLT (CLT master-panel in Fig. 21.11) achieving an average yield of 37% ranging from 7–53%. Overall, the further processing losses of GLT and CLT are similar within different literature sources. This is a result of both EWPs having similar processing steps and by-products during manufacturing (see Fig. 21.11).

For total GLT [142–145], the average losses during production amount to about 16% in a range of 12–18% based on timber input. In CLT production [85, 93, 145–150], roughly 16% of input material is transformed into by-products, ranging from

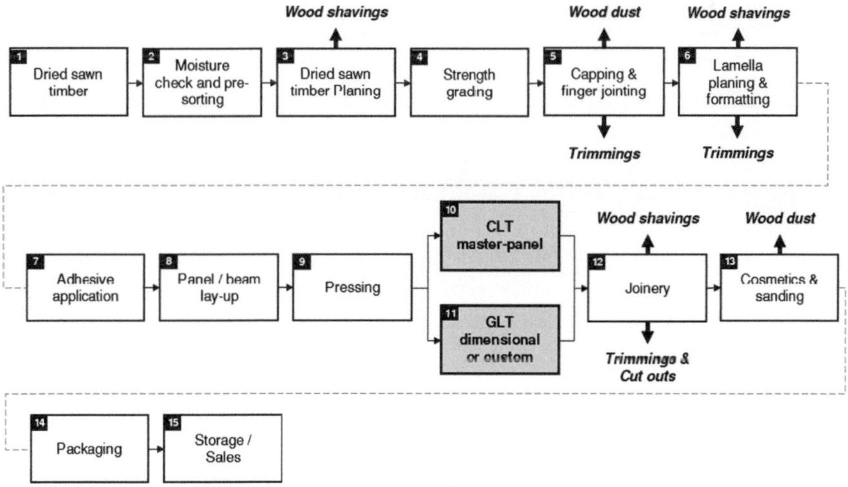

Fig. 21.11 Typical operations and by-products during the manufacturing of GLT and CLT

12% up to 20%. From this, it can be argued that the main reason for the high variability in yield can be found in the base efficiency of the input timber. In both cases the main steps causing additional losses are lumber trimming and finger jointing, planing of the lamellas and joinery 1 in the case of CLT and cosmetics & sanding in the case of GLT [27, 102]. However, it is important to note that especially planer shavings, which are a big loss factor [57, 85], are probably counted double in some cases for GLT as well as CLT as the process may also be started with rough-sawn lumber [147]. For CLT, this is countered by the fact that the losses caused by the joinery 2 (e.g. cut-outs for windows and doors) are not yet included. Depending on the end-use of the panel this could further bring down the final yield significantly. While floor panels do not require many cut-outs, outer and inner walls require openings for windows or doors, with the total area of these openings mainly driven by design aspects as well as building physics (lighting and ventilation) [151, 152].

21.2.5.3 Dowel Laminated Timber (DLT) and Nail Laminated Timber (NLT)

Currently, no publicly available literature deals with the resource efficiency of DLT or NLT. Both products are manufactured from dimensional lumber and instead of an adhesive either wooden dowels (DLT) or nails (NLT) are used to form the finished EWP (see Fig. 21.4). Depending on the producer the typical processing operations (Fig. 21.12) start, similar to GLT and CLT, with dried sawn timber which is in some cases finger jointed before a planing and profiling step to create the final raw lamella. Therefore, similar by-products to GLT and CLT occur. In the case of DLT, the lamellas are then laid up to either a beam or a panel (similar to CLT) and holes are drilled for subsequent dowel insertion. This creates further wood shavings as a by-product. In the case of NLT, the lamellas are directly laid up and nailed together in a repeating pattern. The resulting beam or panel is then further processed in a joiner. However, in the case of NLT, the options for subsequent joining are limited due to the nails within the material. The final EWP can also move through a final cosmetics and sanding step before being either stored or sold directly.

Based on these similarities with GLT and CLT it can be assumed that the overall material yield of DLT and NLT also falls in the same range. More specifically, the material yield tends to be higher compared to CLT and GLT due to the lower requirements on the surface quality of the lamellae. Another potential advantage of DLT and NLT lies in the potential recyclability of beams and panels at the end of their life-cycle due the absence of hard to dissolve adhesive bonds.

21.2.5.4 Laminated Veneer Lumber (LVL) and Plywood (PLY)

Taking a look at the veneer-based products, LVL and PLY achieve higher average yields compared to the lumber-based EWPs. Total LVL yield ranges from 27 to 72% with an average of 52% and an average further processing loss of 9% [137, 153,

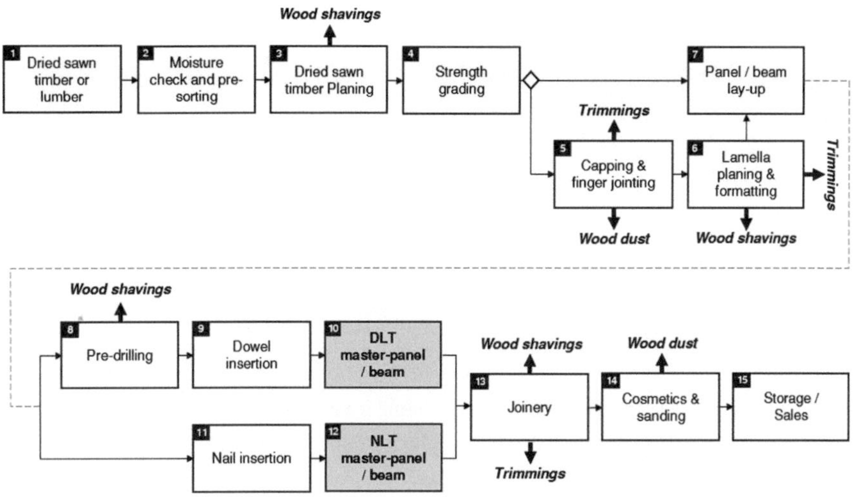

Fig. 21.12 Typical operations and by-products during the manufacturing of DLT and NLT

154]. Total PLY typically achieves an average yield of 52%, in the range of 33–62% according to literature. As depicted in Fig. 21.13, the losses from the input green veneers are caused by waste veneer, lay-up scrap, panel trimming and sanding as well as material testing [33, 64, 137, 138, 153], with improved material utilization of roughly 15% compared to the lumber-based EWPs. The higher maximum yield of LVL compared to PLY could be explained by fewer cross-cuts and fewer surfaces for sanding as usually thicker billets are produced [155].

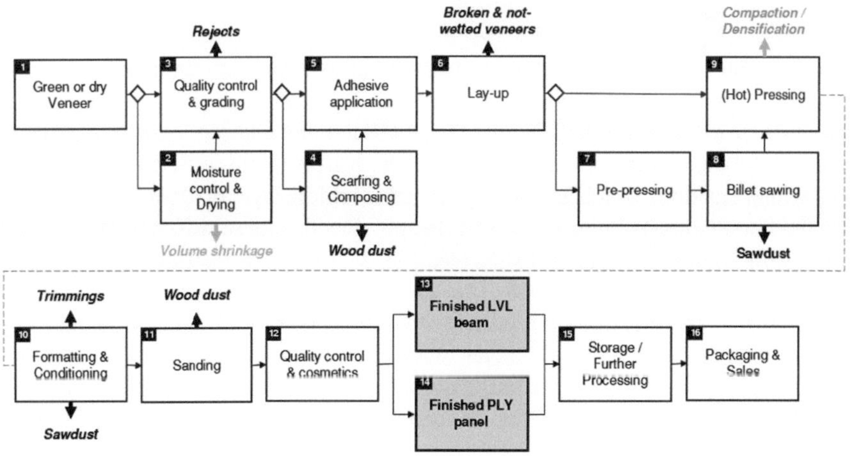

Fig. 21.13 Typical operations and by-products during the manufacturing of LVL and PLY

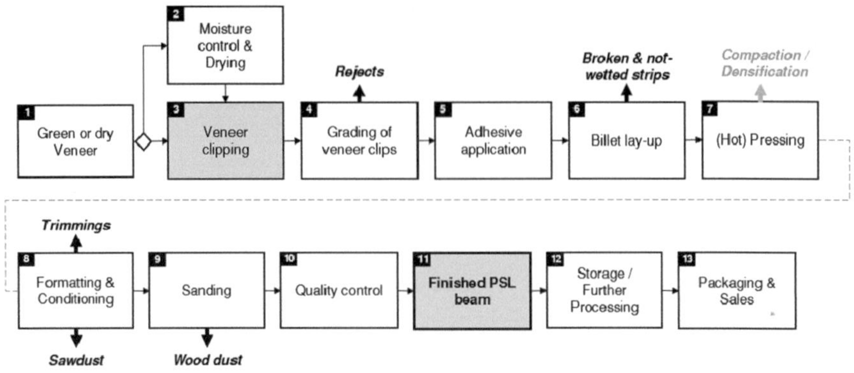

Fig. 21.14 Typical operations and by-products during the manufacturing of PSL

21.2.5.5 Parallel Strand Lumber (PSL)

PSL reaches an average material yield of 64%, ranging from 62 to 65%. Within the group of veneer-based products, it can be seen as a special case of EWP. In comparison to LVL and PLY, the production of PSL includes a secondary breakdown of the veneers into veneer clips (see Fig. 21.14). This enables a higher material yield compared to LVL and PLY, as veneer pieces that are not a full-width veneer sheet (roundups, fishtail, etc.) can be utilized [156]. The resulting clips are continuously feed into a hot press to form an endless beam, which is then formatted to the final dimensions.

21.2.5.6 Laminated Strand Lumber (LSL)

The material yield of LSL ranges from 55% up to 76%, with an average of 69%. The lower yield compared to OSB (see next section) is due to the unidirectional structure over the entire thickness of the material, demanding generally bigger (see Sect. 21.2.4) and more uniform strands, with high aspect ratios [156]. Therefore, part of the finer fractions which are usually put in the middle layer of OSB cannot be used for the homogeneous build-up of LSL beams. Besides these differences, the main drivers for additional material losses (see Fig. 21.15) are formatting and sanding of the finished panels and beams [27, 138, 157].

In general, the log stranding process does not require high-quality, mature logs and is geared towards small-diameter roundwood of lower quality [27]. Therefore, the process allows for a bigger range in roundwood quality and supports more log diameters and shapes. The factors influencing strand yield partly differ from the previous two first disintegration processes. Increasing log moisture content [139] and ambient temperature [158] significantly decreases the produced fines in stranding and therefore increases strand yield. In regards to machine set-up, the knife angle is reported [158] to be an important factor influencing the share of fines. In addition,

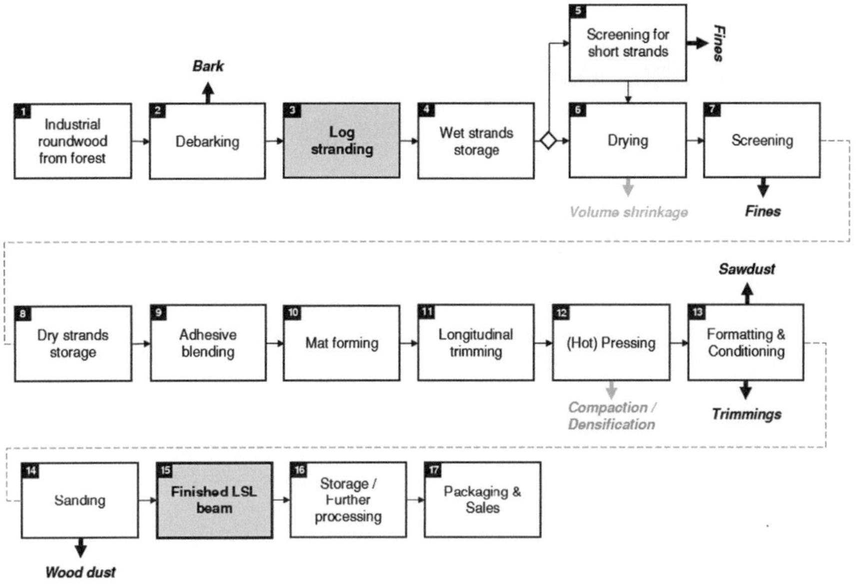

Fig. 21.15 Typical operations and by-products during the manufacturing of LSL

the positioning of the logs against the rotating knife using of clamps or pushers partly dictates the strand geometry and therefore also influences the strand yield [28]. The fines are normally separated from production and are typically used for the provision of energy as they reduce the strength of the final board and increase the amount of adhesive required. Furthermore, strands that do not fulfil the required sizes are also used for energy, but can also be used for PB.

The final target moisture content of the strands is in the range of 2–5% [27, 28] usually dried from above fibre saturation [159]. As mentioned previously, this could further affect the final material yield depending on the reported input and output unit. This is especially critical when comparing the different processes, as volumetric shrinkage is usually considered a loss in volume-based lumber and veneer production. However, as LSL and OSB production is commonly monitored using (dry) weight, this could lead to inaccuracies when comparing the processes.

21.2.5.7 Oriented Strand Board (OSB)

The average material yield of OSB is 83% and ranges from 66 to 96%. Similar to LSL, the main drivers for by-products besides the fines from stranding industrial roundwood are formatting and sanding of the finished panels (see Fig. 21.16). OSB is typically made of three layers, two unidirectionally oriented surface layers and one randomly or perpendicularly oriented middle layer. This was inspired by the orientation of plywood and combines two stiffness-providing surface layers and a

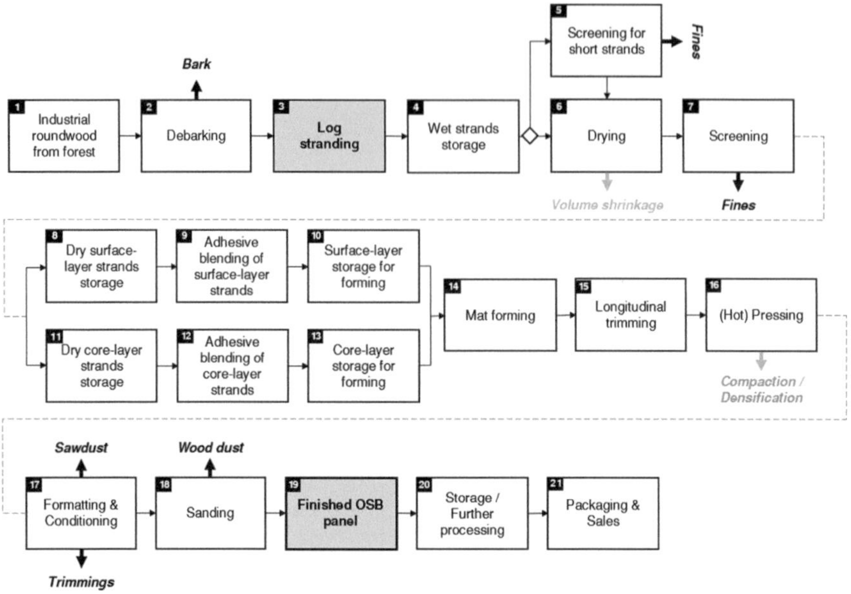

Fig. 21.16 Typical operations and by-products during the manufacturing of OSB

middle layer which gives additional dimensional stability while also improving shear stiffness. This allows for a bigger range of strand sizes to be used in the production compared to LSL, which explains the higher resource efficiency compared to LSL.

21.2.5.8 Scrimber (SCR)

The average material yield of SCR is 86%, ranging from 84 to 90%. These values are partly based on the company that originally produced SCR under the product name TimTek® [160]. Based on the available literature, by-products occur during, the formatting and finishing of the final panels (see Fig. 21.17). Additionally, it should be mentioned that it can be assumed that there is a certain share of unusable intermediaries (fines after scrimming and after drying). However, this is not discussed in the available literature.

As of 2024, SCR is the only EWP discussed here that is not produced on an industrial scale. Nevertheless, the process of splitting small-diameter logs has recently gained renewed interest and should therefore also be mentioned here. Especially as it could be one viable way to process the potentially increasing share of non-sawable and- peelable assortments in the future. Furthermore, it could also be used to add value to currently undervalued assortments like branches and crown tops, which were recently found to offer similar technological potential to stem wood [161].

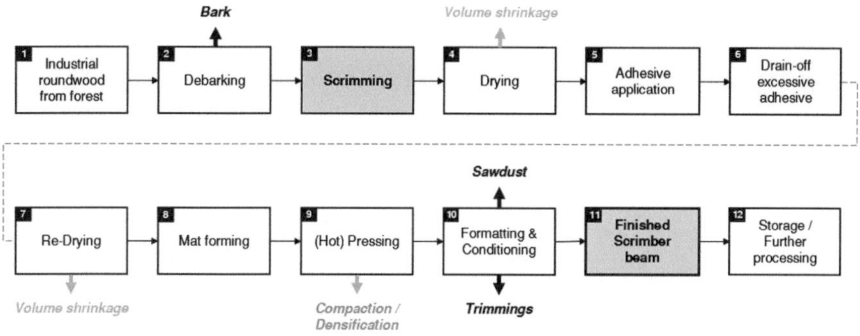

Fig. 21.17 Typical operations and by-products during the manufacturing of SCR

21.2.5.9 Particle Board (PB)

In comparison to the other processes, the production of PB can be seen as having the most flexible input raw material and offers a common utilisation strategy for by-products from sawmilling but also from stranding and possibly also from veneer peeling.

The average material yield for PB is 76% and ranges from 49% up to 87%. The lower end of this spectrum should be taken with a grain of salt as this low value of material yield is not typical for an industrial-scale facility. However, the reported numbers stem from a facility which utilizes mainly low-quality raw materials. According to the study [129], the raw material includes branches and wood from garden pruning, poplar and eucalyptus roundwood, other low-quality deciduous roundwood, forest firewood and industrial by-products, underlining the raw material flexibility of this process. Overall, the production of PB leads to by-products during screening and finishing of the final panel (see Fig. 21.18).

21.2.5.10 Fibre Board (FB)

Completing this comparison of different EWPs, the average material yield for FB is 89%, ranging from 75% up to 98%. This also represents the highest potentially achievable resource efficiency of the presented products. However, this, in turn, requires the highest degree of disintegration of roundwood in order to obtain single wood fibres or fibre-bundles as intermediaries. This is typically achieved by combining two major breakdown processes, chipping of roundwood and subsequent refining of the obtained chips into fibres (see Fig. 21.19).

While this has the potential to yield a 100% useable intermediary material for FB production, it also requires the highest total primary energy input of all the processes. With this, the environmental performance, typically evaluated by Life Cycle Assessment (LCA), as well as the mechanical performance needs to be brought into the comparison, as resource efficiency alone does not provide the complete

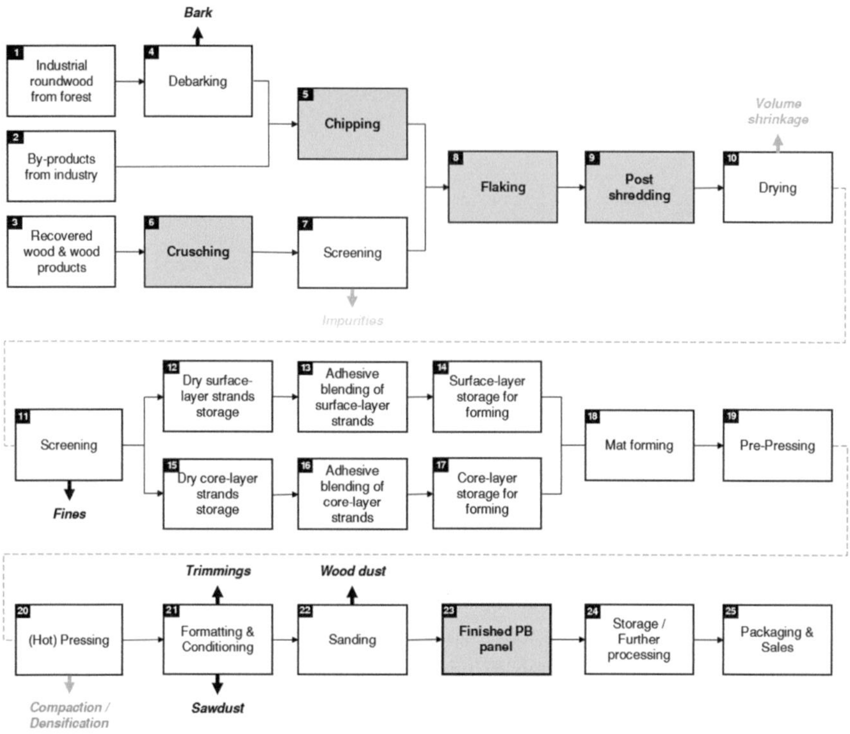

Fig. 21.18 Typical operations and by-products during the manufacturing of PB

picture in regards to sustainability of wood-based building materials in a broader sense.

21.3 Mechanical Performance

The architectural freedom [162] and design phase [163] of a building demands a certain mechanical capacity of the suitable construction materials [164]. Therefore, the potential mechanical performance of the different EWPs is another important aspect. In addition to the strength, stiffness and long-term behaviour already discussed in Pramreiter et al. [24], the specific modulus of elasticity (spec. $E_{0,mean}$) [GPa/kg/m^3] in the main load direction is compared with the material yield in Fig. 21.20.

It should become clear that the lower material yield of the lumber-(GLT, CLT, DLT and NLT) and veneer-based (LVL, PLY and PSL) EWPs, can be balanced by the higher attainable stiffness per kg EWP in contrast to the particle-based products [165, 166]. However, this only covers the serviceability side and one load case (bending) of

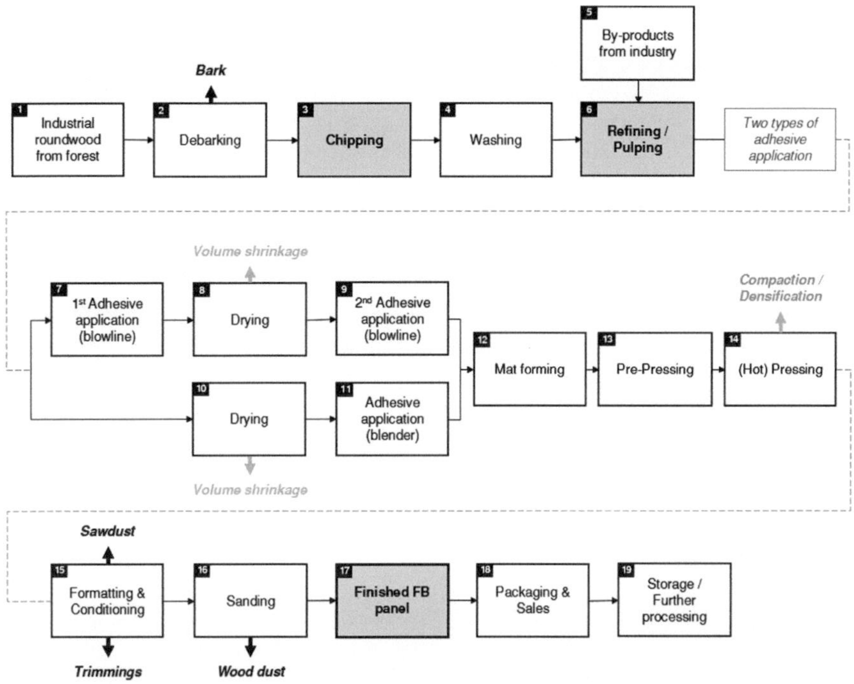

Fig. 21.19 Typical operations and by-products during the manufacturing of FB

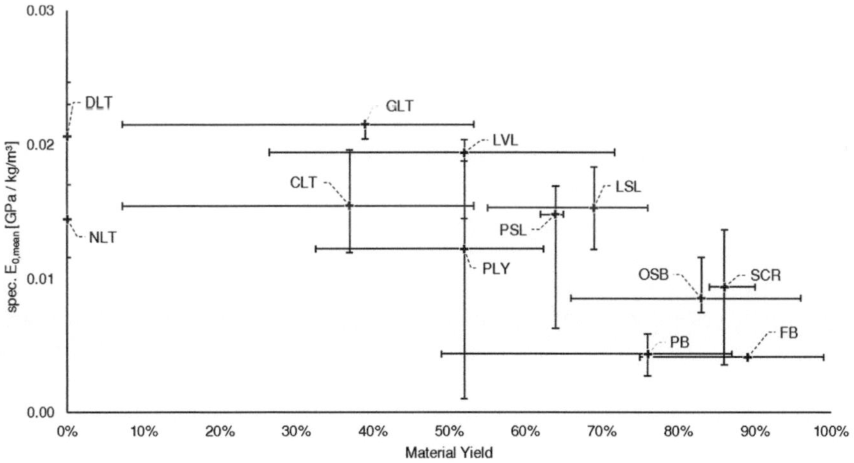

Fig. 21.20 Specific mean Modulus of Elasticity (spec. $E_{0,mean}$) [GPa/kg/m^3] versus respective material yield [%]. Based on values in Table 21.3

the design process. Therefore, it is highly recommended to also consider section four in Pramreiter et al. [24] for a more complete picture of the mechanical performance.

21.4 Environmental Performance

The evaluation of the environmental performance of materials and products is a very complex topic including different possible approaches and indicators [167]. Therefore, the following LCA indicators [168, 169] (functional unit is "1 m^3 of the respective EWP)" were selected to provide an overview of the environmental performance of the established EWPs:

- Released carbon
- Stored carbon
- Net carbon storage potential (difference between "Stored carbon" and "Released carbon")
- Global warming potential (GWP)
- Total primary energy.

21.4.1 Selected LCA Indicators

Considering remarks by Oneil et al. [170] it needs to be mentioned that the presented values in this section are subject to a certain degree of uncertainty for two main reasons. First, the values include two allocation principles (economic and mass allocation), with economic allocation usually leading to a higher environmental impact of the higher-value product [170]. This significantly influences the outcome, as market prices of the main and co-products can fluctuate greatly over time. Second, the number of values found in the literature varied greatly between the different products. In the case of NLT, PSL and SCR no literature could be found and for DLT only values from one environmental product declaration (EPD) are currently available [171]. A similar issue is present in regards to LSL, with only one available publication [157]. This is especially critical when different production regions (e.g. transport distances and energy mixes) and actuality of the data (e.g. age of the facility) are considered. This underlines the need to study the environmental performance of EWPs in more detail, broadening the investigated regions and raw materials. Furthermore, the underlying LCAs need to be kept up to date much more regularly, as factors like environmental regulations, energy provision and efficiency, data quality and availability are subject to more frequent changes than, for example, mechanical performance and can significantly change the outcome of the assessment. Nevertheless, Table 21.4 summarizes the values found in the literature for the discussed products.

Overall, one m^3 of EWP (cradle-to-gate) offers a net carbon storage potential of 705 kg CO_2 equiv. (192 kg C), by temporarily storing 1019 kg CO_2 equiv. (277 kg C) and releasing 314 kg CO_2 equiv. (85 kg C) into the air during production (including provision of raw materials). These numbers are in line with assumptions made by Churkina et al. [1], which underlines the potential of EWPs to decarbonise the building sector. More specifically, PB (889 kg CO_2 equiv.), GLT (827 kg CO_2 equiv.) and DLT (823 kg CO_2 equiv.) offer the highest potential of the available EWPs, followed by LVL (781 kg CO_2 equiv.), LSL (751 kg CO_2 equiv.) and CLT (739 kg CO_2 equiv.). FB, OSB and PLY complete the available literature with 671 kg CO_2 equiv., 576 kg CO_2 equiv. and 533 kg CO_2 equiv. respectively. Currently, there are no values available for PSL and SCR. For material selection, the material with the highest net carbon storage potential should be selected.

In regards to GWP, the established EWPs, except for FB, show a rather homogenous GWP with an average of 232 kg CO_2 equiv. per m^3 material. According to the available literature [131, 172], the high GWP of FB (487 kg CO_2 equiv.) is caused during the manufacturing stage (A3). They don't give a specific cause for this in the discussion, but it is safe to say that the heat required for the drying and (hot) pressing steps as well as the necessary adhesive play a key role. This is also supported by the total primary energy demand (see Table 21.4). While this value may seem very high, it corresponds to the general trend that the GWP increases with decreasing size of the intermediary. In contrary to net carbon storage potential, the goal ought to be to choose the EWP with the lowest GWP [kg CO_2 equiv.]. It should be noted here that GWP is one of the most discussed and accepted indicators in LCA, yet for a full picture, especially when materials from different processes are compared, all the evaluated LCA indicators need to be considered.

A closer look at the total primary energy demand [MJ] in Table 21.4 shows a significant difference on two levels. First, the provision of raw material has a significant influence on the reported total primary energy. On average, the raw material supply and transport stages, in LCA are typically referred to as A1 and A2 [168, 169], cause 49% (4037 MJ of 8178 MJ) of the total primary energy demand in EWPs. For the lumber-based products (GLT, CLT and DLT), these stages represent 66% (3320 MJ of 4997 MJ) of the total primary energy demand. In the veneer- (LVL and PLY) and strand-based products (OSB), they constitute 51% (3885 MJ of 7577 MJ) and 45% (3614 MJ of 8114 MJ) respectively and they account for 50% (3585 MJ of 7118 MJ) in PB and 28% (5540 MJ of 20,043 MJ) in FB. Second, the total primary energy demand for the manufacturing stage, in LCA typically referred to as A3, significantly increases with decreasing size of the intermediary. This is a well-known correlation in wood technology [25] and can be explained by the fact that the disintegration of roundwood into ever smaller intermediaries and the subsequent re-assembly to composites typically causes greater expenditure of energy and requires more energy-intensive auxiliary materials (e.g. synthetic adhesives and hot pressing instead of pressing at room temperature). It also explains the decreasing share of total primary energy demand for the provision of raw materials.

21.4.2 Specific Environmental Performance

Considering the typical functional unit of "1 m^3 EWP" and the different densities of the EWPs, a comparison based on a specific environmental performance similar to the specific mechanical performance is also of interest. Therefore, Fig. 21.21 shows the specific net carbon storage potential (Fig. 21.21a), specific GWP (Fig. 21.21b) and specific total primary energy demand (Fig. 21.21c) in relation to the respective material yield from Sect. 21.2. It needs to be mentioned that the density was taken from Table 21.3 instead of the respective LCA literature. This was done due to the fact, that not all references on LCA had a clear description of the corresponding density of the final product.

Taking a look at Fig. 21.21, an influence of the intermediary size is still visible, with smaller intermediaries leading to a higher environmental impact. LSL and FB seem to also have a particularly high total primary energy demand compared to the other EWPs. In regards to LSL, this high demand is caused by the log storage in ponds to heat the logs before processing and the subsequent stranding as well as the drying process. Furthermore, the typical panel thickness of 75 mm and above requires the use of a steam injection press in order to maintain suitable cycle times. However, as noted earlier this is only based on one publication [157]. Considering FB, the high energy demand is caused by the refining and drying process, which requires significant amounts of energy [126, 131].

21.4.3 A Three-Dimensional Comparison of Established EWPs

Together with the previous comparisons, it should become even clearer now that there is no such thing as one ideal EWP but rather a balance of factors that needs to be achieved. Therefore, the end of this section presents a three-dimensional comparison of established EWPs (see Fig. 21.22). This comparison serves as blueprint for a material selection process which incorporates the discussed material yield, mechanical performance as well as net carbon storage potential. For future investigations it should be extended with other mechanical properties, LCA indicators as well as not yet discussed indicators (e.g. production costs or recyclability) in mind. Especially, aspects like recyclability and different cost factors might further support a more balanced material selection. While it might be difficult to discuss the cost factor using reliable values from industry (not commonly accessible), the aspect of recyclability or circularity has gained significant interest in recent years. Therefore, the last section of this chapter will give a brief comment on the circularity aspects of the discussed EWPs.

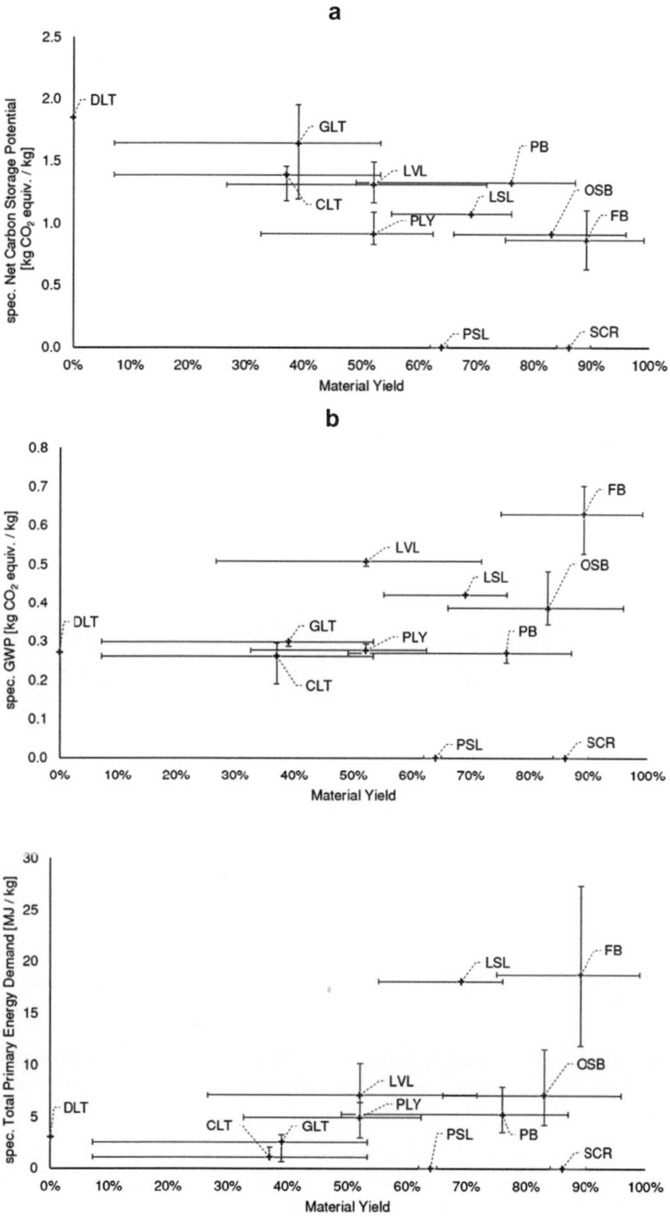

Fig. 21.21 Specific environmental performance of established EWPs versus respective material yield [%]. **a** Specific net carbon storage potential [kg CO_2 equiv./kg]. Numbers are based on LCA containing results on stored as well as released carbon. **b** Specific global warming potential (GWP) [kg CO_2 equiv./kg]. **c** Specific total primary energy demand [MJ/kg]. Based on values in Table 21.4 and densities from Table 21.3. Convert [kg CO_2 equiv.] to [C] using the molecular weight ratio of 44/12 CO_2/C

Table 21.3 Mean modulus of elasticity ($E_{0,mean}$) and raw density (density at a moisture content of typically 12–15%) [kg/m³] of established EWPs

EWP	$E_{0,\ mean}$ [GPa]			Raw density [kg/m³]		
	Min.	Avg.	Max.	Min.	Avg.	Max.
GLT	8.3	12.3	16.8	370	521	828
CLT	5.5	9	15	420	532	698
DLT	5.4	7	11.6	425	441	459
NLT	6.9	10.1	13.8	370	446	510
LVL	7	12.7	17.9	440	596	800
PLY	0.5	7.8	14	459	582	680
PSL	4.2	11.8	15.2	614	727	820
LSL	9	11.7	14.5	673	698	722
OSB	4.9	5.9	8.4	600	631	660
SCR	2.1	8.5	15.9	540	826	1063
PB	1.8	3.2	5	600	670	780
FB	2.7	3.5	5	600	775	1080

Table 21.4 Selected indicators for the environmental performance of 1 m³ EWP

Indicator	Unit	GLT	CLT	DLT	LVL	PLY	LSL	OSB	PB	FB
Released carbon (cradle-to-gate)	[kg CO_2 equiv.]	104	147	155	212	374	295	429	392	714
	[kg C]	28	40	42	58	102	80	117	107	195
Stored carbon (cradle-to-gate)	[kg CO_2 equiv.]	931	886	978	952	866	1046	1033	1145	1334
	[kg C]	254	242	262	260	236	285	282	312	364
GWP (cradle-to-gate)	[kg CO_2 equiv.]	156	139	121	303	161	295	244	181	487
Total primary energy (cradle-to-gate)	[MJ/m³]	5560	4588	4844	9275	5878	–	8114	7118	20,043
Total primary energy (gate-to-gate)	[MJ/m³]	1310	580	1342	4250	3133	13,386	4473	3533	14,503

The numbers represent an average based on literature (taken from [24] and extended with literature from [126, 128, 131, 133, 138, 171, 172]) dealing with Life Cycle Assessment. In the case of stored carbon, a molecular weight ratio of 44/12 CO_2/C was assumed

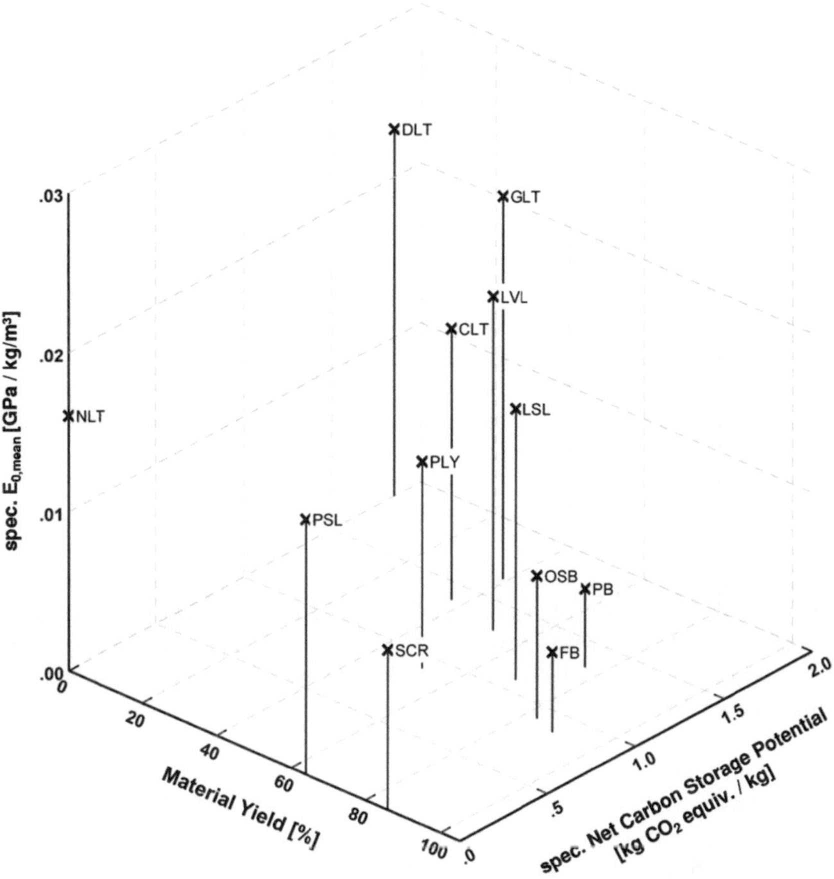

Fig. 21.22 A three-dimensional comparison of Material yield [%], specific Modulus of Elasticity in the main load direction (spec. $E_{0,mean}$) [GPa/kg/m^3] and specific Net Carbon Storage Potential [kg CO_2 equiv./kg] of established EWPs. The values are based on average values from the previous sections. A value of zero indicates that there is no available literature for the respective property

21.5 A Comment on Circularity

Circularity in engineering and architecture has become an important consideration for sustainable design in recent years [173]. Beyond the initial embodied emissions in new products, an increased life span and future reuse of building materials and components can greatly increase the carbon and emission efficiency of new products [174]. Building designs or material systems conceived for an end-of-life scenario with planned deconstruction or reuse (DfD—Design for Deconstruction) have greater chances of being reused or recycled in the future. The selection of EWPs for a new building design has a large influence on the future resource, process and environmental efficiency of their reuse or recycling, as they all have different potentiasl for reuse of the raw materials they are made from [175].

Ideally, a building is designed for a longer life cycle than typically provided by building codes (50–100 years) in order to extend the carbon storage and best utilize the spent carbon and emissions for the construction of the building. Next, adaptive reuse of the building can further extend its life cycle, by bringing upgrades that can serve its new performance requirements (such as structural integrity or thermal performance) and accommodate for the new use of the space. If the building has to be taken down, building materials or components can be reused or recycled to feed into new buildings. Otherwise, the product's intermediary can be reused in new EWPs as salvaged intermediary (e.g. salvaged lumber).

As shown in the previous sections, the selection of suitable EWPs balances a few parameters, such as structural versus environmental performance and material yield. Circularity and reuse of EWPs are currently less quantifiable and hard to include in the analysis but predictions are emerging from current research and references from the past [176, 177]. EWPs made with current adhesive technology might have a lower potential for efficient reuse of the intermediary when compared to EWPs assembled with mechanical fasteners like DLT or NLT. Denailing NLT panels or removing dowels from DLT panels allows for a simple recovery of the product's lumber. Glued products, if not reused as a whole component, will need to be resawn and resurfaced for the production of new EWPs, bringing them back into the same process flow with associated processing losses. Furthermore, information about the durability of the performance of glue over many building life cycles has not been demonstrated. Historical buildings built out of wood demonstrates the durability of the wood fiber itself, as a long-term structural material, like the Hōryū-ji temple in Nara, Japan that has been in use for the last 1300 years.

These predictions don't consider future developments in adhesives and disassembly technologies. However, given the increase appeal for wood-based construction such as mass timber around the world in its current technological state, these considerations should be weighted in to increase the long-term resources efficiency of EWPs.

21.6 Concluding Remarks

The widespread application of Engineered Wood Products (EWPs) for not only small residential but also medium to high rise buildings will be key in order to decarbonize the global construction sector. The recent completion of ever higher building projects (see Fig. 21.23) utilizing selected EWPs as one of the main structural components has already shown the advantages of wood-based products along with existing materials like steel and concrete.

However, the potential demand for built environments in the future could put increasing pressure on the world's forests and may jeopardise the sustainable supply

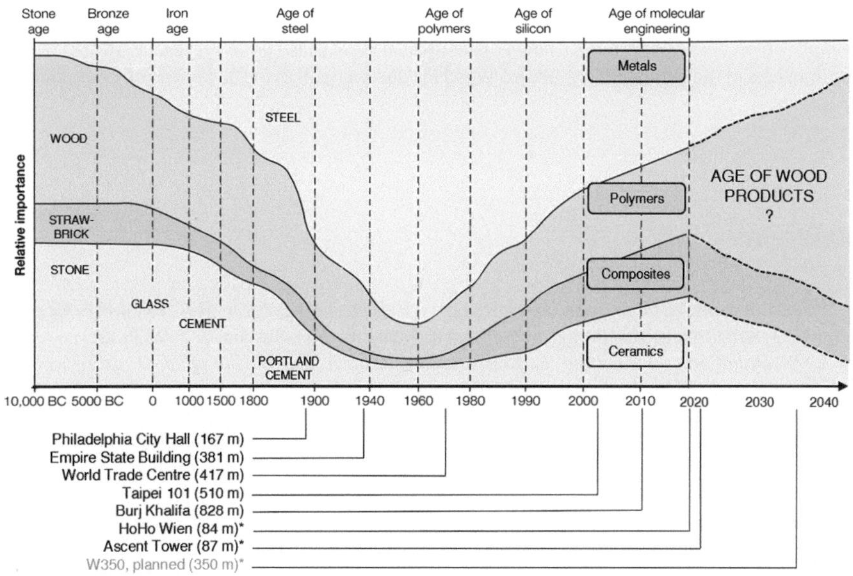

Fig. 21.23 Past, present and potential future importance of common material categories used in different disciplines of engineering (adapted with permission from [7], copyright by Elsevier). *Engineered Wood Products as structural components

of sufficient raw materials. Therefore, the construction industry needs to consider the full variety of EWPs and incorporate aspects like resource efficiency, environmental and mechanical performance as well as circularity into their decision process. This chapter summarized this information and shined a light on the currently occurring by-products during production and provided base-values for material selection, not only in construction, but also for other potential applications in the future. It is important to emphasise here that the comparisons should not be seen as a ranking, but rather a demonstration of a potential material-first approach to wood as a raw material and how the by-products of one EWP can be and partly already are the raw material for another. While there are certain aspects of the wood-based value chain that need critical discussion and further research, the authors firmly believe that the shift towards a climate-resilient society requires the sustainably harvested resources of our forests, which are then carefully handled and utilized in long-lasting products. After a century characterised by mainly non-renewable building materials (see Fig. 21.23), the twenty-first century seems to mark the beginning of the age of wood-based building products.

Acknowledgements The results presented in this chapter are part of the research project "Strong Hardwood—A new resource-efficient material concept based on hardwood branches". The financial support by Amt der Niederösterreichischen Landesregierung (FTI20-003) is gratefully acknowledged. Additionally, the results are part of the research project

"UniStrand". This project is funded by the Forest Fund, an initiative of the Federal Ministry of Agriculture, Forestry, Regions and Water Management, and is carried out as part of the Think.

Wood program of the Austrian Wood Initiative (FFG 893351). The financial contribution by the company partners is gratefully acknowledged. Special thanks also go to Frederic Lam for creating the drawings of the presented Engineered Wood Products as well as the flowcharts of the production processes.

References

1. G. Churkina, A. Organschi, C.P.O. Reyer, A. Ruff, K. Vinke, Z. Liu, B.K. Reck, T.E. Graedel, H.J. Schellnhuber, Buildings as a global carbon sink. Nat. Sustain. **3**, 269 (2020)
2. M. Pramreiter, T. Nenning, L. Malzl, J. Konnerth, *A plea for the efficient use of wood in construction*. Nat. Rev. Mater. (2023)
3. L. Barham et al., Evidence for the earliest structural use of wood at least 476,000 years ago. Nature **622**, 107 (2023)
4. OECD, *Global Material Resources Outlook to 2060: Economic Drivers and Environmental Consequences* (2019)
5. A.L.S. Bais, C. Lauk, T. Kastner, K. Erb, Global patterns and trends of wood harvest and use between 1990 and 2010. Ecol. Econ. **119**, 326 (2015)
6. A.R. Martin, M. Doraisami, S.C. Thomas, Global patterns in wood carbon concentration across the world's trees and forests. Nat. Geosci. **11**, 915 (2018)
7. M.F. Ashby, *Materials Selection in Mechanical Design* (Butterworth-Heinemann, Oxford, 2011)
8. T.D. Searchinger, T. Beringer, B. Holtsmark, D.M. Kammen, E.F. Lambin, W. Lucht, P. Raven, J.P. van Ypersele, Europe's renewable energy directive poised to harm global forests. Nat. Commun. **9**, 10 (2018)
9. U. Mantau, *Wood Flows in Europe (EU 27)*. Project Report Celle 2012 24 (2012)
10. European Commission, *Sankey Diagrams of Woody Biomass Flows in the EU—2021 Release* (2021)
11. I. Alberdi et al., Assessing forest availability for wood supply in Europe. For. Policy Econ. **111** (2020)
12. B.J.W. Lerink et al., *How much wood can we expect from European forests in the near future?* For.: Int. J. For. Res. **96**, 434 (2023)
13. Portions of this document include intellectual property of Esri and its licensors and are used under license. Copyright ©2017 Esri and its licensors. All rights reserved., Sources: ESA. Land Cover CCI Product User Guide Version 2. Tech. Rep. (2017)
14. Portions of this document include intellectual property of Esri and its licensors and are used under license. Copyright ©2014 Esri and its licensors. All rights reserved., Sources: Esri; CIA World Factbook; GMI; NIMA; UN Data; UN Habitat; US Census Bureau (2014)
15. A. Mishra, F. Humpenöder, G. Churkina, C.P.O. Reyer, F. Beier, B.L. Bodirsky, H.J. Schellnhuber, H. Lotze-campen, A. Popp, Land use change and carbon emissions of a transformation to timber cities. Nat. Commun. **13** (2022)
16. V. Augiseau, S. Barles, Studying construction materials flows and stock: a review. Resour. Conserv. Recycl. **123**, 153 (2017)
17. E. Leszczyszyn et al., The future of wood construction: opportunities and barriers based on surveys in Europe and Chile. Sustainability **14** (2022)
18. L. Orozco, H. Svatoš-Ražnjević, H.J. Wagner, M. Abdelaal, F. Amtsberg, D. Weiskopf, A. Menges, Advanced timber construction industry: a quantitative review of 646 global design and construction stakeholders. Buildings **13** (2023)
19. C. Mergel, K. Menrad, T. Decker, Which factors influence consumers' selection of wood as a building material for houses? Can. J. For. Res. (2024)

20. M. Riala, L. Ilola, Multi-storey timber construction and bioeconomy—barriers and opportunities. Scand. J. For. Res. **29**, 367 (2014)
21. P. Blanchet, C. Perez, M.R. Cabral, Wood building construction: trends and opportunities in structural and envelope systems. Curr. For. Rep.
22. FAO, FAOSTAT: Forest Production and Trade. https://www.fao.org/faostat/en/#compare
23. FAO, Global Production and Trade in Forest Products in 2020. https://www.fao.org/forestry/statistics/80938/en/
24. M. Pramreiter, T. Nenning, C. Huber, U. Müller, B. Kromoser, P. Mayencourt, J. Konnerth, A review of the resource efficiency and mechanical performance of commercial wood-based building materials. Sustain. Mater. Technol. **38** (2023)
25. P. Niemz, A. Teischinger, D. Sandberg, *Springer Handbook of Wood Science and Technology* (Springer International Publishing AG, 2023)
26. M. Dunky, P. Niemz, *Holzwerkstoffe Und Leime* (Springer, Berlin Heidelberg, 2002)
27. M.P. Ansell, *Wood Composites* (Woodhead Publishing, Cambridge, 2015)
28. J.C.F. Walker, *Primary Wood Processing—Principles and Practice* (Springer, Dordrecht, 2006)
29. A. Wagenführ, F. Scholz, *Taschenbuch der Holztechnik, 2, aktual* (Carl Hanser Verlag GmbH & Co, KG, 2012)
30. T.G. Williamson, *APA Engineered Wood Handbook* (McGraw-Hill, 2002)
31. Forests Products Laboratory, *Wood Handbook—Wood as Engineering Material* (U.S. Department of Agriculture, Forest Service, Forest Products Laboratory, Madison, WI, 2021)
32. R. Sathre, L. Gustavsson, Process-based analysis of added value in forest product industries. For. Policy Econ. **11**, 65 (2009)
33. M.A. Fonseca, *The Measurement of Roundwood Methodologies and Conversion Ratios* (CABI Publishing, 2005)
34. E.R. Thomas, N.D. Bennett, An analysis of the differences among log scaling methods and actual log volume. For. Prod. J. **67**, 250 (2017)
35. FAO, *Forest Product Conversion Factors* (2020)
36. R. Wagenführ, A. Wagenführ, *Holzatlas* (Carl Hanser Verlag, München, 2022)
37. FAO, *Forest Products Definitions* (2016)
38. ISO 24294, *Timber—Round and Sawn Timber—Vocabulary*
39. EN 844, *Round and Sawn Timber—Terminology*
40. ISO 13059, *Round Timber—Requirements for the Measurement of Dimensions and Methods for the Determination of Volume* (2011)
41. EN 1309-2, *Round and Sawn Timber—Method of Measurement of Dimensions. Part 2: Round Timber—Requirements for Measurement and Volume Calculation Rules* (2006)
42. EN 1927-1 *Qualitative Classification of Softwood Round Timber. Part 1: Spruces and Firs*
43. EN 1927-2, *Qualitative Classification of Softwood Round Timber. Part 2: Pines*
44. EN 1927-3, *Qualitative Classification of Softwood Round Timber. Part 3: Larches and Douglas Fir*
45. EN 1316-2, *Hardwood Round Timber—Qualitative Classification. Part 2: Poplar*
46. EN 1316-1, *Hardwood Round Timber—Qualitative Classification. Part 1: Oak and Beech*
47. EN 1309-3, *Round and Sawn Timber—Methods of Measurements. Part 3: Features and Biological Degradations*
48. J. Hildebrandt, N. Hagemann, D. Thrän, The contribution of wood-based construction materials for leveraging a low carbon building sector in Europe. Sustain. Cities Soc. **34**, 405 (2017)
49. R. Sikkema, D. Styles, R. Jonsson, B. Tobin, K.A. Byrne, A market inventory of construction wood for residential building in Europe—in the light of the green deal and new circular economy ambitions. Sustain. Cities Soc. **90** (2023)
50. Wood Products Council, *Snapshot of Mass Timber Projects in the U.S.* https://www.woodworks.org/resources/mapping-mass-timber/
51. M. Hanewinkel, D.A. Cullmann, M.J. Schelhaas, G.J. Nabuurs, N.E. Zimmermann, Climate change may cause severe loss in the economic value of European forest land—supplementary information. Nat. Clim. Chang. **3**, 203 (2013)

52. W. Thuiller et al., Predicting global change impacts on plant species' distributions: future challenges. Perspect. Plant Ecol. Evol. Syst. **9**, 137 (2008)
53. T. Gschwantner et al., Harmonisation of stem volume estimates in European National Forest Inventories. Ann. For. Sci. **76** (2019)
54. H. Spelter, Conversion of board foot scaled logs to cubic meters in Washington State, 1970–1998. For. Prod. Lab. **FPL-GTR-13** (2002)
55. P. Yang, B.M. Jenkins, Wood residues from Sawmills in California. Biomass Bioenergy **32**, 101 (2008)
56. C.E. Keegan III, T.A. Morgan, K.A. Blatner, J.M. Daniels, Trends in lumber processing in the Western United States. Part II: overrun and lumber recovery factors. For. Prod. J. **60**, 140 (2010)
57. K.A. Blatner, C.E. Keegan, J.M. Daniels, T.A. Morgan, Trends in lumber processing in the Western United States. Part III: residue recovered versus lumber produced. For. Prod. J. **62**, 429 (2012)
58. P.H. Steele, Factors determining lumber recovery in sawmilling. For. Prod. Lab. **FPL-39**, (1984)
59. S. Torno, M. Knorz, J. van de Kuilen, Supply of beech lamellas for the production of glued laminated timber, in *International Scientific Conference on Hardwood Processing, ISCHP* 210 (2013)
60. P.H. Steele, F.G. Wagner, K.E. Skog, Regional softwood sawmill processing variables as influenced by productive capacity. For. Prod. Lab. **FPL-RP-504** (1991)
61. K.A. Kilborn, Lumber recovery studies of Alaska Sawmills, 1997 to 1999. USDA For. Serv. **PNW-GTR-54** (2002)
62. D. Smorfitt, S. Harrison, J. Herbohn, Portable sawmills in a high-value rainforest cabinet timber industry in North Queensland. Small-Scale For. Econ., Manag. Policy **2**, 21 (2003)
63. M.W. Wade, S.H. Bullard, P.H. Steele, P.A. Araman, Estimating hardwood sawmill conversion efficiency based on sawing machine and log characteristics. For. Prod. J. **42**, 21 (1992)
64. T.D. Fahey, R.O. Woodfin, The cubics are coming: predicting product recovery from cubic volume. J. For. **74** (1976)
65. A. Clark III., Sawmill residue yields from yellow-poplar saw logs. For. Prod. J. **26**, 23 (1976)
66. G.Y. Jeong, J.J. Lee, H. Yeo, S.S. Lee, Predicting lamina yield from logs of different diameters for cross laminated timber production. J. Korean Wood Sci. Technol. **44**, 809 (2016)
67. G.R. Middleton, B.D. Munri, *Second-Growth Western Hemlock: Product Yields & Attributes Related to Stand Density* (Forintek Canada Corp., 2001)
68. P.A. Shenga, P. Bomark, O. Broman, D. Sandberg, Simulation of tropical hardwood processing—sawing methods, log positioning, and outer shape. Bioresources **10**, 7640 (2015)
69. S.-Y. Zhang, Q.-J. Tong, Modeling lumber recovery in relation to selected tree characteristics in jack pine using sawing simulator Optitek. Ann. For. Sci. **62**, 219 (2005)
70. I. Pinto, A. Usenius, T. Song, H. Pereira, Sawing simulation of maritime pine (Pinus Pinaster Ait.) stems for production of heartwood containing components. For. Prod. J. **55**, 88 (2005)
71. B. Perkins, R.L. Smith, B. Bond, Solid wood and residue yield analysis of small-diameter red oak logs. For. Prod. J. **58**, 97 (2008)
72. F.P. Vergara, C.D. Palma, H. Sepúlveda, A comparison of optimization models for lumber production planning. Bosque **36**, 239 (2015)
73. P.C. Kersavage, M. Savsar, R.J. Meimban, A mathematical analysis to determine the volume of residues and lumber produced in the sawmill edging process. Wood Fiber Sci. **22**, 271 (1990)
74. C.G. Lundahl, A. Grönlund, Increased yield in sawmills by applying alternate rotation and lateral positioning. For. Prod. J. **60**, 331 (2010)
75. P. Hamner, M. White, P. Araman, The effect of curve sawing two-sided cants from small-diameter hardwood sawlogs on lumber and pallet part yields. For. Prod. J. **56**, 80 (2006)
76. U. Buehlmann and R. E. Thomas, Estimating component yield for CLT production, in *IEEE International Conference on Industrial Engineering and Engineering Management* (2017), pp. 470–474

77. G.Y. Jeong, J.J. Lee, H. Yeo, J.P. Hong, H.K. Kim, W.T. So, W.Y. Chung, Optimized lamina size maximizing yield for cross laminated timber using domestic trees. J. Korean Wood Sci. Technol. **41**, 141 (2013)
78. P. Schlotzhauer, A. Kovryga, L. Emmerich, S. Bollmus, J.W. Van de Kuilen, H. Militz, Analysis of economic feasibility of ash and maple lamella production for glued laminated timber. Forests **10** (2019)
79. J. Edlund, H. Lindström, F. Nilsson, M. Reale, Modulus of elasticity of Norway Spruce saw logs vs. structural lumber grade. Holz Als Roh - Und Werkstoff **64**, 273 (2006)
80. C. Fischer, G.I. Vestøl, A. Øvrum, O.A. Høibø, Pre-sorting of Norway Spruce structural timber using acoustic measurements combined with site-, tree- and log characteristics. Eur. J. Wood Wood Prod.**73**, 819 (2015)
81. A. Rais, E. Ursella, E. Vicario, F. Giudiceandrea, The use of the first industrial X-ray CT scanner increases the lumber recovery value: case study on visually strength-graded Douglas-fir timber. Ann. For. Sci. **74**, 1 (2017)
82. A. Ranta-Maunus, *Strength of European Timber—Part 1. Analysis of Growth Areas Based on Existing Test Results* (VTT Publications, 2009)
83. A. Ranta-Maunus, J.K. Denzler, P. Stapel, *Strength of European Timber—Part 2. Properties of Spruce and Pine Tested in Gradewood Project* (VTT Publications, 2011)
84. M. Green, P. Lister, F. Brüchert, F. Giudiceandrea, A. Katsevich, J. Oja, K.-J. Johansson, *From Imagining to Imaging—High Speed 3D Scanning of Logs*, in *Symposium 2016: The Marcus Wallenberg Prize* (2016)
85. K. Lan, S.S. Kelley, P. Nepal, Y. Yao, Dynamic life cycle carbon and energy analysis for cross-laminated timber in the Southeastern United States. Environ. Res. Lett. **15**, (2020)
86. R.D. Bergman, Gate-to-gate life-cycle inventory on hardwood sawmills in the Northeastern Region of the United States By, Master Thesis, University of Wisconsin (2007)
87. R.D. Bergman, S.A. Bowe, Environmental impact of manufacturing softwood lumber in Northeastern and North Central United States. Wood Fiber Sci. **42**, 67 (2010)
88. R.D. Bergman, S.A. Bowe, Life-cycle inventory of manufacturing hardwood lumber in Southeastern US. Wood Fiber Sci. **44**, 71 (2012)
89. ASMI, *A Cradle-to-Gate Life Cycle Assessment of Canadian Surfaced Dry Softwood Lumber* (2018)
90. ASMI, *A Cradle-to-Gate Life Cycle Assessment of Eastern Canadian Surfaced Dry Softwood Lumber* (2018)
91. P. Döring, S. Glasenapp, U. Mantau, *Rohstoffmonitoring Holz - Sägeindustrie 2015 Einschnitt- Und Produktionsvolumen* (2017)
92. M.R. Milota, C.D. West, I.D. Hartley, Gate-to-gate life-cycle inventory of softwood lumber production. Wood Fiber Sci. **37**, 47 (2005)
93. Y. Han, J.H. Park, Y.S. Chang, Y. Park, J.K. Oh, J.P. Hong, J.J. Lee, H. Yeo, The effect of controlling the drying distortion of laminas on the production yield of cross-laminated timber (CLT) using Larix Kaempferi Wood. Eur. J. Wood Wood Prod. **74**, 519 (2016)
94. R.O. Woodfin, Wood losses in plywood production—four species. For. Prod. J. **23**, 98 (1973)
95. Raute Group, *Data-Driven Veneer, Plywood and LVL Production—Improving the Production Through Intelligence*
96. J.W. Funck, M.G. Babb, Green veneer recovery as influenced by clipping strategies. For. Prod. J. **37**, 23 (1987)
97. Raute Group, *5 Ways to Improve Yield.*
98. M. Hanewinkel, D.A. Cullmann, M.J. Schelhaas, G.J. Nabuurs, N.E. Zimmermann, Climate change may cause severe loss in the economic value of European Forest Land. Nat. Clim. Chang. **3**, 203 (2013)
99. UNECE and FAO, *The European Forest Sector Outlook Study II* (2011)
100. A.A. Marra, *Technology of Wood Bonding: Principles in Practice* (Van Nostrand Reinhold, New York, 1992)
101. ÖNORM EN 14080, *Holzbauwerke - Brettschichtholz Und Balkenschichtholz - Anforderungen*

102. R. Brandner, G. Flatscher, A. Ringhofer, G. Schickhofer, A. Thiel, Cross laminated timber (CLT): overview and development. European Journal of Wood and Wood Products **74**, 331 (2016)
103. ÖNORM EN 16351, *Holzbauwerke - Brettsperrholz - Anforderungen*
104. ESR-4069, *ICC-ES Evaluation Report: Dowel Laminated Timber (DLT)* (2022)
105. ETA-13/0785, *Solid Wood Slab Element—Element of Dowel Jointed Timber to Be Used as a Structural Element in Buildings* (2018)
106. ETA-18/0960, *Vorgefertigte Holzbauelemente- Elemente Aus Mechanisch Verbundenen Kanthölzern Für Tragende Bauteile in Gebäuden* (2018)
107. T. Bell, *A Detailed Investigation into the Engineering Properties and Challenges Affecting the Potential Introduction of a UK Grown Dowel-Laminated Timber Floor Panel into the Domestic Construction Market*, Ph.D. Thesis (2018)
108. ETA-11/0338, *Solid Wood Slab Element—Element of Dowel Jointed Timber Boards to Be Used as a Structural Element in Buildings* (2017)
109. BSLC, *The Nail-Laminated Timber Design and Construction Guide* (2017)
110. BSLC, *Nail-Laminated Timber Design & Constr Uction Guide I v1.0* (2017)
111. PR-L325(C), *FRERES Mass Ply Panel (MPP), and Mass Ply Lam (MPL) Beams and Columns*, (2023)
112. ÖNORM EN 313-2, *Plywood Classification and Terminology—Part 2: Terminology* (2000)
113. Finnish Forest Industries Federation, *Handbook of Finnish Plywood* (2022)
114. ESR-1387, *Structural Composite Lumber and Oriented Strand Board*
115. EN 12369-1, *Holzwerkstoffe - Charakteristische Werte Für Die Berechnung Und Bemessung von Holzbauwerken - Teil 1: OSB, Spanplatten Und Faserplatten*
116. ETA-13/0784, *Swiss Krono Magnumboard*, Deutsches Institut für Bautechnik
117. S.C. Kitchens, T.L. Amburgey, H. Michael Barnes, R.D. Seale, Mechanical and durability properties of steam-pressed scrim lumber. Bioresources **11**, 5343 (2016)
118. Mississippi State University, *TimTek Process*. https://www.cfr.msstate.edu/timtek/process.asp
119. T. Joščák, A. Teischinger, U. Müller, R. Mauritz, *Production and Material Performance of Long—Strand Wood Composites*. Wood Research
120. H. Michael, B.R. Dan, S.J.M. Linton, Steam-pressed scrim lumber (SPSL), in *Proceedings of the International Convention of Society of Wood Science and Technology* (2010)
121. Swedish Wood, *Swedish Glulam—New Possibilities for Halls, Arenas, Commercial Buildings and Bridges* (2012)
122. Studiengemeinschaft Holzleimbau e.V., Sternstunden Des Ingenieurholzbaus, 2017
123. Studiengemeinschaft Holzleimbau e.V., *Aufbauen, Umbauen, Nachverdichten: Neuer Holzbau in Bestand Und Fläche*, (2023)
124. V. Çavuş, A.Ç. Konukçu, Innovative approach for engineered wood products: Scrimber, in *Academic Research and Reviews in Agriculture, Forestry and Aquaculture Sciences*, edited by A. Atik (Duvar Publishing, 2021)
125. ETA-18/0960, *Brettstapel(Element), Dübelholz/Element), Dowel Laminated Timber, DLT, DowelLam*
126. S.K. Diederichs, 2010 Status Quo for life cycle inventory and environmental impact assessment of wood-based panel products in Germany. Wood Fiber Sci. **46**, 340 (2014)
127. T. Enters, *Trash or Treasure? Logging and Mill Residues in Asia and the Pacific*. FAO
128. ASMI, *A Cradle-to-Gate Life Cycle Assessment of Canadian Particleboard- 2013 Update Final*, (2018)
129. H. Kouchaki-Penchah, M. Sharifi, H. Mousazadeh, H. Zarea-Hosseinabadi, A. Nabavi-Pelesaraei, Gate to gate life cycle assessment of flat pressed particleboard production in Islamic Republic of Iran. J. Clean. Prod. **112**, 343 (2016)
130. B. Rivela, A. Hospido, M.T. Moreira, G. Feijoo, Life cycle inventory of particleboard: a case study in the wood sector. Int. J. Life Cycle Assess. **11**, 106 (2006)
131. R. Bergman E. Oneil, *Cradle-To-Gate Life Cycle Assessment of North American Hardboard and Engineered Wood Siding and Trim Production* (2016)

132. U. Saal, H. Weimar, U. Mantau, Wood processing residues. Adv. Biochem. Eng. Biotechnol. **166**, 27 (2019)
133 . ASMI, A cradle-to-gate life cycle assessment of Canadian medium density fibreboard (mdf), in *Review of 3D Printing and Potential Red Meat Applications*, vol. 1 (2013)
134. H. Kouchaki-Penchah, M. Sharifi, H. Mousazadeh, H. Zarea-Hosseinabadi, Life cycle assessment of medium-density fiberboard manufacturing process in Islamic Republic of Iran. J. Clean. Prod. **112**, 351 (2016)
135. B. Rivela, M.T. Moreira, G. Feijoo, Life cycle inventory of medium density fibreboard. Int. J. Life Cycle Assess. **12**, 143 (2007)
136. R.E. Thomas, U. Buehlmann, Using low-grade hardwoods for CLT production: a yield analysis, in *6th International Scientific Conference on Hardwood Processing (ISCHP2017)*, 199 (2017)
137. R.D. Bergman, S. Alanya-Rosenbaum, Cradle-to-gate life-cycle assessment of laminated veneer lumber production in the United States. For. Prod. J. **67**, 343 (2016)
138. D. Kaestner, *Compile a Life Cycle Assessment of the OSB and Plywood Industries in the U.S. & Analysis of the OSB and Plywood Industries in the U.S. Based on a Life Cycle Assessment* (2017)
139. M.W. Feng, R.M. Knudson, Effect of Log Rehydration on quality of OSB Strands manufactured from Beetle-Killed Lodgepole Pine. For. Prod. J. **57**, 35 (2007)
140. R.J. Ross, *Wood Handbook—Wood as an Engineering Material* (Forest Products Laboratory, Madison, WI, 2010)
141. F.F.P. Kollmann, W.A. Cote, *Principles of Wood Science and Technology* (Springer-Verlag, Berlin, Heidelberg, 1968)
142. ASMI, *A Cradle-to-Gate Life Cycle Assessment of Canadian Glulam Manufacture* (2018)
143. M.E. Puettmann, J.B. Wilson, Gate-to-gate life-cycle inventory of glued-laminated timbers production. Wood Fiber Sci. **37**, 99 (2005)
144. T. Bowers, M.E. Puettmann, I. Ganguly, I. Eastin, Cradle-to-gate life-cycle impact analysis of glued-laminated (Glulam) timber: environmental impacts from Glulam produced in the US Pacific Northwest and Southeast. For. Prod. J. **67**, 368 (2017)
145. K. Brandt, A. Wilson, D. Bender, J.D. Dolan, M.P. Wolcott, Techno-economic analysis for manufacturing cross-laminated timber. BioResources **14**, 7790 (2019)
146. M. Fredriksson, P. Bomark, O. Broman, A. Grönlund, Using small diameter logs for cross-laminated timber production. BioResources **10**, 1477 (2015)
147. ASMI, *A Life Cycle Assessment of Cross-Laminated Timber Produced in Canada*
148. G. Jauk, *CLT | Special*, Holzkurier (2020)
149. M. Puettmann, A. Sinha, I. Ganguly, *CORRIM Report—Life Cycle Assessment of Cross Laminated Timber Produced in Oregon* (2017)
150. C.X. Chen, F. Pierobon, I. Ganguly, Life cycle assessment (LCA) of cross-laminated timber (CLT) produced in Western Washington: the role of logistics and wood species mix. Sustainability (Switzerland) **11** (2019)
151. D. Mumovic, M. Santamouris, *A Handbook of Sustainable Building Design and Engineering: An Integrated Approach to Energy, Health and Operational Performance*, vol. 10 (2009)
152. J. Cremers, *Building openings construction manual—windows vents exterior doors* (2016)
153. R.D. Bergman, S. Alanya-Rosenbaum, Assessment of Laminated Veneer Lumber (LVL) Produced in the Pacific Northwest Region of the United States
154. J.B. Wilson, E.R. Dancer, Gate-to-gate life-cycle inventory of laminated veneer lumber production. Wood Fiber Sci. **37**, 114 (2005)
155. T. L. Laufenberg, Parallel-laminated Veneer: processing and performance research review. For. Prod. J. **33**, 21 (1983)
156. S. Smulski, *Engineered Wood Products: A Guide for Specifiers, Designers & Users* (Pfs Research Foundation, 1997)
157. CORRIM and WoodLife Environmental Consultants, *Cradle to Gate Life Cycle Assessment of North American Laminated Strand Lumber Production* (2015)

158. P. Stiglbauer, T.E. Conners, S. Banerjee, Influence of knife angle and ambient temperature on fines generation from flakers. For. Prod. J. **56**, 86 (2006)
159. E. Benetto, M. Becker, J. Welfring, Life cycle assessment of oriented strand boards (OSB): from process innovation to ecodesign. Environ. Sci. Technol. **43**, 6003 (2009)
160. TimTek LLC, *TimTek: The Process*. http://timtekllc.com/process.html (Webpage not available anymore)
161. T. Nenning, A. Tockner, J. Konnerth, W. Gindl-Altmutter, M. Grabner, C. Hansmann, S. Lux, M. Pramreiter, Variability of mechanical properties of hardwood branches according to their position and inclination in the tree. Constr. Build. Mater. **419** (2024)
162. D. Parker, A. Wood, *The Tall Buildings Reference Book* (Routledge, Oxon, 2013)
163. ÖNORM EN 1995-1-1, *Eurocode 5: Design of Timber Structures Part 1: General—Common Rules and Rules for Buildings*
164. M. Sarkisian, *Designing Tall Buildings: Structure as Architecture* (2016)
165. EN 338, *Bauholz Für Tragende Zwecke - Festigkeitsklassen*
166. EN 14374, *Holzbauwerke - Furnierschichtholz Für Tragende Zwecke - Anforderungen*
167. B.G. Hermann, C. Kroeze, W. Jawjit (2007) Assessing environmental performance by combining life cycle assessment, multi-criteria analysis and environmental performance indicators. J. Clean Prod. **15**, 1787 (2007)
168. ISO 14044, *Environmental Management—Life Cycle Assessment—Requirements and Guidelines* (2021)
169. ISO 14067, *Greenhouse Gases—Carbon Footprint of Products—Requirements and Guidelines for Quantification* (2019)
170. E. Oneil, R.D. Bergman, M.E. Puettmann, CORRIM: Forest Products Life-Cycle Analysis Update Overview. For. Prod. J. **67**, 308 (2017)
171. EPD157, *EPD for Dowel Laminated Timber Produced by StructureCraft in Abbotsford*. BC Canada (2020)
172. M. Puettmann, J. Salazar, *Cradle to Gate Life Cycle Assessment of North American Medium Density Fiberboard Production* (2019)
173. E. Stricker, G. Brandi, A. Sonderegger, Baubüro in Situ AG, Zirkular GmbH, M. Angst, B. Buser, and M. Massmünster, *Reuse in Construction: A Compendium of Circular Architecture* (Park Books, 2022).
174. D. Densley Tingley, B. Davison, *Developing an LCA methodology to account for the environmental benefits of design for deconstruction*. Build. Environ. **57**, 387 (2012)
175. V. Nicholson, A. Miatto, *Architects as catalysts of reuse in construction*. Nat. Cities 1:3 **1**, 179 (2024)
176. C. De Wolf, E. Hoxha, C. Fivet, Comparison of environmental assessment methods when reusing building components: a case study. Sustain. Cities Soc. **61**, 102322 (2020)
177. C. Totman, *The Green Archipelago: Forestry in Pre-industrial Japan* (University of California Press, 1989).

Chapter 22
Life Cycle Assessment and Circularity of Bio-Based Insulation Materials: An Overview for the Construction Sector

Rosa Agliata, Nouman Khadim, and Elena Lucchi

Abstract The growing interest in low-energy and sustainable architecture has driven a demand for buildings with minimized environmental footprints. Within this context, building insulation materials serve as key players in curbing operational energy usage within the construction sector. However, their production and use have been identified as significant contributors to carbon emissions and overall environmental impacts. Notably, the embodied carbon in insulation products varies significantly, ranging from materials derived from plastics (e.g., EPS, XPS) or mineral fibers (e.g., glass wool, rockwool) to those based on biological components. Remarkably, the latter often exhibit a carbon-negative balance, presenting a promising solution in the quest for sustainable construction materials. Despite this inherent advantage, the global utilization of bio-based insulating materials still hovers below 10%. Wood fiber materials account for nearly 60% of this usage, followed by cellulose fibers constituting approximately 30%, leaving the remaining 10% to encompass all other bio-based insulation products. This chapter aims to provide a comprehensive analysis of the characteristics of both traditional and innovative bio-based insulation materials to understand: (i) how bio-based insulation materials behave in terms of circularity and sustainability; (ii) what CE–LCA considerations are essential for material selection; and (iii) how to make a trade-off between circularity, LCA and physical properties (energy use, insulation, acoustics).

R. Agliata (✉)
Dipartimento di Ingegneria, Università della Campania "Vanvitelli", Aversa (CE), Italy
e-mail: rosa.agliata@unicampania.it

N. Khadim
Vrije Universiteit Amsterdam, Amsterdam, Netherlands
e-mail: n.khadim@vu.nl

E. Lucchi
Università di Pavia, Pavia, Italy
e-mail: elena.lucchi@unipv.it

D. A. H. Hanaor (ed.), *Sustainable Materials for the Built Environment*, Springer Series in Materials Science 352, https://doi.org/10.1007/978-3-031-97818-0_22

22.1 Introduction

Thermal insulation materials play a crucial role in enhancing the sustainability of buildings. By improving energy efficiency, reducing environmental impact, and promoting healthier indoor environments, these materials contribute significantly to sustainable building and energy efficiency practices. They reduce the heat transfer through building envelopes (e.g., walls, roofs, and floors), lowering the energy consumption for heating and cooling. Besides, they maintain stable indoor temperatures, enhancing occupant comfort and reducing thermal fluctuations. Also, proper insulation reduces the risk of condensation of building structures, preventing mold growth and associated health issues. Many insulation materials also offer soundproofing benefits for reducing noise pollution from outside and between indoor environments. However, their production and use have been identified as significant contributors to carbon emissions and overall environmental impacts. Manufacturing insulation typically involves energy-intensive processes, which often rely on fossil fuels, leading to substantial greenhouse gas emissions. The extraction and processing of raw materials, such as petrochemicals for foam insulation or minerals for fiberglass, also have environmental repercussions, including habitat destruction and resource depletion. Furthermore, some insulation materials, like foamed plastics, release potent greenhouse gases during production and disposal.

This dual nature of insulation underscores the importance of sustainable production practices and the selection of eco-friendly materials to minimize negative impacts while maximizing energy savings. Adopting a Life Cycle Assessment (LCA)-based approach for the selection of the most appropriate insulation materials can help highlighting the carbon footprint of the latter throughout their life stages, including production, use, and disposal phases. Besides, focusing on the circularity aspect of the material and the component, permits to identify opportunities for recycling and reusing materials, reducing waste, and conserving resources. This philosophy encourages the development of insulation products designed for longevity and end-of-life recovery, ultimately supporting a more sustainable and resilient built environment.

A thermal insulating material is characterized by the reduced ability to conduct heat flows. This property is quantified for homogeneous materials by "*thermal conductivity*" (λ), which represents the material's intrinsic ability to conduct heat. For non-homogeneous materials, "*thermal conductance*" (C) is used, which accounts for the material's overall capacity to transmit heat, considering its composite nature. Conventionally, thermal insulation materials have a λ value below 0.065 W/mK. They can be broadly classified into organic and inorganic, based on their chemical composition and origin. Inorganic insulation materials do not contain carbon-hydrogen bonds and are composed of mineral or metallic elements. They can be divided into mineral-based (e.g., rockwool, fiberglass, slag wool, cellular glass), and aerogel (e.g., silica and polymer aerogel) [1].

Conversely, organic materials are primarily composed of carbon-based compounds. They can be further divided into petrochemical-derived and bio-based materials. Synthetic materials are derived from petroleum or natural gas through chemical processes and are usually in the form of polymers. Examples are expanded polystyrene (EPS), extruded polystyrene (XPS), polyurethane (PUR), phenolic foam, polyisocyanurate foam (PIR). According to the EU Commission (2020), bio-based materials «[…] *are fully or partially made from biological resources, rather than fossil raw materials*» [2]. Bio-based materials are derived from renewable substances of natural origin, either vegetable or animal [3]. These products offer sustainable alternatives to traditional, petroleum-based items across a wide range of industries, including packaging, agriculture, construction, personal care, and textiles [4]. They are often considered more sustainable and environmentally friendly compared to other categories. Below, the chemical composition, manufacturing processes, and performance of these materials are described to better understand their applicability in buildings.

22.2 Bio-Based Thermal Insulation Materials

Bio-based thermal insulation materials are *fully or partially made from substances derived from living organisms and biological resources* [4]. These materials can be broadly classified into categories based on their origin and processing methods:

- Plant-based materials that derived from crops like cork, wood, natural textile fibers (e.g., hemp, kenaf, coconut, flax, corn, jute, etc.), and cellulose fiber.
- Animal-based materials that sourced from animal products such as sheep wool, horse mane, and feathers.

Most bio-based insulation materials are fibrous due to the natural fiber content in plants or animal wool. Historically, some of these materials were the most widely used in the construction sector. Recently, there has been a renewed interest in these materials to develop more adaptable and sustainable products. In general, their production and installation processes are not harmful to the environment or human health [5]. Furthermore, they are typically recyclable and biodegradable, which significantly contributes to sustainable construction practices. These properties help reduce the carbon footprint of buildings and support the circular economy by enabling the reuse of materials. Additionally, their natural composition often results in lower emissions of volatile organic compounds (VOCs), improving indoor air quality and promoting healthier living environments.

Below, the properties of the various bio-based insulation materials are described, divided into:

- Plant-based materials:
  - Cork (Sect. 22.2.1).
  - Wood fiber (Sect. 22.2.2).
  - Cellulose fiber (Sect. 22.2.3).
  - Natural textile fibers (Sect. 22.2.4).

- Animal-based materials:
  - Sheep wool (Sect. 22.2.5).

22.2.1 Cork

Cork is a vegetable-based material composed primarily of suberin, a hydrophobic substance, along with lignin, polysaccharides, and various extractives. It is harvested from the bark of cork oak trees, processed into granules, and then compressed to form blocks or sheets without the need for synthetic adhesives. Thermal insulation materials are divided into:

- Natural (or pure) cork.
- Expanded cork.

Natural cork (Fig. 22.1) is obtained from the processing of the bark of the cork oak, which grows in a Mediterranean climate. The bark can be harvested once every ten years, respecting the natural reproduction cycles of the plant. The bark is removed from the tree, stripped of its outermost layer, and then crushed. This yields a product in coarse granules, which are ground and left to age for approximately one year. Subsequently, they are boiled to soften and disinfect them from any parasites. The thickness of natural cork insulation panels is typically between 0.2 and 0.10 m. It is available on the market as panels, strips, loose material, and sheets. Natural cork is distinguished by its discreet insulation properties, excellent thermal storage performance, hygiene, impermeability, durability, non-combustibility, and recyclability [5].

Expanded cork is obtained by expanding cork granules in an autoclave with high-temperature steam (350 °C). The heat causes a 20–30% increase in volume and results in the release of resin (suberin), which serves as the natural binder of the conglomerate. Subsequently, the material is subjected to compression, thereby enhancing its compactness. In addition to its thermal and acoustic insulation properties, expanded cork has high resistance to chemical and biological agents, is elastic, mold-resistant, rot-resistant, and stable over time. It is a breathable material, permeable to vapor, hydrophobic, and waterproof.

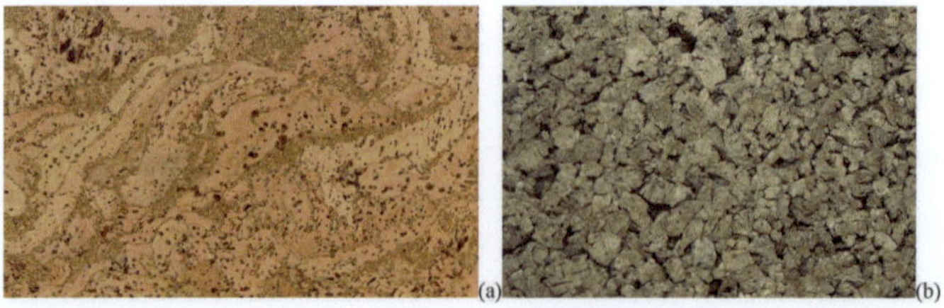

Fig. 22.1 **a** Natural cork; **b** expanded cork. *Photo* Authors

The characteristics of the material are reported considering composition, fabrication process, building applications, performance of density (ρ), λ-value, specific heat capacity (c), and moisture vapor transmission rate (μ) (Table 22.1).

22.2.2 *Wood Fiber*

Wood fiber is composed of cellulose, hemicellulose, and lignin, with small amounts of extractives and minerals. It is produced by breaking down wood into fibers through mechanical or chemical processes, then pressing into mats or boards with the help of natural or synthetic binders. Thermal insulation fibers are divided into:

- Wood fiber, classified according to their density in rigid and flexible.
- Mineralized wood panels.

Wood fiber derives from the processing waste of resinous wood from sawmills. The waste is then subjected to a series of thermomechanical processes, including cutting, grinding and fibrillation. Thermal insulating wood fibers (Fig. 22.2a) are categorized as rigid or flexible, depending on their wood density. In both cases, wood is combined with water and additives (e.g., aluminum sulfate, boron salts) to enhance fire resistance and protection from pests, insects, and mold. Otherwise, wood could be treated with hydrophobic solutions (e.g., bitumen, latex, or beeswax) to improve impermeability and water resistance. The high specific heat of wood ensures high thermal capacity and performance. The fibrous nature of wood provides good acoustic resistance, with sound absorption values dependent on fiber size. The open-pore structure allows the passage of water vapor, and breathability. The material possesses good mechanical resistance, dimensional stability, and damage resistance. However, it is susceptible to attacks from rodents, insects, parasites, and fire [6].

Mineralized wood panels (Fig. 22.2b) are manufactured using timber that has been cultivated rapidly and waste materials derived from the wood processing industry.

The shredded fibers are impregnated with magnesite, Portland cement, or other mineral binders, which renders them compact and rigid, more resistant to biological attack and fire. It has high mechanical, chemical, biological, and fire resistance performance, dimensional stability, and durability. Thermal performance is reduce compared to wood fiber, but it has similar thermal capacity, and higher damping of temperature fluctuations. The material's acoustic performance is ensured by the presence of voids. The characteristics of the material are reported below (Table 22.2).

22.2.3 *Cellulose Fiber*

Cellulose fiber (Fig. 22.3) primarily consists of cellulose, a complex carbohydrate or polysaccharide. It derives from the processing of newspapers and magazines, which are ground and fibrillated to facilitate recycling.

Table 22.1 Characteristics of cork insulation materials considering composition, fabrication process, building applications, ρ, λ-value, c-value, and μ coefficient

Material	Composition	Fabrication process	Building applications	Type	Performances			
					ρ (kg/m^3)	λ (W/mK)	c (J/kgK)	μ (g/m^2/h)
Pure cork	Cork derived from the bark of *Quercus suber* and binders (e.g., polyurethane. synthetic resins)	1. Cork harvesting 2. Drying to stabilize its moisture content 3. Boiling in water to increase the volume 4. Cooling and flattening 5. Granulation 6. Mixing with binders 7. Molding and pressing 8. Cutting and shaping	Insulation of walls, roofs, floors Light-weighting of screeds and plasters	Loose natural cork	80–120	0.043–0.049	1700–2000	5–11
				Mixed natural cork	200–450	0.070–0.10	1700–2000	5–11
Expanded cork	Cork	1–5. Same process of pure cork granules 6. Heating to expand the suberin for binding the granules into solid boards 7. Pressing and cooling 8. Cutting and shaping	Insulation of walls, roofs, floors, ceilings, and casings	Loose expanded cork	65–75	0.034–0.030	1600–2000	9–19
				Mixed expanded cork	200–450	0.048–0.053	1600–2000	9–19

Source Authors' elaboration

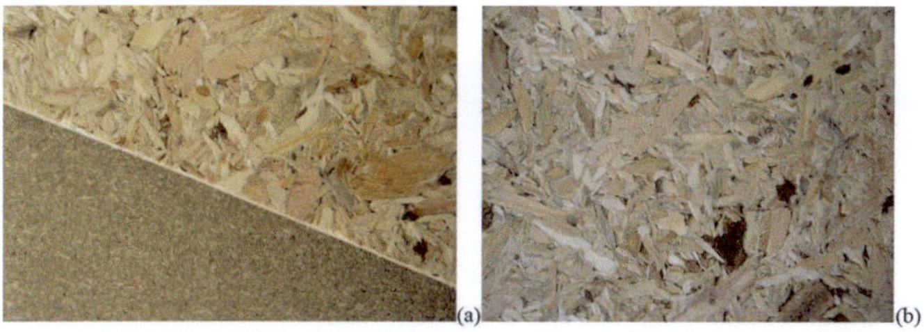

Fig. 22.2 **a** Flexible and rigid wood fiber; **b** mineralized wood panels. *Photo* Authors

The flakes are subjected to boron salts to enhance fire and biological resistance. The material is breathable, hygroscopic, and durable, as well as has elastic properties, minimal deformation, and high resistance to compression and biological agents. Its hygroscopic properties permit it to absorb more than 15% of its weight in water vapor. Nevertheless, prolonged contact with water results in the deterioration of cellulose, leading to rot and degradation. Thermal and acoustic performance are high thanks to the presence of air and fibers oriented in all directions. Furthermore, it has good thermal phase shift and attenuation characteristics. Nevertheless, its fire resistance is diminished, necessitating the application of fire-retardant surface finishes. The material is available in bulk form or as semi-rigid panels, which are produced by weaving cellulose with polyester or jute fibers. Rigid panels are created through the hot processing of cellulose with binders and support fibers. The characteristics of the material are reported below (Table 22.3).

22.2.4 Natural Textile Fibers

Natural textile fibers are a complex of products cropped-by secondary materials left after the primary agricultural products that are processed into insulation materials. These fibers are harvested, cleaned, and processed into fibers through mechanical means, and then spun and formed into insulation batts or loose-fill. Thermal insulation fibers are divided into:

- Hemp fiber
- Kenaf fiber
- Flax (or linen) fiber
- Coconut fiber
- Corn fiber
- Jute fiber
- Rice fiber
- Mixed textiles fiber.

Hemp fiber (Fig. 22.4a) derives from the processing of the hemp plant, which is cultivated in temperate climates to produce oilseeds, textiles, and fodder. The fiber

Table 22.2 Characteristics of wood fibre insulation materials considering composition, fabrication process, building applications, ρ, λ-value, c-value, and μ coefficient

Material	Composition	Fabrication process	Building application	Type	Performances			
					ρ (kg/m^3)	λ (W/mK)	c (J/kgK)	M (g/m^2/h)
Rigid wood fiber	Wood waste from softwood trees (e.g., pine, spruce, fir) and binders (e.g., lignin)	1. Collection and preparation of raw wood 2. Thermomechanical pulping for mechanical fiberizing 3. Bonding with binders 4. Forming and/or pressing e to achieve the desired density and thickness 5. Drying in kilns to remove excess moisture and stabilize the material 6. Cutting and finishing	Insulation of walls, roofs, ceilings, and floors	–	45–55	0.038	1600–2300	1–2
Flexible wood fiber	Wood waste from softwood trees and binders	1–3. Same process of rigid wood fiber 4. Forming and cutting 5. Drying and packaging		–	150–300	0.038–0.055	1600–2300	5–10
Mineralized wood panels	Wood fibers, Portland cement binder, and additives	1. Wood wool production 2. Mixing with cement to create a slurry 3. Molding and forming 4. Compression to remove excess water and compact the mixture 5. Curing at controlled humidity and temperature to harden the cement 6. Cutting and finishing with treatment or colors		Mineralized wood with magnesite	300–625	0.086–0.107	1700–1900	4–10
				Mineralized wood with cement	360–1000	0.060–0.090	1800–2100	5–13

Source Authors' elaboration

Fig. 22.3 Cellulose fiber. *Source* Authors

is subjected to chemical treatments to improve the fire resistance, and addition of adhesive polyester fibers, and potato starch to enhance strength and cohesion of felts and panels. The material has good thermal and acoustic insulation, thermal inertia, breathability, hygroscopicity, and resistance to mold and biological agents [7]. Nevertheless, it has reduced compressive strength especially for wall or ceiling applications that requires reinforcement systems to guarantee mechanical stability [5].

Like hemp, Kenaf fiber (Fig. 22.4b) derives from the processing of a specific plant belonging to the hemp family, cultivated in temperate climates to produce sacks. It has good thermal and acoustic insulation properties, high thermal inertia, breathability, hygroscopicity, mechanical strength, dimensional stability and resistance to mold and other biological agents. Additional additives are required to enhance its fire resistant properties.

Flax fiber (Fig. 22.4c) derives from the processing of a plant from the *Linaceae* family that is cultivated to produce oilseeds and textiles. Insulating materials are produced from fibers discarded by the textile industry with the addition of polyester fibers to enhance their strength. The material is low degradable, resistant to biological attack, stable, and durable. Additional additives are required to improve fire resistance. It has good hygroscopic properties, although it becomes susceptible to decay with prolonged contact with water. Furthermore, it has good thermal and acoustic insulation properties, as well as thermal inertia [8].

Coconut fiber (Fig. 22.4d) is obtained from the fruit of the coconut palm, cultivated in tropical climates to produce textile fibers for furnishings and mats. It has good durability, mechanical resistance to compression, friction, and biological attack, thermal and acoustic performance, breathability, and thermal inertia. It is fire-sensitive but burns very slowly.

Corn fiber derives from the processing of maize kernels. It has good thermal and acoustic performance, breathability, biodegradability as well as low toxic emissions, and smoke emission during combustion.

Table 22.3 Characteristics of cellulose fibre insulation materials considering composition, fabrication process, building applications, ρ, λ-value, c-value, and μ coefficient

Material	Composition	Fabrication process	Building applications	Type	Performances			
					ρ (kg/m^3)	λ (W/mK)	c (J/kgK)	M (g/m^2/h)
Cellulose fiber	Paper waste (e.g., newspapers, cardboard) and fire retardants (e.g., borates or ammonium sulfate)	1. Removing of inks and other contaminants 2. Mechanical or chemical shredding into fibers 3. Treatment with fire retardants 4. Production of loose-fill or dense-packed batts	Blown into attics and wall cavities	Cellulose fiber in flakes	25–500	0.037–0.053	1600–2000	1–3
				Cellulose fiber in panels	65–90	0.090	1600–2000	1–3

Source Authors' elaboration

Fig. 22.4 **a** Hemp fiber; **b** kenaf fiber; **c** flax fiber; **d** coconut. *Source* Authors

Jute fiber is obtained from the processing of jute, a plant cultivated in Southeast Asian countries for textile production. It has good durability, hygroscopicity, breathability, acoustic performances, electrostatically neutrality, and recyclability [6].

Rice husk insulation is produced from the agricultural waste of rice hulls. It has medium thermal and acoustic performance, moisture resistance, durability, and stability. It should be protected from prolonged exposure to high humidity or water to prevent degradation [8]. The characteristics of these materials are reported below (Table 22.4).

22.2.5 Sheep Wool

Sheep wool (Fig. 22.5) consists mainly of keratin, a fibrous structural protein, with minor amounts of lipids and other organic materials. It is obtained from the shearing of the animal's fleece that is treated with natural soaps and sodium carbonate to remove lanolin and impurities, and then carded into fibers.

These fibers are treated with non-toxic agents, such as insecticides or boron salts, to enhance fire resistance and prevent pest infestation and moth, before being formed into insulation batts or rolls. Furthermore. the addition of hemp and polyester fibers can enhance the rigidity of the material. Sheep wool has medium thermal and acoustic performances, thermal capacity, hygroscopicity, and durability. although it can be susceptible to attacks by moths and parasites. The characteristics of the material are reported below (Table 22.5).

Table 22.4 Characteristics of natural textile fibre insulation materials considering composition, fabrication process, building applications, ρ, λ-value, c-value, and μ coefficient

Material	Composition	Fabrication process	Building applications	Type	Performances			
					ρ (kg/m^3)	λ (W/mK)	c (J/kgK)	M (g/m^2/h)
Hemp fiber	Cannabis sativa waste	1. Extraction of fibers from the stalks of the plant 2. Removing of impurities 3. Cleaning and carding of fibers 4. Possible treatment with polyester or other binder 5. Production of batts or rolls	Insulation of cavities, roofs, ventilated coats, partition walls, subfloors, and ceilings	–	25–115	0.039–0.060	1300–1900	1–2
Kenaf fiber	Hibiscus Cannibus waste	1–3. Same process of hemp fiber 4. Production of batts or rolls	Insulation of cavities, roofs, ventilated coats, ceilings, partition walls, floors	–	20–120	0.033–0.050	1600	1–2
Hemp-kenaf	Mixing of hemp and kenaf fibers	1–3. Same process of hemp and kenaf fiber 4. Mixing of hemp and kenaf fibers 5. Production of batts or rolls	Insulation of roofs and walls	–	25–115	0.033–0.050	1600	1–2

(continued)

Table 22.4 (continued)

Material	Composition	Fabrication process	Building applications	Type	Performances			
					ρ (kg/m^3)	λ (W/mK)	c (J/kgK)	M (g/m^2/h)
Flax fiber	Linum Usitatissimum waste and recycled linen	1. Mechanical procession of flax and waste fibers to remove impurities 2–5. Same process of hemp fiber	Insulation of cavities, Roofs, ceilings, floors, holes in walls	–	20–35	0.038–0.044	1300–1700	1–2
Hemp-linen	Mixing of hemp and linen fibers	1–3. Same process of hemp fiber 4. Mixing of hemp and linen fibers 5. Production of batts or rolls	Insulation of roofs and walls	–	25–35	0.038–0.044	1300–1600	1–2

(continued)

Table 22.4 (continued)

Material	Composition	Fabrication process	Building applications	Type	Performances			
					ρ (kg/m^3)	λ (W/mK)	c (J/kgK)	M (g/m^2/h)
Coconut fiber	Cocos Nucifera waste	1. Carding to obtain a uniform direction 2. Mat or sheets formation with mechanical or chemical bonding techniques 3. Mechanical needling to improve cohesion and strength 4. Heat treatment to improve thermal insulation properties and stability 5. Finishing to improve durability and environmental resistance	Insulation of walls and floors not in direct contact with the environment Drainage of green roofs and reinforcement of steep terrain	–	50–150	0.043–0.047	1300–1600	1–2

(continued)

Table 22.4 (continued)

Material	Composition	Fabrication process	Building applications	Type	Performances			
					ρ (kg/m^3)	λ (W/mK)	c (J/kgK)	M (g/m^2/h)
Corn fiber	Corn food waste	1. Extraction of fibers from the corn husks and stalk 2. Cleaning to remove impurities or remaining plant material 3. Drying to reduce moisture 4. Carding to align in a uniform direction 5. Mat or sheets formation using mechanical or chemical bonding 6. Compression to increase density 7. Finishing to improve durability and environmental resistance	Insulation of cavities, ventilated coats, ventilated roofs, partition walls, and ceilings	Corn fiber	10–80	0.038–0.053	1900	3
				Corn husk	119–128	0.047		

(continued)

Table 22.4 (continued)

Material	Composition	Fabrication process	Building applications	Type	Performances			
					ρ (kg/m^3)	λ (W/mK)	c (J/kgK)	M (g/m^2/h)
Jute fiber	Jute waste	1. Mechanical procession of fibers to remove impurities 2. Cleaning of fibers 3. Treatment with borates or other fire retardants 4. Production of batts or rolls	Insulation of cavities and joints between walls and windows, slabs, and floors Plaster base mesh for plastering walls or radiant wall systems	–	100	0.050–0.055	–	1–2
Rice	Rice plant waste + binders (e.g., lime, cement, organic resin)	1. Collection of rice after the rice milling process 2. Cleaning to remove impurities and contaminants 3. Grinding and binding to improve the performance 4. Compression into boards 5. Drying to remove excess moisture 6. Cutting and shaping	Insulation of walls, and roofs	Untreated Rice straw particle board	500	0.086	524	–
				Rice husk composite	378	0.08	524	–

(continued)

Table 22.4 (continued)

Material	Composition	Fabrication process	Building applications	Type	Performances			
					ρ (kg/m^3)	λ (W/mK)	c (J/kgK)	M (g/m^2/h)
Rye straw	Rye plant waste + binders (e.g., casein, liquid glass)	1. Collection of rye straw 2. Straw chopping 3. Straw tubes sifting 4. Mixing straws and binder 5. Shaping in panels 5. Curing	Padding for wooden walls (additional) Insulation plates	Tubes	215	0.059	–	–
Sunflower	Sunflower stalks + binder (e.g., chitosan)	1. Grinding of sunflower stalks (e.g., cutting mill) 2. Sieving of the obtained particles 3. Mixing particles with the binder 4. Compression (shaping) and drying	Vegetal concrete	Fibres	150–200	0.056	–	–
Wheat	Wheat plant waste + binders (e.g., lime, plaster, liquid glass)	1. Collection of wheat waste 2. Chopping 3. Tubes sifting 4. Mixing with binder 5. Shaping in panels 5. Curing	Insulating plates	Wheat husk	448	0.1	718	–
				Wheat straw	220	0.056	–	–

(continued)

Table 22.4 (continued)

Material	Composition	Fabrication process	Building applications	Type	Performances			
					ρ (kg/m^3)	λ (W/mK)	c (J/kgK)	M (g/m^2/h)
Eucalyptus	Waste from eucalyptus short-rotation plantation	1. Wood chipping 2. Chips classification and sorting 3. Crushing and drying 4. Second milling 5. Mixing 6. Panel manufacturing	Insulation panels	Bark	25–100	0.045–0.049	2198	–
				Leaves	217	0.049–0.065		
Onion skins and peanut shells	Waste from onions and peanuts + binder (e.g., cement, gypsum)	1. Oven-drying 2. Pulping 3. Mixing onions shells and peanuts skins 4. Sifting 5. Soaking in water for softening 6. Filtering 7. Panels shaping	Radiation absorbing concrete	–	1230–3330	0.165–0.313	–	–

(continued)

Table 22.4 (continued)

Material	Composition	Fabrication process	Building applications	Type	Performances			
					ρ (kg/m^3)	λ (W/mK)	c (J/kgK)	M ($g/m^2/h$)
Date palm	Waste from dates production (with or without binder)	1. Washing 2. Drying 3. Cutting 4. Adding binder 5. Board manufacturing 6. Loading in a plexiglass box (binderless type) 7. Addition of more fibres (binderless type)	Insulation panels Palm based concrete	Fibres	121 (no binder)—1217 (with binder)	0.033	–	–
Peat	Peatland	Boards are produced by air-lay technology with utilization of bicomponent PES fibers	Building shells, attics, upper floors, and intermediate floors	Boards	64	0.038	1408	–
				Loose-fill insulations	131	0.043		
Moss	*Sphagnum* moss	Under development (unmarketed)		Loose-fill insulations	89.9	0.037	1289	–

(continued)

Table 22.4 (continued)

Material	Composition	Fabrication process	Building applications	Type	Performances			
					ρ (kg/m^3)	λ (W/mK)	c (J/kgK)	M ($g/m^2/h$)
Pineapple	Pineapple by-products from agroindustry + clay	Not specified (unmarketed)	Insulation panels Slabs	–	1098	0.048	–	–
Recycled textile fiber	Textile waste (e.g., discarded denim, clothing)	1. Mechanical procession of textile to remove impurities 2. Cleaning of fibers 3. Treatment with borates or other fire retardants 4. Production of batts or rolls	Insulation of cavities and joints between walls and windows, slabs, and floors	Performances are different for each material				

Source Authors' elaboration

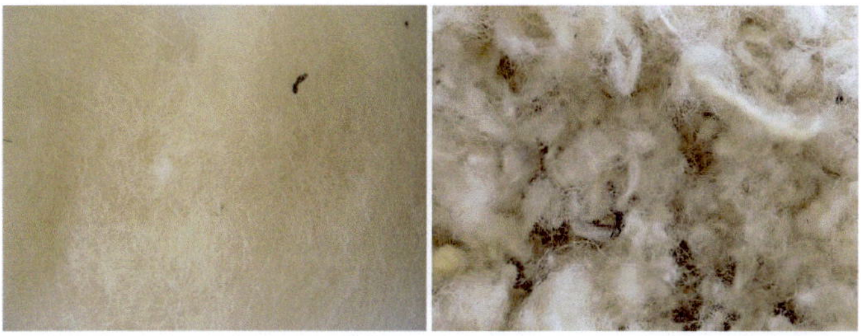

Fig. 22.5 Sheep wool. *Source* Authors

22.3 Environmental Impacts and Carbon Footprints of Insulation Materials

Environmental impact means «[…] *any change to the environment wholly or partially resulting from a product during its life cycle*» [8] while Life cycle means «[…] *the consecutive and interlinked stages of a product from raw material use to final disposal*» [9].

22.3.1 Cork

Cork is derived from renewable materials, specifically the bark of cork oak trees, which are limited in availability and confined to specific geographic regions. Additionally, cork trees have very long regeneration times, requiring several years for the bark to be harvested sustainably. Despite these constraints, the production and processing phases of cork have relatively low environmental impacts. The primary environmental concern in its life cycle is related to the transportation phase, which can generate significant emissions due to the distances involved in moving the raw and finished products. The use of synthetic adhesives may result in the emission of VOCs. Upon reaching the end of its useful life, the material can be reused or composted by simple grinding of the finished panels. The use of synthetic adhesives in cork products can lead to the emission of VOCs, which may affect indoor air quality. However, advancements in eco-friendly adhesives are helping to mitigate this issue. At the end of its useful life, cork materials can be reused or composted, promoting a circular economy. Simple grinding of finished panels allows the material to be repurposed for other applications or composted, returning valuable nutrients to the soil. Overall, cork's life cycle analysis highlights its potential as a sustainable material, with benefits such as low production impacts and recyclability. Addressing transportation emissions and VOCs from adhesives can further enhance its environmental profile.

Table 22.5 Characteristics of sheep wool insulation materials considering composition, fabrication process, building applications, ρ λ-value, c-value, and μ coefficient

Material	Composition	Fabrication process	Building applications	Type	Performances			
					ρ (kg/m^3)	λ (W/mK)	c (J/kgK)	M (g/m^2/h)
Sheep wool	Sheep wool fibers	1. Shearing of sheep 2. Removing dirt. grease (lanolin). and other impurities 3. Washing in in a series of baths 4. Drying to remove excess moisture 5. Carding for fiber separation and continuous web of wool creation 6. Blending to achieve uniformity 7. Felting and possible bonding to improve cohesion. fire resistance. structural integrity. durability 8. Forming and cutting 9. Packaging	Insulation of walls, roofs, floors	Sheep wool batts	20–35	0.035–0.040	1000–1800	1–2
			Insulation of cavity wall, roofs, and irregular spaces	Sheep wool rolls	13–20	0.040–0.045	1000–1800	1–2
				Sheep wool soft mats	30	0.0318	1000–1800	1–2
				Sheep wool loose-fill fibers	75–100	0.035	1000–1800	1–2
Horse mane	Horse's hair from the upper neck part	1. Collection of horse mane 2. Unlocking by hands and carding 3. Wet felting, rolling, and felting 4. Washing and drying	Thermal and acoustic applications in buildings	Binderless	200–250	0.084–0.087		–
Feathers	Chicken feathers	1. Crushing 2. Washing	Building shells, attics, upper floors, and intermediate floors	Loose-fill insulations	38.3	0.033	1331	–

Source Authors' elaboration

22.3.2 Wood Fiber

Fiber-based wood insulation, derived from natural wood fibers, offers excellent sustainability credentials. Wood fiber has a positive energy balance, with reduced energy consumption during production, installation, and recycling phases. However, the environmental impact of transportation from production areas to application sites must be evaluated, particularly for long-distance transport, as it can diminish the material's eco-compatibility. Pollutant emissions during production are primarily linked to additives, which are used in minimal concentrations. Upon reaching the end of its useful life, the material can be recycled and reintroduced into the processing cycle. Alternatively, if it has not been treated with bitumen or adhesives, it can be composted. Mineralized wood fiber, although originating from natural wood fibers, undergoes a mineralization process involving the infusion of minerals or binders. The presence of binders increases emissions during combustion as well as poses challenges for reuse. One potential reuse option involves crushing the panels for use as inert material in concrete construction, such as lightweight subfloors, composite slabs, and prefabricated elements. Both fiber-based and mineralized wood insulation offer sustainable solutions for thermal insulation in buildings, with fiber-based variants characterized by better biodegradability, and reuse. Efforts to optimize transportation logistics and reduce emissions from adhesives and binders can further enhance their environmental profiles.

22.3.3 Cellulose Fiber

The environmental balance of cellulose fiber is positive, particularly for flakes derived from recycled material. This results in decreased energy consumption across the production, transportation, installation, and recycling stages. Upon reaching the end of its useful life, the material can be reintroduced into the processing cycle or disposed of in waste-to-energy plants. During production, pollutant emissions are primarily attributed to the additives used, which are present in minimal concentrations. In contrast, panels require greater energy expenditure during processing and may result in emissions associated with additives and binders. Similarly, recycling is a considerably more complex and costly process. Thus, while its production and recycling phases boast minimal environmental impacts, attention must be directed towards optimizing processing methods and recycling techniques for panels to align with stringent eco-standards.

22.3.4 Natural Textile Fibers

The environmental balance of hemp is positive, as it is a renewable raw material with a relatively short growth cycle and is easily adaptable to areas with very different climates. If cultivated correctly. hemp does not require the use of fertilizers, chemical additives, or herbicides. thereby negating any potential ecological risks. Conversely, it is capable of soil remediation and improvement. The primary source of pollutant emissions is the production of synthetic fibers used as mechanical support. The environmental balance of kenaf is positive, as it is a renewable raw material with a relatively short growth cycle and is widely distributed across European territories. The cultivation of kenaf does not require the use of fertilizers or pesticides, which has the beneficial effect of improving soil quality. The problem is the high water required for irrigation, particularly during the initial stages of germination. Pollutant emissions result from the production of synthetic fibers used as mechanical support. Linen is a renewable raw material with a positive environmental balance. It has a short growth cycle. It is widely spread across European territories, and its cultivation does not require the use of fertilizers, herbicides, or pesticides. The emissions that result from the production of synthetic fibers used as mechanical support are those that are inherent to the manufacturing process. The environmental balance of coconut fiber is influenced by transportation, which heavily impacts emissions and the final energy consumption of the material.

22.3.5 Sheep Wool

Sheep wool is a renewable material that is widely available in areas where sheep are raised. The quantity of material obtained from shearing exceeds the demand of the textile market. The production and processing stages have a minimal environmental impact, particularly in the case of loose wool. The production of mixed panels is associated with a higher environmental cost. Once the material has reached the end of its life cycle, it can be recycled; however, stabilizing treatments must be repeated. Mixed materials must be sent to landfills and waste-to-energy plants.

A synthesis of environmental performance of different bio-based insulation materials is reported below (Table 22.6).

22.4 Circular Principles for Insulation Materials

Circular economy is an emerging concept that contrasts with the traditional take-make-dispose pattern of consumption by emphasizing closed material loops (take-make-reuse) through regeneration, reuse, remanufacturing, refurbishment, and recycling. While there are various definitions of circularity, it can be broadly defined as

Table 22.6 Comparison of environmental performance of different bio-based insulation materials

Material	Renewability	Recyclability	Carbon sequestration	Embodied energy	Production process	Healthiness
Cork	Harvesting: every 9–12 year	Very easily	Very high (in harvested cork this capacity is 5 times higher than in non-harvested one)	Very low	Boiling and cutting the bark + No need for harmful chemicals or additives	Does not emit harmful chemicals or VOCs + Antimicrobial properties
Wood fibres	Harvesting time depends on the species. Renewable practices possible	Easily repurposed	High (it stores CO_2 that the plants absorbed during its growth)	Low	Few chemicals and additives required during production	Reduced potential for indoor air quality issues
Cellulose fibres	Made from recycled paper products	Easily	Average (derived from plant material)	Low	Reduces deforestation and the environmental impacts of logging	Does not off-gas harmful chemicals + Does not contain irritants
Natural textile fibres	Harvested and regrown	Generally possible	High	Low	Requires minimal chemical processing	Non-toxic + Antimicrobial properties
Sheep wool	Shoring: one per year	Easily reprocessed	High	Very low	Washing, carding, and felting	Non-toxic + Antimicrobial properties

Source Authors' elaboration

«(…) *an economic system that is based on business models which replace the 'end-of-life' concept with reducing, alternatively reusing, recycling and recovering materials in production/distribution and consumption processes, thus operating at the micro level (products, companies, consumers), meso level (eco-industrial parks) and macro level (city, region, nation and beyond), with the aim to accomplish sustainable development, which implies creating environmental quality, economic prosperity and social equity, to the benefit of current and future generations*» [10]. Circular design, also known as Design for Circularity (DfC), translates the principles of the circular economy into practical application. Given the massive environmental footprint and high energy demand of the buildings and construction industry, this sector is at the forefront of the transition to circularity. Construction practitioners are actively incorporating circular thinking into building materials, components, and structures. Insulation materials are a prime example. They significantly influence a building's energy consumption during operation, which is often the largest source of carbon emissions. Additionally, many insulation materials have a high environmental impact during their production and disposal stages. Research on 503 houses found that insulation material production is responsible for 26.1% of total carbon emissions, second largest contributor after concrete production [11]. Thus, implementing circularity in insulation materials is vital. This section briefly discusses the five main circularity principles presented by the Ellen MacArthur Foundation (EMF) and their application to insulation materials.

22.4.1 Design Out Waste [Output Side]

Design phase is known as the most effective phase to eradicate waste and implement circular principles. Waste is eliminated when materials and components are designed for reuse through disassembly and refurbishment, a concept known as "closing the material loops". These loops are categorized into biological and technical loops. Biological nutrients are non-toxic and can be easily composted. Bio-based insulation materials like cork, wood, and cellulose fibers are preferred because they can be returned to the natural environment through non-toxic and restorative loops. This generates a closed-loop system where the materials decompose and become nutrients for new growth. However, some bio-based materials may contain contaminants, such as adhesives, that hinder biodegradability. To address this, binder-less insulation materials using coconut husk and bagasse have been developed [12]. Similarly, technical nutrients include polymers and composite materials, should be designed for disassembly and reuse. They should consume minimum energy for reuse/recycle and retain optimal quality at the end of their lifecycle.

22.4.2 Build Resilience Through Diversity

Insulation materials should be designed for longevity and resilience, adapting to external shocks and efficiently meeting ever-changing requirements throughout their lifecycle. This practice is known as "*slowing down the loops*". Buildings are typically designed to last 50–60 years, but they can often endure for up to 100 years. Given such a long lifespan, the purpose and function of buildings continually evolve, making functional requirements hard to predict. To avoid excessive waste and repair during renovations, insulation systems should be designed to last longer and adapt to future needs. However, achieving this may require trade-offs between various material properties, such as balancing high porosity, which enhances insulation but reduces mechanical strength. Durable insulation materials with high moisture resistance and good UV stability are preferred. Additionally, modular designs for insulation systems facilitate future modifications and easy repairs.

22.4.3 Optimize Energy Use

As discussed in Sect. 2.2, materials fabrication process can be energy intensive. Therefore, materials that require less energy for manufacture, repair and reuse should be preferred. Table 22.6 provides a comparison of embodied energy and renewable capability of various materials. The renewable energy sources should be used in manufacturing, construction, disassembly phase of the materials to reduce the embodied energy. Similarly, the use phase of buildings often accounts for the highest energy consumption and carbon emissions. Insulation materials play a crucial role here. Favor materials with high insulation efficiency, as they can significantly reduce the overall energy demand of a building, lowering its environmental impacts. LCA can be valuable tools for selecting the most suitable materials. These assessments consider the entire life cycle of a product and identify options with the lowest overall energy consumption. Insulation materials made from cork, sheep wool and cellulose are known for having lower embodied energy compared to traditional options.

22.4.4 Thinking in System

A building functions as a complex system, with each element affecting and being affected by others. This necessitates a holistic design approach that considers insulation materials not just in isolation, but in relation to the whole building, environment, and social context. Buildings are viewed as "material banks", with components having varying lifespans and maintenance cycles. Brand divided the building into "Six Layers framework" (site, structure, skin, space plan, services, and stuff) [13]. Insulation materials play important role in skin, structure and space plan. Thermal or acoustic insulation is usually provided in roofs, elevations and external walls, these elements form the skin of the building. Partition walls (space plan) may also have insulation materials. In few cases, insulation may be integrated into structural components such as insulated concrete forms. Since every layer has its own function and lifespan, circular design principles become critical. A 'system thinking' is required. Designers need to consider how insulation interacts with other parts (walls, roofs), how easy it is to access and remove, how to assemble and disassemble it for future reuse or recycling, and finally, how to repair or replace it with minimal damage to surrounding elements. By making insulation easy to access through modular design, repairs become more circular, minimizing harm to connected parts.

22.4.5 *Waste is Food [Input Side]*

In theoretical terms, waste is non-existent in a fully circular system. On input side, the waste should be fed back into system as 'feedstock' instead of virgin raw material. This waste input can come from the same process or even from entirely different industries. The bio-based insulation materials discussed in Sect. 2.2 exemplify this principle. They utilize agricultural waste (corn stalks, wheat straw, etc.), paper waste (newspapers), and wood industry residuals to create insulation materials. This eliminates the need for end-of-life waste treatment in first lifecycle and virgin materials extraction in following lifecycle. Technical waste from textiles, rubber, and plastics can also be upcycled into thermal insulation materials [12]. In circularity certain strategies are preferred over others. For instance, upcycling that prioritizes reusing materials in their original form or for a higher-quality purpose, is favored over traditional recycling which can lead to material degradation. In this regard, 'R-Imperatives' also known as 'R-Ladder' and'10 R framework' provides a hierarchy of various approaches that can be used to implement circularity principles. As shown in Table 22.7, Refuse (R0) is the most preferred option, while Recovery (R9) is the least desired.

22.5 LCA and Circularity Indicators (CI) for Insulation Materials

Implementing circularity and sustainability principles in construction such as insulation materials is a complex process that requires several considerations and transformations in the usual practices. The evolving nature of circular economy and blurriness

Table 22.7 R-Imperatives for circular design of insulation materials

Lifecycle phase	R-imperative	Elaboration
Design (inner circles)	R0 Refuse	• Prevent the use of toxic or scarce substance in manufacturing • Rethink about maintenance and end of life of insulation materials • Decrease the use of raw material (light weight construction) • Reuse the reclaimed materials and enable reuse by design
	R1 Rethink	
	R2 Reduce	
	R3 Reuse	
Use (medium circles)	R4 Repair	• Extend the life through easy repair and maintenance
	R5 Refurbish	
	R6 Remanufacture	
	R7 Repurpose	
End of life (outer circles)	R8 Recycle	• Process waste into new products or input material • Energy recovery from the material
	R9 Recover	

Source Authors' elaboration

regarding its conceptual foundation adds another layer of difficulty. Tracking progress towards circularity and sustainability goals can be challenging. However, assessment methods such LCA and circularity indicators offer objective ways to measure the circular and environmental performance of the materials. These methods provide a common language to communicate between stakeholders and allow the comparison between different alternatives. The analysis can help identify the environmental and resource intensive processes that can be optimized to improve the overall performance of the material. Additionally, they provide a holistic lifecycle perspective of the material loops which is vital aspect of multicycle circular economy. This section explores how LCA and circularity indicators can be applied to evaluate bio-based insulation materials.

22.5.1 Circularity Indicators

Circularity indicators are assessment methods that transform the circular principles into measurable values. Indicator developed for buildings are known as building circularity indicators. Indicators can be nano (material), micro (building), meso (industrial symbiosis) and macro (city and region) level. Among the available indicators for building and materials, the Whole Building Circularity Indicator (WBCI) stands out as a multi-level, multi-cycle innovative framework. It synthesizes existing information and offers a lifecycle perspective to track circularity performance. Based on quantitative material, design, and building data, the WBCI provides circularity ratings on a scale from 0 (fully linear) to 1 (fully circular) for materials, components/elements, systems, and the whole-building. While it compares material and component alternatives, it also sheds light on how choice made at material level can significantly influence the building's circularity performance [14]. Research highlights that the supply chain level circularity of insulation materials is quite low, for instance it stands at 1.5% in Swiss residential building sector. The main reason for low circularity score is the difficulty to deconstruct and collect insulation material as a single trough. The insulation materials are generally glued (chemical connections) to the construction elements that hinders the fast and residual free recovery. Furthermore, dedicated recycling facilities are scarce, currently limited to EPS, stone wool, glass wool, and wood fiberboards [15]. In this regard, WBCI is a guiding tool to improve the circularity performance of insulation materials. Building on the WBCI's mathematical and theoretical framework, Table 22.8 outlines key factors influencing circularity performance of influence materials. It explains these factors, provides relevant examples, and explains their impact on overall circularity scores. The application of material circularity indicator on individual biobased insulation materials, such as flax, reveals their potential to enhance the circularity value of façades or walls. Flax insulation exhibits a low linear flow value and a high material circularity value due to its high biobased content and ease of biodegradability at the end of its life. However, this impact becomes negative or insignificant when seen as an element (i.e., wall, roof), due to disassembly and other technical concerns [16].

Table 22.8 Explanation of factors in circularity indicators for bio-based insulation materials

Factor	Level	Impact on circularity	Explanation	Examples
Input feedstock	Material	Very high	• Avoid use of virgin raw material by using biobased, reuse or recycle content in manufacturing	Cellulose fibers, natural textile fiber that use waste as raw material
Output/waste	Material	Very high	• Materials that can be reuse, recycle or biodegraded at end of life • Disassembly, residual value and avoiding contamination are important criteria	Binder-less natural fibers, horse mane, wood fibers that can be repurposed
Utility and service life	Material	Medium	• Longevity, resilience, and enhanced material utility than average traditional product	Coconut fiber, hempcrete
Efficiency of recyclability	Material	Low	• Recycling process that produces high quality output and produce very less waste	Sheep wool, cork, cellulose fiber, natural textile fiber made by denim
Repair, maintenance and construction loss	Material	Medium	• Materials that produce less waste during manufacturing and installation • Materials that require less repair or maintenance throughout building lifecycle	Cork, natural textile fibres
Disassembly	Component	High	• More relevant to component in which insulation material is installed • Insulation system that are disassemble, have dry connects (screws instead of glues), cause no damage to connected elements	Insulation materials in drywalls, disassemblable roofs assemblies
Adaptability	Building	Medium	• Adaptive to change such as building functions (residential to office), renovation or change in layout	Insulation system designed with future proofing

Source Authors' elaboration

22.5.2 *Lifecycle Assessment for Biobased Insulation Materials*

As discussed earlier, some insulation materials demonstrably reduce building operational energy use, while others, such as bio-based materials, appear environmentally beneficial due to their lower manufacturing and disposal emissions. To choose the most suitable material for a specific case, it is essential to analyze the material from a holistic perspective. LCA bridges this gap by considering emissions across the entire product lifecycle, including raw material extraction, manufacturing, operational energy use in buildings, and end-of-life deconstruction, waste disposal, and recycling. Similar to the circularity indicator, LCA assigns values to different material options. These values and ratings enable the comparison of the environmental performance of materials and the identification of areas for improvement.

Bio-based materials store CO_2, helping to reduce climate impact, while non-renewable materials like EPS contribute to an immediate increase in global warming potential (GWP) due to their energy-intensive production processes. Bio-based materials reduce the burden on depleting resources, have higher specific heat capacity, and passively contribute to the indoor environment. In contrast, non-renewable materials are cost-effective, easy to install, and highly resistant to moisture. Conducting a comparative LCA among various bio-based materials, as well as a cross-comparison between non-renewable and bio-based materials, is vital to understanding their environmental performance. However, the comparison of LCA studies is only possible if analyses are conducted under same conditions such as methodological choices and geological locations. LCA results are sensitive to many factors, such as (but not limited to) goal and scope of study, functional unit, system boundaries, selected impact assessment method and database [17].

Table 22.9 compares the environmental impacts of four common bio-based insulation materials under the same conditions. Wood fiber contributes the least to global warming, producing only 2.72 kg CO_2 eq. per m^2 of external wall insulation, which is crucial given the current climate crisis. In contrast, flax has a high impact on global warming, generating 13.62 kg CO_2 eq. for the same application. Overall, insulation materials made from miscanthus perform best in the other three impact categories. Similarly, bio-based materials particularly wood fiber and miscanthus produce significantly less lifecycle impacts across several impact categories compared to EPS and stone wool insulation materials [16]. Biobased insulation material flax performs similar to stone wool in terms of insulation properties. Netherlands's NIBE platform [18] classified flax insulation as 'best' choice in terms of environmental performance, whereas stone wool is considered as 'bad' choice (class 6c) (Table 22.9).

Another important consideration for bio-based insulation materials is that achieving the same thermal resistance may require a thicker layer compared to traditional materials. Figure 22.6 provides a comparison between different bio-based and nonrenewable insulation materials for 100 m^2 with a thermal resistance of (5 m^2K)/W, based on research conducted on 503 residential houses. XPS has significantly higher carbon emissions compared to bio-based alternatives. Straw bale and hemp

Table 22.9 Environmental impacts produced by insulating material required for insulating 1 m^2 of external wall of a residential building

Material	Global warming kg CO_2 eq	Freshwater ecotoxicity kg. 1.4-DCB	Human carcinogenic toxicity kg. 1.4-DCB	Land use m^2*a crop eq
Wood fiber	2.72	0.09	0.27	10.45
Hemp fiber	11.70	0.28	0.34	44.59
Flax	13.62	0.30	0.38	59.17
Miscanthus	2.95	0.09	0.14	6.44

Source Adapted from [17]

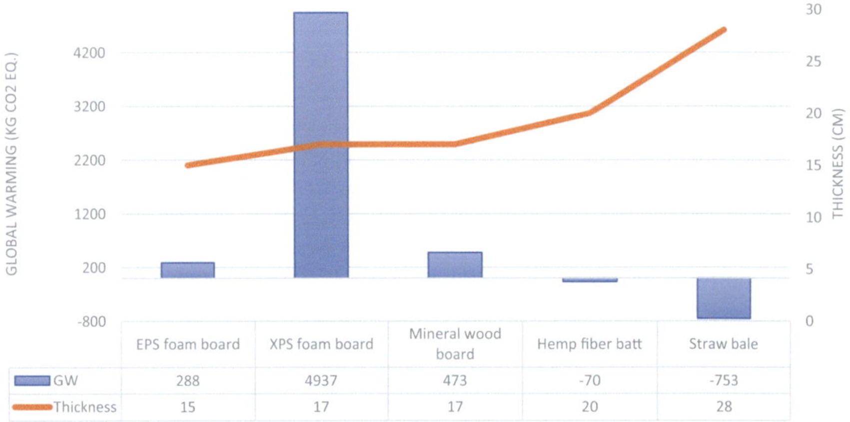

Fig. 22.6 Sheep wool. *Source* Adapted by the Authors from [11]

fiber batt exhibit a negative carbon balance but require significantly larger quantities (20 and 28 cm, respectively) compared to EPS and XPS (15 and 17 cm) [11]. This raises circularity concerns and puts stress on valuable and depleting resources. Therefore, a holistic evaluation of physical properties, sustainability, and circularity performance is recommended to select the most optimal solution. Combining LCA and WBCI, along with various design scenarios, can help objectively make these trade-offs (Fig. 22.6).

22.6 Conclusion

This chapter elucidate the environmental properties of natural and bio-based insulation materials, shading light on the environmental benefits, as well as possible shortcomings, deriving from their application in the building industry. These materials often present a carbon-negative balance compared to traditional plastic or mineral-based insulations, showing a significant potential in advancing sustainability and circularity. Despite their advantages, bio-based insulations currently represent less than 10% of the global market.

However, a thorough evaluation of the environmental pros and cons should rely on quantitative analyses, using tools like LCA and Circularity Indicators, which provide a robust framework for evaluating the environmental performance and sustainability of insulation materials. These tools help identify opportunities for improvement in material selection, usage, and end-of-life management, promoting a more holistic and sustainable approach to building design and construction. This analysis wanted to underscore the importance of adopting sustainable practices and advancing research in bio-based insulation materials to achieve a balanced integration of environmental, economic, and social considerations in the built environment.

References

1. Lakatos A., Lucchi E., Thermal performances of super insulation materials (SIMs): a comprehensive analysis of characteristics. heat transfer mechanisms. laboratory tests. and experimental comparisons. Int. Commun. Heat Mass Transf. **152**, 107293 (2024)
2. European Commission, Bio-based, biodegradable and compostable plastics: The EU will address the sourcing, labelling and use of bio-based plastics, and the use of biodegradable and compostable plastics (2020). https://ec.europa.eu/environment/topics/plastics/bio-based-biodegradable-and-compostable-plastics_en. Accessed on 6/2024
3. L. Cosentino, J. Fernandes, R.A. Mateus, Review of natural bio-based insulation materials. Energies **16**, 4676 (2023)
4. REFLOW, Co-creating circular and regenerative resource flows in cities. EU Horizon 2020 project (2021). https://reflowproject.eu
5. F. Pittau F. Krause G. Lumia G. Habert, Fast-growing bio-based materials as an opportunity for storing carbon in exterior walls. Build. Environ. **129**, 117–129 (2018)
6. Husk. Wheat husk, wood fibers and textile waste fibers: elaboration and performances evaluation. Ind. Crops Prod. **135**, 238–245 (2019)
7. R. Agliata, S. Gianoglio, L. Mollo, Hemp-lime composite for buildings insulation: material properties and regulatory framework. VITRUVIO—Int. J. Archit. Technol. Sustain. **4**, 48 (2019). https://doi.org/10.4995/vitruvio-ijats.2019.11771
8. I. Nardi, E. Lucchi, In situ thermal transmittance assessment of the building envelope: practical advice and outlooks for standard and innovative procedures. Energies **16**(8), 3319 (2023). https://doi.org/10.3390/en16083319
9. R. Muthuraj, C. Lacoste, P. Lacroix, A. Bergeret, Sustainable thermal insulation biocomposites from rice. B. Gaujena, V. Agapovs, A. Borodinecs, K. Strelets, Analysis of thermal parameters of hemp fiber insulation. Energies **13**, 6385 (2020)
10. European Commission, Directive 2009/125/EC ff The European Parliament and of The Council of 21 October 2009 establishing a framework for the setting of ecodesign requirements for energy-related products (recast). https://eur-lex.europa.eu/legal-content/EN/TXT/PDF/?uri=CELEX:02009L0125-20121204&from=EN. Accessed on 6/2024
11. J. Kirchherr, D. Reike, M. Hekkert, Conceptualizing the circular economy: an analysis of 114 definitions. Resour. Conserv. Recycl.. Conserv. Recycl. **127**, 221–232 (2017)
12. C. Magwood, E. Bowden, M. Trottier, *Emissions of Materials Benchmark Assessment for Residential Construction Report* (Efficiency Canada, Ottawa, ON, Canada, 2022)

13. B. Abu-Jdayil, A.H. Mourad, W. Hittini, M. Hassan, S. Hameedi, Traditional, state-of-the-art and renewable thermal building insulation materials: an overview. Constr. Build. Mater. **214**, 709–735 (2019)
14. N. Khadim, R. Agliata, A. Marino, M.J. Thaheem, L. Mollo, Critical review of nano and micro-level building circularity indicators and frameworks. J. Clean. Prod. **357**, 131859 (2022)
15. N. Khadim, R. Agliata, M.J. Thaheem, L. Mollo, Whole building circularity indicator: a circular economy assessment framework for promoting circularity and sustainability in buildings and construction. Build. Environ. **241**, 110498 (2023)
16. M. Wiprächtiger, M. Haupt, N. Heeren, E. Waser, S. Hellweg, A framework for sustainable and circular system design: development and application on thermal insulation materials. Resour. Conserv. Recycl.. Conserv. Recycl. **154**, 104631 (2020)
17. S. Lanting, *Biobased materials in the construction industry: a case study into the effects of using biobased construction materials on life cycle costs* (Master's thesis, University of Twente) (2023)
18. M. Schulte, I. Lewandowski, R. Pude, M. Wagner, Comparative life cycle assessment of bio-based insulation materials: environmental and economic performances. GCB Bioenergy **13**(6), 979–998 (2021)
19. NIBE, NIBE Milieuclassificaties (2023). https://www.nibe-sustainability-experts.com/nl/nibemilieuclassificaties. Accessed on 6/2024

Chapter 23
Analysis of Building Foundation Reconstruction: A Holistic Assessment of Its Economic, Social, and Environmental Impacts

Antonio Ferreira-Sánchez, Mª Desirée Alba-Rodríguez, Valeriano Lucas-Ruiz, and Madelyn Marrero

Abstract The issue regarding rehabilitation versus building demolition remains an open debate. In order to make decisions in a sustainable way, the social, economic, and environmental dimensions all need to be considered. Although rehabilitation is not always the most economical solution, it can, in most cases, offer other benefits beyond solely contributing towards environmental sustainability, including the reduction of project execution times, protection of communities, and the restriction of urban sprawl. A holistic assessment of these factors is therefore essential. In the present work, a multi-family building in Seville, Spain, constructed in 1950 and containing forty social housing dwellings, suffered an accident related to soil subsidence. Consequently, prompt decisions had to be made in order to rehabilitate the property and return the occupants to their dwellings within a year. The building also needed a significant repair to its foundations. In the present work, a methodology is defined to rapidly assess the economic, social, and environmental sustainability of various options of reconstruction work. Four solutions for the reconstruction of the foundations are studied, which give rise to a methodology for the assessment of similar scenarios of repair work.

A. Ferreira-Sánchez
Departamento de Arquitectura, Universidad de Alcalá, Alcalá de Henares, Madrid, España
e-mail: antonio.ferreira@uah.es

M. D. Alba-Rodríguez
Departamento de Ingeniería del Diseño, Escuela Politécnica Superior, Universidad de Sevilla, Sevilla, España
e-mail: malba2@us.es

V. Lucas-Ruiz · M. Marrero (✉)
Departamento de Construcciones Arquitectónicas II, Escuela Técnica Superior de Ingeniería de Edificación, Universidad de Sevilla, Sevilla, España
e-mail: madelyn@us.es

D. A. H. Hanaor (ed.), *Sustainable Materials for the Built Environment*, Springer Series in Materials Science 352, https://doi.org/10.1007/978-3-031-97818-0_23

23.1 Introduction

Approximately 50% of the buildings in Europe are over 50 years old. In comparison to contemporary buildings, these are more vulnerable to catastrophic events, such as floods, earthquakes, and accidents, since their construction predates today's more restrictive construction codes [1]. When a building needs to be repaired under emergency circumstances, the results of the work, the techniques, and the organisation used rarely become available for public scrutiny. This is due not only to the affected stakeholders' avoidance of negative publicity, but also to the related responsibilities and to a general lack of technical knowledge in the background of professionals to deal with the details and complexities of the various techniques available [2]. Another aspect that highlights the need for deeper knowledge regarding buildings affected by an emergency repair is derived from the urgency with which it must be undertaken. This kind of intervention has an insufficient number of previous studies available and hence offers only a limited range of possibilities, and emergency repairs tend to involve insufficient time to find the optimal solution at a certain technical and economic level [3]. Although there are examples of emergency retrofits related to certain infrastructures [4], their solutions cannot be suitable for every type of building repair. It is necessary to increase the publication of case studies, where the origin of the damage, the technical solutions, the results, and the management processes are fully reported [5]. Recent, simulations [6] and studies [7] have determined the feasibility of methodologies using micropiling as a solution for unstable soil. However, economic, environmental, and social aspects must also be considered if a sustainable solution is to be found. Reference [8] revealed that building environmental analysis largely centres on new construction and fails to include the productivity, efficiency, and short- and long-term consequences of the building rehabilitation work. Moreover, it has become necessary to include the environmental aspect since building rehabilitation is not solely an economic issue.

In fact, due to the multifaceted complexity of repairing damaged buildings, a clear methodology is necessary to successfully achieve the goals of the intervention [9]. A method to prioritise the actions to be taken during the rehabilitation is also needed, in order to help in the work planning and to decide on the best solutions to be implemented during execution. Nowadays, limitations on the work plan that include only the technical and economic aspects are becoming increasingly common. The environmental performance of building retrofits is currently promoted by local, regional, and national governments, and will become mandatory in the near future. The European Union encourages stakeholders to use the common framework LEVEL(s) to improve building resilience against climate change, among other major objectives [10]. To maximise the positive effects of environmental building interventions, a wider approach is required that considers the urban context and provides in-depth knowledge regarding the environmental impact related to the work. These constitute key factors in the creation of the strategic measures to be implemented [11].

Certain studies [12] even show that, in comparison with new buildings, higher efficiency standards can be achieved when a life-cycle approach is considered during the rehabilitation.

In the present work, the relationship between environmental, social, and economic factors is analysed. All three factors constitute major aspects of the decision-making process and are proposed while considering the improvements necessary to swiftly render a building habitable subsequent to an accident [13]. In order to consider as many dimensions as possible in this complicated process, a model is proposed that, starting from the project budget perspective, includes the environmental [14], economic (project bill of quantities and budget), and social (health and safety on the construction site) [15] assessment of the foundation underpinning.

For the economic evaluation, the work breakdown system (WBS) of the Andalusia Construction Cost Database, [16] is employed. The WBS can improve the general knowledge by including environmental impact assessment and by allowing an ecological comparison in terms of the carbon footprint (CF) [17], the ecological footprint (EF) [18], and the water footprint (WF) [8, 19].

Social aspects are also relevant since the reconstruction of a building after a disaster carries huge implications for the tenants, and these are also taken into consideration [3]. When a damaged building is under major structural risk and immediate intervention is required, then multiple technical factors and unknown elements become involved, which can influence the final decision [20]. Many social indicators have been employed by researchers, such as occupants comfort, building management, accessibility, etc. The social dimension of sustainable development is difficult to assess mainly due to the subjective aspects in its assessment. Another social aspect involves the construction work itself [14] The breakdown system of construction cost databases can be useful from the perspective of the health & safety risks associated with the foundation repair [21].

Four solutions are considered for the foundation repair work: foundation underpinning, reinforced grouting, cement injection, and injections of expandable polyurethane resin. The generation of information on expertise and tools may help technicians in other, similar situations. The present work focuses on the case study of 40 dwellings damaged during an excavation that was taking place on an adjoining lot and caused the evacuation of all families from said dwellings. This resulted in technical and social problems that needed immediate action. Techniques of psychological first aid for social management are presented. Indirect aspects of the repair work are also assessed using environmental indicators, cost assessment, and occupational-risk evaluation of the project. Lastly, a model of action is proposed for the restoration of buildings under emergency conditions that takes into consideration social, ecological, and economic aspects.

In order to define the methodology, four reparation projects are defined, one for each solution, using the WBS of ACCD. On this basis, the proposed model assesses the feasibility of the repair work of the building foundation, which would otherwise be facing a demolition, by means of a global vision that incorporates environmental, social, and economic evaluation. Furthermore, the risks on the construction site are also considered through the identification of health & safety risks.

23.2 Literature Review

23.2.1 Environmental Assessment

For the environmental assessment of construction projects, identification is made of the carbon emissions and embodied energy (EE) as the most commonly evaluated indicators in the review conducted by [22]. However, other indicators stand out in the assessment of buildings, such as the ecological footprint (EF) and the water footprint (WF), which are normally referred to as part of the footprint family. The term footprint family was coined for the first time by [23] and [24]. Reference [25] consider that this term encompasses the three most widely studied footprints (CF, EF, and WF) due to the intuition of the term footprint. According to these studies, the family of footprints can be considered as an indicator that gathers other indicators, in an effort to obtain basic information from each of the footprints, with special emphasis on comparative aspects [26]. The success of these footprints is due, firstly, to the fact that the results they produce are understandable to non-scientific members of society, and secondly, to their ease of application in environmental policies and decision-making [27, 28]. It is necessary, however, to advance in their standardisation, since research combining multiple footprints remains limited in contrast to single-footprint studies [29]. A good association of these indicators is shown in the evaluation of construction and demolition waste (CDW) in the building life cycle [30], and [31] employ said indicators in the assessment of urbanisation projects. Moreover, the complete building life cycle has also been analysed by simultaneously employing these indicators [14].

The CF indicator constitutes the most widely used indicator from the footprint family in the assessment of buildings. Reviews related to the use of the CF [32] demonstrate its importance in the sector. Other studies strive to establish scales of CO_2 emission intervals in construction processes [33–35]. The ArDiTec group works on the calculation of the CF in buildings [28, 36] and has developed the evaluation tool OERCO2 [37, 38].

Another interesting indicator is that of the EF, which was introduced by [39], who measured the EF of humanity and compared it with the carrying capacity of the planet. The EF is the amount of land that would be required to provide the resources (grain, feed, firewood, fish, and urban land) and absorb the emissions (CO_2) of humanity. Moreover, the ArDiTec research group has successfully adapted the EF indicator to the assessment of construction impacts [40, 41]. Their analysis includes the rehabilitation of buildings [13] and the retrofitting and usage of buildings [8].

The WF considers yet another important asset, that of the water consumption of human processes. Construction materials consume water in their extraction, manufacture, waste management, and reuse. Reference [42] study direct and indirect water consumption in the life cycle of residential buildings and show that the indirect water consumption is greater than the direct household consumption of water. The WF of buildings can be analysed from a global perspective [43] either through an input–output analysis of total consumption in the country or via models that analyse the components in construction projects [44–46]. It is also worth mentioning the research of the ArDiTec group where they incorporate the WF indicator into the assessment of urbanisation projects in which both direct and indirect water consumption are evaluated [47, 48].

23.2.2 Economic Assessment

For the cost evaluation of construction projects, different regions have developed classification methodologies of construction-related activities. These have a long tradition in the sector, such as MasterFormat [49], CI/SfB [50], the Standard Method of Measurement of Civil Engineering [51], Uniformat [52], and Uniclass [53]. In Spain, the classification systems by the Institute of Construction Technology of Catalonia [54] and Generador de Precios [55] appear among those most commonly employed. Over the last three decades, in the south of Spain, the regional government has defined the Andalusian Construction Cost Database (ACCD), which has become mandatory in public bidding [56].

23.2.3 Social Assessment

Direct Impact

There are social impacts by the damaged building that affect tenants directly. Material loss and evacuation generate stress, and tenants' mistrust can arise due to the complexity of the technical solutions or to lack of protocol knowledge established for the intervention [57]. This factor can directly influence the choice of the final solution to be applied due to the desire to reduce the duration of the project.

The effects of a severe structural accident in a building are similar to those caused by earthquakes, floods, and conflicts. The health of the population can be seriously affected if left untreated. In order to prevent such effects, the World Health Organization has developed "Psychological First Aid (PFA)" [58], whose main objective is to provide a human response in support of those suffering and those in need [59]. The conceptual and methodological simplicity of this technique has demonstrated its effectiveness and can be applied by non-specialists [60]. It can be combined with the technical aspects of the repair work of the damaged building [57] from an integral point of view, in a "Model of action in buildings against damage and accidents with social repercussions" [57] where clear commitment and a defined time limit can generate a favourable attitude of the families affected. Situations of post-traumatic stress are prevented and, through the participation of the occupants, a favourable climate can be created both for the execution of the repair work and for its reception.

Indirect Impact: Health & Safety in the Construction

The construction sector presents one of the highest accident rates in the European Union (EU) and accounted for 20% of the fatal accidents occurring at work in 2020 [61]. Furthermore, the types of injuries appear to remain constant over time [62]. The cause-effect of accidents must be recorded in order to propose solutions and interventions towards the reduction and prevention of accidents [63], and the mitigation of risks needs to be delivered at various levels and by different actors [64]. Tools to assess and control H&S risks have been developed over the years, such as the Storybuilder Tool [65], the Risk Assessment Model [66], the Safety, Health, and Environmental Checklist for Contractors in the Netherlands [63], the Construction Site Risk Assessment Tool (CONSRAT) [67], and other similar instruments [68]. However, these are independent of other tools already in place at the construction site regarding cost control, which require additional expertise and/or personnel. In the present work, a methodology is proposed for the control and definition of risks in a systematic way, and this is linked to the economic and environmental control of the project.

23.3 Materials and Methods

23.3.1 Economic Assessment

For the generation of the inventory of resources of the life cycle, the use of construction cost databases and their WBS systems are proposed. The regional database, ACCD, [56] is employed, which has a pyramidal classification. It follows the cost structure established by the Public Procurement Law and its implementing regulations [69], and divides the budget into chapters that group together basic costs (BC), auxiliary costs (AC), and unit costs (UC). Basic costs refer to construction materials, machinery, and labour, while the AC are combinations of the BC that have no entity of a finished work (such as mortars, concrete, work team, etc.). The UC are elements that represent finished elements on the work site such a cubic metre of soil excavated for the foundation, a square metre of terrazzo floor polished and in place, and a window set into the façade. The pyramidal structure continues with the chapters of work that classify the stages of the construction of the buildings and their corresponding tasks, for example, Chap. 1 corresponds to the clearing of the plot and to small demolitions, if necessary. Chapter 2 groups earthwork tasks, while Chap. 3 covers the construction of the foundation. The subsequent chapters of the budget include structure, brickwork, and installations. With the measurement of work units or UC, the amount of each specific activity is then obtained, and, by aggregation, the total budget of the work is attained. The pyramidal structure enables the creation of a database that combines both economic and environmental information (Fig. 23.1).

Fig. 23.1 Cost pyramid

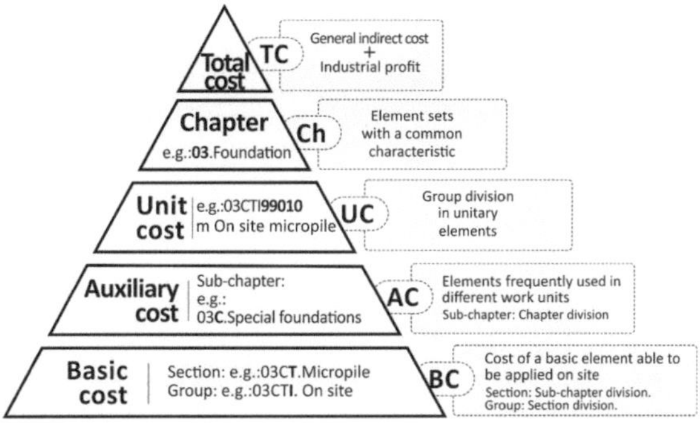

23.3.2 Environmental Assessment

Labour

Of all the footprints only the EF indicator includes the impact of workers through their food and generation of municipal solid waste (MSW). The workers' food is considered their energy source and is obtained from Eq. 1 in Fig. 23.2. as typical adult food intake consisting of meat, fish, cereals, and water [70]. Food also has an energy footprint, due to its transformations.

The footprints according to the food source (meat: EF of pastures; fish: EF of productive sea; cereals: EF of crops) has a natural productivity and equivalence factor between actual productive hectares (ha) and global hectares (gha). The equivalence factors are 2.51 for crops, 0.46 for pastures, 1.26 for forests, 0.37 for sea, and direct occupation is 2.51 gha/ha [71]. The equivalence factors used [72] are implicit in the calculation of EF (gha/person/year) of each country. In Spain, footprints in gha/person/year are 1.45, 0.27, 0.41, and 0.49 in 10^{-3} gha for crops, pastures, fisheries, and fossil fuels, respectively. It has been considered that breakfast and lunch (taking place on-site) represent approximately 60% of the daily food intake of an adult [41]. Equation 3 of Fig. 23.2 is utilised to calculate the impact of municipal solid waste (MSW). An average generation coefficient per worker and the emission factor of its subsequent treatment are used [14]. The impact of the workers mobility to the work site is excluded as per UNE-EN 15978 [73].

Construction Materials

The determination of the impact of construction materials follows the methodology defined by [74]. The first step is to perform a conversion of the original unit of measurement of each basic cost (m^3, m^2, m, t, thousands of units of bricks, etc.) into m^3, and its density is obtained from the Catalogue of Constructive Solutions of the Technical Building Code and the Basic Document of Structural Safety of the Technical Building Code DB-SE AE [1], and hence the weight of each element is attained. This enables the environmental coefficients per kg to be applied as obtained from the LCA database, Ecoinvent LCA [75], which is recognised as being one of the

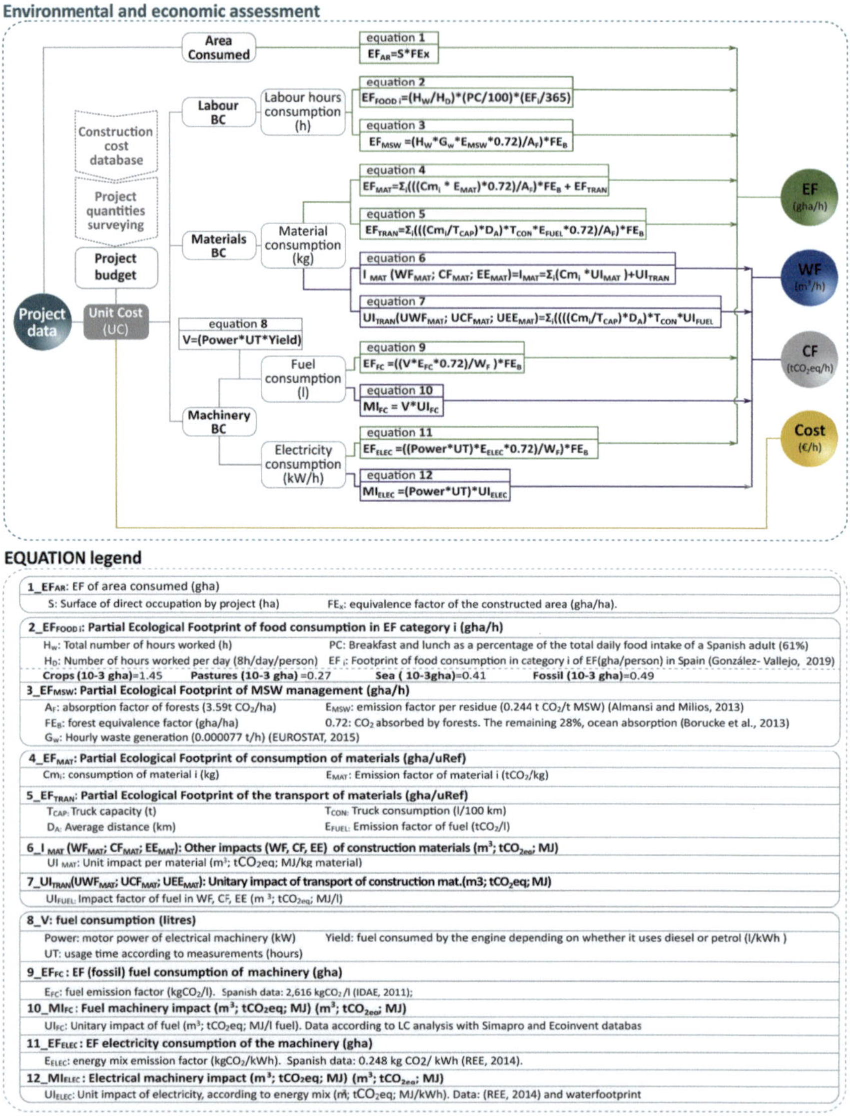

Fig. 23.2 Flow chart and equations for the calculation of the impacts: economic cost, embodied energy (EE), and the ecological (EF), carbon (CF), and water footprints (WF)

most complete databases at European level [76] and known for its integration with the Simapro LCA software [77]. The Life Cycle Inventory of materials using the IPCC 100a methodology is employed to obtain CO_2 emissions of building materials. This methodology isolates CO_2 and other GHG emissions from the LCI inventory.

Equation 4 in Fig. 23.2 is applied to obtain the EF, and Eq. 6 is employed to obtain the remaining indicators.

The work is similar to that conducted for the calculation of the CF with the SOFIAS tool [78], which uses data either from environmental product declarations, OpenDAP, or from the BEDEC platform [79], developed by the Institute of Construction Technology of Catalonia [80], see Fig. 23.2 where the integration of indicators in the budget of the work is shown in a schematic way. The Cmi is obtained by means of the family of materials, measured in weight per ground floor area, by which its environmental impact is multiplied to obtain IUMAT.

In order to evaluate the A4 stage of the life cycle [73], an analysis of the transport of the material is also conducted, thereby establishing approximations of distance travelled, as in the previous section on machinery. The average diesel consumption (26 L/100 km) and its emissions (2.62E−03 tCO_2/litre) are used. The water consumption in its production is 1.26 m^3/litre of diesel and its embodied energy is 57.7 MJ/litre. The capacity of the trucks and travel distance is 2000 and 24,000 kg, and 20 and 250 km, for the transport of concrete and for other materials, respectively [81]. As an example, in Table 23.1, basic costs are analysed, and their costs, transformation coefficients and associated environmental impacts are presented.

Machinery

The construction machinery impact is calculated using the engine power and the on-site hours of use, while the energy consumed is determined in kWh, which will be transformed into CO_2 emissions [81]. The calculation is CDW transport machinery.

To obtain fuel consumption, the "Machinery Manual" prepared by SEOPAN [82] is used, where the technical data of different models and types of machines on the market are collected. From the power of the engine, the litres of fuel consumed are calculated, (Eq. 9 in Fig. 23.2) and the amount of CO_2 generated per litre is applied [83]. The WF per litre of fuel is also obtained from Ecoinvent, and the CF, WF (Eq. 11 in Fig. 23.2), and EF (Eq. 11 in Fig. 23.2) are calculated.

The electrical machinery used on-site, depending on the engine power and the use time, is measured in kWh [81]. The emissions of the production of one kWh by the Spanish electricity system is employed [84], as GHG gases, measured through the global warming potential (GWP) of the various gases emitted into the atmosphere, and through weighting the tons and transforming them into equivalent tons of CO_2. For the WF of the electrical machines, the WF associated with the Spanish energy mix is used. The EF is calculated using Eq. 9 in Fig. 23.2 and, for the remaining indicators, Eq. 12 is applied.

23.3.3 Social Indicators

For the evaluation of the indirect social dimension, which, in the present work, refers to the assessment of the risk to health and safety (H&S) or to occupational risks, the

Table 23.1 Weight calculation of basic construction elements and footprints, where u_r is the measurement unit of the basic construction element

Material	Cost (€)	u_r	Description	Weight (kg/u_r)	CF	WF	EF
					(tCO_2eq/u_r)	m^3/u_r f	gha/u_r
GC00600	98.54	t	cement CEM ii/b-l 32.5R in sacks	1000.00	7.86E−01	3.28E+00	3.75E−01
CA00220	0.68	kg	Steel B400S	1.00	1.46E−03	2.70E−02	6.59E−04
CA01700	1.23	kg	Tying wire	1.00	1.46E−03	2.70E−02	6.59E−04
CA80065	26.46	m	Steel jacket S375JR diam. 60 mm and thickness 3.30 mm	4.98	7.25E−03	1.34E−01	3.28E−03
CM00200	195.18	m^3	Pine wood board	509.85	−4.98E−01	1.34E+00	−2.45E−01
CM00300	225.64	m^3	Pine wood plank	509.85	−4.98E−01	1.34E+00	−2.45E−01
CP00100	0.21	kg	Bentonite	1.00	1.61E−05	1.03E−04	7.35E−06
CW00600	1.72	l	Releaser	0.90	1.71E−03	6.66E−02	7.66E−04
GR00100	0.72	kg	Agglomerate of resins and marble	1.00	6.75E−03	4.03E−01	2.90E−03
GW00100	0.55	m^3	Tap water	1000.00	7.40E−03	1.13E+00	3.61E−03
WW00400	0.30	u	Small material	0.02	1.58E−04	3.32E−03	7.27E−05

project budget is also employed for the definition of the tasks. The methodology is illustrated in Fig. 23.3. The steps are the following:

1. Problem boundaries: collective and individual protection and the delimitation of spaces, in compliance with Spanish law regarding organisational and ergonomic activities necessary to prevent and/or minimise risks [85, 86].
2. Collection of building characteristics, and its bill of quantities is expressed in the work breakdown system of the ACCD.
3. Risk indicator: the "Simplified accident risk assessment system" published by the INSST in the Prevention Technical Note 330 is employed [87]. For ergonomic impacts, either the "Evaluation of working conditions: postural load" is assessed or the REBA method (Rapid Entire Body Assessment) [88, 89], published by the INSST in NTP 601.
4. Work procedures: analysis of each work unit, and of its procedures and/or tasks.
5. Risk assessment: accidents and ergonomic risks of each procedure previously defined are assessed and aggregated into a single indicator.

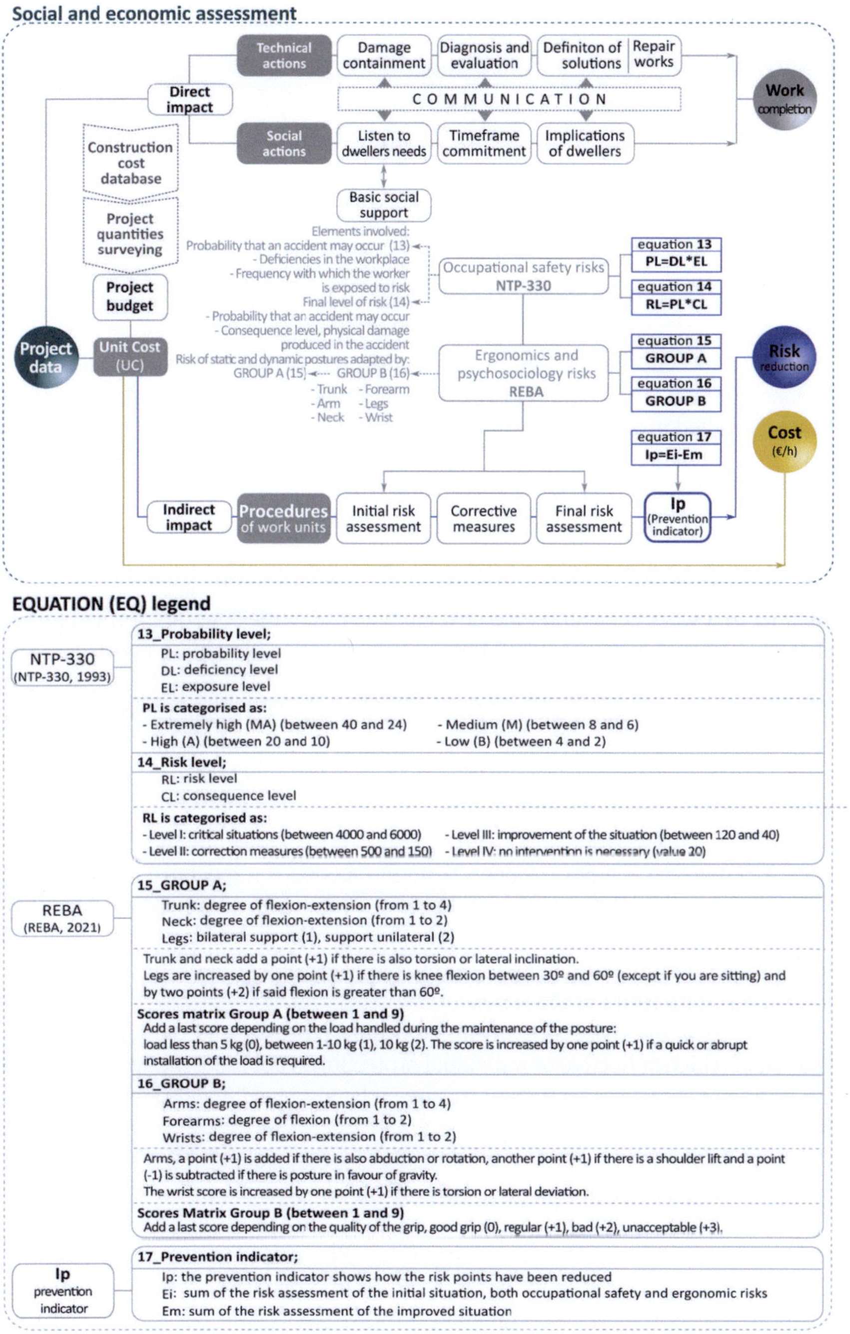

Fig. 23.3 Summary of equations and methodology for the calculation the social indicator

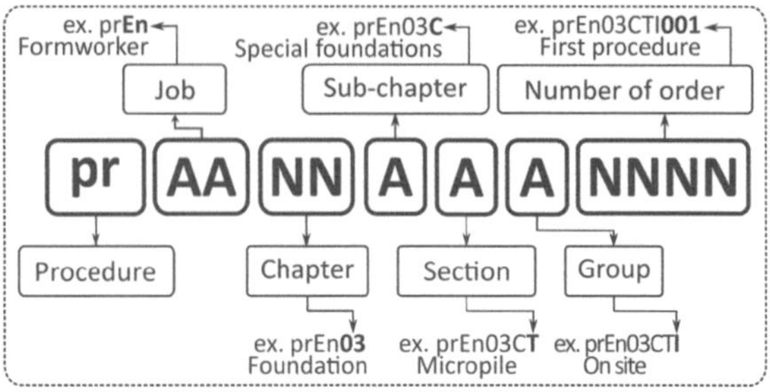

Fig. 23.4 Procedure coding scheme. Example: prEn03CTI001

Based on the unit cost of the project budget, as classified according to ACCD mentioned in Sect. 3.1. Economic assessment, the procedures involved in its execution are identified in each unit cost. It is necessary to generate a coding model for the procedures, for their standardisation within the database. Each worker creates one or more parts of the unit cost that will be defined in the procedure's codification [90], see Fig. 23.4.

The coding of the procedures is formed by seven alphabetical and/or numerical blocks (Fig. 23.4): the first two characters are "pr" as an indicator that it is a code belonging to a work procedure, the next character identifies the job, formed by two alphabetical characters (example: "En" Formwork). The remaining codes are those of the unit cost in the ACCD (Fig. 23.4).

An example of the unit cost 03CTI99010 is shown. This unit cost corresponds to the micropile manufactured on-site with an external diameter of 60 mm. The labour is described in the unit cost description. Their related tasks and construction procedures can be assessed, as shown in Table 23.2.

There are 4 jobs associated with the unit cost shown in the example (see Table 23.2): the first is a Form worker (En), the second is a specialist assistant (Ay), the third is a special labourer (Pe), and finally there is an equipment operator (Ep), all corresponding to the foundation chapter (03), special foundations subchapter (C), micropile section (T) and belonging to the on-site group (I). All these costs are in the first procedure (001) of the series.

In Table 23.2, the risk levels of the procedure represent the most unfavourable situation, in which collective or individual protection is not used, the spaces have not been delimited, and organisational measures related to health and safety (H&S) are not taking place.

The first method to be employed for the risk assessment is NTP-330, which estimates the risk level of the workers [87]. First, deficiencies in the workplace are identified. Second, the probability of an accident and the magnitude of the consequences are determined. The probability level (PL) considers the deficiency level (DL: 10 for very deficient, 6 for deficient, 2 for improbable, and 0 for acceptable), the exposure level (EL: 4 for continuous, 3 for frequent, 2 for occasional, and 1 for sporadic), and the consequence level (CL: 100 for catastrophic or fatal, 60 for very

Table 23.2 Example of methodological development

Definition of the unit cost				
03CTI99010	m	On-site micropile with steel jacket diameter 60 mm		
Micropile manufactured on-site, external diameter is 60 mm, reinforced with a steel jacket whose internal diameter is 53.4 mm and wall thickness is 3.3 mm, up to 15 m deep with toxoid-tropic sludge. Work includes cleaning and transportation of equipment. Mechanical equipment complies with NCSR-02 and Spanish Technical Code				
Measure the length executed from the equipment support platform				
Code	*Concept*	*Quantity*	*Price*	*Cost*
TO00400	h official 1st formwork	0.300	23.17	6.95
TA00200	h specialist assistant	0.300	22.36	6.71
TP00100	h special labourer	0.150	22.01	3.30
GC00200	t cement CEM II/A-L 32.5 N in bags	0.100	161.24	16.12
CP00100	kg bentonite	1.000	0.22	0.22
CA80065	m steel jacket S 375 JR diameter 60 mm	1.000	33.32	33.32
MP00600	u repercussion useful/S, transport, assembly of equipment	1.000	9.39	9.39
MP80000	h micropile drilling rig	0.160	407.83	65.25
		Direct costs		141.27
		10.62% Indirect costs		15.00
		Total euros		156.27
Definition of procedures				
Code	*Procedure*			
PrEn03CTI001	Placement of steel jacket for micropile formwork			
PrAy03CTI001	Support work in the formwork procedure			
PrPe03CTI001	Support work in the micropiling procedure			
PrEp03CTI001	Micropile drilling			
Definition of procedure risks (example of a procedure)				
PrEn03CTI001	Placement of steel jacket for micropile formwork			
Occupational safety risk				
Risk identification: 1. Risk of falls to the same level 2. Risk of blows to upper and lower limbs				
Ergonomics risk				
Risk identification: 3. Risk of trunk injuries due to forced postures 4. Risk of trunk injuries due to the load of the material				

serious, 25 for serious, and 10 for mild). The final risk can be either level I (4000–6000), level II (500–150), level III (120–40), or level IV (40 points) (see Fig. 23.3 for the formulation).

The second method is REBA (Rapid Entire Body Assessment), which enables ergonomic and psychosociology risks to be estimated regarding work-related injuries. The method analyses the positions of arms, forearms, wrists, trunk, neck, and legs. The method was defined by a group of ergonomists, physiotherapists, occupational therapists and nurses, who assessed approximately six hundred working postures [88, 89]. For the definition of body segments, simple tasks with variations in load and movements were analysed (see Fig. 23.3).

Table 23.3 presents the calculation of the risk indicator as 6264.00 m of on-site micropile execution. The quantity of the work unit determines the level of exposure (LE) or the frequency with which the worker has been subjected to said risk (i.e., the operator risk increases with the execution size). The 'Risk number' in the table is defined in Table 23.2. The risk levels (RL) of the NTP-330 method are expressed in Roman numerals from I to IV, while the REBA method uses Arabic numbers. In order to combine these variables, the values I to IV from NTP-330 are taken as corresponding to values 4 to 1 in REBA, and hence the two methods work in the same range, from 1 to 4.

23.4 Case Study

A case study is analysed of the foundation repair of forty apartments located in a building in Seville (Spain) which had been constructed in the 1950s. The building was damaged by a soil movement produced by nearby construction work; the foundation displacement damaged the masonry of the loadbearing walls. It was necessary to evacuate the tenants quickly due to the safety risk. The building presented cracks and deformations that needed major repair work [3], see Fig. 23.5.

For the emergency foundation underpinning, four techniques are considered: foundation underpinning; cement injection; injections of expandable polyurethane resin; and increasing load transmission area of the foundation footings.

The four solutions are briefly described:

- Foundation underpinning: 232 IRS (Repetitive and Selective Injection) 27 m-long micropiles (actual solution), in order to reach the resistant substrate of the ground, located at 25 m below the soil surface, and to transmit the mechanical tensions into solid clay located at that level. These IRS micropiles also enable selective cement injections to be made in selected soil levels.
- Soil cement injection: 300 kg of cement per metre perforated via 360 injections located at each metre of foundation have been considered. Using this solution to increase the soil compaction due to the pressure of the mortar injection. The mortar considered has a water/cement ratio of 0.4 using a cement, EN 197-1-CEM I 42,5 R, with a compression resistance of 42.5 MPa. Its maximum depth

Table 23.3 Example of initial risk assessment

03CTI99010		M	On-site micropile with steel jacket diameter 60 mm							
Code		Procedure								
PrEn03CTI001		Placing steel jacket for micropile formwork								
Safety at work	Risk number	Probability (P)	Consequence (C)	PXC	Initial assessment		Ergonomics	Risk number	Initial assessment	
	1	8	10	80	III	2		3	M-2	2
	2	8	10	80	III	2		4	M-2	2
Σ safety hazards at work						4	Σ ergonomic risks			4
Total										8

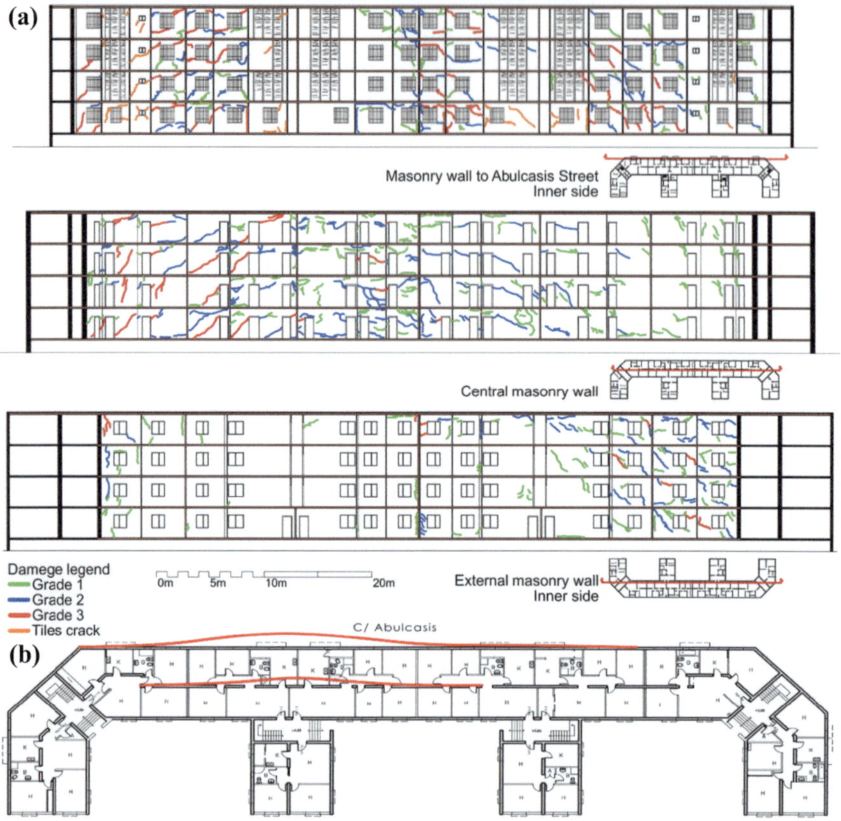

Fig. 23.5 **a** Distribution of cracks in the loadbearing walls, **b** deformation of the load bearing walls

considered is 10 m. This solution improves the resistance of the soil to mechanical loads (Fig. 23.6a).

- Injections of expandable polyurethane resin: perforations are carried out through the original foundation. Their diameter is 26 mm and depth 3 m. There are 240 injections at a maximum depth of 5 m. The technique cannot advance to a greater depth. The use of reinforced injections and polyurethane foam improves the properties of the underlying soil, and stress is not produced on the surface foundation (Fig. 23.6b).
- Increasing foundation footings: This traditional underpinning increases the support surface of the foundation, which in turn reduces pressure on the soil and generates interaction with soil layers underneath the foundation.

In the present case study, micropile underpinning was urgently carried out subsequent to determining the soil characteristics (soft lime and low-resistance clay), which also needed deep structural support, located at 25 m below the surface, see Fig. 23.6. An IRS micropile system, based in repetitive and selective injections along

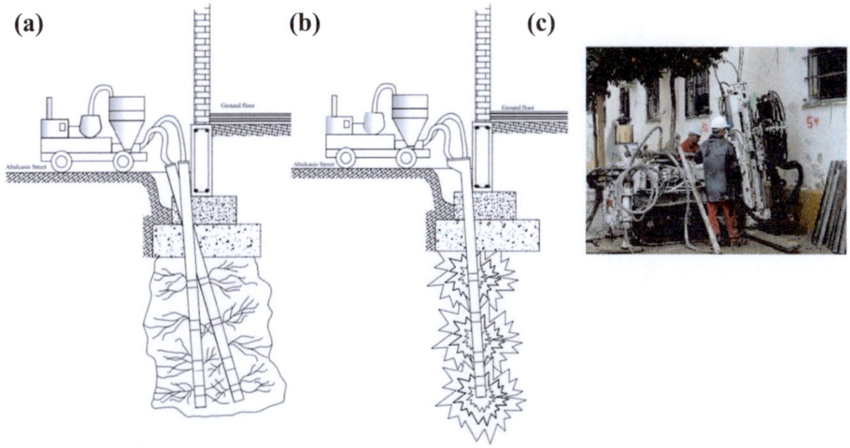

Fig. 23.6 Schema of constructive solutions **a** cement injections (mortar injection), **b** injections of expandable polyurethane resin (polyurethane injection); **c** cement injections works ongoing

it's length, was selected. The diameter was 140 mm and reinforced with a 60.3 mm steel tube. Micropiles, as a small-diameter (typically less than 300 mm), drilled and grouted replacement foundation that is typically reinforced with a steel pipe or bar and the most commonly solution available, were placed along the foundation every 1 to 1.50 m and with a 7° angle to the vertical plane, [57], see Fig. 23.6.

A technical, economic, social, and ecological comparison has been set between foundation underpinning, cement injection, injections of expandable polyurethane and increasing load transmission area of the foundation footings. Technically, there are several differences. The main difference is the depth of the action: in micropiling it is necessary to drill up to a depth of 25 m, while for reinforced grouting it is only necessary to reach the section of soil that needs to be improved (in this case the soil layer that is 6 m deep under the foundation). In the micropiling, a connection between the micropile and the existing foundation is needed, see Fig. 23.7; while in the reinforced grouting, the existing foundations remain untouched and the soil underneath is treated, see Fig. 23.8.

23.5 Results and Discussion

23.5.1 Economic and Technical Comparison

The external side of the building foundation is underpinned with 58 micropiles, of 27 m each in length, resulting in 1566 m in total and a cost of 898,890.78€. In a second hypothetical scenario, the reinforced grouting needs 241 m under the foundation, with 97 grouting points (2 of which are placed every 2–2.5 m) with between 4.5

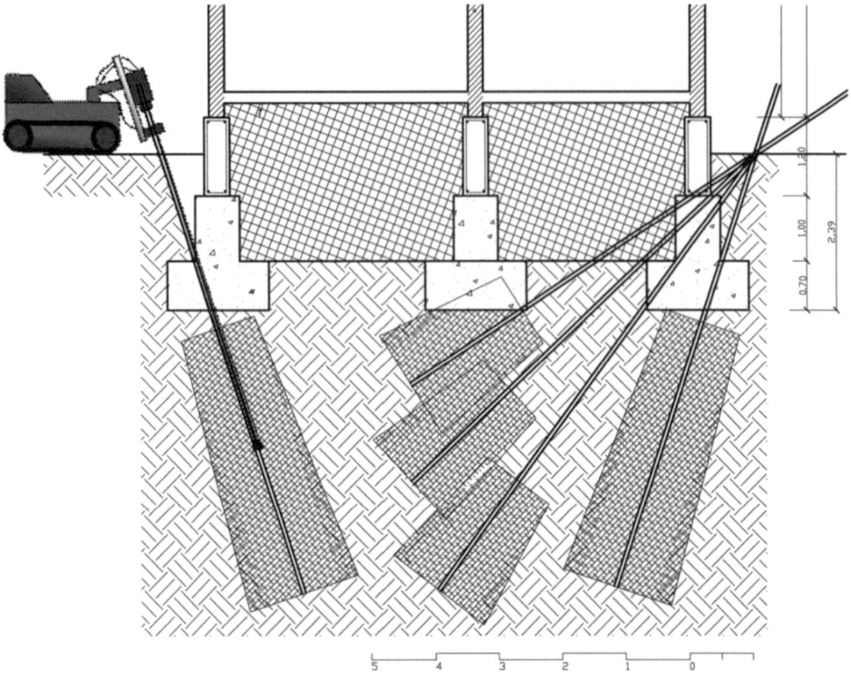

Fig. 23.7 Reinforced grouting [3]

and 6 m distance between each perforation and 1018.5 m in total length. Each unitary cost is described in Table 23.4. The total cost in this case would be 511,117.66€. In the case of using injections of expandable polyurethane resin, the cost would be 544,883€ for 240 points of injection. If the traditional solution were adopted for increasing the foundation support area, the cost would be 324,166€. Although the cost of micropiles is indeed the highest of all the scenarios, the direct transmission of load to a resistant soil layer ensures the success of the solution. Moreover, since this underpinning solution is widely used in construction, it is available for use in emergency situations; in contrast, grout injections, a less known system, despite presenting a more affordable solution, need not only a deeper soil characterisation, but also studies for the design of an optimal solution in terms of effective pressures, deep injections, and materials used. The use of injections of expandable polyurethane resin is a solution that requires no heavy machinery or auxiliary systems (but cannot reach deeper soil layers), which makes it a quick way of dealing with minor subsidence of foundations but, on the other hand, the use of polymers has yet to be proven as a durable solution and the effects of its expansions need rigorous control, which increases its risks in emergency applications. The traditional solution, based on incrementing the foundation support area, uses less machinery but much more manual labour and, for this reason, is the solution with the longest execution time. Moreover,

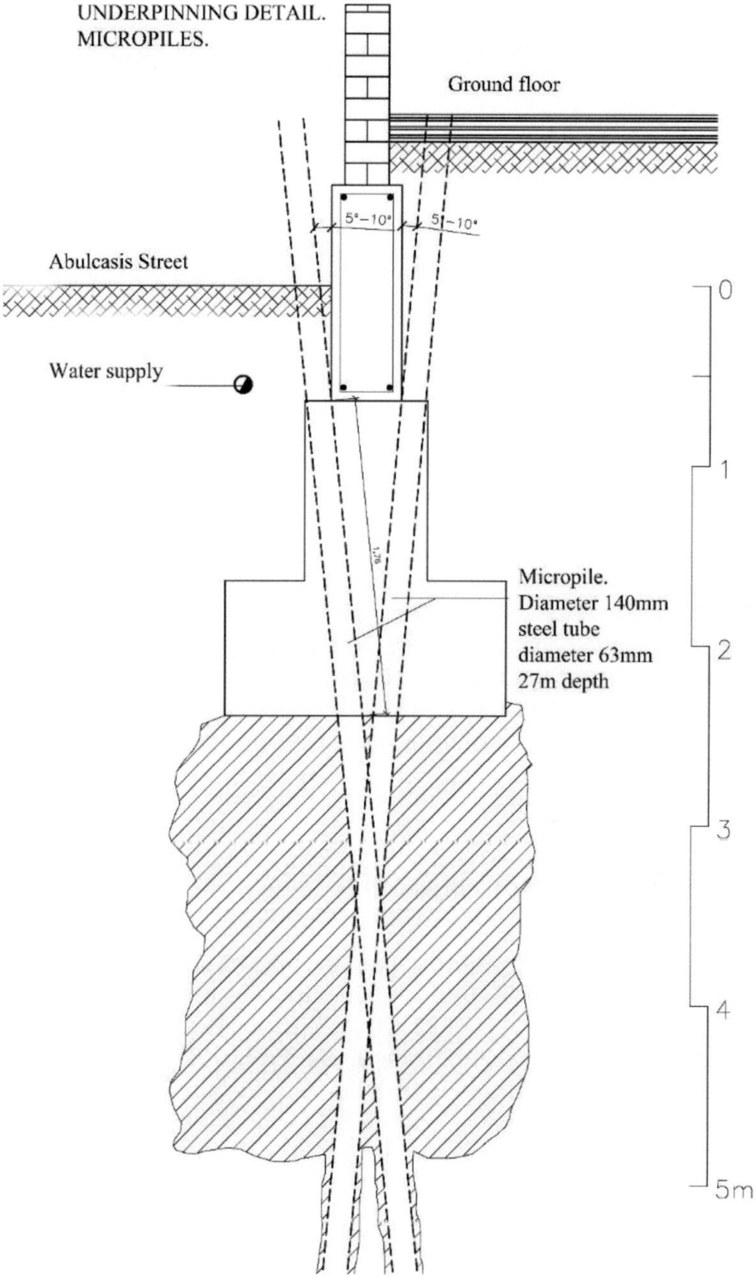

(a)

Fig. 23.8 **a** The building foundation is reinforced with micropiling, **b** Left: micropiling works ongoing; Right: detail of foundation drill

(b)

Fig. 23.8 (continued)

due to the need for a connection to the existing foundation, there is also a technical risk of failure and the involvement of complex planning.

As a result, a foundation recovery solution must be determined not only by means of the economic assessment, but using an approach that takes into consideration all the aspects (solution availability, execution time, and complexity) that contribute towards the success of the solution.

The micropiling solution is the most expensive, at approximately 40% more than the others, and the traditional reconstruction is the cheapest. However, from the technical evaluation of the latter method, it is the most complex to perform and is less reliable in correctly transmitting the loads between the newly built sections and the original foundation.

23.5.2 Environmental Comparison

For comparison with work by other authors, the impacts are calculated per ground floor area of the building, 3504.00 m^2. The CF value of construction of new buildings varies significantly: from 0.128 to 1.35 tCO_2/m^2, as calculated by others [35, 91, 92]. This wide range can be due to the different levels of complexity of the projects under study. In the present work, the repair work can also be expressed in terms of ground floor area. The CF of micropiling is 0.163 $kgCO_2eq/m^2$, while mortar injection is 0.251 $kgCO_2eq/m^2$, polyurethane injection is 0.0359 $kgCO_2eq/m^2$, and traditional restructuring is 0.0629 $kgCO_2eq/m^2$. The CF of the repair work is significantly high: even higher than those found in the construction of complete buildings. Comparing the solutions, the CF impact of mortar injection is substantially greater than in the other solutions (see Fig. 23.9), which is due to the large amount of mortar needed in order to complete just one metre of micropiling, since part of the material is dispersed in the soil and needs 300 kg of cement for every metre in depth thereof, see Table 23.4.

Table 23.4 Ecological footprint per impact source and territory

ACCD code	Description	Quantity	Unit	Cost	CF	WF	EF
				€	tCO_2eq	m^3	(hag)
	Reinforced grouting (micropiling)						
01RSH90002	Mechanical demolition of flooring	188.50	m^2	1264.76	3.55E−02	1.28E−01	3.13E−02
02PMM00002	Shaft excavation by mechanical means	287.10	m^3	2702.44	1.31E+00	4.73E+00	6.55E−01
03CTW00100	Drilling into existing foundation	255.20	m	6302.42	1.47E−01	5.31E−01	9.66E−02
03CTI99010	In-situ micropiling with steel jacket of diameter 60 mm	6264.00	m	884,898.37	5.71E+02	3.02E+03	2.75E+02
03CTW00200	Transport, installation, and removal of complete micropiling equipment	1.00	Unit	3722.78	9.97E−02	3.60E−01	1.24E−04
		Total		898,890.78	5.73E+02	3.02E+03	2.76E+02
	Cement injection (mortar injection)						
01RSH90002	Mechanical demolition of flooring	188.50	m^2	1264.76	3.55E−02	1.28E−01	3.13E−02
02PMM00002	Shaft excavation by mechanical means	287.10	m^3	2702.44	1.31E+00	4.73E+00	6.55E−01
03WSW00001	Consolidated ground column with pressurised cement slurry injections	3600.00	m	505,921.21	8.79E+02	4.06E+03	4.18E+02

(continued)

Table 23.4 (continued)

ACCD code	Description	Quantity	Unit	Cost	CF	WF	EF
				€	tCO_2eq	m^3	(hag)
03WSW00100	Transport, installation, and removal of complete equipment for injections	1.00	Unit	1229.25	1.75E−01	6.33E−01	8.57E−02
		Total		511,117.66	8.81E+02	4.07E+03	4.19E+02
	Injection of expandable polyurethane resin (polyurethane injection)						
01RSH90002	Mechanical demolition of flooring	188.50	m^2	1264.76	3.55E−02	1.28E−01	3.13E−02
02PMM00002	Shaft excavation by mechanical means	287.10	m^3	2702.44	1.31E+00	4.73E+00	6.55E−01
03WSW00010	Injection of expansive resins	1200.00	Unit	539,670.00	1.24E+02	7.38E+03	5.42E+01
03WSW00200	Transport, installation, and removal of complete equipment for injections	1.00	Unit	1245.93	1.76E−01	6.35E−01	8.60E−02
		Total		544,883.13	1.26E+02	7.38E+03	5.49E+01
	Increase of transmission area (traditional restructuring)						
01RSH90002	Mechanical demolition of flooring	673.00	m^2	4515.57	1.27E−01	4.58E−01	1.12E−01
02PMM00002	Shaft digging with mechanical means	1682.50	m^3	15,837.20	7.67E+00	2.77E+01	3.84E+00
03RZK90010	Foundation underpinning	504.75	m^3	265,844.23	2.08E+02	3.34E+03	1.03E+02
02RRB00001	Manual soil backfills	1346.15	m^3	35,525.57	3.99E+00	4.61E+02	2.56E+00

(continued)

Table 23.4 (continued)

ACCD code	Description	Quantity	Unit	Cost	CF	WF	EF
				€	tCO_2eq	m^3	(hag)
02RCM00001	Surface compaction conducted with mechanical tamper	1009.50	m^2	2444.40	5.03E−01	6.88E+01	2.85E−01
		Total		324,166.98	2.20E+02	3.90E+03	1.09E+02

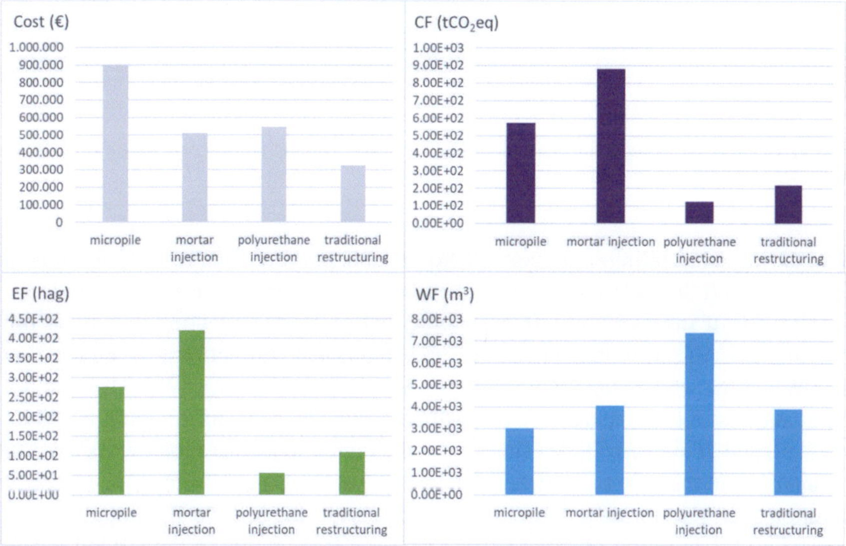

Fig. 23.9 Footprints and cost comparison

The EF determined by other authors is in terms of ground floor area of the construction of new buildings, at 0.6 gha/m^2 [93], and 0.05 gha/m^2 [94]. In the present work, the EF of micropiling is 7.87E−02gha/m^2, mortar injection is 1.20E−01 gha/m^2, polyurethane injection is 1.57E−02 gha/m^2, and traditional restructuring is 3.12E−02 gha/m^2. The results also show that the impact of the repair work has a high EF as compared to new construction. The EF follows a similar trend because the carbon footprint of construction materials is transformed into forest footprint, and is more significant than the impact from machinery or the workforce. If the impact of the materials is separated from other EF categories, see Fig. 23.10, then the traditional solution leaves the greatest footprint in each category. This is largely due to the traditional solution being more labour-intensive and the footprints from workers food is higher in each category. The machinery has a greater impact during injections

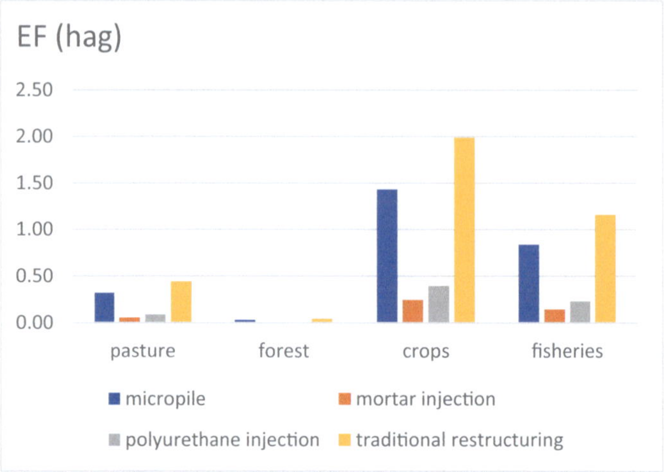

Fig. 23.10 Ecological footprint (EF) of various solutions without considering the CO_2 emissions of construction materials production

than during micropiling, injections of expandable polyurethane resin (polyurethane injection), and the increasing of transmission area (traditional restructuring).

The WF of micropiling is 0.86 m^3/m^2, that of mortar injection is 1.16 m^3/m^2, polyurethane injection is 2.11 m^3/m^2, and traditional restructuring is 1.11 m^3/m^2. The WF of the whole construction of warehouses is 6–12 m^3/m^2 [95]. The results of other authors of the construction of dwellings stand at 20.80 m^3/m^2 [46], 20.83 m^3/m^2 [96], and 26.6 m^3/m^2 [45]. This shows the high impact that repair work exerts in comparison to that of the construction of a new building. From among the suggested solutions, the solution of polyurethane injections has the highest WF impact. This is caused by the injection of expansive resins for consolidation of the ground located under the existing foundation. The procedure is water-intensive: 7380 m^3 of water per metre per column.

23.5.3 Comparison of the Social Indicator

In Fig. 23.11, the risk assessment is presented per unit cost and construction solution. From these results, it is clear that traditional underpinning is the technique that presents the greatest risk in its execution, with 110 risk level points. The highest accumulation of risk for this solution corresponds to unit cost 03RZK90010 (Foundation underpinning) with 87 points, due to both the numerous procedures necessary for its execution, with a total of 26 procedures (compared to the average of 9 procedures required to execute the other constructive solutions), and the risk involved in executing most of them, especially the PrPe03RZK007 procedure (Gallery excavation under foundation) which stands out with 17 points, see Table 23.5. The elevated risk of this procedure is due to the risk of burial of the operator who conducts the gallery excavation. The comparison of the results of the risk assessment broken down into procedures can be consulted in Table 23.5.

The subsequent solution that stands out for its level of risk is the solution solved with micropiling with 67 points, since it requires a greater number of procedures for its execution. In contrast, the unit cost with the highest level of risk is located in the solution with mortar injection, whose unit cost of 03WSW00001 (Consolidated ground column with pressurised cement slurry injections) by itself accumulates 41 risk level points, see Fig. 23.11. This elevated risk is due to the fact that all of its procedures present a high level of risk, between 10 and 11 points each (see scoring of the procedures in Table 23.5).

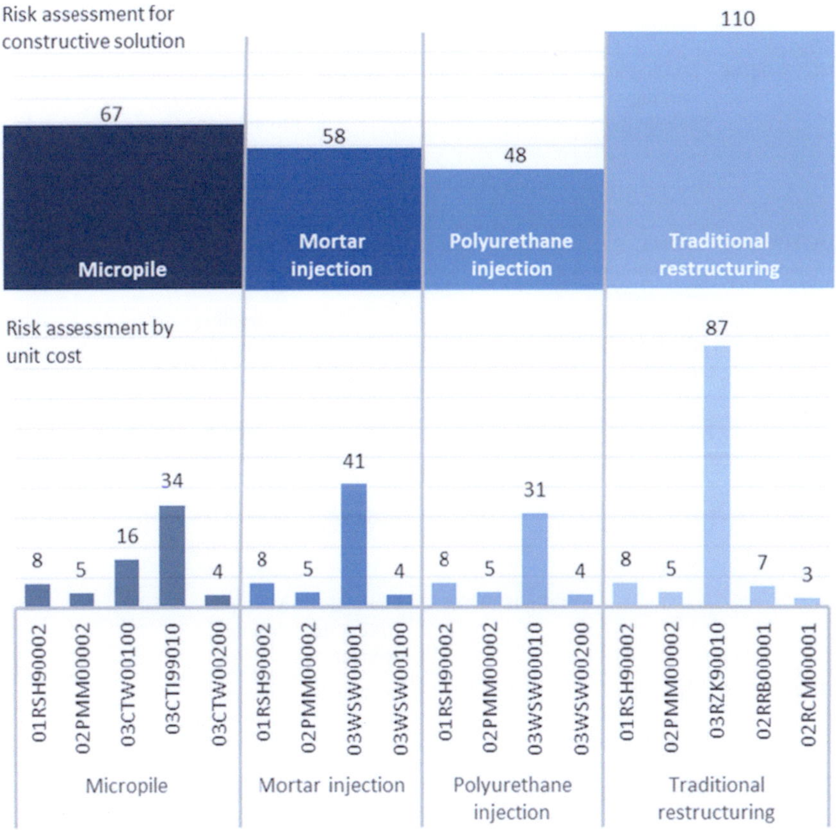

Fig. 23.11 Comparison of risk assessment results: in terms of unit prices and complete construction solution

Table 23.5 Comparison of risk assessment results: per procedure, unit cost, and complete construction solution

ACCD code	ACCD description	Procedure code	Task	Procedure description	Risk assessment Proc.	Unit cost	Total
	Reinforced grouting (micropiling)						
01RSH90002	Mechanical demolition of flooring	PrPe01RSH001	Special labourer	Demolition support work	7	8	67
		PrPc01RSH001	Loader, shovel operator	Demolition work	1		
02PMM00002	Shaft excavation with mechanical means	PrPe02PMM001	Special labourer	Excavation support work	4	5	
		PrPc02PMM001	Backhoe operator	Excavation work	1		
03CTW00100	Drilling into existing foundation	PrAy03CTW001	Specialist assistant	Support work in foundation drilling	8	16	
		PrEp03CTW001	Drilling rig operator	Foundation drilling	8		
03CTI99010	In-situ micropiling with steel jacket of diameter 60 mm	PrEn03CTI001	Official 1st formwork	Placement steel jacket for micropiling formwork	8	34	
		PrAy03CTI001	Specialist assistant	Support work in the formwork procedure	8		
		PrPe03CTI001	Special labourer	Support work in the micropiling procedure	8		
		PrEp03CTI001	Micropile drilling rig operator	Micropile drilling	10		
03CTW00200	Transport, installation, and removal of complete micropiling equipment	PrTr03CTW001	Equipment operator	Transport, installation, and removal of micropiling equipment	4	4	

(continued)

Table 23.5 (continued)

					Risk assessment		
ACCD code	ACCD description	Procedure code	Task	Procedure description	Proc.	Unit cost	Total
	Cement injection (mortar injection)						
01RSH90002	Mechanical demolition of flooring	PrPe01RSH001	Special labourer	Demolition support work	7	8	58
		PrPc01RSH001	Loader, shovel operator	Demolition work	1		
02PMM00002	Shaft excavation with mechanical means	PrPe02PMM001	Special labourer	Excavation support work	4	5	
		PrPc02PMM001	Backhoe operator	Excavation work	1		
03WSW00001	Consolidated ground column with pressurised cement slurry injections	PrEn03WSW001	Official 1st formwork	Placement of injection tubes	10	41.00	
		PrAy03WSW001	Specialist assistant	Support work in tube placement	10		
		PrPe03WSW001	Special labourer	Support work in the injection procedure	10		
		PrEp03WSW001	Injection equipment operator	Cement grout injection	11		
03WSW00100	Transport, installation and removal of complete equipment for injections	PrTr03WSW001	Equipment operator	Transport, installation and removal of equipment for injections	4	4	
	Injection of expandable polyurethane resin (polyurethane injection)						
01RSH90002	Mechanical demolition of flooring	PrPe01RSH001	Special labourer	Demolition support work	7	8	48
		PrPc01RSH001	Loader, shovel operator	Demolition work	1		

(continued)

Table 23.5 (continued)

					Risk assessment		
ACCD code	ACCD description	Procedure code	Task	Procedure description	Proc.	Unit cost	Total
02PMM00002	Shaft excavation with mechanical means	PrPe02PMM001	Special labourer	Excavation support work	4	5	
		PrPc02PMM001	Backhoe operator	Excavation work	1		
03WSW00010	Injection of expansive resins	PrEn03WSW002	Official 1st formwork	Placement of injection formwork	10	31	
		PrAy03WSW002	Specialist assistant	Support work in the injection procedure	10		
		PrEp03WSW002	Injection equipment operator	Expandable polyurethane resin injection	11		
03WSW00200	Transport, installation, and removal of complete equipment for injections	PrTr03WSW002	Equipment operator	Transport, installation, and removal of equipment for injections	4	4	
	Increase in transmission area (traditional restructuring)						
01RSH90002	Mechanical demolition of flooring	PrPe01RSH001	Special labourer	Demolition support work	7	8	110
		PrPc01RSH001	Loader shovel operator	Demolition work	1		
02PMM00002	Shaft excavation with mechanical means	PrPe02PMM001	Special labourer	Excavation support work	4	5	
		PrPc02PMM001	Backhoe operator	Excavation work	1		
03RZK90010	Running and centred shoe underpinning	PrPe03RZK001	Special labourer	Support work in the pumping procedure	4	87	
		PrBa03RZK001	Pumping equipment operator	Pumping of water and sludge	1		

(continued)

Table 23.5 (continued)

					Risk assessment		
ACCD code	ACCD description	Procedure code	Task	Procedure description	Proc.	Unit cost	Total
		PrPe03RZK002	Special labourer	Ground excavation (manual, max. depth 1.5 m)	6		
		PrPe03RZK003	Special labourer	Rubble excavation (manual, max. depth 1.5 m)	6		
		PrPe03RZK004	Special labourer	Ground excavation (manual, depth between 1.5 and 3 m)	7		
		PrPe03RZK005	Special labourer	Support work in ground filling	4		
		PrPi03RZK001	Mechanical compactor operator	Ground compaction	3		
		PrPe03RZK006	Special labourer	Transport of earth by manual means	9		
		PrPc03RZK001	Loader, shovel operator	Load earth by truck	2		
		PrCb03RZK001	Truck operator	Transport earth to landfill	2		
		PrPe03RZK007	Special labourer	Gallery excavation under foundation	17		
		PrFe03RZK001	Official 1st rebar	Placement rebar in the foundation	4		

(continued)

Table 23.5 (continued)

					Risk assessment		
ACCD code	ACCD description	Procedure code	Task	Procedure description	Proc.	Unit cost	Total
		PrEn03RZK001	Official 1st formwork	Placement wooden formwork	5		
		PrPe03RZK008	Special labourer	Support work in the wooden formwork procedure	5		
		PrPe03RZK009	Special labourer	Support work in HA-20 concrete vibration	3		
		PrVi03RZK001	Concrete vibrator operator	Concrete vibration	3		
		PrPe03RZK010	Special labourer	Support work in HA-25 concrete vibration	3		
		PrVi03RZK002	Concrete vibrator operator	Concrete vibration	3		
02RRB00001	Manual soil backfills	PrPe02RRB001	Special labourer	Support work in ground compaction	4	7	
		PrPi02RRB001	Manual compactor operator	Manual ground compaction	3		
02RCM00001	Surface compaction carried out with mechanical tamper	PrPe02RCM001	Special labourer	Support work on surface compaction of the ground	2	3	
		PrPi02RCM001	Mechanical compactor operator	Surface compaction of the ground	1		

23.6 Discussion

The results obtained show that the elements that most significantly affect the ecological impact are the construction materials due to its embodied resources, although machinery fuel consumption and workforce are also major factors. The EF is the only indicator which takes into account the impact of labour.

From the economic perspective, the micropile foundation exerts the highest impact of all the solutions evaluated, whose social indicator is also high, following on the tail of the traditional reconstruction of the foundation. The traditional solution involves an elevated risk of the worker being buried by soil displacement during the repair work. Strategies that reduce the EE of construction materials can reduce CF and EF significantly.

From a technical point of view, the results obtained conclude that, by using IRS micropiles, under emergency conditions, the underpinning of the foundations on soft soils improves the soil characteristics (also using grout injections) and transfers the loads to resistant soil layers.

From its economic dimension, the traditional solution remains the cheapest. The environmental results conclude that despite being the most economical, its environmental cost is higher than that of polyurethane injections due to an intense use of labour. Finally, in this case study, polyurethane injections are considered the optimal solution when all three aspects are considered: social, environmental, and economic.

23.7 Conclusion

The foundation rehabilitation of a 40-dwelling building, having been subject to an accident, is studied from a social, economic, and environmental perspective. The project budget information, more specifically the work breakdown structure facilitates the evaluation of all impacts.

For the first time, the footprint family indicators are employed to analyse a foundation repair in order to identify the materials, machinery, and labour that exert high environmental impact. The indicators are sensitive to the differences between the four solutions studied herein. The application of the proposed procedure has proved to be effective to address technical, economic, social, and environmental aspects simultaneously, when a work breakdown system is applied.

From the solutions evaluated, the cheapest solution is the traditional restructuring. Its environmental indicators are also on the lower side, but its social indicator is the highest due to the risk of suffocation during the excavation. The micropilling is the most expensive and its social indicator is also high. Also, it CF is the second highest. But is the most commonly employed due to technical aspects and availability of the equipment. The underpinning of the foundations on soft soils improves the soil characteristics and transfers the loads to resistant soil layers. For a better understanding, interrelations between variables and indicators need to be established in a unified indicator.

In future research, a database of repair work and its classification is to be defined in order to assess all activities taking place in any building reconstruction. To this end, social, economic, and environmental impacts must accompany each construction unit selected in the project.

References

1. CTE, Spain MH (Ministry of Housing). Código Técnico de la Edificación (Building Technical Code). (Madrid. Spain, 2006)
2. F. Da Casa, E. Echeverría, F. Celis, La intervención bajo rasante, la importancia de su conocimiento. La técnica de la armada de inyección. 'La intervención bajo tierra, la importancia de su conocimiento. La técnica de la inyección armada. nformes de la Construcción. **59**, 21–35 (2007)
3. A. Ferreira-Sánchez, M. Alba-Rodríguez, M. Marrero, Technical-social tandem in restoration of buildings under emergency conditions: 40 homes in Spain. J. Perform. Constr. Facil. **33**, 04019031 (2019)
4. D.E. Splitstone, S.A. Stonecheck, R.L. Dodson, J.A. Fuller, Birmingham Bridge emergency repairs: micropile foundation retrofit. *Advances in Analysis, Modeling & Design*, 1479–1487 (2010)
5. A.A. Sezer, Environmental assessment tools and efficiency in housing and office refurbishment, in *28th Annual Conference on Association of Researchers in Construction, 2012*, pp. 1331–1341
6. Z.S. Liu, W. Zou, S. Deng, Y. Fu, S. Tang, D. Gou et al., A simulation study of the emergency double-row micropile reinforcement of the Xituo Landslide. IOP Conf. Ser. Earth Environ. Sci. **570**, 022039 (2020)
7. A. Gutiérrez-Martín, J. Pedro Millán-Martín, R. Castedo, J. Ignacio Yenes, A. Gutierrez-Martın, J. Pedro Millan-Martın, et al., Calculation of micropiles and anchors to reinforce a slope in emergency situations: application in Málaga, Spain, vol. 12 (Taylor & Francis, 2021), pp. 716–740
8. M.D. Alba-Rodríguez, J. Solís-Guzmán, M. Marrero, Evaluation model of the economic-environmental impact on housing recovery. Application in the city of Seville, Spain. Sustain. Cities Soc. **83**, 103940 (2022)
9. B. Nowogońska, A methodology for determining the rehabilitation needs of buildings. Appl. Sci. (Switzerland) **10** (2020)
10. European_Commission, LEVEL(S) Taking action on the TOTAL impact of the construction sector (2019). https://doi.org/10.2779/458570
11. M.M. Santos, J.C.G. Lanzinha, A.V. Ferreira, Proposal for a methodology for sustainable rehabilitation strategies of the existing building stock—the ponte gêa neighborhood. Des. (Basel), **5** (2021)
12. T. McGrath, S. Nanukuttan, K. Owens, M. Basheer, P. Keig, Retrofit versus new-build house using life-cycle assessment. Eng. Sustain. **166**, 122–137 (2015)
13. M.D. Alba-Rodríguez, A. Martínez-Rocamora, P. González-Vallejo, A. Ferreira-Sánchez, M. Marrero, Building rehabilitation versus demolition and new construction: economic and environmental assessment. Environ. Impact Assess. Rev. **66**, 115–126 (2017)
14. C. Rivero-Camacho, J.J. Martín-del-Río, M. Marrero-Meléndez, Evolution of the life cycle of residential buildings in Andalusia: economic and environmental evaluation of their direct and indirect impacts. Sustain. Cities Soc. **93**, 104507 (2023)
15. M. Marrero, C. Rivero-Camacho, A. Martínez-Rocamora, D. Alba-Rodríguez, V. Lucas-Ruiz, Holistic assessment of the economic, environmental, and social impact of building construction. Application to housing construction in Andalusia. J Clean Prod. **434**, 140170 (2024)

16. ACCD. Andalusia Construction Cost Database (ACCD). Andalusia Government. [Internet]. 2017. Available from: https://www.juntadeandalucia.es/organismos/fomentoyvivienda/areas/vivienda-rehabilitacion/planes-instrumentos/paginas/bcca-sept-2017.html
17. M. Ruiz-Pérez, C. Rivero-Camacho, M. Alba-Rodríguez, M. Marrero, Evaluation of carbon footprint of the renovation of urban spaces, in *Environmental Footprints and Eco-Design of Products and Processes* (Springer, 2022)
18. C. Rivero-Camacho, J.J. Martín-Del-Río, J. Solís-Guzmán, M. Marrero, Ecological footprint of the life cycle of buildings, in *Environmental Footprints and Eco-Design of Products and Processes* (Springer, 2021), pp. 1–39
19. M. Ruiz-Pérez, M. Alba-Rodríguez, M. Marrero Evaluation of water footprint of urban renewal projects. Case study in Seville, Andalusia. Water Res. **221**, 118715 (2022)
20. M.Y. Kaltakci, M. Kamanli, M. Ozturk, M.H. Arslan, H.H. Korkmaz, Sudden complete collapse of Zumrut apartment building and the causes. J. Perform. Constr. Facil. **27**, 381–390 (2013)
21. L.E. Yamin, J.F. Correal, M. Asce, J.C. Reyes, F. Ramirez, R. Rincón, et al., Sudden Collapse of the 27-Story Space Building in Medellin, Colombia. J. Perform. Constr. Facil. **32**, 04018010 (2018)
22. M. Bahramian, K. Yetilmezsoy Life cycle assessment of the building industry: an overview of two decades of research (1995–2018). Energy Build. 219 (2020)
23. S. Giljum, E. Burger, F. Hinterberger, S. Lutter, M. Bruckner, A comprehensive set of resource use indicators from the micro to the macro level. Resour. Conserv. Recycl. **55**, 300–308 (2011)
24. G. Stoeglehner, M. Narodoslawsky, Implementing ecological footprinting in decision-making processes. Land Use Policy **25**, 421–431 (2008)
25. J. Matuštík, V. Kočí, What is a footprint? A conceptual analysis of environmental footprint indicators. J. Clean. Prod. 285 (2021)
26. K. Fang, R. Heijungs, G.R. de Snoo, Theoretical exploration for the combination of the ecological, energy, carbon, and water footprints: overview of a footprint family. Ecol. Indic. **36**, 508–518 (2014)
27. J.C. Bare, P. Hofstetter, D.W. Pennington, H.A.U. de Haes, Midpoints versus endpoints: the sacrifices and benefits. Int. J. Life Cycle Assess. **5**, 319 (2000)
28. J. Solís-Guzmán, A. Martínez-Rocamora, M. Marrero, *Methodology for determining the carbon footprint of the construction of residential buildings Assessment of Carbon Footprint in Different Industrial Sectors* (Springer, 2014), pp. 49–83
29. A. Laurent, M. Owsianiak, Potentials and limitations of footprints for gauging environmental sustainability. Curr. Opin. Environ. Sustain. **25**, 20–27 (2017)
30. M. Marrero, C. Rivero-Camacho, M.D. Alba-Rodríguez, What are we discarding during the life cycle of a building? Case studies of social housing in Andalusia, Spain. Waste Manag. **102**, 391–403 (2020)
31. M. Marrero, M. Wojtasiewicz, A. Martínez-Rocamora, J. Solís-Guzmán, M.D. Alba-Rodríguez, BIM-LCA integration for the environmental impact assessment of the urbanization process. Sustainability (Switzerland) **12**, 1–24 (2020)
32. S. Geng, Y. Wang, J. Zuo, Z. Zhou, H. Du, G. Mao, Building life cycle assessment research: a review by bibliometric analysis. Renew. Sustain. Energy Rev. **76**, 176–184 (2017)
33. C. Dossche, V. Boel, W. De Corte, Use of life cycle assessments in the construction sector: critical review. Procedia Eng. **171**, 302–311 (2017)
34. R. Geng, S.A. Mansouri, E. Aktas, The relationship between green supply chain management and performance: a meta-analysis of empirical evidences in Asian emerging economies. Int. J. Prod. Econ. **183**, 245–258 (2017)
35. P. Chastas, T. Theodosiou, K.J. Kontoleon, D. Bikas, Normalising and assessing carbon emissions in the building sector: a review on the embodied CO_2 emissions of residential buildings. Build. Environ. **130**, 212–226 (2018)
36. A. Freire, M. Marrero, Analysis of the ecological footprint produced by machinery in construction. World Sustainable Building 14. Barcelona (2014)
37. J. Solís-Guzmán, C. Rivero-Camacho, M. Tristancho, A. Martínez-Rocamora, M. Marrero Software for calculation of carbon footprint for residential buildings, in *Environmental Footprints and Eco-Design of Products and Processes* (Springer, 2020), pp. 55–79

38. J. Solís-Guzmán, C. Rivero-Camacho, D. Alba-Rodríguez, A. Martínez-Rocamora, Carbon footprint estimation tool for residential buildings for non-specialized users: OERCO2 project. Sustainability **10**, 1359 (2018)
39. M. Wackernagel, W. Rees, Our Ecological Footprint: Reducing Human Impact on the Earth. New Catalyst Bioregional Series (1997)
40. J. Solís-Guzmán, M. Marrero, The ecological footprint of building construction: a Spanish case study. Bentham eBooks (2015)
41. P. González-Vallejo, C. Muñoz-Sanguinetti, M. Marrero, Environmental and economic assessment of dwelling construction in Spain and Chile. A comparative analysis of two representative case studies. J Clean Prod. **208**, 621–635 (2019)
42. R. Crawford, S. Pullen, Life cycle water analysis of a residential building and its occupants. Build. Res. Inf. 589–602 (2011)
43. Y. Chang, Z. Huang, R.J. Ries, E. Masanet, The embodied air pollutant emissions and water footprints of buildings in China: a quantification using disaggregated input-output life cycle inventory model. J. Clean. Prod. **113**, 274–284 (2016)
44. S. Bardhan, Assessment of water resource consumption in building construction in India. Ecosyst. Sustain. Dev. **144**, 1743–3541 (2011)
45. M.Y. Han, G.Q. Chen, J. Meng, X.D. Wu, A. Alsaedi, B. Ahmad, Virtual water accounting for a building construction engineering project with nine sub-projects: a case in E-town, Beijing. J. Clean. Prod. **112**, 4691–4700 (2016)
46. G. Heravi, M.M. Abdolvand, Assessment of water consumption during production of material and construction phases of residential building projects. Sustain. Cities Soc. **51**, 101785 (2019)
47. M.R. Ruiz-Pérez, M.D. Alba Rodríguez, M. Marrero, *Systems of Water Supply and Sanitation for Domestic Use. Water Footprint and Carbon Footprint Evaluation: First Results*. In: Research IIC on C and B, editor. Santa Cruz de Tenerife, España (2017)
48. M.R. Ruiz-Pérez, M.D. Alba-Rodríguez, M. Marrero, The water footprint of city naturalisation. Evaluation of the water balance of city gardens, in *The 22nd Biennial Conference of The International Society for Ecological Modelling (ISEM)*. Salzburg, Austria, 2019
49. CSI/CSC, Construction Specifications Institute/Construction Specifications Canada. MasterFormat 2016 Edition: Numbers and Titles (2016)
50. A.R. Jones, *CI/SfB Construction Indexing Manual* (Royal Institute of British Architects RIBA, London, UK., 1987)
51. T. Telford, *Civil Engineering Standard Method of Measurement*, 3rd edn. U.K., pp. 4–39 (1991)
52. UniFormatTM, The Construction Specifications Institute. A Uniform Classification of Construction Systems and Assemblies. Alexandria, VA (1998)
53. Omniclass, A strategy for classifying the built environment - Table 13: Spaces by function (2012)
54. ITeC BEDEC, Instituto Tecnológico de Cataluña (2019)
55. CYPE Ingenieros SA, Construction cost generator. Spain. http://www.generadordeprecios.info/#gsc.tab=0 (2022)
56. M. Marrero, A. Ramirez-de-Arellano, The building cost system in Andalusia: application to construction and demolition waste management. Constr. Manag. Econ. **28**, 495–507 (2010)
57. A. Ferreira Sánchez, Recuperación de viviendas en situación de emergencia. Estudio de caso: Viviendas RENFE, Sevilla (Dwelling rehabilitation under emergency conditions. Case study: RENFE Dwellings, Seville) (University of Seville, 2015)
58. WHO, Psychological first aid [Internet]. Worl Health Organization. 2024. Cited 2024 Mar 27. Available from: https://www.who.int/publications/i/item/psychological-first-aid
59. J.H. Fox, F.M. Burkle, J. Bass, F.A. Pia, J.L. Epstein, D. Markenson, The effectiveness of psychological first aid as a disaster intervention tool: research analysis of peer-reviewed literature from 1990–2010. Disaster Med. Public Health Prep. **6**, 247–252 (2012)
60. J.I. Ruzek, M.J. Brymer, A.K. Jacobs, C.M. Layne, E.M. Vernberg, P.J. Watson, Psychological first aid. J. Ment. Health Couns. **29**, 17–49 (2007)
61. INSST, Observatorio Condiciones de Trabajo Prevención de Riesgos Laborales [Internet]. Instituto Nacional de Seguridad y Salud en el Trabajo. 2022. Cited 2023 Mar 12. Available from: https://www.insst.es/el-observatorio

62. F. Lander, K.J. Nielsen, J. Lauritsen, Work injury trends during the last three decades in the construction industry. Saf. Sci. **85**, 60–66 (2016)
63. P. Swuste, A. Frijters, F. Guldenmund, Is it possible to influence safety in the building sector? A literature review extending from 1980 until the present. Saf. Sci. **50**, 1333–1343 (2012)
64. S. Winge, E. Albrechtsen, J. Arnesen, A comparative analysis of safety management and safety performance in twelve construction projects. J. Safety Res. **71**, 139–152 (2019)
65. L.J. Bellamy, B.J.M. Ale, J.Y. Whiston, M.L. Mud, H. Baksteen, A.R. Hale et al., The software tool storybuilder and the analysis of the horrible stories of occupational accidents. Saf. Sci. **46**, 186–197 (2008)
66. I.W.H. Fung, V.W.Y. Tam, T.Y. Lo, L.L.H. Lu, Developing a risk assessment model for construction safety. Int. J. Project Manage. **28**, 593–600 (2010)
67. F.J. Forteza, A. Sesé, J.M. Carretero-Gómez, CONSRAT. Construction sites risk assessment tool. Saf Sci. **89**, 338–354 (2016)
68. S. Winge, E. Albrechtsen, B.A. Mostue, Causal factors and connections in construction accidents. Saf. Sci. **112**, 130–141 (2019)
69. LPSC_9/2017, Law 9/2017 of 8 November, on Public Sector Contracts, which transposes into Spanish law the Directives of the European Parliament and Council 2014/23/EU and 2014/24/EU, of 26 February 2014. Government of Spain. Spain (2017)
70. N. Grunewald, A. Galli, I. Katsunori, M. Halle, M. Gressot, The Ecological Footprint of Mediterranean Diets (2015)
71. WWF, *The Built Environment Sector, Pre-Seminar Report* (Council for Environmental Education WWF, Department of Environment, De Monfort University Leicester, 1993)
72. GFN, *Learning Package of National Footprint Accounts*, 2014 edn. (Global Footprint Network, 2014)
73. UNE-EN 15978, *Sustainability of Construction Works. Assessment of Environmental Performance of Buildings. Calculation Method* (2012)
74. A. Freire-Guerrero, M.D. Alba-Rodríguez, M. Marrero, A budget for the ecological footprint of buildings is possible: a case study using the dwelling construction cost database of Andalusia. Sustain. Cities Soc. **51**, 101737 (2019)
75. Ecoinvent Centre, Ecoinvent database v3 ecoinvent report (2013)
76. A. Martínez-Rocamora, J. Solís-Guzmán, M. Marrero-Meléndez, LCA databases focused on construction materials: a review. Renew. Sustain. Energy Rev. **58**, 565–573 (2016)
77. PRé Sustainability. SimaPro 8 (2016)
78. X. Oregi, J. Tenorio, C. Gazulla, I. Zabalza, D. Cambra, S. Leão, et al., SOFIAS—Software for life-cycle assessment and environmental rating of buildings. Inf. Constr. **68** (2016)
79. BEDEC, Construction cost bases. Institute of construction technology of catalonia ITeC. ITeC (2019)
80. ITeC, Institute of Construction Technology of Catalonia, Barcelona (2012)
81. A. Freire Guerrero, M. Marrero, Evaluation of the embodied energy of a construction project using the budget. Habitat Sustentable. **5**, 54–63 (2015)
82. SEOPAN, Machinery costs manual (in Spanish: Manual de costes de maquinaria) (2008). Available: http://www.concretonline.com/pdf/07construcciones/art_tec/SeopanManualCostes.pdf. Accessed 01.07.16
83. IDEA, Factores de conversión a Energía Primaria y factores de emisión de CO_2/CO_2 emission factors (2011)
84. REE, *El sistema eléctrico español* (The Spanish Electric System, 2014)
85. Law_31/95, Law of November 8, Prevention of Occupational Risks. BOE.Es–BOE-A-1995-24292, 1999). Jefatura del Estado, «BOE» núm. 269, de 10/11/1995. Jefatura del Estado; 1999
86. RD_2015/97, Royal Decree 1215/1997, of July 18, which establishes the minimum health and safety provisions for the use of work equipment by workers. [Internet]. BOE-A-1997-17824. 1997. Cited 2023 Mar 12. Available from: https://www.boe.es/buscar/act.php?id=BOE-A-1997-17824

87. NTP-330, *NTP 330: Simplified Accident Risk Assessment System* (National Institute of Safety and Hygiene at Work. 1993)
88. S. Hignett, L. McAtamney, Rapid entire body assessment (REBA). Appl. Ergon. **31**, 201–205 (2000)
89. REBA, *NTP 601: Método REBA (Rapid Entire Body Assessment)* (National Institute of Safety and Hygiene at Work, 2001)
90. V. Zamorano Cardoso, Clasificación sistemática de los procedimientos de trabajo. Universidad de Sevilla Trabajo Fin de Máster en Seguridad Integral en Edificación (2010), pp. 1–405
91. D. Clark, *What Colour is Your Building? Measuring and Reducing the Energy and Carbon Footprint of Buildings* (RIBA Publishing, 2019)
92. C. De Wolf, *Material Quantities in Building Structures and Their Environmental Impact* (Massachusetts Institute of Technology, Boston, 2014)
93. D. Husain, R. Prakash, Life cycle ecological footprint assessment of an academic building. J. Inst. Eng. (India): Ser A **A100**, 97–110 (2019)
94. D. Husain, R. Prakash, A. Ahmad, Life cycle ecological footprint reduction for a tropical building. Adv. Civ. Eng. **2022** (2022)
95. D. Lopes, R. Silva, C. Rivero-Camacho, D. Rusu, M. Marrero, Methodology for improving the sustainability of industrial buildings via matrix of combinations water and carbon footprint assessment. Sustainability **14** (2022)
96. J. Meng, G.Q. Chen, L. Shao, J.S. Li, H.S. Tang, T. Hayat et al., Virtual water accounting for building: case study for E-town, Beijing. J. Clean. Prod. **68**, 7–15 (2014)

Chapter 24
LCA Analysis and Strategic Scenarios for the Reduction of Gas Emissions from Concrete Constructions: A Case Study

Federica Cucchiella, Chiara Marchionni, and Marianna Rotilio

Abstract The construction industry plays a key role in future sustainability strategies, as it is responsible for 39% of global carbon dioxide emissions. To counteract the impact generated by the construction sector, it is necessary to direct efforts to make more informed design choices. Although many studies have shown that dry construction systems are able to reduce emissions during the entire life cycle compared to traditional on-site construction, there are circumstances in which the latter must necessarily be used. In the present research, after having identified a gap in the literature between studies that investigate how to reduce the environmental impact of such constructions, the Life Cycle Assessment analysis of a case study is proposed. In particular, the case of a multi-stored building which uses the cast-in-situ reinforced concrete construction system as an executive choice is analyzed. For the LCA analysis the SimaPro software was used and the data obtained lead to identifying the impact categories in the respective pre-use, use and end-of-life phases, through the ReCiPe 2016 Endpoint (H) v.1.05 method, to determining which of these is more impactful and worthy of attention. The study therefore proposes strategic scenarios to mitigate polluting emissions, particularly in the phase of greatest impact, i.e. the pre-use phase.

F. Cucchiella
Department of Industrial and Information Engineering and Economics, University of L'Aquila, L'Aquila, Italy

C. Marchionni (✉) · M. Rotilio
Department of Civil, Construction-Architectural and Environmental Engineering, University of L'Aquila, L'Aquila, Italy
e-mail: chiara.marchionni@univaq.it

D. A. H. Hanaor (ed.), *Sustainable Materials for the Built Environment*, Springer Series in Materials Science 352, https://doi.org/10.1007/978-3-031-97818-0_24

24.1 Introduction

The construction sector is responsible for 40% of global energy consumption [1], causing approximately 39% of global greenhouse gas emissions (GHG), 12% of global drinking water use: for these reasons it is considered the most responsible for the changes climate [2–4]. Regarding these, the construction sector is also called upon to reduce the quantity of construction and demolition waste, which represents 33.5% of that produced in Europe [5]. Achieving the decarbonization objectives of the Paris Agreement (COP21) on climate change, i.e. zero net carbon emissions in the construction sector by 2050, requires that strong attention be paid to the construction front by adopting development concepts capable of taking natural resources into account, from the early stages of the project and the surrounding environment. The current critical environmental issue requires the construction sector, therefore, to identify design solutions capable of reducing the consumption of resources and the production of waste with the assumption of the time horizon of the entire life cycle, starting from the consideration of energy and matter flows [6]. In fact, while the linear economy focuses on the limited life span of resources without considering the products end of life [7], the circular economy (CE) promotes a sustainable built environment with increased efficiency for construction resources and waste minimization at construction stage and end of life [8]. Reference [9] assert as this is why the CE has attracted increased attention as one of the most powerful moves towards sustainability. However [10] underlines as the concept of CE is still relatively new in the building construction industry (BCI): [11, 12] study as the CE concept for building construction and demolition waste (BCDW) emerged in 2014. Núñez-Cacho et al. [13] show as CE is regarded as an important alternative to the linear economy that define unsustainable: the need for a shift from a linear approach to a CE clarify the recent attention given to it, also determined by the most recent European directives [14, 15], which aim to close the circle of the material cycle. The CE takes part in the objectives of the green economy, understood as an economy "with low carbon content, with efficient use of resources and socially inclusive" [16] and is outlined as a subcategory of the same, acting as the main center of interest is the efficiency in the use of resources and the reduction of waste production [17, 18]. Also managing the built environment according to the principles of the CE with the aim of limiting energy consumption and environmental pollution is a highly topical issue and digitization can play a crucial role in promoting virtuous scenarios [19]. However, the concept of circularity applied to concrete cases has struggled to take hold, due to the complexity of the method: [20] study that more than 100 hundred articles were published on the topic in 2016, compared to only about 30 articles in 2014. Life cycle Assessment analysis (LCA) is a systematic methodology that evaluates the environmental footprint of a construction, from the extraction phases of the raw materials that constitute it, to its construction, to its use, up to its final disposal. By quantifying the environmental impacts associated with its life cycle, it has a strategic role in identifying critical aspects. Nußholz et al. [21] in their research investigates 65 real-life circular building cases of new build, renovation, and demolition to the understanding of the application and decarbonisation potential of circular strategies in the building industry. Cai et al. [22] studies the lifecycle GHG and reduction potential of traditional cast-in-place building and prefabricated light-steel building using LCA method. Aye et al. [23] assesses the embodied greenhouse gas emissions of circular building and conventional building elements, to demonstrate with environmental data as the business model has the potential to reduce gas emissions. Also [24, 25], using the LCA methodology, prove as prefabricated constructions provides improved environmental performance over traditional in-situ concrete construction. In the Australian context [26] evaluate the environmental impacts of various concrete types, including innovative alternatives, using a computational LCA model. Results show as recycling 60% of concrete demolition waste further decreases environmental impacts by over 20%, aligning with CE principles and supporting resource recovery.

The main bibliography previously discussed is shown in Table 24.1. However, in cases where the use of traditional concrete construction is essential, it is necessary to develop strategies to evaluate the environmental impact of this material, improve its performance and adopt construction strategies that help to increasing the life of a concrete structure [27].

This study proposes a methodology based on the application of LCA analysis divided into four phases, in compliance with the international standards EN ISO 14040 and ISO 14044, and the subsequent analysis of the most impactful phases, to which strategies are applied for reduction of GHG emissions. The study is expressed through the evaluation of the environmental impact produced by a specific case, a newly built three-level underground concrete car park with a garden roof, in order to direct the design process towards more sustainable choices. The following paragraphs illustrate the results of the four phases on which an LCA analysis is based, the conclusions aim to summarize the results obtained, to identify mitigation strategies and to identify future research developments.

24.2 Materials and Methods

The circular perspective imposes a design that considers the entire cycle of materials, from their origin (extraction of the raw material) to their disposal [28]. The LCA analysis allows us to consider the effects of the transformation actions of the built environment by considering the flows of resources and environmental impacts along all phases of the life cycle, from the extraction of raw materials to their transformation, from production to assembly and installation, from marketing to use, from maintenance to end-of-life management. The evaluation of the environmental sustainability of the project is linked to the definition of environmental reference thresholds (benchmarks) which allow measuring the level of sustainability of a material, a system and a building designed within the building processes considered [29]. In accordance with the international standards UN EN ISO 14040 [30] and UN EN ISO 14044 [29] the current LCA study was divided into four phases, as shown in the Fig. 24.1:

Table 24.1 Essential bibliography relating to the main case studies analyzed

References	Aim of the research	Case study analized	Keyword selection for the sustainable design criteria
Aye et al. [24]	Quantify the embodied energy of modular prefabricated steel and timber multi-residential buildings to determine whether this form of construction provides improved environmental performance over conventional concrete construction methods	Multi-residential building in Melbourne	– Prefabricated construction in steel and timber – Comparison with concrete construction method
Andersen et al. [23]	Circular solutions are compared with conventional solutions to ascertain whether the business model has the potential to reduce GHG emissions	Five circular building and three conventional building elements/ components	– LCA – GHG emissions
Jayawardana et al. [25]	Investigate the cradle-to-gate environmental performance of prefabricated construction methods compared to traditional in-situ construction in buildings using LCA methodology	Office building constructed in Sri Lanka using traditional and prefabrication technologies	– Prefabricated construction – LCA
Cai et al. [22]	Study the reduction potential of light-steel building and traditional cast-in-place building (TCB) using LCA method	A prefabricated light-steel building (PLB) and a traditional cast-in-place building (TCB)), with almost identical architectural features	– GHG emissions – Prefabricated light-steel building (PLB) – Traditional cast-in-place building (TCB) – LCA

(continued)

Table 24.1 (continued)

References	Aim of the research	Case study analized	Keyword selection for the sustainable design criteria
Nußholz et al. [21]	Analyse cases regarding the type of CE strategy—narrowing, slowing, closing, and regenerating resource loops—, level of application in the building, and their decarbonisation potential	65 novel real-world cases of new build, renovation, and demolition projects in Europe compiled from academic and grey literature	Application of the circular strategies Decarbonisation potential
Tran et al. [26]	Develop a computational model to evaluate the environmental impacts of various concrete types commonly used in civil construction projects. The study focuses on key environmental indicators such as global warming potential and life cycle impacts	Eenvironmental impacts of various concrete types in Australian	– LCA – GHG emissions – Decarbonisation potential

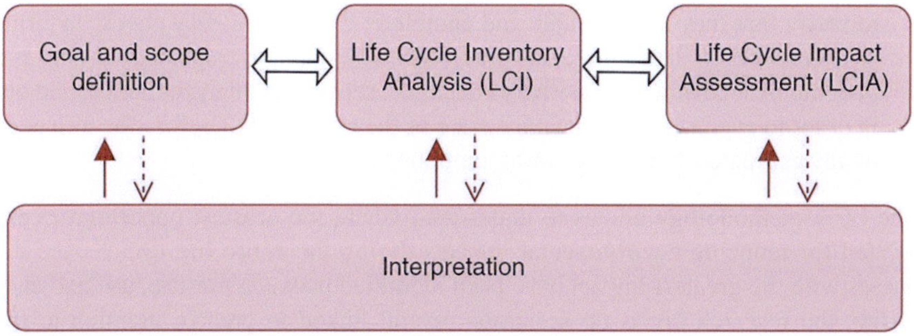

Fig. 24.1 The four phases of a life cycle impact assessment

1. Goal and scope definition of the LCA: it is the phase in which the objective of the study is defined, the functional unit (i.e. the reference unit with respect to which all energy consumption and resources and the related environmental impacts will be accounted for) the boundaries of the system (i.e. the processes and aspects to be included in the study based on its rationale for the study and any constraints induced by the data) the categories of data to be collected and analysed.
2. Life Cycle Inventory Analysis (LCI) phase: it is the most delicate and demanding phase, fundamental for the subsequent life cycle impact assessment (LCIA) phase. It involves the quantification of input and output data relating to the system

under study; in it the data necessary to achieve the objectives of the study are collected through archiving in an inventory table. The data collected can have different origins: primary source, i.e. data collected directly at the production sites; secondary source, literature data whose source must be reported (publications, manuals, previous studies, etc.); tertiary source, data obtained from estimates or technical coefficients, for which the calculation method must be indicated. The data collected and processed constitute the basis of calculation for the assessment of the impact of the life cycle.

3. Life Cycle Impact Assessment (LCIA) phase: the EN ISO 14044 standard defines the LCIA as the "LCA phase aimed at understanding and evaluating the extent and significance of the potential environmental impact of a product system". The purpose of the impact assessment phase is to use the data relating to energy and resource consumption, coming from the LCI phase, to build some environmental impact indicators. The purpose of this phase is to determine the system's potential effects on the environment by linking inventory data to specific impact categories, providing additional information to help evaluate a product system's LCI results to better understand their environmental significance.
4. Life Cycle Interpretation phase: life cycle interpretation is the final stage of the LCA procedure, in which all results are summarized and discussed as the basis for conclusions and decisions in accordance with the objective. In particular, in this phase, in addition to analyzing the results and therefore identifying any improvement actions for both the study and the production processes analyzed, a series of checks are carried out: completeness check on the data and information to verify that they are available and complete; data consistency check, in terms of differences in data source, accuracy, technological coverage, age of data and geographical coverage; sensitivity check, i.e. sensitivity analyzes are carried out in order to evaluate the percentage error of the study and therefore the influence of any estimates, hypotheses or assumptions.

The LCA methodology allows to understand where the greatest opportunities are located for reducing environmental impacts during the entire life cycle: once the phases with the greatest impact have been identified between pre-use, use and end-of-life, the research opens up scenarios several linked to greater attention to the performance of the material, to its supply, to the identification of the most suitable materials depending on their origin and production.

24.2.1 Goal and Scope Definition

The object of interest of this paper is the possibility of reducing the environmental impacts of traditional reinforced concrete construction. To this end, for the case study a project was selected relating to the construction of a three-storey underground car park with a garden roof serving the Coppito Campus of the University of L'Aquila, which was sized to have a gross floor area equal to 5155.38 m^2, a total height of

9.50 m and a total volume of 48,976.11 m^3. The project was evaluated through an LCA analysis performed with SimaPro software version 9.3.0.2, which allowed the environmental performance of the project to be monitored.

The processes that were considered in the case study can be traced back to the following phases:

- extraction of raw materials;
- production of the materials making up the car park;
- transport of materials to the construction site;
- construction of parking lot;
- parking use phase;
- demolition of the car park;
- recycling/disposal of materials.

The functional unit to which the LCA results were referred is the entire car park. In reference to the Technical Standards for Construction (Ministerial Decree 17 January 2018), a useful life of 50 years has been envisaged.

24.2.2 *Inventory Analysis (LCI)*

In the second phase of the study, the LCI was carried out, quantifying all the materials making up the car park and the related processes for their creation. The datasets were obtained mainly from the Ecoinvent v3 database, integrated with input data obtained from the literature and manufacturing companies. The model was divided into the following components: foundation slab, pillars, bulkhead, walls, intermediate floors and ramp, roof slab. The choice of construction materials was conducted with a "transformation" approach, i.e. using data inclusive of production inputs and not of transport on site, to obtain clear information on the exact origin of the raw materials involved in the creation of each product.

Pre-use Phase

Tables 24.2, 24.3, 24.4, 24.5, 24.6 and 24.7 show the materials related to the pre-use phase.

Listed in Fig. 24.2, for each material used in the construction of the car park, is the corresponding process implemented in the SimaPro model, with reference to: the category, the unit of measurement and the type of waste generated at the end of its life by disposal.

Table 24.2 Foundation slab materials

Materials	Weight [T]	Volume [Mc]
Concrete	8352.52	3608.00
Steel	721.60	91.92
Polyethylene	3.48	2.58

Table 24.3 Pillar materials

Materials	Weight [T]	Volume [Mc]
Concrete	883.64	381.70
Steel	241.99	30.83

Table 24.4 Bulkhead materials

Materials	Weight [T]	Volume [Mc]
Concrete	5150.88	2225.00
Steel	378.25	48.18
Polyethylene	1.89	1.40

Table 24.5 Wall materials

Materials	Weight [T]	Volume [Mc]
Concrete	714.18	308.50
Steel	46.28	5.89
Brick	32.80	41.00
Mineral wool	4.60	51.10

Table 24.6 Intermediate floor and ramp materials

Materials	Weight [T]	Volume [Mc]
Concrete	4114.24	1777.21
Steel	328.92	41.90
Polyethylene	70.11	2337.09
Polyurethane resin	14.02	233.70

In the pre-use phase, consumption relating to the construction phase was considered (Tables 24.8 and 24.9):

- Diesel fuel for excavators (48,976.11 m^3);
- Electrical energy necessary to power machines, tools and for lighting the construction site.

Table 24.7 Roofing materials

Materials	Weight [T]	Volume [Mc]
Concrete	1903.62	822.30
Steel	164.46	20.95
Organic soil	0.50	1731.18
Polyethylene	87.63	64.91
Gravel	2856.44	1904.29

	SIMAPRO CATEGORY	SIMAPRO PROCESS	UNIT	REFUSAL
		FOUNDATION STAGE		
CONCRETE	Materials/Construction/ Binders/Transformation	Cement, pozzolana and fly ash 36-55% {Europe without Switzerland}| cement production, pozzolana and fly ash 36-55% | Cut-off, S	ton	Cement
STEEL	Materials/Metals/Ferro/ Transformation	Reinforcing steel {Europe without Austria}| reinforcing steel production | Cut-off, S	ton	Steel
POLYETHYLENE	Materials/Plastic/ Thermoplats/ Transformation	Polyethylene terephthalate, granulate, amorphous, recycled {Europe without Switzerland}| polyethylene terephthalate production, granulate, amorphous, recycled | Cut-off, S	ton	PE
		PILLARS		
CONCRETE	Materials/Construction/ Binders/Transformation	Cement, pozzolana and fly ash 36-55% {Europe without Switzerland}| cement production, pozzolana and fly ash 36-55% | Cut-off, S	ton	Cement
STEEL	Materials/Metals/Ferro/ Transformation	Reinforcing steel {Europe without Austria}| reinforcing steel production | Cut-off, S	ton	Steel
		BULKHEAD		
CONCRETE	Materials/Construction/ Binders/Transformation	Cement, pozzolana and fly ash 36-55% {Europe without Switzerland}| cement production, pozzolana and fly ash 36-55% | Cut-off, S	ton	Cement
STEEL	Materials/Metals/Ferro/ Transformation	Reinforcing steel {Europe without Austria}| reinforcing steel production | Cut-off, S	ton	Steel
POLYETHYLENE	Materials/Plastic/ Thermoplats/ Transformation	Polyethylene terephthalate, granulate, amorphous, recycled {Europe without Switzerland}| polyethylene terephthalate production, granulate, amorphous, recycled | Cut-off, S	ton	PE
		WALLS		
CONCRETE	Materials/Construction/ Binders/Transformation	Cement, pozzolana and fly ash 36-55% {Europe without Switzerland}| cement production, pozzolana and fly ash 36-55% | Cut-off, S	ton	Cement
STEEL	Materials/Metals/Ferro/ Transformation	Reinforcing steel {Europe without Austria}| reinforcing steel production | Cut-off, S	ton	Steel
BRICK	Materials/Construction/ Bricks/Transformation	Clay brick {RER}| production | Cut-off, S	ton	Brick
MINERAL WOOL	Materials/Construction/ Insulation/Transformation	Stone wool {CH}| stone wool production | Cut-off, S	ton	Others
		INTERMEDIATE SLABS AND RAMP		
CONCRETE	Materials/Construction/ Binders/Transformation	Cement, pozzolana and fly ash 36-55% {Europe without Switzerland}| cement production, pozzolana and fly ash 36-55% | Cut-off, S	ton	Cement
STEEL	Materials/Metals/Ferro/ Transformation	Reinforcing steel {Europe without Austria}| reinforcing steel production | Cut-off, S	ton	Steel
POLYSTYRENE	Materials/Construction/ Insulation/Transformation	Polystyrene foam slab {CH}| production, 100% recycled | Cut-off, S	ton	PS
RESIN POLYURETHANE	Materials/Plastic/ Thermosets/Transformation	Polyurethane, flexible foam {RER}| production | Cut-off, S	ton	PUR
		ROOF SLAB		
CONCRETE	Materials/Construction/ Binders/Transformation	Cement, pozzolana and fly ash 36-55% {Europe without Switzerland}| cement production, pozzolana and fly ash 36-55% | Cut-off, S	ton	Cement
STEEL	Materials/Metals/Ferro/ Transformation	Reinforcing steel {Europe without Austria}| reinforcing steel production | Cut-off, S	ton	Steel
GROUND ORGANIC	Materials/Agricultural/ Animal feed/Transformation	Grass, organic {CH}| grass production, permanent grassland, organic, extensive | Cut-off, S	ton	Compost
POLYETHYLENE	Materials/Plastic/ Thermoplats/ Transformation	Polyethylene terephthalate, granulate, amorphous, recycled {Europe without Switzerland}| polyethylene terephthalate production, granulate, amorphous, recycled | Cut-off, S	ton	PE
GRAVEL	Materials/Minerals/ Transformation	Gravel, crushed {RoW}| production | Cut-off, S	ton	Others

Fig. 24.2 Correspondence to categories, processes, units and type of waste materials for underground parking

Table 24.8 Consumption during the car park construction phase

Use	Amount	Unit
Powering excavation machines	48.976,11	m^3
Electric energy	120.739,72	MJ

Table 24.9 Correspondence categories, processes and consumption units construction phase

SimaPro category	SimaPro process	Unit
Powering excavation machines		
Transport/Building equipment/Market	Excavation, hydraulic digger {RER}| market for | Cut-off, S	m^3
Electricity machinery, equipment and lighting construction phase		
Electricity/Electricity country mix/Low voltage/Market	Electricity, low voltage {IT}| market for | Cut-off, S	MJ

During the construction phase, the transport of materials from the production sites to the construction site was also considered. An average distance of approximately 50 km from the construction site was assumed. For each material, a value expressed in tkm (tons per km) was calculated, i.e. mass transported per km of distance, as explain Table 24.10.

Use Phase

In the use phase, the energy needs linked to the lighting of the underground car park were considered. For the modeling in SimaPro, watertight LED lighting fixtures with a power of 26 W were chosen. The installation of 50 lighting fixtures for each floor integrated with presence sensors with a daily consumption of 60 kWh was hypothesized. All consumption was reported in the model with reference to the assumed useful life of 50 years for a total of 1103.76 MWh (Table 24.11).

End of Life

According to what was reported in the pre-use phase, for the end-of-life phase of the car park, a complete demolition of the same was hypothesized, divided into three phases: selective demolition, controlled demolition and treatment of the rubble.

Table 24.10 Construction phase: correspondence to construction phase categories, processes and consumption units

SimaPro category	SimaPro process	Unit
Transporting materials to the construction site		tkm
Transport/Road/Market	Transport, freight, lorry 16–32 metric ton, euro4 {RER}| market for transport, freight, lorry 16–32 metric ton, EURO4 | Cut-off, S	

Table 24.11 Use phase: correspondence to construction phase categories, processes and consumption units

SimaPro category	SimaPro process	Unit
Electricity/Electricity country mix/Low voltage/Market	Electricity, low voltage {IT}| market for | Cut-off, S	MJ

- Selective demolition: before dismantling the car park, some parts of the building will be removed with special disassembly equipment. The consumption linked to this phase is linked to the use of specific equipment and the transport of materials to the treatment sites, as is reported in Table 24.12.

- Controlled demolition: in this phase the demolition of the car park using an excavator with a hydraulic demolition hammer was considered. A machine power of 210 kWh was assumed (Table 24.13).

- Treatment of rubble: following the demolition of the car park, the rubble will be loaded by mechanical shovel onto an articulated lorry and transported to the treatment sites or landfill. The energy consumption of this phase is mainly due to the handling and loading of the rubble using a mechanical shovel and its transport, as shown in Table 24.14.

Table 24.12 End of life: correspondence to categories, processes and consumption units of selective demolition phase

SimaPro category	SimaPro process	Unit
Transport/Building equipment/Market	Excavation, skid-steer loader {GLO}| market for | Cut-off, S	m^3
Electricity/Electricity country mix/Low voltage/Market	Electricity, low voltage {IT}| market for | Cut-off, S	MJ
Transport/Road/Market	Transport, freight, lorry 16–32 metric ton, euro4 {RER}| market for transport, freight, lorry 16–32 metric ton, EURO4 | Cut-off, S	tkm

Table 24.13 End of life: correspondence to categories, processes and consumption units of controlled demolition phase

SimaPro category	SimaPro process	Unit
Transport/Building equipment/Market	Excavation, hydraulic digger {RER}| market for | Cut-off, S	m^3
Electricity/Electricity country mix/Low voltage/Market	Electricity, low voltage {IT}| market for | Cut-off, S	MJ

Table 24.14 End of life: correspondence to categories, processes and consumption units of treatment of rubble phase

SimaPro category	SimaPro process	Unit
Transport/Building equipment/Market	Excavation, skid-steer loader {GLO}| market for | Cut-off, S	m^3
Electricity/Electricity country mix/Low voltage/ Market	Electricity, low voltage {IT}| market for | Cut-off, S	MJ
Transport/Road/Market	Transport, freight, lorry 16–32 metric ton, euro4 {RER}| market for transport, freight, lorry 16–32 metric ton, EURO4 | Cut-off, S	tkm

Listed in Table 24.15, for each material used in the car park, is the corresponding recycling and reuse process implemented in the SimaPro model, with reference to the category, unit of measurement and type of waste.

24.2.3 Life Cycle Impact Assessment and Interpretation (LCIA)

Following the inventory analysis, the LCIA of the underground car park was performed. To construct the life cycle of the artefact, the following phases were considered:

- Assembly: contains a list of all the raw materials or semi-finished products (with relative quantities) necessary for the realization of the project and the processes that define the energy consumption for the assembly itself.
- Life cycle: contains a link to the assembly defined above and to the processes relating to the pre-use, use and end-of-life phases of the product.

There are various LCIA methods for conducting an LCA analysis, such as CML2001, Eco-indicator 99, and ReCiPe [31]. In any case, this phase is complex due to the large number of impact categories to be examined. For the environmental impact assessment, the ReCiPe 2016 Endpoint (H) method in version 1.05, created by RIVM, Radboud University, Norwegian University of Science and Technology and PRé Sustainability [32] and the IPCC 2013 method, developed by Intergovernmental Panel on Climate Change, which contains IPCC climate change drivers with a 100-year time horizon.

ReCiPe 2016 Endpoint (H) v1.05 Method

In this paper the ReCiPe 2016 Endpoint method [33] was used, chosen for some fundamental aspects: as underlined [34] "*it contains the broadest set of midpoint impact categories; enables impact categories to implement characterisation factors*

Table 24.15 Correspondence between material and recycling and reuse process

Materials	SimaPro category	SimaPro process	Landfill	Recycling
Concrete	Waste treatment/ Recycling/ Transformation	Concrete (waste treatment) {GLO}| recycling of concrete | Cut-off, S	Re	0%
	Waste treatment/ Landfill/Inert material landfill/Transformation	Waste concrete {Europe without Switzerland}| treatment of waste concrete, inert material landfill | Cut-off, S		
Steel	Waste treatment/ Recycling/ Transformation	Steel and iron (waste treatment) {GLO}| recycling of steel and iron | Cut-off, S	80%	20%
	Waste treatment/ Landfill/Inert material landfill/Transformation	Scrap steel {Europe without Switzerland}| treatment of scrap steel, inert material landfill | Cut-off, S		
Polyethylene	Waste treatment/ Recycling/ Transformation	PE (waste treatment) {GLO}| recycling of PE | Cut-off, S	50%	50%
	Waste treatment/ Landfill/Sanitarlandfill/ Transformation	Waste polyethylene {CH}| treatment of, sanitary landfill | Cut-off, S		
Brick	Waste treatment/ Construction waste/ Market	Waste brick {Europe without Switzerland}| market for waste brick | Cut-off, S	80%	0.20
	Waste treatment/ Landfill/Sanitary landfill/Transformation	Municipal solid waste {CH}| treatment of, sanitary landfill | Cut-off, S		
Mineral wool	Waste treatment/ Construction waste/ Market	Waste mineral wool {Europe without Switzerland}| market for waste mineral wool | Cut-off, S	90%	0.10
	Waste treatment/ Landfill/Inert material landfill/Transformation	Waste mineral wool, for final disposal {Europe without Switzerland}| treatment of waste mineral wool, inert material landfill | Cut-off, S		

(continued)

Table 24.15 (continued)

Materials	SimaPro category	SimaPro process	Landfill	Recycling
Polystyrene	Waste treatment/ Recycling/ Transformation	PS (waste treatment) {GLO}| recycling of PS | Cut-off, S	50%	50%
	Waste treatment/ Landfill/Sanitary landfill/Transformation	Waste polystyrene {CH}| treatment of, sanitary landfill | Cut-off, S		
Polyurethane resin	Waste treatment/ Plastics/Market	Waste polyurethane {IT}| market for waste polyurethane | Cut-off, S	50%	50%
	Waste treatment/ Landfill/Inert material landfill/Transformation	Waste polyurethane {CH}| treatment of, inert material landfill | Cut-off, S		

at an international scale; and does not include potential impacts from future extractions in the impact assessment and assumes that they have been included in the inventory analysis". The chosen method includes two levels for evaluation as shown in Fig. 24.3:

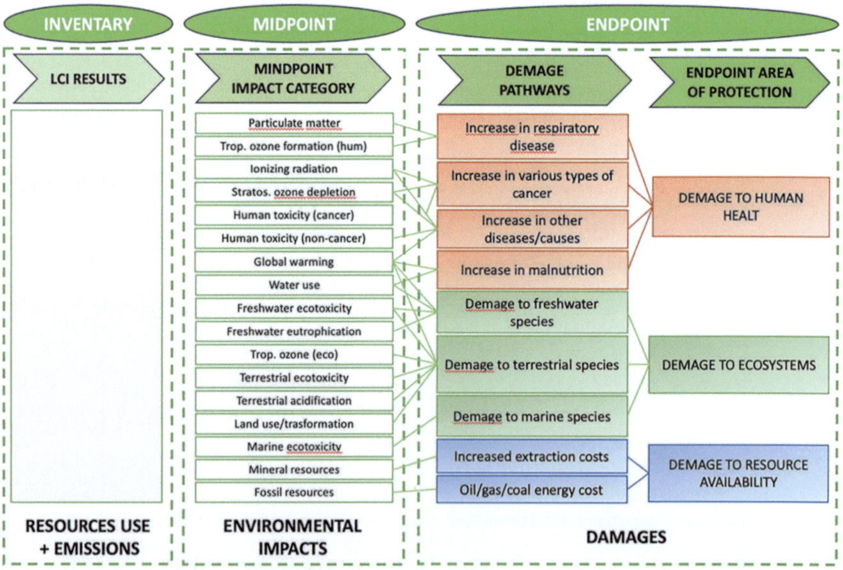

Fig. 24.3 The Endpoint method phases: it defines impact categories disaggregated at midpoint level and aggregate impact categories at the endpoint level

- a problem-oriented approach, which defines impact categories disaggregated at midpoint level;
- a harm-oriented approach, which defines aggregate impact categories at the endpoint level.

This method is considered the most recent and advanced, combining the approach based on the midpoint of the CML-IA and the endpoint of the Eco-indicator 99, methods recognized worldwide. At an intermediate level, 22 impact categories are considered, such as global warming, particulate formation, ecotoxicity, etc. At the endpoint level, these categories are multiplied by damage factors and aggregated into three endpoint categories:

- Human Health;
- Ecosystems;
- Resources.

The ReCiPe method outlines five steps in calculating life cycle impact: characterization, damage assessment, normalization, weighting, and single scoring. According to ISO 14040 standards, the only mandatory phase is characterization, while the others are optional. For greater clarity, the initial characterization phase and the final single scoring phase are reported in the results, leaving out the intermediate phases of damage assessment, normalization and weighting.

The results relating to the characterization phase are listed in Fig. 24.4, divided by impact category, expressed in the respective units of measurement and in percentage terms.

By analyzing the data relating to impacts, it is possible to observe in the Fig. 24.5:

- in the pre-use phase, the most relevant impact categories are "Human carcinogenic toxicity" (99.3%) and "Mineral resource scarcity" (99.3%), both mainly linked to the extraction of raw materials necessary for construction of the parking lot.
- in the use phase, the most relevant impact categories are "Stratospheric ozone depletion" (12.54%) and "Water consumption, human health" (10.81%), linked to the production of electricity.
- in the end-of-life phase, the most important category is "Terrestrial acidification" (18.48%) mainly caused by the transport of materials to waste treatment sites.

The results are reported in Fig. 24.6, in the form of tables and graphs, in terms of scores.

It can be seen that in all damage categories, the most impactful phase is the pre-use phase with a total damage equal to 452.14 kPt, as show Fig. 24.7. Within the pre-use phase, the most impactful damage category is "Human health" with a score of 434.6 kPt. The pre-use phase contributes 94.7% to the overall damage, the use phase 2.9% and the end-of-life phase 2.34%. Below is the impact assessment relating to the pre-use phase only.

Analyzing the pre-use phase, the greatest impact is attributed to the foundation slab with approximately 172 kPt (38.1%), followed by the bulkhead with 99.1 kPt (21.9%) and the intermediate floors and ramp with 85.1 kPt (18.8%); while the minor

IMPACT CATEGORY	UNIT	TOTAL	PRE-USE PHASE	USE PHASE	END OF LIFE
Global warming, Human health	DALY	14,15889643	13,40142607	0,434535391	0,32293497
Stratospheric ozone depletion	DALY	0,001232241	0,000969695	0,000154557	0,000107989
Ionizing radiation	DALY	0,000487777	0,000435238	2,94325E-05	2,31067E-05
Ozone formation, Human health	DALY	0,026509483	0,024504658	0,000728943	0,001275882
Fine particulate matter formation	DALY	7,98954057	7,416844517	0,306288477	0,266407576
Human carcinogenic toxicity	DALY	4,860913438	4,826770439	0,020885724	0,013257276
Human non-carcinogenic toxicity	DALY	0,327703692	0,29407767	0,017634527	0,015991494
Water consumption, Human health	DALY	0,105484195	0,089830695	0,011400973	0,004252527
Global warming, Terrestrial ecosystems	species.yr	0,042724471	0,040438698	0,001311295	0,000974478
Global warming, Freshwater ecosystems	species.yr	1,16713E-06	1,10469E-06	3,58164E-08	2,66177E-08
Ozone formation, Terrestrial ecosystems	species.yr	0,003876469	0,003586942	0,000105262	0,000184265
Terrestrial acidification	species.yr	0,006007293	0,005483182	0,000301848	0,000222262
Freshwater eutrophication	species.yr	0,000218879	0,000207383	8,63806E-06	2,85744E-06
Marine eutrophication	species.yr	2,59931E-07	2,4876E-07	2,13594E-09	9,03505E-09
Terrestrial ecotoxicity	species.yr	0,000215534	0,000169845	5,84993E-06	3,98389E-05
Freshwater ecotoxicity	species.yr	1,7418E-05	1,68632E-05	1,13336E-07	4,41416E-07
Marine ecotoxicity	species.yr	5,03369E-06	4,66804E-06	7,21744E-08	2,93477E-07
Land use	species.yr	0,001608999	0,001379025	0,000104043	0,000125931
Water consumption, Terrestrial ecosystem	species.yr	0,000658044	0,000609509	3,06468E-05	1,78887E-05
Water consumption, Aquatic ecosystems	species.yr	6,03308E-08	5,72659E-08	1,94755E-09	1,11741E-09
Mineral resource scarcity	USD2013	42789,18177	42488,54967	190,9121907	109,7199084
Fossil resource scarcity	USD2013	537211,0376	448939,4373	42330,45178	45941,14854

Fig. 24.4 Life cycle analysis –ReCiPe 2016 method Endpoint (H) V1.05- characterization: numerical results

impact is associated with the walls with 13.5 kPt (3%) and the pillars with 36.1 kPt (8%).

The impact categories and substances relating to the foundation slab that cause the greatest environmental damage are:

- "Global warming, Human health" with 85.9 kPt (49.8%) mainly due to 95.7% carbon dioxide, a fossil present for 76.3% in concrete processes and for 22.4% in 'steel; and 3.87% of methane, a fossil present for 64.8% in steel processes and 34.3% in concrete;
- "Fine particulate matter formation" with 46.6 kPt (27%) mainly due to the 38.1% sulfur dioxide present 56.9% in the concrete and 41.3% in the steel; by 36.8% of particulates, < 2.5 um present for 74.4% in steel processes; and from 29.3% nitrogen oxides to 66% in concrete processes;
- "Human carcinogenic toxicity" with 30.8 kPt (17.8%) due to the 99.2% of chromium VI present in 97.4% of the steel processes;
- "Global warming, terrestrial ecosystems" with 4.2 kPt (2.44%) due to 95.7% carbon dioxide, fossil present for 76.3% from concrete processes and 22.3% from steel.

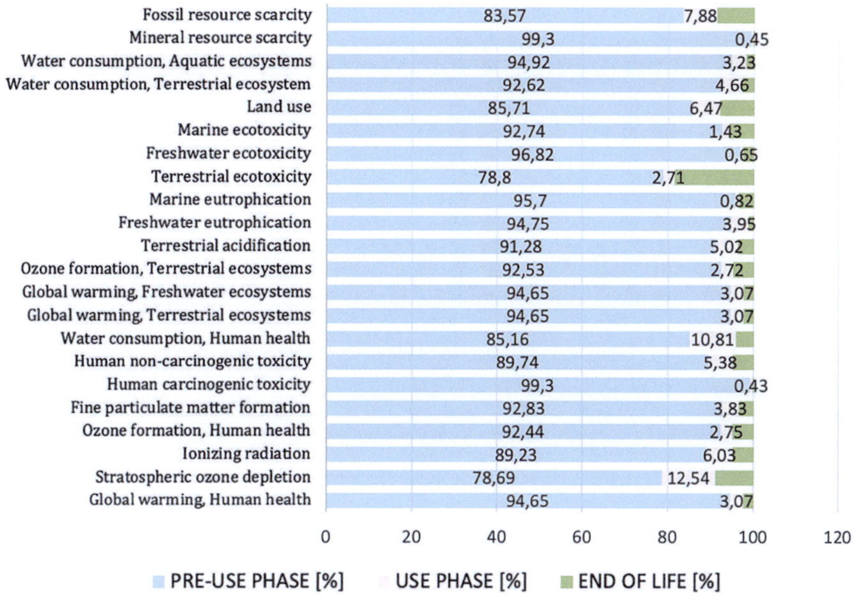

Fig. 24.5 Life cycle analysis –ReCiPe 2016 method Endpoint (H) V1.05- characterization: graphical results

The impact categories and substances relating to the bulkhead that cause the greatest environmental damage are:

- "Global warming, Human health" with 51.1 kPt (51.5%) mainly due to 96% carbon dioxide, a fossil present for 79% in concrete processes and 20% insteel; and 3.62% methane, a fossil present for 61.1% in steel processes and 38% in concrete;
- "Fine particulate matter formation" with 26.5 kPt (26.7%) mainly due to the 38.1% sulfur dioxide present 61% in the concrete and 38% in the steel; by 35.4% of particulates, < 2.5 um present for 71% in steel processes and 27% in concrete processes; and from 24.7% nitrogen oxides to 69.2% in concrete processes;
- "Human carcinogenic toxicity" with 16.2 kPt (16.3%) due to the 99.2% of chromium VI present in 96.9% of the steel processes.

Figures 24.8 and 24.9 show the single scores of Pre-use phase analysis and of End of life phase analysis.

IPCC 2013 GWP 100° V1.03 Method

The IPCC 2013 GWP method was also used for a more detailed analysis of the underground car park's environmental impacts. This method exclusively evaluates the environmental impact linked to the greenhouse effect and, therefore, to equivalent CO_2 emissions. Below in Tables 24.16 and 24.17 is the analysis of the pre-use phase and Fig. 24.10 the contribution of the singles materials.

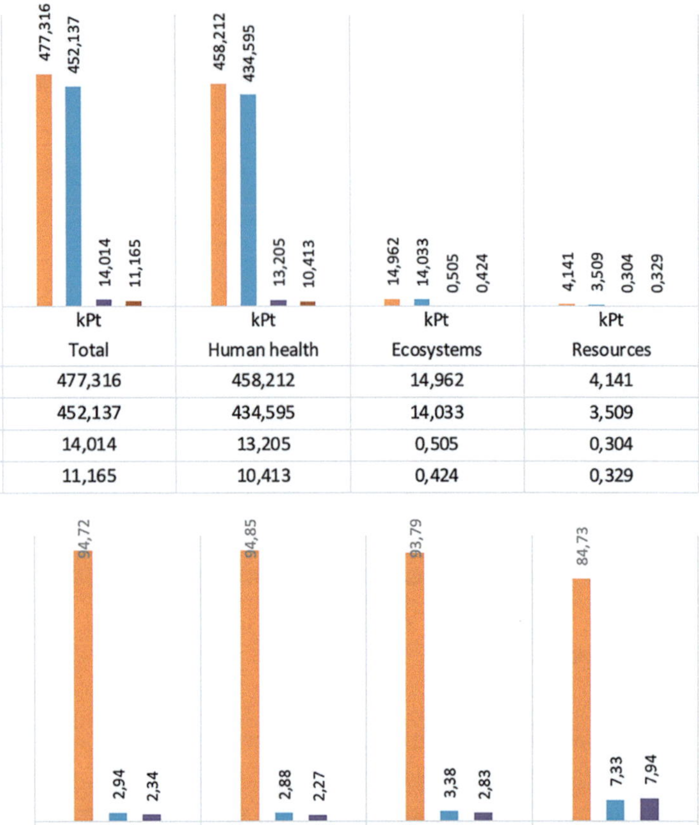

	kPt Total	kPt Human health	kPt Ecosystems	kPt Resources
Total	477,316	458,212	14,962	4,141
PRE-USE PHASE	452,137	434,595	14,033	3,509
USE PHASE	14,014	13,205	0,505	0,304
END OF LIFE	11,165	10,413	0,424	0,329

	Total	Human health	Ecosystems	Resources
PRE-USE PHASE [%]	94,72	94,85	93,79	84,73
USE PHASE [%]	2,94	2,88	3,38	7,33
END OF LIFE [%]	2,34	2,27	2,83	7,94

Fig. 24.6 Life cycle analysis—ReCiPe method 2016 Endpoint (H) V1.05—single score

Analysing the pre-use phase, the Fig. 24.11 illustrate as the greatest impact comes from the foundation slab for 38.5% (5.52E6 kg CO2-eq), followed by the bulkhead for 22.9% (3.28E6 kg CO2-eq) and from the processes of the floors and ramp for 19.3% (2.77E6 kg CO2-eq).

24.3 Results and Discussion

The concrete production process appears to have the most impact from the results of the impacts processed, both with the ReCiPe 2016 Endpoint (H) method and with the IPCC 2013 GWP 100° V1.03 method.

IMPACT CATEGORY	UNIT	TOTAL	FOUNDATION STAGE	PILLARS	BULKHEAD	WALLS	INTERMEDIATE SLABS AND RAMP	ROOF SLAB
Total	**kPt**	**452,10**	**172,40**	**36,07**	**99,14**	**13,51**	**85,15**	**44,15**
Global warming, Human health	kPt	223,50	85,93	13,86	51,06	7,10	43,12	21,84
Global warming, Terrestrial	kPt	10,93	4,20	0,68	2,50	0,35	2,11	1,07
Global warming, Freshwater	kPt	0,00	0,00	0,00	0,00	0,00	0,00	0,00
Stratospheric ozone depletion	kPt	0,02	0,01	0,00	0,00	0,00	0,00	0,00
Ionizing radiation	kPt	0,01	0,00	0,00	0,00	0,00	0,00	0,00
Ozone formation, Human health	kPt	0,41	0,15	0,03	0,09	0,01	0,08	0,04
Fine particulate matter formation	kPt	123,70	46,56	10,36	26,52	3,66	23,23	12,49
Ozone formation, Terrestrial	kPt	0,97	0,37	0,07	0,21	0,03	0,18	0,10
Terrestrial acidification	kPt	1,48	0,56	0,11	0,33	0,05	0,28	0,15
Freshwater eutrophication	kPt	0,06	0,02	0,00	0,01	0,00	0,01	0,01
Marine eutrophication	kPt	0,00	0,00	0,00	0,00	0,00	0,00	0,00
Terrestrial ecotoxicity	kPt	0,05	0,02	0,00	0,01	0,00	0,01	0,01
Freshwater ecotoxicity	kPt	0,00	0,00	0,00	0,00	0,00	0,00	0,00
Marine ecotoxicity	kPt	0,00	0,00	0,00	0,00	0,00	0,00	0,00
Human carcinogenic toxicity	kPt	80,51	30,75	10,10	16,20	2,01	14,10	7,30
Human non-carcinogenic toxicity	kPt	4,91	1,85	0,44	1,04	0,14	0,89	0,54
Land use	kPt	0,37	0,14	0,03	0,08	0,01	0,07	0,05
Mineral resource scarcity	kPt	0,30	0,12	0,03	0,06	0,01	0,05	0,03
Fossil resource scarcity	kPt	3,21	1,17	0,22	0,68	0,10	0,65	0,36
Water consumption, Human health	kPt	1,50	0,53	0,12	0,30	0,04	0,32	0,17
Water consumption, Terrestrial	kPt	0,16	0,06	0,01	0,03	0,00	0,03	0,02
Water consumption, Aquatic	kPt	0,00	0,00	0,00	0,00	0,00	0,00	0,00

Fig. 24.7 Analysis of the pre-use phase—ReCiPe 2016 Endpoint (H) V1.05 method—single score

In fact, analyzing the impacts due to the pre-use phase with the ReCiPe 2016 Endpoint (H) method, it is clear that concrete production is the predominant one with 233 kPt (51.6%). This is due on the one hand to the high environmental load associated with the production of the material, and on the other to the quantity of concrete used in the car park. While in the use phase the impact is due only to the use of electricity with 14 kPt (100%) to power the lighting of the underground car park. In the end-of-life phase, however, the greatest impact is attributed to the transport (55.4%) of the materials to waste treatment sites, followed by the electricity (24.9%) used for the recycling and reuse of the materials after the decommissioning of the car park. Even with the examination of the results of the analysis performed with the IPCC GWP method, the greatest impact is attributed to the pre-use phase with 1.43E7 kg CO2-eq (94.7%). In particular, from the contribution analysis, the concrete process is also the most impactful for this method, with 73%, corresponding to 1.04E7 kg CO2-eq.

In order to reduce impacts and improve the green footprint of the concrete industry, the scientific community is making a contribution through the study of design choices

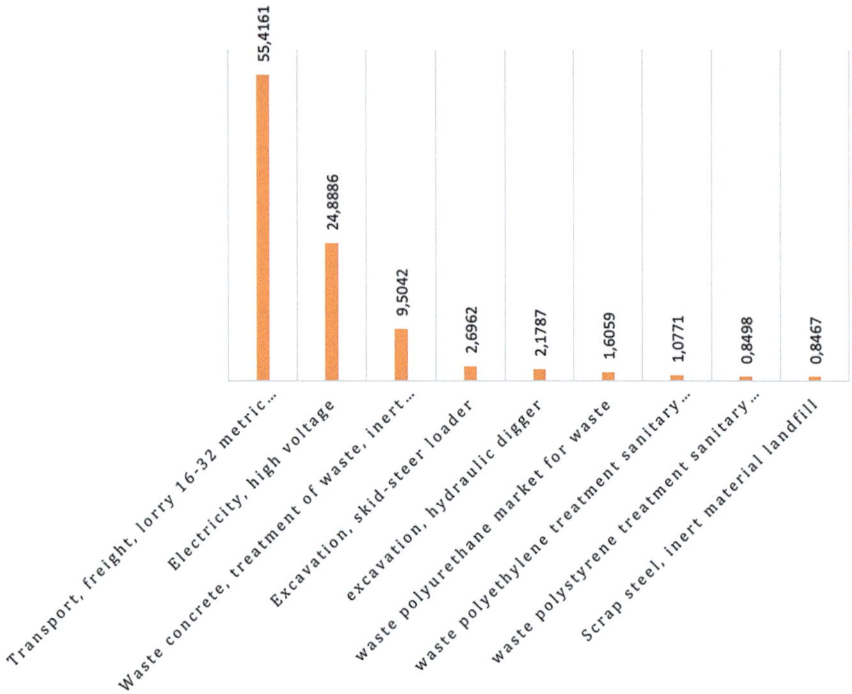

Fig. 24.8 End of life phase analysis—ReCiPe 2016 method Endpoint (H) V1.05—single score

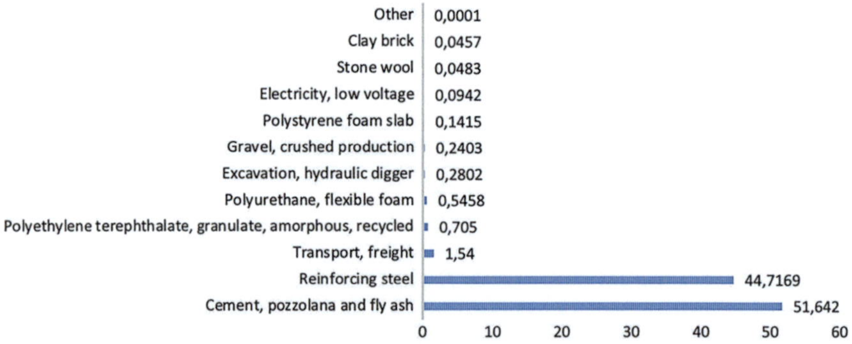

Fig. 24.9 Pre-use phase analysis—ReCiPe 2016 method Endpoint (H) V1.05—single score

Table 24.16 Life cycle analysis—IPCC 2013 GWP 100° V1.03 method—characterization

	Unit	Total	Pre-use phase	Use phase	End of life
IPCC GWP 100a	Kg CO2 eq	1.51E7	1.43E7	4.6E5	3.43E5

Table 24.17 Life cycle analysis—method IPCC 2013 GWP 100° V1.03—characterization

Impact category	Unit	Total	Pre-use phase	Use phase	End of life
IPCC GWP 100a	%	100	94.7	3.04	2.26

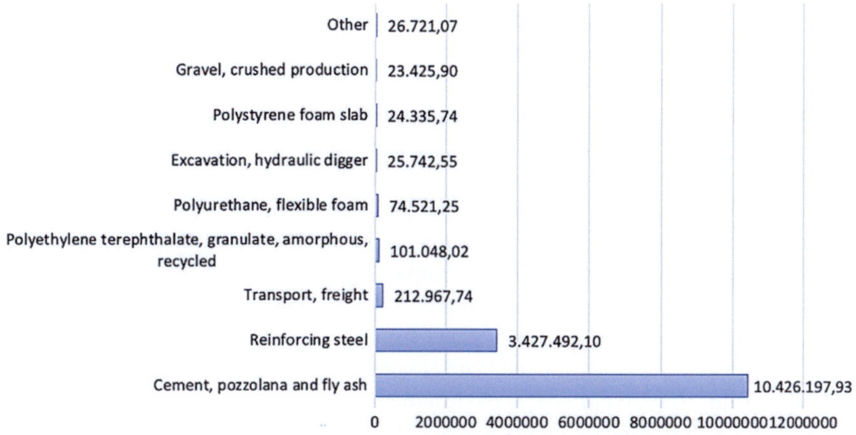

Fig. 24.10 Contribution to the pre-use phase process—IPCC 2013 GWP 100° V1.03 method—characterization

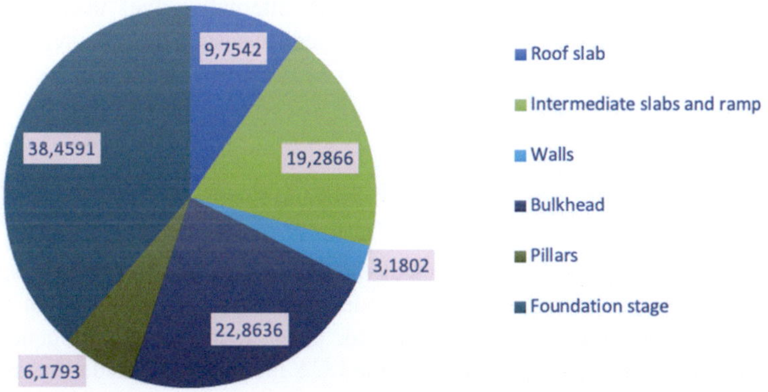

Fig. 24.11 Pre-use phase analysis—IPCC 2013 GWP 100° V1.03 method—characterization

and alternative solutions to be put in synergy during the processing phase of the main material of traditional constructions. There are numerous strategies that can be used to decarbonise cement production. To date, those on which researchers agree and which are also starting to make headway in the industrial world are the replacement of clinker, the use of alternative fuels, the improvement of production kilns, the capture and storage of carbon [35]. Mikulčić et al. [36] in their studies, through the LCA methodology, demonstrate that in 2050 cement can be produced with significant reductions in CO2-eq emissions compared to 2020 through the replacement of clinker (42%), the use of alternative fuels (25%) or the improvement of kiln efficiency (12%); if combined, these measures could reduce CO2-eq emissions. of cement production by approximately 58%.

To reduce the negative impacts of concrete production while maintaining the engineering properties, alternating raw materials such as recycled aggregates and supplementary cementitious materials are encouraging [37, 38]. For example, some studies have evaluated the use of rubberised concretes: [39] and [40] show that there is a promising future for the use of waste tire rubber as a partial replacement for aggregate in cement concrete for the purpose of environmental sustainability and in the production of pervious concrete [41]. Others with the addition of textile materials [42], or recycled [43, 44]. For example, [45] design a concrete panel with integrated mineral wool produced with recycled material deriving from glass, silica, which allows the reduction of environmental impact through the reduction of embodied energy and carbon as well as the production of waste. Also [46] developed a methodological approach to produce a new panel for use in construction and use the LCA methodology to make assessments related to the reduction of environmental impacts.

[47] presents the comparison, in environmental and economic terms, of the cradle-to-gate life cycle of concrete with coarse natural and recycled concrete aggregates. By analyzing the environmental impact of 216 concrete mixes from 24 references, the results show how using recycled coarse aggregates from concrete can significantly reduce environmental impacts and costs.

Furthermore, promoting a greater share of alternative fuel use, leading to cleaner cement production, could significantly reduce sustainable concrete's carbon emissions and energy consumption.

24.4 Conclusions

This study aims to provide awareness of the environmental impact deriving from a specific project, the construction of a underground car park with garden roof, to highlight how the choice of materials and the construction process influence environmental emissions. The application of the LCA methodology made it possible to highlight the greatest impact in the pre-use phase, which includes the processes for the extraction of raw materials, for the construction of the product and for the energy necessary to power means and equipment of worksite. Although significant impacts have emerged in the concrete and steel production processes, in the end-of-life phase the impacts appear to decrease drastically through recycling treatments and disposal of waste in landfill for subsequent reuse.

The research leads to reflections regarding design choices, the use of innovative materials (such as concrete with the addition of aggregates) and the reduction of consumption in concrete production factories, to reduce consumption and improve environmental sustainability. Life cycle analysis has proven to be a reliable tool for assessing the real sustainability of the product, allowing environmental critical issues to be highlighted in the choice of materials, avoiding moving from one phase of the cycle to another. Finally, it is underlined how the availability of an accessible and updated database of building materials relating to the Italian context would allow to increase the reliability and significance of the results obtained.

Acknowledgements Part of the study is funded by the European Union - NextGenerationEU under the Italian Ministry of University and Research (MUR) National Innovation Ecosystem grant ECS00000041 - VITALITY - CUP E13C22001060006.

Author Contributions Conceptualization, M.R. and F.C.; methodology, M.R. and F.C.; software, F.C. and C.M..; validation, F.C.; data curation, F.C., M.R. and C.M.; writing—original draft preparation, F.C. (24.2, 24.4), M.R. (24.2, 24.3) and C.M. (24.1, 24.3, 24.4); writing—review and editing, C.M.; supervision, F.C. Authors in alphabetic order. All authors have read and agreed to the published version of the manuscript.

Informed Consent Statement "Not applicable".

Conflicts of Interest "The authors declare no conflict of interest".

References

1. A. Darko, A.P.C. Chan, Critical analysis of green building research trend in construction journals. Habitat Int. **57**, 53–63 (2016)
2. T. Abergel, B. Dean, J. Dulac, I. Hamilton, T. Wheeler, *Global Status Report—Towards a Zero-Emission. Efficient and Resilient Buildings and Construction Sector* (2018)
3. S. Zhao, Q. Song, H. Duan, Z. Wen, C. Wang, Uncovering the lifecycle GHG emissions and its reduction opportunities from the urban buildings: a case study of Macau. Resour. Conserv. Recycl. **147**, 214–226 (2019)
4. M.H.A. Nasir, A. Genovese, A.A. Acquaye, S.C.L. Koh, F. Yamoah, Comparing linear and circular supply chains: a case study from the construction industry. Int. J. Prod. Econ. **183**, 443–457 (2017)
5. Eurostat, *Key figures on Europe* (Key Figures on Europe, 2016)
6. M. Lavagna, Life cycle assessment in Edilizia, in *Progettare e Costruire in Una Prospettiva di Sostenibilità Ambientale* (Hoepli, Milano, 2008)
7. N.B. Jacobsen, Industrial symbiosis in Kalundborg, Denmark: a quantitative assessment of economic and environmental aspects. J. Ind. Ecol. **10**, 239–255 (2008)
8. H.P. Tserng, C.-M. Chou, Y.-T. Chang, The key strategies to implement circular economy in building projects—a case study of Taiwan. Sustainability **13** (2021)
9. J.T. Scott, The sustainable business: a practitioner's guide to aaschieving long-term profitability and competitiveness, in *The Sustainable Business a Practitioner's Guide to Achieving Long-Term Profitability and Competitiveness*, 2nd edn. (2013)
10. E. Leising, J. Quist, N. Bocken, Circular economy in the building sector: three cases and a collaboration tool. J. Clean. Prod. **176**, 976–989 (2018)
11. C. Marlet, GtoG Life + Project: first step towards gypsum circular economy. ZKG Int. **67**, 48–55 (2014)
12. M. Sansom, N. Avery, Briefing: reuse and recycling rates of UK steel demolition arisings. Proc. Inst. Civ. Eng.: Eng. Sustain. **167**, 89–94 (2014)
13. P. Núñez-Cacho, J. Górecki, V. Molina, F.A. Corpas-Iglesias, New measures of circular economy thinking in construction companies. J. EU Res. Bus. **2018**, 1–16 (2018)

14. Communication from the Commission to the European Parliament, the Council, the European Economic and Social Committee and the Committee of the Regions, A new Circular Economy Action Plan For a Cleaner and More Competitive Europe, COM(2020) 98 (2020)
15. Communication from the Commission to the European Parliament, the Council, the European Economic and Social Committee and the Committee of the Regions, The European Economic and Social Committee and the Committee of the Regions. A Farm to Fork Strategy for a fair, healthy and environmentally-friendly food system. COM(2020) 381 (2020)
16. United Nations Environment Programme, Towards A Green Economy, Pathways to Sustainable Development and Poverty Eradication - A Synthesis for Policy Makers (2011)
17. B.I. Oluleye, D.W.M. Chan, A.B. Saka, T.O. Olawumi, Circular economy research on building construction and demolition waste: a review of current trends and future research directions. J. Clean. Prod. **357** (2022)
18. A.E.d.M. Ambient, A. Reichel, E.E. Agency, M. De Schoenmakere, J. Gillabel, *Circular Economy in Europe: Developing the Knowledge Base* (Publications Office of the European Union, 2016)
19. C. Marchionni, A. Corneli, Circular economy in the built environment management supported by digital twin. A review, in *Proceedings of 2024 IEEE International Workshop on Metrology for Living Environment, MetroLivEnv 2024 IEEE International Workshop on Metrology for Living Environment, 2024*, pp. 546–551
20. M. Geissdoerfer, P. Savaget, N.M.P. Bocken, E.J. Hultink, The circular economy—a new sustainability paradigm? J. Clean. Prod. **143**, 757–768 (2017)
21. J. Nußholz, S. Çetin, L. Eberhardt, C. De Wolf, N. Bocken, From circular strategies to actions: 65 European circular building cases and their decarbonisation potential. Resour., Conserv. Recycl. Adv. **17** (2023)
22. K. Cai, H. Wang, J. Wang, J. Bai, J. Zuo, K. Chan, et al., Mitigating lifecycle GHG emissions of building sector through prefabricated light-steel buildings in comparison with traditional cast-in-place buildings. Resour., Conserv. Recycl. **194** (2023)
23. C.E. Andersen, K. Kanafani, R.K. Zimmermann, F.N. Rasmussen, H. Birgisdóttir, Comparison of GHG emissions from circular and conventional building components. Build. Cities (1) (2020)
24. L. Aye, T. Ngo, R.H. Crawford, R. Gammampila, P. Mendis, Life cycle greenhouse gas emissions and energy analysis of prefabricated reusable building modules. Energy Build. **47**, 159–168 (2012)
25. J. Jayawardana, M. Sandanayake, J.A.S.C. Jayasinghe, A.K. Kulatunga, G. Zhang, A comparative life cycle assessment of prefabricated and traditional construction—a case of a developing country. J. Build. Eng. **72** (2023)
26. C.N.N. Tran, I.M.C.S. Illankoon, V.W.Y. Tam, Decoding concrete's environmental impact: a path toward sustainable construction. Buildings **15**(3), 442 (2025)
27. H.S. Müller, M. Haist, M. Vogel, Assessment of the sustainability potential of concrete and concrete structures considering their environmental impact, performance and lifetime. Constr. Build. Mater. **67**, 321–337 (2014)
28. A. Campioli, A. Dalla Valle, S. Ganassali, S. Giorgi, Designing the life cycle of materials: new trends in environmental perspective. TECHNE **16**, 86–95 (2018)
29. S. Ganassali, M. Lavagna, A. Campioli, Benchmark LCA e uso di EPD nei Green building rating system, in *XI Convegno della Rete Italiana LCA–Resource efficiency e sustainable development goals: Il ruolo del life cycle thinking* (2017), pp. 312–320
30. UNI EN ISO 14044:2021, Gestione ambientale-Valutazione del ciclo di vita-Requisiti e Linee guida
31. UNI EN ISO 14040:2006+A1:2020, Gestione Ambientale-Valutazione del Ciclo di Vita-Principi e Quadro di Riferimento
32. A.P. Acero, C. Rodríguez, A. Ciroth, Impact assessment methods in life cycle assessment and their impact categories, in *LCIA Methods—Impact Assessment Methods in Life Cycle Assessment and Their Impact Categories, 2014*
33. M.A.J. Huijbregts, Z.J.N. Steinmann, P.M.F. Elshout, G. Stam, F. Verones, M. Vieira et al., ReCiPe2016: a harmonised life cycle impact assessment method at midpoint and endpoint level. Int. J. Life Cycle Assess. **22**, 138–147 (2016)

34. M. Goedkoop, R. Heijungs, M. Huijbregts, A. De Schryver, J. Struijs, R. Van Zelm, ReCiPe 2008—a life cycle impact assessment method which comprises harmonised category indicators at the midpoint and the endpoint level, 2009
35. H. Feng, J. Zhao, A. Hollberg, G. Habert, Where to focus? Developing a LCA impact category selection tool for manufacturers of building materials. J. Clean. Prod. **405** (2023)
36. H. Mikulčić, J.J. Klemeš, M. Vujanović, K. Urbaniec, N. Duić, Reducing greenhouse gasses emissions by fostering the deployment of alternative raw materials and energy sources in the cleaner cement manufacturing process. J. Clean. Prod. **136**, 119–132 (2016)
37. M. Georgiades, I.H. Shah, B. Steubing, C. Cheeseman, R.J. Myers, Prospective life cycle assessment of European cement production. Resour., Conserv. Recycl. **194** (2023)
38. W. Xing, V.W.Y. Tam, K.N. Le, J.L. Hao, J. Wang, Life cycle assessment of sustainable concrete with recycled aggregate and supplementary cementitious materials. Resour., Conserv. Recycl. **193** (2023)
39. B. Bennett, P. Visintin, T. Xie, Global warming potential of recycled aggregate concrete with supplementary cementitious materials. J. Build. Eng. **52** (2022)
40. A.B. Fraj, R. Idir, Concrete based on recycled aggregates—recycling and environmental analysis: a case study of Paris' region. Constr. Build. Mater. **157**, 952–964 (2017)
41. B.S. Thomas, R.C. Gupta, A comprehensive review on the applications of waste tire rubber in cement concrete. Renew. Sustain. Energy Rev. **54**, 1323–1333 (2016)
42. S. Surehali, A. Singh, K.P. Biligiri, A state-of-the-art review on recycling rubber in concrete: sustainability aspects, specialty mixtures, and treatment methods. Dev. Built Environ. **14** (2023)
43. A. Mohajerani, L. Burnett, J.V. Smith, S. Markovski, G. Rodwell, M.T. Rahman, et al., Recycling waste rubber tyres in construction materials and associated environmental considerations: a review. Resour. Conserv. Recycl. **155** (2020)
44. X. Zhu, Z. Jiang, Reuse of waste rubber in pervious concrete: experiment and DEM simulation. J. Build. Eng. **71** (2023)
45. A.S.M.A. Awal, H. Mohammadhosseini, Green concrete production incorporating waste carpet fiber and palm oil fuel ash. J. Clean. Prod. **137**, 157–166 (2016)
46. B. Milovanovic, N. Stirmer, I. Milicevic, The sustainable prefabricated wall panel system made of recycled aggregates, in *International Symposium on Life Cycle Assessment and Construction* (2012)
47. A.M. Braga, J.D. Silvestre, J. de Brito, Compared environmental and economic impact from cradle to gate of concrete with natural and recycled coarse aggregates. J. Clean. Prod. **162**, 529–543 (2017)

Chapter 25
Life Cycle Assessment of Hemp-Lime Concrete: Analyzing Methodology and Environmental Impacts

Ilija Bošković

Abstract The construction industry is responsible for significant greenhouse gas emissions, leading to climate change. Utilising more environmentally sustainable building materials is a key strategy to improving the overall sustainability of the construction industry. Bio-based building materials have been identified as potentially more sustainable alternatives to conventional materials as they can make use of renewable raw materials and can sequester carbon dioxide during the life cycle of the material. Hemp-lime concrete is a composite bio-based material formed by combining the chopped woody core (shives/hurds) of the industrial hemp plant with a lime-based binder and water. This study analyzes the approaches taken by prominent studies on the life cycle assessment of hemp-lime concrete. An overview of the life cycle of a typical in situ hemp-lime concrete wall is presented. Key factors including functional unit properties, carbon sequestration factors and end of life scenarios are discussed. The environmental impacts of the life cycle of the material and the relative impacts of various phases and processes are analyzed and discussed. Finally, key uncertainties in the life cycle of hemp-lime concrete are identified and suggestions for improving the accuracy of hemp-lime concrete life cycle assessments are presented.

25.1 Introduction

The construction industry is responsible for significant energy use, resource depletion and greenhouse gas emissions. Utilising more sustainable building materials is one of the strategies, the construction industry can implement to reduce the negative environmental impacts associated with its activities. With the pressing need to reduce the negative impacts of climate change, utilising building materials with a low life cycle global warming potential is crucial. Novel bio-based building materials can be a more sustainable alternative to the traditionally used mineral building materials, as

I. Bošković (✉)
Faculty of Architecture, University of Belgrade, Belgrade, Serbia
e-mail: ilija.boskovic@arh.bg.ac.rs

D. A. H. Hanaor (ed.), *Sustainable Materials for the Built Environment*, Springer Series in Materials Science 352, https://doi.org/10.1007/978-3-031-97818-0_25

they utilise renewable raw materials that can sequester carbon dioxide (CO_2) from the atmosphere through photosynthesis. However it is also important to consider the negative impacts on other environmental phenomena such as eutrophication, acidification, stratospheric ozone depletion, photochemical ozone formation and the depletion of mineral and fossil resources and water.

Hemp-lime concrete or hempcrete is a prominent composite bio-based building material formed by combining the chopped woody core (shives/hurds) of the industrial hemp plant with a lime-based binder and water. It is a porous and vapour permeable non-loadbearing thermal insulation material most often used for the construction of exterior walls. One of the advantages of hemp-lime concrete is its versatility, as it can be applied in situ by casting it around a timber frame and it can also be prefabricated into blocks or panels. In addition, it can also be used for the construction of floors and roof insulation. It was developed in the 1980s as a vapour permeable replacement for deteriorated wattle and daub in medieval timber buildings [1]. Thus, it can be used for historical restoration as well as in new construction and the energy renovation of existing modern buildings. The industrial hemp plant is a fast growing crop that can be used to create a variety of products including textiles, biocomposites, cosmetics, building materials, food, paper, biofuels and health supplements. Hemp shives are the byproduct of hemp fibre extraction and are a readily available renewable biological resource with few competing applications, as the other primary use is for animal bedding. The key advantage of the material is that it can proved long-term storage of the CO_2 sequestered by the industrial hemp plant during a short growth cycle of several months. Finally, studies [2–10] have shown that hemp-lime concrete exhibits positive hygrothermal performance and can delay the effects of oscillations in the exterior environment, thus maintaining interior comfort and reducing the need for active climate control.

The environmental sustainability of hemp-lime concrete has so far been investigated through published life cycle assessment studies [11–20] and commercial environmental product declarations (EPD) [21–25]. As such the environmental impacts of various hemp-lime concrete products have been examined in several different geographical locations. In general, studies have analysed the life cycle of exterior hemp-lime concrete wall constructions. This study aims to analyse the life cycle of hemp-lime concrete wall constructions through a comparative analysis of the approaches taken by prominent published hemp-lime concrete life cycle assessment studies. An overview of the life cycle of a typical in situ hemp-lime wall construction is presented and key stages and unit processes are discussed in the context of the methodologies applied in existing studies. The relative impacts of unit processes and uncertainties in the life cycle of the material are identified and suggestions for improving the accuracy of hemp-lime concrete life cycle assessments are presented.

25.2 Method

The framework for the study is comprised of key published life cycle assessments that examine the environmental impacts of functional units defined as a 1m^2 hemp-lime concrete wall construction.

25.2.1 *Functional Units and Key Scenarios*

Table 25.1 shows the characteristics of the primary functional units or key scenarios in the chosen studies. For most studies, the primary functional units was chosen for the comparison. For the scenarios from Sinka et al. [18] and Arehart et al. [20], where a variety of different scenarios were defined with no primary functional unit the scenarios with the most similar physical characteristics and binder compositions are presented. In Heidari et al. [19] the functional unit is defined as 1 kg of hemp shiv, thus the wall represented in the comparison is an additional scenario and not the primary functional unit in the study.

All of the studies followed the life cycle assessment methodology defined in ISO Standards 14040 [26] and 14044 [27]. The life cycle assessments were therefore carried out in four major phases: goal and scope, inventory analysis, environmental impact assessment and interpretation. The initial step of defining the goal of the life cycle assessment entails the definition of the aim and reasons for carrying out the life cycle assessment. The scope primarily entails the definition of the product system to be analysed, the functional unit, system boundary and environmental impact categories through which the results of the assessment will be represented. The system boundary includes all the unit processes that will be analysed, while the functional unit is the reference unit connected to all the inputs and outputs of the product system [26]. Outside of the studies presented in Table 25.1, most functional units are also defined as 1 m^2 hemp-lime concrete wall constructions. Several EPDs [22, 24, 25] used a functional unit defined as a 1 m^2 hemp-lime concrete block wall constructions, while others took a different approach analysing the life cycle of 1 m^3 of in situ hemp-lime concrete [20] and 1 m^3 of prefabricated blocks [23] respectively. Additionally, Rivas-Aybar, John and Biswas [28], analysed the life cycle of a different type of hemp-based material, by defining a functional unit of 1 m^2 of novel hemp-based prefabricated board.

While all of the scenarios presented in Table 25.1 have the same height and length their thicknesses vary due to the definition of different thermal conductivities and U-values. The density of hemp-lime concrete can vary due to differences in hemp shiv bulk density, the ratio of hemp shiv to binder used when preparing the composition and the level of compaction applied when applying the material, thus variations in the defined densities are to be expected. Variations in density can also be assumed to be one of the primary reasons for variations in the defined thermal conductivity. Studies have shown that increasing the density of the material negatively impacts its thermal

Table 25.1 1 m^2 hemp-lime concrete wall constructions

Study	Functional unit/scenario		Life span	Hemp-lime concrete		Wall U-value (W/ m^2K)	System boundary
				ρ (kg/m^3)	λ (W/mK)		
Bošković and Radivojević [11]	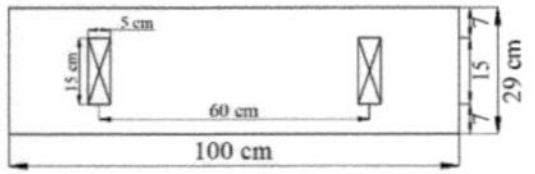	A	100 years	350	0.0894	0.30	Cradle to grave
Bošković [12]	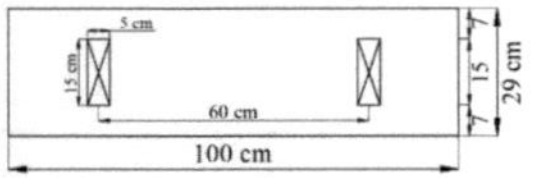	A^2	100 years	350	0.0894	0.30	Cradle to grave
Bošković and Radivojević [13]	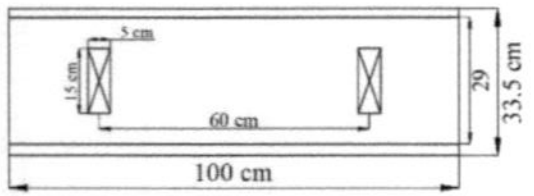	A+Flp	100 years	350	0.0894	0.296	Cradle to grave
Boutin et al. [14]	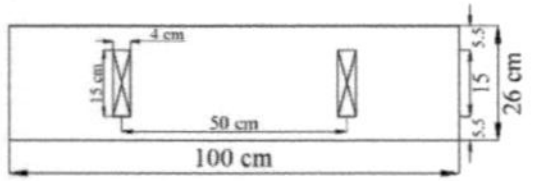	FU	100 years	330	0.118	0.42	Cradle to grave
Ip and Miller [15]	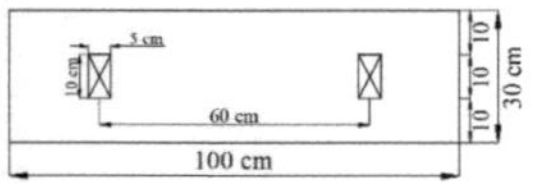	FU	100 years	275	0.06	0.19	Cradle to use

(continued)

Table 25.1 (continued)

Study	Functional unit/scenario		Life span	Hemp-lime concrete		Wall U-value (W/m^{2}K)	System boundary
				ρ (kg/m^3)	λ (W/mK)		
Pretot et al. [16]	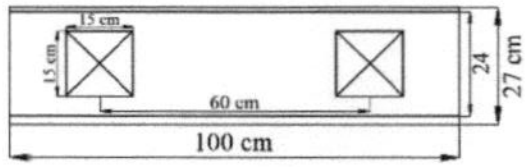	FU	100 years	/	0.087	0.36	Cradle to grave
Arrigioni et al. [17]	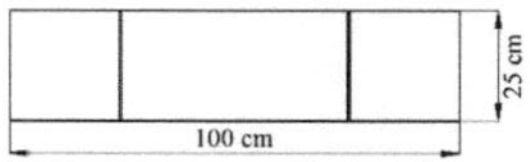	FU	100 years	330	0.07	0.27	Cradle to use
Sinka et al. [18]	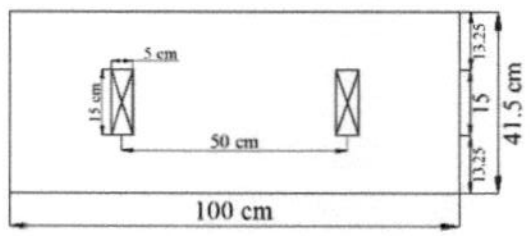	HL (0.15)	100 years	310	0.075	0.18	Cradle to use

(continued)

Table 25.1 (continued)

Study	Functional unit/scenario		Life span	Hemp-lime concrete		Wall U-value (W/m²K)	System boundary
				ρ (kg/m³)	λ (W/mK)		
Heidara et al. [19]	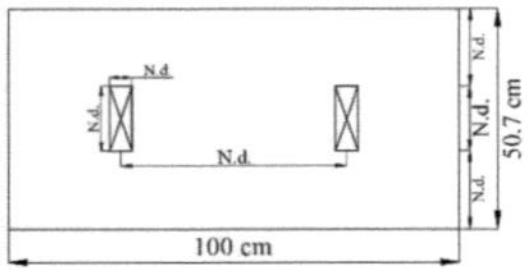	Untreat. MA/EA	60 years	/	/	0.15	Cradle to grave
Arehart et al. [20]	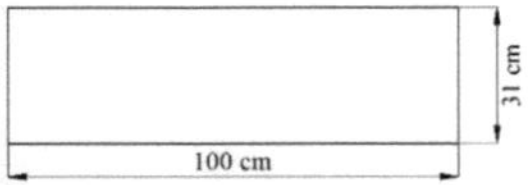	Mix 20	60–100 years	310	0.085	0.27	Cradle to gate + use stage

conductivity [29, 30]. Prefabricated blocks [31], also typically have lower thermal conductivities, as unlike with in situ applications, the material can be compacted in the same direction as thermal loading [31, 32]. In Bošković [11] and Bošković and Radivojević [9, 11] the thermal conductivity of the hemp lime concrete was defined using the density-dependent formula defined in Cerezo [29], while Arehart et al. [20] used the formula defined in Collet and Pretot [30]. The varying thicknesses in the wall constructions ultimately therefore result in the scenarios having different inputs.

Most of the analysed studies define the life span of the material as 100 years, though a lower value of 60 years is also used. Hemp-lime concrete buildings are still relatively young, with the first use widely reported to be in 1986 in the renovation of an XVI century building (Maison de la Turque) in Nogent sur Seine. However, it is assumed that hemp/lime concrete wall construction can have an equivalent life span to conventional wall constructions [16]. Leonard and Groslambert [24] define a life cycle of 60 years according to the Belgian standard for EPDs, but they state that hemp-lime concrete can have a life span of up to 150 years or the life span of the building. Commercial producers also advertise a life span that is as long as the service life of the building [33]. As it is a vapour permeable material, hemp-lime concrete is able to absorb and desorb moisture preventing the excess build-up of moisture in the wall. The high alkalinity of the lime-based binder protects the hemp shives from pests and microbial deterioration and increases the fire resistance of the material.

However, some microorganisms can grow in alkaline conditions and the pH of the binder will fall as a result of carbonation [34, 35]. Experimental research has shown that if the material is produced properly, with an adequate binder, microorganisms may form only under extreme conditions. Walker et al. [36] kept samples of hemp-lime concrete in hot and humid conditions (30 °C, RH = 80%) for 7 months and directly inoculated them with microorganisms (Aspergillus and Penicillium fungi and Bacillus bacterium) on several occasions. The samples experienced no physical deterioration and microorganisms didn't develop. In Marceau et al. [37] microorganisms didn't develop in samples that were subjected to 8 wetting and drying cycles (over 75 days), but did develop in samples that were kept under very hot and humid conditions (30 °C, RH = 98%) for several months and the pH dropped under 10. Obuka et al. [38] found similar results. Directly inoculated samples kept for 45 days under very humid conditions (20 °C, RH = 99%) showed no sign of microorganism growth. However, samples with a lower ratio of binder in the mix with a pH under 9 showed microbial growth. Delannoy et al. [39] concluded that though viable fungal cells may remain on the hemp shiv following retting and carbonation may reduce the pH of the binder, the relative humidity inside a coated wall is unlikely to be high enough for the period of time necessary for fungal growth to begin, and that no fungal growth should occur "without wrong manufacturing or accidental damages [40]." Experimental research has therefore shown that microorganisms cannot develop in natural conditions in which the material can dry after reaching a higher moisture content.

In terms of durability, research has shown that the material can withstand 10 [41] and 25 [42] freeze–thaw cycles without showing physical damage and significant negative impacts to the compressive strength of the material. It is important to note that all of these studies were undertaken with uncoated hemp-lime concrete. In practice the material will either be coated by a lime-based render or covered by a ventilated façade, providing the material with additional protection. Though it is important that an adequate coating is used as coatings that allow excess liquid water to pass into the material or prevent water vapour diffusion are likely to increase the relative humidity of the hemp-concrete wall and increase the risk of fungal growth [43, 44]. Applying the coating before the material has sufficiently dried can inhibit the drying of the wall [1, 9, 39]. The results of existing research and experiences from practice do not go against the assumption that the material, if formulated and applied correctly can have a life cycle of up to and possibly even over 100 years, though further published research testing the long-term durability of the material to confirm this further would be welcome.

25.2.2 System Boundary

As can be seen in Table 25.1, the studies set varying system boundaries. As a minimum, all of the studies took into account production of the raw materials (cradle to gate) and carbonation of the binder in the use stage [20]. Uncertainty regarding the end of life of the material is the primary reason behind the exclusion of the end of life stage in some of the studies. As a minimum, the studies that considered the end of life stage took into account transport of the wall construction materials to the disposal site [16]. All of the studies only took into account embodied environmental impacts, thus the impact of the material on operational energy and water use in the use phase was not considered. Almost all of the studies considered several sensitivity analyses within the system boundary. Table 25.2 shows all of the different sensitivity analyses undertaken so far.

Additional sensitivity analyses aim to analyse the environmental impacts of variation in the life cycle of hemp-lime wall constructions. In Bošković [12] and Bošković and Radivojević [11, 13] an average (A), pessimistic (P) and optimistic scenario (O) is defined by varying key variables that include the hemp sequestration factor, transport distance, construction waste formation rate and carbonation factor. In Bošković and Radivojević [13] and Pretot et al. [16] the environmental impacts of applying various finishes to the wall construction are examined, as well as the impacts of various renewal regimes in the use phase. These variables were also examined in Bošković [12], as well as comparisons between in situ and prefabricated block applications and comparisons with typical construction practices. Among other sensitivity analyses, Arrigioni et al. [17] experimentally measured the degree of carbonation in a hemp-lime concrete block and examined the impact of varying the binder carbonation factor. Sinka et al. [18] primarily examined the environmental impacts of using different types of binder. Arehart, Nelson and Srubar [20] also examined the impacts

Table 25.2 Sensity analyses analysed in existing studies

Study	Sensitivity analysis/variable
Bošković and Radivojević [11]	– Hemp shiv sequestration factor – Raw material transport distance to site – Raw material wastage rate during construction – Binder carbonation factor – End of life scenario
Bošković [12]	– Hemp straw yield – Hemp shiv sequestration factor – Raw material transport distance to site – Raw material wastage rate during construction – Binder carbonation factor – End of life scenario, – Comparing in situ and prefabrication – Wall finish type and number of renewals during the use stage – Wall thickness –– Comparison with conventional wall constructions
Bošković and Radivojević [13]	– Wall finish type – Number of coating/finish renewals during the use stage
Boutin et al. [14]	– Allocation in hemp shiv production
Ip and Miller [15]	– Application of coating to wall – Wall thickness – Raw material transport distance to site
Pretot et al. [16]	– Wall finish type – Number of coating/finish renewals during the use stage – Wall thickness – Comparison with conventional wall constructions
Arrigioni et al. [17]	– Binder carbonation factor – Binder composition – Raw material transport distance to site – Hemp shiv to binder ratio
Sinka et al. [18]	– Binder composition – Wall thickness
Heidara et al. [19]	– Hemp shiv treatment – Allocation in hemp shiv production – End of life scenario – Comparison with conventional wall construction
Arehart et al. [20]	– Binder composition and binder carbonation factor – Hemp shiv to binder ratio – Wall thickness

of applying different types of binder as a means of analysing the impacts of varying the binder carbonation factor. Heidari et al. [19] and Boutin et al. [14] examined the impacts of applying different allocation methods in hemp shiv production. Heidari et al. [19] also analysed the impact of applying a sol–gel coating to hemp shiv. A wide variety of variables have been analysed, providing insight into the relative impact of unit processes and phases in the life cycle of hemp-lime concrete wall constructions.

They provide an insight into general principles that can be applied to improve the embodied environmental impacts of hemp-lime concrete constructions.

25.2.3 Inventory Analysis

The inventory analysis (LCI) is an iterative process that entails the data collection and calculation procedures quantifying the inputs and outputs associated with the product system [26, 27]. The four primary stages in the life cycle of the product system are the production, construction, use and end-of-life stages. The lifecycle inventory of a typical in situ hemp-lime concrete wall from cradle to grave is shown in Fig. 25.1.

As all of the studies didn't use the same life cycle impact assessment (LCIA) methodology and didn't uniformly evaluate the same environmental impact categories, the relative impacts of stages and processes in the life cycle of hemp-lime concrete wall constructions will be primarily discussed in the context of scenarios A [11, 13] and A+Flp [13] presented in Table 25.1. The studies by Bošković and Radivojević [11, 13] utilised the cut-off by classification system model and the environmental impact categories were calculated using the Environmental Footprint impact assessment method conforming to EN15804:A2+2019 [28]. The first study [11] evaluated the four global warming potential categories (GWPtotal, GWPbiogenic, GWPfossil and GWPluluc), while the second study [13] also analysed ozone depletion potential (ODP), acidification potential (AP), freshwater eutrophication potential (EPfresh), marine eutrophication potential (EPmarine), terrestrial eutrophication potential (EPter), photochemical ozone formation potential (POCP), abiotic depletion potential (ADP) fossil depletion potential (ADPfossil) and water depletion potential (WDP). Figure 25.2 shows the relative environmental impacts of the average uncoated hemp-lime concrete wall (scenario A) shown in Table 25.1.

Figure 25.3 shows the relative environmental impacts of a coated hemp-lime concrete wall (scenario A+Flp) shown in Table 25.1. Scenario A+Flp is constructed by applying a lime putty and sand finish (+Flp) to both sides of scenario A [13].

25.2.3.1 The Production Stage

The production stage includes the extraction of raw materials (A1), transport of the raw materials (A2) and manufacturing processes (A3). The production stage primarily includes the production of the hemp shiv and binder, as well as the timber frame and coating materials, where they are present. As can be seen in Figs. 25.2 and 25.3, the production stage is responsible for most of the emissions and negative environmental impacts in both a coated and uncoated hemp-lime concrete wall. In the case of hemp-lime concrete blocks the production stage would also include the prefabrication processes, but production of the mortar and steel fixings is likely to be attributed to the construction stage.

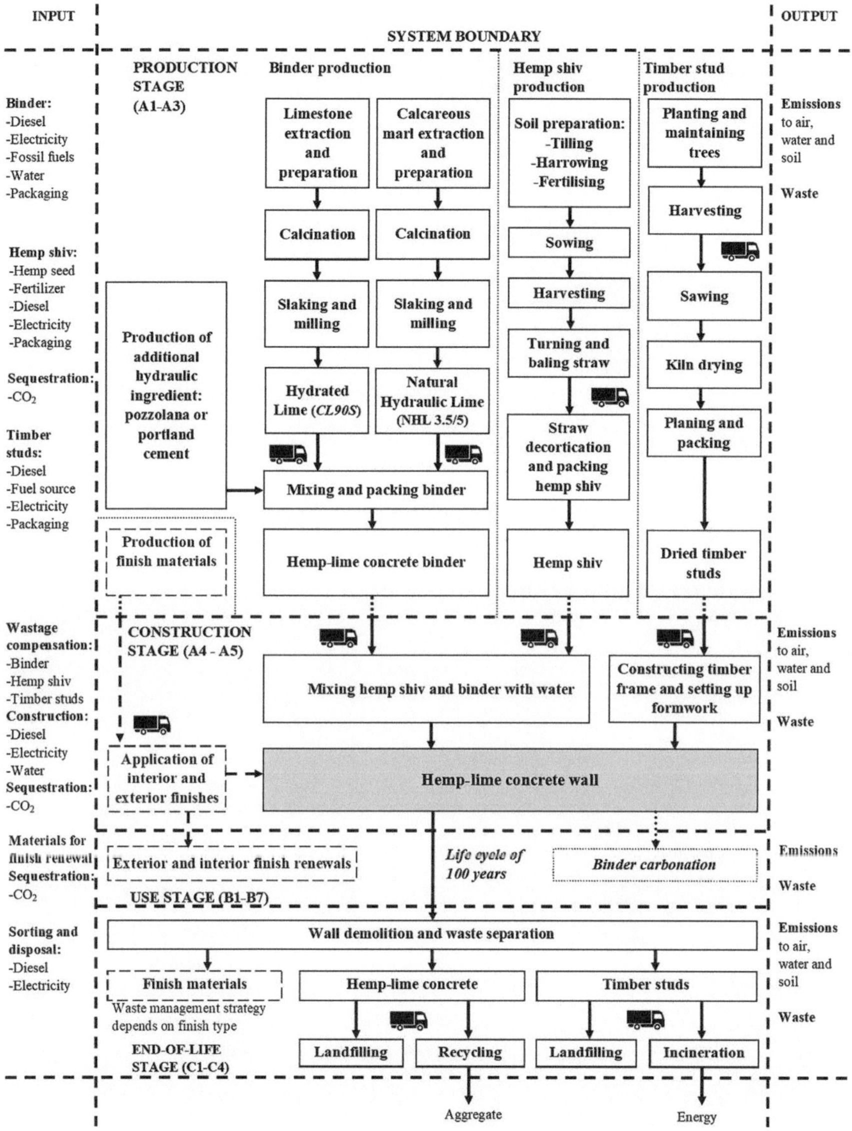

Fig. 25.1 The typical life cycle of an in situ hemp-lime concrete wall construction [12]

Hemp Shiv Production

The primary material inputs for industrial hemp production are industrial hemp seed, diesel for the agricultural machinery and fertiliser. Table 25.3 shows some of the primary parameters in the inventory analysis of hemp shiv production. Values not

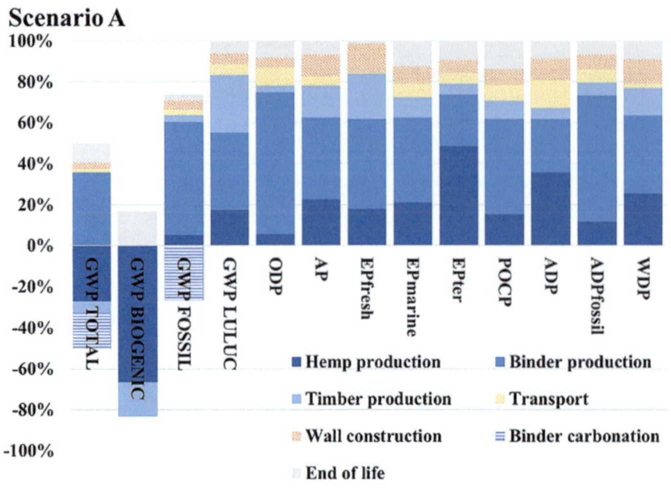

Fig. 25.2 The relative environmental impacts of the life cycle of scenario A [13]

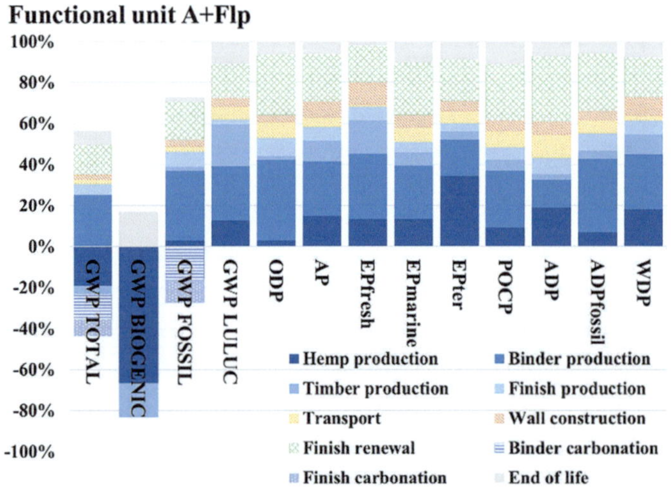

Fig. 25.3 The relative environmental impacts of the life cycle of scenario A+Flp [13]

directly quoted in the studies, which have been inferred based on the data presented in the text, tables or graphs, have been marked with an approximation symbol (≈).

Before sowing, the soil is prepared by ploughing and harrowing. Fertilisers are typically applied after sowing, though they can also be applied partially beforehand. The fertilization regime will vary according to the context and soil requirements, though according to van der Werf [42] in typical farming practice in France, industrial hemp production required lower inputs of nitrogen (N) and phosphate (P) than crops such as rapeseed, wheat, maize, potato and sugarbeet. However only potato

Table 25.3 Key inputs parameters of hemp shiv production

Study	Allocation method	Impacts allocated to hemp shiv production (%)	Hemp straw yield (t/ha)	Fertiliser use (kg/ha)			Global warming potential of hemp shiv production (kg CO_{2eq}/kg shiv)		
				N	*P*	*K*	Emissions	Uptake	Net
Bošković and Radivojević [11, 13]	Mass	60	10	80	60	60	0.167	1.349–1.547	– 1.182 to – 1.38
Bošković [12]	Mass	60	7.5–12.5	80	60	60	0.139–0.173	1.349–1.547	– 1.176 to – 1.408
Boutin et al. [14]	Mass	60	7.75	69	49	21	≈ 0.243	– 2.1	– 1.446
	Economic	32	7.75	69	49	21	≈ 0.123	– 1.1	– 0.97
Ip and Miller [15]	Mass	52	7.5	93	100	125	≈ 0.192	1.527	≈ – 1.335
Pretot et al. [16]	Mass	60	8	/	/	/	/	1.7	/
Arrigoni et al. [17]	Mass	75	15	/	/	/	0.104	1.84	– 1.736
Sinka et al. [18]	Economic	27	16.5	80	70	147	/	1.84	/
Heidari el al. [19]	Mass	47	8	85	65	125	0.29	1.84	– 1.55
	Economic	32	8	85	65	125	0.21	1.84	– 1.63
Arehart et al. [20]	Mass	75	/	/	/	/	0.104	1.84	– 1.736

and sugarbeet required higher inputs of potassium (K). As a result of fertilization, ammonia (NH_3) and nitrous oxide (N_2O) are emitted to the atmosphere, while nitrates (NO_3) and phosphates (PO_4) are leached and discharged into water systems. At harvest, the plants are cut and left in the field for roughly two to four weeks for the dew retting process to take place. Retting is a microbial process by which the chemical bonds holding the stem of the industrial hemp plant together are broken down, allowing for the separation of bast fibres from the woody core. During this period the straw should be turned several times to allow for even retting. Once the straw has been sufficiently broken down, it is baled and transported to the fibre processing plant. At the processing plant, the fibres are separated from the woody core through the decortication process. The primary product of the decortication process is hemp fibre, while hemp shiv and dust are the byproducts. Based on the existing studies, hemp shiv is assumed to make up 47–75% of hemp straw (Table 25.3).

As hemp shiv is a byproduct of industrial hemp fibre production, it is appropriate to allocate only a certain percentage of the environmental impacts associated with growing and processing industrial hemp, to hemp shiv production. Environmental impacts can be allocated by mass, meaning most of the environmental impacts are attributed to hemp shiv production. Almost all of the existing studies took into account mass allocation, though some also considered economic allocation. As the fibre is the primary and more valuable product, applying economic allocation reduces the environmental burden of hemp shiv production. In applying economic allocation 27–32% of the environmental impacts of hemp straw production and processing were ascribed to hemp shiv production [14, 18, 19]. Dust is typically considered waste, but it can be used to make fuel briquettes [15]. Existing studies therefore did not consider dust to be a waste product and didn't allocate any of the environmental impacts of producing and potentially disposing of the dust to hemp shiv production. As the difference between mass and economic allocation can vary from 20 to 48% of the environmental impacts of hemp straw production and processing the choice of allocation method can have a significant effect on the environmental impacts of hemp shiv production. Though it is worth noting that with economic allocation Boutin et al. [14] also assigned a lower CO_2 sequestration factor. Whichever allocation method is used, the accuracy of the life cycle assessment will depend on using an accurate hemp shiv mass value for the industrial hemp cultivar(s) in question or following up-to-date market trends.

Though the environmental impacts of hemp shiv production will be context-specific it can be argued that it can have a relatively low impact on the overall life cycle of a hemp-lime concrete wall. In Bošković and Radivojević [11] hemp shiv production was responsible for 6.35% of greenhouse gas emissions (GWPtotal) in the uncoated hemp-lime concrete wall (A), which was reduced to 4.15% when the same wall was coated with a hemp-lime putty render/plaster (A+Flp) [13]. Though in some studies hemp shiv production had a higher global warming potential (GWP), the relative impact was still far less when compared to binder production. Hemp shiv production had a higher relative impact on most other environmental impact categories. For scenario A in Bošković & Radivojević [13] hemp shiv production had a higher relative impact on other environmental impact categories such as acidification potential (AP = 22.47%), water depletion potential (WDP = 25.84%) abiotic depletion potential (ADP = 36.00%) and terrestrial eutrophication potential (Epter = 48.67%), primarily due to fertiliser production and use and the processing and packaging of hemp shiv. The environmental impacts of hemp shiv production will therefore primarily depend on the amount of fertiliser used and the efficiency of the decortication process.

The hemp straw yield varies depending on the cultivar, growing conditions and agricultural practices. In Bošković [12] the environmental impacts of varying hemp straw yield were examined by varying an average hemp straw yield of 10t per hectare (ha) by ± 2.5 t (total yield of 7.5 t to 12.5 t per hectare) with the same fuel and fertiliser inputs. The study utilised the CML-IA (v4.8) impact assessment methodology. Global warming potential (GWP) had the highest relative variation and the 5 t/ha yield differential accounted for 1.079kg$CO_{2eq.}$ The remaining environmental impact categories were affected to a lesser extent. The 5t/ha yield increase reduced acidification potential (AP), photochemical ozone formation (POCP), ozone depletion potential (ODP) and fossil resources depletion (ADPf) by around 2%. Eutrophication potential (EP) was reduced by 4.24%, while abiotic resource depletion (ADP) decreased by 13.79%. Though the benefits are less pronounced when comparing the pessimistic and optimistic yields to the average scenario, it is clear that using higher yielding industrial hemp varieties that don't require additional fertiliser inputs can have a positive environmental impact. However, as climate conditions that can affect yield can't be controlled it's also important to note that reducing the environmental impacts of hemp shiv production through consistent yields may not always be possible.

Carbon dioxide (CO_2) sequestration is a crucial parameter to take into account for the life cycle assessment of hemp shiv production. The existing studies in Table 25.3 utilising mass allocation, use a range of values for the hemp shiv CO_2 sequestration factor. The values range from 1.349 kgCO_2/kg of hemp shiv [11–13] to 2.105 kgCO_2/kg of hemp shiv [14]. The carbon sequestration factor can be calculated with Formula (25.1), where M_{DM} represents the mass of dry matter, C_{DM} the carbon content, MM_{CO_2} the molar mass of carbon dioxide and MM_C the molar mass of carbon [11, 12]:

$$Sq_{CO_2} = M_{DM} \cdot C_{DM} \cdot MM_{CO_2}/MM_C \quad (25.1)$$

The CO_2 sequestration factor can therefore be determined based on the carbon and moisture content in the hemp shiv. The carbon content will vary based on cultivar and growing conditions, as Sausserde and Adamovicz [43] measured carbon content values ranging from 41.62 to 45.38% and 41.33 to 43.94% for the same selection of 10 cultivars in two consecutive years. Amziane et al. [44] examined the initial water content of hemp shiv and determined an average value of 10.9% with a standard deviation of 0.6%. Bošković [12] and Bošković and Radivojević [11, 13] examined the environmental impacts of varying the hemp shiv carbon sequestration factor. An average, pessimistic and optimistic scenario were defined based on the lowest

(41.33% [43]) and highest (47.4% [45]) carbon content values reported in the literature and an average moisture content of 11% based on the results from Amziane et al. [44]. With an average hemp shiv carbon content of 44.37% a CO_2 sequestration factor of 1.448 $kgCO_{2eq}$/kg hemp shiv at 11% moisture content was defined for the functional unit (A). Based on the highest and lowest carbon content values from literature a hemp shiv a CO_2 sequestration factor of 1.349 $kgCO_{2eq}$/kg hemp shiv at 11% moisture content was defined for the pessimistic scenario (P) and 1.547$kgCO_{2eq}$/kg hemp shiv for the optimistic scenario (O). The results from Bošković and Radivojević [11] indicated that variation of the CO_2 sequestration factor had a significant impact on global warming potential (GWP), as the 6.07% carbon content difference accounted for the emission of 5.943$kgCO_{2eq.}$ (0.198$kgCO_{2eq.}$ per kg of hemp shiv). Though carbon content is also likely to depend on climate conditions identifying and promoting industrial hemp cultivars that can sequester more CO_2 (have a higher hemp shiv carbon content) can have environmental benefits.

The ability of the hemp plant to sequester CO_2 is a key environmental benefit of using hemp-lime concrete, but the variability of the CO_2 sequestration factor, due to cultivar specificity and growing conditions, introduces additional uncertainty into the life cycle assessment of hemp-lime concrete that can have a large impact on the reported global warming potential value. Care should be taken not to overestimate the ability of the plant to sequester CO_2. The most typically used CO_2 sequestration factor of 1.84 $kgCO_2$/kg of hemp shiv used in some existing studies [17–20] would have to assume a carbon content of around 50% and a moisture content of 0%. Based on currently available data for hemp shiv carbon content [43, 45–48] this appears to be an overestimation. Additional context-specific experimental data for the carbon content of hemp shiv would be valuable to show which cultivars have can sequester more biogenic carbon. However, the most crucial parameter is to accurately determine for how long the sequestered CO_2 can be stored in the material. This primarily depends on the life span of the wall construction, whether the material can be successfully recycled at end of life and organic carbon degradation at final disposal.

Binder Production

Hemp-lime concrete generally utilises a lime-based binder which is primarily made up of hydrated lime (CL90s), (60–80%) that provide a slow set over time through carbonation, and hydraulic components that provide a faster initial set through hydration (20–40%). The hydraulic component is typically comprised of naturally hydraulic lime (NHL3.5 or NHL5) and various pozzolans (metakaolin, GGBS etc.), though some scientific studies have also used Portland cement [17, 20, 34, 49, 50] and "natural cement binder is also available commercially [51]. Most of the studies used commercial binder Tradical PF70 or a binder based on it in either the base case [11–16] or sensitivity analyses [17, 18]. The full composition of Tradical PF70 is a commercial secret, but its composition is defined as 75% hydrated lime (CL90s), 15% natural hydraulic lime and 10% pozzolana. In Bošković [12] and Bošković and Radivojević [11, 13] the pozzolan was defined as metakaolin. Sinka et al. [18] and

Arehart et al. [20] also examined binders that contained metakaolin. The primary purpose of the study by Sinka et al. [18] was to examine the environmental impacts of magnesium phosphate cement (case MPC 0.15/0.50) and magnesium oxychloride cement binders (case MOC 0.15/0.50). Arrigioni et al. [17] also defined an alternative binder for the base case, as a binder containing 80% dolomitic lime and 20% portland cement was used, while Heidari et al. [19] used an alternative commercial binder (Tradical Thermo). Arehart et al. [20] examined the largest number of binder variations, by comparing the environmental impacts of 36 different mixes comprised of hydrated lime (50–80%) and either natural hydraulic lime (20–50%), portland cement (10–25%) or metakaolin (40–50%). Mix 20 defined in Table 25.1 used a binder composed of 65% hydrated lime and 35% natural hydraulic lime. Regardless of binder composition, all studies found that binder production was the single most environmentally damaging process in the the life cycle of hemp-lime concrete.

Hydrated lime production starts with the extraction and preparation of limestone. The crushed limestone is then burned in a kiln at temperatures of 900–1200 °C typically using fossil fuels as the energy source. During calcination calcite ($CaCO_3$) in the limestone is turned to quicklime (CaO) which releases large amounts of CO_2. The quicklime is then milled and slaked with water resulting in hydrated lime ($Ca(OH)_2$). The process of producing natural hydraulic lime is less energy intensive, as temperatures of 950–1050 °C are required for the calcination of calcareous marl and less CO_2 is released due to the lower calcite content in the stone. However, it still involves the burning of fossil fuels and releases CO_2 directly into the atmosphere during calcination. Pozzolans are typically favoured over portland cement as they can be naturally occurring and can require no thermal activation (e.g. pumice) and will almost certainly require much less energy if they do require calcining (e.g. metakaolin). Pozzolans can also be the byproducts of other industrial processes (e.g. GGBS, fly ash etc.), thus allowing for the reuse of waste or less desirable products. The environmental impacts of binder production will be dependent on its composition as well as the efficiency of the lime kiln and the type of fuel used.

In Bošković and Radivojević [13], binder production is the single most impactful in all of the environmental impact categories other than terrestrial eutrophication potential (EPter) and abiotic depletion potential (ADP) in both the coated (A+Flp) and uncoated (A) wall constructions. Binder production was responsible for 64.36% of all greenhouse gas emissions (GWPtotal) in scenario A and 42.07% in scenario A+Flp. However it is important to state that part of the CO_2 emitted during calcination will be sequestered due to carbonation in the use phase. Therefore the overall impact of binder production will be somewhat reduced depending on the rate of carbonation. Carbonation is the process by which the material hardens, as hydrated lime (calcium hydroxide, $Ca(OH)_2$) in the binder reacts with CO_2 from the atmosphere to form calcium carbonate ($CaCO_3$). It is a process that occurs during the use stage, but will be discussed here to cover the overall environmental impact of the binder.

The degree of carbonation depends on ambient conditions (temperature and relative humidity) and is defined as the percentage of the CO_2 released during calcination that is reabsorbed. Studies in general assumed that CO_2 will be able to dissolve through the whole thickness of the wall and the defined degree of carbonation can be achieved during the relatively long life cycle of the material. In addition, if the assumed degree of carbonation is not met during the use stage it is likely to be achieved at the end of life, when the material is crushed and the surface area in direct contact with air is increased [12]. The amount of CO_2 released during calcination depends on the composition of the raw material being calcined and how much of it is transformed. However, the binder carbonation factors used in the studies primarily vary due to differences in binder composition and assumed degrees of carbonation. Studies used a degree of carbonation of 75–100% for hydrated lime and 40–75% for natural hydraulic lime. In an extensive literature review, Despotou et al. [49] found a complete degree of carbonation of 80–92% for hydrated lime. Thus, using a degree of carbonation of 100% may be an overestimation. The degree of carbonation for natural hydraulic lime is generally considered to be lower, as the process is more complex. Natural hydraulic lime has a lower amount of "free lime" ($Ca(OH)_2$), available for carbonation as the calcium oxide (CaO) present in the material is also found in the hydraulic mineral phases (primarily belite). During hydration, the belite reacts with water forming calcium-silicate-hydrate (CSH) and additional $Ca(OH)_2$ available for carbonation.

In Bošković and Radivojević [11, 13] carbonation of the binder in the use stage reduced the total global warming potential (GWPtotal) of the binder from + 50.699 $kgCO_{2eq}$ to between + 23.579 and 30.239 $kgCO_{2eq}$, depending on the binder carbonation factor. Binder carbonation is therefore a key factor in reducing the global warming potential of the material. However, just as with the hemp shiv carbon sequestration factor, overestimating the carbonation factor can skew the results. The long predicted life span of the material and crushing it at the end of life should ensure a high degree of carbonation that can offset some of the greenhouse gas emissions of binder production. In scenario A, binder production also had a particularly high impact on ozone depletion potential (ODP = 69.36%) and the depletion of fossil resources (ADPf = 61.39%) [13]. Due to the high overall environmental impacts of binder production, developing binders with lower emissions in the production stage would be very advantageous and could further improve the environmental sustainability of the material.

Timber Production

As hemp-lime concrete is a nonloadbearing material all in situ hemp-lime walls must be cast around a timber frame. This is typically a timber stud frame utilising untreated softwood studs placed at 40–60 cm centres. Most of the in situ hemp-lime wall constructions shown in Table 25.1 include two timber studs. Hemp-lime concrete blocks are also nonloadbearing, but do not require a timber substructure for support and can thus be used independently to form non-loadbearing walls or used as infill for concrete, steel and timber frame structures. Arrigioni et al. [17] as well as all of the environmental product declarations analysing hemp-lime block constructions therefore don't include any timber frame elements. Sinka et al. [18] and Heidari et al. [19] also included the steel fixings required for constructing the timber stud frame. As timber is also a bio-based material it sequesters CO_2 in the production stage, which can be stored until the end of life. The CO_2 sequestration factor will depend on the type of timber used and can be calculated using Formula 25.1. Usually untreated softwood timber is used. In the life cycle of scenario A [13] timber stud production impacted freshwater eutrophication potential (EPfresh = 22.2%), acidification potential (AP = 15.62%) and water depletion potential (WDP = 13.35%) the most, but had a relatively low impact on greenhouse gas emissions (GWPfossil = 3.99%). In the case of coated wall construction A+Flp, timber production was responsible for 2.61% of greenhouse gas emissions [13]. Thus, while the timber frame is a key element of the in situ wall construction it typically has a low relative environmental impact.

Finishes

Studies have primarily focused on the material itself, analysing the life cycle of uncoated hemp-lime wall constructions. The external surface of a hemp-lime concrete wall needs to be covered with a vapour permeable finish to protect the wall from damage, prevent excess moisture ingress and complement the hygrothermal properties of the material [1]. Lime-based renders and plasters are commonly used as they are vapour permeable and allow infiltrated water to evaporate [52]. Alternatively ventilated façade systems can also be used and they are typically finished with timber cladding. So far studies have examined the environmental impacts of applying lime-based renders and plasters [11, 13, 15, 16] and ventilated façade systems with timber cladding [11, 13]. Production of the finish materials is attributed to the production stage. In Bošković and Radivojević [13] production of the lime putty render and plaster for the initial application has a much lower relative impact than binder production, due to the much lower amount of lime required, but it had a higher relative impact than hemp shiv production in several environmental impact categories (GWPtotal, ODP and ADPfossil). Lime putty render and plaster production for the initial exterior and interior contributed to 8.88% of greenhouse gas emissions, but the overall impact of the finish materials depends on the amount of maintenance and the number of replacements required in the use stage. The durability of the applied finish materials is therefore also an important parameter to consider, alongside the environmental impacts of the initial production.

25.2.3.2 Construction Stage

The construction stage includes the transport of raw materials from gate to site and the installation of the hemp-lime concrete wall construction. Table 25.4 shows the inputs of primary raw materials and the transport distance to site in the analysed studies.

Table 25.4 Raw material inputs and transport distances

Study		Material inputs (kg)					Transport distance gate to site (km)			
		Hemp shiv	Binder	Water	Timber	Other	Hemp	Binder	Timber	Other
Bošković and Radivojević [11]	A	30	60	90	6.75	/	125	125	125	/
Bošković [12]	A^2	30	60	90	6.75	/	125	125	125	
Bošković and Radivojević [13]	A+Flp	30	60	90	6.75	Lime putty: 17.5 Sand: 53.1	125	125	125	Lime: 125 Sand: 145
Boutin et al. [14]	FU	24.8	54.5	37.2	5.5	/	260	921	40/80	/
Ip and Miller [15]	FU	30	50	75	4.6	/	100	100	100	/
Pretot et al. [16]	FU	20.4	45	67	20	Lime: 3.2 Sand: 17.1	300	800	80	N.d
Arrigoni et al. [17]	FU	31.4	44.5	58.6	Mortar: 4.25	/	100 in block	100 in block	/	Mortar: 100
Sinka et al. [18]	HL (0.15)	52.1	75.8	103	N.d	Steel fastening: 0.08	200	200	50	N.d
Heidari el al. [19]	Untreat. MA/EA	50.7	91.2	152	9.3	Steel fastening: 5	N.d	N.d	N.d	N.d
Arehart et al. [20]	Mix 20	20.8	36.4	36.4	/	/	/	/	/	/

Transport to Site

The transport distances defined in the studies vary as they are dependent on geographical context. All of the studies assumed the use of lorries to transport the hemp shiv, binder, timber and finish materials to the site. In the life cycle of an uncoated hemp-lime concrete wall construction (A), transporting the hemp shiv, timber (EURO 5, 7.5-16t lorry) and binder (EURO 5, 16-32t lorry) 125 km impacted abiotic resource depletion potential (ADP = 13.72%), ozone depletion potential (ODP = 8.52%) and photochemical ozone depletion potential (POCP = 8.17%) the most [13]. Transport of the raw materials contributed to only 3%/of total greenhouse gas emissions (GWPtotal). As transport distances are context-specific, defining the overall relative environmental impact of transport can be difficult as they can vary significantly. Increasing the transport distances to 180 km [11, 13] only slightly increased the relative impact of transport in the construction stage on greenhouse gas emissions (GWPtotal = 4.2%). However, with a more significant increase to 500 km for each raw material, the relative impact of transport on global warming potential (GWPtotal) increases to 10.95%. As the binder is a specialized product currently produced by a small number of commercial producers that makes up most of the mass of the wall construction, its transport has the highest potential of negatively impacting the life cycle of hemp-lime concrete. Using the framework of the same study [13], if the transport of the hemp shiv and timber remained 125 km, but the binder is transported 1600 km (EURO 5, > 32 t lorry) from a producer in France, the relative impact of transport on global warming potential increases to 11.59%. Long gate to site transport distances can therefore have a notable negative impact on global warming potential and offset some of the inherent environmental advantages of the material. However if the materials can be sourced locally, transport from gate to site has a relatively low effect on most environmental impact categories.

Installation on Site

Installation of the wall entails erecting the timber stud wall and formwork, mixing the hemp-lime concrete and casting it in the formwork. The mixing process requires inputs of water and electricity. In situ hemp-lime concrete can be cast by hand placement or spraying. Hand application is a more manual process, whereby the hemp shiv, binder and water are usually mixed in a pan mixer and transferred in buckets or other containers and placed in the formwork by hand. Spray application is a mechanized process in which the hemp shiv and binder slurry are mixed at the nozzle of the sprayer and sprayed onto the surface of a permanent shuttering board or wall. The key parameter to consider during construction is the amount of material and packaging waste that will be produced on site. Most existing studies didn't consider construction waste formation. Where construction material waste formation was considered, a wastage rate of 0–10% was examined [11–13]. Thus the construction stage also includes the production and transport of the hemp shiv, binder and timber necessary to replace the materials lost in transport or on site. Following installation the waste materials and packaging are transported to a disposal site. The waste hemp and timber waste will also contribute to the sequestration of CO_2. However, to avoid waste formation having a positive impact on global warming potential it would be appropriate to not take this into account and where possible reuse the biobased waste for another purpose or incinerate it to provide energy. Binder waste will have the greatest impact, though some of the waste hydrated lime will inevitably undergo carbonation to a certain degree, depending on where it is lost or disposed of. At a 5% wastage rate constructing in scenario A impacted freshwater eutrophication (EPfresh = 13.95), water depletion potential (WDP = 11.77%), acidification potential (AP = 10.86%/7.95%) and abiotic resource depletion (ADP = 10.48%). The global warming potential can vary notably as at 0% waste formation construction of the wall had a relative impact of 1.58% on global warming potential (GWPtotal), which rose to 8.97% at a 10% waste formation rate. While achieving 0% waste is highly unlikely in practice improving construction practices and reducing waste can improve the environmental performance of the material.

25.2.3.3 Use Stage

All of the studies assumed that the hemp-lime concrete would not require any maintenance or refurbishment during the life span of the wall construction and that the binder would undergo carbonation [12–20]. The environmental impacts of carbonation were discussed in the Binder Production section. Studies that analysed the life cycle of uncoated hemp-lime concrete wall constructions therefore didn't attribute any negative environmental impacts to the use stage. The studies that analysed wall constructions with applied finishes [12, 13, 16] took into account the replacement of the finish materials at their end of life. The use stage therefore includes the production and transport of the replacement materials and the disposal of the old materials and packaging. The environmental impacts associated with the use stage therefore depend on the life span of the finish materials and the number of exterior and interior replacements required during the life cycle of the hemp-lime concrete wall construction. In practice, the life span of finish materials will vary due to varying environmental conditions, material quality and user preference. As the life span of lime render is typically 25–40 years [53, 54] the studies [12, 13, 16] took into account various renewal scenarios. In Bošković and Radivojević [13] varying renewal regimes resulted in the use stage having an impact of 18.99–27.81% on global warming potential (GWPtotal) in scenario A+Flp. When the initial production of the finish materials is also factored in, it was found that the lime coating for the hemp-lime concrete wall construction can be responsible for over a third of greenhouse gas emissions (GWPtotal = 27.87–36.69%). Ultimately applying a lime putty coating increased the global warming potential of the wall construction from 0.135kgCO_2 to between 21.747kgCO_2 and 30.273kgCO_2. Applying and renewing wall finishes therefore has a pronounced effect on the overall life cycle environmental impacts of a hemp-lime concrete wall and should not be ignored when evaluating the environmental performance of the material.

25.2.3.4 End of Life

As the end of life strategy for the material hasn't been widely developed in practice, the end of life stage is still relatively unknown. The key uncertainty that needs to be determined is the carbon storage potential of the material at the end of the life of the hemp-lime concrete wall construction. It is important to determine how the material can be recycled and reused to extend the period of carbon storage. Ip and Miller [15] suggest that at end of life hemp-lime concrete can be reused as a lightweight aggregate, an aggregate for construction blocks or to adjust the pH of soil. It is also crucial to determine the degradation mechanisms at final disposal to determine how much, if any, of the initially sequestered carbon can remain stored in the material if it is landfilled. Pretot et al. [16] assumed that the lime binder would make the hemp shives inert, preventing organic carbon degradation and the emission of greenhouse gases at end of life.

A static or dynamic approach may be taken when accounting for biogenic carbon in the life cycle of building materials. In a static approach, the "0/0" or "− 1/ + 1" approaches are typically used. The "0/0" or "carbon neutral" approach assumes that all sequestered biogenic carbon will be released at end of life, so the movement of biogenic carbon is not considered at all. The "− 1/ + 1" approach tracks the flow of biogenic carbon through the life cycle of the building material, accounting for sequestration in the production stage, but also assuming that all of the sequestered carbon will be released at the end of life stage. A criticism of the static approaches is that they don't take into account the timing of the carbon emissions and the rotation periods as they relate to biomass growth [55]. As a result, dynamic approaches have been developed to factor in the impact of time [56–61]. All of the existing studies defined in Table 25.1 essentially take a static approach, but by not considering the end of life phase or assuming the material remains inert, most studies have essentially taken a "− 1/0" approach [62]. In this approach, all of the carbon sequestered by the hemp and timber remains stored in the material, which can lead to an overestimation of the environmental performance of the material and overall wall construction.

At end of life the hemp-lime wall construction will be demolished and the materials will be transported to final disposal. Studies typically considered that the entire construction will be landfilled with no further emissions. Alternatively, in Bošković [12] in Bošković and Radivojević [11, 13] the hemp-lime concrete and timber construction were assumed to be separated on site and treated separately. The timber was assumed to be incinerated releasing all of the initially sequestered carbon dioxide ("− 1/ + 1" approach) and producing energy to be used in another product system. The hemp-lime concrete was assumed to be recyclable and would therefore be sent to a recycling centre where it would be crushed, sorted and prepared for reuse. In scenarios A and A+Flp the hemp-lime concrete was also assumed to be inert at the end of life ("− 1/0"). In this scenario, 16.57% (A) or 11.64% (A+Flp) of greenhouse gas emissions take place at the end of life stage due to the incineration of the timber. Though the greenhouse gas emissions are higher than in the construction stage, most of these emissions are neutralised by sequestration in the production stage.

In a sensitivity analysis, the hemp-lime concrete was considered to be landfilled at end of life and would release greenhouse gases as a result of organic carbon degradation [11–13]. Data regarding the degradation of organic carbon in the hemp-lime concrete in ambient or landfill conditions at the end of life is not yet available. Environmental product declarations for hemp-lime block products [22, 24] take into account biodegradation after landfilling and the release of CO_2 and methane (CH_4) at the end of life. They used a fraction of degradable organic carbon dissimilated (DOC_F) of 15%, thus 15% of the organic carbon in the hemp shiv was assumed to degrade at the end of life. This value was defined based on French conventions for timber products in general and therefore may not be reflective of the actual behaviour of hemp-lime concrete under landfill conditions. The default DOC_F for wood waste prescribed by the IPCC methodology is 50% [63], though experimental investigations suggest this value is an overestimation.

Through laboratory simulations, Wang et al. [64] determined a DOCF of 0.1–1.8% for softwood lumber and 7.8% for red oak lumber, while DOCF of values of 0–9.5% values were determined for pine branches and 17.1–28.5% for oak branches [65]. Ximenes et al. [66] found similar results when examining samples from various landfills in Australia. Factors such as ambient conditions, the ratio of lignin and holocellulose in the plant, lignin type and the type and amount of extractives influence the degradation of organic carbon [64, 66]. It can be assumed that the DOC_F of hemp shives would be higher than in pine or spruce, as hemp shives have a lower lignin content and higher holocellulose and lignin ratio. However, without extensive experimental data it is difficult to predict how much the lime binder would inhibit degradation. As such Bošković [12] and Bošković and Radivojević [11] analysed the end of life greenhouse gas emissions of various DOC_F values used for wood products in literature (1.5% [67], 10% [68], 15% [22, 24], 23% [69] and 50% [63]). It was assumed that 50% of the DOC_F would be dissimilated as CO_2 and 50% as CH_4 and that the landfill site would have a CH_4 recovery system in place with a collection efficiency (CE) of 75% and a destruction efficiency of 99.9%. Landfill sites need to have an efficient CH_4 recovery system in place, as CH_4 has a much higher global warming potential than CO_2. Burning the captured methane can produce energy to be used in another product system and ensures that the captured CH_4 is emitted as CO_2, reducing the environmental impacts of organic carbon degradation. In the Environmental Footprint impact assessment method, the characterisation factor used for 1 kg of CH_4 is 36.75 $kgCO_{2eq.}$. It was determined that with a low to moderate DOC_F and an efficient methane recovery system not all of the CO_2 sequestered in the production phase would be emitted at the end of life. Assuming a DOC_F of 15% and CE of 75% it was found that 35.86% of the biogenic carbon sequestered in the hemp shiv would be emitted at end of life. This would decrease to 23.91% with a DOC_F of 10% and increase to 54.98% with a DOC_F of 23%. However, with a lower CE of 50%, the amount of sequestered biogenic carbon released at end of life would vary from 37.81 to 86.7% at the same DOC_F values. The degradation of hemp shiv can therefore result in high greenhouse gas emissions at end of life and have a large impact on the global warming potential (GWPtotal) impact category, as biogenic carbon sequestered in the production stage is released. Composting, as suggested by Heidari et al. [19] may be the environmentally preferable option, but would also result in greenhouse gas emissions. Due to the lack of experimental data and the high variability of potential greenhouse gas emissions, the end of life stage is a large factor of uncertainty in the life cycle of hemp-lime concrete.

25.2.3.5 Environmental Impact Assessment and Interpretation

Table 25.5 shows the results of the environmental impact assessments carried out by the studies. Only the results of the global warming potential indicator are shown as it is widely considered to be the most important impact category and is the only impact category that is present in all of the studies. In addition, though the studies used different impact assessment methodologies that use some different characterisation factors, the unit for the impact category is the same ($kgCO_{2eq.}$). Results highlighted as "()*" [11–13] show the results of the scenario with an alternative end of life strategy that assumes the hemp-lime concrete is landfilled and degrades at the end of life with an assumed DOC_F of 15% and CE of 75% ($+DOC_{F=15\%}$). Though the described differences in the functional units/scenarios, system boundaries and impact assessment methods make direct quantitative comparisons of the results non-viable, contextualising them can help identify trends and the impact of variations in the approaches and system boundaries.

It can be seen that the base scenarios and scenarios with similar binder compositions (hydrated lime 65–75%, natural hydraulic lime 15–35% and pozzolans 0–10%) in all of the studies have carbon negative or close to carbon neutral global warming potential. It was determined that global warming potential could be as low as −36.08 $kgCO_{2eq}$. [15]. However, it is important to note that all of the studies with the lowest results analysed uncoated wall constructions and assumed that all of the CO_2 sequestered by the hemp and timber frame would stay stored at the end of life and some also used potentially overestimated CO_2 uptake factors. This can be seen when comparing the results of scenarios P and O [11, 13] that are variations of the same uncoated hemp-concrete wall construction. Both scenarios assume that the hemp shiv stays inert at end of life, but scenario P uses lower CO_2 uptake factors, longer transport distances and takes into account the production of construction material waste in the construction stage. The results of scenario O are carbon negative and in line with many other existing studies, while scenario P emits more greenhouse gases than it sequesters. This demonstrates the potentially large variations in global warming potential that can be caused by external factors.

Other variations in the life cycle of the wall construction can also cause the global warming potential to become positive. Sinka et al. [18] demonstrated that by using alternative binders such as magnesium phosphate cement or magnesium oxychloride cement instead of a commercial hemp-lime concrete binder, the global warming potential becomes positive even within a cradle to use system boundary. Similarly, treating the hemp shiv with sol gel [19] also made the global warming potential indicator positive. Applying finish materials to the wall construction significantly increases global warming potential. Applying lime putty coatings (+Flp) to scenario

Table 25.5 Global warming potential of key scenarios in the analysed studies

Study		System boundary	LCIA method	Hemp shiv ($kgCO_{2eq.}$/kg shiv)		Binder ($kgCO_{2eq.}$/kg binder)		GWPtotal/GWP ($kgCO_{2eq.}$)
				Production	Uptake	Production	Carbonation	
Bošković and Radivojević [11, 13]	A (FU)	Cradle to grave	EF	0.167	1.448	0.845	0.397	0.135 (14.899)*
	P	Cradle to grave	EF	0.167	1.349	0.845	0.341	10.165 (23.872)*
	O	Cradle to grave	EF	0.167	1.547	0.845	0.452	– 9.696 (6.142)*
Bošković and Radivojević [13]	A+Flp (FU)	Cradle to grave	EF	0.167	1.448	0.845	0.397	25.548 (40.322)*
	P+Flp	Cradle to grave	EF	0.167	1.349	0.845	0.341	40.958 (54.665)*
	O+Flp	Cradle to grave	EF	0.167	1.547	0.845	0.452	9.358 (25.196)*
Bošković [12]	A^2 (FU)	Cradle to grave	CML-IA	0.367	1.448	0.838	0.367	– 0.261 (12.983)*
Boutin et al. [14]	FU (Mass allocation)	Cradle to grave	CML-IA	≈ 0.243	– 2.1	≈ 0.606	0.249	– 35.53
	(Economic allocation)	Cradle to grave	CML-IA	≈ 0.123	– 1.1	≈ 0.606	0.249	– 14.1
Ip and Miller [15]	FU	Cradle to use	IPCC 2007	≈ 0.192	1.527	≈ 0.704	0.571	– 36.08
Pretot et al. [16]	FU	Cradle to grave	CML-IA	/	1.7	0.778	≈ 0.462	– 1.6
Arrigoni et al. [17]	FU	Cradle to use	CML-IA	0.104	1.84	/	0.325	– 26.01
Sinka et al. [18]	HL (0.15)	Cradle to use	CML-IA	/	1.84	/	≈ 0.487	– 19.28
	MPC (0.15)	Cradle to use	CML-IA	/	1.84	/	0	26.49

(continued)

Table 25.5 (continued)

Study		System boundary	LCIA method	Hemp shiv ($kgCO_{2eq.}$/kg shiv)		Binder ($kgCO_{2eq.}$/kg binder)		GWPtotal/GWP ($kgCO_{2eq.}$)
				Production	Uptake	Production	Carbonation	
Heidari el al. [19]	Untreated (M/E allocation)	Cradle to grave	IPCC 2013	0.29	1.84	/	/	– 15.27 to – 8.92
	Treated (M/E allocation)	Cradle to grave	IPCC 2013	0.21	1.84	/	/	22.51–28.77
Arehart et al. [20]	Mix 20	Cradle to gate + use	/	0.104	1.84	≈ 1.002	0.47	– 17.00 to – 11.56

A increases the global warming potential by 25.413 $kgCO_{2eq.}$ and the global warming potential of optimistic scenario O became positive (O+Flp). In Pretot et al. global warming potential remains negative despite the applied lime finishes, but thinner coatings and less interior renewals were defined compared to scenario O+Flp [13]. Taking into account alternative end of life scenarios that consider the degradation of hemp shiv further increases global warming potential. In Bošković and Radivojević [13] the global warming potential of a pessimistic coated wall construction construction scenario that assumes hemp shiv degradation at the end of life is 54.665 $kgCO_{2eq}$ (P+Flp+$DOC_{F=15\%.}$). This is 90.745 $kgCO_{2eq.}$ higher than the life cycle global warming potential of the primary functional unit in Ip and Miller [15], demonstrating how much differing unit process inventories and system boundaries can impact the final result and the conclusions regarding the environmental performance of the material.

25.3 Discussion

Analysing the relative impacts of the unit processes in the life cycle of a hemp-lime concrete wall construction it is apparent that binder production overall has the most negative environmental impacts. Binder production also releases the most greenhouse gases in both uncoated and coated scenarios, though part of the CO_2 released during carbonation can be sequestered through carbonation. A high degree of carbonation can be assumed due to the long expected life span of the material and end of life treatment, carbonation may still be variable in practice. Therefore using mixes with a lower binder content where possible and potentially developing binders with lower negative emissions in the production stage can notably improve the environmental performance of the material. However, Sinka et al. [18], demonstrate that some alternative binders can be significantly more environmentally damaging than the typically used commercial binders. Hemp shiv and timber production have a much lower relative impact in most environmental impact categories compared to binder production. Heidari et al. [19] showed that applying treatments to hemp shiv can also have a very adverse effect on the environmental impacts of hemp shiv production. Defining the relative impact of gate to site transport distance can be difficult as transport distances are context dependent, thus the relative impact of transport can vary significantly. With very high transport distances transport to the site can have a more negative impact than hemp shiv and timber production on global warming potential, but it will still have a much lower impact than binder production. The relative impact of installing the wall construction on site can be relatively low, but will primarily depend on the amount of construction material waste formed during transport and installation.

Comparing the results of pessimistic and optimistic scenarios P and O [11, 13] in Table 25.5 it is clear that variations to unit processes in the production, construction and use stages can significantly impact global warming potential. The results vary by 19.861 $kgCO_{2eq.}$ and controlling the variables may not be fully possible in practice. Having a low hemp sequestration factor can be caused by cultivar choice or unfavourable growing conditions, transport distances will be dictated by geographical context and the availability of raw materials, high material wastage can be caused by poor construction site practices or unforeseen events and a low degree of binder carbonation can be caused by unconducive ambient conditions and end of life disposal. The ability of the material to sequester CO_2 as a result of the photosynthesis of the industrial hemp plant and the carbonation of the binder is a significant advantage. However it is important to take into account the potential impact of suboptimal uptake factors. Comparisons with other wall constructions would help contextualize the performance of the material in the other environmental impact categories.

Most studies analysed the life cycle of uncoated hemp-lime concrete wall constructions. As the exterior surface of a hemp-lime concrete wall must be protected in practice, thus current studies have primarily focused on the materials and not on complete wall constructions. Considering the life cycle of coated or covered wall constructions is therefore crucial to fully understanding the environmental impacts of using the material. In Bošković and Radivojević [13], though producing the hemp-lime putty coatings had a relatively low impact, the environmental impacts of the use stage were comparable to binder production in many of the environmental impact categories (Fig. 25.3). Thus, it is important to not only consider the environmental impacts of producing the finish materials but also their durability, as high maintenance and renewals can have a high relative impact on life cycle emissions. In Bošković & Radivojević [13] applying lime putty finishes somewhat changed the environmental perception of the material in the average (A) and optimistic (O) scenarios, as with the added coatings the global warming potential went from almost carbon neutral or carbon negative, to carbon positive. Applying the coatings also had a more negative impact on global warming potential than just applying the defined hemp shiv degradation scenario (+$DOC_{F=15\%}$) [11, 13]. However, with a higher fraction of dissimilated organic carbon or a lower methane collection efficiency the degradation of hemp shiv at the end of life can have an even greater impact.

The sequestration of CO_2 by the hemp plant is considered the primary advantage of hemp-lime concrete. However, the flow of biogenic carbon in the end of life phase is the greatest factor of uncertainty in the life cycle of the material. Typically in traditional LCA with a static approach ("0/0" or "− 1/ + 1") no biogenic carbon can be assumed to be stored after end of life. All of the existing studies however have taken a "− 1/0" or partial approach, assuming that all or some of the biogenic carbon will remain stored in the material after final disposal. Taking a "− 1/ + 1" approach in scenario A+Flp [13] (Table 25.5) and assuming that all of the biogenic carbon sequestered in the production stage is emitted at the end of life, the global warming potential would increase from 25.548 $kgCO_{2eq.}$ to 68.896 $kgCO_{2eq}$. For comparison, in Bošković [12] the global warming potential of a typical masonry wall construction in Serbia (thermo blocks + ETICS façade system), with an equivalent U-value and life span, was determinrd to be 108.845 $kgCO_{2eq.}$. Thus even though the global warming potential of the hemp-lime concrete wall may still be significantly lower than for the conventional masonry wall, the perception of the material is significantly different when compared to the carbon negative global warming potentials presented in existing studies. Applying this static approach somewhat nullifies the advantage of sequestering CO_2 and ignores the stages of the life cycle when the emissions take place. Therefore it may be more appropriate to examine the life cycle of the material using a dynamic, rather than static approach. Even if it is assumed that all of the biogenic carbon is released at the end of life, hemp-lime concrete still has significant advantages when the temporal dimension is also considered. Pittau et al. [70] highlighted that fast growing bio-based materials such as hemp and straw can provide effective mitigation to global warming due to their short growing cycle. The hemp-lime concrete should be able to store the biogenic carbon for 100 years (or more if successfully recycled) and as hemp has a short growth cycle the biomass is replaced quickly.

The " + 1/− 1" approach typically assumes that the biological material will be incinerated at the end of life. This may not be practical in the case of hemp-lime concrete, as incinerating the material may calcinate some of the hardened binder releasing additional CO_2 into the atmosphere. Thus the primary strategy at the end of life should be to recycle as much of the material as possible to extend the period of biogenic carbon storage. Unless new technologies or strategies are developed, at practical end of life composting or disposal in a landfill would likely be preferable to incineration. It is currently unknown to what extent the lime binder may be able to inhibit the degradation of organic carbon in the hemp shiv. To be able to accurately model the end of life phase it is therefore important to experimentally examine the degradation mechanisms of the material when composted and in landfill conditions and find to what extent the degradable organic carbon can be expected to dissimilate and how much biogenic carbon, if any, can be stored beyond the life cycle of the material. Once this data is available the uncertainty factor of the end of life phase can be mitigated and the overall accuracy of hemp-lime concrete life cycle assessment studies can be improved.

25.4 Conclusion

The life cycle of hemp-lime concrete wall constructions was examined by analysing the approaches taken by existing hemp-lime concrete life cycle assessment studies. The studies defined varying system boundaries and only some considered the life cycle from cradle to grave. Most studies considered the material to have a life span of 60–100 years and the durability of the material against microbial growth, freeze–thaw cycles and accelerated aging has been experimentally verified. The relative environmental impacts of the unit processes were examined and it was determined that binder production was the primary hotspot and the most negatively impactful unit process. As a result, the production stage is responsible for most of the environmental impacts, though the relative impacts of the remaining life cycle stages showed potentially high variability. The relative impact on the construction stage depends primarily on gate to site transport distances and construction material waste formation.

Studies primarily focused on the material itself and few analysed the impacts of coating or covering the hemp-lime concrete wall construction. As such the use stage in most studies only included carbonation of the lime binder. However, when a lime putty coating is applied to the wall the environmental impacts of the use stage can be comparable to binder production in many environmental impact categories. The environmental impacts associated with the use stage depend on the environmental impacts of producing and transporting the finish materials and the number of replacements necessary throughout the life cycle of the material. The ability of the material to sequester CO_2 through photosynthesis of the hemp plant and carbonation of the binder is an important advantage of the material that mitigates greenhouse gas emissions in the production and use stages. As such some studies determined the global warming potential of the material to be carbon negative, though care should be taken not to overestimate the CO_2 uptake of the material, by using overoptimistic sequestration and carbonation factors. The end of life stage is the primary factor of uncertainty in the life cycle of the hemp-lime wall constructions, as disposal strategies have not yet been fully developed and data regarding organic carbon degradation of hemp shiv is not yet available. As such it is difficult to estimate the expected greenhouse gas emissions at the end of life and determine how much, if any, of the biogenic carbon can remain stored for a significant period of time after the end of life. To accurately model the end of life stage of hemp-lime concrete wall constructions, future research into the material should include experimental studies examining the degradation of organic carbon in hemp-lime concrete in landfill conditions and when it is composted. In the absence of this data using a dynamic approach for biogenic carbon accounting may be preferable to the static approach utilised in all of the analysed studies.

References

1. W. Stanwix, A. Sparrow, *The Hempcrete Book* (Green Books, Cambridge, 2014), p.23
2. A. Evrard, A. De Herde, J. Building Phys. **34**, 5–25 (2010). https://doi.org/10.1177/1744259109355730
3. A.D. Tran Le, C.Maalouf, T.H. Mai, E. Wurtz, F. Collet, Energy Build. **42**, 1797–1806 (2010). https://doi.org/10.1016/j.enbuild.2010.05.016
4. A. Shea, M. Lawrence, P. Walker, Constr. Build. Mater. **36**, 270–275 (2012). https://doi.org/10.1016/j.conbuildmat.2012.04.123
5. S. Pretot, F. Collet, Constr. Build. Mater. **36**, 58–65 (2012) https://doi.org/10.1016/j.conbuildmat.2012.04.139
6. S. Pretot, F. Collet, Build. Environ. **82**, 459–466 (2014). https://doi.org/10.1016/j.buildenv.2014.09.018
7. E. Latif, M. Lawrence, A. Shea, P. Walker, Build. Environ. **93**, 199–209 (2015). https://doi.org/10.1016/j.buildenv.2015.07.011
8. U. Dhakal, U. Berardi, M. Gorgolewski, R. Richman, J. Clean. Prod. **142**, 3655–3664 (2017). https://doi.org/10.1016/j.jclepro.2016.10.102
9. B. Moujalled, Y. Aït Oumeziane, S. Moissette, M. Bart, C. Lanos, D. Samri, Build. Environ. **136**, 11–27 (2018). https://doi.org/10.1016/j.buildenv.2018.03.025
10. F. Bennai, M. Yacine Ferroukhi, F. Benmahiddine, R. Belarbi, A. Nouviaire, Constr. Build. Mater. **316**, 126007 (2022). https://doi.org/10.1016/j.conbuildmat.2021.126007

11. I. Bošković, A. Radivojević, J. Build. Eng. **66**, 105908 (2023). https://doi.org/10.1016/j.jobe.2023.105908
12. I. Bošković, Doctoral Dissertation, Формирање модела за процену животног циклуса бетона од конопље и креча у контексту Србије (Creating Life Cycle Assessment Models for Hemp Lime Concrete in the Context of Serbia), University of Belgrade (2023)
13. I. Bošković, A. Radivojević, J. Build. Eng. **86**, 108940 (2024). https://doi.org/10.1016/j.jobe.2024.108940
14. M. Boutin, C. Flamin, S. Quinton, G. Gosse, Etude des caracteristiques environmentales du chanvre par l'analyse de son cycle de vie, Ministere de l'agriculture et de la peche, MAP 04 B1 05 01 (2006)
15. K. Ip, A. Miller, Resour. Conserv. Recycl. **69**, 1–9 (2012). https://doi.org/10.1016/j.resconrec.2012.09.0012012
16. S. Pretot, F. Collet, C. Garnier, Build. Environ. **72**, 223–231 (2014). https://doi.org/10.1016/j.buildenv.2013.11.010
17. A. Arrigoni, R. Pelosato, P. Melia, G. Ruggieri, S. Sabbadini, G. Dotelli, J. Clean. Prod. **149**, 1051–1061 (2017). https://doi.org/10.1016/j.jclepro.2017.02.161
18. M. Sinka, P. Van den Heede, N. De Belie, D. Bajare, G. Sahmenko, A. Korjakins, Resour. Conserv. Recycl. **133**, 288–299 (2018). https://doi.org/10.1016/j.resconrec.2018.02.024
19. M.D. Heidari, M. Lawrence, P. Blanchet, B. Amor, Materials **12**(18), 2987 (2019). https://doi.org/10.3390/ma12182987
20. J.H. Arehart, W.S. Nelson, W.V. Srubar III., J. Clean. Prod. **266**, 1–13 (2020). https://doi.org/10.1016/j.jclepro.2020.121846
21. EVEA, (FDES) Construire en Chanvre—1 m^3 de béton de chanvre pour remplissage et isolation de mur ou cloison, Inies Catalogue de la base, (2018)
22. M. Sié, (FDES) VICAT—Bloc de beton de chanvre BIOSYS®, Inies Catalogue de la base (2018)
23. A. Léonard, S. Groslambert, Isohemp hempcrete blocks—life cycle assessment EN 15804+A1 Project Report (Liège Université, 2017)
24. S. Groslambert, A. Léonard, ISOHEMP – PAL36 – Bloc de béton chanvre, B-EPD, (2021)
25. Senini, Hemp and Lime Biocomposite Building Products, EPD Italy, SEN-01_22 (2022)
26. The International Organisation for Standardisation, ISO 14040, *2006 Environmental Management—Life Cycle Assessment - Principles and Framework* (2006)
27. The International Organisation for Standardisation, ISO 14044, *2006 Environmental Management—Life Cycle Assessmentd—Requirements and Guidelines* (2006)
28. D. Rivas-Aybar, M. John, W. Biswas, Materials **16**(22), 7208 (2023). https://doi.org/10.3390/ma16227208
29. V. Cerezo, Doctoral Dissertation, Propriétés mécaniques, thermiques et acoustiques d'un matériau à base de particules végétales:approche expérimentale et modélisation théorique, INSA Lyon
30. F. Collet, S. Pretot, Constr. Build. Mater. **65**, 612–619 (2014). https://doi.org/10.1016/j.conbuildmat.2014.05.039
31. J. Williams, M. Lawrence, P. Walker, Constr. Build. Mater. **159**, 9–17 (2018). https://doi.org/10.1016/j.conbuildmat.2017.10.109
32. T. Pierre, T. Colinart, P. Gloucannec, Int. J. Thermophys. **35**, 1832–1852 (2014). https://doi.org/10.1007/s10765-013-1477-0
33. Isohemp, *FAQ* (2024) https://www.isohemp.com/en/faq
34. R. Walker, S. Pavia, R. Mitchell, Constr. Build. Mater. **61**, 340–348 (2014). https://doi.org/10.1016/j.conbuildmat.2014.02.065
35. S. Marceau, P. Glé, M. Guéguen-Minerbe, E. Gourlay, S. Moscardelli, I. Nour, S. Amziane, Constr. Build. Mater. **139**, 524–530 (2017). https://doi.org/10.1016/j.conbuildmat.2016.11.129
36. V. Obuka, M. Sinka, V. Nikolajeva, S. Kostjukova, R. Ozola-Davidane, M. Klavins, J. Ecol. Eng. **22**(4), 296–313, (2021) https://doi.org/10.12911/22998993/134033

37. G. Delannoy, M. Guéguen-Minerbe, I. Nour, S. Marceau, P. Glé, E. Gourlay, D. Diafi, S. Amziane, F. Farcas, Impacts of alkalinity and environmental ageing on fungal growth in hemp concretes, in *Proceedings of the Final Conference of RILEM 253-MCI Microorganisms-Cementitious Materials Interactions Volume II*, ed. by A. Bertron, H. Jonkers (RILEM Publications, Paris, 2018), pp. 347–353
38. P. Strandberg-de Bruijn, K.H. Jeppson, K. Sandin, C. Nilsson, Biosyst. Eng. **103**(4), 474–479 (2009). https://doi.org/10.1016/j.biosystemseng.2009.02.005
39. G. Delannoy, S. Marceau, P. Gle, E. Gourlay, M. Gueguen Minerbe, S. Amziane, F. Farcas, Constr. Build. Mater. **252**, 119043 (2020). https://doi.org/10.1016/j.conbuildmat.2020.119043
40. A. Piot, T. Béjat, A. Jay, L. Bessette, E. Wurtz, L. Barnes-Davin, Constr. Build. Mater. **139**, 540–550 (2017). https://doi.org/10.1016/j.conbuildmat.2016.12.143
41. European Committee for Standardization, EN 15804:2012+A, *2019 Sustainability of Construction Works—Environmental Product Declarations—Core Rules for the Product Category of Construction Products* (2019)
42. H. Van der Werf, Euphytica **140**, 13–23 (2014). https://doi.org/10.1007/s10681-004-4750-2
43. R. Sausserde, A. Adamovics, J. Agric. Eng. **44**, 619–622 (2013). https://doi.org/10.4081/jae.2013.365
44. S. Amziane, F. Collet, M. Lawrence, C. Magniont, V. Picandet, Round robin for hemp shiv characterization—part 1: evaluation of initial water content and water, absorption, in *Bio-Aggregates Based Building Materials: State-of-the-Art Report of the RILEM Technical Committee 236-BBM* ed. by S. Amziane, F. Collet (RILM and Springer, Dordrecht, 2017), pp. 207–223
45. P. Brazdausks, R. Tupciauskas, M. Andzs, J. Rizhikovs, M. Puke, A. Paze, K. Meile, N. Vedernikovs, Energy Procedia **72**, 34–41 (2015). https://doi.org/10.1016/j.egypro.2015.06.006
46. M.-S. Diakité, H. Lenormand, V. Lequart, S. Arufe, P. Martin, N. Leblanc, Molecules **26**, 6334 (2021). https://doi.org/10.3390/molecules26216334
47. L. Marrot, K. Candelier, J. Valette, C. Lanvin, B. Horvat, L. Legan, D.B. DeVallance, Waste and Biomass Valorization **13**, 2267–2285 (2022). https://doi.org/10.1007/s12649-021-01640-6
48. C. Mongiovi, D. Lacalamita, N. Morin-Crini, X. Gabrion, A. Ivanovska, F. Sala, V. Placet, V. Rizzi, J. Gubitosa, E. Mesto, A.R. Lado Ribeiro, P. Fini, N. De Vietro, E. Schingaro, M. Kostić, C. Cosentino, P. Cosma, C. Bradu, G. Chanet, G. Crini, Molecules **26**(15), 4574 (2021). https://doi.org/10.3390/molecules26154574
49. F. Delhomme, E. Prud'homme, C. Julliot, T. Guillot, S. Amziane, S. Marceau, Results Chem. **4**, 100440 (2022). https://doi.org/10.1016/j.rechem.2022.100440
50. Y. Diquelou, E. Gourlay, L. Arnaud, B. Kurek, Constr. Build. Mater. **112**, 506–517 (2016). https://doi.org/10.1016/j.conbuildmat.2016.02.138
51. Vicat, *Prompt Natural Cement* (2024). https://www.prompt-natural-cement.com/product
52. R. Nogueira, A.P. Ferreira Pinto, A.Gomes, Cem. Concr. Compos. **89**, 192–204 (2018). https://doi.org/10.1016/j.cemconcomp.2018.03.005
53. D. Kellenberger, H.-J. Althaus, Build. Environ. **44**, 818–825 (2009). https://doi.org/10.1016/j.buildenv.2008.06.002
54. S. Sturgis, J. Andersen, P. Astle, C. Bagenal George, L. Bowles, L. Hamot, L. Leston-Jones, D. Qian Li, A. Papakosta, Whole life carbon assessment for the built environment. *Royal Institute of Chartered Engineers* (2023)
55. E. Hoxha, A. Passer, M.R.M. Saade, D. Trigaux, A. Shuttleworth, F. Pittau, K. Allacker, G. Habert, Buildings Cities **1**, 504–524 (2020). https://doi.org/10.5334/bc.46
56. A. Levasseur, P. Lesage, M. Margni, L. Deschênes, R. Samson, Environ. Sci. Technol. **44**(8), 3169–3174 (2010). https://doi.org/10.1021/es9030003
57. F. Cherubini, G.P. Peters, T. Berntsen, A.H. Strømman, E. Hertwich, GCB Bioenergy **3**(5), 413–426 (2011). https://doi.org/10.1111/j.1757-1707.2011.01102.x
58. F. Cherubini, G. Guest, A.H. Strømman, GCB Bioenergy **4**, 784–798 (2012). https://doi.org/10.1111/j.1757-1707.2011.01156.x
59. A. Levasseur, P. Lesage, M. Margni, M. Brandao, R. Samson, Clim. Change **115**, 759–776 (2012). https://doi.org/10.1007/s10584-012-0473-x

60. A. Levasseur, P. Lesage, M. Margni, R. Samson, J. Ind. Ecol. **17**(1), 117–128 (2013). https://doi.org/10.1111/j.1530-9290.2012.00503.x
61. G. Guest, F. Cherubini, A.H. Strømman, J. Ind. Ecol. **17**, 20–30 (2013). https://doi.org/10.1111/j.1530-9290.2012.00507.x
62. J.H. Arehart, J. Hart, F. Pomponi, B. D'Amico, Sustain. Prod. Consumption **27**, 1047–1063 (2021). https://doi.org/10.1016/j.spc.2021.02.028
63. D.R. Gómez, J.D. Watterson, B. B. Americano, C. Ha, G. Marland, E. Matsika, L. Nenge Namayanga, B. Osman-Elasha, J.D. Kalenga Saka, K. Treanton, in *Stationary Combustion*. 2006 IPCC Guidelines for National Greenhouse Gas Inventories—vol. 2: Energy ed. by S. Eggleston, L. Buendia, K. Miwa, T. Ngara, K. Tanabe (IGES, Hayama, 2006), pp. 2.1–2.47
64. X. Wang, J.M. Padgett, F.B. De la Cruz, M.A. Barlaz, Environ. Sci. Technol. **45**, 6864–6871 (2011). https://doi.org/10.1021/es201241g
65. X. Wang, M.A. Barlaz, Sci. Total. Environ. **557–558**, 355–362 (2016). https://doi.org/10.1016/j.scitotenv.2016.03.091
66. F. Ximenes, C. Björdal, A. Cowie, M. Barlaz, Waste Manage. **41**, 101–110 (2015). https://doi.org/10.1016/j.wasman.2015.03.032
67. R. Hischier, *Treatment of Waste Wood, Untreated, Sanitary Landfill, CH, Ecoinvent Database, Version 3.8* (2021)
68. Australian Government Department of Industry, Science, Energy and Resources, National greenhouse accounts factors, Australian Government (2021)
69. M.A. Barlaz, Global Biogeochem. Cycles **12**(2), 373–380 (1998). https://doi.org/10.1029/98GB00350
70. F. Pittau, F. Krause, G. Lumia, G. Habert, Build. Environ. **129**, 117–129 (2018). https://doi.org/10.1016/j.buildenv.2017.12.006

Chapter 26
Lime for the Built Environment

Marco Simoni, Cecilia Pesce, and Theodore Hanein

Abstract This chapter offers a general overview of the lime industry in relation to the built environment. The vital role of lime in cement making is also discussed. The production of lime from limestone calcination is carbon and energy-intensive requiring more than 3000 kJ per kg of lime and with processing temperatures around 1000 °C. In cement making, even higher temperatures (1600 °C) are required for the formation of the targeted calcium-rich clinker phases. More than half of the carbon emissions from lime making arise due to the process chemistry where calcium carbonate is thermally decomposed into calcium oxide and carbon dioxide gas. The use of lime in construction dates back thousands of years; since then, the fundamental principles of lime production, based on thermal calcination, has not changed. This is mainly because, until recent breakthroughs, society believed this to be the only way. Although the calcination process is industrially robust and its efficiency has been improving over the years, process improvements alone are not sufficient to achieve Net Zero and our climate goals. A description of novel approaches that avoid thermal calcination, and which have the potential of abating the carbon emissions from lime and cement manufacturing, are highlighted as they represent genuine solutions for the complete decarbonization of these industries that are crucial to our built environment.

M. Simoni (✉) · C. Pesce · T. Hanein
earth4Earth Technology Ltd., London W1W 7LT, UK
e-mail: marco.simoni@earth4earth.co.uk

C. Pesce
e-mail: cecilia.pesce@earth4earth.co.uk

T. Hanein
e-mail: t.hanein@leeds.ac.uk

T. Hanein
School of Civil Engineering, University of Leeds, Leeds LS2 9JT, UK

D. A. H. Hanaor (ed.), *Sustainable Materials for the Built Environment*, Springer Series in Materials Science 352, https://doi.org/10.1007/978-3-031-97818-0_26

26.1 Lime

The term "lime" is usually referred to all those products derived from the calcination of different types of limestone distinguished by purity degree or rock source. The types of lime obtainable may be classified as: quicklime CaO, hard-burnt lime CaO, slaked/hydrated lime $Ca(OH)_2$, and dolomitic lime (containing significant amounts of Magnesium [1]). Due to the variability of the chemical and physical properties, stemming from the source and processing conditions used for its manufacturing, lime is a very versatile inorganic bulk chemical. Its setting properties were already well known in ancient civilisations, such as the Romans, Greek, and Egyptians, where it was mainly used in construction [2]. It is likely that the fundamental scientific reasons behind its properties were unknown, and arguably not still fully understood today, but that did not stop its widespread use contributing to the development of those civilizations.

CaO is the direct product of the thermal calcination of limestone (mainly $CaCO_3$) and is also a precursor for conventional cement manufacturing [3]. The calcination of $CaCO_3$ (Eq. 26.1) is a highly endothermic reaction (requiring ~ 1800 kJ per kilogram of $CaCO_3$ [4]). In Eq. 26.1, the CO_2 released upon calcination is differentiated between fuel, i.e., arising from the combustion of the mostly fossil fuels used in the process, and raw material CO_2, arising from the carbonate rock that is decomposed [5].

$$CaCO_{3(Solid)} + \text{heat energy} \rightarrow CaO_{(Solid)} + CO_{2(Combustion)} + CO_{2(raw\ material)} \tag{26.1}$$

The physical and chemical properties of CaO are largely influenced by the temperature, residence time, impurities content, and type of fuel used during manufacturing, leading to higher or lower reactivities and an array of potential applications. That is also valid for the derivatives of the quicklime, such as slaked lime, and limes obtained from different natural sources, such as dolomite, or even biogenic calcium carbonate such as chalk. The reactivity of the CaO can be assessed by exploiting the exothermic properties upon contact with water and monitoring the rate of temperature increase [6] or by adopting commonly used reactivity tests [7].

26.2 Lime and Cement Manufacturing

Outside the cement industry, in 2023 about 430 Mt of lime were produced globally [8], of which 20–22 were in the EU. Such production rate indicates its extraordinary physical and chemical properties, which make it suitable for a wide variety of industrial applications.

To truly grasp the vital importance that lime has in modern society, the cement industry must be considered. With over 4 Gt globally produced every year (180 Mt/

y in the EU), cement is the most manufactured commodity worldwide [9]; it was found that the cement production rate of a given country is an accurate indicator of its development degree. In the near future, the global cement production is expected to increase in some underdeveloped/developing areas [10]. Global cement demand forecasting is essential to understand the past, the present, and the future of lime's role in the global market and wider society. Lime is the main ingredient (50–60%) of ordinary Portland cement [11], and a key ingredient of most mineral binders used in construction, which still rely on the fundamental role of calcium in the overall binder chemistry.

Conventional cement and lime manufacturing routes share the same preliminary steps prior to firing such as: quarrying and crushing prior to calcination [12]. The composition of the limestone fed into the high temperature kiln may vary according to the specifics of the desired product but generally requiring ≥ 80 wt% ($CaCO_3$ + $MgCO_3$) [13]. Cement is typically manufactured in the long rotary kiln (LRK) configuration, comprising the counter-current flow of solid feed and combustion flame, generally obtained by burning natural gas, coke, and waste, such as exhausted tires [14]. The solid feed is fed at the top of the 90 m long kiln unit, while both the rotation and inclination of the kiln ensure homogeneous mixing between the reactants and a sufficient residence time allowing for the formation of the main clinker phases, namely: alite (Ca_3SiO_4), belite (Ca_2SiO_5), tetracalcium alumino-ferrite ($Ca_4Al_2Fe_2O_{10}$), and tricalcium aluminate ($Ca_3Al_2O_6$) [15]. Notably, clinker refers to the indurated output products of the firing in the rotary kiln, whereas the term cement applies once gypsum (5%) and other optional mineral additives or admixtures are blended with the clinker at the end of the processing [16]. The Long Rotary Kiln with Pre-heaters (PRK) configuration reflects the most energy-efficient option for clinkering, which was firstly introduced in the 1960s. Depending on the design considered [17], such as air-through or air-separate pre-calciners, up to the 40–95% of the raw feed can be decarbonized prior to entering the rotary kiln. It is essential that the formation of the clinker does not occur in the pre-calciner unit; for this reason, around 70% of the overall amount of fuel can be efficiently used in the pre-calciner unit [18] as the decarbonisation of limestone is much more energy intensive than clinker formation.

Lime is the essential component for clinker making and subsequent cementitious properties, since calcium ions are essential for the formation of the clinker and hydration phases, whilst contributing to the adequate environment in the pore solution of the hardened cement to promote mechanical properties. Upon hydration of Portland cement, the gel-like structure of the main hydration phase, calcium silicate hydrate (known as C–S–H) readily forms, providing early strength and formation of micro-domains within the structure, where $Ca(OH)_2$ precipitates and allows for the achievement of those alkaline conditions required for further reaction and hardening [11]. Like lime hydration, the hydration of cement is also exothermic, mainly because of the formation of $Ca(OH)_2$ [19]. In fact, like for lime, the reactivity of cement upon hydration is also monitored through heat release; the higher and faster the heat release, the more reactive the cement is considered. It must be noted that

Table 26.1 Overview of the ranges of thermal efficiency with respect to the ideal value of ~ 1800 kJ per kilogram of $CaCO_3$ (η, expressed in %), CO_2 emissions factor (EF, expressed in tt_{CO_2}/t_{CaO}), electricity consumption (EC, expressed in kWh/t_{CaO}), and productivity range (PR, expressed in tons/day)

Kiln type	η(%)	EF (tt_{CO_2}/t_{CaO})	EC (kWh/t)	PR (t/d)
PFRK	75–99	1.0–1.2	20–40	100–600
LRK	35–53	1.2–1.8	18–25	160–1500
PRK	41–62	1.1–1.4	17–45	150–1500

other non-Portland or blended binders will have other hydration reactions and phases not covered here.

The favoured design of the kilns used for lime making is generally different from those used for cement making and allows for the achievement of operating efficiencies as high as 99% with the Parallel Flow Regenerative Kiln configuration (PFRK). This design represents the most used type of kiln in the EU, counting 158 units in 2013, for lime making. It consists of two shaft kilns, both composed of a pre-heating (top), burning (middle), and cooling (bottom) zone. At the burning zone, the two shafts are interconnected by a crossover channel that enables very low specific heat consumption. The fuel is injected through a series of lances vertically placed in the material bed of the burning zone at the top of the first shaft, leading to combustion with the counter-current cooling air blown from the bottom. These combustion gases, together with the process CO_2 released from the burned limestone, enter the crossover channel at a temperature of ~ 1050 °C. These streams then flow upwards upon mixing with the cooling air from the bottom of shaft two, heating the limestone in the respective pre-heating zone. The limestone is fed at the top of both shaft kilns and slowly drops by gravity to the bottom of each calcination unit, while converting to CaO. A PFRK is suitable for different types of fuel and produces a high-reactivity and low-CO_2 residual quicklime [20]. Table 26.1 shows the main features of the LRK, PRK, and PFRK, in terms of feed requirements, efficiency and energy consumption, emission factor, and productivity range.

26.3 Lime in Ferrous and Non-ferrous Metallurgy

Steel is a foundational material for the built environment and lime is a crucial component in the steel industry, ensuring the effective refining of the final ferrous products. Steel is mainly produced through the integrated (namely Integrated Steel Making—ISM) and electrical (namely Electric Arc Furnace steel making—EAF) routes, in the EU accounting for the 57% and 43% of the overall production (170 Mt), respectively [21]. Despite the differences in feed requirements (ISM uses natural iron ores, whereas EAF relies on recycled steel scrap), energy input (ISM requires energy input of 6.0 MWh/$t_{Crude\ Steel}$, EAF only 0.6 MWh/$t_{Crude\ Steel}$) [22], industrial units used for pig iron (ISM operation relies on the Blast Furnace—BF, EAF on Electric

Arc Furnace) and steel refining in secondary metallurgy (Blast Oxygen Furnace—BOF—for ISM, vacuum and non-vacuum refining for EAF), and environmental impact (1.85 and 0.4 t_{CO_2} per ton of crude steel produced through the ISM and EAF routes, respectively [23]), both routes must converge on the use of lime as essential component in steel manufacturing. In fact, lime is the main precursor of the so-called slags, essential oxide mixtures which prevent molten steel from oxidation and allow for the refinement to high-quality final products through chemical interaction between the impurities (undesirable chemical elements) in the molten iron and the slag. Given the lower density, the slags float on the molten iron bath; the convective flows in the molten iron bath, obtained by bubbling of inert gas from the bottom of the so-called ladle refining units, ensures for an effective interaction with the floating slag bath. To understand the importance of slags, and therefore lime, in the steelmaking industry, it is sufficient to mention that about 230 kg of slags are generated per ton of steel produced in the EU [24]. The slags from the ISM are fully recycled in the EU for cement making, given that the chemical composition is similar to the target clinker in conventional processing. Again, lime plays the most determining role in exploiting the BFS for recycling (40–50% CaO content [25]). However, the recycling of the slags from EAF is currently not common practice, mainly due to the lower CaO content (25–30%) and the excess FeO (about 30%) [26]. Again, lime content is a main consideration which would allow (or not) for novel recycling practices and synergy between two industrial sectors which affect the economy of any country. Hydrated lime is also used in the steel industry as a lubricant and a neutralization agent for sulfuric acid [27].

Given the large consumption of lime (about 0.15 t CaO per ton of steel manufactured), steel manufacturers usually have on-site limestone mines to cut transportation costs. As a result, the use of lime in the steel industry also accounts for significant indirect CO_2 emissions, in the order of 0.2–0.3 t CO_2 emitted per ton of steel produced, depending on the calcining unit considered. Given the focus of steel making, the long service life, and the lower costs, LRK are usually employed in steelmaking facilities, resulting in higher carbon emissions.

Lime also plays a determining role in the non-ferrous metallurgical sector; these metals are also a crucial part of our built environment whether directly or indirectly. For example, it is used for copper production during the beneficiation process that involves grinding, followed by flotation to improve the purity of the copper concentrate; in the latter step, lime is essential since it provides the necessary alkalinity. Its alkaline properties are also required during the flotation step for nickel, zinc and lead production, and during the extraction of gold, silver, and uranium [28].

26.4 Lime Binders

The high porosity and permeability properties of lime makes it ideal for the restoration of old buildings, a practice referred to as grouting [29], where the application of cement-based layers would instead lead to detachment, cracks, and shrinkage [30].

In fact, the flexibility of lime, the high viscosity, and the enhanced permeability allow for the transpiration of the foundations underneath the grout. In these terms, lime mortars are a well-known type of binder, and feasible for a wide range of applications given the high flexibility, workability, freeze thaw resistance, water retention, higher strength, vapor permeability, and autogenous healing. However, the main drawbacks are represented by a longer setting and hardening time with respect to Portland cement.

Pozzolanic materials are those [alumino] silicate-based non materials which lead to the formation of cementitious phases through reaction with $Ca(OH)_2$ [31, 32]. Examples include: volcanic ash, fly ash, silica fume, metakaolin, and steel slag. Lime-pozzolan cements are alternative binders, that can have lower environmental burden than clinker-based cements. However, the variability of the source of pozzolan might affect strength, durability, and hydration properties; moreover, the kinetics of the pozzolanic reaction is generally lower at the early stages of hydration, when compared to Portland cement [33].

Lime is also used as an additive and filler in asphalt to improve its water resistance, rheological and mechanical properties [34].

26.5 Lime-Stabilised Soil

It is well-known that the addition of small percentages of lime (5–30%) to clay-rich soils can promote the formation of cementitious phases in the mixture under certain conditions [35]. Both hydrated lime and quicklime can be used for this scope. Different outcomes for soil stabilization have been shown [36]; quicklime results are shown to be more effective than slaked lime at parity of replacement rate, but this is simply based on a mass balance, given the higher provision of Ca ions from the former than the latter. In fact, Ca content in quicklime and slaked lime is 71 wt% and 54 wt%, respectively. However, whilst the use of quicklime imposes strict storage and handling conditions, i.e., quicklime readily hydrates to $Ca(OH)_2$ in slightly moist atmosphere releasing heat, the addition of slaked lime appears straightforward and solely depends on the content of active slaked lime, i.e., purity. It has been reported that small additions of NaOH (2–4% of total water), Na_2SO_4 (0.5–1.1% of total water), and Na_2CO_3 (up to 1% of total water) can provide beneficial results in terms of mechanical properties [37]; however, care must be paid to avoid a too high concentration of soluble salt, which could result in, for example, leaching upon hardening and therefore loss of performance [36].

The stabilization of soil through addition of lime is conducted through different methods and for various purposes, but all of them depend on the nature of the soil used [38]. The crushing or mixing (or can be called soil treatment) is normally the first step, with finely ground clay-rich soils generally allowing for better performance due to the higher active surface for reaction with lime. This may be followed by blending with other materials such as sand. The mixing of the lime and the soil is then carried out with the aim of this step to ensure a full homogenization of the components. A following compression step can lead to the final product targeted for certain purposes and needs to be proportioned depending on various factors.

Lime in soil stabilisation has various applications. One type of application includes the stabilisation of weak soils for road or structure construction and is a way to stop structures or vehicles from sinking into the ground by making the soil less porous and easy to pack down. Another type of application is to produce monoliths, such as rammed earth or even soil-based bricks [39]. Preliminary characterization of the soil is strictly required to choose the optimal conditions for each application. Finally, the stabilized soil can be cured at a range of environmental conditions of e.g., temperatures and moisture to achieve desired performance and within a desired timeframe. Overtime, the lime stabilised soil exposed to air will also carbonate through CO_2 sequestration from the air.

26.6 Lime in Other Applications Related to the Built Environment

Lime has been used for wastewater treatment for over a century, with globally recognized beneficial outcomes for the environment. Thanks to its high alkalinity, lime helps preventing the spread of pathogenic microorganisms in municipal wastes [40], whilst precipitating the nitrogen and phosphorous which would otherwise lead to eutrophication. It represents a valid alternative to the management of sludge from wastewater treatment plants, which is either landfilled or incinerated and then used in agriculture; both practices have raised controversial opinions, with lime representing an affordable and effective alternative solution [41]. Lime is also used for the effective removal of carbonates from potable water, diminishing the overall hardness index, and for the removal of target contaminants, i.e., carbon dioxide, sulfur, or chlorine, from industrial exhaust flue gases. The $CaCO_3 \leftrightarrow CaO$ calcium looping cycle represents a valid carbon capture technology which exploits the capacity of lime to adsorb the CO_2 from industrial exhaust gases, followed by the thermal decomposition of the so-formed $CaCO_3$ to produce a pure CO_2 gaseous stream ready for compression and storage [42].

With the constantly increasing global population, and the progressing desertification of more and more extended areas, the protection and wise management of the soil is key to preserve the optimal conditions for crop growth. As an additional factor, the use of biomass for clean energy production has become common practice, and it must be sustained with approximately constant provision to ensure industrial spread as a contribution to the Net zero transition. In these terms, lime is globally recognized as a beneficial component since it provides soil stability, alkalinity to neutralize naturally occurring acidic soils, and to provide nutrients necessary for crop growth, such as calcium and magnesium [43]. Lime also finds applications in farming, as calcium supplement for livestock nutrition, as disinfectant to sanitise barns, as well as for controlling the pH of bodies of water for fish farming [44, 45].

Lime is also largely used in the pulp and paper industry. In the Kraft pulping process, wood chips are dissolved into sodium hydroxide and sodium sulphide solutions, namely liquor; such liquor is heated and then discharged for the bleaching step. Slaked lime is the bleaching agent conventionally used. After bleaching, heat is recovered from the burning of the organic residues in the recovery furnace, upon which sodium sulphide and sodium carbonate are recovered. The following reaction with slaked lime leads to the precipitation of $CaCO_3$, which is then calcined in tail-end to recover lime; Na_2CO_3 is converted back to NaOH through the causticization reaction [46]. Lime is also used to produce high-purity precipitated $CaCO_3$, mainly used in high-quality coated and uncoated papers, paints and plastics [47].

26.7 Lime's Environmental Burden

In both lime and cement making, which both have similar CO_2 composition in the exhaust gases (20–23 Vol.%) [4], most of the total energy input (and cost) is required for the calcination of the carbonate rock [5]. From an environmental perspective, the carbon emissions from factory operations of modern kilns have been abated to a very high extent, especially when compared to other sectors. However, the cement production process has more heat losses because of the LRK and PRK design used in cement making, and the relatively low efficiency arises from the heat dispersion throughout the external walls of the rotating kiln [48]. Such heat losses cannot be easily avoided, and further heat recovery is also not immediately achievable. In any case, the largest share of carbon emissions from both processes arise from the raw material carbonate source used as a feed, being 0.44 t_{CO_2}, for limestone (assumed $CaCO_3$) and 0.48 t_{CO2} for dolomite (assumed $CaMg(CO_3)_2$). However, given that the largest fraction of lime produced through calcination is used for clinker production, the global CO_2 emissions solely linked to lime making amounts to about 1% of the global anthropogenic carbon emitted to the atmosphere, whereas CaO required for the cement industry accounts for about 6%. For slaked lime synthesis, the carbon emissions related to the slaking process, i.e., hydration of the quicklime to $Ca(OH)_2$, might be considered negligible since there is no calcination of any carbonate-bearing minerals and the overall energy consumption ranges between 5 and 30 kWh per ton of $Ca(OH)_2$ [18].

Apart from CO_2, lime (or cement) kiln dust, NO_x, and SO_x, also arise upon calcination, with concentrations ranging between 1.4–2 mg/Nm^3, 100–2000 mg/Nm^3, and 50–400 mg/Nm^3, respectively [12]. While dust may be considered a valuable by-product, if effectively removed via modern techniques, both NO_x and SO_x represent a threat to human health and the environment and arise mainly from the fuel but also from the raw material used. The NO_x emissions may be limited through primary techniques (fuel selection, burner design, air staging) which prevent their formation, or by application of secondary reduction techniques transforming the NOx species produced into N_2 and H_2O [49]. The SO_x emissions may be limited by using fine limestone, injecting lime into the combustion air, sulfur scrubbing, and injecting an absorbent into the exhaust gases. The composition of the exhaust off-gas stream from a lime kiln varies in accordance with the kiln, fuel, and mineral feed used in the process.

26.8 Advances in Lime Making

Despite the increased efficiency of the state-of-the-art kilns for lime making and the increasing share of alternative fuels to provide heating energy and achieve the operating temperatures required for calcination, the CO_2 emissions arising from the thermal calcination cannot be fully abated. This is because CO_2 emissions arise both from the combustion of the fuels and from the stoichiometric release of CO_2 from natural carbonate rocks sources (Eq. 26.1). Such CO_2 source, namely raw material or process CO_2, accounts for up to 0.44 t_{CO2} per ton of $CaCO_3$ processed, or 0.59 t_{CO2} release per ton of lime produced. Apart from representing the major carbon outcome from the lime making process, such emissions are still considered unavoidable, and end-of-pipe carbon capture will be necessary to achieve carbon neutrality.

26.8.1 Carbon Capture

Carbon Capture & Storage (CCS) is the ultimate solution when no other solutions allow for the mitigation of the carbon emissions from a given process. They are distinguished into pre- and post-combustion technologies, depending on whether they are applied prior or upon processing of the emitting carbon source [50]. Post-combustion technologies are normally favourable since it is generally possible to retro-fit existing plants, but challenging to do so for pre-combustion technologies such as oxyfuel combustion. In either case, the application of CCS is accompanied with higher production costs and complex processing [20]. Amongst all, the post-combustion CCS technology with highest readiness level is based on amine adsorption, relying on the adsorption properties of engineered amine solutions for the CO_2 present in the exhaust flue gases [51]. The main drawback of this solution is the additional processing costs for the regeneration of the amine solutions prior to reuse.

Other technologies including calcium looping (post-combustion CCS) and oxyfuel (pre-combustion CCS) have also shown promising results [52]. The calcium looping technology has been successfully tested at pilot scale, showing that high-grade waste heat can be recovered from the steam cycle, therefore offsetting the higher CAPEX and greater consumption of fossil fuel [53]. It is essential to contrast the naturally occurring sintering events upon a certain number of adsorption/heating cycles; in these terms, the progressive feed of fresh $CaCO_3$ has proven successful, as well as the doping of the reactive bed with inert supports or synthetic sorbents [54].

Even considering successful and cost-effective applications of CCS technologies for lime making, the CO_2 captured must be either be utilized or safely stored upon capture, but safe long-term storage is still unproven. The utilization of the CO_2 captured through CCS could represent a viable option for a wide variety of industrial processes relying on carbon sources currently obtained through conventional calcination and natural resources consumption, such as in soda ash (Na_2CO_3) and methanol (CH_3OH) production or even in the production of various types of construction materials. However, the utilization of all the carbon sequestrated through CCS will be challenging both from a technical and operational point of view.

26.8.2 Electrochemical Production of Lime

An electrochemical process has been recently developed for the decarbonisation of limestone [55]. The preliminary investigations focused on the electrochemical calcination of $CaCO_3$, whose balanced stoichiometry is reported in Eq. 26.2. Despite being pioneering, such an approach in this form would still not avoid the emission of the process CO_2 from the decarbonized $CaCO_3$, and therefore CCS would still be required.

$$CaCO_{3(s)} + 2H_2O_{(l)} \rightarrow Ca(OH)_{2(s)} + H_{2(g)} + \frac{1}{2}O_{2(g)} + CO_{2(g)} \tag{26.2}$$

This lime-making electrochemical process now focuses on the treatment of hardened cement paste in demolished concrete (or other calcium silicate) through an electrolyzing unit, conceptually like conventional water electrolyzing systems. In the cathodic chamber of the electrolyser, reactive Ca ions are extracted from inert materials and dissolved, producing a Ca-rich solution and reactive silicates, i.e., the extraction of calcium and separation from silicates occurs in this single step. At the anodic chamber, alkalinity is produced through redox reactions and electricity input, which promotes the precipitation of reactive lime. Finally, the reactive calcium and silicates can be blended in a low-temperature process to make the final cement product. It is undeniable that the technology shows clear advantages; however, it relies on the availability of non-carbonated construction and demolition waste (that can be separated efficiently) or other naturally occurring calcium silicates.

26.8.3 Lime Production Through Anion Exchange

A novel chemical approach towards the decarbonization of lime manufacturing has recently been proposed [56–58]; this method implies the aqueous reaction between $CaCO_3$ and NaOH solutions at atmospheric pressure and temperature. As reported in Eq. 26.3 below, 2 mol of NaOH are required to convert $CaCO_3$ to the products $Ca(OH)_2$ and Na_2CO_3 (soda ash).

$$CaCO_3 + 2NaOH \rightleftharpoons Ca(OH)_2 + Na_2CO_3 \quad (26.3)$$

While producing slaked lime, this approach would allow for the concomitant capture of the process CO_2 through synthesis of Na_2CO_3. This process would avoid the direct combustion CO_2 emissions conventionally arising in $CaCO_3$ calcination, and the raw material CO_2 emissions through their precipitation into Na_2CO_3.

The widespread use of this anion exchange technology could be hindered by the indirect energy requirements for the main reactant NaOH, and the disproportion between the market demand for lime and soda ash. However, more recent work has shown that this technology can immediately be used to co-produce lime and soda ash and reduce global CO_2 emissions by 37 Mt per year in an economically feasible manner [59] with scope to provide all world demand for lime in the future depending on further process advancements, the increased usage of soda ash (for example in energy storage applications) or the economic safe storage of soda ash.

In short, this approach could drastically change the industrial reality for both the lime and soda ash state-of-the-art. However, industrial stakeholders in these sectors are generally reserved to radical change given the relatively long service life of existing capital equipment for lime and soda ash making (~ 50 years) as well as existing R&D or other related project investments. Nonetheless, this process has recently been taken up by industry.

26.9 Concluding Remarks: It's Lime Time

Lime is one of the most versatile, high performing, and cost-competitive bulk chemicals necessary for the built environment. It is successfully applied in a wide and diverse range of applications and generally plays a critical role in all of them. Its conventional production route, although polluting, is robust and well known; this is an advantage and a drawback at the same time, since it hinders the stakeholders to swiftly accept alternative routes for cleaner production. Market demand slows the full potential of applicability of novel production routes, which are nonetheless starting to emerge at a fast pace. Novel production routes described in this chapter are currently being tested and applied at industrial scale, portraying the steadfastness of the research community to achieve climate goals, and strengthened by global investment shifts.

References

1. A. Dowling, J. O'Dwyer, C. Adley, Lime in the limelight. J. Clean. Prod. 13–22 (2015)
2. J. Elsen, K. Van Balen, G. Mertens, Hydraulicity in historic lime mortars: a review, in *Historic Mortars: Characterisation, Assessment and Repair*, pp. 125–139 (2012)
3. M. Schneider, M. Romer, M. Tschudin, H. Bolio, Sustainable cement production—present and future. Cem. Concr. Res. **41**(7), 642–650 (2011)

4. B.R. Stanmore, P. Gilot, Calcination and carbonation of limestone during thermal cycling for CO_2 sequestration. Fuel Process. Technol. **86**(16), 1707–1743 (2005)
5. Oates, J., *Lime and Limestone: Chemistry and Technology, Production and Uses*. Wiley (2008)
6. The European Monitoring and Evaluation Programme (EMEP) & European Environment Agency (EEA), Manufacture of Cement, Lime and Plaster, Lime Production. Date accessed: 30.06.24 https://ww.eea.europa.eu/publications/emep.../2-a-2-lime-production., The EMEP & EEA Emission Inventory Guidebook 2013 (2013)
7. X. Li et al., Reactivity tests for supplementary cementitious materials: RILEM TC 267-TRM phase 1. Mater. Struct. **51**, 1–14 (2018)
8. Online Source: https://www.statista.com/statistics/1006040/production-of-lime-worldwide/#:~:text=In%202023%2C%20the%20world%27s%20total,at%20311%20million%20metric%20tons. Accessed 1 July 2024
9. Global Cement and Concrete Association, Online Source: https://gccassociation.org/key-facts/#:~:text=Manufacture,production%20is%204.1%20billion%20tonnes. Accessed 1 July 2024
10. D. Cheng et al., Projecting future carbon emissions from cement production in developing countries. Nature Commun. **14**(1) (2023)
11. H. Taylor, *Cement Chemistry*, vol. 2 (Thomas Telford, London 1997)
12. European Commission, Reference document on BAT in the cement and lime manufacturing industries. JRC Reference Report, E.-D.G. (2007)
13. T. Oates, *Lime and limestone* (Kirk-Othmer Encyclopedia of Chemical Technology, 2000)
14. A. Rahman et al., Recent development on the uses of alternative fuels in cement manufacturing process. Fuel **145**, 84–89 (2015)
15. L. Xuerun et al., Effect of SO_3 and MgO on Portland cement clinker: formation of clinker phases and alite polymorphism. Fuel Process. Technol. 182–192 (2014)
16. D.N. Huntzinger, T.D. Eatmon, A life-cycle assessment of Portland cement manufacturing: comparing the traditional process with alternqtive technologies. J. Clean. Prod. 668–675 (2009)
17. T. Hanein, F.P. Glasser, M.N. Bannerman, One-dimensional steady-state thermal model for rotary kilns used in the manufacture of cement. Adv. Appl. Ceram. **116**(4), 207–215 (2017)
18. F. Schorcht et al., Best available techniques (BAT) reference document for the production of cement, lime and magnesium oxide, in *European Commission Joint Research Centre Institute for Prospective Technological Studies* (2013)
19. J.W. Bullard et al., Mechanisms of cement hydration, in *Cement and concrete Research*, pp. 1208–1223 (2011)
20. M. Simoni, M.D. Wilkes, S. Brown, J.L. Provis, H. Kinoshita, T. Hanein, Decarbonising the lime industry: state-of-the-art. Renew. Sustain. Energy Rev. **168** (2022)
21. The European Steel Association, Online Source: https://www.eurofer.eu/publications/brochures-booklets-and-factsheets/european-steel-in-figures-2023/. Accessed 1 July 2024
22. J. Suer, M. Traverso, N. Jäger, Review of life cycle assessments for steel and environmental analysis of future steel production scenarios. Sustainability **14**, 14131 (2022)
23. L. Holappa, A general vision for reduction of energy consumption and CO_2 emissions from the steel industry. Metals **10**, 1117 (2020)
24. M. Simoni et al., Towards the circularity of the EU steel industry: modern technologies for the recycling of the dusts and recovery of resources. Metals **14**(2) (2024)
25. R. Remus et al., in *Best Available Techniques (BAT) Reference Document for Iron and Steel Production—Industrial Emissions Directive 2010/75/EU*. Integrated pollution prevention and control, Publications Office
26. The European Steel Association, The European steel in figures (2022). Online Source: https://www.eurofer.eu/as-sets/publications/brochures-booklets-and-factsheets/european-steel-in-figures-2022/European-Steel-in-Figures-2022-v2.pdf. Accessed on 22 Sept. Accessed 1 July 2022
27. J.L. Clifford, R.S. Boynton, Acid neutralization with lime—Bulletin No. 216. For environmental control and manufacturing processes, Online Source: https://www.lime.org/documents/publications/free_downloads/Acid_NeNeutralizati_with_Lime.pdf. Accessed 1 July 2024

28. F. Habashi, *Principles of Extractive Metallurgy* (Gordon and Breach, CRC Press, New York, 1969)
29. L.G. Baltazar, F. Henriques, F. Jorne, M.T. Cidade, Combined effect of superplasticizer, silica fume and temperature in the performance of natural hydraulic lime grouts. Constr. Build. Mater. **50**, 584–597 (2014)
30. S. Holmes, *Building with Lime: A Practical Introduction* (Intermediate Technology Publications, 1997)
31. V. Dodson, Pozzolans and the pozzolanic reaction, in *Concrete Admixtures* (1990), pp. 159–201
32. T. Hanein, A.G. De la Torre, Z. Zhang, J.L. Provis, Alternative non-Portland binders. Elem. Int. Mag. Mineral. Geochem. Petrol. **18**(5), 314–320 (2022)
33. C. Poon, L. Lam, S. Kou, Y. Wong, R. Wong, Rate of pozzolanic reaction of metakaolin in high-performance cement pastes, in *Cement and Concrete Research* (2021), pp. 1301–1306
34. European Lime Association, Online Source: https://eula.eu/application/construction-civil-engineering/. Accessed 12 July 2024
35. A.S. Negi, M. Faizan, S.P. Siddharth, R. Singh, Soil stabilization using lime. Int. J. Innov. Res. Sci., Eng. Technol. 448–453 (2013)
36. H. Houben et al., *Earth Construction: A Comprehensive Guide* (Theo schilderman, ITDG, Marseille, 1989)
37. H. Houben et al., *Earth Construction a Comprehensive Guide* (Intermediate Technology Publications, 1994)
38. P. Maniatidis et al. A review of rammed earth construction, in *Innovation Project "Developing Rammed Earth for UK Housing", Natural Building Technology Group, Department of Architecture & Civil Engineering, University of Bath*, 12 (2003)
39. Y. Luo, B. Yin, X. Peng, Y. Xu, L. Zhang, Wind-rain erosion of Fujian Tulou Hakka earth buildings. Sustain. Cities Soc. **50** (2019)
40. W. Grabow, I. Middendorff, N. Basson, Role of lime treatment in the removal of bacteria, enteric viruses, and coliphages in a wastewater reclamation plant. Appl. Environ. Microbiol. 663–669 (1978)
41. R. Renou, J.S. Thomas, E. Aoustin, M.N. Pons, Influence of impact assessment methods in wastewater treatment. J. Clean. Prod. 1098–1105 (2008)
42. M. Voldsund et al., Comparison of technologies for CO_2 capture from cement production—part 1: technical evaluation. Energies **12**(3), 559 (2019)
43. U.S. Environmental Protection Agency, Lime Production: Industry Profile (2000). Online Source: http://www.epa.gov/ttnecas1/regdata/IPs/Lime%20Manufacturing_IP.pdf. Accessed 1 July 2024
44. European Lime Association A pathway to negative CO_2 emissions by 2050, The contribution of the lime industry to a carbon-neutral Europe. Online Source: https://eula.eu/wp-content/uploads/2023/09/WEB_EULA-2030-Climate-Roadmap_Lime-Acts_A4_v19.pdf. Accessed 19 Dec 2024
45. European Lime Association, Agriculture, Forestry & Fish Farming. Online Source: https://eula.eu/application/agriculture-forestry-fish-farming/. Accessed 19 Dec 2024
46. J. Shen, P. Fatehi, P. Soleimani, Y. Ni, Recovery of lignocelluloses from pre-hydrolysis liquor in the lime kiln of kraft-based dissolving pulp production process by adsorption to lime mud. Bioresour. Technol. 10035–10039 (2011)
47. J. Kemperl, J. Maček, Precipitation of calcium carbonate from hydrated lime of variable reactivity, granulation and optical properties. Int. J. Miner. Process. 84–88 (2009
48. A. Boateng, in *Rotary Kilns: Transport Phenomena and Transport Processes* (Butterworth-Heinemann, 2015)
49. R. Schreiber, C. Russell, J. Evers, Evaluation of suitability of selective catalytic reduction and selective non-catalytic reduction for use in Portland cement industry, in *Portland Cement Association to the Ozone Transport Commission*
50. D. Leeson et al., A Techno-economic analysis and systematic review of carbon capture and storage (CCS) applied to the iron and steel, cement, oil refining and pulp and paper industries, as well as other high purity sources. Int. J. Greenh. Gas Control. 71–84 (2017)

51. M. Bui et al., Carbon capture and storage (CCS): the way forward. Energy Environ. Sci. **11**(5), 1062–1176 (2018)
52. D. Granados, F. Chejne, J. Mejía, Oxy-fuel combustion as an alternative for increasing lime production in rotary kilns. Appl. Energy 107–117 (2015)
53. A. Coppola et al., Performance of natural sorbents during calcium looping cycles: a comparison between fluidized bed and thermo-gravimetric tests. Energy Fuels **27**(10), 6048–6054 (2013)
54. H. Pawlak-Kruczek, M. Baranowski, Effectiveness of CO_2 capture by calcium looping with regenerated calcium sorbents–last step calcination. Energy Procedia 4499–4512 (2017)
55. L. Ellis, A. Badel, M. Chiang, R. Park, Y. Chiang, Toward electrochemical synthesis of cement—an electrolyzer-based process for decarbonating $CaCO_3$ while producing useful gas streams. Proc. Natl. Acad. Sci. **117**(23), 12584–12591 (2020)
56. T. Hanein, M. Simoni, C.L. Woo, J.L. Provis, H. Kinoshita, Decarbonisation of calcium carbonate at atmospheric temperatures and pressures, with simultaneous CO2 capture, through production of sodium carbonate. Energy Environ. Sci. 6595–6604, 2021.
57. Simoni, M., Hanein, T., Woo, C., L., Provis, J., H. Kinoshita, Effect of impurities on the decarbonisation of calcium carbonate using acqueous sodium hydroxide. ACS Sustain. Chem. 11913–11925 (2022)
58. M. Simoni et al., Decarbonisation of calcium carbonate in sodium hydroxide solutions under ambient conditions: effect of residence time and mixing rates. Phys. Chem. Chem. Phys. 16125–16138 (2022)
59. A. Baral, J.L. Galvez-Martos, T. Hanein, Realizing CO2 emission reduction in lime and soda ash manufacturing through anion exchange. Green Chem. (2025)

Chapter 27
Cross-Disciplinary Integration Towards Sustainable Construction

Dorian A. H. Hanaor

Abstract The built environment is the basis for human experience in the world, yet its current practices contribute significantly to global greenhouse gas emissions, resource depletion, and environmental degradation. Addressing the unsustainable practices of our contemporary construction practices requires the combined efforts of professionals across multiple fields. This chapter explores the critical role of cross-disciplinary integration in achieving sustainable construction, emphasising the need to bridge the gaps between material science, architecture, engineering, environmental studies, and social sciences. By examining past and present examples and trends we explore how holistic, systems-oriented approaches can drive innovation and sustainability. We examine how the siloed nature of modern disciplines has hindered progress, leading to fragmented outcomes and missed opportunities for circularity and resource efficiency. We present the case for a paradigm shift towards cross-disciplinarity, where stakeholders across sectors and disciplines collaborate to create cohesive solutions that balance technical and economic performance, environmental stewardship, and social equity. This chapter addresses the economic and regulatory barriers that impede the adoption of sustainable materials and practices, advocating for structural changes in education, industry, and policy and highlighting the tools that are now enhancing this integration of knowledge. Ultimately, this work underscores the transformative potential of cross-disciplinary integration in redefining the built environment, offering a roadmap for achieving sustainability through collaboration, innovation, and a shared commitment to planetary well-being.

D. A. H. Hanaor (✉)
Faculty of Science and Engineering, Southern Cross University, East Lismore, NSW 2480, Australia
e-mail: dorian@materialeyes.net
URL: https://www.materialeyes.net

Materialeyes, Materials Innovation Consulting, Berlin, Germany

D. A. H. Hanaor (ed.), *Sustainable Materials for the Built Environment*, Springer Series in Materials Science 352, https://doi.org/10.1007/978-3-031-97818-0_27

27.1 Introduction

During the Imperial period of Rome (27 B.C.—4th Century A.D) the production of building materials underwent tremendous advancement [1]. Through the well informed extraction and combination of suitable raw materials and their physical and thermal processing, the industrial and centralised production of bricks, glass and concrete allowed structural achievements in Rome itself, and throughout the empire, surpassing anything seen in comparable civilisations at the time [2]. Roman concrete is to this day an object of wonder and inspiration, enabling structures of unprecedented complexity and size. The processing—microstructure—performance relationships of roman concrete are to this day the subject of investigations, and findings have inspired a range of recent innovative ventures including among others DMAT [3] and Silica-X [4]. While perhaps lesser known, industrial roman brickmaking also allowed the expansion of efficiently constructed and well-built structures around the empire, using masonry units of consistent and standardised dimensions [5]. The centralised industrial production of materials for European built environments largely diminished with the decline of the Roman Empire.

Between the fall of the Roman Empire and the emergence of industrial societies in the early nineteenth century, vernacular construction prevailed, with construction of dwellings, places of worship, castles and fortifications being conducted as projects within communities. Typically, materials for construction were based on locally available resources. Wood, stone, lime-based cements and bricks were produced from regionally varying earth, rocks and biomass [6]. With the industrial revolution a paradigm shift occurred and the production of bricks, cement, timber elements, iron and later steel became the domain of centralised industrial production. Bricks in particular are typical of this change. In vernacular construction, hand-made loam based bricks were often used unfired, while fired clay bricks were produced locally in temporary clamp-kilns or simple intermittently fired bottle-shaped kilns [7]. Bricks produced in this manner are often inconsistent in their dimensions, and are not conducive to mass construction of structures. Unlike roman bricks and later industrial bricks that dominate the architectural landscape of much of Victorian London.

The shift to industrial production of materials in general, and building materials in particular, like many other trends of the nineteenth century, gave rise to increasingly narrowly defined disciplines within the realm of engineering and materials science and a diminishment of the role of local trades and craftspeople in shaping the material aspects of built environments. The producers of construction materials became increasingly disconnected from the broader contexts of the sourcing of raw materials and their implementation in built environments, which were often distant from the location of production. Standardised products have long since become the norm, and this has naturally provided advantages for architects and engineers seeking to design structures based on known quantities of material performance. However the disconnect between environmental custodians, materials producers and construction professionals has made it difficult to improve lifecycle outcomes and human experience in the production and use of materials.

A stark example of the emergence of new disciplines and the resulting disconnect from vernacular construction can be found in the field of cement chemistry. In the most typical production of cement clinker, limestone and clay rich minerals are milled and fired at very high temperatures in clinker kiln. The resulting carbon emissions, at least 60% of which arise due to the calcination of limestone, are a major obstacle towards sustainability of our built environments worldwide. Using a multitude of honed analytical tools such as the Bogue calculation, the Lea and Parker equations and numerous others, the cement chemist knows how to tailor raw materials and processing towards the achievement of desired microstructures, phase assemblages and cement performance [8–10]. However, the isolation of the discipline from geologists, environmentalists and environmentally-conscious architects and stonemasons means that the use of quarried limestone directly as masonry material is not considered by cement chemists and does not feature widely in the modern construction landscape, despite the improved sustainability outcomes offered by the use of quarried stone and the aesthetics and value experienced by inhabitants of stone dwellings [11–13].

Addressing climate change and biodiversity loss is emerging as a rallying point for human activity, as the awareness of the gravity of these challenges grows. Even with complete elimination of carbon emissions, it has become clear that carbon dioxide removal (CDR) is further essential to avoid catastrophic environmental change. To effectively align human activities with our environmental objectives, disciplines need to operate in synergy with one another to develop a systems thinking oriented approach to achieving human development goals in a sustainable manner [14].

Systems thinking enables a holistic understanding of the many interconnected factors influencing sustainability in the built environment. By bringing together the diverse stakeholders in this sector and viewing the built environment as a dynamic system, we can get a better view of how materials choices impact resource extraction, manufacturing, energy consumption, structural performance and lifecycle impacts [15, 16].

27.2 Disciplinary Silos

The breadth of our economy means that much of the economic activity that takes place occurs in insulated silos. More specifically, producers of goods and suppliers of services are less aware in general of the broader context in which their economic activity takes place and are primarily occupied with optimising performance within in a limited field of focus. In the analysis of sustainable economics, the integration of disciplines and bridging of scales have been determined as key factors in delivering outcomes relating to industrial symbiosis, and in particular circularity, as well as enhancing the effectiveness of climate action and sustainable development [17–19]. Methodological and technological tools, including IoT, big data, game theory, network modelling and numerous others have been developed to expand system boundaries, bring together numerous input and output metrics from diverse stakeholders and provide disciplinary-integrating insights towards improved holistic industrial and environmental outcomes in many sectors of industry [19–21]. The study of industrial ecologies as complex systems exhibiting emergent properties underpins many lifecycle assessments and material flow analyses [22, 23].

Expanding cross disciplinary integration in sustainable construction to extend beyond metrics of Greenhouse gas (GHG) emissions and create truly inclusive decision making processes, is rendered complex by the diversity of materials performance indicators, the complex environmental and social contexts in which the value chain of construction industry takes place and the consequent subjectivity in defining an effective functional unit in LCAs relating to materials and built environments [24–27]. The fuzzy nature of many metrics, and the difficulty in weighting human experience and social impacts has motivated the adoption of concepts of the built environment as social-ecological systems, [28–30]. These studies highlight the need to integrate cultural and natural elements across spatial scales from the national to the local scale to achieve improved social-LCA outcomes and identify the intricacies and assumptions that need to be considered when addressing a people-focused approach to sustainability and the built environment.

The more complex social intricacies of built environments notwithstanding, even within the confines of the industrial realm, the optimisation of processing of materials towards use in the built environment is also carried out within silo-constrained closed loops, sensitive to performance and cost-effectiveness measured by discipline-specific metrics. The lack of transparency between supply chains in the construction industry and the absence of systems thinking are seen as barriers to disciplinary integration and efficient outcomes in the construction materials sector [15, 31]. As with other industries, expanding the scope of symbiosis is achieved by developing holistic understandings of interconnected industries and identifying opportunities for circularity and process optimisation by integration across primary and secondary industries. For example, greater opportunities for industrial symbiosis and resulting circularity exist when processes in metallurgy and cement production are viewed holistically as opposed to separate loops of production and consumption. In particular tailings, metallurgical slags and overburdens from mining and metallurgy offer a rich menu of ingredients to produce low-carbon cements, Supplementary Cementing Materials (SCMs) and alternative aggregates [32–36].

Taking a broader view further inclusivity can be achieved by integration with silos beyond primary and secondary industries, extending to the disciplines such as architecture and engineering that drive the use of materials in structures, and community planners who help shape the role of structures in built environments, and further afield to other stakeholders influencing the interface between humans and the built world. Some disciplinary silos are illustrated in Fig. 27.1, showing the stakeholders and fields of focus within these silos. Naturally one can define silos with a higher degree of granularity and around the world a certain degree of integration is generally present between actors in these silos, which are naturally not totally isolated from one another.

Fig. 27.1 Disciplinary silos, showing fields of focus and stakeholders within a set of definable disciplines related to the contemporary built environment

Architecture in particular serves a prominent role in acting as a nexus point for disciplinary integration and indeed interdisciplinary design features in architectural education [37, 38]. It is in the realm of architecture where the technical world is brought into harmony with human-experience at the structure-level. Waste management too sits at a strategically advantageous point, being a key enabler of the integration of industries and the built environment. For example, the use of waste products from agriculture, food processing and metallurgy as components in construction materials including concrete presents inspiring examples of industrial symbiosis that can be achieved when diverse silos are brought together. [39, 40]. Also shown in Fig. 27.1, Venture Capital, despite exhibiting its own finance-centric silo-mentality, also serves as a nexus point and acts as a conduit for economic catalysis of collaborative ventures, provided these produce a realisable financial advantage.

As mentioned, acknowledging the interconnectedness and complex-system nature of production and consumption of resources in the materials sector is a key step towards improving outcomes in the built environment and its impact on spheres of the environment, which themselves are highly interconnected. A systems thinking approach to building materials supply is accepted is being important towards robust and sustainable built environment value chains [41, 42]. A systems thinking approach to building materials supply and use recognises that every stage—from extraction and production to transportation and disposal—is interconnected, and therefore decisions made at one point ripple through the entire value chain. This perspective, built upon the interconnection of the silos shown in Fig. 27.1, allows a better understanding of the complex trade-offs involved in material selection, design, and construction processes.

Transcending the disciplinary silos and enabling systems thinking is also valuable towards addressing other twenty-first century challenges relating to the built environment in addition to sustainability. Thus for example heritage conservation, mass migration, conflict and disaster response may pose together conflated challenges towards the design of sustainable built environments that necessitate integrative systems thinking [43–46]. Moreover, the traditional separation of disciplines often fails to account for cultural, historical, and geographical contexts that influence material use. For example, vernacular construction practices, which often embodied deep local knowledge about material sourcing and environmental adaptation, have been largely sidelined in favour of standardised industrial materials. This disconnect limits opportunities to integrate traditional knowledge with modern advancements to achieve both sustainability and cultural relevance.

To break free from the confinement of these disciplinary silos and address the complex challenges in the construction material ecosystem, there is a pressing need for continued improvement in cross-disciplinary collaboration and systems thinking. As we have seen, the interdependencies between material production, design, policy, and environmental impact demand integrated solutions. Research has shown that siloed approaches often lead to fragmented outcomes, whereas cross-disciplinary collaboration frameworks and competencies in academia and industry foster effective innovation and sustainability [47, 48]

By fostering dialogue and partnerships across the spectrum of stakeholders in the construction materials ecosystem, including raw material suppliers, manufacturers, architects, urban planners, policymakers, and environmental advocates, we can create a shared vision for sustainable innovation. This requires not only structural changes within organisations and academic institutions but also the cultivation of a mindset that values interdisciplinary learning and co-creation. Educational programs, research initiatives, and industry practices must prioritise holistic thinking, encouraging professionals to understand the downstream and upstream implications of their decisions. This mentality is gaining increasing traction in educational domains, where programs that integrate engineering, environmental science, and social sciences have been shown to better prepare students for the complexities of sustainable development [49, 50]. However, when it comes to the industrial and governmental implementation of integrated strategies for the production and usage of sustainable construction materials, significant barriers still remain.

27.3 Cross Disciplinary Integration

The concept of bringing multiple disciplines together in the context of engineering challenges is well known and is a key feature of Systems Thinking and contemporary design philosophies. The need to bring multiple disciplines together in the identification and implementation of sustainable materials and practices is now well recognised towards various objectives, including materials circularity [51], heritage management [52], architectural design [53], energy efficiency [54] and many others. To a certain extent, nearly all projects and studies involve some integration of multiple disciplines and therefore it is important to highlight the uniqueness of the built environment in the demands it places on the breadth and depth of the fields that need to be brought under one roof. As the economic activities associated with construction and the usage of structures account for around 40% of global greenhouse gas emissions [55] and are a main driver of resource usage [56], the need to expand the integration of disciplines is particularly pressing in this sector.

In academic discourse, distinctions are often made between multi-disciplinarity, inter-disciplinarity, and trans-disciplinarity. Multi-disciplinary approaches involve parallel efforts by experts from different fields, while inter-disciplinary collaboration seeks to blend insights from these fields to address shared challenges. Trans-disciplinarity, however, goes a step further, creating a unified framework that transcends traditional disciplinary boundaries to foster co-creation and innovation. In the context of construction technologies, cross-disciplinary integration aligns closely with the principles of trans-disciplinarity, emphasising the synthesis of knowledge rather than its parallel application [57]. This distinction highlights the shift from merely involving multiple disciplines to truly integrating them into a cohesive, systems-oriented approach [58].

27.4 Applied Disciplinary Integration in the Built Environment

Cross-disciplinary integration has historically been a ubiquitous aspect of construction endeavours, evident even in early human societies. However, the degree to which this integration influences the nature and outcomes of construction activities has varied significantly over time. In today's construction landscape, characterised by siloed industry practices, fragmented planning processes, misaligned economic incentives, and the predominance of consumer-centric paradigms, the need for enhanced cross-disciplinary collaboration is increasingly apparent. Addressing these systemic challenges requires fostering integration across technical, economic, and operational domains to unlock efficiencies, drive innovation, and achieve sustainability objectives. Examples from our past, present and future can be found, highlighting the advantages that cross-disciplinary integration can bring to the realm of sustainable construction and inspiring the expansion of this integration.

Pullman, a region in Chicago built rapidly as a centrally planned industrial community, provides excellent examples for the integration across multiple disciplines towards industrial symbiosis and effective outcomes in the built environment [59–61]. Initiated by George Pullman in the nineteenth century, the construction of this community featured the integration of materials engineering, town planning, energy systems, residential construction and manufacturing to name but a few. The Clay used as the raw materials for the high quality bricks used to construct Pullman was sourced from the waste from the dredging of nearby lake Calumet, done to facilitate the transport of iron ore to the nearby steel mills, an early example of circular economy, while the factory operations were used to provide heat to the adjacent planned housing developments and waste water from residential communities was utilised as fertiliser in agricultural farms located near the brickyards. The efficiency implemented in the design, construction and operation of this community was made possible through cross-disciplinary integration, enabled by the centralised planning of this project.

Similar centrally planned projects including Masdar City in Abu Dhabi and Songdo Business District in Korea also are distinguished by an integration of fields to promote sustainable urban environments, with building sustainability metrics being certified in both cases by according to the LEED rating system (Leadership in Energy and Environmental Design), corresponding to aspects of lifecycle impacts in construction, operation and maintenance (Fig. 27.2).

Holistic approaches, often enabled by centralised planning, reduce barriers to systems thinking and foster disciplinary integration. Centralised planning allows for the alignment of diverse stakeholders, the harmonisation of goals, and the implementation of large-scale, innovative solutions that might otherwise be fragmented or constrained by competing interests. However, these barriers are harder to overcome in smaller-scale development projects, where the operational environment often prioritises cost-efficiency, risk mitigation, and adherence to established practices. This

Fig. 27.2 Centrally planned urban and urban/industrial developments. (left) photograph of Masdar City, in Abu Dhabi, UAE and (right) A watercolour painting of Pullman, Illinois, USA. Two examples of centralised planning across multiple disciplines enabling sustainable use of materials. Images available for use under the CC BY 4.0 license [62, 63]

can lead to more conservative decision-making in the selection and implementation of construction materials and methods, limiting the potential for transformative innovation.

Despite these challenges, several pioneering projects and initiatives have demonstrated how holistic, systems-oriented approaches can be successfully applied across different scales and contexts. These examples illustrate the potential for integrating sustainability, innovation, and interdisciplinary collaboration into the built environment:

BedZED (Beddington Zero Energy Development), a residential development led by the Peabody trust and completed in 2002 in Sutton, south London integrates energy efficiency, renewable energy, and sustainable materials into a cohesive urban community that aims to offer all-round sustainability. By integrating professionals from a diverse range of fields the project stands out in enabling the use of local materials (over half materials were sourced within 50km of the project), and reclaimed materials (reportedly 15% of the total used) [64, 65]. The integrative multi-level planning of this development enable also sustainable lifecycle outcomes in terms of energy usage, housing affordability, water consumption, transport needs and community experiences [66, 67]. The success of this development demonstrates the feasibility of reconciling creative and ambitious architecture with excellent social and environmental outcomes through cross-disciplinary integration. The total project cost including research and development and cost overruns, was £15 Million. As this project includes 100 homes and ~ 1000 m^2 of office space, the project demonstrates that sustainable outcomes are economically scalable [68] (Fig. 27.3).

The Edge in Amsterdam: designed by PLP Architecture and completed in 2014, the Edge is a commercial building that has been ranked by BREEAM as one the most efficient and sustainable commercial buildings worldwide [70]. In this development he integration of advanced smart materials, recycled materials, smart sensors and

Fig. 27.3 BedZed (Beddington Zero Energy Development) Eco Village in the southern outskirts of London, England. Image available for use under the CC BY 4.0 license [69]

systems for energy generation and storage [71]. Through a sophisticated building management system (BMS) that draws on human feedback to optimise energy usage, and the installation of extensive renewable energy generation on the roof and façade, the building is able to generate more energy than it is consumed, with onsite storage further enhancing the utility of the energy generated. The demonstrated sustainability outcomes of this development show that by addressing challenges at multiple levels and integrating feedback from both humans and sensors, excellent outcomes can be achieved. This project is seen as a source of inspiration for office buildings of the future.

Circularity Networks, such as the Ellen MacArthur Foundation stands as a prominent example within a growing global network of organisations dedicated to accelerating the transition to a circular economy. Circularity networks such as this catalyse the transparency, collaboration, and systems thinking that are essential for achieving effective circularity across industries and the built environment. They act as conveners, knowledge hubs, and advocates, driving change at policy making, industry and consumer levels. In particular the Ellen MacArthur Foundation and other circularity networks work with cities and municipal waste handlers to implement circular economy strategies in the built environment, including promoting the reuse of building materials, modular construction and creating closed-loop systems for construction waste [72–74]. The New European Bauhaus (NEB) is a similar network that engages primarily on the community level and serves the role of bringing together stakeholders to reduce the barriers to industrial symbiosis and social, cultural and environmental sustainability outcomes.

Whether at small or large scales, it is clear that one of the main roles to be played by catalysts of cross-disciplinary integration in the built environment is lowering the barriers to communication and cooperation between disciplinary silos as well as addressing the contextual constraints imposed by the organisation of human activity—notably economic and regulatory.

27.5 Economic and Regulatory Constraints

The tendency for professionals across the various disciplinary silos illustrated in the preceding section to focus on their particular set of stakeholders, clients, and challenges stems largely from the structural and systemic pressures inherent in the economic and regulatory landscape. These constraints often incentivise short-term thinking, discipline-specific optimisation, and resistance to innovation—all of which can act as significant barriers to achieving holistic sustainability goals in the built environment.

27.5.1 Economic Pressures and Market Dynamics

In many cases, the economic incentives driving decisions in the built environment prioritise cost minimization and return on investment over long-term sustainability. Developers, for instance, often operate within tight budgets and timelines, which may lead to the selection of less expensive but environmentally detrimental materials. Similarly, materials producers may prioritise volume and cost efficiency over developing and scaling innovative, sustainable alternatives, given the high upfront costs and uncertain demand for such products. These economic pressures create a fragmented value chain where each stakeholder seeks to optimise for their immediate economic priorities rather than broader systemic goals.

The dominance of cost-driven decision-making presents a significant economic constraint to the adoption of sustainable materials in construction. This isn't simply a matter of clients aiming to save money; it's a complex interplay of market dynamics, risk aversion, and entrenched practices. Developers, facing pressure to maximise returns, may seek to minimise upfront costs, to maintain competitiveness in the market. This short-term focus may result in the selection of widely available materials with poor sustainability outcomes, and tends to overshadow the long-term benefits of sustainable materials, for which supply chains may not be as well-established. The widespread use of steel and OPC based concrete is particularly exemplary of this; despite the multitude of alternative materials with better LCA outcomes, including Cross Laminated Timber, Bamboo [75], Hempcrete [76] and Geopolymer concrete [77], steel and OPC based concrete prevail even in low-rise construction in many construction markets around the world. This is partly a result of the highly developed understandings of the processing and application of these materials. Metallurgy and cement chemistry have been intensively studied and developed through the lens of science and engineering for well over a century now. Within the confines of our industrial supply chain, research has shown that even minor adjustments to the selection process for materials, and giving further consideration to the application specific demands of the materials, can result in significant reduction in embodied carbon, without significantly higher costs [78, 79]. Many of the poor practices in selecting and applying materials in the built environment can be attributed to an entanglement of education, labour experience and material availability (lead times). A feedback loop emerges in the market dynamics of construction sectors, whereby siloed practices reinforce one another, and create inertia in the technological landscape, which becomes resistant to the adoption of sustainable materials and practices.

This cost-centric approach is deeply ingrained in the fragmented structure of the construction industry. Multiple stakeholders including architects, engineers, contractors, and subcontractors operate under separate contracts with often conflicting incentives. The aforementioned siloed approach hinders collaboration and innovation. For instance, a contractor might be incentivised to minimise material costs within their specific scope, even if a slightly more expensive, sustainable material would lead to significant long-term savings for the building owner. The lack of integrated project delivery models and lifecycle cost analysis further exacerbates this issue. There's often no mechanism to capture and reward the long-term benefits of sustainable materials within the existing contractual frameworks, resulting in a so-called "Green Premium".

Supply chains for sustainable materials, while growing, often lack the aforementioned scale and maturity of those for conventional materials like concrete and steel. As long as sustainable alternatives remain sidelined it can be difficult to overcome barriers of longer lead times, and concerns about consistent availability. For example, while the demand for engineered wood products (EWPs) and the technology and understandings required for producing highly functional structural components from materials like Cross-laminated Timber (CLT) are developing, the relatively limited number of production facilities can create bottlenecks and price volatility [80–82]. This market immaturity creates a similar feedback loop in market dynamics, similar to the proverbial "chicken and egg" problem: greater demand is needed to drive economies of scale and reduce costs, but widespread adoption is hindered by current price premiums. It remains to be seen how the changing of economic, social and ethical incentives will alter these feedback loops. In particular scope-3 emissions reporting requirements, a rising carbon offset costs and more stable emissions trading markets are poised to tip the scales in favour of market growth in the realm of sustainable materials for the built environment.

Motivated by the prospect of future profitability through emissions reduction incentives, venture capital (VC) investments are increasingly driving the development and growth of sustainable materials ventures and their associated innovations. However, as we have seen above, venture capital flows are themselves subjected to silo-confinement, being obliged to deliver short-term returns on investments (ROI) in terms of portfolio growth, revenue and successful exits, thus neglecting integrative ventures that add community value. Moreover, while defined investment sectors of "Green-Tech", "ConTech", and similar labels have shown growing interest, investment in the sustainable construction sector is still relatively small compared to other areas of technology, with much of the venture capital entering this sector coming from venture arms of large corporate entities in the construction materials sector, such as Cemex Ventures, one of the leading investors in construction material startups. This can be attributed to barriers of higher capital expenditure and longer development times that sustainable construction ventures and other "hard-tech" enterprises face, when compared to those in sectors such as data analytics, software, or informatics.

As we have seen, due to feedback loops in market dynamics, sustainable material startups face difficulties scaling up production and competing with established materials producers. However, there are numerous focal points of growth and positive developments in the world of materials-oriented construction technologies. Motivated by integrative projects and networks that emphasise sustainability and systems thinking, such as those we have seen in the preceding section, materials innovators have been inspired to create various products and processes that produce more favourable sustainability outcomes, as indexed by schemes such as LEED, BREEAM, C2C and other sustainability certification schemes. There are several prominent trends in recent materials innovation aimed at improved sustainability metrics that are targeted or followed by startups and their VC backers. Some of these trends includes:

(1) **Biobased polymers**: the derivation of polymers such as polyurethane, polyhydroxyalkanoates (PHAs), polylactic acid (PLA) from waste biomass, or through the metabolism of micro-organisms, provides a pathway for the production of materials analogous or identical to those derived from fossil-hydrocarbons, with improved carbon-balance. By addressing particular niche applications of these materials, startups are able to gain market entry, through the provision of focused solutions in sustainability conscious construction projects. An illustrative example of this trend can be found in Klima-Pur windows, produced by Indresmat© in Spain. This is a bio-based polyurethane foam window frame with good recycling potential that provides high performance in thermal insulation, fostering energy-efficient buildings while providing a carbon-sink, as the carbon in the material come from the atmosphere carbon rather than fossil carbon [83, 84].

(2) **Composites based on natural materials**: Novel engineered structural materials based on the adaptation of geomaterials or bio-based lignocellulosic materials incorporated into various composite systems provide a pathway towards high performance with reduced environmental impacts. Examples include the use of cut stone laminated with carbon fibres, as demonstrated by Pierre-Carbone of Technocarbon and the use of high strength natural fibres (e.g. flax, jute or hemp) in composites for structural components. Hempcrete, a widely studied lime-based concrete reinforced with hemp fibres, and EWPs are also part of this trend, whereby the natural attributes of wood are harnessed in ways that are conducive to application-optimised mechanical performance. EWPs are also combined with other natural materials to create hybrid composites with improved sustainability and mechanical performance.

(3) **Robotic construction and 3D printing of materials**: The use of mechatronic approaches for production of structures is drawing increasing interest. A large part of this interest comes from the motivation of VC funds and innovators to apply technological solutions to housing shortages, rather than vernacular or community based solutions. The attractiveness of mechatronic approaches

stems from economic factors, namely the possible reduction in labour costs and delivery times, as well as efficiency, as material wastage is reduced. Robotic fabrication and 3D printing enable precision and speed, facilitating the creation of intricate geometries that would be difficult to achieve with traditional methods. Ventures in this field seek to integrate innovations into the equipment, formulations and incorporation of waste feedstocks.

(4) **Advanced nanomaterials**: Despite the seemingly stark contrast between the costs of nanoparticles and conventional structural materials, the integration of advanced nanoparticles in concrete has been demonstrated as a pathway towards synergetic outcomes. Of particular note is the use of graphene, or more accurately graphene oxide, as an additive in concrete [85–89]. By catalysing hydration, nucleating high density calcium silicate hydrates and by bridging cracks, small levels of these additives can produce dramatic improvements in the performance of the cementing phase in concrete, enabling the use of lower amounts of cement and providing a pathway to lower embodied emissions in the material. In general, the integration of advancements in nanotechnology in materials for the built environment is an area of high growth with numerous ventures active in this sector.

(5) **Biotechnology**: Around the globe, biotechnology ventures have seen rapid growth in recent years. Towards sustainable materials for the built environment, the integration of disciplines of molecular biology and genetic engineering has allowed the harnessing of various microscopic organisms including yeast, plankton, algae and fungi towards the production of improved materials based on mycelium, bio-based polymers, and biocementation of minerals.

(6) **Green concrete**: The realisation of opportunities for industrial symbiosis underpins many emerging enterprises in the field of sustainable construction materials. Bringing the right waste streams and processes into place to facilitate high performance concrete with better sustainability outcomes has enabled a plethora of ventures broadly categorised as Green Concrete. This includes cements and SCMs based on slags, fly ash and other industrial by-products.

The representative trends described above are by no means an exhaustive list of the numerous fields of innovation supported by capital investments. It can be seen that these trends are all in some way examples of cross-disciplinary integration, bridging fields of biotechnology, mineralogy, agriculture and many more. Despite the numerous fields of innovation, there are some significant limitations to the ability of VC funds to promote sustainable built environments. VCs often seek rapid returns on investment, which can be challenging in the construction industry where project cycles are long and adoption of new materials can be slow. Moreover, the profitability and high expected internal rates of return (IRRs) that are required for recipient ventures excludes enterprises seeking to develop holistic solutions that do not have pathways to high revenues. Thus for example earth and stone construction, vernacular building techniques, regenerative agriculture and other approaches that are not conducive to patenting, or establishment of exclusive supply chains may not gain momentum under such funding models.

Venture capital and corporate venture capital (CVC) clearly play a critical role in funding innovative and sustainable approaches in construction materials, enabling agile responses to emissions regulations and growing demand for eco-friendly alternatives. However, the profit-driven nature of these ecosystems may hinder cross-disciplinary integration. Venture capital prioritises investments that promise high returns and market exclusivity, which can limit the scope of integration of disciplines. As a result, cross-disciplinary integration is typically fostered by investors only when it offers a unique, proprietary advantage that ensures competitive differentiation. This focus on exclusivity and economic efficiency can inadvertently stifle broader, more inclusive innovation, as projects that require open collaboration or lack immediate commercial potential may struggle to secure funding. Consequently, while VC and CVC ecosystems drive progress in sustainable construction materials, their inherent constraints can hinder the full potential of cross-disciplinary solutions needed to address complex global challenges.

Economic and financial constraints to cross disciplinary integration are further compounded by regulatory frameworks that may lack the flexibility to support approaches that span multiple disciplines or industries.

27.5.2 Regulatory Frameworks and Their Limitations

Regulatory frameworks in the built environment are traditionally developed within specific disciplinary boundaries, focusing on core objectives such as safety, structural integrity, and basic environmental compliance. While these regulations are needed to ensure baseline performance, they often lag behind the pace of innovation and fail to incentivise cross-disciplinary collaboration, which is critical for advancing sustainable material use. For example, while building codes frequently mandate minimum energy efficiency standards, these may not sufficiently reward materials or processes that exceed these standards [90, 91].

Regulatory environments tend to be rigid, lacking adaptability and often prioritise compliance over innovation or excellence in sustainability. This is rarely conducive to the exploration of disciplinary integration towards outstanding sustainability outcomes. Building codes and standards are generally slow to incorporate advancements in sustainable materials, such as bio-based composites, mycelium derived materials or other innovations in construction materials. This can create uncertainty for architects, engineers, and developers who may be hesitant to adopt these materials due to concerns about compliance, delays or liability [92, 93]. Additionally, the absence of unified global protocols for measuring and certifying environmental performance, such as embodied carbon or circularity potential, further complicates efforts to align regulatory frameworks with sustainability goals. Without clear guidelines, professionals are often left to navigate a patchwork of local, national, and international regulations, which can stifle innovation and collaboration across silo and national borders.

The regulatory framework may pose particular obstacles with regard to the use of natural materials, which often exhibit variable material properties. In particular, the use of earthen construction techniques, such as rammed earth, adobe, and cob, can face significant hurdles due to building code compliance requirements [94–96]. These materials, while offering pathways to high performance in terms of hygroscopic temperature regulation, thermal mass, and low embodied carbon [97–102], may face challenges meeting the standardised performance metrics required by building codes and guidelines. This disconnect between the inherent benefits of natural materials and rigid regulatory standards can stifle innovation and limit their adoption, despite their potential to contribute to sustainable, resilient, and energy-efficient built environments. Addressing these challenges will require updated regulatory frameworks that account for the unique properties and performance of natural materials, fostering their integration into mainstream construction practices.

Gradually, regulatory frameworks are beginning to encourage cross-disciplinary integration. For instance, the European Union's Circular Economy Action Plan from 2020 includes measures to promote the use of recycled materials in construction and mandates building deconstruction over demolition, fostering collaboration between material scientists, designers, and contractors [103].

27.6 Tools for Disciplinary Integration

Many modelling and analytical tools developed for design in the built environment embody some measure of cross-disciplinary integration; Finite element analysis (FEA) can simulate the structural and thermal performance in materials, integrating multi-physics understandings in the context of architecture and engineering [104]. Importantly, for the analysis of lifecycle impacts of applied materials in the built environment, LCA tools including software like GaBi and OpenLCA help bring together inputs from the silos of diverse stakeholders and combine them to produce actionable insights into the environmental impacts and sustainability outcomes [105, 106]. A new emerging tool to enhance the integration of industries and disciplines are dynamic LCAs, and real-time LCAs, that take into account the variation of lifecycle inventories over time, and can be updated to reflect changes across various nodes in the supply chain.

Emerging digital tools, including neural networks and artificial intelligence (AI), are revolutionizing cross-disciplinary integration in the built environment by enabling seamless collaboration and data-driven decision-making. AI-powered platforms, such as generative design tools, allow architects, engineers, and material scientists to explore countless design iterations optimised for sustainability, performance, and cost. These tools leverage machine learning to analyse complex datasets, identifying patterns and solutions that might be overlooked in traditional workflows [107, 108]. For example, AI can optimise material selection by balancing factors like embodied carbon, structural performance, and availability, fostering collaboration between disciplines. Moreover AI tools are fostering cross-disciplinary collaboration by accelerating research and knowledge acquisition by individuals, enabling the rapid exploration of many dimensions of engineering challenges, empowering architects, engineers, and material scientists to holistically evaluate sustainable material design, integration, and performance across the entire lifecycle of built environments.

A particularly important tool facilitating the integration of disciplines is Building Information Modeling (BIM), which provides a shareable digital representation, often in a format compatible across multiple platforms that enables stakeholders to collaborate in real time to improve outcomes regarding the construction and use of structures. BIMs bring together the various elements of a building and its materials in a standardised format that can be used to optimise design and use of materials and structures throughout the lifetime of the structure. Construction, Energy use, maintenance, utilities, renovation and other aspects of a building can be included in the information of BIMs. In recent years, BIMs are becoming a mandatory requirement for projects over a certain size [109]. Integrating BIMs with artificial intelligence and Internet of things (IoT) offers pathways to create integrative outcomes in construction and urban development [110, 111].

27.7 Cross-Disciplinary Integration and Systems Thinking in the Context of Vernacular Construction

In the context of vernacular construction, cross-disciplinary integration is playing a growing role. In earth based construction, disciplines of granular physics, mineralogy, agriculture, architecture, materials science and microbiology are brought together through the manual labour of craftsmen to form sustainable vernacular structures [94]. The use of local materials in construction, including earth, stone, wood and waste biomass, requires the integration of insights from local experts as well as discipline professionals. Improving and expanding the field of vernacular construction may benefit from further platforms and channels for discourse and collaboration.

3D printing and digitisation have been applied to earthen construction, posing an inspiring example of cross-disciplinary integration [112–115]. By combining advanced fabrication technologies with traditional earth-based techniques, researchers and practitioners are creating innovative solutions that enhance structural performance, design flexibility, and scalability. For instance, 3D-printed earthen structures leverage computational design to optimise material use and structural integrity, while maintaining the sustainability and cultural relevance of vernacular methods. This fusion of ancient wisdom and modern technology not only preserves the ecological benefits of natural materials but also opens new avenues for affordable, resilient, and aesthetically rich construction. Such advancements underscore the transformative potential of cross-disciplinary collaboration in redefining sustainable building practices for the future.

27.8 Conclusions and Outlook

As we have seen, the integration of diverse disciplines is essential for advancing sustainable construction practices addressing the complex challenges of resource efficiency, environmental impact, and human well-being. Sustainable construction requires the collaboration of architects, engineers, material scientists, environmentalists, educators, and policymakers to ensure that every stage of the building lifecycle—from material extraction and production to construction, operation, and end-of-life disposal—is optimised for sustainability. By breaking down the silos that traditionally separate these fields, cross-disciplinary integration can foster innovative solutions that balance technical performance, environmental stewardship, and social equity. The incorporation of LCA tools into design processes allows architects and engineers to evaluate the environmental impacts of material choices, while input from environmental scientists can guide the selection of low-carbon alternatives. This collaborative approach may reduce the carbon footprint of buildings and further enhance their resilience and adaptability to changing environmental conditions.

Towards sustainable outcomes the integration of disciplines must extend beyond technical fields to include social sciences, economics, and community engagement, ensuring that construction practices are culturally relevant and socially inclusive. For example, vernacular construction techniques, which have historically relied on local materials and traditional knowledge, can be revitalised through the integration of modern technologies such as 3D printing and digital fabrication. This fusion of ancient wisdom and contemporary innovation not only preserves cultural heritage but also promotes the use of sustainable, locally sourced materials. By fostering systems thinking and shared visions among diverse stakeholders, discipline integration can drive the transition towards a more sustainable and equitable built environment, where innovation and tradition coexist to meet the needs of present and future generations.

References

1. M. Fulford, J.-P. Adam, *Roman Building: Materials and Techniques* (Routledge, 2005)
2. M.H. Strickland, *Clemson University* (2010)
3. L.M. Seymour, J. Maragh, P. Sabatini, M. Di Tommaso, J.C. Weaver, A. Masic, Sci. Adv. **9**, eadd1602 (2023)
4. The Secret to Making Concrete That Lasts 1000 Years, https://www.wired.com/story/secret-roman-self-repairing-concrete/ Accessed 20 Jan 2025
5. E.M. Pérez-Monserrat, M.-A. Causarano, L. Maritan, A. Chavarria, G.P. Brogiolo, G. Cultrone, J. Cult. Herit. **54**, 12 (2022)
6. R. Langenbach, in *Nonconventional and Vernacular Construction Materials* (Elsevier, 2020), pp. 1
7. H. Valdes, J. Vilches, G. Felmer, M. Hurtado, J. Figueroa, Sustainability **12**, 7724 (2020)
8. H. Taylor (Thomas Telford, 1997)
9. R.H. Bogue, *Some Recent Advances in Our Understanding of the Chemistry of Portland Cement* (Portland Cement Association Fellowship, 1947)
10. H.A. Mboya, C.K. King'ondu, K.N. Njau, A.L. Mrema, Adv. Civ. Eng. **2017**, 6952645 (2017)

11. D. Dernie, *New Stone Architecture* (Laurence King Publishing, 2003)
12. D. Ioannidou, S. Zerbi, G. Habert, Build. Environ. **82**, 628 (2014)
13. J. Kerr, S. Rayburg, M. Neave, J. Rodwell, Sustainability **14**, 9019 (2022)
14. P. Healey, T. Kruger, J. Lezaun, Front. Clim. **5**, 1293650 (2024)
15. E. Iacovidou, J.N. Hahladakis, P. Purnell, Environ. Sci. Pollut. Res. **28**, 24785 (2021)
16. A.M. Bassi, M. Bianchi, M. Guzzetti, G. Pallaske, C. Tapia, Sustain. Prod. Consum. **27**, 128 (2021)
17. A. Nikas et al., Sustain. Prod. Consum. **30**, 269 (2022)
18. N.T. Yap, J.F. Devlin, J. Ind. Ecol. **21**, 6 (2017)
19. J.S. Hofstetter et al., Circ. Econ. Sustain. **1**, 21 (2021)
20. F. Bento, M. Tagliabue, F. Lorenzo, Societies **10**, 56 (2020)
21. M. Lawal, S.R.W. Alwi, Z.A. Manan, W.S. Ho, J. Clean. Prod. **280**, 124327 (2021)
22. M. Demartini, F. Tonelli, K. Govindan, Clean. Logist. Supply Chain. **3**, 100020 (2022)
23. D.A. DeLaurentis, S. Ayyalasomayajula, J. Ind. Ecol. **13**, 247 (2009)
24. J.G. Backes, R. Hinkle-Johnson, M. Traverso, Case Stud. Constr. Mater. **18**, e01966 (2023)
25. D.K. Panesar, K.E. Seto, C.J. Churchill, Int. J. Life Cycle Assess. **22**, 1969 (2017)
26. H.H. de Simone Souza, P.P. de Abreu Evangelista, D.L. Medeiros, J. Albertí, P. Fullana-i-Palmer, M.Á. Boncz, A. Kiperstok, J.P. Gonçalves, Int. J. Life Cycle Assess. **26**, 435 (2021)
27. I. Arzoumanidis, M. D'Eusanio, A. Raggi, L. Petti, in *Perspectives on Social LCA: Contributions from the 6th International Conference* (Springer, 2020), pp. 1.
28. S. Moffatt, N. Kohler, Build. Res. Inf. **36**, 248 (2008)
29. E. Conte, V. Monno, Environ. Impact Assess. Rev. **34**, 31 (2012)
30. C. Herrero-Jáuregui et al., Sustainability **10**, 2950 (2018)
31. J. Lohne, K. Mohn, in *ECPPM 2022-eWork and eBusiness in Architecture, Engineering and Construction 2022* (CRC Press, 2023), pp. 622
32. R.I. Iacobescu, G.N. Angelopoulos, P.T. Jones, B. Blanpain, Y. Pontikes, J. Clean. Prod. **112**, 872 (2016)
33. M. Tanguler-Bayramtan, S. Turk, I.O. Yaman, Constr. Build. Mater. **447**, 138090 (2024)
34. G. Biava, A. Zacco, A. Zanoletti, G.P. Sorrentino, C. Capone, A. Princigallo, L.E. Depero, E. Bontempi, Materials **16**, 4055 (2023)
35. M.A. Sellitto, F.K. Murakami, Chem. Eng. Trans. **70**, 211 (2018)
36. D. Adiguzel, S. Tuylu, H. Eker, Constr. Build. Mater. **360**, 129574 (2022)
37. M. Phocas, A. Michael, P. Fokaides, J. Renew. Energy Power Qual. **9**, 501 (2011)
38. G. Schneider-Skalska, World Trans. Eng. Technol. Educ. **16**, 356 (2018)
39. S. Luhar, T.-W. Cheng, I. Luhar, Compos. B Eng. **175**, 107076 (2019)
40. S. Tedesco, E. Montacchini, J. Andreotti, Acta Polytech. CTU Proc. **38**, 216 (2022)
41. M. Ghufran, K.I.A. Khan, F. Ullah, W.S. Alaloul, M.A. Musarat, Sustainability **14**, 11815 (2022)
42. C. Shrubsole et al., Indoor Built Environ. **28**, 100 (2019)
43. M. KeeGAN, N. Nguyen, Migr. Aust. **1**, 50 (2011)
44. E.C. Nabong, L. Hocking, A. Opdyke, J.P. Walters, Wiley Interdiscip. Rev.: Clim. Chang. **14**, e828 (2023)
45. M.S. Uddin, J.K. Routray, P. Warnitchai, Resilient Struct. Infrastruct. **379** (2019)
46. J.M. Anderies, Build. Res. Inf. **42**, 130 (2014)
47. A. Wiek, L. Withycombe, C.L. Redman, Sustain. Sci. **6**, 203 (2011)
48. R.W. Scholz, C.R. Binder (2011)
49. J. Segalàs, D. Ferrer-Balas, K.F. Mulder, J. Clean. Prod. **18**, 275 (2010)
50. J.A. Vargas-Merino, C.A. Rios-Lama, M.H. Panez-Bendezu, Int. J. Manag. Educ. **22**, 100904 (2024)
51. F. Pomponi, A. Moncaster, J. Clean. Prod. **143**, 710 (2017)
52. A.M. Colavitti, A.M. Colavitti, *Urban Heritage Management: Planning with History*, vol. 55 (2018)
53. N. Mamajonova, M. Oydin, T. Usmonali, A. Olimjon, A. Madina, M. Marg'uba, Holders REASON **2**, 327 (2024)

54. N. Soares et al., Renew. Sustain. Energy Rev. **77**, 845 (2017)
55. M. Röck, M.R.M. Saade, M. Balouktsi, F.N. Rasmussen, H. Birgisdottir, R. Frischknecht, G. Habert, T. Lützkendorf, A. Passer, Appl. Energy **258**, 114107 (2020)
56. S. Steger, R. Bleischwitz, J. Clean. Prod. **19**, 816 (2011)
57. A.N. Butt, B. Dimitrijević, Sustainability **14**, 10339 (2022)
58. B.V. Ikejiaku, *The Capability Approach and the Sustainable Development Goals: Inter/multi/trans Disciplinary Perspectives* (Taylor & Francis, 2024)
59. T. Scarlett, S.A. Walton (2017)
60. R.M. Lillibridge, J. Soc. Archit. Hist. **12**, 17 (1953)
61. J.E. Baxter, Int. J. Hist. Archaeol. **16**, 651 (2012)
62. W. Commons (2024)
63. G. Yellich (1895)
64. C. Twinn, Arup J. **38**, 10 (2003)
65. Case-studies: Bedzed-the-uks-first-large-scale-eco-village, https://www.bioregional.com/projects-and-services/case-studies/bedzed-the-uks-first-large-scale-eco-village
66. A. Suwanchaiskul, M. Sambandaraksa, A. Chaladdee, D. Jareemit, B. Limmeechokchai, K. Aramraks, in *2024 International Conference on Sustainable Energy: Energy Transition and Net-Zero Climate Future (ICUE)* (IEEE, 2024), pp. 1.
67. R. Horne, A. Nelson, L. Dorignon, in *Post-carbon Inclusion* (Bristol University Press, 2024), pp. 84
68. N. Schoon, *Bioregional Development Group*. Available online: https://storage.googleapis.com/www.bioregional.com/downloads/The-BedZED-Story_Bioregional_2017.pdf. Accessed on 10 Apr 2023) (2016)
69. T. Chance, Wikimedia Commons, 2007).
70. J. Knikker, in *How To Win Work* (RIBA Publishing, 2021), pp. 137
71. A. Jalia, R. Bakker, M. Ramage, (Centre for Digital Built Britain, University of Cambridge, Cambridge, 2019)
72. E. MacArthur, J. Ind. Ecol. **2**, 23 (2013)
73. E. Macarthur, H. Heading, Ellen MacArthur Found. **1**, 1 (2019)
74. A. Petit-Boix, S. Leipold, J. Clean. Prod. **195**, 1270 (2018)
75. E. Zea Escamilla, G. Habert, J.F. Correal Daza, H.F. Archilla, J.S. Echeverry Fernández, D. Trujillo, Sustainability **10**, 3096 (2018)
76. L. Essaghouri, R. Mao, X. Li, Environ. Impact Assess. Rev. **100**, 107085 (2023)
77. D.A. Salas, A.D. Ramirez, N. Ulloa, H. Baykara, A.J. Boero, Constr. Build. Mater. **190**, 170 (2018)
78. R. Hingorani, N. Dittrich, J. Köhler, D.B. Müller, Build. Environ. **245**, 110830 (2023)
79. J. de Brito, R.V. Silva, (International Association for Bridge and Structural Engineering, 2024)
80. K. Jones, J. Stegemann, J. Sykes, P. Winslow, Constr. Build. Mater. **125**, 690 (2016)
81. M. Pramreiter, M. Grabner, Forests **14**, 1419 (2023)
82. M. Pramreiter, T. Nenning, C. Huber, U. Müller, B. Kromoser, P. Mayencourt, J. Konnerth, Sustain. Mater. Technol. e00728 (2023)
83. Indresmat Launches Eco-Friendly KLIMA-PUR Windows: A Milestone in Sustainable Building Solutions, https://www.mezeroe.eu/news/indresmat-launches-eco-friendly-klima-pur-windows-milestone-sustainable-building-solutions
84. KLIMA-PUR® BioPolyurethane Windows, https://www.indresmat.com/KLIMA-PUR
85. B.A. Salami, F. Mukhtar, S.A. Ganiyu, S. Adekunle, T.A. Saleh, Constr. Build. Mater. **396**, 132296 (2023)
86. Z. Pan et al., Cement Concr. Compos. **58**, 140 (2015)
87. S. Lv, Y. Ma, C. Qiu, Q. Zhou, Mag. Concr. Res. **65**, 1246 (2013)
88. S. Lv, Y. Ma, C. Qiu, T. Sun, J. Liu, Q. Zhou, Constr. Build. Mater. **49**, 121 (2013)
89. C.S.R. Indukuri, R. Nerella, S.R.C. Madduru, Constr. Build. Mater. **229**, 116863 (2019)
90. W. O'Brien et al., Build. Environ. **179**, 106906 (2020)
91. M.d.M. Solà, A. de Ayala, I. Galarraga, M. Escapa, Energy Effi. **14**, 6 (2021)
92. K. Krueger, A. Stoker, G. Gaustad, Smart Sustain. Built Environ. **8**, 270 (2019)

93. K.T. Adams, M. Osmani, T. Thorpe, J. Thornback, in *Proceedings of the Institution of Civil Engineers-Waste and Resource Management* (Thomas Telford Ltd, 2017), pp. 15
94. G. Giuffrida, R. Caponetto, F. Nocera, M. Cuomo, Sustainability **13**, 11948 (2021)
95. G. Giuffrida, V. Costanzo, F. Nocera, M. Cuomo, R. Caponetto, in *International Conference on Sustainability in Energy and Buildings* (Springer, 2022), pp. 164.
96. G. Giuffrida, L. Dipasquale, R.M. Pulselli, R. Caponetto, Sustainability **16**, 1367 (2024)
97. G. Giuffrida, L. Ibos, A. Boudenne, H. Allam, J. Build. Eng. **90**, 109370 (2024)
98. M. Gomaa, J. Carfrae, S. Goodhew, W. Jabi, A. Veliz Reyes, Architectural Sci. Rev. **62**, 230 (2019)
99. K. Harries, L. Ben Alon, and B. Sharma, Characterisation, properties and applications (2019).
100. L. Ben-Alon, V. Loftness, K. Harries, E. Cochran Hameen, in *IOP Conference Series: Earth and Environmental Science* (IOP Publishing, 2019), p. 012139.
101. L. Ben-Alon, V. Loftness, K.A. Harries, E.C. Hameen, M. Bridges, J. Green Build. **15**, 87 (2020)
102. L. Ben-Alon, A.R. Rempel, Build. Environ. **238**, 110339 (2023)
103. M.C. Friant, W.J. Vermeulen, R. Salomone, Sustain. Prod. Consumption **27**, 337 (2021)
104. W. B. Zimmerman, *Multiphysics Modeling With Finite Element Methods* (World Scientific Publishing Company, 2006), Vol. 18.
105. N. Khadim, R. Agliata, Q. Han, L. Mollo, Build. Environ. **269**, 112413 (2025)
106. M. Rotilio, C. Marchionni, P. De Berardinis, F. Cucchiella, in *International Conference on Innovation in Urban and Regional Planning* (Springer, 2023), pp. 599.
107. E. Rodrigues, Á. Gomes, A.R. Gaspar, C.H. Antunes, Renew. Sustain. Energy Rev. **94**, 959 (2018)
108. R. Oxman, Des. Stud. **27**, 229 (2006)
109. A. Borrmann, C. Forster, T. Liebich, M. König, J. Tulke, in *Proceedings of the 18th International Conference on Computing in Civil and Building Engineering: ICCCBE 2020* (Springer, 2021), pp. 452.
110. R.O. Ajirotutu, A.B. Adeyemi, G.-O. Ifechukwu, T.C. Ohakawa, O. Iwuanyanwu, B.M.P. Garba, J. Adv. Infrastruct. Stud. (2024)
111. F. Saleh, A. Elhendawi, A.S. Darwish, P. Farrell, J. Intell. Syst. Internet Things **11**, 75 (2024)
112. M. Gomaa, W. Jabi, A.V. Reyes, V. Soebarto, Autom. Constr. **124**, 103577 (2021)
113. M. Gomaa, S. Schade, D.W. Bao, Y.M. Xie, J. Clean. Prod. **398**, 136569 (2023)
114. F. Faleschini, D. Trento, M. Masoomi, C. Pellegrino, M.A. Zanini, Constr. Build. Mater. **398**, 132496 (2023)
115. D. Trento, F. Faleschini, M. Masoomi, C. Pellegrino, and M. A. Zanini, in *International Conference on Earthen Construction* (Springer, 2024), pp. 3

Chapter 28
Circular Bio-Based Building Materials for a Sustainable Built Environment: Categorisation, Adoption Barriers and Sustainability Assessment

Dinh Linh Le, Roberta Salomone, and Quan T. Nguyen

Abstract Adopting circular bio-based building materials provides opportunities to mitigate greenhouse gas emissions and other environmental impacts of the construction industry. However, as these materials are mostly new and emerging, their contribution to achieving a sustainable built environment needs to be thoroughly investigated. To this end, this chapter first categorises these materials into three types, with specific examples for each category, to give an overview of the potential utilisation of these materials in the construction industry. Furthermore, it is reported that the adoption of circular bio-based building materials in both developed and developing economies is hindered by various barriers, which need to be investigated and addressed to promote these materials' successful and widespread utilisation. To shed light on this aspect, the adoption barriers of these materials are investigated in the cases of Flanders in Belgium and Vietnam. A comparative study shows that while cost and risk-related barriers are the most critical barriers in Flanders, the government's role-related barriers are deemed the biggest hindrance to these materials' utilisation in Vietnam. Despite these differences, the government's roles are essential to addressing barriers in both developed and developing economies. Finally, to unravel the sustainability performance of these materials, this chapter synthesises,

Supplementary Information The online version contains supplementary material available at https://doi.org/10.1007/978-3-031-97818-0_28.

D. L. Le (✉)
University School for Advanced Studies IUSS Pavia, Pavia, Italy
e-mail: dinh.le@iusspavia.it; linhld@huce.edu.vn

R. Salomone
Department of Economics, University of Messina, Messina, Italy
e-mail: roberta.salomone@unime.it

Q. T. Nguyen
Faculty of Construction Economics and Management, Hanoi University of Civil Engineering, Hanoi, Vietnam
e-mail: quannt@huce.edu.vn

D. A. H. Hanaor (ed.), *Sustainable Materials for the Built Environment*, Springer Series in Materials Science 352, https://doi.org/10.1007/978-3-031-97818-0_28

analyses, and discusses findings from relevant studies identified in the existing literature. The results indicate that most existing studies are dedicated to assessing the environmental sustainability of specific materials, lacking those that evaluate the comprehensive sustainability of buildings. The sustainability assessment methods used in these studies, mostly life cycle assessments, have methodological issues that need to be addressed in future studies. Moreover, although circular bio-based building materials outperform their traditional counterparts in mitigating climate change and reducing production costs, their use does not always offer better economic and environmental sustainability.

28.1 Introduction

It is widely accepted that the construction sector has severe impacts on the environment, particularly climate change and natural resource scarcity. This is because, firstly, only buildings contribute to nearly 35% of energy consumption, more than half (55%) of total electricity use, and nearly two-fifths (38%) of greenhouse gas (GHG) emissions globally [1]. Secondly, the construction of only infrastructure and buildings in developing economies consumes up to half of the total amount of raw natural resources extracted for material purposes worldwide [2]. The construction industry is also well-known for the large amount of materials wasted, with around 15% annually, and this figure is forecast to continue rising [1].

To address the aforementioned impacts attributed to the construction industry, many different solutions have been implemented. For example, to cut off GHG emissions, scholars and practitioners have proposed and applied different technologies to improve buildings' energy efficiency, such as air tightness, active shading systems, efficient lighting, etc., and transition to renewable energy sources [3]. However, it is noted that most of the current solutions are focused only on reducing the operational impacts of buildings, even though the contribution of embodied GHG associated with the use of materials can be half of the total buildings' life-cycle emissions [4]. Therefore, to achieve a sustainable built environment, it is necessary to address the embodied impacts of building materials.

Among the solutions to reduce energy consumption, GHG emissions, and virgin natural resource dependence in the construction industry, using circular bio-based building materials emerges as a coherent approach. These materials are defined as "*materials wholly or partly derived from renewable biological origins, or by-products and biowaste of plant and/or animal biomass that can be used as raw building materials and decorating items in construction, in their original forms or after being reprocessed*" [5, 6]. There are strong links between these materials and circular economy strategies, for instance, "recycle" and "reduce", according to the 9Rs framework [5, 7]. The definition suggests that the input sources of these materials should be circular and be retrieved from secondary sources. Consequently, using these materials can alleviate the demand for virgin materials in the construction industry. Furthermore, adopting these materials can not only dramatically reduce GHG emissions associated with the construction industry and bring economic benefits [5, 8], but can also offer great opportunities for transitioning the sector from the current linear model to a circular one.

Although the term "*circular bio-based building materials*" and the definition of these materials have just been proposed recently [5], many kinds of them have been developed in previous practice and discussed in previous scientific studies. However, these studies are fragmented, and the materials themselves are diversified in size, composition, shape, function, and biological origins. This may confuse the adopters when considering the use of these materials, highlighting the need to clearly categorise and also indicate the potential use of these materials. Furthermore, since most of these materials are new and emerging, their adoption encounters various barriers and challenges. This is because the introduction of new materials to any sector of the economy has to face barriers, especially in a highly conservative and cost-driven sector like construction [9]. These barriers have to be investigated and addressed to promote the successful and widespread adoption of these materials. Alongside addressing the barriers, a better understanding of the pros and cons in terms of sustainability performance is essential to convincing stakeholders in the construction industry to uptake these materials. This information should be supported by sustainability assessments regarding the use of these materials. Despite this importance, there is still a lack of studies assessing the sustainability of circular bio-based building materials [6], since most studies on this topic have been dedicated to the development and performance testing of these materials. This highlights the need to figure out the status of assessing the sustainable performance of circular bio-based building materials. Because of these motivations, it is necessary to conduct a study to shed light on the categorisation of CBBMs, investigate their adoption barriers in both developed and developing economies, and evaluate the state of the art of sustainability assessment of these materials.

In order to address the aforementioned gaps, this chapter first summarises the findings from deep analyses of 97 relevant studies identified from the existing literature to provide insights into the categorisation, potential use, and sustainability assessments of these materials. In particular, the relevant articles included in the literature review are identified from the Scopus and Web of Science databases and screened using inclusion and exclusion criteria. Only peer-reviewed articles written in English that focus on assessing the sustainability of circular bio-based building materials are selected. The literature review is carried out using systematic and critical approaches. Secondly, questionnaire surveys are adopted to investigate the adoption barriers of these materials in both developed and developing economies.

To provide the research findings, the chapter is organised into five sections, including the introduction. Section 28.2 presents the categorisation and potential use of circular bio-based building materials, while Sect. 28.3 discusses the adoption barriers of these materials, including a comparative study between developed and developing economies. Meanwhile, the sustainability assessment regarding the use of these materials is presented in Sect. 28.4, and finally, Sect. 28.5 concludes the chapter.

28.2 Categorisation and the Potential Use of Circular Bio-Based Building Materials

28.2.1 Categorisation

It is highlighted that circular bio-based building materials are heterogeneous in size, composition, shape, function, and biological origins [5]. They can be insulation board made from cork waste [10], pineapple by-products [11], or recycled cellulose and sheep wool waste [12]; hemp-based materials like hempcrete [13, 14] and hemp hurds [15]; and concrete or bricks partly retrieved from bio-wastes like coconut shell [16], rice husk [17–21], wheat straw and olive stone flour [22], etc. Despite these heterogeneities, these materials can be grouped into three macro categories [5]: (i) circular bio-based insulation materials; (ii) circular bio-based cementitious materials; and (iii) indoor-used circular bio-based building materials (e.g., ceiling, flooring, and furniture).

28.2.1.1 Circular Bio-Based Insulation Materials

This category is the most dominant of the three aforementioned categories, highlighting that circular bio-based building materials are better suited to be used as insulation materials than for other purposes [5]. Within this category, materials are also diversified in forms and origins. Among them, hemp-based, cork-based, and straw-based materials have emerged as the most popular ones [5, 23]. A list of example materials in this category is presented in Table 28.1.

Circular bio-based insulation materials emerge as promising alternatives to be installed in building envelopes to improve in-use energy efficiency [5, 79]. This is because of their distinct qualities regarding hygrothermal performance, for example, the sorption and desorption of water vapour in their porous structure in dynamic equilibrium with their surroundings, constantly adjusting their hygric and thermal characteristics while causing energy transfers themselves [79]. Hence, using these materials can reduce dramatic operational energy consumption. For example, a building using agglomerated cork and hempcrete walls can save up to 75% [62] and 16% [52] of its operational energy consumption compared to a brick-wall building, respectively. Furthermore, as these materials are sourced from renewable biosources and have a low carbon footprint compared to petrol-based insulation materials, rock and glass wool, etc., their use offers great opportunities to mitigate the climate impacts of the construction industry [5].

Table 28.1 Circular bio-based insulation materials and their origins [5]

Name of materials	Origins	Sources
Straw-based materials (panels, straw bales, straw-clay blocks, light straw clay brick, straw chips, prefabricated panel, pressed straw, straw board)	Straw	[24–44]
Hemp-based materials (hurds, hempcrete, hemp-lime, composites)	Hemp waste and by-products	[13–15, 28, 42, 45–57]
Cork-based materials (expanded cork slab and granules, corkboards, composite)	Cork waste	[10, 32, 58–65]
Coconut-based materials (coconut shell concrete, coir fibre insulation, coir compressed boards, biocomposites)	Coconut shell and coconut husk	[16, 66, 67]
Bricks with olive core flour	Olive core flour	[22, 42]
Cob (corn-pith)	Cob	[30, 63, 65]
Animals' waste-based materials (lightweight brick made from oyster shells, panels made from sheep wool, bio-epoxy composite board)	Oyster shells and sheep wool waste	[12, 31, 43]
Wood-waste-based materials (composite, bio-based panels, mud bricks made with walnut shells)	Sawdust, residual coniferous bark, and walnut shells	[68–70]
Rice-husk-based materials (rice husk panels and waste-based bricks)	Rice husk	[71, 72]
Paper-waste-based materials (bio-epoxy composite board, insulation panels, and cardboard-based panels, bricks)	Recycled cellulose, waste paper, old newsprint, recovered paper and cardboard	[12, 72–76]
Waste-based bricks (miscanthus blocks and agrostone)	Miscanthus shiv; water hyacinth bagasse, and grass	[77, 78]
Bio-based panel	Pineapple by-products	[11]

28.2.1.2 Circular Bio-Based Cementitious Materials

It is worth noting that materials in this category are rarely discussed in the literature as bio-based building materials [5]. However, since these materials fulfil the circular bio-based building material definition and their environmental benefits are considerable, they should be considered an important group of circular bio-based building materials [5]. A list of example materials in this category is presented in Table 28.2.

Table 28.2 indicates that most of the materials in this group relate to the use of biowaste-based inputs to partially replace cement or, in some instances, sand in concrete and mortar. This is because the production of cement has detrimental impacts on the environment. Indeed, the cement industry consumes around 12%-15% of the total available industrial energy and contributes about 5% of the total

Table 28.2 Circular bio-based cementitious materials and their origins [5]

Name of materials	Origins	Sources
Sustainable concrete	Coconut shell	[16]
	Rice husk ash (RHA)	[17–21, 80–82]
Mortars	Jacaranda seed pods	[83]
	Wood waste	[84]
	Spent coffee ground and coffee chaff	[85–87]
Filler or binder for concrete and mortar	Waste wood	[88]
	Bran products	[89]
	Hemp shives	[14, 15, 57]

global greenhouse gas emissions [90]. Therefore, the more cement consumption is reduced, the less harm this material poses to the environment. Furthermore, using biowaste-based ingredients to replace traditional ones in concrete and mortar can also bring not only environmental but also economic and social benefits from the valorisation of these waste streams. For instance, the majority of RHA is currently landfilled which pollutes soil and water environments and therefore, affects the health of residents in the surrounding areas [91]. Therefore, using RHA in the construction industry can contribute to overcoming these issues. Besides, since RHA is very low-cost or even no-cost in some countries like Cambodia [91], using RHA to replace cement can not only save on disposal costs but also reduce the production cost of concrete.

28.2.1.3 Indoor-Used Circular Bio-Based Building Materials

Indoor-used circular bio-based building materials can be used for finishing ceilings, flooring, and producing furniture. For instance, circular bio-based linoleum can be used as a no-joint flooring material, especially in high-traffic areas like sports and office facilities and also can be used in residential buildings [92]. Meanwhile, circular bio-based composites have potential uses in framing, wallboards, window frames, doors, flooring, decorative panelling, cubicle walls, and ceiling panels [93]. These highlight the varied potential applications of indoor-used circular bio-based building materials within the construction industry. A list of example materials in this category is presented in Table 28.3.

Table 28.3 Indoor-used circular bio-based building materials and their origins [5]

Name of materials	Origins	Sources
Bio-flexi HDF fibreboards	Straw	[94]
Bio-based wood flooring coatings	Corn stover	[95]
Bio-based biocomposite	Oil palm fibre	[93, 96]
	Recycled cellulose	[97]
Ultralight particleboards	Wood residues from sawmills	[98]
Composite boards	Shredded cotton fibres	[99]
Linoleum	Plants, pine resin, linseed oil, wood, and cork flour	[92]
Cork floating floor	Cork	[100]
Particleboard	Residues from sugarcane bagasse	[101]
	Wood residues/waste	[102–109]
	Pinus and eucalyptus wastes	[110]
Clay ceramics	Agro-wastes	[111]

28.2.2 Potential Use

Generally, circular bio-based building materials can be used for three main purposes associated with their three categories. Particularly, materials within the first category can mainly be used as thermal [42, 50, 66, 70] and acoustic insulation [42, 52] to be installed mostly in walls and also ceilings [5]. In this context, hemp-based, cork-based, and straw-based are the most commonly used circular bio-based insulation materials [5, 23]. Meanwhile, materials in the second category can be used for diversified applications, from masonry works [84–87] to constructing load-bearing structures [80, 81, 84], and even for insulation purposes [85–87]. This suggests that circular bio-based building materials can be used either indoors or outdoors, as well as for constructing substructures like foundations, upper structures such as columns, beams, slabs, etc., or infrastructure like bridges, sewers, etc. Finally, materials in the third category can be used for various purposes, such as flooring, ceilings, wallboards to divide space, decoration, windows and doors, and furniture [92, 93].

With the aforementioned potential uses, circular bio-based building materials can be adopted at different scales, such as components, buildings, and building stock scales [5]. Generally, these materials can be adopted in any building type, including residential, commercial, and industrial ones. These buildings can be new [54] or existing ones [34, 112]. In developed countries, especially in Europe, where around four-fifths of residential buildings were built before 1990 [113] and 75% of the building stock is energy-inefficient [112]. This implies that in these countries, energy renovation of existing buildings is the main market for circular bio-based building materials, especially for insulation purposes. Meanwhile, in developing countries, with a high need for new buildings to meet the requirements of the population and economic growth, circular bio-based building materials are more likely to be used not only for insulation and finishing purposes but also for constructing load-bearing structures using circular bio-based cementitious materials, for example, RHA concrete and mortar [114].

28.3 Barriers to Circular Bio-Based Building Material Adoption

28.3.1 Potential Barriers

According to Le et al. [115], the adoption of circular bio-based building materials in both developed and developing economies is faced with various barriers. These barriers are attributed to the cost-driven and conservative nature of the construction industry [9]. Indeed, cost-related barriers such as fear of high initial costs and extra costs [116–119] and the sector's culture-related factors like reluctance to change from easy-procurement-traditional materials to new materials [116, 117, 120] are identified as the main barriers to the adoption of sustainable materials in the construction industry. As most circular bio-based building materials are new and emerging, their adoption has to face market-related barriers regarding their availability in the market [116, 121], the challenges in supplying [117, 122], etc., as well as barriers linked to the lack of data and information [118, 123]. Finally, these materials' use is affected by barriers related to the government's role, such as the lack of public incentives and promotion [116, 121] and the inadequacy of building codes and regulations [116, 118], for instance. A list of 23 potential barriers to the adoption of these materials is identified from the existing literature and presented in Table 28.4.

28.3.2 A Comparative Study of Adoption Barriers in Developed and Developing Economies

To conduct a comparative study of the barriers to the adoption of circular bio-based building materials in developed and developing economies, Flanders in Belgium and Vietnam are chosen as case studies, respectively. First, the criticality of the 23 identified barriers is investigated based on the perspectives of building professionals in the Flemish and Vietnamese construction industries. To collect data, a questionnaire survey, which is a technique commonly used to gather specialist perceptions on factors affecting the implementation of sustainable practices in the construction industry [116, 124, 125], is employed. The questionnaire was developed based on the potential barriers identified in the literature (Table 28.4). Particularly, the research aims and contact details are presented in the first section, while the second section is designed to gather background information about respondents (e.g., position, area of expertise, working experience, etc.). Meanwhile, the rank of respondents

Table 28.4 The comparison of barriers' ranking for the two cases of Flanders and Vietnam

Code	Barriers	Sources	Developed economy (Flanders, Belgium)				Developing economy (Vietnam) [115]			
			Mean	SD[a]	Rank	NV[b]	Mean	SD[a]	Rank	NV[b]
B01	Building code restriction and lack of regulation	[116, 118, 121, 123, 144, 145]	3.65	0.89	7	0.77*	3.95	0.871	1	1.00*
B02	Lack of government promotion and incentives	[116, 121, 146, 147]	3.54	0.93	9	0.68*	3.69	0.990	3[c]	0.69*
B03	Lack of design guides and tools	[117, 122]	3.19	0.88	18	0.42	3.69	0.883	5	0.69*
B04	Lack of information (e.g., environmental impacts, performance)	[116, 118, 121–123, 148]	3.27	1.02	16	0.48	3.71	1.026	2	0.71*
B05	Lack of comprehensive tools and data to compare material alternatives	[116, 118]	3.49	1.04	12	0.65*	3.10	0.900	23	0.00
B06	Maintenance concern	[116, 118, 144, 145, 149]	3.27	0.87	17	0.48	3.65	0.934	6	0.64*
B07	Lack of case studies and demonstration projects	[117, 122, 148]	3.16	0.83	19	0.39	3.41	1.004	18	0.37
B08	Lack of design knowledge and skills	[116, 121, 123, 147, 150]	3.51	0.84	10	0.66*	3.49	1.007	13	0.46
B09	Shortage of skilled labourers	[116, 117, 122, 123, 151]	3.62	0.86	8	0.75*	3.45	1.119	14	0.41
B10	Contractors lack technical knowledge and experience	[117, 122, 150, 152, 153]	3.92	0.64	4	0.98*	3.53	0.966	12	0.51*

(continued)

Table 28.4 (continued)

Code	Barriers	Sources	Developed economy (Flanders, Belgium)				Developing economy (Vietnam) [115]			
			Mean	SD[a]	Rank	NV[b]	Mean	SD[a]	Rank	NV[b]
B11	Unwillingness to change from conventional materials to CBBMs	[116, 118, 122, 153]	3.76	0.98	6	0.85*	3.57	1.025	9	0.55*
B12	The perception that CBBMs are low in quality (fire and moisture resistance, low durability, etc.)	[117, 118, 122, 144, 145, 154, 155]	3.49	1.07	11	0.65*	3.43	0.855	17	0.39
B13	Perception of the extra cost being incurred	[116–118, 122, 123, 155]	3.92	0.76	3	0.98*	3.40	0.852	19	0.35
B14	Clients worrying about profitability	[148, 156]	3.84	0.8	5	0.92*	3.53	1.007	10[c]	0.51*
B15	Unavailability of materials in the local market	[116, 121, 122, 145, 153]	3.3	0.97	14	0.50*	3.63	0.979	7	0.62*
B16	Challenging in finding a reliable supply of materials	[118, 121–123]	3.14	1.03	20	0.38	3.27	0.918	21	0.21
B17	Lack of effective marketing from material producers	[117, 122, 157]	3.35	1.03	13	0.54*	3.39	0.850	20	0.35
B18	Low flexibility of alternatives or substitutes	[118]	2.65	0.95	23	0.00	3.57	1.044	8	0.55*

(continued)

Table 28.4 (continued)

Code	Barriers	Sources	Developed economy (Flanders, Belgium)				Developing economy (Vietnam) [115]			
			Mean	SD[a]	Rank	NV[b]	Mean	SD[a]	Rank	NV[b]
B19	Clients with less empathy for sustainability-related goals	[123, 148, 158]	3.14	0.89	21	0.38	3.43	0.944	16	0.39
B20	Unstable prices of materials	[148, 158]	2.76	0.68	22	0.08	3.22	0.966	22	0.14
B21	Concern about the high initial cost	[122, 144, 152, 153, 158, 159]	3.95	0.78	1	1.00*	3.53	1.007	10[c]	0.51*
B22	Time and liability constraints of architects and contractors	[118, 122, 123]	3.3	1.13	15	0.50*	3.45	0.923	15	0.41
B23	Risks and uncertainties involved in adopting new materials	[116, 118, 122, 144]	3.92	0.89	2	0.98*	3.69	0.990	3[c]	0.69*

Note [a] Standard deviation; [b] Normalised value, which is calculated by the Eq. (28.1); [c] Equal rankings in which the subsequent rank is omitted; * Critical barriers ($NV \geq 0.5$)

to barriers to circular bio-based building material utilisation is gathered in another section using a five-point Likert scale (1 = extremely low and 5 = extremely high). This is because using the Likert scale can speed up the data collection and provide clear and simple-to-understand results [126]. Before carrying out the mass survey, a pilot study involving three experts in the building sector was conducted to evaluate and refine the questionnaire.

The data collection was conducted simultaneously in the cases of Flanders and Vietnam, using a combination of purposive and snowball sampling techniques. Although these two methods are non-probability sampling techniques, their use is appropriate. This is because the probability-based random sampling technique is rarely feasible in construction management research [116, 127]. In this context, non-probability sampling techniques, especially purposive and snowball samplings, are most commonly used in construction management research [116, 128, 129]. Therefore, the combination of these two methods is appropriate to be used to gather data in this study. The questionnaire was presented in Google Forms and distributed to potential respondents via personal emails. Besides, in line with the snowball sampling technique, potential respondents were asked to circulate the questionnaire to other experts in their network that suited the survey.

To analyse the data, different methods are applied. Firstly, Cronbach's alpha is used to evaluate the reliability of the data collected using the five-point scale. The reliability is evaluated via the alpha coefficient (α), which ranges from 0 to 1 [129, 130]. It is widely accepted that this coefficient must be greater than 0.7 to ensure the acceptable reliability of the scale [129, 130]. Next, the mean score technique, a typical quantitative analysis technique to rank the criticality or relative importance of factors [129], is adopted to rank the level of importance of barriers based on the votes of respondents. The rank of different barriers is decided by relying on their mean scores and standard deviation (SD). In cases where more than two barriers have equal mean scores, the one with the highest SD is ranked lowest. Finally, to identify the critical barriers, the normalised value (NV) of the mean scores is calculated according to Eq. (28.1) [129]. Barriers with a value equal to or higher than 0.5 are deemed to be critical [129].

$$\text{NV} = \frac{(\text{Mean score} - \text{lowest mean score})}{(\text{Highest mean score} - \text{lowest mean score})} \tag{28.1}$$

28.3.2.1 The Case of the Developed Economy of Flanders

Flanders is the most populated region in Belgium [131], where the government has chosen the construction industry as a key sector for the transition to a circular economy [132]. Therefore, solutions and initiatives to transform the current linear to circular construction industry are the priority in the region. Besides, within the Flemish construction industry, research shows that strategies to address the embodied impacts associated with the production of energy and building materials are still lacking [133]. To contribute to the transition of the construction industry and provide a solution to address the embodied impacts of the sector, promoting the use of circular bio-based building materials is one of the most potential approaches [5, 6].

To investigate the impact and interaction of adoption barriers of circular bio-based building materials, the data collection targeted specialists, including researchers and industry practitioners, as well as firms that are frontrunners in circular and sustainable construction practices in the region. The data collection was taken placed from mid-April to mid-November 2023. The combination of purposive and snowball sampling techniques helped collect 37 valid responses. Among these 37 respondents, 11 are architects, accounting for nearly 30% of the sample, followed by eight green building development consultants (21.6%), academic staff (n = 4), contractors, and material producers (n = 3 for each). The remaining respondents are working as authorities, owners, staff of industry associations or civic organisations, engineers, educators, advisers for natural building development, and facilitators of the circular economy. Regarding working experience, nearly four-fifths of them (73%) have at least five years of experience in the construction industry, and more than half (54.05%) have hands-on experience in using circular bio-based building materials.

28.3.2.2 The Case of the Developing Economy of Vietnam

Vietnam is one of the most vulnerable countries to climate change, especially sea level rise and its consequences [134]. The economy in Vietnam relies on the agricultural sector as a backbone, as the country is the sixth-largest and second-largest rice [135] and coffee producer [136], respectively. It implies that there are diversified biowaste and byproducts generated in the country, which may pose negative impacts on the environment, economics and also society. For example, the majority of rice straw in Vietnam currently has been burned or mixed with the soil, and both of these methods harm the air, water and soil environment, affecting human health in the neighbourhood area [137]. Hence, there is an urgent need for other approaches to valorising these waste and by-product streams to mitigate their impacts and turn them into valuable materials. In this context, developing and promoting circular bio-based building materials in the construction industry is one of the best solutions. The use of these materials also offers a coherent solution to mitigate climate change and other negative impacts of the construction industry on the environment, economy, and society [5, 6]. As a result, the impact of sea level rise and other consequences of climate change on the country may be mitigated.

Since the development and utilisation of circular bio-based building materials in Vietnam are only in a very early stage, the number of experts with hands-on experience in using these materials is very limited. Hence, the data collection targeted building professionals and companies that are working on or have knowledge of sustainable construction practices. This is because the adoption of sustainable materials like circular bio-based materials is one important aspect of sustainable construction [138]; therefore, these targeted respondents are assumed to be able to provide valuable insights into the barriers affecting the adoption of these materials in Vietnam.

The questionnaire was open from February 2023 to mid-May 2023, which helped to gather 51 responses [115].

Of these respondents, 17 are civil engineers (33.33%), followed by 13 academics (25.49%), nine employees of housing developers (17.65%), three architects (5.88%), three green building consultants (5.88%), and those working for other stakeholders, including materials producers, quantity surveyors and authorities (n = 6, accounting for 11.77%) [115]. Nearly four-fifths of the respondents have at least 5 years of experience working in the construction sector. However, it should be highlighted that only 5.88% of respondents have firsthand experience in developing and utilising circular bio-based building materials [115]. This indicates that the use of these materials is in its infancy in Vietnam, highlighting the need to promote their adoption in the country.

28.3.2.3 Comparison Barriers in Developed and Developing Economies

The collected data is first evaluated for the reliability of the five-point Likert scale by using SPSS 29.0 statistical software [139]. The α coefficient for the 23 barriers is computed to be 0.83 and 0.95 for the cases of Flanders and Vietnam, respectively. These values are far higher than the threshold of 0.7, indicating the appropriate reliability of the five-point scale used is at the 5% significance level, and, therefore, the data collected is suitable for further analyses. In the next step, mean scores are calculated to rank barriers in both cases. The calculation of normalised values helps to identify 15 and 12 critical barriers to the adoption of circular bio-based building materials in Flanders and Vietnam, respectively. The mean score, rank, SD, normalised value of each barrier, and comparison of the two cases are presented in Table 28.4.

It can be seen that the rank of almost all barriers is significantly different between the two cases. In Flanders, cost- and risk-related barriers, such as B21 (concern about the high initial cost), B23 (risks and uncertainties involved in adopting new materials), and B13 (perception of the extra cost being incurred), arise as the three most critical ones hindering the use of circular bio-based building materials. Meanwhile, these barriers are ranked only 10th, 3rd, and 19th in the case of Vietnam. Except for the similarity in consideration of the risks and uncertainties associated with the use of these materials (B23), the perceptions about the cost of building professionals in the case of a developed economy are totally different from those in a developing economy. It should be highlighted that the results in the developed economy of Flanders are indeed in line with the cost-driven nature of the construction industry [9]. Otherwise, cost-related barriers are not the most significant obstacles in the case of the developing economy of Vietnam, suggesting that there may be other factors that surpass costs in hindering the adoption of new materials in the country. Indeed, the government's role-related barriers are deemed to be the most critical to circular bio-based building materials in Vietnam. As presented in Table 28.4, B01 (building code restriction and lack of regulation), B04 (lack of information on CBBMs), B02 (lack of government promotion and incentives), and B03 (lack of design guides and tools) are in the top five most significant barriers in Vietnam, while these are ranked only 7th, 16th, 9th, and 18th in the case of Flanders. This implies that without the involvement of the government, it is hard to achieve sustainability in the construction industry in developed economies like Vietnam. This is echoed with the previous studies on barriers to the implementation of green building, which is a sustainable practice in the Vietnamese construction industry [140]. Furthermore, this finding also highlights that the adoption of CBBMs in Vietnam is hindered by inadequate information regarding the environmental impacts, performance, available types, how to install these materials, etc. It suggests that the collection and dissemination of these data sets can facilitate the wide adoption of CBBMs in the context of Vietnam.

The aforementioned differences indicate that developed economies like Flanders may have better preparation for the implementation of sustainable construction practices, including the adoption of new materials like circular bio-based building materials, than developing economies like Vietnam. Indeed, the Flemish government already took action more than a decade ago to promote circular bio-based building materials [141], and these materials already have a specific share in the construction market. Although less than 5% of the total building projects are partly made of circular bio-based building materials [142], these buildings can perfectly represent either case studies or demonstration projects. Conversely, the development of circular bio-based building materials was officially mentioned in the Vietnamese government's document less than two years ago [143], and the use of these materials is just in its infancy and not yet well documented. That is why Vietnamese building professionals highlight the need for guidance from the government through issuing building codes and regulations, as well as building demonstration projects, forming technical databases, and introducing tools to compare and select circular bio-based building materials over their traditional counterparts. Therefore, positive involvement in issuing regulations and taking needed actions by the government can also be considered lessons learned from a developed economy to be applied in a developing economy for promoting circular bio-based building materials.

28.4 Sustainability Assessment of Circular Bio-Based Building Materials

28.4.1 Scales of Case Studies and Assessed Dimensions

To better analyse the state-of-the-art sustainability assessment studies of circular bio-based building materials, Le et al. [5, 6] categorised the research subjects of the relevant studies into four scales. These include the material, component, building, and building stock scales. Studies on the first scale take into account the sustainable performance of certain materials, while those on the second scale evaluate the sustainability of these materials when they are incorporated into components like walls. Meanwhile, the third scale comprises studies that take into account the influence of circular bio-based building materials on the sustainability of buildings. The last scale refers to studies assessing the effects of utilising these materials to build houses in a city, region, country, or beyond. This categorisation is vital because the research aims and objectives differ at each scale, leading to the differences in methods and data needed for carrying out sustainability assessments [5, 6]. Indeed, while the operational phase is mostly excluded from sustainability assessments at the material scale [10, 95], this phase is essentially included in assessments of buildings [32, 52]. Since sustainability comprises three dimensions: social, economic, and environmental [160], sustainability assessments can cover either a single dimension, simultaneously two dimensions, or all three dimensions (comprehensive sustainability). The share of assessed dimensions at each scale is presented in Fig. 28.1

It can be seen from Fig. 28.1 that the environment is the most assessed dimension, while most of the relevant studies are distributed to the material scale. Indeed, among the 97 studies identified from the existing literature, 74 evaluate only the environmental impacts of circular bio-based building materials, accounting for more than 76%. Nine papers consider only the economic sustainability of these materials, which is equal to the number of studies that simultaneously consider the environmental and

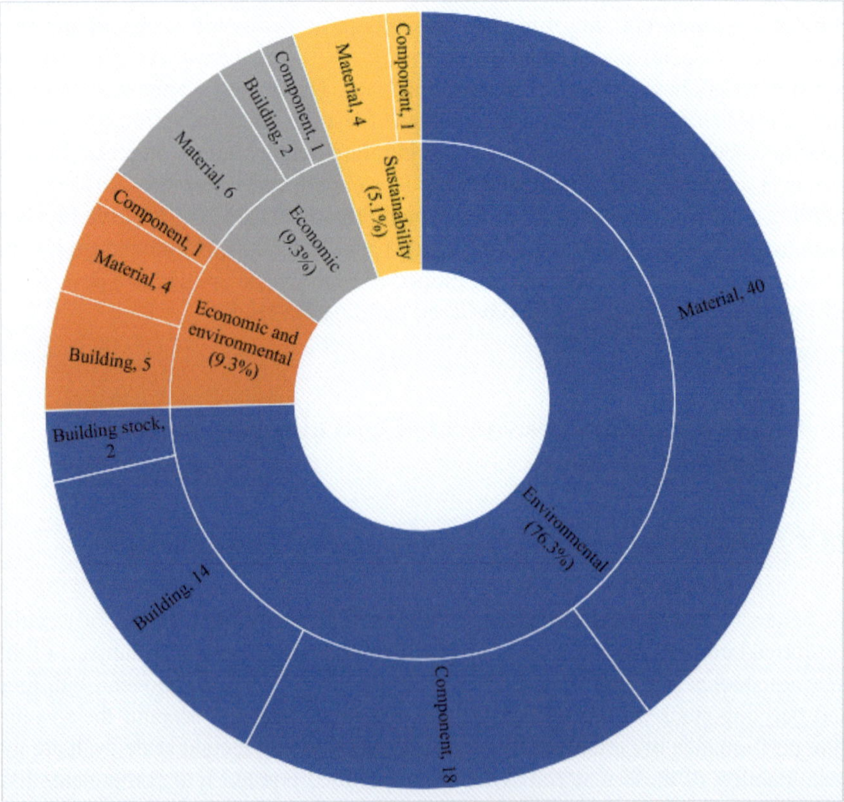

Fig. 28.1 Sustainable dimensions assessed in each case study scale

economic impacts of these materials. Meanwhile, only five studies take into account all three dimensions of sustainability, despite these pillars interacting and relying on each other to survive [161]. Furthermore, no study that either evaluates solely the social aspect or simultaneously with another dimension can be found. These highlight the need for more studies evaluating the social impacts of these materials as well as comprehensive sustainability assessments of circular bio-based building materials.

Regarding the scales of case studies, more than half of the 97 papers (n = 53) consider only the sustainable performance of materials themselves (material scale) rather than considering the interaction and influence of these materials in the sustainable performance of a system. Therefore, the results of these studies may not comprehensively reflect the pros and cons of these materials compared to their traditional counterparts regarding sustainability. The building and component scales are the second and third most evaluated, with 24 and 18 papers, respectively. Meanwhile, only two papers consider the impacts of these materials when they are used in the residential building stock of a city.

28.4.2 Sustainability Assessment Methods

Generally, the methods used to evaluate the sustainability performance of these materials at all four scales are diversified, which are presented in Fig. 28.2.

(Note: BiSA = Bio-Inspired sustainability assessment; CFA = Carbon footprint analysis; EEA = Embodied energy analysis; EFA = Ecological footprint analysis; EPA = Energy performance analysis; LCC = Life cycle costing; LCSA = Life cycle sustainability assessment; MCDA = Multi-criteria decision analysis; MDM = Multi-criteria decision-making; MFA = Material flow analysis; MIVES = Modelo Integrado de Valor para Estructuras Sostenibles; SCA = Simple cost analysis; SDSS = Sustainable Decision Support System) [6].

It can be seen that life cycle assessment (LCA) is the most commonly applied. LCA is an approach that identifies and assesses the potential environmental impacts of a product thoroughly at all stages of its life cycle [162, 163]. In the context of circular bio-based building materials, LCA is employed to evaluate the environmental performance in 58 studies and is either integrated or combined with other methods in environmental assessments (n = 2), environmental and economic assessments (n = 4), and comprehensive sustainability assessments (n = 2). It means that LCA is adopted in 66 out of 97 studies, accounting for nearly 70% [5, 6].

It is highlighted that most LCA-related assessments focus on the material scale (n = 37), followed by 20 studies on the component scales [6]. Concerning the functional unit, the most commonly adopted is area-related, followed by mass-related and volume-related. This is an unsurprising result since insulation materials are the most dominant among the three groups of circular materials discussed in Sect. 28.2, and those materials are mostly in the form of panels or boards and are installed on walls, which are typically measured by area [6]. The dominance of studies at the material

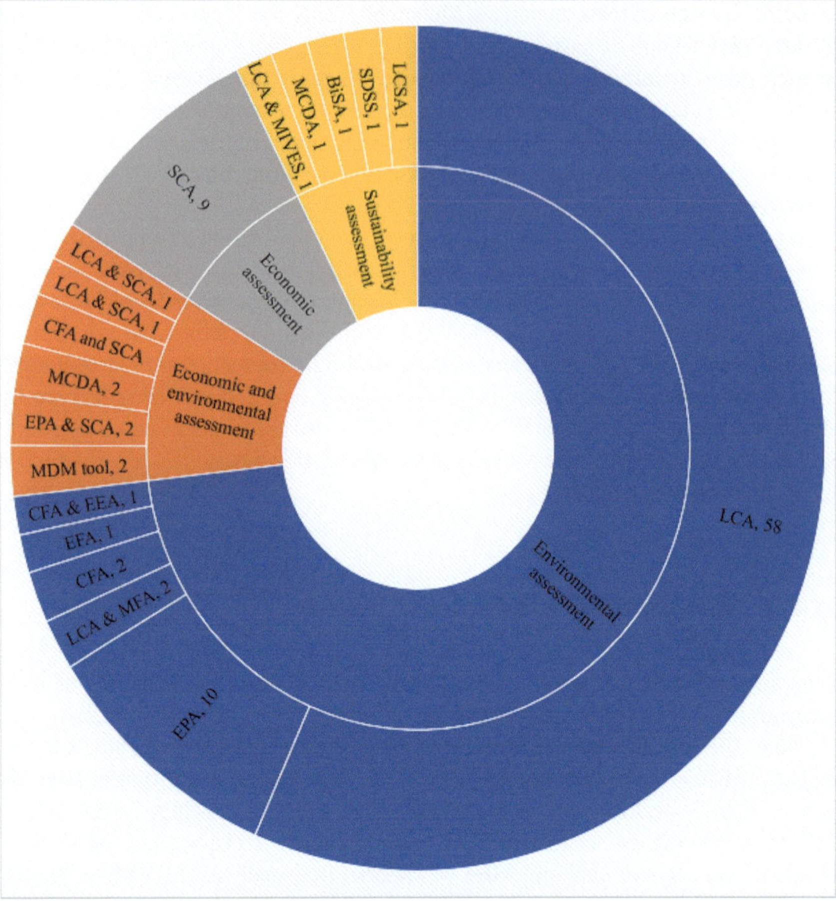

Fig. 28.2 Sustainability assessment methods used

scale leads to "cradle-to-gate" being the most applied research boundary (n = 28), where only production impacts are considered, excluding impacts from other life cycle stages. Regarding the end-of-life phase, only 18 studies specifically propose scenarios, yet most of them are unsustainable or have a low level of circularity, such as landfilling (n = 14), incineration without energy recovery and recycling (n = 9 for each), and energy recovery (n = 4). These limitations highlight the necessity of conducting more holistic assessments such as "cradle-to-grave" or "cradle-to-cradle" with more sustainable end-of-life scenarios.

Additionally, there are other methodological issues associated with these studies that need to be highlighted. Firstly, concerning the life cycle inventory, data quality directly determines the reliability and accuracy of LCA results, and, therefore, this aspect should be clearly described to avoid biases due to data variations and transparency issues [6]. However, up to 21 studies do not precisely provide detailed information about material inventory data. Secondly, to conduct the life cycle impact assessment, most studies consider multiple categories [6], primarily climate change, i.e., global warming potential, carbon footprint, greenhouse gas emissions, etc., followed by acidification potential, eutrophication, etc. Conversely, despite GHG emissions being attributed to energy consumption, the cumulative energy demand indicator is evaluated in only 28 studies. Similarly, carbon sequestration presents additional pros of these materials in addressing climate change impacts but is considered in only 25 studies, with no consensus on assessment methods. This aspect affects the comparability of the results, which may not accurately reflect the benefits of these materials. Lastly, in the life cycle interpretation phase, since most of these materials are new and emerging, and many of them are even just under research at a lab scale, not all the needed data for LCA studies are available. This limitation necessitates making assumptions for LCA analyses, potentially affecting the accuracy of results and decision-making. To address this shortcoming, sensitivity analysis is needed to assess the impacts of uncertain parameters on the results [6]. Nonetheless, only 25 studies consider sensitivity analyses. Likewise, only six studies investigate the effects of uncertainty data on the environmental sustainability of these materials, and most of them do not specify the parameters considered [6]. Future LCA studies in the context of circular bio-based building materials should address these issues to ensure the appropriateness of decisions and strategies to promote the widespread adoption of these materials.

The second most employed method is energy performance analysis (EPA, n = 10), which is an approach to evaluating the energy performance of buildings. That is why all ten relevant studies are attributed to the building scale. However, since this method can only take into account the energy consumption during the operation phase, it cannot holistically reflect the sustainable performance of a building. It is recommended to combine it with other methods, especially life cycle thinking tools, to provide accuracy and comprehensive results on the influence of circular bio-based building materials on the sustainability of buildings. The same recommendation is made for simple cost analysis, which is used in economic assessment as well as economic and environmental assessments, especially at the building scale. As this method is only a simple calculation of expenses that occurred during the production or construction phase, it cannot cover the whole life cycle of products and the time value of money [5, 6], affecting the accuracy of results. For the methods applied in five comprehensive sustainability studies, i.e., assessing all three sustainable pillars, it is highlighted that most of them are tailored tools rather than commonly used methods, affecting the comparability of these studies. For instance, Valtere et al. [89] and Garas et al. [18] develop tailored sustainability assessment frameworks based on multi-criteria decision analysis (MCDA) approaches, yet the specific MCDA methods used are distinctive. Indeed, while the former study employs the TOPSIS method, the latter adopts a combination of the AHP, Entropy, and TOPSIS approaches. Besides, the criteria used in these studies are also different, challenging the comparison of these studies' results.

28.4.3 Economic Sustainability

The findings from nine economic sustainability assessments of circular bio-based building materials indicate that these materials are generally more economical than their traditional counterparts. For instance, with the same use, concrete that uses coconut shells to supplement 30% of fine aggregates can reduce the production cost by nearly 6% [16]. A similar result can be observed in the case of replacing 5% of cement with rice husk ash [17], etc. This is unsurprising as most of these biowaste-based materials are either free or low-cost, leading to a reduction in the production costs of circular bio-based building materials.

However, does the use of circular bio-based building materials always provide better economic performance compared to traditional materials? To answer this question, it is essential to take into account the comprehensiveness of the assessment methods used. Indeed, if using the SCA method, the economic performance of these materials surpasses that of conventional materials [5]. Conversely, when it comes to more comprehensive methods like life cycle costing, the result shows that using circular bio-based building materials is economically weaker than traditional materials [69]. However, since the number of economic assessments is limited, this result may not be true for all types of circular bio-based building materials that have not been assessed yet. To shed light on the economic performance of circular bio-based building materials, it is needed to conduct more studies and synthesise the results to evaluate if the findings are different from those reported in this chapter.

28.4.4 Environmental Sustainability

To provide insights into the environmental sustainability of circular bio-based building materials, the findings from studies that compare their environmental performance with traditional materials are collected and discussed. Firstly, results from ten studies using the EPA approach at the building scale indicate that these materials offer great opportunities to reduce the operational energy consumption of buildings compared to conventional materials. For example, using giant reed residues and agglomerated cork to replace bricks in walls reduces more than half and three-fourths of the total amount of energy that a building consumes in the operation phase [62]. The same results can also be observed when comparing the energy performance of a hemp concrete building to brick and aerated concrete buildings [52]. It is noted that the greater the reduction in energy consumption, the lower the climate change impact of a building. It means that at the building scale, the use of circular bio-based building materials significantly reduces the climate change impact, especially when assessed by the EPA method. These findings are consistent with those from LCA-related studies, which are synthesised and presented in Table 28.5 [5]. It can be seen that circular bio-based building materials offer better environmental performance compared to traditional ones in almost all impact categories, especially climate change and, to a lesser extent, abiotic depletion, human toxicity, etc.

Nonetheless, does the use of circular bio-based building materials always bring better environmental sustainability? Answering this question is crucial to providing a concrete basis for scholars, industry practitioners, and, especially, policymakers, to take appropriate actions towards the use of these materials. As presented in Table 28.5, several impact categories of these materials are worse than those of their traditional counterparts. For example, although reducing the climate impact, expanded cork agglomerate causes higher ozone depletion, acidification potential, and eutrophication impacts [59] compared to polystyrene, etc. This highlights that, in some cases, using circular bio-based building materials to mitigate climate change impacts may have to be a tradeoff with other impact indicators. To address this shortcoming, research needs to identify and address the hotspots among the life cycle stages of these materials to reduce their harm to the environment. Prioritising the development of materials based on local biowaste and by-products can reduce the impact of logistics and, therefore, reduce their general environmental impacts. Furthermore, according to Table 28.5, materials made from straw, recycled papers, or wood waste, have excellent environmental sustainability compared to traditional materials [5]; therefore, these materials should be prioritised. Finally, the decision-making process regarding the use of circular bio-based building materials should also consider the benefits of valorising biowaste or residues used to produce these materials.

As discussed in Sect. 4.1., only five studies evaluate the comprehensive sustainability, i.e., all three sustainable dimensions, of circular bio-based building materials. The findings of these studies indicate that circular bio-based building materials have better sustainability than traditional materials. For instance, RHA concrete [18] and agro-waste-based bricks [72] have better general sustainability than benchmarking materials. However, in these studies, not all environmental impact categories of circular bio-based building materials outperform those of conventional materials, underscoring again that using these materials does not always imply improved environmental sustainability [6]. Furthermore, it should be noted that most of these five studies exclude the impacts of the use, maintenance, and end-of-life phases of these materials, so the results may not be comprehensive and accurate [6]. Therefore, future studies should evaluate the comprehensive sustainability of circular bio-based building materials from a whole-lifecycle perspective to better understand their pros and cons and assist decision-making processes towards their use.

28.5 Conclusion

Although circular bio-based building materials offer promising opportunities to combat climate change, their contribution to a sustainable built environment needs to be better investigated. Plus, since these materials are mostly new and emerging, they have not yet gained enough attention and acknowledgement from scholars, practitioners, and policymakers. To convince stakeholders to use circular bio-based

Table 28.5 The environmental sustainability of circular bio-based building materials [5]

References	Circular bio-based building materials	Biological origins	Comparing the environmental impact categories of circular bio-based building materials to their traditional counterparts																		
			CC	OD	HT	PMF	POF	WT	TE	AP	EP	LU	WU	MD	FD	ARU	IR	CED	HH	ES	RS
Group 1: Circular bio-based insulation materials																					
[11]	Insulation panel	Pineapple by-product	–	–			+			+	+					-					
[12]	Sheep wool	Sheep wool	+	+	+	-	+	–	+	+	+	+	-	+	+		+				
[13]	Hempcrete	Hemp shiv	+																		
[14]	Binders	Hemp shives	+																		
[22]	Bricks	Wheat straw, olive stone flour	+	+	+	+	+	+	+	+	+	+		+	+		+				
[24]	Straw panel	Straw	+	+	+	+	+	+		+	+	–	–	+	+		+				
[25]	Straw bales	Straw	+															+			
[26]	Straw bale	Straw	+							+								+			
[27]	Straw wall	Straw	+		+		+	+	+	+	+			+	+						
[29]	Straw bales	Straw	+																		
[30]	Light straw clay	Straw	+			+				+								+			
[31]	Lightweight brick	Oyster shells, sorghum	+																		
[33]	Straw bales	Straw	+																		
[38]	Compressed straw	Straw																+			

(continued)

Table 28.5 (continued)

References	Circular bio-based building materials	Biological origins	Comparing the environmental impact categories of circular bio-based building materials to their traditional counterparts																		
			CC	OD	HT	PMF	POF	WT	TE	AP	EP	LU	WU	MD	FD	ARU	IR	CED	HH	ES	RS
[43]	Circular bio-based panels	Straw board, Sheep wool	–																		
[44]	Straw bales	Straw	+																		
[46]	Hemp lime	Hemp shiv	+															+			
[48]	Hempcrete	Hemp shiv	+				+			+			–	–	–			+			
[49]	Hempcrete	Hemp shiv	+																+	–	+
[53]	Hemp lime	Hemp shiv	+															+			
[54]	Hempcrete	Hemp shiv	+																		
[59]	Expanded cork agglomerate	Cork	+	-			+			–	–					+		+			
[60]	Cork slab	Cork	+										+					+			
[64]	Eco-sandwich material	Cork	–		+		–	–	–	–	–					–					
[66]	Wall assemblies	Coconut	+							+	+		+					+			
[69]	Multilayer panel	Coniferous bark	+	+			+			+	–							+			
[71]	Rice husk panels	Rice husk	+															+			
	Cork panel	Cork	+															+			
	Coffee chaff	Coffee chaff	+															+			

(continued)

Table 28.5 (continued)

References	Circular bio-based building materials	Biological origins	Comparing the environmental impact categories of circular bio-based building materials to their traditional counterparts																		
			CC	OD	HT	PMF	POF	WT	TE	AP	EP	LU	WU	MD	FD	ARU	IR	CED	HH	ES	RS
[72]	Waste based bricks	Cotton waste	+							+					+						
	Waste based bricks	Paper waste	+							+					+						
	Waste based bricks	Rice husk ash	+							+					+						
[73]	Panels	Waste paper	+															+			
[74, 75]	Paper wool	Old newsprint	+				+			+	+		+			+					
[76]	Cardboard panel	Recycled cardboard	+		+				+		+	–		–	–			–			
Group 2: Circular bio-based cementitious materials																					
[14]	Binders	Hemp shives	+																		
[83]	Lime mortars	Jacaranda seed pods	+																		
Group 3: Indoor-used circular bio-based building materials																					
[92]	Bio-based pigments	Wild fast-growing trees	+	+	+	+	+	+	+	+	+	+	+	+	+		+				
[93]	Bio-based biocomposite	Oil palm fibres																	+	–	+

(continued)

Table 28.5 (continued)

References	Circular bio-based building materials	Biological origins	Comparing the environmental impact categories of circular bio-based building materials to their traditional counterparts																		
			CC	OD	HT	PMF	POF	WT	TE	AP	EP	LU	WU	MD	FD	ARU	IR	CED	HH	ES	RS
[95]	Bio-coating	Corn stover	+	+	+	–	–		+	-	-				+						
[96]	Biocomposite	Oil palm fibres	+	-	+	+	+	+	-	+		–		–	+		+				
[98]	Bio-particleboards	Sawmill	+						-												
[99]	Shredded cotton board	Shredded cotton fibre	+	+	–	+	–	–	–	–	+	+		+	+		+				
[110]	Particleboards	Recycled wood waste	+		+		+			+	+					+					
[164]	Bio-concrete	Invasive alien wood chips	+																		
[165]	Hempcrete	Hemp shiv	+																		
	Straw	Straw	+																		

Note (+): Better; (-): worse; CC: Climate change; OD: Ozone depletion; HT: Human toxicity; PMF: Particulate matter formation; POF: Photochemical ozone formation; WT: Water ecotoxicity; TE: Terrestrial ecotoxicity; AP: Acidification potential; EP: Eutrophication; LU: Land use; WU: Water use; MD: Metal depletion; FD: Fossil depletion; ARU: Abiotic resource use; IR: Ionising radiation; CED: Cumulative energy demand; HH: Human health; ES: Ecosystem; RS: Resources. 4.5. Comprehensive sustainability

building materials, this chapter first provides a precise categorisation, highlights the potential utilisation, and analyses state-of-the-art sustainable assessments of these materials to identify their pros and cons regarding sustainability. This chapter also sheds light on the barriers to the adoption of these materials in both developed and developing economies, with the specific case of Flanders in Belgium and Vietnam. A comparative study is conducted to identify the differences between these two cases and collect lessons learned in Flanders that can be applied in Vietnam.

Particularly, circular bio-based building materials are grouped into three main groups: circular bio-based insulation materials, circular bio-based cementitious materials, and indoor-used circular bio-based building materials, with various potential uses ranging from insulation to masonry, finishing works, structural purposes, etc. These materials can be used in both new and existing buildings. Regarding the sustainability assessment, the existing studies mostly evaluate the environmental pillar of certain materials associated with the production phase, whatever their influence on the sustainability performance of a building in other phases, such as the use, maintenance, and end-of-life phases. The results indicate a lack of studies that deeply evaluate the social impacts and the comprehensive sustainability of these materials, especially at the building scale. Concerning the assessment methods, LCA is the most commonly used in the existing papers, yet most LCA studies consider only the cradle-to-gate impacts of materials. Furthermore, assessment methods used, mainly LCA, also have methodological issues, such as the lack of a clear description of material inventory data, shortcomings in the life cycle interpretation phase, end-of-life scenarios having low levels of sustainability and circularity, etc. These limitations affect the comparability, accuracy, and comprehensiveness of results and, therefore, need to be addressed in future studies.

The sustainability assessments indicate that, compared to traditional materials, circular bio-based building materials have better sustainable performance because they have lower production costs and fewer impacts on climate change, human toxicity, etc. The comprehensive sustainability scores of these materials are also higher than those of conventional materials. Nonetheless, the utilisation of these materials does not always bring better sustainability performance. When assessed by comprehensive methods like life cycle costing, the economic performance of these materials is slightly weaker than that of standard materials. LCA results also reveal that various impact categories of these materials are higher than those of benchmark materials. Therefore, when considering adopting these materials, comprehensive methods should be applied to evaluate their sustainability to ensure the appropriateness of the selection.

Next, barriers to the adoption of these materials exist in both developed and developing economies. 23 potential barriers identified from the existing literature are investigated in the cases of Flanders (a developed economy) and Vietnam (a developing economy) to identify the most significant barriers in each case and also the differences between these two cases. The investigation identifies 15 and 12 critical barriers to the adoption of these materials in Flanders and Vietnam, respectively. The results also show that cost- and risk-related barriers are deemed the most critical in Flanders, while the government's role-related barriers are the biggest hindrances to the adoption of circular bio-based building materials in Vietnam. These findings suggest that in the developing economy of Vietnam, circular bio-based materials cannot be successfully adopted without the involvement of the government in issuing building codes, regulations, and design guides. Besides, it seems that a lack of government promotion and incentives may also prevent the takeoff of these materials in the Vietnamese construction industry. To encourage adopters, the government needs to offer them appropriate promotions and incentives. This action is also essential to offset the additional costs and contribute to alleviating the most critical barriers in Flanders.

Generally, this chapter may be a good reference for not only scholars but also industry practitioners and especially policymakers in the construction industry. This is because its findings may assist these stakeholders in taking appropriate actions to promote the adoption of circular bio-based building materials, contributing to achieving a sustainable built environment. However, it should be noted that this chapter also has limitations. For instance, the sample size in both cases is relatively small, which may affect the comprehensiveness of the results on the adoption barriers of these materials. Furthermore, the characteristics of the sample in the two case studies are quite distinctive, which may affect the appropriateness of the comparative study. Future studies should increase the sample size in both cases and include more respondents with firsthand experience in using circular bio-based building materials in Vietnam to investigate if the results differ from those reported in this chapter.

Acknowledgements This chapter and related research have been conducted during and with the support of the Italian inter-university PhD course in Sustainable Development and Climate Change. The authors would like to acknowledge Arch. Alexis Versele and Prof. Chiara Piccardo at the Building Physics and Sustainable Design Group, Department of Civil Engineering, Ghent Campus, KU Leuven in Belgium, as well as Mr. Nguyen Trung Kien, Ms. Ly Thi Thuy Huong, and other staff of the Viet Land Investment Consulting and Construction Corporation (VILANDCO) in Vietnam for their support in collecting data. We also appreciate the survey respondents and blind reviewers for their time and support during the data collection and peer review of this chapter.

References

1. United Nations Environment Programme, *2020 Global Status Report for Buildings and Construction: Towards a Zero-emission, Efficient and Resilient Buildings and Construction Sector.* Nairobi (2020)
2. M. De Wit, et al., The circularity gap report: an analysis of the circular state of the global economy. Circ. Econ.: Amst. Neth. 1–36 (2018)
3. L.F. Cabeza, M. Chàfer Technological options and strategies towards zero energy buildings contributing to climate change mitigation: a systematic review. Energy Build. **219** (2020)
4. P. Enkvist, et al., The circular economy–a powerful force for climate mitigation: transformative innovation for prosperous and low-carbon industry, in Material Economics Sverige AB.: Stockholm, Sweden (2018)
5. D.L. Le, R. Salomone, Q.T. Nguyen Circular bio-based building materials: a literature review of case studies and sustainability assessment methods. Build. Environ. (2023)
6. D.L. Le, R. Salomone, Q.T. Nguyen Sustainability assessment methods for circular bio-based building materials: a literature review. J. Environ. Manag. **352** (2024)

7. J. Kirchherr, L. Piscicelli Towards an education for the circular economy (ECE): five teaching principles and a case study. Resour. Conserv. Recycl. **150** (2019)
8. C. Rabbat, et al., Sustainability of biomass-based insulation materials in buildings: current status in France, end-of-life projections and energy recovery potentials. Renew. Sustain. Energy Rev. **156** (2022)
9. C. Zhang, L. Canning, Application of non-conventional materials in construction. Proc. Inst. Civ. Eng.-Constr. Mater. **164**(4), 165–172 (2011)
10. J. Sierra-Perez et al., Environmental implications of the use of agglomerated cork as thermal insulation in buildings. J. Clean. Prod. **126**, 97–107 (2016)
11. D.A. Arellano-Vazquez, et al., Use of EPD system for designing new building materials: the case study of a bio-based thermal insulation panel from the pineapple industry by-product. Sustainability **12**(17) (2020)
12. A. Quintana-Gallardo et al., Life-cycle assessment and acoustic simulation of drywall building partitions with bio-based materials. Polymers (Basel) **12**(9), 1–16 (2020)
13. J.H. Arehart, W.S. Nelson, W.V. Srubar On the theoretical carbon storage and carbon sequestration potential of hempcrete. J. Clean. Prod. **266** (2020)
14. G. Bumanis, et al., Gypsum, geopolymers, and starch—alternative binders for bio-based building materials: a review and life-cycle assessment. Sustainability **12**(14) (2020)
15. F. Scrucca, et al., Energy and carbon footprint assessment of production of hemp hurds for application in buildings. Environ. Impact Assess. Rev. **84** (2020)
16. S.U. Azunna, et al., Characterization of lightweight cement concrete with partial replacement of coconut shell fine aggregate. SN Appl. Sci. **1**(6) (2019)
17. A.L.G. Gastaldini et al., Total shrinkage, chloride penetration, and compressive strength of concretes that contain clear-colored rice husk ash. Constr. Build. Mater. **54**, 369–377 (2014)
18. G. Garas, A.M. Sayed, E.S.H. Bakhoum, Application of nano waste particles in concrete for sustainable construction: a comparative study. Int. J. Sustain. Eng. **14**(6), 2041–2047 (2021)
19. A.L.G. Gastaldini et al., Influence of curing time on the chloride penetration resistance of concrete containing rice husk ash: a technical and economical feasibility study. Cement Concr. Compos. **32**(10), 783–793 (2010)
20. O. Gencel, et al., Properties of eco-friendly foam concrete containing PCM impregnated rice husk ash for thermal management of buildings. J. Build. Eng. **58** (2022)
21. R. Khan et al., Reduction in environmental problems using rice-husk ash in concrete. Constr. Build. Mater. **30**, 360–365 (2012)
22. C. Bories et al., Development of porous fired clay bricks with bio-based additives: study of the environmental impacts by life cycle assessment (LCA). Constr. Build. Mater. **125**, 1142–1151 (2016)
23. L. Mouton, K. Allacker, M. Röck Bio-based building material solutions for environmental benefits over conventional construction products—Life cycle assessment of regenerative design strategies (1/2). Energy Build. **282** (2023)
24. A. Quintana-Gallardo, et al., Waste valorization of rice straw as a building material in Valencia and its implications for local and global ecosystems. J. Clean. Prod. **318** (2021)
25. A.D. Gonzalez, Energy and carbon embodied in straw and clay wall blocks produced locally in the Andean Patagonia. Energy Build. **70**, 15–22 (2014)
26. L. Brojan, A. Petric, P.L. Clouston, Comparative study of brick and straw bale wall systems from environmental, economical and energy perspectives. ARPN J. Eng. Appl. Sci. **8**(11), 920–926 (2013)
27. J.L. Menet, I.C. Gruescu, A comparative life cycle assessment of exterior walls constructed using natural insulation materials. Environ. Eng. Sustain. Dev. Entrep. **1**(4), 14 (2012)
28. F. Pittau et al., Fast-growing bio-based materials as an opportunity for storing carbon in exterior walls. Build. Environ. **129**, 117–129 (2018)
29. V. Zieger, T. Lecompte, A. Hellouin de Menibus Impact of GHGs temporal dynamics on the GWP assessment of building materials: a case study on bio-based and non-bio-based walls. Build. Environ. **185** (2020)

30. L. Ben-Alon, et al., Life cycle assessment (LCA) of natural vs conventional building assemblies. Renew. Sustain. Energy Rev. **144** (2021)
31. H.-Y. Liu, H.-S. Wu, C.-P. Chou Study on engineering and thermal properties of environment-friendly lightweight brick made from Kinmen oyster shells & sorghum waste. Constr. Build. Mater. **246** (2020)
32. G. Mutani, et al., Straw buildings: a good compromise between environmental sustainability and energy-economic savings. Appl. Sci.-Basel **10**(8) (2020)
33. B. Sodagar et al., The carbon-reduction potential of straw-bale housing. Build. Res. Inf. **39**(1), 51–65 (2011)
34. F. Rosso et al., A sustainable approach towards the retrofit of the public housing building stock: energy-architectural experimental and numerical analysis. Sustainability **13**(5), 1–19 (2021)
35. C.H. Koh, D. Kraniotis Hygrothermal performance, energy use and embodied emissions in straw bale buildings. Energy Build. **245** (2021)
36. X. Yin, et al., Energy-saving potential of applying prefabricated straw bale construction (PSBC) in domestic buildings in Northern China. Sustainability **12**(8) (2020)
37. H. Guo, et al., Energy sustainability of bio-based building materials in the cold and severe cold regions of china—a case study of residential buildings. Appl. Sci. **10**(5) (2020)
38. S. Cascone et al., Energy performance and environmental and economic assessment of the platform frame system with compressed straw. Energy Build. **166**, 83–92 (2018)
39. C. Cornaro, et al., An innovative straw bale wall package for sustainable buildings: experimental characterization, energy and environmental performance assessment. Energy Build. **208** (2020)
40. F. D'Alessandro et al., Straw bale constructions: laboratory, in field and numerical assessment of energy and environmental performance. J. Build. Eng. **11**, 56–68 (2017)
41. G. Garas, M. Allam, R. El Dessuky, Straw bale construction as an economic environmental building alternative-a case study. ARPN J. Eng. Appl. Sci. **4**(9), 54–59 (2009)
42. A. Vilaboa Díaz, A. Francisco López, P.M. Bello Bugallo Analysis of biowaste-based materials in the construction sector: evaluation of thermal behaviour and life cycle assessment (LCA). Waste Biomass Valorization **13**(12), 4983–5004 (2022)
43. V. Cascione, et al., Integration of life cycle assessments (LCA) in circular bio-based wall panel design. J. Clean. Prod. **344** (2022)
44. E. Krasny, S. Klarić, A. Korjenić, Analysis and comparison of environmental impacts and cost of bio-based house versus concrete house. J. Clean. Prod. **161**, 968–976 (2017)
45. K. Ip, A. Miller, Life cycle greenhouse gas emissions of hemp–lime wall constructions in the UK. Resour. Conserv. Recycl. **69**, 1–9 (2012)
46. A.S. Nordby, A.D. Shea, Building materials in the operational phase: Impacts of direct carbon exchanges and hygrothermal effects. J. Ind. Ecol. **17**(5), 763–776 (2013)
47. M. Sinka et al., Comparative life cycle assessment of magnesium binders as an alternative for hemp concrete. Resour. Conserv. Recycl. **133**, 288–299 (2018)
48. S. Pretot, F. Collet, C. Garnier, Life cycle assessment of a hemp concrete wall: impact of thickness and coating. Build. Environ. **72**, 223–231 (2014)
49. M.D. Heidari, et al., Regionalised life cycle assessment of bio-based materials in construction; the case of hemp shiv treated with sol-gel coatings. Materials (Basel) **12**(18) (2019)
50. M. Sinka et al., Hempcrete—CO2 neutral wall solutions for 3D printing. Environ. Clim. Technol. **26**(1), 742–753 (2022)
51. A. Arrigoni et al., Life cycle assessment of natural building materials: the role of carbonation, mixture components and transport in the environmental impacts of hempcrete blocks. J. Clean. Prod. **149**, 1051–1061 (2017)
52. F. Bennai, et al., Assessment of hygrothermal performance of hemp concrete compared to conventional building materials at overall building scale. Constr. Build. Mater. **316** (2022)
53. Y. Florentin et al., A life-cycle energy and carbon analysis of hemp-lime bio-composite building materials. Energy Build. **156**, 293–305 (2017)

54. Y. Shang, Tariku F Hempcrete building performance in mild and cold climates: Integrated analysis of carbon footprint, energy, and indoor thermal and moisture buffering. Build. Environ. **206** (2021)
55. T. Moussa et al., Bio-based and recycled-waste materials in buildings: a study of energy performance of hemp-lime concrete and recycled-polyethylene terephthalate façades for office facilities in France and Italy. Sci. Technol. Built Environ. **24**(5), 492–501 (2018)
56. C. Maalouf et al., An energy and carbon footprint assessment upon the usage of hemp-lime concrete and recycled-PET façades for office facilities in France and Italy. J. Clean. Prod. **170**, 1640–1653 (2018)
57. M. Sinka et al., Fast setting binders for application in 3D printing of bio-based building materials. Sustainability **12**(21), 1–12 (2020)
58. M. Demertzi et al., Environmental performance of expanded cork slab and granules through life cycle assessment. J. Clean. Prod. **145**, 294–302 (2017)
59. N. Pargana et al., Comparative environmental life cycle assessment of thermal insulation materials of buildings. Energy and Buildings **82**, 466–481 (2014)
60. I. Zabalza Bribián, A. Valero Capilla, A. Aranda Usón Life cycle assessment of building materials: Comparative analysis of energy and environmental impacts and evaluation of the eco-efficiency improvement potential. Build. Environ. **46**(5), 1133–1140 (2011)
61. J.D. Silvestre, et al., Insulation cork boards-environmental life cycle assessment of an organic construction material. Materials (Basel) **9**(5) (2016)
62. F. Barreca et al., Innovative use of giant reed and cork residues for panels of buildings in Mediterranean area. Resour. Conserv. Recycl. **140**, 259–266 (2019)
63. A. Torres-Rivas, et al., Systematic combination of insulation biomaterials to enhance energy and environmental efficiency in buildings. Constr. Build. Mater. **267** (2021)
64. A.D. La Rosa et al., Environmental impacts and thermal insulation performance of innovative composite solutions for building applications. Constr. Build. Mater. **55**, 406–414 (2014)
65. A. Torres-Rivas et al., Multi-objective optimisation of bio-based thermal insulation materials in building envelopes considering condensation risk. Appl. Energy **224**, 602–614 (2018)
66. N. Keena, et al., A life-cycle approach to investigate the potential of novel biobased construction materials toward a circular built environment. Energies **15**(19) (2022)
67. J. Jeon et al., Latent heat storage biocomposites of phase change material-biochar as feasible eco-friendly building materials. Environ. Res. **172**, 637–648 (2019)
68. R. Opoku, et al., Minimizing heat transmission loads and improving energy efficiency of building envelopes in sub-Saharan Africa using bio-based composite materials. Sci. Afr. **8** (2020)
69. A. Barrio, et al., life cycle sustainability assessment of a novel bio-based multilayer panel for construction applications. Resources, **10**(10) (2021)
70. A. El hammouti, et al., Thermal insulation of mud bricks made with walnut shells: Characterization and simulation study. Mater. Today: Proc. **62**, 4545–4550 (2022)
71. C. Buratti et al., Rice husk panels for building applications: thermal, acoustic and environmental characterization and comparison with other innovative recycled waste materials. Constr. Build. Mater. **171**, 338–349 (2018)
72. S.N. Joglekar et al., Sustainability assessment of brick work for low-cost housing: a comparison between waste based bricks and burnt clay bricks. Sustain. Cities Soc. **37**, 396–406 (2018)
73. P. Ricciardi, E. Belloni, F. Cotana, Innovative panels with recycled materials: thermal and acoustic performance and life cycle assessment. Appl. Energy **134**, 150–162 (2014)
74. A.C. Schmidt, A.A. Jensen, A.U. Clausen, O. Kamstrup, D. Postlethwaite, A comparative life cycle assessment of building insulation products made of stone wool, paper wool and flax. Int. J. Life Cycle Assess. **9**, 53–66 (2004)
75. A.C. Schmidt, A.A. Jensen, A.U. Clausen, O. Kamstrup, D. Postlethwaite, A comparative life cycle assessment of building insulation products made of stone wool, paper wool and flax. Int. J. Life Cycle Assess. **9**(2), 122–129 (2004)

76. S. Secchi et al., Experimental and environmental analysis of new sound-absorbing and insulating elements in recycled cardboard. J. Build. Eng. **5**, 1–12 (2016)
77. F. Ntimugura, et al., Environmental performance of miscanthus-lime lightweight concrete using life cycle assessment: application in external wall assemblies. Sustain. Mater. Technol. **28** (2021)
78. Y.M. Degu, Effect of curing time on compressive strength of agrostone building material with water hyacinth, bagasse and grass bio-fillers—a comparative study. J. Eng. Sci. Technol. **16**(1), 792–806 (2021)
79. M. Palumbo et al., Bio-based insulation materials and their hygrothermal performance in a building envelope system (ETICS). Energy Build. **174**, 147–155 (2018)
80. N. Van Tuan et al., The study of using rice husk ash to produce ultra high performance concrete. Constr. Build. Mater. **25**(4), 2030–2035 (2011)
81. N. Van Tuan et al., Hydration and microstructure of ultra high performance concrete incorporating rice husk ash. Cem. Concr. Res. **41**(11), 1104–1111 (2011)
82. K.T. Tong, R. Vinai, M.N. Soutsos, Use of Vietnamese rice husk ash for the production of sodium silicate as the activator for alkali-activated binders. J. Clean. Prod. **201**, 272–286 (2018)
83. D. Ayasgil, et al., The long-term engineering properties and sustainability indices of dewatering hydrated lime mortars through Jacaranda seed pods. Sustain. Mater. Technol. **32** (2022)
84. C. Ince, S. Tayançlı, S. Derogar Recycling waste wood in cement mortars towards the regeneration of sustainable environment. Constr. Build. Mater. **299** (2021)
85. M. Saeli, et al., Architectural technologies for life environment: spent coffee ground reuse in lime-based mortars. a preliminary assessment for innovative green thermo-plasters. Constr. Build. Mater. **319** (2022)
86. G. La Scalia et al., Coffee biowaste valorization within circular economy: an evaluation method of spent coffee grounds potentials for mortar production. Int. J. Life Cycle Assess. **26**(9), 1–11 (2021)
87. P. Ricciardi et al., Environmental characterisation of coffee chaff, a new recycled material for building applications. Constr. Build. Mater. **147**, 185–193 (2017)
88. G.C.H. Doudart de la Gree, et al., Contaminated biomass fly ashes--Characterization and treatment optimization for reuse as building materials. Waste Manag. **49**, 96–109 (2016)
89. M. Valtere, et al., The Versatility of the Bioeconomy. Sustainability Aspects of the Use of Bran. Environ. Clim. Technol. **26**(1), 658–669 (2022)
90. M. Ahmed et al., An overview of Asian cement industry: environmental impacts, research methodologies and mitigation measures. Sustain. Prod. Consum. **28**, 1018–1039 (2021)
91. R. Pode, Potential applications of rice husk ash waste from rice husk biomass power plant. Renew. Sustain. Energy Rev. **53**, 1468–1485 (2016)
92. F. Rosso, et al., Natural, bio-based, colored linoleum: design, preparation, characteristics and preliminary life cycle assessment. J. Clean. Prod. **267** (2020)
93. S.M. Khoshnava, et al., The role of green building materials in reducing environmental and human health impacts. Int. J. Environ. Res. Public Health **17**(7) (2020)
94. R. Horn, et al., Bio-inspired sustainability assessment for building product development—concept and case study. Sustainability **10**(2) (2018)
95. M. Montazeri, M.J. Eckelman, Life cycle assessment of UV-Curable bio-based wood flooring coatings. J. Clean. Prod. **192**, 932–939 (2018)
96. S.M. Khoshnava et al., A cradle-to-gate based life cycle impact assessment comparing the KBF w EFB hybrid reinforced poly hydroxybutyrate biocomposite and common petroleum-based composites as building materials. Environ. Impact Assess. Rev. **70**, 11–21 (2018)
97. O. Platnieks, et al., Sustainable tetra pak recycled cellulose / Poly(Butylene succinate) based woody-like composites for a circular economy. J. Clean. Prod. **270** (2020)
98. C. Ganne-Chédeville, S. Diederichs, Potential environmental benefits of ultralight particleboards with biobased foam cores. Int. J. Polym. Sci. **2015**, 1–14 (2015)

99. A. Quintana et al., Comparative life cycle assessment of gypsum plasterboard and a new kind of bio-based epoxy composite containing different natural fibers. J. Clean. Prod. **185**, 408–420 (2018)
100. M. Demertzi et al., Environmental performance of a cork floating floor. Mater. Des. **82**, 317–325 (2015)
101. M.F.N. dos Santos, et al., Comparative study of the life cycle assessment of particleboards made of residues from sugarcane bagasse (Saccharum spp.) and pine wood shavings (Pinus elliottii). J. Clean. Prod. **64**, 345–355 (2014)
102. R. Garcia, F. Freire, Carbon footprint of particleboard: a comparison between ISO/TS 14067, GHG Protocol, PAS 2050 and climate declaration. J. Clean. Prod. **66**, 199–209 (2014)
103. H. Kouchaki-Penchah et al., Gate to gate life cycle assessment of flat pressed particleboard production in Islamic Republic of Iran. J. Clean. Prod. **112**, 343–350 (2016)
104. H. Merrild, T.H. Christensen, Recycling of wood for particle board production: accounting of greenhouse gases and global warming contributions. Waste Manag Res **27**(8), 781–788 (2009)
105. B. Rivela et al., Life cycle inventory of particleboard: a case study in the wood sector. Int. J. Life Cycle Assess. **11**(2), 106–113 (2006)
106. B. Rivela et al., Life cycle assessment of wood wastes: a case study of ephemeral architecture. Sci. Total. Environ. **357**(1–3), 1–11 (2006)
107. A.M. Saravia-Cortez et al., Assessing environmental sustainability of particleboard production process by ecological footprint. J. Clean. Prod. **52**, 301–308 (2013)
108. J.B. Wilson, Life-cycle inventory of particleboard in terms of resources, emissions, energy and carbon. Wood Fiber Sci. 90–106 (2010)
109. M.U. Hossain et al., Environmental and technical feasibility study of upcycling wood waste into cement-bonded particleboard. Constr. Build. Mater. **173**, 474–480 (2018)
110. V. Uemura Silva, et al., Circular vs. linear economy of building materials: a case study for particleboards made of recycled wood and biopolymer vs. conventional particleboards. Constr. Build. Mater. **285** (2021)
111. R. Sani, A. Nzihou, Production of clay ceramics using agricultural wastes: study of properties, energy savings and environmental indicators. Appl. Clay Sci. **146**, 106–114 (2017)
112. A. Galimshina, et al., Bio-based materials as a robust solution for building renovation: a case study. Appl. Energy **316** (2022)
113. R. Volk, J. Stengel, F. Schultmann, Building information modeling (bim) for existing buildings—literature review and future needs. Autom. Constr. **38**, 109–127 (2014)
114. D. Safarik, et al., Effect of rice husk ash and fly ash on the workability of concrete mixture in the high-rise construction. E3S Web Conf. **33** (2018)
115. D.L. Le, et al., Factors Influencing the Implementation of Circular Bio-Based Building Materials in Vietnam. in *Proceedings of the 29th International Sustainable Development Research Society (ISDRS) Conference, Half-way through Agenda 2030: Assessing the 5Ps of SDGs (People, Planet, Prosperity, Peace, and Partnership)*. Malaysia: Universiti Kebangsaan Malaysia (UKM) (2023)
116. S. Gounder et al., Barriers to the use of sustainable materials in Australian building projects. Eng. Constr. Archit. Manag. **30**(1), 189–209 (2021)
117. J. Giesekam, J.R. Barrett, P. Taylor, Construction sector views on low carbon building materials. Build. Res. & Inf. **44**(4), 423–444 (2015)
118. P.O. Akadiri, Understanding barriers affecting the selection of sustainable materials in building projects. J. Build. Eng. **4**, 86–93 (2015)
119. S. Durdyev, et al., Sustainable construction industry in Cambodia: awareness, drivers and barriers. Sustainability **10**(2) (2018)
120. S. AlSanad, Awareness, drivers, actions, and barriers of sustainable construction in Kuwait. Procedia Eng. **118**, 969–983 (2015)
121. J. Ameh, A. Soyingbe, O. Oyediran Acceptability and use of innovative bamboo products for the construction of residential buildings in Nigeria. Int. J. Technol. **10**(4) (2019)

122. J. Giesekam et al., The greenhouse gas emissions and mitigation options for materials used in UK construction. Energy Build. **78**, 202–214 (2014)
123. K. Krueger, A. Stoker, G. Gaustad, "Alternative" materials in the green building and construction sector. Smart Sustain. Built Environ. **8**(4), 270–291 (2019)
124. A. Darko et al., Examining issues influencing green building technologies adoption: the United States green building experts' perspectives. Energy Build. **144**, 320–332 (2017)
125. B. Cease et al., Barriers and incentives for sustainable urban development: an analysis of the adoption of LEED-ND projects. J. Environ. Manage. **244**, 304–312 (2019)
126. T. Nemoto, D. Beglar Likert-scale questionnaires. in *JALT 2013 conference proceedings.* (2014)
127. D.A. Abowitz, T.M. Toole, Mixed method research: fundamental issues of design, validity, and reliability in construction research. J. Constr. Eng. Manag. **136**(1), 108–116 (2010)
128. S.P. Low, S. Gao, S.K. Ng The adoption of mass-engineered timber (MET) in the Singapore construction industry: Barriers and drivers. J. Clean. Prod. **327** (2021)
129. A.P.C. Chan et al., Critical barriers to green building technologies adoption in developing countries: the case of Ghana. J. Clean. Prod. **172**, 1067–1079 (2018)
130. J.C Nunnally, I.H. Bernstein *Psychometric theory (3rd Ed.).* New York: McGraw-Hill (1994)
131. T. D'Haenens, *Flanders, Belgium*
132. C. Charlotte, et al., *Final report on the Green Deal for Circular Construction.* (2023)
133. L. Mouton, et al. Preliminary study on the use of Big Data for environmental benchmarks of residential buildings in Flanders. in *IOP Conference Series: Earth and Environmental Science.* (IOP Publishing) (2023)
134. S.A. Kulp, B.H. Strauss, New elevation data triple estimates of global vulnerability to sea-level rise and coastal flooding. Nat. Commun. **10**(1), 4844 (2019)
135. US Department of Agriculture *Global leading rice producing countries based on area harvested in 2021/2022 (in million hectares)* [Graph].* Statista (2023)
136. ICO, *Coffee production worldwide from 2003/04 to 2020/21 (in million 60 kilogram bags)* [Graph].* (2022)
137. P.T. Duong, H. Yoshiro Current situation and possibilities of rice straw management in Vietnam. University of Tsukuba, (2015)
138. G.Y. Qi et al., The drivers for contractors' green innovation: an industry perspective. J. Clean. Prod. **18**(14), 1358–1365 (2010)
139. Statistics IBM SPSS, *SPSS Statistics for Windows, Version 29.0.* IBM Corp. Released, 2022
140. H.-T. Nguyen et al., Will green building development take off? An exploratory study of barriers to green building in Vietnam. Resour. Conserv. Recycl. **127**, 8–20 (2017)
141. G. Scialpi, D. Perrotti, Circular economy in the valorisation of food and other biowaste: case studies in small and medium-sized enterprises in the Belgian construction sector. Environ. Sci. Pollut. Res. Int. **31**(12), 17914–17931 (2023)
142. D.L. Le, et al., Status and barriers to circular bio-based building material adoption in developed economies: the case of flanders, belgium. unpublished manuscript, (2024)
143. The Prime Minister of Vietnam, *Decision No. 882/QĐ-TTg, Approving the national action plan on green growth for the 2021–2030 period.* (2022)
144. E. Markström, et al., Drivers and barriers for an increased use of bio-based building materials in Sweden. in *12th meeting of the Northern European Network for Wood Science and Engineering (WSE),* Riga 12–13 September 2016. (Latvian State Institute of Wood Chemistry) (2016)
145. E. Markström, et al., Use of bio-based building materials: perceptions of Swedish architects and contractors. in *Forest Products Society International Convention:* 26–29 June 2016. Forest Products Society (2016)
146. M. Sodangi, Z.A. Kazmi Integrated evaluation of the impediments to the adoption of coconut palm wood as a sustainable material for building construction. Sustainability **12**(18) (2020)
147. E. Hurmekoski, R. Jonsson, T. Nord, Context, drivers, and future potential for wood-frame multi-story construction in Europe. Technol. Forecast. Soc. Chang. **99**, 181–196 (2015)
148. K. Ernest, S. Mohammed Abdulai, Y. Daniel Drivers militating against the pricing of sustainable construction materials: the Ghanaian quantity surveyors perspective. Case Stud. Constr. Mater. **8**, 507–516 (2018)

149. M. Riala, L. Ilola, Multi-storey timber construction and bioeconomy—barriers and opportunities. Scand. J. For. Res. **29**(4), 367–377 (2014)
150. A. Viļuma, U. Bratuškins, Barriers for use of wood in architecture: the Latvian case. Arch. Urban Plan. **13**(1), 43–47 (2017)
151. O.S. Dosumu, C. Aigbavboa, An investigation of the barriers to the uptake of local materials in Africa: a literature review approach. Afr. J. Sci. Technol. Innov. Dev. **12**(4), 365–371 (2019)
152. J. O'Connor, et al., Wood use in nonresidential buildings: opportunities and barriers. For. Prod. J. **54**(3) (2004)
153. A. Gosselin et al., Main motivations and barriers for using wood in multi-story and non-residential construction projects. BioResources **12**(1), 546–570 (2017)
154. B.K. Baiden, E. Badu, F.S. Menz, Exploring the barriers to the use and potential of timber for housing construction in Ghana. Constr. Build. Mater. **19**(5), 347–352 (2005)
155. V. Enjily, J. Bregulla Barriers to the enhanced use of wood. CEI Bois: Rue Montoyer, Brussels, (2013)
156. O. Ametepey, C. Aigbavboa, K. Ansah, Barriers to successful implementation of sustainable construction in the Ghanaian construction industry. Procedia Manuf. **3**, 1682–1689 (2015)
157. F. Franzini, R. Toivonen, A. Toppinen Why Not Wood? Benefits and barriers of wood as a multistory construction material: perceptions of municipal civil servants from Finland. Buildings **8**(11) (2018)
158. E. Leszczyszyn, et al., The future of wood construction: opportunities and barriers based on surveys in Europe and Chile. Sustainability **14**(7) (2022)
159. M.F. Laguarda Mallo, O. Espinoza awareness, perceptions and willingness to adopt cross-laminated timber by the architecture community in the United States. J. Clean. Prod. **94**, 198–210 (2015)
160. A.M. Walker, et al., What Is the Relation between Circular Economy and Sustainability? Answers from Frontrunner Companies Engaged with Circular Economy Practices. Circ Econ Sustain 1–28 (2021)
161. M. Ranjbari et al., Three pillars of sustainability in the wake of COVID-19: a systematic review and future research agenda for sustainable development. J. Clean. Prod. **297**, 126660 (2021)
162. R. Salomone et al., Environmental impact of food waste bioconversion by insects: application of life cycle assessment to process using Hermetia illucens. J. Clean. Prod. **140**, 890–905 (2017)
163. ISO, ISO 14040: Environmental management—Life cycle assessment—Principles and framework. (2006)
164. V. Göswein, et al., Invasive alien plants as an alternative resource for concrete production – multi-scale optimization including carbon compensation, cleared land and saved water runoff in South Africa. Resour. Conserv. Recycl. **167** (2021)
165. F. Pittau et al., Retrofit as a carbon sink: the carbon storage potentials of the EU housing stock. J. Clean. Prod. **214**, 365–376 (2019)

Chapter 29
Cradle-To-Grave LCA of a Bio-Based Retrofitting of an Old Italian Dwelling

Nicolas Bertoldo, Sergio Renato Maria Sabbadini, Gianluca Ruggieri, and Giovanni Dotelli

Abstract The construction industry stands as a primary energy consumer and a significant emitter of Greenhouse gases, making a substantial environmental impact. The industry's extensive material usage and inadequate building energy efficiency compound the detrimental environmental effects. Notably, the energy performance of buildings within the Italian built-stock registers particularly low figures. A closer examination of energy certifications issued in 2021 reveals that 34% of these certifications fall into the G class. This study explores the potential environmental advantages of using bio-based materials for building retrofitting. To achieve this, a comparative cradle-to-grave Life Cycle Assessment (LCA) will be conducted on a real case retrofit project. The LCA will compare two design scenarios: one utilizing bio-based materials and the other using conventional synthetic materials. Results indicate that material production has the greatest impact across most categories, but building energy use dominates global warming, abiotic depletion, and ozone depletion. Bio-based materials offer a potential environmental benefit, particularly for global warming potential where a 96% reduction is observed compared to conventional materials. This advantage is attributed to the biogenic carbon sequestration of bio based materials.

29.1 Introduction

The construction industry is a critical sector for global development, providing essential infrastructure and shaping our built environment. However, this progress comes at a significant environmental cost. The industry is a major consumer of energy,

N. Bertoldo · G. Dotelli (✉)
Department of Chemistry, Materials and Chemical Engineering "Giulio Natta", Politecnico Di Milano, Milan, Italy
e-mail: giovanni.dotelli@polimi.it

S. R. M. Sabbadini
A.N Studio Di Architettura Disstudio.It, a.N.a.B., Milan, Italy

G. Ruggieri
Dipartimento Di Scienze Teoriche E Applicate, Università Dell'Insubria, Varese, Italy

D. A. H. Hanaor (ed.), *Sustainable Materials for the Built Environment*, Springer Series in Materials Science 352, https://doi.org/10.1007/978-3-031-97818-0_29

accounting for 34% of global energy consumption, and a leading emitter of greenhouse gases, contributing 37% of global emissions [1]. This substantial environmental impact arises from two key factors: the extensive use of resource-intensive materials and the often inadequate energy efficiency of buildings [2].

Material usage throughout the construction lifecycle, from extraction and processing to transportation and on-site construction, creates a significant environmental burden. Energy-intensive processes like cement production contribute heavily to greenhouse gas emissions [3]. Additionally, the construction of buildings often fails to prioritize energy efficiency, leading to buildings that require excessive energy for heating, cooling, and lighting. This is particularly concerning in Italy, where a recent analysis of energy certifications in 2023 revealed a disproportionate share (31.3%) of buildings falling into the lowest energy efficiency class (G), with only a small percentage (11.2%) achieving the highest efficiency (A classes) [4]. This highlights the urgent need for strategies to improve the energy performance of the existing building stock and future constructions.

By carefully selecting building materials, we can minimize the environmental impact of construction [5, 6]. Bio-based materials, for example, offer a sustainable alternative to traditional materials. Their production can even act as a form of carbon sequestration, trapping the CO_2 absorbed by plants during their growth phase [5, 7–9]. Additionally, many natural materials offer comparable thermal performance to conventional materials. This means that by embracing new construction paradigms that prioritize the use of bio-based and other sustainable materials, we can achieve good building thermal insulation and enhance indoor air quality and comfort, all while significantly reducing a building's energy consumption [10].

Retrofitting existing buildings offers a key approach to achieving sustainability goals [11]. Compared to constructing new buildings, it presents significant environmental, social, and economic advantages [11, 12]. Constructing a new building necessitates the use of vast quantities of raw materials, each with its environmental footprint. Moreover, building new structures consumes land, leading to habitat loss, ecosystem disruption, and a strain on natural resources [13–15]. Retrofitting breathes new life into existing structures, minimizing the need for virgin materials and the associated environmental damage from their extraction and processing [12]. Furthermore, retrofitting an existing building is often cheaper than demolishing and rebuilding a new structure [11]. Retrofitting can significantly reduce a building's operational energy consumption, leading to a reduction in operational costs. By extending the lifespan of buildings, retrofits contribute to the preservation of valuable natural resources like timber, steel, and concrete [14].

Retrofitting can play a vital role in preserving the character and history embodied in existing structures [16, 17]. Improving the building envelope and ventilation systems within a retrofit can significantly enhance indoor environmental quality for occupants. This translates to healthier and more comfortable living and working environments, potentially leading to improved well-being and productivity [16].

The European Union (EU) has established ambitious goals for energy efficiency, aiming to reduce its overall energy consumption and greenhouse gas emissions [18–20]. Achieving these targets demands a multi-purpose approach, and building retrofitting plays a key role in this strategy. The huge volume of existing buildings in the EU necessitates a focus on improving their performance as well. Buildings are responsible for a significant portion of the overall energy consumption [1, 20], and a large percentage of these structures were built before stringent energy efficiency standards were in place [4, 21]. Retrofitting these existing buildings offers a much faster and more cost-effective path towards achieving the EU's energy efficiency goals compared to solely relying on new construction [11, 22].

While building retrofits offer a promising strategy for achieving energy efficiency goals, a comprehensive assessment of their environmental impact across their entire lifecycle is essential to ensure their potential. Life Cycle Assessment (LCA) methodology provides a robust framework for such an evaluation. LCA studies allow quantifying the environmental burdens associated with a retrofit project, encompassing all stages from the cradle (e.g., extraction of raw materials for new building components) to the grave (e.g., demolition and disposal of the renovated structure), including the operational energy use of the retrofitted building.

A critical challenge in LCA studies lies in acquiring reliable and consistent data on the environmental impact of materials and processes. This is where Environmental Product Declarations (EPDs) emerge as a valuable resource. Environmental Product Declarations (EPD) are type III environmental declarations defined by ISO 14025 [23]. An EPD is a technical document that provides a standardized and verified assessment of a product's environmental footprint throughout its life cycle [24]. To ensure consistent comparisons between products, EPDs establish common criteria for each category. For what concerns construction products, the category rules are defined by the standard EN 15,804 [25].

This study investigates the environmental benefits of utilizing bio-based construction materials in building retrofitting. We aim to achieve this by conducting a comparative LCA of a real case retrofit project. The LCA will assess two design scenarios: one employing bio-based materials and the other using conventional synthetic materials.

Through this comparative approach, the study seeks to address the following research questions:

Question 1. To what extent does the choice of materials during a building retrofitting design influence the overall environmental footprint of the building across its entire lifecycle?

Question 2. Can bio-based construction materials offer a significant reduction in the environmental impact of buildings compared to conventional synthetic materials?

29.2 Materials and Methods

This study aims to assess the potential environmental impact of a retrofitting project employing bio-based construction materials in the restoration of a historic Italian dwelling. To achieve this goal, we conducted a cradle-to-grave LCA in a practical case study involving the retrofitting of an old farmstead in Italy, utilizing bio-based materials, following the attributional approach [26, 27]. Through this methodology, we assessed the environmental impact associated with material production, and carbon sequestration within bio-based materials, and compared it to a conventional scenario employing synthetic materials.

In the following sections, first, the case study is introduced, followed by a detailed description of the LCA phases.

29.2.1 Case Study

The object of this study is a retrofitting project of an historic Italian farmstead, located in the town of Casatenovo (LC). When the renovation concerns, as in this case, a historic farmstead, the dialectic comes into play between the preservation of architectural, constructional and material characteristics and the issue of improving energy performance required by legislation [28]. In this case, the complexity is accentuated by the fact that the intervention concerned only one portion of the entire farmstead (Fig. 29.1), a closed courtyard farmstead once used for livestock fairs and then agricultural equipment. The dwelling develops over two floors, with an overall useful surface of 104 m^2. The original plasterwork of both the interior and exterior façades consisting of lime mortar and sand not left rough as workmanship and not completed by finishing or painting, characterised the portion of the farmstead undergoing the intervention, making it possible to justify choices compatible with the existing building while differentiating it from the modern cement plasters of the neighbouring portions. The energy recovery choices concerned both the envelope and the building systems that previously consisted only of the heating of a chimney on the ground floor. Constrained by the façade alignments, the insulation of the external walls was carried out for the thickness allowed on the façade and for the rest by acting on the internal part of the masonry. The new ventilated crawl space, which also involved reinforcing the masonry, was insulated with a cork and vitrified screed. The plant engineering involved radiant heating systems positioned mainly on the ceiling. The heating circuits are powered by the condensing boiler assisted by a heat pump, with the possibility of a photovoltaic supply, to produce domestic hot water. For the mid-seasons, there is also a thermo-fireplace that replaces the original, less-performing one. Summer air conditioning has not been entrusted to the plant engineering but to the bioclimatic operation summer operation of the building with large wall masses that contribute to thermal inertia, the existing and new openings that guarantee excellent air feedback, to the solar protection given by the eaves overhangs and the existing south-facing loggia. The parameters of the thermal improvements were always compared with the architectural choices, respecting the historical building and the healthiness of the building. The materials and techniques applied had to guarantee the breathability of the walls and low environmental impact, favouring natural insulators of vegetable origin: panels in mineralised wood wool, mineralised wood fibre, hemp fibre and hemp-lime.

Fig. 29.1 Picture of the farmstead with the portion involved in the retrofitting project highlighted in green

The environmental impact principle of the materials also involved the use of raw earth; the same red earth rich in clay and iron oxides that were excavated on-site to create the aerated crawl space. The compounds for plasters and finishes were formulated and fine-tuned on-site differentiated according to the use and function they were to perform. An example of the intervention carried out on the ground floor of the farmstead is shown in Fig. 29.2, comparing the situation before and after the renovation.

This paper explores this retrofitting project in detail, focusing on the potential of bio-based materials and improved building design to reduce the environmental impact of the construction industry, particularly within the context of the Italian built environment, strongly vernacular. We explored the environmental benefits of bio-based materials, and their thermal and indoor air quality properties, comparing them to an alternative scenario employing conventional synthetic materials.

29.2.2 Goal and Scope Definition

In this study, the goal is to assess the environmental impact of a retrofitting project employing bio-based materials and in order to understand the environmental implications of such design, we compared it with a conventional one using synthetic materials. With this study we want to understand the importance of materials selection in a retrofitting project, considering the entire life cycle of the building, from the extraction of raw materials to the end of life after the demolition stage. We would like first to understand the contribution of the production of the materials over the whole life cycle. The outcomes of this study could be used to support the materials selection in the design stage of a retrofitting project, presenting a real case study.

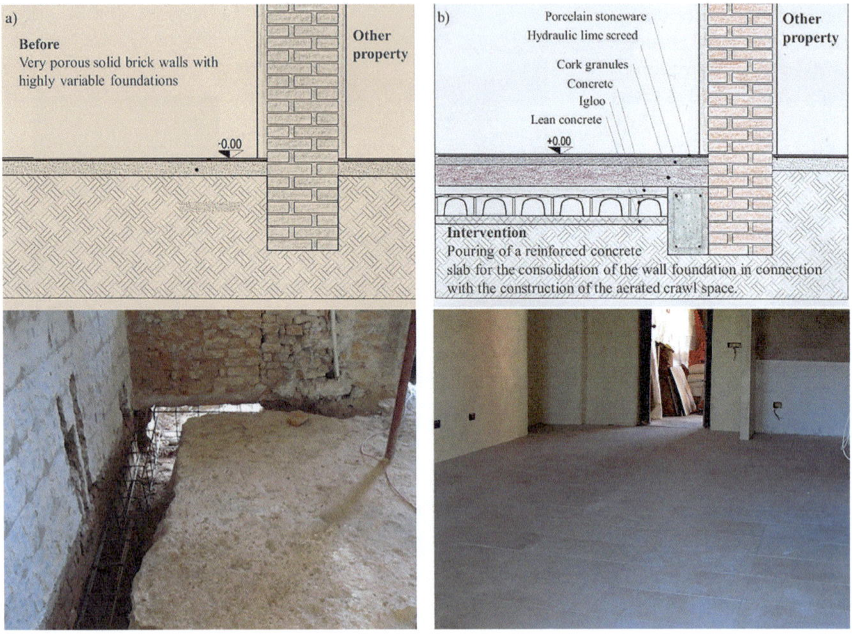

Fig. 29.2 Details of the renovations carried out on the ground floor of the farmstead. Situation before, **a** and after, **b** the intervention

29.2.2.1 System Boundary

The system boundary of the study covers from the extraction and processing of raw materials needed to manufacture the products involved in the retrofitting project to the end of life of such products after the demolition of the building. The scheme follows the one defined in the standard EN 15,804 including the core rules for the product category of construction products [25]. The life cycle stages are subdivided into modules: the product stage (from A1 to A3) is related to the production of the building materials, including the extraction and processing of raw materials, the transport of raw materials to the manufacturing site and the production of the building materials for both the bio-based and synthetic scenarios. The retrofitting stage is subdivided into the transportation of building materials to the building site (A4) and the installation into the building (A5). The use stage is subdivided into various modules and includes the use of the material (B1), maintenance (B2), repair (B3), replacement (B4), refurbishment (B5), operational energy use (B6) and operational water use (B7). Then, the end-of-life stage includes the demolition of the building (C1), the transportation of the waste to the processing facility (C2), the processing of the waste for reuse, recovery and recycling (C3) and the disposal of the waste (C4) [25]. In this study, not all modules were considered, but only those that were relevant with respect to the goal and for which detailed information was available. For example, regarding the use stage, the module B3 and B7 were not considered.

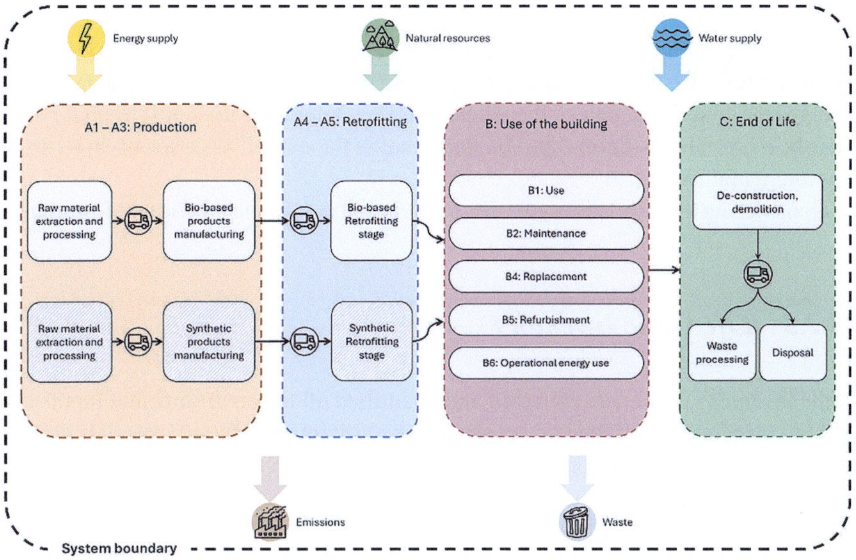

Fig. 29.3 The system boundary of the LCA study following the scheme from the EN 15,804:2012 + A2:2019 [25]

The system boundary is reported in Fig. 29.1. The geographical scope considered is Italy for the construction and use phase, while Europe is considered for the material production phase (Fig. 29.3).

29.2.2.2 Functional Unit

To effectively compare the two retrofitting scenarios, employing bio-based and synthetic materials, a proper functional unit must be defined. First, we must define the function to consider. In this case, we are dealing with a retrofitting project whose main goal is to upgrade the energy performance of the building. Therefore, the function is related to the insulating properties of the materials. This function is considered to properly define the alternative scenario with synthetic materials with similar characteristics to the bio-based one, in order to have the same thermal resistance and the same energy performance of the bio-based scenario. This means that it is necessary to define the thickness in relation to thermal conductivity and the amount of material in terms of weight, in relation to density, that are required to fulfil the expected performance.

Taking an overall balance, accounting for both energy and impacts caused by the production and construction of the building, and energy and impacts generated during the use phase of the building, it is necessary to define the building's lifetime. The choice generally made is to estimate a scenario of 50–100 years, which is the expected lifetime for a building [29]. In this study, we considered an expected lifetime of 80 years. Given the significant influence of this parameter on the outcome of our first research question, we conducted a sensitivity analysis to ensure a robust understanding. We are aware that performing a sensitivity analysis on this parameter changes the functional unit of the whole building, but this is done just to understand how the materials selection could be important in the overall assessment, in terms of environmental impact.

Considering the entire life cycle, the reference flow is related to the whole building.

29.2.3 Life Cycle Inventory

In the inventory phase, we collected and quantified all the environmental inputs and outputs associated with the two scenarios: one utilizing bio-based materials and the other employing conventional synthetic materials. The complete inventory for the whole building life cycle is available in the supplementary material. The software used to perform the LCA study is the open-source framework brightway [30].

29.2.3.1 Materials Production and Retrofitting Phase

The construction materials employed in the retrofitting project are mainly of natural origin. Natural insulating materials such as hemp-derived building materials, in particular hempcrete and hemp fibre mats [31], wood wool and wood fibre panels, and cork granules, as well as earthen plasters [32] and bio-based thermal plasters. For such materials is of particular importance the assessment of biogenic carbon [7]. To comprehensively track all biogenic carbon fluxes, we align with the EN 15,804 standard, and thus also with the results of EPDs, where the $-1/+1$ approach is used, considering both CO^2 uptake and release [7, 9].

Starting from the bio-based scenario, the inventories to produce the construction materials were both primary, provided directly by some of the producers, and secondary, considering activities from ecoinvent 3.10 cut-off, and from building materials EPD. The entire Casatenovo retrofitting was modelled by taking all the drawings of the various walls and floors involved in the intervention and their materials composition, in order to obtain the materials requirement for the entire building.

The Synthetic scenario represents a more traditional approach. It utilizes widely employed construction materials, primarily of synthetic origin, commonly encountered in conventional renovation projects. This scenario was developed by analysing the materials employed in the actual renovation of the dwelling in Casatenovo, Italy. Some of these materials were then replaced with alternatives commonly used in traditional construction methods. The replacement materials were carefully selected to ensure they possessed similar thermal conductivity and density to the original materials. Furthermore, a comprehensive evaluation of their specific characteristics was conducted to confirm their suitability as replacements. This evaluation process began with the formula for thermal resistance, which demonstrates the inverse relationship between thermal resistance and conductivity:

$$R_i = \frac{d_i}{\lambda_i}$$

where, R_i is the thermal resistance [m^2 K W^{-1}], d_i is the thickness of the material [m] and λ_i is the thermal conductivity [W m^{-1} K^{-1}]. To keep the thermal resistance between the synthetic material and the original bio-based material unchanged, once the conductivity has been set, it is necessary to vary the thickness of the material. Many of the substitution synthetic materials, were considered from available EPDs, where the required parameters, i.e. density and thermal conductivity, were reported. In Fig. 29.4 it is reported, as an example, the stratigraphy of a structure of the dwelling, with the materials employed in the bio-based scenario and the selected substitution materials considered for the synthetic scenario. The complete list of materials employed in the bio-based scenario and the substitute materials included in the synthetic scenario with their data source is shown in Table 29.1. For further details refer to the data available in the supplementary materials.

Following material selection, the next step involves transporting them to the construction site. We aimed to maintain consistency in transportation distances and means of transport whenever possible. However, in cases where primary data on specific materials was unavailable, certain assumptions were necessary to fill these knowledge gaps.

The retrofitting phase includes the installation of the materials in the building structures. This phase relies on the information reported in the selected EPDs, including the possible use of machinery and tools for the laying.

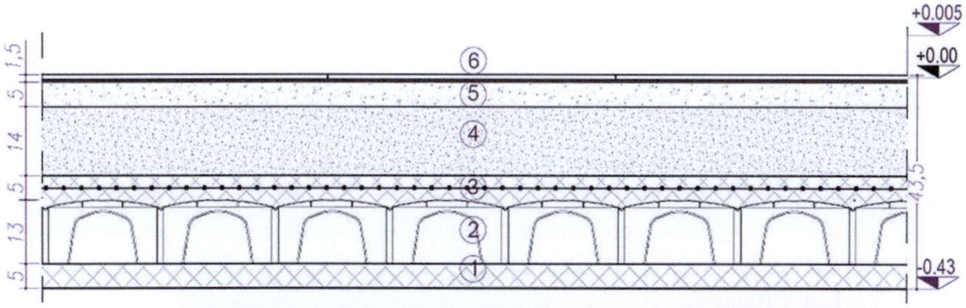

Layer	Bio-based material	Synthetic material
1	lean concrete	lean concrete
2	igloo	igloo
3	natural hydraulic lime fibre-reinforced screed	Cement-based Floor Screeds
4	cork granulate	expanded perlite
5	natural hydraulic lime fibre-reinforced screed	Cement-based Floor Screeds
6	Porcelain Stoneware Surfaces	Porcelain Stoneware Surfaces

Fig. 29.4 Details of the composition of a structure, reporting the materials employed in the bio-based scenario and the substitution materials selected for the synthetic scenario

Table 29.1 list of materials employed in the bio-based scenario and the substitute materials included in the synthetic scenario, with their data source

Bio-based scenario	Source	Synthetic scenario	Source
Natural hydraulic lime fibre-reinforced screed	Primary	Cement-based floor screeds	EPD [33]
Rasocalce	Primary	Base plaster	Ecoinvent
igloo	Primary	igloo	Primary
Decorcalce	Primary	Decorcalce	Primary
Earthen plaster B1	Primary	Earthen plaster B1	Primary
Earthen plaster B2	Primary	Earthen plaster B2	Primary
Earthen plaster R1	Primary	Earthen plaster R1	Primary
Cork granulate	Primary	Expanded perlite	Ecoinvent
Base plaster	Ecoinvent	Base plaster	Ecoinvent
Gypsum fibreboard	Ecoinvent	Gypsum fibreboard	Ecoinvent
Sand	Ecoinvent	Sand	Ecoinvent
Fibreboard, soft	Ecoinvent	Glass wool mat	Ecoinvent
Lean concrete	Ecoinvent	Lean concrete	Ecoinvent
Porcelain stoneware surfaces	EPD [34]	Porcelain stoneware surfaces	EPD [34]
Multi-layer parquet	EPD [35]	Multi-layer parquet	EPD [35]
Ekolution hemp fiber insulation	EPD [36]	Extruded polystyrene (XPS) foam board	EPD [37]
Foamglas	EPD [38]	Foamglas	EPD [38]
Blocco ambiente	EPD [39]	Solid precast concrete insulated wall	EPD [40]
Bio beton pronto	EPD [39]	Solid precast concrete insulated wall	EPD [40]
Bio beton jet	EPD [39]	CT300 thermal plaster	EPD [41]
Celenit N insulating panel	EPD [42]	Rockwool stone wool thermal insulation	EPD [43]
Benesserebio Kerakoll thermo plaster	EPD [44]	CT300 thermal plaster	EPD [41]

29.2.3.2 Use Phase

For the use phase, we considered the modules related to the use of the material (B1), maintenance (B2), replacement (B4), refurbishment (B5) and operational energy (B6).

The modules B1, B2, B4 and B5 rely on the information reported in the EPD of the considered materials. Information on the impacts of module B1 is available only for the hempcrete materials, namely Blocco Ambiente, Bio Beton jet and Bio Beton pronto produced by the company Senini S.r.l. For these materials, in this module is considered the carbon uptake due to carbonation [45]. Carbonation consists of the reaction of the calcium hydroxide contained in the lime with the carbon dioxide present in the atmosphere, resulting in the formation of calcium carbonate; this phenomenon involves the removal of CO_2 from the atmosphere and the subsequent storage of carbon in the material in the form of calcium carbonate [46]. In the EPD, a lifespan of 100 years is assumed for these hemp-derived materials, so the CO_2 absorption is reduced to take into account the lifespan of the building.

Module B2 is considered only for multi-layer parquet and Porcelain stoneware. This module includes the cleaning of these materials. However, adjustments were necessary to account for the discrepancy between the reference service lives (RSL) used in the EPDs and the actual lifespan considered for this study. The EPD for multi-layer parquet references a service life of 50 years. To align with the 80-year lifespan of the building, the cleaning impacts from module B2 were proportionally increased. Similarly, the EPD for porcelain stoneware references a one-year maintenance cycle. In this case, the cleaning impacts from module B2 were multiplied by 80 to reflect the 80-year lifespan of the building.

According to the standard EN 15,804, module B4 includes the replacement of a whole construction element [25]. Therefore, RSL must be considered for this module too, in order to cover the whole building lifespan. RSL of multi-layer parquet is 50 years, hence, to cover the building lifespan, we must consider that this element has to be replaced one time. This also applies to porcelain stoneware, as they have an expected RSL of 50–60 years. Replacement includes the impacts of replacing old materials and their disposal, and the impacts of new material production, transport to site and installation.

On the other hand, module B5 includes all the interventions to restore a construction element to perform its required function. This module is disclosed only for multi-layer parquet and involves 2 renovations over its RSL. Therefore, the impacts related to this module were properly increased.

The operational energy use module (B6) typically involves monitoring energy consumption for heating, cooling, and hot water. However, due to the nature of the Casatenovo farmstead—a holiday home used seasonally—daily consumption data wasn't obtainable. To address this limitation, we leveraged the issued energy certification, which classified the renovated dwelling in the A1 class. Only the thermal energy consumption data for winter heating computed in the certification was considered since the energy consumption due to summer air-conditioning and domestic hot water was considered unreliable. The dwelling does not have any air-conditioning system and does not use an electric storage tank for domestic hot water heating, but a heat pump. Therefore, the primary data of the condensing boiler installed in the dwelling were used to make the inventory concerning the use phase of the thermal system for winter heating. According to the type of boiler, a proper activity within the ecoinvent database was selected and modified with the specific boiler efficiency. The value relative to the annual thermal energy demand for winter heating (ETh) was used for the calculation of the total annual thermal energy consumption, multiplied then for the considered building lifespan (80 years). Furthermore, since the synthetic scenario was defined to have the same thermal resistance as the bio-based one, the same annual energy consumption was considered for both scenarios.

29.2.3.3 End-Of-Life Stage

This stage analyses the environmental impact associated with dismantling the elements added during the building's retrofit (C1). The total amount of these materials is considered to assess the impact of demolition. The materials are then transported (C2) assuming an average distance of 100 km by truck to a processing facility (C3). Existing data (EPDs) are used to determine the environmental impact of processing these materials. In cases where multiple end-of-life scenarios are presented in the EPDs, the most conservative option is chosen for C3 and disposal of any remaining waste (C4). Landfill disposal or incineration is assumed for any remaining materials. For further information about the end-of-life modelling refer to table 19 in the supplementary materials.

The end-of-life scenario considered is specifically relevant for the bio-based materials since it determines whether the stored biogenic carbon is released back into the environment [7, 9].

29.2.4 Life Cycle Impact Assessment

To guarantee consistency with the results taken from EPDs, we considered the same impact assessment method to compute the overall results for the entire life cycle. The method considered is the version of the method Environmental Footprint (EF) 3.1 [47] associated with the standard EN 15,804, named EF v3.1 EN15804 implemented in brightway. For the same reason, the impact categories considered are the ones included in the EPDs, following the same standard.

Building lifespan is a critical parameter for understanding the environmental impact of construction materials throughout the building's life cycle. To assess its significance, we conducted a sensitivity analysis evaluating the cradle-to-grave effects for lifespans of 50 and 100 years.

29.3 Results and Discussion

In this section, the cradle-to-grave results of the two retrofitting scenarios are presented. In Fig. 29.5 the results for the considered impact categories are presented, comparing the bio-based scenario with the synthetic one, highlighted with a diagonal pattern in the figure. These results allow us to investigate our research questions. The first question focuses on the influence of material selection on the entire building life cycle's environmental performance. Material production (A1–A3) emerges as the primary contributor across most impact categories. However, for global warming, abiotic depletion, and ozone depletion, the dominant contributor is energy consumption during the building's use phase (B6).

Regarding the second research question, we can see if the retrofitting solution employing bio-based materials allows for the reduction of the environmental impacts when compared to the conventional solution employing synthetic materials. If we compare the two scenarios, we can spot that the bio-based one has lower environmental impacts for most of the impact categories, being worse only for acidification potential, eutrophication potential and abiotic depletion potential related to elements and materials.

Another module which results to be relevant seems to be A4, related to the transportation of the materials to the construction site. The difference between the two scenarios for this module is related to the higher weight of the materials selected for the synthetic scenario, due to their higher density.

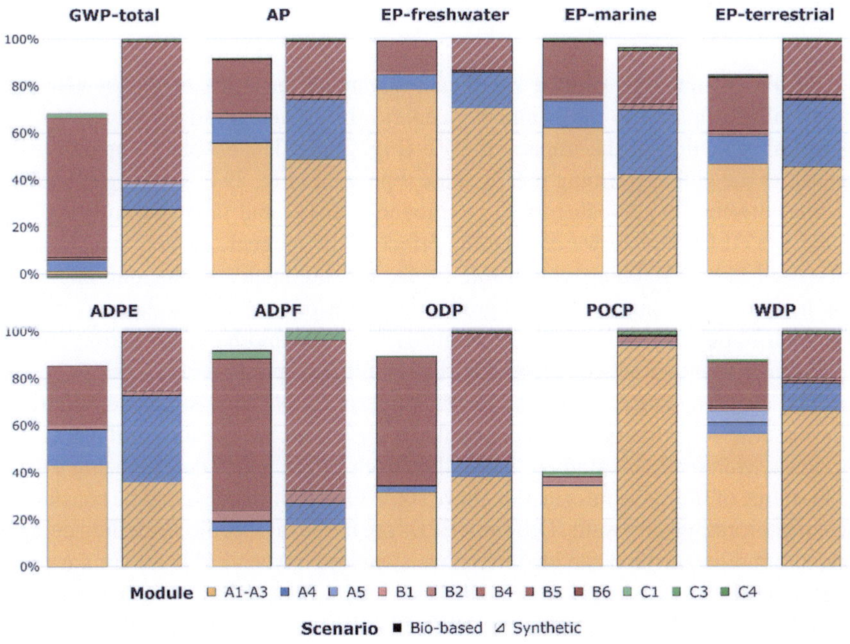

Fig. 29.5 Cradle-to-grave relative results for the retrofitting project employing bio-based materials compared with the alternative scenario with synthetic materials (highlighted with a pattern). The scenario with the highest impact for each category is set to 100%. GWP = Global Warming Potential, AP = Acidification Potential, EP = Eutrophication Potential, ADPE = Abiotic resource Depletion Potential, minerals and element, ADPF = Abiotic resource Depletion Potential, fossils, ODP = Ozone Depletion Potential, POCP = Photochemical Ozone formation Potential, WDP = Water user Deprivation Potential

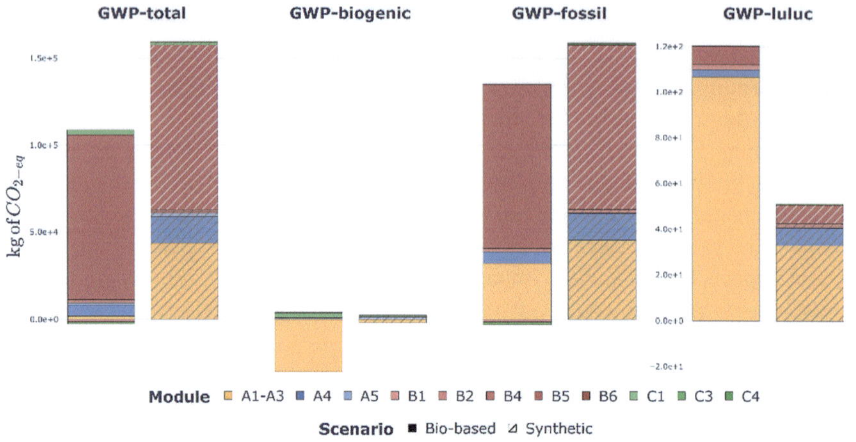

Fig. 29.6 Cradle-to-grave absolute results for the bio-based and synthetic scenarios, focusing on global warming potential impact category and its sub-categories. LULUC = Land Use and Land Use Change

Bio-based materials offer a significant advantage in terms of global warming potential. Compared to the synthetic scenario, bio-based solutions demonstrate a potential reduction of the impact of 96% (Fig. 29.5). The absolute impact results related to the global warming potential are reported in Fig. 29.6, showing the subdivision between its sub-categories, i.e. biogenic, fossil and land use and land use change (LULUC). Here we can see the effect of the biogenic carbon uptake of the bio-based materials, being responsible for the great reduction of the total impact. If for the synthetic scenario, material production contributes 27.3% of the total impact, the contribution of this module is equal to 1.7% for the bio-based scenario.

It's important to acknowledge that the results are particularly sensitive to the defined scenarios, especially the alternative scenario utilizing synthetic materials. The specific selection of materials within this scenario significantly influences our findings. Moreover, this study relies on data from EPDs, which despite being a valuable source of life cycle information, have some limitations that should be considered when interpreting the results. Different EPDs are based on specific modelling choices and assumptions, which can lead to inconsistencies between EPDs for similar products. Additionally, the quality of an EPD can be influenced by the chosen system boundaries and allocation methods, potentially affecting the comparability of results across different products. Therefore, careful consideration should be given to these limitations when drawing conclusions from this study.

29.3.1 Sensitivity Analysis on Building Lifetime

The use phase significantly impacts the overall life cycle of the retrofit project. A key parameter influencing this stage is the assumed building lifespan. To assess the importance of this parameter and gain a more comprehensive understanding of material production's contribution, we conducted a sensitivity analysis. We evaluated two alternative lifespans: 50 and 100 years, representing the lower and upper bounds commonly found in building life cycle studies [29]. The use phase impacts are directly tied to the assumed building lifespan. For a 100-year lifespan, Module B6 (energy consumption) simply increases by 20 years, reflecting additional daily energy use. Conversely, a 50-year lifespan reduces impacts in B6. Similarly, modules B1 (material use—including hempcrete materials CO_2 uptake through carbonation), B2 (maintenance), and B5 (refurbishment) are proportionally adjusted based on lifespan. Module B4 (component replacement) behaves differently. With a 50-year lifespan, no replacements are needed since materials like multi-layer parquet and porcelain stoneware last as long as the building. However, replacements become necessary for a 100-year lifespan, leading to increased impacts in B4.

The sensitivity analysis results for the bio-based scenario are reported in Fig. 29.7. From these results, we can notice that the greatest variation is recorded for the global warming impact category since the use phase is the main contributor to its overall impact. Moreover, as expected, changing the assumed building lifespan affects the relative importance of material selection. Indeed, if we consider a building lifespan of 50 years, the materials production phase becomes equally relevant concerning module B6, also for the ozone depletion impact category.

29.4 Conclusions

Through this study, we aimed to understand the importance of materials selection on the whole building life cycle and to assess if a retrofitting project employing bio-based materials has the potential to reduce the environmental impacts compared to a conventional retrofitting solution employing synthetic materials.

From our findings, material production (A1–A3) results are the main contributor to most impact categories. However, building use phase energy consumption (B6) dominates global warming, abiotic depletion, and ozone depletion. The bio-based scenario shows lower environmental impact for most categories, apart from acidification, eutrophication, and abiotic depletion related to elements/materials. Bio-based materials have the potential to significantly reduce global warming compared to synthetics (96% reduction). This is mainly attributed to the biogenic carbon uptake of bio-based materials.

While the study offers valuable insights, the results are influenced by specific scenario choices and limitations of EPD data. Material selection within the synthetic

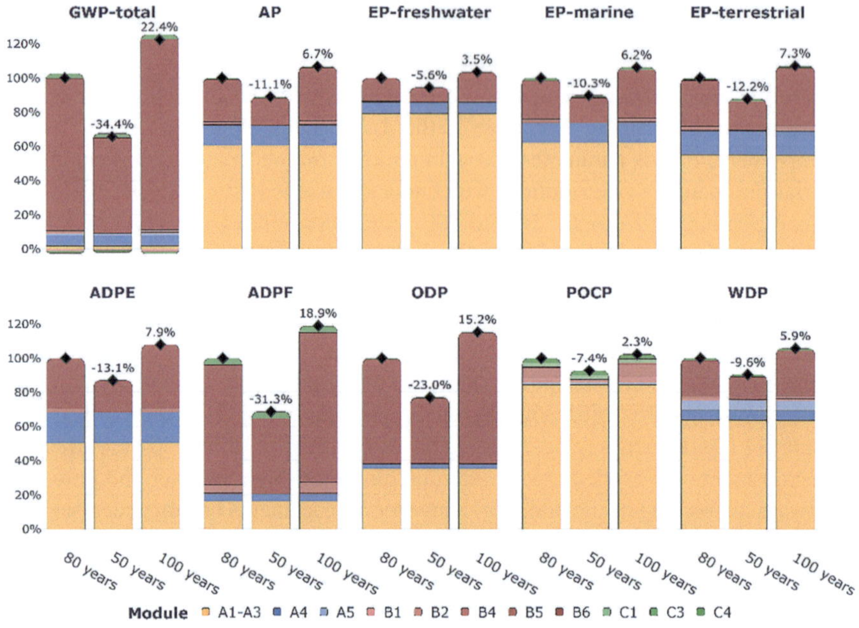

Fig. 29.7 Sensitivity analysis results for the bio-based scenario, comparing the base case (80 years) with two variations of the building lifespan parameter equal to 50 years and 100 years

scenario significantly impacts the findings. Additionally, EPDs have inherent limitations due to varying modelling choices and approaches, potentially affecting comparability across products. In future studies, a broader assessment considering other construction materials might be needed to have deeper scenario definitions and greater control on these issues.

Given the long lifespan of a building, a prospective LCA might be more appropriate. This approach would allow us to incorporate potential future scenarios, particularly regarding energy mix projections, and accounting for future climate change effects, which might change the energy consumption requirements for heating and cooling [9, 48]. By including projections, we can gain a more accurate understanding of the long-term environmental impact of the retrofit project.

29.5 Competing Interests

The authors declare that they have no known competing financial interests or personal relationships that could have appeared to influence the work reported in this paper.

Acknowledgements This work was carried out thanks to the financial support from the Italian National Recovery and Resilience Plan (PNRR).

References

1. United Nations Environment Programme, in *2022 Global Status Report for Buildings and Construction: Towards a Zero-emission, Efficient and Resilient Buildings and Construction Sector* (Nairobi, 2022)
2. I. Artola, K. Rademaekers, R. Williams, J. Yearwood, Boosting Building Renovation: What potential and value for Europe? Study for the ITRE Committee. Dir. Gen. Intern. Policies Policy Dep. A: Econ. Sci. Policy PE 587.326, 1–72 (2016)
3. O. Cavalett, M.D.B. Watanabe, M. Voldsund et al., Paving the way for sustainable decarbonization of the European cement industry. Nat. Sustain. (2024). https://doi.org/10.1038/s41893-024-01320-y
4. ENEA, Rapporto Annuale Sulla Certificazione Energetica Degli Edifici (2023)
5. D. Peñaloza, M. Erlandsson, A. Falk, Exploring the climate impact effects of increased use of bio-based materials in buildings. Constr. Build. Mater. **125**, 219–226 (2016). https://doi.org/10.1016/j.conbuildmat.2016.08.041
6. N. Heeren, C.L. Mutel, B. Steubing et al., Environmental impact of buildings—what matters? Environ. Sci. Technol. **49**, 9832–9841 (2015). https://doi.org/10.1021/acs.est.5b01735
7. E. Hoxha, A. Passer, M.R.M. Saade et al., Biogenic carbon in buildings: a critical overview of LCA methods. Build. Cities **1**, 504–524 (2020). https://doi.org/10.5334/bc.46
8. F. Pittau, F. Krause, G. Lumia, G. Habert, Fast-growing bio-based materials as an opportunity for storing carbon in exterior walls. Build. Environ. **129**, 117–129 (2018). https://doi.org/10.1016/j.buildenv.2017.12.006
9. S. Cucurachi, B. Steubing, F. Siebler, N. Navarre, *Prospective LCA Methodology for Novel and Emerging Technologies for BIO-based products—The PLANET BIO project* (Publications Office of the European Union, Luxembourg, 2022)
10. M. Schulte, I. Lewandowski, R. Pude, M. Wagner, Comparative life cycle assessment of bio-based insulation materials: environmental and economic performances. GCB Bioenergy **13**, 979–998 (2021). https://doi.org/10.1111/gcbb.12825
11. R. Jagarajan, M.N. Abdullah Mohd Asmoni, A.H. Mohammed et al., Green retrofitting—a review of current status, implementations and challenges. Renew. Sustain. Energy Rev. **67**, 1360–1368 (2017)
12. T. Konstantinou, U. Knaack, Refurbishment of residential buildings: a design approach to energy-efficiency upgrades, in *Procedia Engineering* (Elsevier Ltd., (2011), pp. 666–675
13. P.S. Roy, R.M. Ramachandran, O. Paul et al., Anthropogenic land use and land cover changes—a review on its environmental consequences and climate change. J. Indian Soc. Remote. Sens. **50**, 1615–1640 (2022)
14. H.S. Koren, C.D. Butler, The interconnection between the built environment ecology and health (2006)
15. D. Beillouin, R. Cardinael, D. Berre et al., A global overview of studies about land management, land-use change, and climate change effects on soil organic carbon. Glob. Chang. Biol. **28**, 1690–1702 (2022). https://doi.org/10.1111/gcb.15998
16. C. Passoni, A. Marini, A. Belleri, C. Menna, Redefining the concept of sustainable renovation of buildings: state of the art and an LCT-based design framework. Sustain. Cities Soc. **64** (2021). https://doi.org/10.1016/j.scs.2020.102519
17. E. Piaia, F. Maietti, R. Di Giulio et al., BIM-based cultural heritage asset management tool. Innovative solution to orient the preservation and valorization of historic buildings. Int. J. Arch. Herit. **15**, 897–920 (2021). https://doi.org/10.1080/15583058.2020.1734686
18. The European Parliament and the Council of the European Union, Directive (EU) 2023/1791 of the European Parliament and of the Council of 13 September 2023 on energy efficiency and amending regulation (EU) 2023/955 (2023)
19. The European Parliament and the Council of the European Union, Directive (EU) 2024/1275 of the European Parliament and of the Council of 24 April 2024 on the energy performance of buildings(2024)

20. K.N. Streicher, S. Mennel, J. Chambers et al., Cost-effectiveness of large-scale deep energy retrofit packages for residential buildings under different economic assessment approaches. Energy Build. **215** (2020). https://doi.org/10.1016/j.enbuild.2020.109870
21. S. Pezzutto, F. Haas, D. Exner, S. Zambotti, in *Europe's Building Stock and Its Energy Demand: A Comparison Between Austria and Italy* (Springer International Publishing, 2018)
22. G. Ruggieri, G. Dotelli, P. Melià, S. Sabbadini, Life cycle assessment of refurbishment strategies for historic buildings (2013)
23. British Standards Institution, in *Environmental Labels and Declarations—Type III Environmental Declarations : Principles and Procedures* (British Standards Institution, 2010)
24. C. Strazza, A. Del Borghi, F. Magrassi, M. Gallo, Using environmental product declaration as source of data for life cycle assessment: a case study. J. Clean. Prod. **112**, 333–342 (2016). https://doi.org/10.1016/J.JCLEPRO.2015.07.058
25. BSI British Standards Institution, in BS EN 15804:2012+A2:2019, vol. 70 (British Standards Institution, 2013)
26. B.P. Weidema, M. Pizzol, J. Schmidt, G. Thoma, Attributional or consequential life cycle assessment: a matter of social responsibility. J. Clean. Prod. **174**, 305–314 (2018). https://doi.org/10.1016/J.JCLEPRO.2017.10.340
27. M.A. Curran, M. Mann, G. Norris, The international workshop on electricity data for life cycle inventories. J. Clean. Prod. **13**, 853–862 (2005). https://doi.org/10.1016/J.JCLEPRO.2002.03.001
28. V. Annibaldi, F. Cucchiella, P. De Berardinis et al., An integrated sustainable and profitable approach of energy efficiency in heritage buildings. J. Clean Prod. **251** (2020). https://doi.org/10.1016/j.jclepro.2019.119516
29. M. Bahramian, K. Yetilmezsoy, Life cycle assessment of the building industry: an overview of two decades of research (1995–2018). Energy Build. **219** (2020)
30. C. Mutel, Brightway: an open source framework for life cycle assessment. J. Open Source Softw. **2**, 236 (2017). https://doi.org/10.21105/joss.00236
31. C. Ingrao, A. Lo Giudice, J. Bacenetti et al., Energy and environmental assessment of industrial hemp for building applications: a review. Renew. Sustain. Energy Rev. **51**, 29–42 (2015). https://doi.org/10.1016/J.RSER.2015.06.002
32. P. Melià, G. Ruggieri, S. Sabbadini, G. Dotelli, Environmental impacts of natural and conventional building materials: a case study on earth plasters. J. Clean. Prod. **80**, 179–186 (2014). https://doi.org/10.1016/j.jclepro.2014.05.073
33. S.A. Nordia, in *Environmental Product Declaration* (The International EPD System, 2023)
34. S.P.A. Florim Ceramiche, in *Environmental Product Declaration* (EPD Italy, 2023)
35. VdP—Verband der Deutschen Parkettindustrie e.V., in *Environmental Product Declaration* (Institut Bauen und Umwelt e.V. (IBU), 2022)
36. A.B. Ekolution, in *Environmental Product Declaration* (The International EPD System, 2023)
37. K.G. Karl Bachl Kunststoffverarbeitung GmbH & Co., in *Environmental Product Declaration* (Institut Bauen und Umwelt e.V. (IBU), 2022)
38. N.V. Pittsburgh Corning Europe, in *Environmental Product Declaration* (Institut Bauen und Umwelt e.V. (IBU), 2021)
39. S.R.L Senini, in *Environmental Product Declaration* (EPD Italy, 2022)
40. A.B. Gripen Betongelement, in *Environmental Product Declaration* (The International EPD System, 2023)
41. Prima Ciment, in *Environmental Product Declaration* (The International EPD System, 2023)
42. S.P.A Celenit, in *Environmental Product Declaration* (The International EPD System, 2021)
43. S.P.A. Rockwool Italia, in *Environmental Product Declaration* (The International EPD System, 2023)
44. Kerakoll, in *Environmental Product Declaration* (The International EPD System, 2021)
45. A. Arrigoni, R. Pelosato, P. Melià et al., Life cycle assessment of natural building materials: the role of carbonation, mixture components and transport in the environmental impacts of hempcrete blocks. J. Clean. Prod. **149**, 1051–1061 (2017). https://doi.org/10.1016/j.jclepro.2017.02.161

46. Z. Liu, W. Meng, Fundamental understanding of carbonation curing and durability of carbonation-cured cement-based composites: a review. J. CO2 Util. **44** (2021)
47. Environmental Footprint, in *European Commission, Joint Research Centre (JRC)*. https://eplca.jrc.ec.europa.eu/EnvironmentalFootprint.html. Accessed 1 Jul 2024
48. R. Sacchi, T. Terlouw, K. Siala et al., Prospective environmental impact assement (premise): a streamlined approach to producing databases for prospective life cycle assessment using integrated assessment models. Renew. Sustain. Energy Rev. **160** (2022). https://doi.org/10.1016/j.rser.2022.112311

Chapter 30
A New Framework for the Automated Generation of Discontinuum-Based Structural Analysis Models from Images

Andrei Farcasiu, Peter Griesbach, Qipei Mei, Sinan Acikgoz, and Bora Pulatsu

Abstract Masonry walls, arches, and other structural elements have been primary components of buildings from ancient to modern times; residential and public masonry buildings are widespread in North America and Europe. These buildings often have distinctive architectural and cultural character which needs to be preserved. To assess these buildings' response to hazards, such as seismic action and settlements, high-fidelity computational models are needed. Discontinuum-based structural analysis tools, where the contact of each individual unit (*e.g.*, brick or stone blocks) to other units is modeled, can be used for this purpose. However, manually generating the computational models with thousands of units per building is arduous and time-consuming. To this end, this chapter introduces a new data-driven framework where AI-assisted object detection and instance segmentation algorithms are used to generate discontinuum-based models from images of the building. The modeling pipeline is illustrated with an application to a stone masonry building with a nonperiodic masonry wall morphology. It is envisioned that the developed tools will enable reliable structural analysis of masonry buildings and support future activities related to their sustainable rehabilitation and/or reuse.

30.1 Introduction

The carbon embodied in new construction materials makes up 11% of the annual greenhouse gas emissions. To meet Net Zero targets and ensure a sustainable built environment, we need to reduce these emissions. Rehabilitation and/or reuse of our

A. Farcasiu · P. Griesbach · B. Pulatsu (✉)
Department of Civil and Environmental Engineering, Carleton University, Ottawa, Canada
e-mail: bora.pulatsu@carleton.ca

Q. Mei
Department of Civil and Environmental Engineering, University of Alberta, Edmonton, Canada

S. Acikgoz
Department of Engineering Science, University of Oxford, Oxford, UK

D. A. H. Hanaor (ed.), *Sustainable Materials for the Built Environment*, Springer Series in Materials Science 352, https://doi.org/10.1007/978-3-031-97818-0_30

existing building stock can play a significant role in this effort. A significant part of the building stock that makes up urban centers worldwide is constructed of masonry.

Structural analysis is used to support the rehabilitation and reuse of existing masonry buildings. The first objective of these analyses is to assess whether the buildings are structurally sound under gravity loads. Furthermore, since masonry buildings are constructed of brittle materials that possess limited resistance against tensile stresses, the impact of likely hazards that may induce significant deformation demands on the structure must also be evaluated. In seismically active countries, earthquakes pose the biggest risk to masonry buildings. Structural analyses are required to prevent collapses and ensure life safety. Another important risk faced by buildings is ground movements, particularly in congested urban areas. Nearby deep excavation activities, e.g., for the construction of other buildings or infrastructure such as tunnels, inevitably induce ground movements, which alter the equilibrium state of the building. While collapse is unlikely in this case, structural analyses are needed to accurately predict crack development. If moderate and above structural damage is predicted, preventative measures need to be taken.

Structural analysis techniques which can be used to assess masonry buildings are introduced in Sect. 30.2. This section highlights that reliable analysis of failures requires explicit modeling of units (*e.g.*, bricks or stones) with discontinuum-based analyses. However, generating high-fidelity discontinuum models that represent real buildings is a labour- and time-intensive task; it requires identifying thousands of units one by one and idealizing them for structural analysis purposes. Whilst this is possible for research activities, the time and resource commitment required for these activities renders discontinuum-based analysis impractical for practical consultancy settings, except for special buildings.

To this end, Sect. 30.3 explores the automated generation of structural analysis models, harnessing the excellent performance of AI algorithms in various image processing tasks, such as object detection and instance segmentation. Different from several previous studies on segmentation [1] and automated model generation [2] for masonry buildings, object detection and segmentation are carried out with the cutting edge convolutional neural network model. This approach allows the use of robust image processing algorithms, already pre-trained on large datasets, to be fine-tuned with single-shot specialized training to conduct the required processing. The adopted strategy is known to reduce the volume of data required for conducting specialized image processing tasks and offers the possibility to benefit from state-of-the-art image processing algorithms. Another novel aspect of this study relates to the morphological analysis of masonry texture (*e.g.*, related to the size and relative arrangement of units) following image processing. This information is used to generate artificial masonry textures that are compatible with the observations; these textures are then used to complete the model in areas image processing has failed. Section 30.4 demonstrates the application of the automated model generation technique to a two-storey masonry building from Ontario, Canada. The generated model is subjected to lateral loads and settlements to demonstrate how it may be used for assessments and how the new '*as-is*' representation of masonry texture in the model delivers new insight into structural response.

30.2 Discontinuum-Based Anaylsis of Urm Structures: Concept, Theory and Applications

Computational structural mechanics of unreinforced masonry (URM) constructions can be grouped under the continuum and discontinuum-based analysis depending on how stone or brickwork assemblages are represented within the adopted modeling framework. Specifically, continuum-based analysis, also called *macro-modeling*, denotes composite material characteristics of masonry as a homogeneous isotropic or orthotropic medium with no distinction between masonry constituents (*i.e.*, masonry units, mortar, and unit-mortar interfaces). The continuous finite element analysis (FEA) is the most used macro-modeling strategy to simulate masonry buildings where nonlinear material behaviour follows plasticity or other macro-scale constitutive relationships [3–7]. Typically, the damage is smeared out through a continuous medium unless special crack-tracking algorithms are utilized (*e.g.*, [8]). Alternatively, the discontinuum-based analyses explicitly represent the morphological features of masonry structures using discrete elastic or rigid polyhedral blocks that can interact with each other along their contact planes [9]. Discontinuous FEA, Discrete element method (DEM), Non-Smooth Contact Dynamics (NSCD) and Applied Element Method (AEM) are the main discontinuum-based analysis approaches employed in the literature, which differ based on their implementation of contact constitutive relationships and mechanics as well as numerical solution schemes (*e.g.*, [10–17]).

To illustrate the outcome of different computational modeling techniques, a typical diagonal tension failure mechanism of a URM wall subjected to in-plane loading is utilized (Fig. 30.1a), where the associated damage pattern is predicted via continuum and discontinuum-based analyses. The result of continuum analysis is shown in Fig. 30.1b, where the diagonal tension failure is smeared out through the continuous medium without necessarily addressing the interaction between the masonry unit and mortar. On the other hand, the explicit representation of local failure mechanisms developing at the unit-mortar interfaces and masonry units is obtained from the block-based representation of the analyzed URM wall (Fig. 30.1c, d). It is important to note that the presented discontinuum-based models follow the *simplified micro-modeling* strategy where the masonry units are expanded up to half-thickness of the mortar joint and nonlinear bond behaviour lumped at the contact surfaces among the adjacent block ignoring the presence of mortar, suggested by Lourenco and Rots [18]. The adopted simplification offers an acceptable compromise between computational cost and accuracy. However, it should be noted that the masonry units can be simulated with different levels of detail within the simplified micro-modeling framework, depending on the expected accuracy in fracture resolution. The most common technique is representing each masonry unit via two discrete blocks (rigid or elastic), considering a single potential crack surface (*e.g.*, [19, 20]), which can be depicted in Fig. 30.1c. Alternatively, masonry units can be represented via multiple randomly generated tetrahedral blocks (or polyhedral blocks, *e.g.*, [21, 22]) to better represent spatially varying crack patterns in bricks and stone blocks, which offers

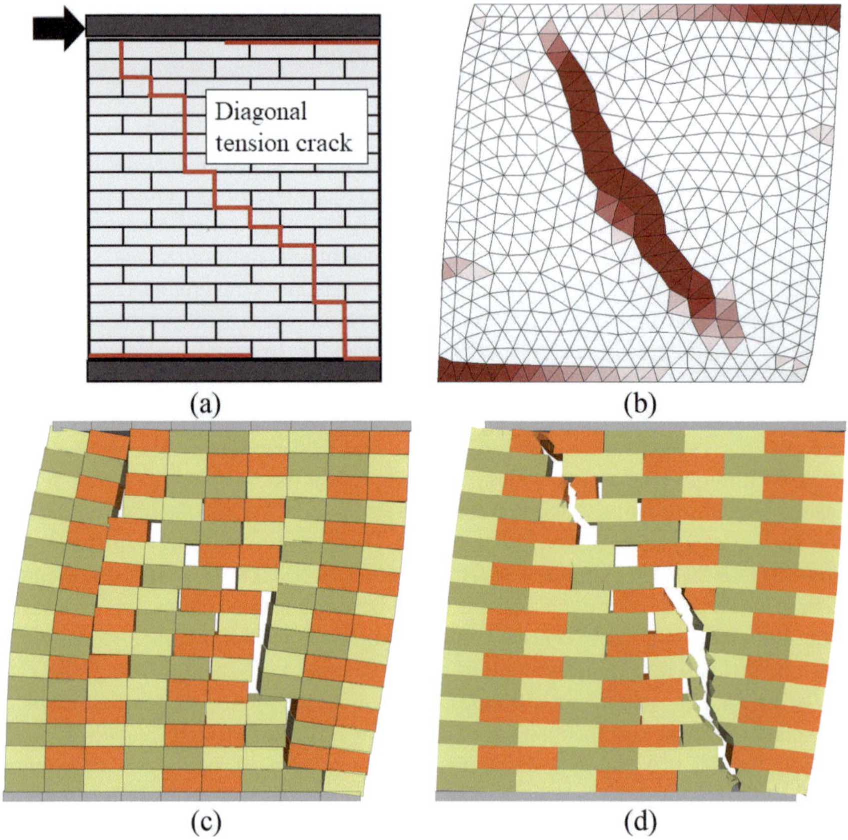

Fig. 30.1 Illustration of continuum (macro-modeling) and discontinuum-based analysis of a URM wall: **a** representation of URM wall subjected to in-plane loading; **b** simulating diagonal tension failure using macro-modeling, **c** discontinuum-based analysis and **d** mixed discrete-continuum approach

less restraint on the fracture mechanism developing within the masonry units and yield more realistic failure mechanism, as shown in Fig. 30.1d.

30.2.1 A Brief Overview of the Discrete Element Method

In this chapter, we use discontinuum-based analysis to assess the structural behaviour of an old stone masonry building using DEM, developed by Cundall [23]. The adopted modeling strategy, following the simplified micro-modeling approach, represents the stone masonry morphology as a system of discrete rigid blocks, as shown in Fig. 30.2a. The geometrical properties of the defined block pattern can be obtained from vision-based data (*i.e.*, images, points cloud, etc.), or a fictitious representation of the analyzed URM wall section is utilized that may or may not be informed by the existing construction characteristics of masonry. The block interactions are computed at the sub-contacts (defined along each contact plane), where three orthogonal springs are assigned, and inter-block penetration is controlled by the pre-defined sub-contact stiffness in the normal (k_n) and shear (k_s) directions, illustrated in Fig. 30.2b. The cohesive bond behaviour between the discrete blocks is described using elasto-softening contact constitutive laws, which require bond tensile strength (f_T), cohesion (c) and friction angle (θ..) as input parameters.

The rigid block movements are computed by solving the governing equations of motion using the central difference method to predict translational and rotational velocities given in Eqs. 30.1 and 30.2, respectively. The adopted explicit solution scheme is formulated by calculated block velocities at the mid-time step (i.e., $t^+ = t + \Delta t/2; t^- = t - \Delta t/2$, Δt: time step) and quasi-static solutions are obtained using artificial damping.

$$\dot{u}_i^{t+} = \dot{u}_i^{t-} + \frac{\Delta t}{m}\left(\Sigma F_i^t - \lambda\left|\Sigma F_i^t\right| sgn\left(\dot{u}_i^{t-}\right)\right) \tag{30.1}$$

$$\omega_i^{t+} = \omega_i^{t-} + \frac{\Delta t}{I}\left(\Sigma M_i^t - \lambda\left|M_i^t\right| sgn\left(\omega_i^{t-}\right)\right) \tag{30.2}$$

where $\dot{u}_i$, ssss ω_i., are the translational and rotational velocities of the block centroid, respectively. The sign function is indicated as $sgn(\Delta)$ which impose the following condition: $if\ \zeta \geq 0,\ sgn(\zeta) = 1;\ if\ \zeta < 0.,\ sgn(\zeta) = -1$. The sum of forces and moments acting on a rigid block is obtained via ΣF_i^t and ΣM_i^t, also called unbalanced force and moment. Furthermore, m and I denote a block mass and moment of inertia about the block center of gravity. The employed numerical procedure offers conditionally stable solions that necessitate time steps less than the critical value (Δt_c) that can be predicted in terms of the minimum block mass (m_{min}) and the maximum

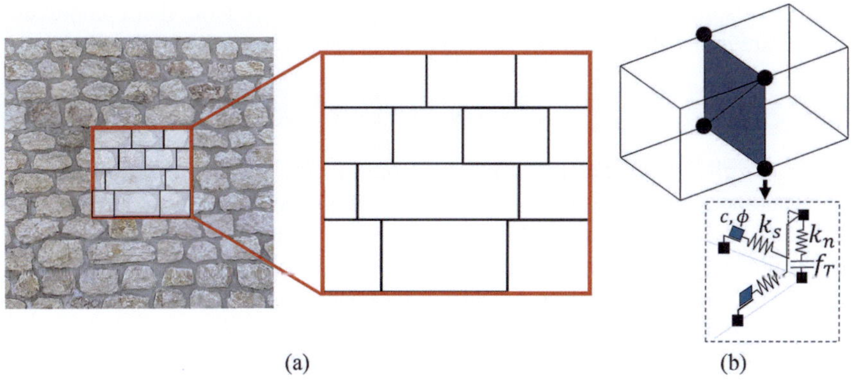

Fig. 30.2 **a** Discrete block representation of a stone masonry wall, **b** point contact defined at the corner of the two adjacent blocks

contact stiffness ($k_{n,max}$) in the system ($\Delta t_c = \sqrt{m_{min}/k_{n,max}}$). An artificial damping force and moment is used to obtain quasi-static solutions, adopting Cundall's local damping formulation ($F_d = \lambda |F_i^t| sgn(\dot{u}^{t-})$, $M_d = \lambda |\Sigma M_i^t| sgn(\omega^{t-})$) opposing motion proportional to the local unbalanced force [24]. Note that the givedamping scheme uses a non-dimensional damping constant (λ), which is taken as 0.8. Once the block velocities are predicted, their new position is computed together with the relative normal (Δu_n^c) and shear (Δu_s^c) sub-contact displacements. Accordingly, the normal and shear elastic sub-contact stress increments are calculated (*i.e.*, $\Delta\sigma = k_n\ \Delta u_n^c$, $\Delta\tau = k_s\ \Delta u_s^c$) and added to old ones (σ^t, τ^t) to obtain the new sub-contact stresses, written in Eq. 30.3 and 30.4, respectively. Then, contact stress update routine is executed to correct the sub-contact stresses based on the pre-determined failure criteria (unless they are defined as fully elastic).

$$\sigma^{t+} = \sigma^t + \Delta\sigma \tag{30.3}$$

$$\tau^{t+} = \tau^t + \Delta\tau \tag{30.4}$$

A recently proposed elasto-softening contact model adopting a linear softening branch for tension, shear, and compression regimes is utilized in this research [25]. As shown in Fig. 30.3a, the ultimate relative sub-contact displacement in tension ($\Delta u_{n_t}^c$) and compression ($\Delta u_{n_c}^c$) is controlled by the defined mode-I (G_f^I) and compression (G_c) fracture energies. Sub-contact stiffness degradation is considered in the normal direction in the unloading and reloading cycles, which can be depicted in Fig. 30.3a. The maximum tension and compression sub-contact stresses are limited by the defined bond tensile strength (f_T) and masonry compression strength (f_m), respectively. Moreover, a standard Mohr–Coulomb failure criterion is used to compute ultimate (τ_0) and residual (τ_{res}) sub-contact shear stress where the maximum shear strength is associated with the initial bond cohesion (c_0) and friction angle (θ_0) while residual one simply relies on the frictional resistance with no cohesive bond considering the residual friction angle (θ_{res}), written in Eq. 30.5. Note that compression stresses are indicated as negative while tension is positive.

$$\tau_0 = c_0 - tan(\theta_0)\sigma \tag{30.5}$$

It is worth noting that the simulated compression behaviour corresponds to the phenomenological response of masonry composite, while bond behaviour is described in tension and shear. Similarly, the ultimate sub-contact displacement in the shear direction is calculated based on the assigned mode-II (G_f^{II}) fracture energy (Fig. 30.3b). A global damage parameter, two-way coupling of tension and cohesive bond strength, is used to relate the tension failure to cohesion and vice-versa. Overall, a composite failure envelope described by the Mohr–Coulomb failure criterion with a compression and tension cut-off is utilized, as shown in Fig. 30.3c.

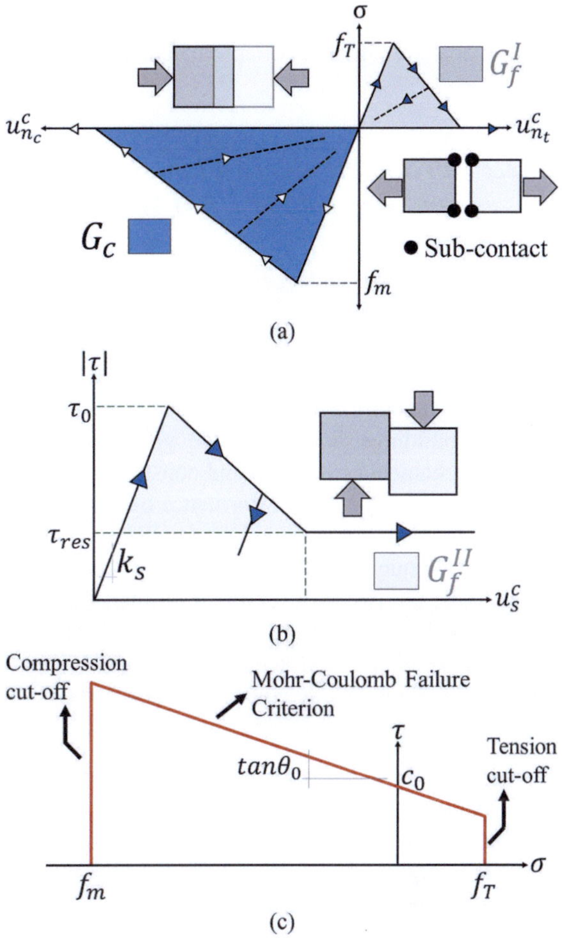

Fig. 30.3 Stress-displacement behavior in the (**a**) normal, (**b**) shear directions, and (**c**) composite failure envelope

Readers are referred to the reference study for further details regarding the mathematical background and implementation of the proposed elastic-softening contact model [25].

Once the updated sub-contact stresses are obtained, they are multiplied by their associated sub-contact area to get sub-contact forces and further utilized in the equations of motion discussed earlier. The explained numerical procedure is continued until a quasi-static solution (if any) is reached under the prescribed external loading condition. The proposed discontinuum-based analysis is performed using a commercial three-dimensional discrete element code, 3DEC, developed by Itasca Consulting Group [23], and the discussed elasto-softening contact model is implemented in C + + and compiled as DLL (dynamic link library) into 3DEC via the user-defined constitutive model option.[1] Next, the comprehensive application of DEM to analyze the in-plane strength and collapse mechanism of URM walls is presented.

30.2.2 Predicting the In-Plane Failure Mechanism and Capacity of URM Walls via DEM

As witnessed in past and recent seismic actions, URM buildings are severely affected by strong ground motions and exhibit highly complex and nonlinear behaviour influenced by various factors. The construction quality of the load-bearing masonry walls and the implementation of effective connections between floor-wall systems are the primary factors affecting the local and global structural behaviour of URM masonry buildings. Moreover, the geometrical properties, vertical compression stress, mechanical properties, and construction quality are the critical factors associated with the structural performance of URM walls [26]. In-plane failure mechanisms are usually observed in URM buildings once the local failure mechanisms are prevented, frequently noted during post-earthquake site visits [27, 28]. To this end, we simulate the two most common in-plane mechanisms, flexural/rocking failure, and diagonal shear cracking, using the proposed computational modeling strategy to demonstrate its potential and applicability. Furthermore, the results obtained from computational models are compared against frequently used strength formulas based on the fundamental concepts of static and strength of the material, summarized below.

Flexural/Rocking Failure: The flexural failure mode is observed once the lateral load yields an overturning mechanism due to violating the static moment equilibrium condition of the URM wall, typically accompanied by toe-crushing. The ultimate lateral load associated with the rocking failure can be obtained simply by considering moment equilibrium and neglecting the bed-joint tensile strength. Therefore, the in-plane maximum lateral load for rocking failure (V_F) can be calculated using the Eq. 30.6. Note that the geometrical characteristics, masonry compression strength (related to the toe-crushing) and vertical pre-compression stress (σ_0) strongly influence the flexural capacity of the URM wall, which can also be depicted in Eq. 30.6.

$$V_F = \frac{\sigma_0 L^2 t}{2\psi H}\left(1 - \frac{\sigma_0}{\kappa f_m}\right) \tag{30.6}$$

where σ_0, ψ, κ and t are the uniformly distributed vertical pre-compression stress, a coefficient adjusting the effective height based on the boundary conditions (i.e., $\psi = 1$ for cantilever and $\psi = 0.5$ fixed at both ends), masonry compression strength reduced factor (typically assumed as 0.85) and wall thickness (taken as 0.25 m, to

[1] Elasto-softening contact model can be accessed through the link: https://www.itasca.ca/software/udm-library/coupled-elastic-softening-joint-model.

correspond to a single brick thick wall for this analysis), respectively. H and L refer to the height and length of the wall.

Diagonal Shear Cracking: The shear failure of URM walls, corresponding to diagonal cracking passing through mortar joints and masonry units, demonstrates a complex behaviour, including different local failure mechanisms, discussed in [20]. The widely used strength formula, proposed by Turnsek and Cacovic [29], assumes that masonry is a homogenous and isotropic material and the diagonal shear failure occurs when the principal stress at the wall panel center reaches the critical value (i.e., reference tensile strength of the masonry). Accordingly, the following expression, comprehensively discussed in the reference study [30], is derived that is provided in Eq. 30.7.

$$V_{DSC} = Lt\left(\frac{f_t}{b}\right)\sqrt{1 + \frac{\sigma_0}{f_t}} \tag{30.7}$$

where f_t and b corresponds to masonry tensile strength (not the bond) and unitless shape factor, which is determined based on the wall/pier aspect ratio (i.e., $b = 1$, for $H/L \leq 1$, $b = H/L$, for $1 < H/L < 1.5$, and $b = 1.5$, for $H/L \geq 1$). Furthermore, diagonal shear cracking yields limited deformability compared to rocking failure due to abrupt failure in tension developing at the mortar joint and bricks.

It is worth noting that both strength formulations, presented in Eqs. 30.6 and 30.7, are frequently used in practice. Therefore, a comparison study is performed to test the performance and accuracy of the adopted modeling strategy. The in-plane load-carrying capacity of URM walls and associated failure mechanisms, mentioned earlier, are predicted considering different wall aspect ratios, denoted by ζ (defined as the height-to-length ratio (H/L)) and vertical pre-compression stress (σ_0) using DEM. The computational investigations are performed on a full-scale mass masonry wall with a height of 2.8 m, and a cantilever boundary condition (fixed at the bottom and free at the top) is considered during the analyses. Different aspect ratios, ranging from 0.5 to 2.0, are considered to capture squat and slender wall behaviour that are subjected to combined shear-compression loading. The contact properties are determined by considering typical strength values for a weak-bond strong-unit combination, following recommended values presented in the literature [31, 32]. Recall that the adopted modeling strategy represents each masonry unit via two identical rigid blocks with a potential crack surface in between, while the cohesive bond action and deformability of the discrete block system are lumped at the contact planes among the masonry units. The contact stiffness (corresponding to the composite action of masonry) is predicted based on the ratio of masonry elastic modulus (E_m—taken as 2.0 GPa), and horizontal joint spacing (h–0.1 m for the analyzed URM wall), i.e., $k_n = E_m/h$.. The compressive strength of masonry is determined as 8.0 MPa. Note that the predicted elastic masonry stiffness is in line with the suggested values in the literature for soft and weak mortar joint brickwork assemblages ($E_m \approx 200 - 350 f_m$, see [33, 34]).

To capture stiff masonry unit action relative to the unit-mortar interface, ten times stiffer contact properties are defined at the sub-contact between blocks associated with each masonry unit, as suggested in the reference study [12]. The assigned contact properties for the bond and masonry units are given in Table 30.1.

Figure 30.4 shows the predicted strength values (peak load) per incrementally applied lateral loads obtained from the computational model and analytical formulations for flexural failure and diagonal shear cracking for different wall aspect ratios and vertical pre-compression stress. The results indicate that discrete element models are in good agreement with commonly used strength formulas for various geometrical properties of URM wall and pre-compression vertical loads (see Fig. 30.4a, b, c and d). It is important to underline that when the collapse mechanism indicates a pure-rocking failure (e.g., Fig. 30.4d), the DEM-based simulations provide almost identical results to the theoretical strength formulation. On the other hand, slight differences are observed when interactions occur between the flexural and diagonal shear cracking collapse mechanisms, as expected. Still, the predicted lateral load capacity aligns with the analytical formulation presented earlier. Additionally, the effect of wall aspect ratio on the in-plane response of the analyzed URM walls is demonstrated in Fig. 30.5, considering a vertical pre-compression stress of 0.4 MPa. The rocking dominant response can be observed for slender walls ($\zeta > 1$) that can be also noticed from the sub-contact failures computed along the contact planes, as shown in Fig. 30.5.

The above results show that the proposed discontinuum-based analyses indicate satisfactory agreement with well-established strength criteria. The presented computational investigations serve as the validation of the adopted modeling approach, which will be further applied to explore the mechanics of an existing stone masonry building, including its as-is construction morphology, in the following section.

Table 30.1 Defined contact properties corresponding to the mechanical characteristics of the bond

Bond properties			
k_n, k_s (GPa/m)	f_T, c, f_m (MPa)	$tan\theta_0$, $tan\theta_{res}$ (°)	G_f^I, G_f^{II}, G_c $\left({}^{N}/{}_{m}\right)$
20, 8	0.1, 0.1, 8.0	0.6, 0.6	3, 30, 25,600
Masonry unit properties			
k_n, k_s (GPa/m)	f_t, c (MPa)	$tan\theta_0$, $tan\theta_{res}$ (°)	G_f^I, G_f^{II} (N/m)
200, 80	1.0, 2.0	1.4, 1.4	30, 300

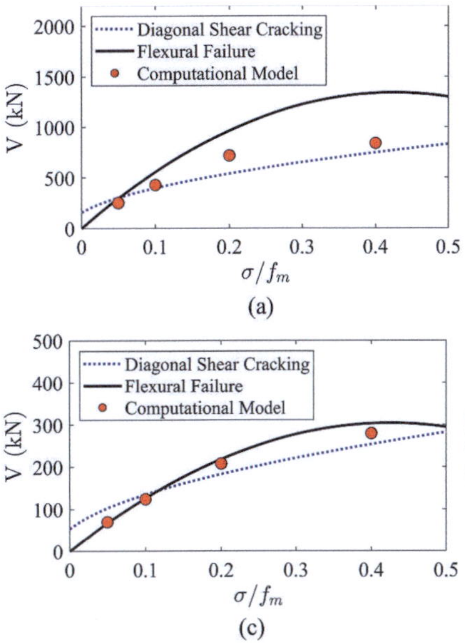

Fig. 30.4 Comparison of in-plane strength prediction between analytical formula and adopted computational model considering different wall aspect ratios: **a** 0.5, **b** 1.0, **c** 1.5 and **d** 2.0

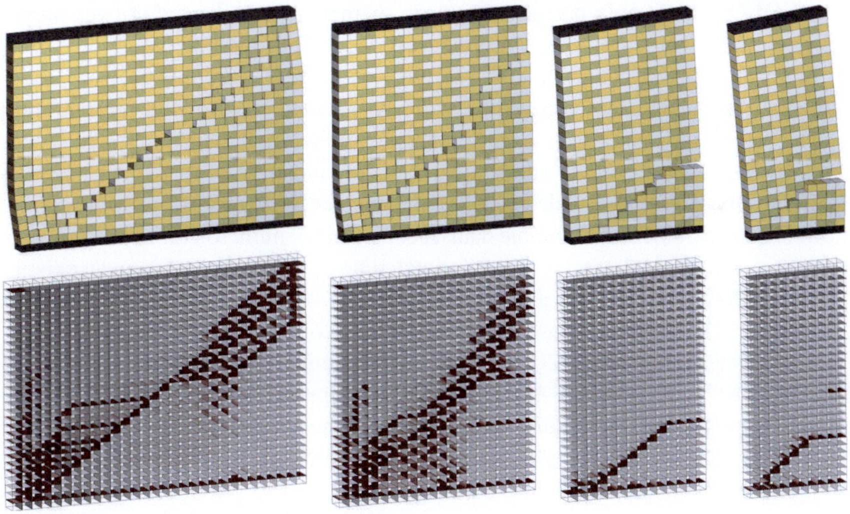

Fig. 30.5 Predicted failure mechanism via discontinuum-based analysis for URM walls with different aspect ratios—from left to right: $\zeta = 0.5, 1.0, 1.5$ and 2.0 ($\sigma_0 = 0.4$ MPa). The sub-contact failures in tension-shear are denoted as red-colored surfaces

30.3 AI-Assisted Data Collection for Documentation and Structural Analysis

Performing discontinuum-based analyses of unreinforced stone and/or brickwork masonry assemblages begins with generating discrete block-based geometries, which may vary in complexity depending on the analyzed assembly and expectation from the computational analysis. As URM buildings contain numerous units, generating such models by hand results in a tedious, time-consuming effort. Advances in machine learning technology have provided tools for streamlining vision-based tasks via the available open-source object detection and instance segmentation frameworks. Leveraging such frameworks can be advantageous for documenting masonry structures and eliminating the need to prepare computational models through manual processes [35]. This study primarily employs YOLOv8, an advanced computer vision model for object detection and instance segmentation. YOLOv8 performs real-time bounding box detection and instance segmentation of multiple objects within images via trained convolutional neural networks [36]. YOLOv8 used in this study allows for the segmentation of high-resolution images by performing detections on different sections of the image separated into grids. An additional segmentation model, Segment Anything (SAM), is used to streamline training data preparation for the YOLOv8. SAM identifies and delineates objects in images directed to different prompting formats, such as text queries and bounding boxes [37]. This model differs from YOLOv8, as it only performs instance segmentation tasks and utilizes proprietary neural networks trained on vast datasets. While it is possible to fine-tune SAM for more specific segmentation tasks, original network checkpoints must be loaded with large file sizes. The generation of an AI-assisted tool for the present study is achieved using Ultralytics, a comprehensive toolset for developing and testing convolutional neural networks developed by an organization of the same name [38].

The ability of YOLOv8 to detect and segment objects relevant to this study requires the model to be trained on a suitable dataset consisting of numerous annotated images. Therefore, a sizeable dataset of 3768 annotated images is compiled from an existing dataset in the literature [39], and numerous royalty-free images are obtained from the internet. The original dataset, "MCrack1300", comprises 1300 images obtained from a crack detection dataset, online sources, and numerous photographs of masonry buildings near the University of Birmingham Campus [39]. While this dataset was initially designed for fine-tuning the SAM model, the annotations are converted into the YOLO format using Roboflow, a web-based suite of dataset preparation tools [39]. A preliminary model is trained using this dataset to provide a checkpoint to further expand the dataset. The images, a selection provided in Fig. 30.6, are split for training and testing. In this case, 1000 images were used to train the neural network, 150 were used to validate accuracy and performance, and the remaining 150 were set aside for automated testing. This dataset contains numerous annotations for bricks of various types and cracks of different sizes and severities. In addition, there is also a separate annotation type for damaged bricks, such as those split apart by cracks. YOLOv8 is trained for one hundred epochs, the number of pass-throughs of the

Fig. 30.6 Annotations included in the "MCrack1300" dataset

entire dataset during the training process. In addition, the boilerplate "yolov8n-seg" checkpoint, trained on the COCO dataset from Microsoft [40], is used as the starting point for training.

The YOLOv8 model trained using just the "MCrack1300" dataset is not meant to be the primary detector of masonry units in this study but rather to perform detections on additional smaller images to efficiently generate additional training and validation annotations for a larger, more diverse dataset. These additional images originate from royalty-free photographs of masonry façades subdivided and resized into smaller 640 × 640 images. Ultralytics provides an automatic annotation workflow that uses YOLOv8 to perform bounding-box detections and SAM to generate segmentation masks of the objects within the bounding boxes. This procedure has resulted in the instant preparation of an additional 2468 annotated masonry images of various types, colours, and sizes. Figure 30.7 provides a selection of the annotations obtained from this workflow, showing that it effectively prepares annotations suitable for training and validation. As the sole purpose of this dataset is to detect masonry units, the crack annotations from the original dataset were discarded, and the broken brick annotations were merged with the brick annotations.

While datasets can be expanded using new images, they can also be expanded significantly via data augmentation to increase the model's performance by creating variations of each image with various manipulations such as rotations, colour adjustments, and mirroring [41]. This procedure was completed using Roboflow, where the combined images from the "MCrack1300" dataset and the new annotated images were mirrored horizontally and vertically, rotated 90 degrees in both directions and rotated further slightly. Hence, the dataset was enlarged to 16,275 images by incorporating these augmentations and removing redundant and poorly annotated images. While Roboflow provides many different augmentation options, rotations

Fig. 30.7 Annotations generated using Ultralytics' automatic annotation algorithm

were primarily used to provide sufficient augmentations to accurately detect vertically placed masonry units, such as those making up lintels above openings. Upon completion of the augmentations, the dataset was exported from Roboflow for training using Ultralytics. Since this dataset already incorporates the photos from the original dataset, the new model is trained again using the COCO-based boilerplate checkpoint rather than the previously trained model as a starting point.

The performance of the newly trained YOLOv8 model is assessed during the training process. Multiple validation metrics exist for convolutional neural networks, such as precision-recall and the mean average precisions at different intersection-over-union (IoU) thresholds. The precision metric provides an outlook into the accuracy of optimistic predictions compared to the ground truth validation data [42]. On the other hand, the recall metric is the ratio of true positives to the sum of both true and false positives. Higher recall values signify the neural network's effectiveness in correctly identifying objects. During the validation process, the IoU of a mask is determined by the overlap between the predicted mask and the ground truth mask. The mean average precision is quantified for different IoU thresholds, which provides insight into the model performance at different levels of accuracy, either at a specific threshold or a range between two thresholds. As shown in Fig. 30.8a, b. c and d, these metrics are obtained across each consecutive epoch during the training process.

The above results indicate that after being trained for 100 epochs, the model effectively detects the bricks found within the validation with an average precision of 94%. Similarly, a high recall ratio of around 91% shows that most positive detections are true positives, meaning that the detected objects match up with the ground truth validation data. Even though the mean annual precision at the 50% IoU threshold is at 97%, the model is precise, albeit with predicted masks that weakly overlap with the ground truth masks. While observing the same metric within an IoU threshold

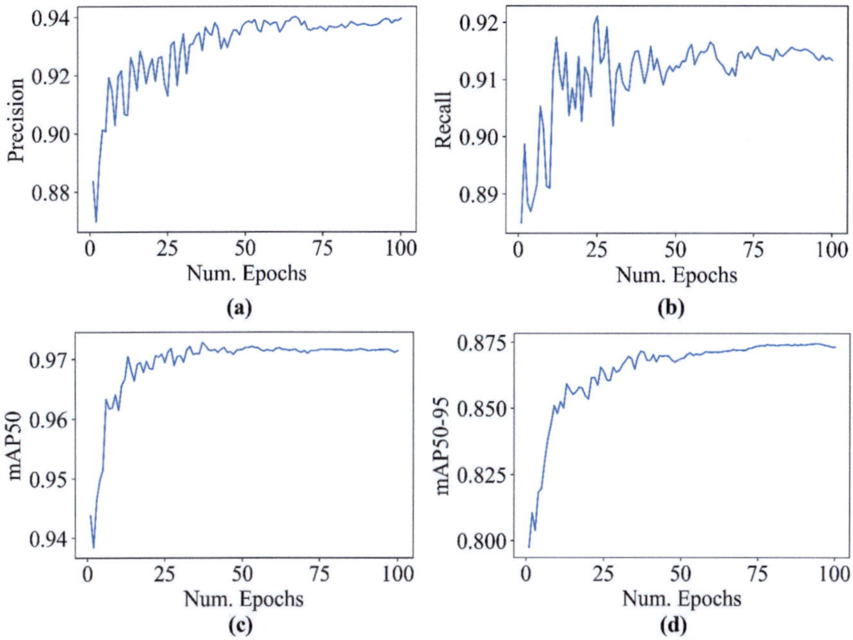

Fig. 30.8 Validation metrics per epoch: **a** Precision, **b** recall, **c** mean Average Precision at 50% IoU threshold, and **d** mean Average Precision between 50 and 95% IoU thresholds

range between 50 and 95%, the mean annual precision decreases to approximately 87.5%.

The direct outputs of YOLOv8 comprise ordered lists of object descriptions, confidence ratios, and areas and vector coordinates. As shown in Fig. 30.9, presenting the implemented workflow, the data obtained from the training of the neural network and processing of input images is exported into spreadsheets and visual formats such as PDFs, images, and vectors graphics. The primary uses of this data are to aid in generating discrete element models and to automatically provide quantitative assessments of the morphological features of stone masonry walls. As such, all of the brick masks are exported into a raster image which is then used alongside a simplified CAD drawing of the building as data input for a block generation algorithm. The direct output from YOLO inferencing can be used for additional structural analyses related tasks, such as identifying and enumerating the geometry and severity of cracks (e.g., [43]). This represents a potential future improvement to enable the use of images to estimate the severity of pre-existing damage sustained by the wall, which may be used to alter the bond properties during the automated DEM model generation workflow.

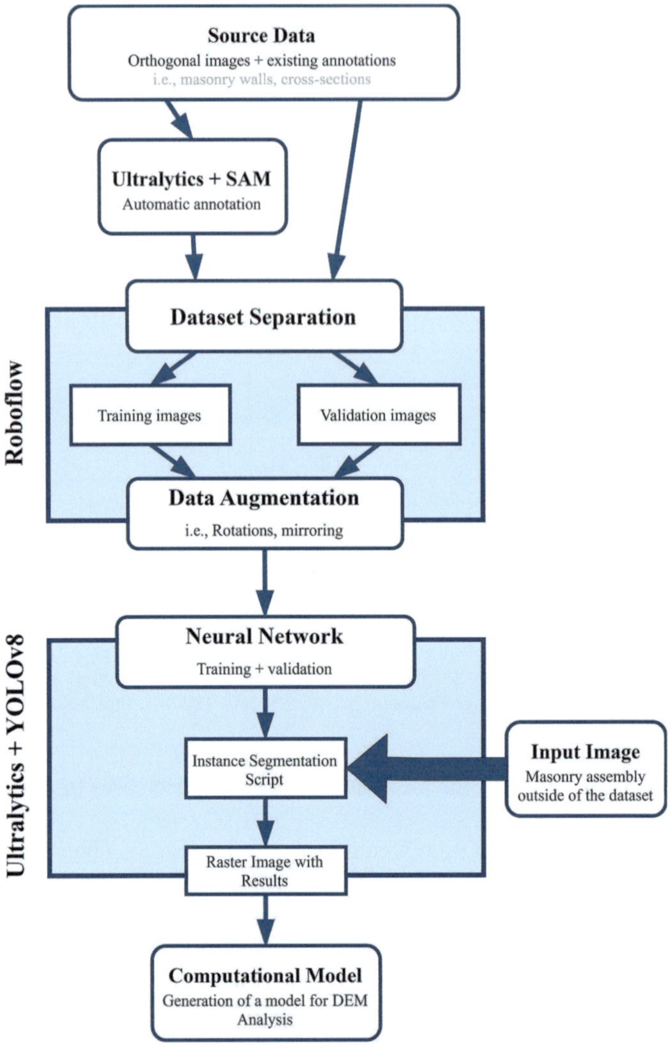

Fig. 30.9 Masonry documentation and model generation workflow

30.4 Application of the Proposed Data-Driven Computational Modeling Approach on a Stone Masonry Building

This section shows a potential application of the proposed data-driven computational modeling framework on a historic stone masonry building (The Buchanan Block located in Kemptville, Ontario, Canada), with a load-bearing wall system comprising stone masonry units visually distinct from the outside. The Buchanan Block is a

Fig. 30.10 Analyzed URM building (Buchanan Block)

commercial space built in the 1860s. It consists of squared ashlar stone masonry with flat arched lintels above its openings along its western elevation, as shown in Fig. 30.10. The building's architectural style can be characterized as Georgian, in keeping with the large number of buildings built in this style during the late 18th and early nineteenth centuries [44]. The simplicity of the building allows for a preliminary evaluation of the proposed model generation framework. The Buchanan Block is located along a busy intersection, making it arduous to take orthogonally correct photographs. The orthophotos of the building's western elevation are created from a photogrammetric 3D model generated using Agisoft Metashape from a collection of photographs taken by a DJI Mini 3 Pro drone. The creation of orthophotos via photogrammetry aims to provide accurate two-dimensional representations of the building despite the lack of space to take orthophotos with a conventional camera. Upon acquiring photographs of the building in multiple positions and angles, they are processed through Agisoft Metashape's algorithms to create point clouds and 3D models.

30.4.1 AI-Assisted Block Generation

The proposed workflow includes a series of Python scripts utilizing different libraries necessary to generate a group of discrete blocks replicating the load-bearing stone masonry wall based on the collected vision-based data for subsequent discontinuum-based analysis. The orthophoto in Fig. 30.10 serves as the input image for the YOLOv8 object detection script, leveraging the masonry detection neural network mentioned in Sect. 30.3. The model processes the orthophoto and then superimposes masks with randomly selected colours to ensure they are visually distinct. Figure 30.11a shows the output masks obtained above a confidence ratio threshold of 80%. This metric is similar to precision and recall as it measures the model's confidence that the detected object is indeed what the model is trained to detect. As the results appear incomplete, it denotes that the model cannot detect all of the masonry units with a high level of confidence. While obtaining more masks by lowering the confidence ratio threshold is possible, they will be of poorer quality. Even though the adopted methodology successfully detects most masonry units at the analyzed stone masonry façade, it struggles particularly with the lintels and discoloured sections, potentially due to a further lack or imbalance of training material, which may signify the need for further training with an improved dataset. Nonetheless, during the generation of the discrete block model for discontinued-based analysis, the algorithm samples the sizes and staggering ratios of the contours to create missing (or undetected) masonry units in line with the as-is construction quality and construction features that will be discussed in the following section. Even at a confidence threshold of 80%, there are still poor-quality masks, such as those that incompletely capture the bricks from the input image. Therefore, as shown Fig. 30.11b, many of these masks were discarded to eliminate the generation of inaccurate geometry by the algorithm.

30.4.2 From Vision-Based Data to a Discrete Block Model

Once masonry units of the analyzed stone masonry wall have been detected and defined by a set of masks, they are further utilized to create a discrete block system for computational modeling based on the simplified micro-modeling approach. It is important to recall that in the adopted modeling approach, the masonry units are expanded up to half-thickness of the surrounding mortar joints, and they directly contact their neighbours at the zero-thickness interfaces. Throughout this research, the expanded units are defined as rigid, and the bond behaviour is simulated at the interfaces of adjacent blocks. Therefore, the proposed approach requires converting detected masonry units, discussed in the previous section, into "expanded" blocks compared to their actual dimensions. It is worth noting that this conversion requires both geometric continuity and maintaining existing mechanical interactions without introducing new ones. This section will detail the implemented approach regarding the conversion process and discuss the implemented steps.

The proposed workflow begins with analyzing the geometrical features of horizontal bed joints. Typically, the bed joints of well-maintained and periodic (or quasi-periodic) masonry walls continuously span the entire wall and remain at an approximately constant height. A skeletonization procedure is implemented to determine the position vectors of bed joints for each masonry course. The proposed skeletonization procedure starts with a high-resolution binary image of the analyzed wall where block masks and the regions they encompass are true, and all other regions are false. The head joints separating horizontally adjacent blocks in the binary mask are removed. This is done by repeatedly applying a kernel to the image, which extends blocks horizontally without modifying bed joints. When only the bed joints remain

(a)

(b)

Fig. 30.11 Buchanan block: **a** Raw segmentation output, and **b** edited version discarding poor detections

in the image, an inverse Boolean operation is applied in which ‘blocks’ correspond to false, and the mortar bed joint corresponds to true. Results can be improved by setting large empty regions (missed detections or windows) currently considered mortar bed joints to be represented instead as blocks (false). This approach reduces white (or true) regions into minimum thickness (single pixel) lines that follow the centerline of the initially provided mortar bed joint shapes. Then, white pixels in the skeletonized image can be used to determine the average vertical location of bed joints. If a large portion of the masks has been missing, the bed joint detection tool can fail to detect the bed joints. In this case, an additional check is executed to find spacings between detected bed joints greater than twice that of the average distance between bed joints. These regions are then assigned additional bed joints spaced as close to the average distance between bed joints as possible. The initial block mask data drawn on the binary image is presented in white, and window data is denoted using gray, as shown in Fig. 30.12a. Furthermore, the determined bed joints based on the described methodology are illustrated explicitly in Fig. 30.12b. Note that the proposed algorithm does not consider significant irregularities along the bed joints; instead, it provides a continuous and approximate bed-joint height for each course of masonry (see Fig. 30.12b).

The wall geometry is stored in an array where each row represents a single course of masonry. Within a row, each entry or 'cell' represents a ~ 1 cm length of the masonry wall. These cells are assigned integers based on which block they are a part of. Each block in the wall has a unique integer associated with it. Therefore,

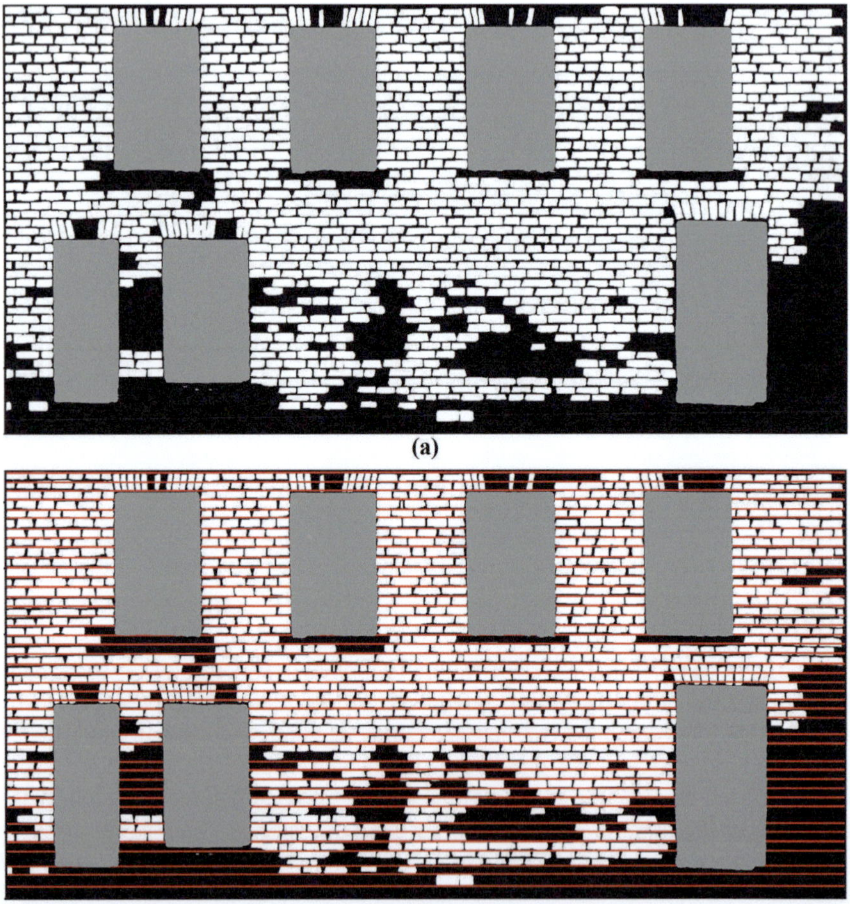

(a)

(b)

Fig. 30.12 Binary mask of the URM stone building with gray windows superimposed: **a** initial input and **b** prescribed bed joints (continuous red line)

multiple cells horizontally adjacent with identical assigned integers define a single masonry unit. It should be noted that the ~ 1 cm length represented by a single cell is an adjustable parameter. This can be reduced if the wall comprises very small blocks or increased if a lower fidelity model is preferred. An initial area-based detection is performed to fill the discrete block array. Each entry in the discrete block array is assigned either a block integer or zero (corresponding to empty). This is accomplished by taking the wall described through masks (as seen in Fig. 30.12b and Fig. 30.13a) and superimposing a grid system onto it, shown in Fig. 30.13b). Each cell in this superimposed grid system corresponds to an entry in the discrete block array. The grid's horizontal lines correspond to bed joints in the wall, and vertical lines are spaced at the same ~ 1 cm length represented in the discrete array. Examining which block mask is present in each of the grid system's cells, a block integer is assigned to the corresponding entry in the discrete block array. If no block is present in one of the grid system's cells, it is considered empty. If two blocks are present in a cell of the superimposed grid, then the one with the largest area in the cell is transcribed to the discrete block array. The blocks assigned to each grid cell (along with those considered empty) can be seen in Fig. 30.13c.

All block masks are now represented in the discrete block array by clusters of cells assigned the same block ID. However, many cells within the discrete block array still have no block ID or window identifier assigned to them. These two cases must be remedied with separate methods. The adjacent blocks are expanded to fill the gap for small clusters of unassigned cells (usually associated with head joints). This increase is consistent with the expanded block geometry and zero-thickness interface representation of analyzed masonry walls prescribed by the simplified micro-modeling

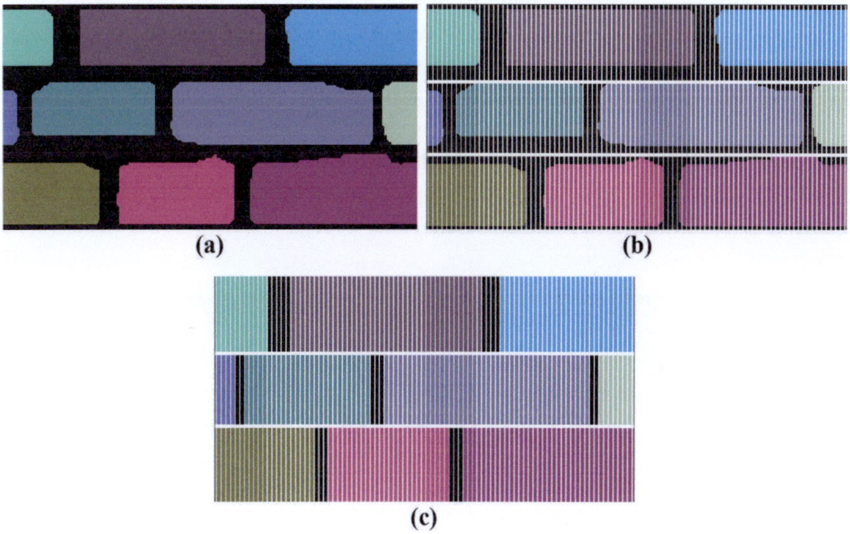

Fig. 30.13 Converting block masks into discrete block geometry: **a** initial masks, **b** masks with discrete array cells, and **c** generated distinct block geometry

approach. The array operation to achieve this gap-filling procedure starts by flagging unassigned cells that are horizontally contiguous with a certain number of other unassigned cells. Having less contiguous cells, the smallest block in the wall is the threshold for gaps to be flagged. These flagged cells are then assigned the same block ID as the nearest block (already assigned cell) in the same row. The result of this operation is showcased in Fig. 30.14. Moreover, a novel custom generation tool is used to remedy larger sections of the array where no blocks were assigned to cells. It matches the statistical properties of newly generated regions with those of already generated sections, discussed in the following section.

Once block IDs are assigned to all remaining portions of the discrete block array, a deficiency remains in the wall's lintels. Lintel blocks above the openings in the wall often do not conform to the assumptions made about blocks represented in the discrete block array. They rarely have perfectly vertical head joints and usually have different heights from blocks in adjacent layers. Discrete block representations are created based on a Voronoi transform utilizing the initial lintel masks to address this issue. Initially, lintel blocks are detected and associated with a window by scanning which blocks neighbour a given window and have a centroid higher than its top frame. A Voronoi transformation is applied to the block masks, as seen in Fig. 30.15a, b. This transformation expands the masks until the entirety of the wall is encompassed by a mask. The space initially outside of a mask becomes part of the nearest adjacent mask. Each window is then individually processed by excluding all blocks and windows that do not belong to the window and the associated lintels under consideration. Figure 30.15c shows an image that only contains the expanded (via Voronoi transformation) window shapes and their corresponding lintels. Points defining the lintel blocks are created wherever three Voronoi cells meet (seen in Fig. 30.15d); these Voronoi cells can correspond to blocks, the window, or the rest of the wall. Additional points must be added for the corners furthest from the window centroid in the leftmost and rightmost lintels blocks (Fig. 15e). Finally, all the vertical coordinates of the lintel points are normalized to the nearest bed joint height and utilized to generate discrete blocks for the lintel, as shown in Fig. 30.15f. It should be noted that there is a substantial variance in the character of window lintels frequently seen in historic masonry. In this study, we proposed a solution primarily for flat arches where it is appropriate to normalize the lintel block's vertical coordinates to the

Fig. 30.14 Expanding discrete blocks to fill head joint gaps **a** unexpanded blocks **b** expanded blocks

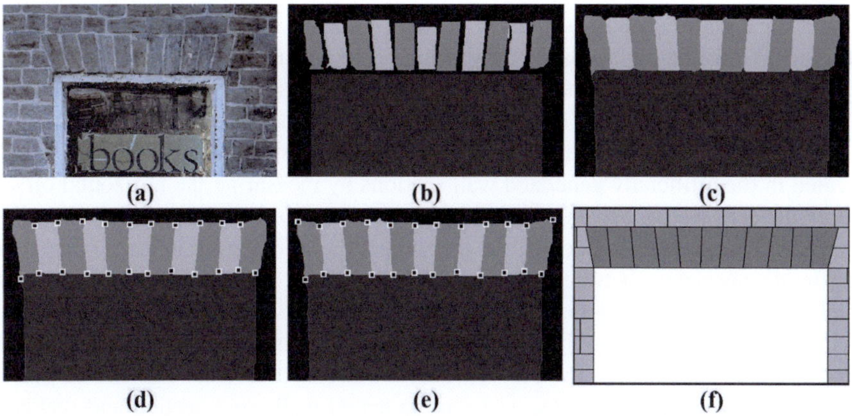

Fig. 30.15 Process to create the points defining a lintel, **a** Image of window and lintel, **b** AI detected masks, **c** voronoi transformation, **d** points defined by Voronoi cell intersection, **e** points defined by extremities of Voronoi cells, **f** discrete block geometry

nearest bed joint. However, the implemented method would not be suitable for arch lintels. Therefore, the readers are referred to the reference study exploring the other techniques to handle various lintel forms [45]. Lastly, a three-dimensional model is created using the coordinate data stored in the discrete block array and the separately detected lintel points. Since the points corresponding to each masonry unit are in two dimensions, they are extruded by considering the actual wall thickness, and three-dimensional geometry is obtained. If the quality and number of documented masonry units for lintels are not satisfactory, as in the case of several lintel sections here, they can be generated externally considering the essential construction features noted from the reference geometry, which has been previously demonstrated by the authors in [45].

30.4.3 Generating the Missing Section of the Masonry Wall

A portion of a given masonry wall may not be fully detected through AI-assisted or manual block segmentation methods. The primary reasons may include material degradation, staining, efflorescence, or no clear lines of sight. Additionally, an entire wall may be undetectable due to the neighbouring building with no clear vision. In these situations, generating a "*geometric twin*" becomes impossible. Therefore, we propose an alternative solution in which undetected or invisible wall sections are created utilizing the collected information from the already documented wall morphology, including the size distribution of masonry units and the staggering ratio of vertical head joints. The two parameters adopted for characterizing the wall morphology can be directly obtained from the detected wall cross-sections. However, it must be highlighted that the present study only considers single-layer walls; hence, additional information needs to be collected for multi-layer walls. In Fig. 30.16, the block size distribution corresponding to the case study is presented considering only the length of masonry units since no considerable variation is noted in height.

The nonperiodic bond pattern used in the analyzed stone masonry wall is represented in the artificially generated wall sections by measuring the horizontal offset ratio, denoted by M_l, between head joints in adjacent rows in the actual wall geometry. As demonstrated in Fig. 30.17a, the staggering ratio of head joints is computed as the distance ratio between the stepped line (shown in red colour), the shortest path starting from an arbitrary head joint to the bottom section of the representative window, divided by the straight line (blue), the direct distance between the start and end of the stepped line. Thus, a representative masonry wall can be generated based on these two parameters collected from the actual wall cross-section and realized via the proposed AI-assisted discrete block-generation methodology. At the artificially generated wall sections, blocks are selected at a rate corresponding to their frequency in the sample of block sizes. The M_l ratio of the generated wall is modified by considering multiple candidate blocks during placement in an unoccupied space. Each candidate block is selected using the previously described method, following the block size frequency. Note that the one creating a new head joint with the greatest offset from the nearest head joints in the adjacent row is placed. By modifying the number of candidate blocks, the M_l ratio of the wall can be adjusted to match that of the already detected region. As shown in Fig. 30.17b, M_l ratio is not a constant value for nonperiodic or quasi-periodic masonry walls through its length. The distribution of M_l, presented in Fig. 30.17c, can be found by measuring the head joint staggering ratio in multiple wall regions. This approach provides a more realistic characterization of the construction quality of the analyzed masonry wall. The adopted parameters are part of the Masonry Quality Index (MQI) method introduced by Borri et al. [46], developed to perform preliminary seismic analysis and predict the material strength of existing unreinforced masonry buildings relying solely on visual inspection.

Hence, the discrete block model is generated utilizing the detected stone units (or blocks) based on the implemented AI-assisted framework, and missing ones are created following actual morphological features of the masonry wall by taking into

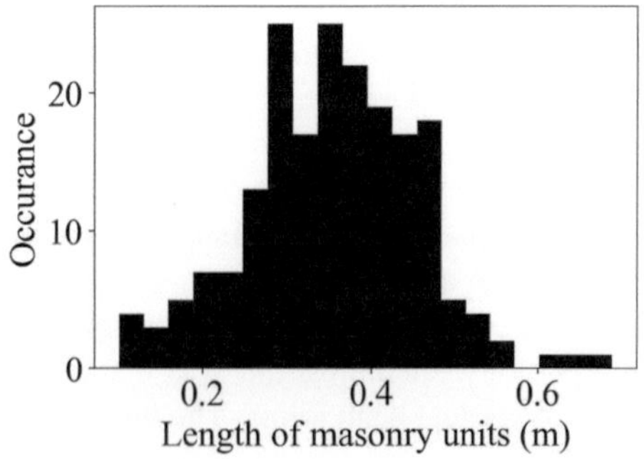

Fig. 30.16 Block size distribution of the analyzed stone masonry façade detected through the proposed AI-assisted approach

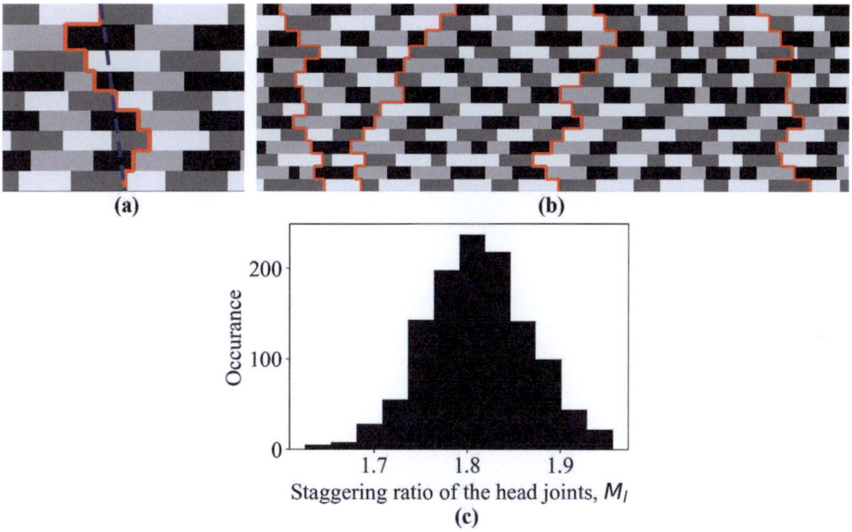

Fig. 30.17 **a** Illustration of the M_l ratio, **b** multiple paths used to create M_l ratio distribution and **c** obtained distribution of M_l ratio for the analyzed stone masonry building

account the block size distribution and M_l ratio, discussed earlier. It is worth noting that the distribution of M_l ratios in the artificially generated wall section align with the documented parts of the façade. The final discrete block model is shown in Fig. 30.18, where the red coloured blocks associated to load-bearing masonry walls represent those generated artificially, whereas the white blocks correspond to actual units. It is important to underline that most undetected blocks suffer from significant deterioration and damage observed at the bottom of the building together with the most right slender pier, which can be depicted in Fig. 30.18. Overall, the proposed approach offers high fidelity in terms of representation of the discontinuous medium by taking advantage of AI-assisted tools and the most relevant wall construction features. Next, the generated discrete block medium is utilized to perform the structural analysis of the stone masonry façade.

30.4.4 Data-Driven Computational Modeling of a Stone Masonry Façade

In this section, the proposed data-driven computational modeling strategy is applied to the reference stone masonry building, where the analyzed section of the building (façade) is documented using AI-assisted tools mentioned earlier. The discrete block model, consisting of roughly 1600 masonry units, is utilized to predict the structural behaviour of the stone masonry façade subjected to lateral loading and differential

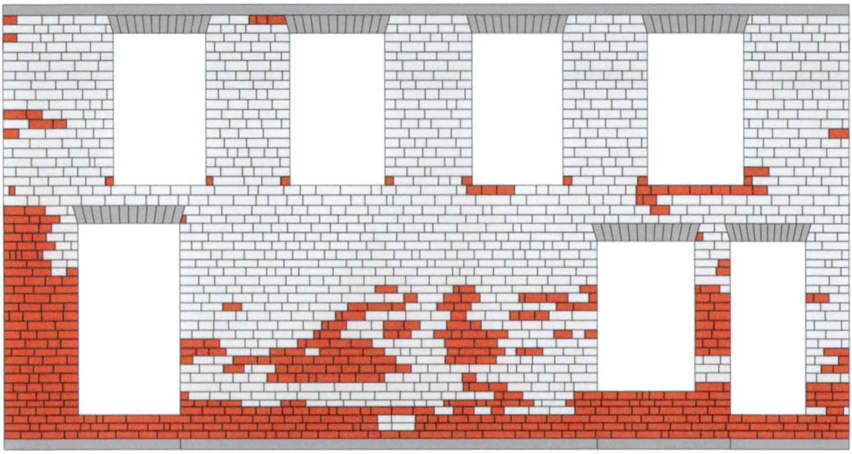

Fig. 30.18 Discrete block model of the analyzed stone masonry façade

support settlement. The primary objective is to demonstrate the potential use of the adopted methodology in the structural assessment of the analyzed building.

The identical mechanical properties utilized in Sect. 30.2 to predict the in-plane capacity and behaviour of masonry walls with different aspect ratios are used for the large-scale façade simulation without considering any potential crack surface for the masonry units following the assumption of weak mortar joints with strong stone blocks. Moderate vertical stress (i.e., 0.1 MPa) is considered to address roof and floor loads, and lateral loads are applied at the floor slab and roof location as point loads. First, the simulated façade is brought into equilibrium under its self-weight, and then lateral loads are uniformly applied along its height. During the analysis, the cantilever boundary condition is considered, and the lateral displacement is monitored at the top corner of the building. Figure 30.19 shows the predicted damage progression and associated force (base shear) versus drift curves under the increasing horizontal displacement where the lateral load is applied from right to left (denoted as L1). Note that the drift is calculated by dividing the lateral displacement traced at the monitoring point to the height of the building. The result indicates that the in-plane load-carrying capacity of the analyzed stone façade is limited by the diagonal shear cracking capacity of the large panel at the first storey (see Fig. 30.19a). Also, cracks at the spandrel walls and rocking mechanism at the slender masonry piers are noted. After the severe diagonal crack, an excessive sliding mechanism is noted along the bed joint with further bond failures among the stone units (Fig. 30.19b-c). Overall, the macro behaviour of the analyzed stone masonry building can be idealized as a bi-linear elasto-plastic response (see Fig. 30.19d). The damage progression associated to bond failure can be noticed along the bed and head joints indicated using red color at the contact planes in Fig. 30.19.

Next, the same loading scheme but in the opposite direction (L2) is utilized to assess the structural behaviour and capacity of the analyzed stone masonry building.

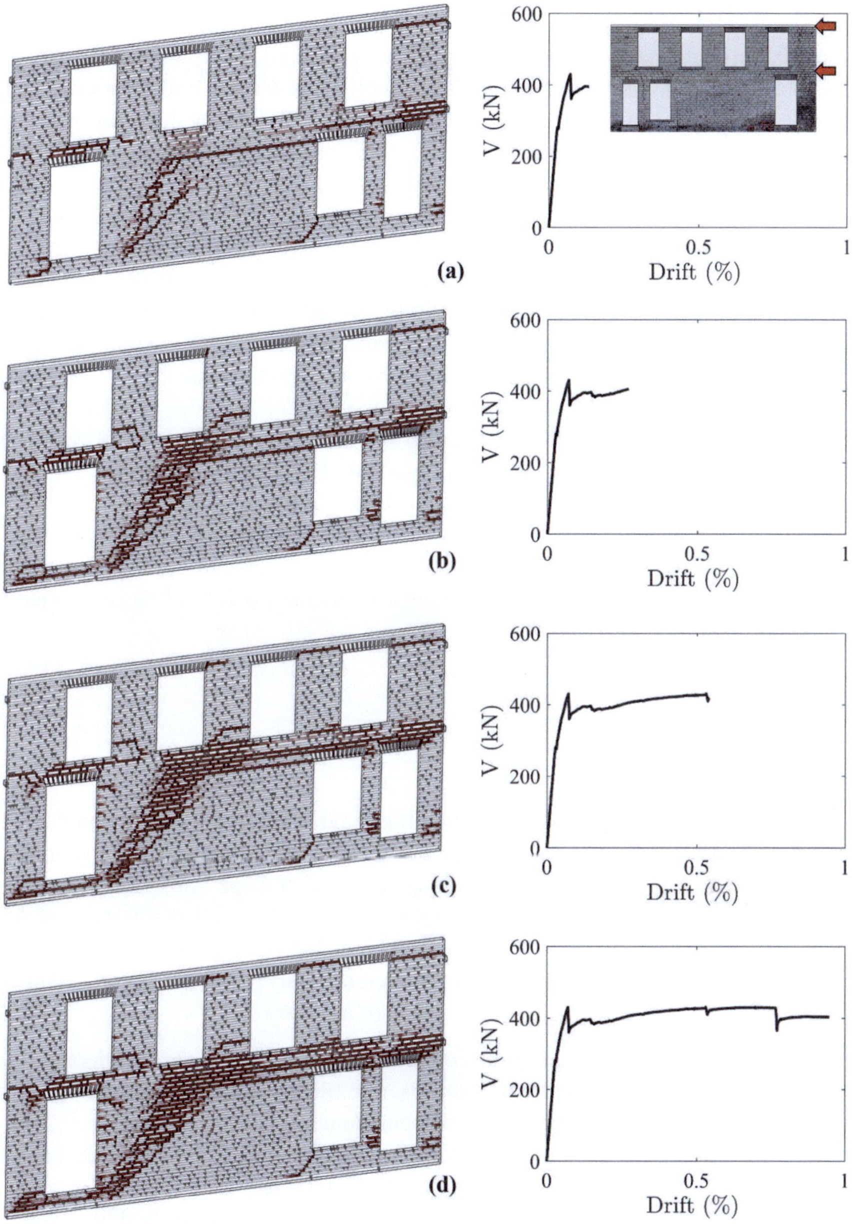

Fig. 30.19 Predicted damage progression in the analyzed stone masonry façade subjected to lateral loading scenario L1 and the associated force-drift curve

The results align with the previous analysis (L1) regarding maximum lateral load (Fig. 30.20a), whereas a less ductile response is predicted from the second analysis. The localized cracks at the sides of flat-arch lintels are noticed in both structural analyses. It is important to note that the analyzed building comprises irregular openings and has a non-periodic bond pattern, which yields considerable variations regarding the macro-mechanics of the discontinuum under different loading directions. As shown in Fig. 30.20a, b, the diagonal shear crack and several vertical cracks at the flat-arch lintels are the primary failure mechanisms. In line with the previous computational investigation, the structure has approximately kept its strength by relying mainly on the sliding resistance along the bed joints, which can be depicted in Fig. 30.20c, d. Furthermore, localized toe-crushing phenomena are observed once the structure has reached the ultimate in-plane load-carrying capacity.

Finally, the structural behaviour of the stone masonry façade is predicted under different differential soil settlement scenarios. The support blocks, placed under the large openings, are displaced 5 mm separately (Fig. 30.21). The computational investigations show that even relatively small displacement (Fig. 30.21a) or variations of boundary conditions may yield widespread crack distributions in low-bond stone masonry buildings (Fig. 30.21b). Similar to the lateral behaviour of the building, the vulnerable condition of the flat arches against even minor deformations is explicitly noted.

30.5 Conclusions

With the advances in computational modeling and computer vision technologies, data-driven simulations offer more precise inferences regarding structural behaviour and performance of masonry buildings and cultural heritage subjected to various loading scenarios. Through this study, we demonstrate a novel application of AI-assisted visual inspection tools employed to document the geometrical characteristics of a stone masonry building and nonlinear structural analysis via the discrete element method. The primary outcomes are summarized as follows:

- The implementation of AI in discontinuum-based simulations accelerates the conversion of vision-based data into discrete block representations of existing stone masonry buildings, making the documentation process efficient.
- Overall, the employed image segmentation and classification tools provide reasonable accuracy in detecting masonry units and offer a source of information regarding the construction quality of the investigated masonry building, which may be further utilized to generate the block-based representation of the structure.
- The undetected sections of the masonry walls can be later generated by utilizing the knowledge obtained through the documented sections of the building, including but not limited to block size distribution and staggering ratio of the vertical joints.

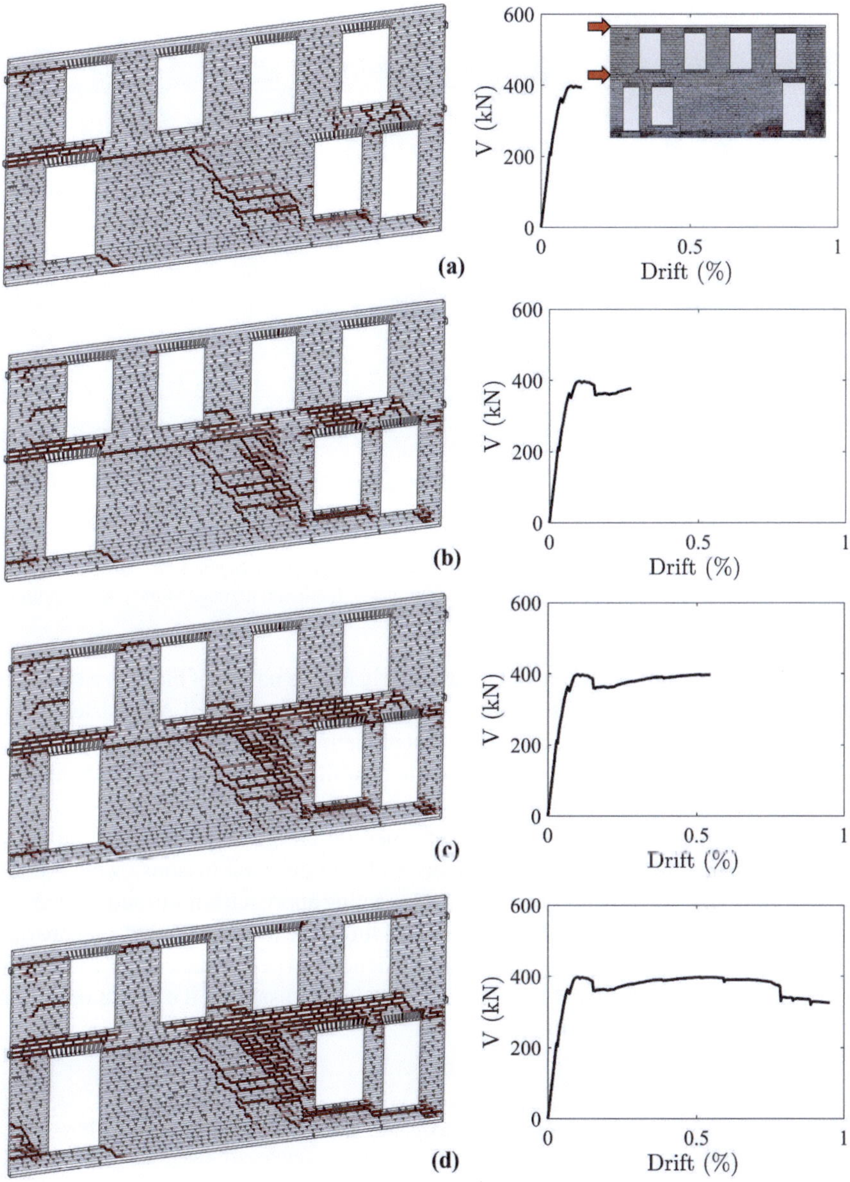

Fig. 30.20 Predicted damage progression in the analyzed stone masonry façade subjected to lateral loading scenario L2 and the associated force-drift curve

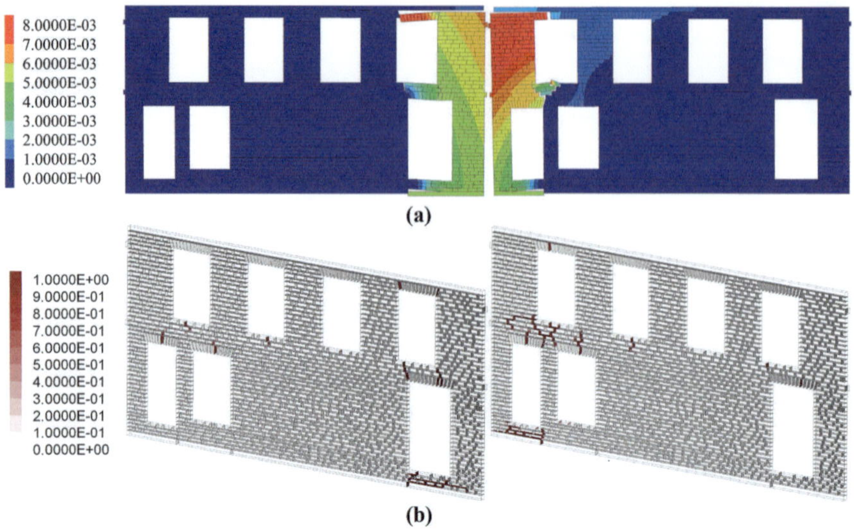

Fig. 30.21 Structural behaviour of the stone masonry façade subjected to vertical support settlement: **a** Displacement contours (units are in meters), **b** tension damage at the contact points

- The discontinuum-based analysis particularly fits the nature of block-based documentation of the existing stone masonry buildings via adopted AI-assisted tools. The proposed DEM-based computational modeling framework takes advantage of the documented morphology of the masonry walls and utilizes it in the structural analysis.
- The computational investigations performed on an old stone masonry building highlight the complex failure mechanism of unreinforced masonry walls. Since the proposed data-driven computational modeling approach aims to address the *as-is* construction features of the masonry buildings, the presented *what-if* scenarios offer critical information regarding the resilience and performance of the masonry buildings that can be used as a step for sustainable usage or the implementation of the optimum intervention plans.

Acknowledgements The first two authors have equally contributed to the study. The support of the Natural Sciences and Engineering Research Council of Canada (NSERC) grant RGPIN- 2022-04938 is gratefully acknowledged. Moreover, the first author is supported through the Undergraduate Student Research Awards (USRA) program administered by NSERC.

References

1. D. Loverdos, V. Sarhosis, Automatic image-based brick segmentation and crack detection of masonry walls using machine learning. Autom. Constr.. Constr. **140**, 104389 (2022). https://doi.org/10.1016/j.autcon.2022.104389
2. N. Kassotakis, V. Sarhosis, M.-V. Peppa, J.P. Mills, Semi-automated discrete-element modelling of arch structures incorporating SfM photogrammetry. Proc Inst Civ Eng—Eng Hist Herit **176**, 3–17 (2023). https://doi.org/10.1680/jenhh.21.00007
3. P.B. Lourenço, J.G. Rots, J. Blaauwendraad, Continuum model for masonry: Parameter estimation and validation. J.Struct. Eng. **124**, 642–652 (1998). https://doi.org/10.1061/(ASCE)0733-9445(1998)124:6(642)
4. P. Roca, M. Cervera, G. Gariup, L. Pelà, Structural analysis of masonry historical constructions. Classical and advanced approaches. Arch. Comput. Methods Eng. **17**, 299–325 (2010). https://doi.org/10.1007/s11831-010-9046-1
5. N. Mendes, P.B. Lourenço, Seismic assessment of masonry "Gaioleiro" buildings in Lisbon. Portugal. J Earthq Eng **14**, 80–101 (2009). https://doi.org/10.1080/13632460902977474
6. N. Mendes, P.B. Lourenço, Sensitivity analysis of the seismic performance of existing masonry buildings. Eng. Struct.Struct. **80**, 137–146 (2014). https://doi.org/10.1016/j.engstruct.2014.09.005
7. S. Saloustros, L. Pelà, M. Cervera, P. Roca, Finite element modelling of internal and multiple localized cracks. Comput. Mech.. Mech. **59**, 299–316 (2017). https://doi.org/10.1007/s00466-016-1351-6
8. S. Saloustros, M. Cervera, L. Pelà, Tracking multi-directional intersecting cracks in numerical modelling of masonry shear walls under cyclic loading. Meccanica **53**, 1757–1776 (2018). https://doi.org/10.1007/s11012-017-0712-3
9. D. Malomo, B. Pulatsu, Discontinuum models for the structural and seismic assessment of unreinforced masonry structures: a critical appraisal. Structures **62**, 106108 (2024). https://doi.org/10.1016/j.istruc.2024.106108
10. R.K. Adhikari, A. Parammal Vatteri, D. D'Ayala, Seismic performance assessment of low-rise unreinforced and confined brick masonry school buildings using the applied element method. Buildings **13** (2023). https://doi.org/10.3390/buildings13010159
11. D. Malomo, M.J. DeJong, A. Penna, Influence of bond pattern on the in-plane behavior of URM Piers. Int. J. Archit. Herit. **15**, 1492–1511 (2021). https://doi.org/10.1080/15583058.2019.1702738
12. B. Pulatsu, E. Erdogmus, P.B. Lourenço, J.V. Lemos, K. Tuncay, Simulation of the in-plane structural behavior of unreinforced masonry walls and buildings using DEM. Structures **27**, 2274–2287 (2020). https://doi.org/10.1016/j.istruc.2020.08.026
13. B. Pulatsu, E.M. Bretas, P.B. Lourenço, Discrete element modeling of masonry structures: Validation and application. Earthq. Struct. **11** 563–82 (2016). https://doi.org/10.12989/eas.2016.11.4.563
14. J.V. Lemos, Discrete element modeling of masonry structures. Int. J. Archit. Herit. **1**, 190–213 (2007). https://doi.org/10.1080/15583050601176868
15. B. Chetouane, F. Dubois, M. Vinches, C. Bohatier, NSCD discrete element method for modelling masonry structures. Int. J. Numer. Methods Eng. **64**, 65–94 (2005). https://doi.org/10.1002/nme.1358
16. S. Gonen, B. Pulatsu, E. Erdogmus, E. Karaesmen, Quasi static nonlinear seismic assessment of a fourth century A.D. Roman Aqueduct in Istanbul, Turkey. Heritage **4**, 401–421 (2021). https://doi.org/10.3390/heritage4010025
17. V. Giamundo, V. Sarhosis, G.P. Lignola, Y. Sheng, G. Manfredi, Evaluation of different computational modelling strategies for the analysis of low strength masonry structures. Eng. Struct.Struct. **73**, 160–169 (2014). https://doi.org/10.1016/j.engstruct.2014.05.007
18. P.B. Lourenço, J.G. Rots, Multisurface interface model for analysis of masonry structures. J. Eng. Mech. **123**, 660–668 (1997)

19. B. Pulatsu, R. Wilson, J.V. Lemos, N. Mojsilović, Exploring the Cyclic behaviour of URM walls with and without damp-proof course (DPC) membranes through discrete element method. Infrastructures **9**, 11 (2024). https://doi.org/10.3390/infrastructures9010011
20. B. Pulatsu, K. Tuncay, Computational modeling of damage progression in unreinforced masonry walls via DEM. Turkish J. Civ. Eng.35 (2024). https://doi.org/10.18400/tjce.1323977
21. B. Pulatsu, E. Erdogmus, P.B. Lourenço, J.V. Lemos, J. Hazzard, Discontinuum analysis of the fracture mechanism in masonry prisms and wallettes via discrete element method. Meccanica **55**, 505–523 (2020). https://doi.org/10.1007/s11012-020-01133-1
22. B. Pulatsu, S. Gonen, E. Erdogmus, P.B. Lourenço, J.V. Lemos, J. Hazzard, Tensile fracture mechanism of masonry wallettes parallel to bed joints: a stochastic discontinuum analysis. Model. Open Access J Model. Eng Sci **1**, 78–93 (2020). https://doi.org/10.3390/modelling1020006
23. P.A. Cundall. A computer model for simulating progressive, large-scale movements in blocky rock systems. *International Symposium on Rock Mechanics*, vol. 2, (Nancy: 1971), pp. 47-65
24. P.A. Cundall, C. Detournay, Dynamic relaxation applied to continuum and discontinuum numerical models in geomechanics. (1st Edition. Boca Raton, FL: CRC Press; 2017)
25. B. Pulatsu, Coupled elasto-softening contact models in DEM to predict the in-plane response of masonry walls. Comput Part Mech **10**, 1759–1770 (2023). https://doi.org/10.1007/s40571-023-00586-x
26. G. Vlachakis, E. Vlachaki, P.B. Lourenço, Learning from failure: Damage and failure of masonry structures, after the 2017 Lesvos earthquake (Greece). Eng. Fail. Anal. **117**, 104803 (2020). https://doi.org/10.1016/j.engfailanal.2020.104803
27. Reconnaissance Report on February 6, 2023 Kahramanmaraş Earthquakes in (Türkiye. Vancouver: 2024)
28. M. Bruneau, State-of-the-Art report on seismic performance of unreinforced masonry buildings. J. Struct. Eng.Struct. Eng. **120**, 230–251 (1994). https://doi.org/10.1061/(asce)0733-9445(1994)120:1(230)
29. V. Turnšek, F. Čačovič, Some experimental results on the strength of brick masonry walls. In *Proceedings of the 2nd International Brick Masonry Conference*, (Stoke-on-Trent, UK: British Ceramic Research Association; 1971) pp. 149–156
30. M. Tomaževič, Shear resistance of masonry walls and Eurocode 6: Shear versus tensile strength of masonry. Mater Struct Constr **42**, 889–907 (2009). https://doi.org/10.1617/s11527-008-9430-6
31. S. Jafari, J.G. Rots, R. Esposito, A correlation study to support material characterisation of typical Dutch masonry structures. J Build Eng **45**, 103450 (2022). https://doi.org/10.1016/j.jobe.2021.103450
32. P.B. Lourenço, A. Gaetani, Recommended properties for advanced numerical analysis. Finite Elem. Anal. Build. Assess., New York: Routledge 209–320 (2022). https://doi.org/10.1201/9780429341564-4
33. J. Segura, L. Pelà, P. Roca, Monotonic and cyclic testing of clay brick and lime mortar masonry in compression. Constr. Build. Mater. **193**, 453–466 (2018). https://doi.org/10.1016/j.conbuildmat.2018.10.198
34. J. Segura, L. Pelà, P. Roca, A. Cabané, Experimental analysis of the size effect on the compressive behaviour of cylindrical samples core-drilled from existing brick masonry. Constr. Build. Mater. **228**, 116759 (2019). https://doi.org/10.1016/j.conbuildmat.2019.116759
35. H.T. Thai, Machine learning for structural engineering: A state-of-the-art review. Structures **38**, 448–491 (2022). https://doi.org/10.1016/j.istruc.2022.02.003
36. J. Redmon, S. Divvala, R. Girshick, A. Farhadi, You Only Look Once: Unified, Real-Time Object Detection. In *Proceedings of the IEEE Computer Society Conference Computer Vision and Pattern Recognition*, (2015) December 2016, pp. 779–88. https://doi.org/10.48550/arXiv.1506.02640
37. A. Kirillov, E. Mintun, N. Ravi, H. Mao, C. Rolland, L. Gustafson, T. Xiao, S. Whitehead, A.C. Berg, W.Y. Lo, P. Dollar, Segment anything. In *2023 IEEE/CVF International Conference on Computer Vision*, (IEEE, 2023), pp. 3992–4003

38. M. Sohan, T. Sai Ram, C.V. Rami Reddy, A Review on YOLOv8 and Its Advancements. *Data Intelligence and Cognitive Informatics*, (Springer Nature Singapore; 2024), pp. 529–545. https://doi.org/10.1007/978-981-99-7962-2_39
39. Z. Ye, L. Lovell, A. Faramarzi, J. Ninić, Sam-based instance segmentation models for the automation of structural damage detection. Adv. Eng. Informatics **62** (2024). https://doi.org/10.1016/j.aei.2024.102826
40. T.Y. Lin, M. Maire, S. Belongie, L. Bourdev, R.Girshick, J. Hays, P. Perona, D. Ramanan, C.L. Zitnick, Microsoft COCO: common objects in context. *13th European Conference on Computer Vision*, (Zurich: Springer; 2014)
41. C.T. Bati, G. Ser, Effects of data augmentation methods on YOLO v5s: application of deep learning with pytorch for individual cattle identification. Yuz. Yil. Univ. J. Agric. Sci. **33**, 363–376 (2023). https://doi.org/10.29133/yyutbd.1246901
42. C. Goutte, E. Gaussier, In A Probabilistic Interpretation of Precision, Recall And F-Score, With Implication For Evaluation. *European Conference on Information Retrieval*, 2005, pp. 345–359. https://doi.org/10.1007/978-3-540-31865-1_25
43. V. Sarhosis, D. Dais, E. Smyrou, I.E. Bal, A. Drougkas, Quantification of damage evolution in masonry walls subjected to induced seismicity. Eng. Struct. 243 (2021). https://doi.org/10.1016/j.engstruct.2021.112529
44. R. Mikel, *Ontario house styles: the distinctive architecture of the province's 18th and 19th century homes* (J. Lorimer, Toronto, 2004)
45. P. Griesbach, R. Wilson, B. Karakus, B. Pulatsu, Transferring vision-based data to discontinuum analysis for the assessment of URM walls. Eur. J. Environ. Civ. Eng. **0**, 1–16 (2023). https://doi.org/10.1080/19648189.2023.2254376
46. A. Borri, M. Corradi, G. Castori, A method for the analysis and classification of historic masonry. Bull. Earthq. Eng. 2647–2665 (2015). https://doi.org/10.1007/s10518-015-9731-4

Chapter 31
LCA of Self-Compacting Concrete Incorporating Waste Plastic Fibre, Recycled Aggregate and SCMs

Zhanggen Guo, Ling Zhou, Qiansen Sun, Zhiwei Gao, Qinglong Miao, Conglin Chen, and Yujie Chen

Abstract This study intends to conduct an in-depth life cycle assessment (LCA) and life cost analysis (LCCA) of waste plastic fiber (WPF) reinforced sustainable recycled aggregate self-compacting concrete (WPF-RSCC). A sustainable self-compacting concrete (SCC) was manufactured by recycling of waste concrete, industrial waste, and waste plastic. This paper assessed the advantages of the addition of waste plastic fiber (WPF), recycled coarse aggregate (RCA) and supplementary cementitious materials (SCMs) on the environmental impact and economic cost of SCC. 10 WPF reinforced recycled aggregate SCC (WPF-RA-SCC) mixtures were prepared by substantially replacing natural aggregates and cement with RCAs and SCMs, respectively. The environmental and economic impacts of WPF-RA-SCC were quantified using Life cycle assessment and life cycle cost analysis. The VIKOR MCDA method was deployed to identify the optimal concrete mixture by integrating fresh and hardened properties, environmental and economic performance. The research findings indicate that combined addition of RCAs, SCMs and WPF offered considerable advantages on environmental and economic performance, by reducing the environmental load and economic cost up to 73.9% and 34.7%, respectively. Quaternary RA-SCC comprising 1.0% WPF and 75% SCMs using a combination type was recognized by the MCDA as the optimized mix, which exhibited superior mechanical properties, low environmental impacts and economic cost. This study proposed a novel approach for prioritizing sustainable materials selection, thus facilitating the recycling of waste concrete, waste plastic and industrial by-products.

Z. Guo (✉) · L. Zhou · Q. Sun · Z. Gao · Q. Miao · C. Chen · Y. Chen
School of Civil Engineering, Nanjing Tech University, Nanjing, China
e-mail: zhgguo@njtech.edu.cn

D. A. H. Hanaor (ed.), *Sustainable Materials for the Built Environment*, Springer Series in Materials Science 352, https://doi.org/10.1007/978-3-031-97818-0_31

31.1 Introduction

In China, Accelerated urbanization and the processes of old city reconstruction have led to a significant increase in the generation of construction and demolition waste (CDW) [1]. According to statistics, China recently generated about 2.4 billion tons of CDW annually [2], of which nearly 35% ended up in landfills [3]. Landfilling CDW not only requires vast land and human resources but also emits pollutant emissions, resulting in severe environmental pollution. To address this issue, recycling CDW to produce recycled coarse aggregate (RCA) as a substitute for natural aggregate has become a feasible option [4, 5].

Over the past two decades, the inclusion of recycled coarse and fine aggregates into self-compacting concrete (SCC) has attained extensive research attentions and has been systematically studied [6–9]. On the other hand, the substituting cement with supplementary cementitious materials (SCMs), such as fly ash (FA), silica fume (SF), and ground granulated blast furnace slag (GGBFS), for the production of SCC is significantly promoted recently, which resulted in the benefit of declining CO_2 and embodied energy of concrete [10–12]. Portland cement, a key component of concrete, is one of the most energy-intensive materials, releasing about 0.87 tons of CO_2 per ton of Portland cement clinker. Cement production is estimated to account for 5–8% of annual CO_2 emissions [13–15]. Therefore, the inclusion of SCMs into recycled aggregate SCC (RA-SCC) can mitigate the negative effect of the recycled aggregates on the fresh and hardened properties of SCC, thus improving the mechanical and durability properties of RA-SCC. By incorporating of recycled aggregate and SCMs into the SCC, a sustainable and cost-effective concrete material cab be developed [4, 5].

In recent years, plastics have become an indispensable part of industries such as packaging, agriculture, construction, and electronics due to their high durability, low density, relative lightweight, and low cost [16, 17]. Over the past several years, 350 million tons of plastics are produced around the world every year [18]. The extensive use of plastics has led to a severe environmental crisis. Majority of waste plastics persist in the environment for decades or even centuries without decomposing, contributing to significant environment pollution [19]. Among various types of plastics, polyethylene terephthalate (PET) is recognized as a major polymer. Although PET is a recyclable material, it is frequently discarded into the environment improperly, leading to microplastic pollution. It is estimated that people ingest 74,000–121,000 micro-plastic particles annually through food consumption and inhalation [20], which generate various immune disorders and potentially damage the kidneys, digestive system, and nervous system in humans [21]. Research implemented by Prata et al. [22] suggested that microplastic pollution could elevate the risk of cancer. Addressing this issue necessitates enhancing the efficiency of plastic waste collection and exploring innovative recycling methods. An effective method involves incorporating PET waste to concrete production, which not only enhances the properties of the concrete but also effectively recycles the PET waste, thereby reducing the environmental footprint [23–25].

By recycling PET waste into concrete, an eco-efficient and cost-effective construction material was developed [23, 26, 27]. In recent years, a considerable amount of research has have been conducted on the workability, mechanical properties and durability of waste plastic fiber (WPF) reinforced recycled aggregate concrete (RAC). Ngoc et al. [26] evaluated the effect of WPFs on the mechanical properties and durability of RAC and found that RAC containing SF and recycled PET fibers showed various degrees of improvement in compressive strength, modulus of elasticity, and splitting tensile strength compared to RAC without fiber. Guo et al. [28] found that combined addition of SCMs and WPF improved the fresh and hardened properties of RA-SCC. Ahmad et al. [29] also found that the addition of WPFs can significantly improve the mechanical properties and durability of concrete. Guo et al. [30] studied the effect of adding an appropriate content of WPFs on the mechanical properties of RAC and found that the addition of WPFs significantly improved the compressive and flexural tensile strength, with the maximum improvement was observed with the addition of WPFs with a length of 30 mm and a volume ratio of 1%. Pereira et al. [31] and Al-Hadithi et al. [32] both obtained that the incorporation of waste PET fibers effectively improved the hardened properties of concrete. Alani et al. [33] concluded that the mechanical properties of concrete with oil fuel ash and silica fume were enhanced after being reinforced with recycled PET fibers. Nikbin et al. [23] explored the plain high strength concrete properties containing different percentages of PET particles which could be accounted as a partial volumetric replacement for sand and found that replacing up to 15% of sand with PET reduces tensile strength and modulus of elasticity by about 20%. Additionally, concrete with PET particles is more environmentally friendly due to lower CO_2 emissions.

More recently, several studies have explored the environmental and economic impacts of WPF reinforced recycled concrete. Tahir, Sbahieh and Al-Ghamdi [25] reviewed the application and environmental impact of recycled plastics in concrete aggregates and concluded that recycling plastics not only significantly reduces environmental burdens such as land eutrophication, global warming potential, and acidification potential compared to landfilling and incineration, but also offers significant economic benefits. Borhan et al. [34] found that the combined inclusion of waste fibers and recycled concrete aggregates can not only conserve huge amounts of non-renewable natural resources, but also decrease the carbon footprints of concrete. Bahmani and Mostofinejad [24] studied the impact of steel, basalt, and WPF on various aspects of high-performance concrete with calcium oxide-activated slag and found that replacing steel or basalt fibers with WPFs, obtained from discarded plastic bottles, in high-performance concrete is a sustainable and cost-effective alternative. Rispoli et al. [35] conducted a comparative life cycle assessment of traditional pavement systems, demonstrating that sustainable road pavement mixes using recycled plastic and carbonated aggregates significantly reduced environmental impacts. Javadabadi [36] conducted a cradle-to-gate analysis to evaluate the environmental effects of using recycled PET as a fine aggregate in concrete applications and revealed that utilizing recycled-based aggregates significantly reduces the impact on the environment in terms of global warming potential and human toxicity potential. Ersan et al. [37] explored the environmental impacts of incorporating FA and WPFs for the production of lightweight concrete and found that lightweight concrete containing 30% WPFs and 20% FA had a 13% less greenhouse gas emissions when compared to conventional concrete.

Majority of existing research on WPF reinforced sustainable recycled aggregate SCC (WPF-RA-SCC) focused on specific individual criteria, such as technical performance, environmental or economic aspects. Life Cycle Assessment (LCA) has been widely used to evaluate environmental issues, such as climate change and resource depletion. Life Cycle Cost Analysis (LCCA) has been extensively employed in economic studies in the field of concrete to evaluate the total cost of concrete structures over their entire lifecycle, including initial construction costs, maintenance and repair expenses, and eventual demolition and disposal costs. However, the concept of sustainability should involve a comprehensive evaluation of concrete considering environmental, technical, and economic aspects. This essentially requires taking multiple factors into account, often involving complex trade-offs. Therefore, Multi-Criteria Decision Analysis (MCDA) is needed to address conflicts arising from multiple attributes. MCDA encompasses a set of methods and procedures designed to rank various alternative solutions based on different criteria, ultimately comparing and selecting the optimal alternative [38]. Nowadays, MCDA methods have become a popular tool for selecting the best solution, with main methods including TOPSIS, VIseKriterijumska Optimizacija I Kompromisno Resenje (VIKOR), Data Envelopment Analysis, Analytic Hierarchy Process, ELECTRE algorithms, etc. [39–42]. Furthermore, to the best knowledge of the authors, there is limited relevant research in the open literatures which focused specifically on the application of MCDA employing VIKOR method to evaluate the environmental and economic advantages of recycling waste concrete, industrial by-products and waste plastic for the production of sustainable high-performance SCC.

This paper is part of a comprehensive study aimed at proposing a high-performance and sustainable SCC, which was produced by substantially replacing cement with SCMs and natural aggregates with RCA, as well as reinforced with WPFs. Previous studies implemented by the authors have investigated the workability, mechanical properties, and durability of WPF-RA-SCC mixtures. The test results indicated that the combined use of SCMs and WPF produced a synergistic effect, substantially improving the mechanical properties and durability of RA-SCC [10, 28]. This paper intends to conduct an in-depth LCA and LCCA of WPF-RA-SCC, combining the diverse of the performance keys in terms of engineering, environmental, and economic perspectives. LCA and LCCA were performed to quantify the environmental and economic impacts of RA-SCC containing different percentages of WPF and SCMs. In addition, MCDA was carried out to compare the production of SCC made with WPF, SCMs and RCAs with the production of conventional concrete made with natural aggregates and ordinary Portland cement. Considering environmental, economic, and technical criteria, the VIKOR MCDA method was adopted to determine the optimal mix proportion. Five key performance indicators were identified, and WPF-RA-SCC mixtures with excellent mechanical properties, low environmental impact, and economic cost were determined, optimizing the proportions of SCMs and WPF.

31.2 Experimental Program and Test Results

31.2.1 Experimental Program

The materials used in this study include water, cement, natural coarse aggregate (NCA), RCA, river sand, water-reducing agent, and SCMs such as FA, GGBFS, and SF. The compressive strength of cement is 42.5 MPa. The river sand had a maximum particle size of 5 mm, a water absorption capacity of 1.1%, a fineness modulus of 2.42, and an apparent density of 2610 kg/m^3, as determined according to the Chinese standard JGJ 52–2006 [43]. The NCA was crushed limestone sourced from a local plant in Nanjing. RCA is obtained by crushing concrete components with a compressive strength of 35 MPa, as shown in Fig. 31.1. The WPFs utilized were derived from waste PET bottles. The bottles were manually cut into small pieces with a dimension of 0.3 mm in thickness, 3 mm in width, and 20 mm in length. The WPFs exhibited a specific gravity of 1.12, a density of 1250 kg/m^3, and a tensile strength of 57 MPa.

Table 31.1 illustrates the mix proportions of WPF-RA-SCC utilized in this study, with each specimen designated as A (B, C, D)-X-Y-2 (3). In this notation, A, B, C, and D represent the volume content of WPFs, namely 0%, 0.5%, 1.0%, and 1.5%, respectively; X indicates the substitution rate of RCA, and Y indicates the substitution rate of mineral admixture. 2 represents a ternary mix, incorporating FA and GGBS to partially replace cement; 3 indicates a quaternary mix, involving FA, GGBS, and SF to partially replace cement. For example, specimen B-100-75-3 indicates a quaternary mix containing 100% RCA and 75% SCM (25% FA + 25% GGBFS + 25% SF) as well as strengthened with 0.5% WPFs. The detailed information of concrete proportions can be obtained in the Guo et al. [28].

The fresh tests of all WPF-RA-SCC mixes were performed align with SCC Specification and Guidelines [44]. The slump-flow value, T500 slump flow time, J-Ring

Fig. 31.1 Recycled aggregates

Table 31.1 Mixture proportions (kg/m^3)

Mix code	Binder				Water	NCA	RCA	Sand	SP	WPF	Additional water
	Cement	FA	GGBFS	SF							
0-0	542	0	0	0	189	806	0	834	8.13	0	0
100-0	542	0	0	0	189	0	768	834	8.13	0	23.04
A-100-50-2	271	135.5	135.5	0	189	0	768	834	8.13	0	32.26
A-100-75-3	135.5	162.6	162.6	81.3	189	0	768	834	8.13	0	32.26
B-100-50-2	271	135.5	135.5	0	189	0	768	834	8.13	6.25	32.26
B-100-75-3	135.5	162.6	162.6	81.3	189	0	768	834	8.13	6.25	32.26
C-100-50-2	271	135.5	135.5	0	189	0	768	834	8.13	12.5	32.26
C-100-75-3	135.5	162.6	162.6	81.3	189	0	768	834	8.13	12.5	32.26
D-100-50-2	271	135.5	135.5	0	189	0	768	834	8.13	18.75	32.26
D-100-75-3	135.5	162.6	162.6	81.3	189	0	768	834	8.13	18.75	32.26

flow and segregation rate were measured and the filling ability, viscosity, passing ability and segregation resistance were evaluated. The detailed fresh properties of all WPF-RA-SCC mixes were presented and analyzed in a preceding paper [10]. Some of the fresh properties are presented in Table 31.2. As shown in Table 31.2, the addition of WPFs consistently impaired the workability of RA-SCC mixes. The slump-flow value decreased and the SF-JF value and T500 slump flow time increased as WPF content increased. Nevertheless, the segregation rate was reduced with the inclusion of WPFs. Although the inclusion of WPFs had a negative effect on the fresh properties, the vast majority of WPF-RA-SCC mixes had excellent fresh properties and met the limitation described by EFNARC [44]. This is ascribed to the inclusion of various SCMs considerably enhanced the workability of WPF-RA-SCC mixes.

31.2.2 Test Results

The cubic and axial compressive strength, splitting tensile and flexural strength, of all specimens are summarized in Table 31.2. The performance of WPF-RA-SCC mixes with various WPFs and SCM is comparatively analyzed in detail, as well as the coupling effect of WPFs and various SCM combinations.

As shown in Table 31.2, the addition of RCAs declined the cubic compressive strength, which is aligned with the prior test results [34]. This is mainly ascribed to inferior performances, such as porosity, high water absorption and residual old mortars of RCAs. Finally, it is obvious that the incorporation of SCMs using different combinations significantly affected the cubic compressive strength. Quaternary mixes containing of FA, GGBFS, and SF had the highest compressive strength. The least strength was obtained for binary mixes with FA. This is ascribed to the pozzolanic reaction and the filling effect, as well as the concurrent effect of these mineral admixtures [45].

Table 31.2 Fresh and hardened properties of WPF-RA-SCC mixes

Mix no	Fresh properties				Hardened properties					
	Slump flow value (mm)	T500 Slump flow time (sec)	SF-JF (mm)	Segregation rate	f_{cu} (MPa)			f_f (MPa)	f_{cp} (MPa)	f_{st} (MPa)
					7d	28d	90d	28d	28d	28d
0-0	755	4.0	25	20	32.71	46.28	51.19	4.82	37.82	4.27
100-0	765	3.1	20	20	33.33	43.67	48.14	4.53	–	3.58
A-100-50-2	840	3.2	30	23	30.21	36.28	–	4.56	–	–
A-100-75-3	720	4.5	40	18	30.4	42.51	49.18	5.00	–	3.91
B-100-50-2	770	3.8	42	19	32.05	40.07	–	6.32	–	–
B-100-75-3	690	5.2	50	15	35.1	42.63	50.11	7.02	–	4.31
C-100-50-2	740	4.6	44	16	40.92	51.26	–	6.86	–	–
C-100-75-3	650	6.5	55	15	42.65	52.07	62	7.26	–	4.68
D-100-50-2	710	6.7	50	14	35.39	46.65	–	6.94	–	–
D-100-75-3	640	7.6	60	15	36.83	48.50	57.03	7.71	–	4.45

The relationship between WPF content and the compressive strength of WPF-RA-SCC mixes can be clearly obtained from Table 31.2, which demonstrates that the incorporation of WPFs is effective in enhancing the compressive strength of RA-SCC. The highest increase in compressive strength was achieved with the addition of 1% WPFs. The addition of 0.5% WPFs had the least enhancement. According to the fresh tests, WPF-RA-SCC mixes with 0.5% and 1.0% WPFs had superior workability, resulting in the uniform distribution of WPFs within the concrete matrix, which improved the homogeneity and reduced the internal voids. Additionally, the incorporation of WPFs hindered the formation and expansion of cracks and reduced the stress concentration at crack tips. Similar findings were obtained in previous studies [46, 47].

31.3 Research Methodology

31.3.1 Life Cycle Assessment

Recently, the environmental performance of RAC has been given substantial attention and extensive studies have been conducted by employing the LCA method, which systematically investigates the environmental burdens and calculates resource inputs and environmental outputs throughout a product's life cycle, thus assessing the environmental performance and proposing the potential improvement options [48]. Some research has addressed the positive environmental effects of using recycled materials in concrete [49, 50]. In this paper, LCA was employed to study the environmental performance of WPF-RA-SCC mixes. A comprehensive introduction of the LCA methodology is given in international standards ISO14040 [51] and ISO14044 [52] and the LCA methodology commonly includes four stages (Goal and Scope Definition, Life Cycle Inventory (LCI), Life Cycle Impact Assessment (LCIA), and Interpretation).

The study aims at quantifying and comparing the environmental performance of WPF-RA-SCC with various WPFs, RCAs, and SCMs. The functional unit used is 1 m3 concrete mix, which was commonly adopted in prior research. Additionally, the cradle-to-gate system boundary was employed in this study, which consists of material production and transportation and the production of concrete in the plant [53].

The total LCI inventory for 1 m^3 WPF-RA-SCC production is shown in Table 31.3, which was obtained from literatures, monitoring analysis, interviews and field investigation with local suppliers and databases, including the Ecoinvent database in Simapro 9.0. The inputs include comprise energy (i.e., electricity and fuels) and raw materials, and outputs include the emissions and pollutants (i.e., CO_2, NO_X, CH_4, SO_2, CO, etc.). NCA was obtained from a quarry situated in Anhui Province and transported to Nanjing by trucks over a distance of 500 km. RCAs were obtained from a demolished concrete structural element in Nanjing. The concrete components were carried to a nearby recycling plant by truck over a distance of 20 km and were crushed to produce RCAs. Natural sand was purchased from a supplier near the concrete plant and transported by truck; The distance is 50 km. Barjoveanu et al. [54] concluded that the environmental pollution caused by landfilling or incinerating WPF is significantly higher than that of recycling WPF. WPFs used in this test is produced by manually cutting PET bottles, which produces much little CO_2 during the hand-cutting process. Therefore, the environmental impact of WPF preprocessing adopted in this study was ignored. In addition, Foolmaun and Ramjeeawon [55] investigated and compared the environmental and social impacts of four disposal methods of waste PET bottles and concluded that the landfilling WPF will produce CO_2 of 1.19 per kg, which was adopted as the WPF inventory (a negative values of 1.19 per kg) in this study. The cementitious binders such as cement, FA, GGBFS, and SF were all obtained from local suppliers and transported to the concrete plant by truck over 50 km. Electricity and fuel were used to produce concrete in the concrete plant, and the energy depletion and related criteria air pollutants were determined from the field statistic investigation at the concrete factory.

In this study, several impact categories described in CML2001, such as global warming potential (GWP [kg CO_2 eq.]), acidification potential (AP [kg SO_2 eq.]), eutrophication potential (EP [kg PO_4^{3-} eq.]), human toxicity potential (HTP inf. [kg 1,4-DCB eq.]), and photochemical ozone creation potential (POCP [kg Ethylene eq.]) were employed and calculated align with CML mid-point method.

31.3.2 Life Cycle Cost Analysis (LCCA)

This study employs Environmental Life Cycle Cost analysis (E-LCCA) proposed by international standard ISO 15686 [56] to evaluate the economic feasibility of WPF-RA-SCC. According to the framework proposed by the Society of Environmental Toxicology and Chemistry, E-LCCA is a common economic tool used in conjunction with LCA [57], which translates the environmental impacts into economic costs. Because environmental emissions represent abstract hidden costs, the calculation of these hidden costs is both complex and challenging. Since monetary units serve as the sole measure for LCCA, Willingness to Pay (WTP) was adopted in this study to convert environmental impacts into monetary units, facilitating the calculation of the E-LCCA for WPF-RA-SCC. WTP is now mostly used internationally to capture the costs caused by environmental impacts, which has been widely used in environmental and economic impact assessment [58 61].

Considering the system boundary of this study is cradle to gate, the effects of operating and end-of-life costs are not considered, assuming the entire life cycle is completed within a relatively short timeframe. In the E-LCCA analysis, the total life cycle costs are divided into fixed costs and variable costs. In this study, variable costs were calculated, which include direct production costs, environmental emission costs, and so forth.

Table 31.3 LCI of WPF-RA-SCC

	Serial	Cement	Water	NCA	RCA	Sand	SP	WPF	Transportation(kg/tkm)	
									Truck_18t	Rail
Resource input	Limestone	9.03E−02	6.17E−05	1.34E−01	0.00E+00	0.00E+00	0.00E+00	0.00E+00	1.35E−04	9.12E−05
	Raw coal	5.91E−02	5.56E−05	1.00E−03	1.08E−03	1.78E−04	8.13E−02	0.00E+00	1.94E−03	5.99E−04
	Fossil oil	3.44E−02	9.33E−06	6.97E−04	6.69E−04	3.30E−04	7.65E−02	0.00E+00	2.62E−02	1.19E−02
	Natural gas	6.76E−04	2.39E−05	1.10E−03	0.00E+00	1.52E−04	6.40E−03	0.00E+00	6.98E−04	4.35E−05
Emission output	CO_2	7.67E−01	5.84E−05	1.81E−03	5.71E−03	2.34E−03	6.90E−01	− 1.19E+00	1.03E−01	3.38E−02
	NO_x	1.09E−03	3.26E−08	1.82E−05	2.40E−05	1.52E−05	3.50E−03	0.00E+00	1.94E−03	5.89E−04
	CH_4	8.29E−06	8.87E−10	8.40E−08	1.00E−06	3.72E−06	1.20E−03	− 1.31E−03	3.64E−04	1.40E−06
	SO_2	3.26E−04	5.00E−06	1.04E−06	5.00E−06	9.49E−06	0.00E+00	− 1.62E−03	9.18E−05	6.57E−06
	CO	5.17E−04	3.28E−07	1.38E−05	1.00E−06	4.19E−06	2.10E−03	0.00E+00	7.74E−04	1.72E−04
	NMVOC	6.66E−05	2.00E−08	2.26E−06	3.00E−06	2.13E−07	4.16E−04	0.00E+00	4.69E−04	5.43E−05
	N_2O	2.22E−06	4.12E−09	2.08E−08	8.45E−08	3.81E−08	0.00E+00	0.00E+00	5.07E−06	6.36E−08
	NH_3	3.91E−05	2.18E−13	1.02E−08	1.00E−08	7.24E−09	0.00E+00	0.00E+00	1.43E−07	2.28E−08
	PM_{10}	2.37E−06	1.94E−07	4.21E−07	1.70E−07	1.20E−07	0.00E+00	− 1.13E−04	4.99E−05	4.25E−05

In this study, the total E-LCCA was calculated by following Eq. (31.1)–(31.3).

$$C_{E-LCCA} = C_D + C_E \tag{31.1}$$

$$C_D = C_{mat} + C_{tra} + C_{ene} \tag{31.2}$$

$$C_E = C_{dis} + C_{lan} + C_{eni} \tag{31.3}$$

where C_{E-LCCA} is the total cost; C_D, C_E represent direct cost, environmental cost, respectively; C_{mat} is the cost of the raw material; C_{tra} is the cost of the transportation phase; C_{ene} is the cost of energy consumption during concrete production; C_{dis} is the avoidable cost of waste concrete disposal; C_{lan} is the avoidable cost of land-filling waste plastics; $C_{eni}$is the cost of the environmental loads from raw material production.

31.3.2.1 Direct Costs

(1) Raw material cost.

The cost of raw materials was calculated by following Eq. (31.4):

$$C_{mat} = \sum (G_i * CF_i) \tag{31.4}$$

where i is the type of raw material, G_i is the amount of the ith material used (unit kg), CF_i is the unit direct cost of the ith material; Table 31.4 presents the cost of the raw materials per unit weight in RMB.

Table 31.4 Unit cost of raw materials for WPF-RA-SCC mixture

Cost item	Unit cost	Source
Cement	¥0.49/kg	[62]
FA	¥0.484/kg	[63]
GGBFS	¥0.381/kg	[64]
SF	¥0.35/kg	Average of several Nanjing providers
Water	¥0.0057/kg	[63]
NCA	¥0.106/kg	[65]
RCA	¥0.0782/kg	[65]
Sand	¥0.14/kg	[62]
SP	¥0.03022/m^3	[65]
WPF	¥0.23/kg	[66]

(2) Transportation cost

The transportation stage cost was calculated from Eq. (31.5) [67]:

$$C_{tra} = \sum(M_i * X_i * D_i * C_d) \quad (31.5)$$

where M_i is the amount of the ith material used (in kg); D_i is the distance of the ith material transported (in km); X_i is the diesel consumption per unit cost (in L/kg/km); C_d is the price of diesel fuel (RMB/L). Statistically, the price of diesel fuel is 7 RMB/L [58].

(3) Energy consumption cost

The energy consumption cost was calculated from Eq. (31.6):

$$C_{ene} = V_e * U_e \quad (31.6)$$

where V_e is the electricity consumption; U_e is the unit price of electricity, which is 1 RMB/kWh [58].

31.3.2.2 Environmental Costs

(1) Landfill costs for waste concrete

Currently, waste concrete in China is primarily disposed of through landfilling, accompanied by a prescribed pollution fee. This study proposes for the reuse of coarse aggregates, thereby circumventing the landfilling of waste concrete. Consequently, the avoidable cost of waste concrete disposal was calculated from Eq. (31.7):

$$C_{dis} = m_1 * C_s \quad (31.7)$$

where C_s is the sewage charges per ton of waste concrete. According to China's pollution fee collection standards [68], a fee of 25 RMB per ton of waste concrete is charged. With China's escalating focus on environmental conservation, it is projected that pollution fee rates will increase.

(2) Cost of landfilling of plastic waste

According to statistics, the plastic waste recycling rate in China is approximately 17% [66], while the landfilling rate is about 80% on average [25]. Therefore, assuming that the WPF recovered in this study was originally intended for landfilling, the avoidable landfilling processing fee was calculated using Eq. (31.8).

$$C_{lan} = m_2 * C_i \quad (31.8)$$

where m_2 is the amount of WPF used and C_i is the cost of landfilling plastic waste, which is 973 RMB/tonne [69].

Table 31.5 Unit environmental costs

Impact category	WTP	Unit	Source
GWP	0.22	¥/kg CO_2	[70]
AP	0.67	¥/kg SO_2	[58]
EP	0.63	¥/kg $PO_4{}^{-3}$	[58]
HTP	0.50	¥/kg 1,4-DB	[59]
ODP	12.23	¥/kg CFC-11	[70]

(3) Environmental loads from raw material production

This study specifically refers to the WTP to pay for environmental changes, and applies the WTP to calculate the environmental costs of RA-SCC. According to WTP, the environment stage cost was calculated as following:

$$C_{eni} = \sum \left(H_j * CF_j\right) + \sum \left(K_j * CF_j\right) \quad (31.9)$$

where $\sum \left(H_j * CF_j\right)$ is the environmental cost at the raw material stage; $\sum \left(K_j * CF_j\right)$ is the environmental cost of the transportation phase; j is the type of environmental impact subcategory; H_j is the total emissions at the raw material stage for the jth environmental subcategory K_j is the total emissions in the transportation stage of the jth environmental subcategory; and CF_j is the unit environmental cost of the jth environmental subcategory. Table 31.5 is the environmental cost of the five environmental impacts based on WTP.

31.3.3 Multi-Criteria Decision-Making Analysis

The feasibility of using RCAs, SCMs and WPF to produce sustainable SCC must be analyzed with respect to technical performance, environmental impacts, and economic benefit to improve the recycling efficiency of these waste materials and waste plastic management, which is a multifaceted issue involving a range of considerations. Therefore, the most appropriate approach to address this concern is MCDA, which is a set of methods and process that can be applied to the sustainable assessment of complex issues involving multiple interests and viewpoints [71].

In this study, the VIKOR MCDA method was employed to assess the environmental, economic, and technical criteria of the production of WPF-RA-SCC as well as seek the optimal mixtures. The fresh and hardened properties, environmental and economic performance was integrated to evaluated and identify the optimal concrete mixture. By considering decision makers preferences and integrating evaluation indicators, VIKOR method aims to maximize group benefits and minimize individual regrets. Through the assessment, comparison, and ranking of alternative solutions based on stability criteria of the optimal solution and the selection of one or more compromise solutions, the VIKOR method aims to strike a reasonable balance between different decision variables and possibilities, minimizing subjectivity [72, 73]. The process begins by determining the relevant weights for each criterion. Then, a specific method must be selected to rank and choose the alternative solutions [39, 74]. The VIKOR MCDA steps used in this study include defining the problem, identifying the criteria, assigning weights, evaluating alternatives, and ranking the solutions. [72, 73].

This study employed the VIKOR method to calculate and rank 10 alternatives of WPF-RA-SCC based on four standardized functions for workability, mechanical properties, environmental impact, and economic impact, namely F_1, F_2, F_3, and F_4, respectively. Index F_1 was related to the workability, index F_2 was associated with compressive strength and flexural strength, index F_3 corresponded to five environmental impact categories (GWP, AP, EP, HTP, ODP), and index F_4 was related to economic cost. Firstly, F_1, F_2, F_3 and F_4 were normalized using the aggregation function Eq. (31.10):

$$\mathrm{f} = \sum\nolimits_{ic=1}^{n} w \frac{\left(f_{ic}^{*} - f_{icj}\right)}{\left(f_{ic}^{*} - f_{ic}^{-}\right)} \tag{31.10}$$

where, w = weighting factor; f_{icj} = impact category value; f_{ic}^{*} = ideal value of the impact category; f_{ic}^{-} = anti-ideal value of the impact category.

The ideal value is the minimum value of the impact category, while the anti-ideal value is the maximum value of the impact category. Based on Tošić et al. [71], the weighting factor associated with the environmental impact category was 0.2, the weighting factor associated with costs and workability were 1, and the weighting factor associated with mechanical properties was 0.33.

Then the best (f_i^{*}) and worst (f_i^{-}) benefits for each criterion are determined using the Eq. (31.11):

$$f_j^{*} = \mathrm{Max}_i f_{ij}, f_j^{-} = \mathrm{Min}_i f_{ij}; j = 1, 2, \ldots, n \tag{31.11}$$

The next and very important step is to calculate the S_i and R_i values, as in Eq. (31.12) and Eq. (31.13):

$$S_i = \sum\nolimits_{j=1}^{n} w_j \frac{\left(f_j^{*} - f_{ij}\right)}{\left(f_j^{*} - f_j^{-}\right)} \tag{31.12}$$

$$R_i = Max_j \left[w_j \frac{\left(f_j^{*} - f_{ij}\right)}{\left(f_j^{*} - f_j^{-}\right)} \right] \tag{31.13}$$

In which, in order to perform multi-criteria optimization, the weights of the criteria (w_j) need to be assigned to indicate their relative importance. Five prioritized criteria were established for this calculation, each emphasizing a different criterion function, as shown in Table 31.6.

The value Q_i is calculated by Eq. (31.14), which denotes the linear interpolation of the values of S_j and R_j.

$$Q_i = \gamma \frac{(S_i - S^*)}{(S^- - S^*)} + (1 - \gamma) \frac{(R_i - R^*)}{(R^- - R^*)} \tag{31.14}$$

where $S^* = \text{Min}_i\{S_i\}$, $S^- = \text{Max}_i\{S_i\}$, $R^* = \text{Min}_i\{R_i\}$, $R^- = \text{Max}_i\{R_i\}$.

The coefficient γ can be any value between 0 and 1. In this study, $\gamma = 0.5$ was chosen to prove that S_j is as important as R_j, avoiding favoritism to either side.

Based on the results of the calculations, the values of S, R, and Q are first arranged in descending order, and in general the optimal solution is the one closest to 0, and the worst solution is the one closest to 1. However, the final optimal compromise solution (a') needs to satisfy the following two conditions, and if both conditions are satisfied, the solution is the best solution to rank the Q:

C1: $Q(a'') - Q(a') \geq DQ$

Where, a'' is the second position of Q and $\text{DQ} = \frac{1}{m-1} = 0.111$.

C2: a' must also be the best ranked by S or/and R.

If one of the conditions is not satisfied, then a set of compromise solutions is proposed:

Solution a: Alternatives $a', a'', \ldots, a^M$ if condition 1 is not satisfied.

where, a^{M} is determined by $Q(a^{\text{M}}) - Q(a') < 1/(m-1) = 0.111$ for maximum M.

Solution b: Alternatives a' and a'' if only condition 2 is not satisfied.

Table 31.6 Weight of each criterion by different scenarios

Ranking criteria	W_j (F_1)	W_j (F_2)	W_j (F_3)	W_j (F_4)
Equal importance	0.25	0.25	0.25	0.25
Workability advantage	0.4	0.2	0.2	0.2
Mechanical properties advantage	0.2	0.4	0.2	0.2
Environmental advantage	0.2	0.2	0.4	0.2
Economic advantage	0.2	0.2	0.2	0.4

31.4 Results and Discussions

31.4.1 LCA Results and Discussions

Table 31.7 summarizes the calculated impact indicators resulting from the production of 1 m^3 concrete. The calculated GWP (kg of CO_2-eq/m^3) from the production of 1 m^3 WPF-RA-SCC is shown in Table 31.8, as well as the contributions to the GWP from various concrete constituents. It is obvious from Table 31.8, that normal SCC A-0-0 had the highest GWP, which is mainly due to cement production emitting the highest GHG emissions (66% to 95% of GWP) [11]. It is apparent from Tables 31.7 and 31.8 that the replacement cement with SCMs significantly reduced the GWP. For example, when cement was 50% and 75% replaced with SCMs, GWP decreased by 42.8% and 65.0%, respectively. Teixeira et al. [75] had similar results. At last, it is clear from Table 31.8 that the inclusion of WPFs consistently reduced the GWP, which is ascribed to the negative value of WPF on the GWP emissions. For example, the inclusion of 0.5%, 1.0% and 1.5% WPFs reduced the GWP by 7.7%, 15.3% and 23.0%, respectively. This is primarily attributed to the incineration and landfill of WPFs generating huge amounts of pollutant gases.

The effect of RCAs, various SCM combinations and various WPF content on the AP, EP, HTP, and POCP is similar to that on the GWP. It can be seen from Table 31.7 that the AP, EP, HTP, and POCP reduced when NCAs were 50% and 100% replaced by RCAs, which is ascribed to the much shorter transportation of RCAs. Moreover, as shown in Table 31.7, replacement cement with SCMs had notable potential to decrease AP, EP, HTP, and POCP.

To search for an optimum content of RCAs, WPFs, and SCM combinations, the CO_2-eq intensity, which is the ratio of calculated GWP to the compressive strength, was utilized to analyze the relationship between strength and environmental

Table 31.7 LCA results of 1 m^3 WPF-RA-SCC mixture considered in the study

Mix code	GWP	AP	EP	HTP	ODP
0-0	5.25E+02	1.28E+00	3.64E−01	9.71E+01	2.13E−05
100-0	5.10E+02	1.13E+00	3.30E−01	9.27E+01	1.96E−05
A-100-50-2	2.91E+02	7.66E−01	2.14E−01	7.04E+01	1.44E−05
A-100-75-3	1.78E+02	5.52E−01	1.48E−01	5.60E+01	1.13E−05
B-100-50-2	2.78E+02	7.56E−01	2.14E−01	7.01E+01	1.30E−05
B-100-75-3	1.64E+02	5.42E−01	1.48E−01	5.58E+01	9.83E−06
C-100-50-2	2.64E+02	7.45E−01	2.13E−01	6.99E+01	1.15E−05
C-100-75-3	1.51E+02	5.32E−01	1.48E−01	5.55E+01	8.39E−06
D-100-50-2	2.50E+02	7.35E−01	2.13E−01	6.97E+01	1.01E−05
D-100-75-3	1.37E+02	5.21E−01	1.48E−01	5.53E+01	6.95E−06

Table 31.8 Comparison of GWP for WPF-RA-SCC production (kg of CO_2-eq/m^3 of concrete)

Mix code	Cement	SCMs	Water	NCA	RCA	Sand	SP	WPF	Transportation
0-0	467	0	0.061	4.51	0	9.73	10.4	0	33.9
100-0	467	0	0.061	0	4.38	9.73	10.4	0	18.8
A-100-50-2	233	16.35	0.061	0	4.38	9.73	10.4	0	17.6
A-100-75-3	117	19.87	0.061	0	4.38	9.73	10.4	0	17.2
B-100-50-2	233	16.35	0.061	0	4.38	9.73	10.4	– 13.7	17.7
B-100-75-3	117	19.87	0.061	0	4.38	9.73	10.4	– 13.7	17.3
C-100-50-2	233	16.35	0.061	0	4.38	9.73	10.4	– 27.4	17.7
C-100-75-3	117	19.87	0.061	0	4.38	9.73	10.4	– 27.4	17.3
D-100-50-2	233	16.35	0.061	0	4.38	9.73	10.4	– 41.1	17.7
D-100-75-3	117	19.87	0.061	0	4.38	9.73	10.4	– 41.1	17.3

impacts [76]. The calculated CO_2-eq intensity per compressive strength of WPF-RA-SCC mixtures is given in Table 31.9. The CO_2-eq strength significantly decreased with SCM content increasing (i.e., cement content decreasing) (Table 31.9). At a specified compressive strength, the CO_2-eq intensities can be substantially reduced by replacing cement with SCMs, especially at 90 days. For example, at a 90-day compressive strength of approximately 48.14 MPa, specimen A-100-0 with 100% cement produced 10.6 kg CO_2-eq/MPa, while quaternary mixes A-100-75-3 with 75% SCMs produced 3.63 kg CO_2-eq/MPa. A reduction of about 65.7% in CO_2-eq intensity was obtained by replacement cement with SCMs (GGBFS, SF, and FA), which demonstrates that substituting cement with mineral admixtures (GGBFS, SF, and FA) considerably declined the CO_2-eq intensities. It can also be seen from Table 31.9 that the inclusion of WPFs effectively reduced the CO_2-eq intensity. For instance, the CO_2-eq intensity of mix D-100-75-3 with 1.0% WPFs is 33.6% less than that of mix A-100-75-3 without WPFs at 90 days. Quaternary mixes D-100-75-3 with 100% RCAs and 75% SCMs (FA + GGBFS + SF) and 1.5% WPFs had the lowest 90-day CO_2-eq intensity (2.41 kg CO_2-eq/MPa). Therefore, for a given performance, such as strength, the combined incorporation of WPFs and SCM combination (FA, GGBFS, and SF) significantly reduced the CO_2-eq intensity of WPF-RA-SCC mixes.

The experimental result reveals that the combined incorporation of WPFs and various mineral admixtures produced a significant concurrent effect, which resulted in an additional enhancement in the net gain of hardened properties and dramatically enhanced the mechanical properties of WPF-RA-SCC. This is commonly ascribed to several reasons. Firstly, incorporating different types of SCMs, such as FA, and GGBFS, can effectively improve the fresh properties. Therefore, WPFs can be uniformly distributed in the concrete matrix. Secondly, the crack-bridge effect of WPFs led to a load transfer from cement paste to fibers, thus lowering the stress concentration and slowing down the occurrence and propagation of cracks. At last, the pozzolanic reaction and hydraulic activity, as well as the remarkable synergistic effect due to the inclusion of various mineral additives such as FA, GGBFS,

Table 31.9 Normalized GWP with respect to compressive strength of WPF-RA-SCC mixes

Mix code	Normalized GWP with respect to compressive strength		
	7-day	28-day	90-day
0-0	16.06	11.35	10.26
100-0	15.31	11.68	10.60
A-100-50-2	9.66	8.05	–
A-100-75-3	5.87	4.20	3.63
B-100-50-2	8.68	6.95	–
B-100-75-3	4.69	3.86	3.29
C-100-50-2	6.47	5.16	–
C-100-75-3	3.54	2.90	2.44
D-100-50-2	7.09	5.38	–
D-100-75-3	3.73	2.83	2.41

and SF enhanced the microstructure. Additionally, the bonding strength between WPFs and cement paste was enhanced with the high packing density and compact microstructure. Ali et al. [12] obtained similar conclusion. LCA results indicate that the combined addition of WPFs and various SCM combinations remarkably reduced the environmental impacts, and WPF-RA-SCC mixes had a much lower carbon footprint along with superior mechanical properties and durability.

31.4.2 LCCA Results and Discussions

The total cost $C_{E\text{-}LCCA}$ of each WPF-RA-SCC mixture is presented in Table 31.10. It is clear from Table 31.10 that C_{mat} (raw material cost) accounted for the largest proportion in E-LCCA, ranging from 67.3% to 91.8%. The second largest proportion was C_{eni} (Environmental loads from raw material production), which accounted for 13–27% of the total E-LCCA. In addition, C_{dis} (avoidable cost of waste concrete disposal) and C_{lan} (avoidable cost of landfilling waste plastics) exhibited a net cost benefit with the inclusion of RCA and WPF, stemming from the avoidance of disposal costs associated with landfilling of waste concrete and incinerating of plastic waste, indicating that recycling RCA and WPF to produce RA-SCC not only effectively reduces the disposal costs of waste concrete, but also increases the recycling rate of waste plastics, contributing to a more sustainable construction practice. This was consistent with the findings of Zhang et al. [58] that the production of recycled aggregate concrete can effectively reduce the CDW disposal costs and the associated transportation costs.

It also can be seen from Table 31.10, the substituting of 75% cement with SCMs resulted in a 31.3% reduction in the total cost $C_{E\text{-}LCCA}$. In addition, the total cost $C_{E\text{-}LCCA}$ was further reduced with the addition of WPF. For example, the total cost

Table 31.10 Total cost of WPF-RA-SCC

Mix code	C_D				C_E		Total cost
	C_{mat}	C_{tra}	C_{ene}	C_{dis}	C_{lan}	C_{eni}	
0-0	469.10	61.10	1.13	0.00	0	165.24	696.57
100-0	443.85	10.90	1.12	− 19.20	0	159.44	596.11
A-100-50-2	428.32	13.00	1.12	− 19.20	0	99.95	523.20
A-100-75-3	413.82	15.00	1.12	− 19.20	0	67.58	478.33
B-100-50-2	429.76	13.03	1.13	− 19.20	− 6.06	96.83	515.46
B-100-75-3	415.26	15.03	1.13	− 19.20	− 6.06	64.46	470.60
C-100-50-2	431.20	13.05	1.13	− 19.20	− 12.16	93.71	507.73
C-100-75-3	416.70	15.05	1.13	− 19.20	− 12.16	61.34	462.86
D-100-50-2	432.63	13.08	1.13	− 19.20	− 18.24	90.59	499.99
D-100-75-3	418.14	15.08	1.13	− 19.20	− 18.24	58.21	455.12

of the quaternary mix D-100-75-3 containing 1.5% WPF was 4.9% lower than that of the mix A-100-75-3 without WPF. This is mainly due to the synergistic effect of the SCMs and the WPFs that resulted in lower environmental impacts, thus reducing the environmental costs of the RA-SCC. Saha et al. [66] obtained similar results that the total cost of concrete decreased with increasing of plastic waste content. This is mainly due to the reduced environmental and social costs resulted from the recycling of plastic waste. At last, as shown in Table 31.10, when NCA was fully replaced with RCA, C_{tra} (transportation cost) decreased by 82%, which was attributable to transportation distance for RCA being significantly shorter compared to NCA, thus reducing economic costs and highlighting the logistical and cost efficiency of using recycled concrete aggregates.

Tables 31.7 and 31.10 presents a comparison of LCA and LCCA for the production of WPF-RA-SCC. The five environmental impacts categories and the corresponding costs related to WPF-RA-SCC production exhibited similar trends. When cement was replaced with SCMs, both environmental impacts and E-LCCA showed an overall decreasing trend. In addition, the inclusion of WPF led to a slight reduction in environmental impacts and E-LCCA. In addition, it is clear from Tables 31.7 and 31.10 that quaternary mix D-100-75-3 with 1.5% WPF and 75% SCMs had the lowest environmental impacts and economic costs, with E-LCCA that was 35% less than that of conventional concrete (0-0) and GWP value that was 73.9% less than that of conventional concrete (0-0).

31.4.3 Multi-Criteria Decision-Making Analysis

The MCDA using VIKOR was conducted to qualitative and quantitative assess the workability, mechanical properties, environmental impact, and economic cost of the production of WPF-RA-SCC, thus ranking the various alternatives and identifying the optimal solution. Table 31.11 presents the technical performance (segregation rate and compressive strength) of WPF-RA-SCC mixtures, which were integrated with environmental impacts and economic cost to evaluate the sustainability aspect of WPF-RA-SCC. Detailed information in terms of fresh and hardened properties are provided in the study conducted by Dong et al. and Guo et al. [10, 28]. From Table 31.11, it is clear that segregation rate consistently reduced as the content of WPF increased. It can also be seen from Table 31.11 that quaternary mix C-100-75-3 with 1.0% WPF achieved the highest compressive strength at both 7 and 28 days, which was 30.4% and 12.5% higher than that of conventional concrete 0-0. Meanwhile, quaternary mix D-100-75-3 with 1.5% WPF had the highest flexural strength at 28 days, which was 60% higher than that of conventional concrete 0-0.

Table 31.12 presents the results of the VIKOR MCDA method of all WPF-RA-SCC alternatives. Considering Q refers to a linear interpolation of S and R, the results of Q (Q values and the ranking lists) are calculated to select the optimal solution and the final list of hierarchical alternatives in this study. As shown in Table 31.12, when all five criterias are assigned to equal importance, the two compromise solutions were D-100-75-3 and C-100-75-3. The advantage of C-100-75-3 over D-100-75-3 was minimal (only 0.067), falling short of the 0.111 threshold and not meeting condition C_1 specified in Sect. 3.3. Additionally, as shown in Table 31.12, alternative C-100-50-2 was in third and D-100-50-2 was in fourth in the list of Q. When priority is given to the workability aspect, the optimal alternatives proposed by the MCDA results were also mix D-100-75-3 and C-100-75-3. Alternative D-100-50-2 was in third and B-100-75-3 was in fourth in the list of Q. Under "environmental advantage" and "economic advantage" criteria, the order of the optimal alternatives in Q ranking lists was both similar to the order under "workability advantage" criteria. When an

Table 31.11 Workability and mechanical properties of WPF-RA-SCC mixes

Mix code	Workability	Mechanical properties		
	Segregation rate	Compressive strength		Flexural strength
		7 days	28 days	28 days
	%	MPa	MPa	MPa
0-0	20	32.71	46.28	4.82
100-0	20	33.33	43.67	4.53
A-100-50-2	23	30.21	36.28	4.56
A-100-75-3	18	30.4	42.51	5
B-100-50-2	19	32.05	40.07	6.32
B-100-75-3	15	35.1	42.63	7.02
C-100-50-2	16	40.92	51.26	6.86
C-100-75-3	15	42.65	52.07	7.26
D-100-50-2	14	35.39	46.65	6.94
D-100-75-3	15	36.83	48.5	7.71

advantage is given to the mechanical properties, the optimal solution was C-100-75-3, which had a 0.119 advantage over alternative D-100-75-3 (in the second in the list of Q). This indicates that when mechanical properties are prioritized, quaternary mix C-100-75-3 incorporating 1.0% WPF and 75% SCMs is the single best compromise solution. This is mainly attributed to the synergistic effect of WPF and SCMs led to a considerably increment on the mechanical properties of WPF-RA-SCC [28].

Table 31.12 presents the radar chart of Q values for all WPF-RA-SCC alternatives under five preference criteria, revealing that all alternatives show similar trends in Q values. In addition, from Table 31.12, it can be seen that quaternary mix C-100-75-3 was the compromise solution for all five preference criteria. At last, for all five preference criteria, normal SCC is recognized as the worst alternative all five preference criteria, consistently ranked the lowest in all priority lists. The trended for the best and worst alternatives across various preference criteria was nearly identical, as validated by the VIKOR ranking method. This further corroborates the methodology of this study, demonstrating that the proposed framework can effectively optimize mixture selection to meet target performance requirements.

According to VIKOR MCDA results, quaternary mix WPF-RA-SCC comprising RCA, WPF and SCMs (FA, GGBS and SF) was ranked as the best solutions for all five classifications criteria in the list of Q, indicating that the inclusion of RCA, SCMs and WPF offered considerable benefits for environmental performance and economic cost in addition to the improvement on the fresh and hardened properties. This is primarily attribute to these reasons. First of all, inclusion of various SCMs (FA, and GGBFS) mitigated the detrimental effect of RCA and WPF, improved the workability of WPF-RA-SCC. Secondly, the remarkable synergistic effect of WPF and various mineral additives led to a strong microstructure and significant bridging effect, which substantially enhanced the mechanical properties. At last, the inclusion of the waste concrete, plastic waste and industrial waste effectively addressed the

Table 31.12 Q values and rank for WPF-RA-SCC alternatives

Mix code	Equal importance		Workability advantage		Mechanical properties advantage		Environmental advantage		Economic advantage	
	Q	Rank	Q	Rank	Q	Rank	Q	Rank	Q	Rank
0-0	1.000	10	0.813	9	0.874	9	1.000	10	1.000	10
100-0	0.916	9	0.761	8	0.868	8	0.927	9	0.683	9
A-100-50-2	0.897	8	0.957	10	0.997	10	0.585	8	0.570	8
A-100-75-3	0.591	7	0.402	6	0.691	7	0.345	6	0.345	6
B-100-50-2	0.579	6	0.535	7	0.657	6	0.417	7	0.377	7
B-100-75-3	0.271	5	0.133	4	0.344	5	0.155	3	0.155	3
C-100-50-2	0.249	3	0.175	5	0.182	3	0.282	5	0.196	5
C-100-75-3	0.000	1	0.000	1	0.000	1	0.000	1	0.000	1
D-100-50-2	0.250	4	0.117	3	0.310	4	0.258	4	0.171	4
D-100-75-3	0.067	1	0.000	1	0.119	2	0.034	1	0.034	1

disposal of these waste materials and the consumption of natural resources, thus dramatically reduced the carbon footprint and the economic cost WPF-RA-SCC.

31.5 Conclusions

A novel methodology for VIKOR MCDA for WPF-RA-SCC was presented in this paper, by integrating workability, mechanical properties, environmental and economic performance to identify optimal concrete mixture with superior mechanical properties, lower carbon footprint and economic cost. The main research findings were listed as following:

1. LCA results indicated that the substitution of cement and natural aggregates with RCA and SCMs, respectively, and the inclusion of WPFs in SCC production resulted in a substantial reduction (up to 73.9%) in the environmental load. In addition, the type of concrete with the lowest environmental impacts of all the categories evaluated was quaternary mix D-100-75-3 with WPFs and various SCM combinations, indicating that the combined addition of WPFs and various SCM combinations substantially decreased the carbon footprint, in addition to the improvement on the mechanical properties.
2. E-LCCA results indicated that the substitution of cement and natural aggregates with RCA and SCMs, respectively, significantly reduced the total economic cost $C_{E\text{-}LCCA}$. In addition, the total cost $C_{E\text{-}LCCA}$ was further reduced with the addition of WPF. Higher $C_{E\text{-}LCCA}$ were obtained for the conventional SCC than for the WPF-RA-SCC. The lowest $C_{E\text{-}LCCA}$ were determined for quaternary mix D-100-75-3 with 100% RCAs, 1.5% WPF and 75% SCM, which is 34.7% less than that of normal SCC.
3. MCDA results indicate that the best option, considering technical performance, environmental impact, and economic cost, was quaternary mix C-100-75-3 containing 100% RCA, 75% SCMs, and 1.0% WPF for all five classifications criteria in the list of Q, indicating that the inclusion of RCA, SCMs and WPF offered considerable benefits for environmental performance and economic cost in addition to the improvement on the engineering properties. Conventional SCC was identified as the worst option in all classifications of the criteria.

By conducting LCA, LCCA and MCDA, the technical, economic and environmental benefits of recycling waste plastic, waste concrete and industrial waste into SCC production was highlighted in this study. In addition, for the first time, this study specifically addressed the application of VIKOR MCDA to identify the optimal solution to manufacture sustainable concrete by integrating technical, environmental and economic performances of WPF-RA-SCC and proposed a novel approach for prospective decision-making as well as presented new method for prioritizing materials selection, thus promoting the sustainability of recycling of waste plastic and waste concrete. Further research is recommended to consider more factors and variables, such as further evaluation of economic impacts at the societal dimension.

Moreover, the above conclusions are valid only in the cradle-to-gate system boundary adopted in this study, and further studies in different systems and extended boundaries are needed to expand the technical, environmental, and economic feasibility of WPF-RA-SCC application.

Acknowledgements This work was supported by National Natural Science Foundation of China (Grant No. 52178144), Postgraduate Research & Practice Innovation Program of Jiangsu Province (Grant No. SJCX24_0515). The authors wish to gratefully acknowledge the support of these organizations for this study.

References

1. X. Hu, Y. Zhou, S. Vanhullebusch, R. Mestdagh, Z. Cui, J. Li, Smart building demolition and waste management frame with image-to-BIM. J. Build. Eng. **49**, 104058 (2022). https://doi.org/10.1016/j.jobe.2022.104058
2. H. Wang, D. Chen, H. Duan, F. Yin, Y. Niu, Characterizing urban building metabolism with a 4D-GIS model: a case study in China. J. Clean. Prod. **228**, 1446–1454 (2019). https://doi.org/10.1016/j.jclepro.2019.04.341
3. Z. Chen, Q. Feng, R. Yue, Z. Chen, O. Moselhi, A. Soliman, A. Hammad, C. An, Construction, renovation, and demolition waste in landfill: a review of waste characteristics, environmental impacts, and mitigation measures. Environ. Sci. Pollut. R. **29**, 46509–46526 (2022). https://doi.org/10.1007/s11356-022-20479-5
4. V. Corinaldesi, G. Moriconi, Influence of mineral additions on the performance of 100% recycled aggregate concrete. Constr. Build. Mater. **23**, 2869–2876 (2009). https://doi.org/10.1016/j.conbuildmat.2009.02.004
5. N. Serres, S. Braymand, F. Feugeas, Environmental evaluation of concrete made from recycled concrete aggregate implementing life cycle assessment. J. Build. Eng. **5**, 24–33 (2016). https://doi.org/10.1016/j.jobe.2015.11.004
6. B.S. Saini, S.P. Singh, Flexure fatigue strength and failure probability of self compacting concrete made with RCA and blended cements. Structures **45**, 299–314 (2022). https://doi.org/10.1016/j.istruc.2022.09.031
7. Y.J. Zrar, K.H. Younis, A.F.H. Sherwani, Properties of sustainable self-compacted concrete with recycled concrete and waste tire crumb rubber aggregates. Constr. Build. Mater. **407**, 133524 (2023). https://doi.org/10.1016/j.conbuildmat.2023.133524
8. S. Malazdrewicz, K. Adam Ostrowski, Ł Sadowski, Self-compacting concrete with recycled coarse aggregates from concrete construction and demolition waste—Current state-of-the art and perspectives. Constr. Build. Mater. **370**, 130702 (2023). https://doi.org/10.1016/j.conbuildmat.2023.130702
9. A. Simalti, A.P. Singh, Comparative study on performance of manufactured steel fiber and shredded tire recycled steel fiber reinforced self-consolidating concrete. Constr. Build. Mater. **266**, 121102 (2021). https://doi.org/10.1016/j.conbuildmat.2020.121102
10. C. Dong, Q. Zhang, C. Chen, T. Jiang, Z. Guo, Y. Liu, S. Lin, Fresh and hardened properties of recycled plastic fiber reinforced self-compacting concrete made with recycled concrete aggregate and fly ash, slag, silica fume. J. Build. Eng. **62** (2022). https://doi.org/10.1016/j.jobe.2022.105384
11. R. Kurda, J.D. Silvestre, J. de Brito, Life cycle assessment of concrete made with high volume of recycled concrete aggregates and fly ash. Resour. Conserv. Recycl. **139**, 407–417 (2018). https://doi.org/10.1016/j.resconrec.2018.07.004

12. B. Ali, L.A. Qureshi, S.H.A. Shah, S.U. Rehman, I. Hussain, M. Iqbal, A step towards durable, ductile and sustainable concrete: simultaneous incorporation of recycled aggregates, glass fiber and fly ash. Constr. Build. Mater. 251 (2020). https://doi.org/10.1016/j.conbuildmat.2020.118980
13. A.P. Gursel, in *Life-Cycle Assessment of Concrete: Decision-Support Tool and Case Study Application* (2014)
14. R.M. Andrew, Global CO_2 emissions from cement production. Earth Syst. Sci. Data **10**, 195–217 (2018). https://doi.org/10.5194/essd-10-195-2018
15. H. Li, M. Liebscher, J. Yang, M. Davoodabadi, L. Li, Y. Du, B. Yang, S. Hempel, V. Mechtcherine, Electrochemical oxidation of recycled carbon fibers for an improved interaction toward alkali-activated composites. J. Clean. Prod. **368**, 133093 (2022). https://doi.org/10.1016/j.jclepro.2022.133093
16. R.J. Gravina, T. Xie, F. Giustozzi, X. Zhao, P. Visintin, Assessment of the variability and uncertainty of using post-customer plastics as natural aggregate replacement in concrete. Constr. Build. Mater. **273**, 121747 (2021). https://doi.org/10.1016/j.conbuildmat.2020.121747
17. L. Gu, T. Ozbakkaloglu, Use of recycled plastics in concrete: a critical review. Waste Manage. **51**, 19–42 (2016). https://doi.org/10.1016/j.wasman.2016.03.005
18. N. Gandhi, N. Farfaras, N.-H.L. Wang, W.-T. Chen, Life cycle assessment of recycling high-density polyethylene plastic waste. J. Renew. Mater. **9**, 1463–1483 (2021). https://doi.org/10.32604/jrm.2021.015529
19. A. Alsabri, F. Tahir, S.G. Al-Ghamdi, Life-cycle assessment of polypropylene production in the gulf cooperation council (GCC) region. Polymers **13** (2021). https://doi.org/10.3390/polym13213793
20. K.D. Cox, G.A. Covernton, H.L. Davies, J.F. Dower, F. Juanes, S.E. Dudas, Human consumption of microplastics. Environ. Sci. Technol. **53**, 7068–7074 (2019). https://doi.org/10.1021/acs.est.9b01517
21. C. Campanale, C. Massarelli, I. Savino, V. Locaputo, V.F. Uricchio, A detailed review study on potential effects of microplastics and additives of concern on human health. Int. J. Env. Res. Pub. He. **17** (2020). https://doi.org/10.3390/ijerph17041212
22. J.C. Prata, J.P. da Costa, I. Lopes, A.C. Duarte, T. Rocha-Santos, Environmental exposure to microplastics: an overview on possible human health effects. Sci. Total. Environ. **702**, 134455 (2020). https://doi.org/10.1016/j.scitotenv.2019.134455
23. I.M. Nikbin, S. Dezhampanah, S. Charkhtab, S. Mehdipour, I. Shahvareh, M. Ebrahimi, A. Pournasir, H. Pourghorban, Life cycle assessment and mechanical properties of high strength steel fiber reinforced concrete containing waste PET bottle. Constr. Build. Mater. **337**, 127553 (2022). https://doi.org/10.1016/j.conbuildmat.2022.127553
24. H. Bahmani, D. Mostofinejad, Comparative analysis of environmental, social, and mechanical aspects of high-performance concrete with calcium oxide-activated slag reinforced with basalt, and recycled PET fibers. Case Stud. Constr. Mat. **20**, e02895 (2024). https://doi.org/10.1016/j.cscm.2024.e02895
25. F. Tahir, S. Sbahieh, S.G. Al-Ghamdi, Environmental impacts of using recycled plastics in concrete. Mater. Today Proc. **62**, 4013–4017 (2022). https://doi.org/10.1016/j.matpr.2022.04.593
26. B. Ngoc Kien, T. Satomi, H. Takahashi, Recycling woven plastic sack waste and PET bottle waste as fiber in recycled aggregate concrete: an experimental study. Waste Manage. **78**, 79–93 (2018). https://doi.org/10.1016/j.wasman.2018.05.035
27. Z. Duan, Q. Deng, C. Liang, Z. Ma, H. Wu, Upcycling of recycled plastic fiber for sustainable cementitious composites: a critical review and new perspective. Cem. Concr. Compos. **142** (2023). https://doi.org/10.1016/j.cemconcomp.2023.105192
28. Z. Guo, Q. Sun, L. Zhou, T. Jiang, C. Dong, Q. Zhang, Mechanical properties, durability and life-cycle assessment of waste plastic fiber reinforced sustainable recycled aggregate self-compacting concrete. J. Build. Eng. **91**, 109683 (2024). https://doi.org/10.1016/j.jobe.2024.109683

29. J. Ahmad, A. Majdi, A.B. Elhag, A.F. Deifalla, M. Soomro, H.F. Isleem, S.A. Qaidi, Step towards sustainable concrete with substitution of plastic waste in concrete: overview on mechanical, durability and microstructure analysis. Crystals **12** (2022). https://doi.org/10.3390/cryst12070944
30. S. Guo, D. Zheng, L. Zhao, Q. Lu, X. Liu, Mechanical test and constitutive model of recycled plastic fiber reinforced recycled concrete. Constr. Build. Mater. **348** (2022). https://doi.org/10.1016/j.conbuildmat.2022.128578
31. E.L. Pereira, A.L. de Oliveira Junior, A.G. Fineza, Optimization of mechanical properties in concrete reinforced with fibers from solid urban wastes (PET bottles) for the production of ecological concrete. Constr. Build. Mater. **149**, 837–848 (2017). https://doi.org/10.1016/j.conbuildmat.2017.05.148
32. A.I. Al-Hadithi, A.T. Noaman, W.K. Mosleh, Mechanical properties and impact behavior of PET fiber reinforced self-compacting concrete (SCC). Compos. Struct. **224** (2019). https://doi.org/10.1016/j.compstruct.2019.111021
33. A.H. Alani, M.A.M. Johari, A.T. Noaman, N.M. Bunnori, T.A. Majid, Effect of the incorporation of PET fiber and ternary blended binder on the flexural and tensile behaviour of ultra-high performance green concrete. Constr. Build. Mater. **331** (2022). https://doi.org/10.1016/j.conbuildmat.2022.127306
34. T.M. Borhan, M.S.A. Dhaheer, Z.A. Mandi, Characteristics of sustainable self-compacting concrete reinforced by fibres from waste materials. Arab. J. Sci. Eng. **45**, 4359–4367 (2020). https://doi.org/10.1007/s13369-020-04460-3
35. O. Rispoli, O.O. Ajibade, Comparative life cycle assessment of a novel sustainable road pavement system adopting recycled plastic from PET bottles and carbonated aggregate. Heliyon **10**, e24354 (2024). https://doi.org/10.1016/j.heliyon.2024.e24354
36. M. Tahanpour Javadabadi, Comparative life cycle assessment of incorporating recycled PET aggregates into concrete. NTNU (2019)
37. Y.C. Ersan, S. Gulcimen, T.N. Imis, O. Saygin, N. Uzal, Life cycle assessment of lightweight concrete containing recycled plastics and fly ash. Eur. J. Environ. Civ. En. **26**, 2722–2735 (2022). https://doi.org/10.1080/19648189.2020.1767216
38. P.G. Palafox-Alcantar, D.V.L. Hunt, C.D.F. Rogers, The complementary use of game theory for the circular economy: a review of waste management decision-making methods in civil engineering. Waste Manage. **102**, 598–612 (2020). https://doi.org/10.1016/j.wasman.2019.11.014
39. B. Kiani, R.Y. Liang, J. Gross, Material selection for repair of structural concrete using VIKOR method. Case Stud. Constr. Mat. **8**, 489–497 (2018). https://doi.org/10.1016/j.cscm.2018.03.008
40. A. Nestico, P. Somma, Comparative analysis of multi-criteria methods for the enhancement of historical buildings. Sustainability **11** (2019). https://doi.org/10.3390/su11174526
41. L. Alberto López Ruiz, X. Roca Ramon, C. Melissa Lara Mercedes, S. Gasso Domingo, Multicriteria analysis of the environmental and economic performance of circularity strategies for concrete waste recycling in Spain. Waste Manage. **144**, 387–400 (2022). https://doi.org/10.1016/j.wasman.2022.04.008
42. J.-J. Wang, Y.-Y. Jing, C.-F. Zhang, J.-H. Zhao, Review on multi-criteria decision analysis aid in sustainable energy decision-making. Renew. Sust. Energ. Rev. **13**, 2263–2278 (2009). https://doi.org/10.1016/j.rser.2009.06.021
43. JGJ52–2006, in *Standard for technical requirements and test method of sand and crushed stone (or gravel) for ordinary concrete*. Chinese Standard (2006)
44. S.-C.J.B. Concrete et al. The European guidelines for self-compacting concrete. BIBM **22**, 563 (2005)
45. Z. Guo, T. Jiang, J. Zhang, X. Kong, C. Chen, D.E. Lehman, Mechanical and durability properties of sustainable self-compacting concrete with recycled concrete aggregate and fly ash, slag and silica fume. Constr. Build. Mater. **231** (2020). https://doi.org/10.1016/j.conbuildmat.2019.117115

46. S. Kakooei, H.M. Akil, M. Jamshidi, J. Rouhi, The effects of polypropylene fibers on the properties of reinforced concrete structures. Constr. Build. Mater. **27**, 73–77 (2012). https://doi.org/10.1016/j.conbuildmat.2011.08.015
47. F. Fraternali, V. Ciancia, R. Chechile, G. Rizzano, L. Feo, L. Incarnato, Experimental study of the thermo-mechanical properties of recycled PET fiber-reinforced concrete. Compos. Struct. **93**, 2368–2374 (2011). https://doi.org/10.1016/j.compstruct.2011.03.025
48. A. Laurent, I. Bakas, J. Clavreul, A. Bernstad, M. Niero, E. Gentil, M.Z. Hauschild, T.H. Christensen, Review of LCA studies of solid waste management systems—Part I: lessons learned and perspectives. Waste Manage. **34**, 573–588 (2014). https://doi.org/10.1016/j.wasman.2013.10.045
49. S. Dezhampanah, I. Nikbin, S. Charkhtab, F. Fakhimi, S.M. Bazkiaei, R. Mohebbi, Environmental performance and durability of concrete incorporating waste tire rubber and steel fiber subjected to acid attack. J. Clean Prod. **268** (2020). https://doi.org/10.1016/j.jclepro.2020.122216
50. S. Mehdipour, I.M. Nikbin, S. Dezhampanah, R. Mohebbi, H. Moghadam, S. Charkhtab, A. Moradi, Mechanical properties, durability and environmental evaluation of rubberized concrete incorporating steel fiber and metakaolin at elevated temperatures. J. Clean Prod. **254** (2020). https://doi.org/10.1016/j.jclepro.2020.120126
51. ISO. ISO 14040 International Standard, in *Environmental Management—Life Cycle Assessment—Principles and Framework* (2006)
52. ISO. ISO 14044 International Standard, in *Environmental Management—Life Cycle Assessment—Requirements and Guidelines* (2006)
53. R. Kurda, J. de Brito, J.D. Silvestre, A comparative study of the mechanical and life cycle assessment of high-content fly ash and recycled aggregates concrete. J. Build. Eng. **29** (2020). https://doi.org/10.1016/j.jobe.2020.101173
54. G. Barjoveanu, G. Gadaleta, G. Santomasi, S. De Gisi, M. Notarnicola, C. Teodosiu, Does PET trays sorting affect the sustainability of plastic waste? An LCA and cost-revenue approach. Sci. Total Environ. **897** (2023). https://doi.org/10.1016/j.scitotenv.2023.165222
55. R.K. Foolmaun, T. Ramjeeawon, Comparative life cycle assessment and social life cycle assessment of used polyethylene terephthalate (PET) bottles in Mauritius. Int. J. Life Cycle Ass. **18**, 155–171 (2013). https://doi.org/10.1007/s11367-012-0447-2
56. ISO. ISO 15686–5, in *Buildings and Constructed Assets e Service—Life Planning e Part 5: Life-cycle Costing* (2017)
57. T.E. Swarr, D. Hunkeler, W. Kloepffer, H.-L. Pesonen, A. Ciroth, A.C. Brent, R. Pagan, Environmental life-cycle costing: a code of practice. Int. J. Life Cycle Ass. **16**, 389–391 (2011). https://doi.org/10.1007/s11367-011-0287-5
58. M. Zhang, X. Liu, L. Kong, Evaluation of carbon and economic benefits of producing recycled aggregates from construction and demolition waste. J. Clean Prod. **425** (2023). https://doi.org/10.1016/j.jclepro.2023.138946
59. T. Zhang, Q. Wen, S. Gao, J. Tang, Comparative study on mechanical and environmental properties of coal gangue sand concrete. Constr. Build. Mater. **400**, 132646 (2023). https://doi.org/10.1016/j.conbuildmat.2023.132646
60. T. Wang, J. Wang, P. Wu, J. Wang, Q. He, X. Wang, Estimating the environmental costs and benefits of demolition waste using life cycle assessment and willingness-to-pay: a case study in Shenzhen. J. Clean. Prod. **172**, 14–26 (2018). https://doi.org/10.1016/j.jclepro.2017.10.168
61. X. Cao, X. Li, Y. Zhu, Z. Zhang, A comparative study of environmental performance between prefabricated and traditional residential buildings in China. J. Clean. Prod. **109**, 131–143 (2015). https://doi.org/10.1016/j.jclepro.2015.04.120
62. M. Chen, H. Zhong, L. Chen, Y. Zhang, M. Zhang, Engineering properties and sustainability assessment of recycled fibre reinforced rubberised cementitious composite. J. Clean Prod. **278** (2021). https://doi.org/10.1016/j.jclepro.2020.123996
63. X. Shi, A. Mukhopadhyay, D. Zollinger, Z. Grasley, Economic input-output life cycle assessment of concrete pavement containing recycled concrete aggregate. J. Clean. Prod. **225**, 414–425 (2019). https://doi.org/10.1016/j.jclepro.2019.03.288

64. C. Moro, Comparative analysis of multi-criteria decision making and life cycle assessment methods for sustainable evaluation of concrete mixtures. Sustainability **15** (2023). https://doi.org/10.3390/su151712746
65. A. Younis, U. Ebead, S. Judd, Life cycle cost analysis of structural concrete using seawater, recycled concrete aggregate, and GFRP reinforcement. Constr. Build. Mater. **175**, 152–160 (2018). https://doi.org/10.1016/j.conbuildmat.2018.04.183
66. S. Saha, D. Sau, T. Hazra, Economic viability analysis of recycling waste plastic as aggregates in green sustainable concrete. Waste Manage. **169**, 289–300 (2023). https://doi.org/10.1016/j.wasman.2023.07.023
67. D.J. Thomas, P.M. Griffin, Coordinated supply chain management. Eur. J. Oper. Res. **94**, 1–15 (1996). https://doi.org/10.1016/0377-2217(96)00098-7
68. Y. Li, Economic Analysis of Recycled Concrete Application Based on LCA, Thesis (2008)
69. R.H.J.M. Gradus, P.H.L. Nillesen, E. Dijkgraaf, R.J. van Koppen, A cost-effectiveness analysis for incineration or recycling of Dutch household plastic waste. Ecol. Econ. **135**, 22–28 (2017). https://doi.org/10.1016/j.ecolecon.2016.12.021
70. Q. Wang, H. Tang, Q. Ma, R. Mu, X. Yuan, J. Hong, J. Zhang, J. Zuo, Z. Mu, S. Cao et al., Life cycle assessment and the willingness to pay of waste polyester recycling. J. Clean. Prod. **234**, 275–284 (2019). https://doi.org/10.1016/j.jclepro.2019.06.123
71. N. Tosic, S. Marinkovic, T. Dasic, M. Stanic, Multicriteria optimization of natural and recycled aggregate concrete for structural use. J. Clean. Prod. **87**, 766–776 (2015). https://doi.org/10.1016/j.jclepro.2014.10.070
72. M.A. Mir, P.T. Ghazvinei, N.M.N. Sulaiman, N.E.A. Basri, S. Saheri, N.Z. Mahmood, A. Jahan, R.A. Begum, N. Aghamohammadi, Application of TOPSIS and VIKOR improved versions in a multi criteria decision analysis to develop an optimized municipal solid waste management model. J. Environ. Manage. **166**, 109–115 (2016). https://doi.org/10.1016/j.jenvman.2015.09.028
73. A.J. Sánchez-Garrido, I.J. Navarro, V. Yepes, Neutrosophic multi-criteria evaluation of sustainable alternatives for the structure of single-family homes. Environ. Impact. Asses. **89**, 106572 (2021). https://doi.org/10.1016/j.eiar.2021.106572
74. J.M.C. Ongpeng, B.I.B. Rabe, L.F. Razon, K.B. Aviso, R.R. Tan, A multi-criterion decision analysis framework for sustainable energy retrofit in buildings. Energy **239**, 122315 (2022). https://doi.org/10.1016/j.energy.2021.122315
75. E.R. Teixeira, R. Mateus, A.F. Camoes, L. Braganca, F.G. Branco, Comparative environmental life-cycle analysis of concretes using biomass and coal fly ashes as partial cement replacement material. J. Clean. Prod. **112**, 2221–2230 (2016). https://doi.org/10.1016/j.jclepro.2015.09.124
76. A.P. Gursel, H. Maryman, C. Ostertag, A life-cycle approach to environmental, mechanical, and durability properties of "green" concrete mixes with rice husk ash. J. Clean. Prod. **112**, 823–836 (2016). https://doi.org/10.1016/j.jclepro.2015.06.029

Chapter 32
CO_2 Sequestration and Storage in Recycled Concrete Waste: An Emerging Solution for Sustainable Construction

Hamideh Mehdizadeh and Mohammad Hajmohammadian Baghban

Abstract Our world faces critical environmental challenges, such as greenhouse gas emissions, natural resources depletion, and solid waste disposal. Concrete, the second most consumed material on Earth after water, is responsible for 8–10% of global CO_2 emissions, primarily stemming from clinker production. Meanwhile, the substantial volume of concrete waste (CW) generated throughout its lifecycle, along with the rising global demand for concrete due to rapid urbanization, exacerbates these pressing concerns. Accelerated carbonation of CW is an effective method to consume CW and sequester long-lasting CO_2. Upcycling carbonated CW to substitute aggregate and cement in sustainable construction materials can also benefit in reducing CO_2 emissions and conserving natural resources. However, bridging the gap between research and practice by promoting the effective use of carbonated concrete waste in the construction sector is the key for achieving this goal. This chapter thus discusses advancements in the carbonation of concrete waste, focusing on the carbonation mechanism and its strategic application to various recycled CW fractions, namely recycled coarse and fine aggregates, as well as recycled concrete powders.

32.1 Introduction

The cement and concrete industry is one of the major contributors to CO_2 emissions (8–10%) and heavily depends on natural resources. Recycling concrete-based components in construction and demolition (C&D) wastes at the end of the life of concrete structures is considered an ideal strategy to address the environmental challenges faced by the industry [1]. Billions of tons of concrete waste enter the global

H. Mehdizadeh (✉) · M. Hajmohammadian Baghban
Department of Manufacturing and Civil Engineering, Norwegian University of Science and Technology (NTNU), Gjøvik, Norway
e-mail: hamideh.mehdizadeh@ntnu.no

D. A. H. Hanaor (ed.), *Sustainable Materials for the Built Environment*, Springer Series in Materials Science 352, https://doi.org/10.1007/978-3-031-97818-0_32

C&D waste stream annually, with amounts rising due to rapid urbanization. To mitigate the environmental impacts of concrete production and save natural resources, technologies have recently been developed to recycle concrete waste into aggregates with two size fractions: below 4.75 mm (recycled concrete fines, RCFs), and above 4.75 mm (recycled coarse aggregates, RCAs) [2, 3]. Despite the potential to recycle 50–90% of concrete waste as aggregates, the inferior properties of recycled aggregates compared to natural aggregates result in their down cycling or use in applications like road construction, rather than reintroducing them into concrete production. For example, the hydrated cement paste residue on the surface of RCAs after crushing increases their porosity and water absorption, leading to the formation of weak interfacial transition zones (ITZ) among old cement paste, RCA, and new cement paste, which lowers the performance of new concrete [4]. In comparison, RCFs, with a higher content of residual cement pastes due to their finer particle size, have more challenges regarding the workability, drying shrinkage, and transport properties of concrete [5]. Consequently, only very low amounts of untreated RCAs and RCFs are used in concrete production.

Alongside the interest in recycling concrete waste into aggregates, there has been growing attention on re-utilizing the waste powder generated during the recycling process, particularly those with size fractions below 0.15 mm [2, 3]. Studies have shown that recycled concrete powders (RCPs) contain 80–90% hardened paste, which is significantly higher than that found in RCFs, making them more like hydrated cement [6]. Despite efforts to reuse RCPs as a cement substitute, the significant amount of gel pores in RCPs and their high-water absorption properties hinder the direct use of untreated RCPs in cementitious binders. Moreover, RCPs have poor pozzolanic activity compared to cement particles, resulting in the poor mechanical and durability performance of cementitious materials [7–9]. For instance, Mehdizadeh et al. [7] have reported that replacing 10–30% of cement with untreated RCPs causes a 12–35% reduction in the 28-day compressive strength of cement paste. Hence, it is crucial to enhance the properties of recycled concrete waste (i.e. RCA, RCF, and RCP) to facilitate their effective utilization in cement-based materials.

Accelerated mineral carbonation, also known as CO_2 mineralization, involves the reaction of oxides such as CaO or MgO in cementitious materials with dissolved CO_2 to produce Ca/Mg carbonate minerals. These carbonate phases exhibit thermodynamic stability under atmospheric conditions, providing a permanent CO_2 storage solution [2]. In recent years, extensive research has been conducted on the carbonation of concrete waste, aiming to strengthen the RCAs and RCFs, as well as enhance the reactivity of RCPs [2–9]. It has also been reported that the carbonation of concrete waste can effectively capture about 7.9 kg CO_2/t of RCA, 20 kg CO_2/t of RCF, and 190 kg CO_2/t of RCP, which is significantly higher than that achieved by chemical absorption methods (15–54 kg/t) [10].

To achieve a comprehensive recycling strategy for concrete waste in sustainable construction, this chapter discusses the advances in the carbonation of concrete waste. It will cover the carbonation strategies and their effects on the properties of

RCAs, RCFs, and RCPs. Additionally, the application of carbonated products in new cementitious materials will be described.

32.2 Accelerated Carbonation of Recycled Concrete Waste

Due to the high content of Ca-based minerals and alkalinity, various fraction sizes of concrete waste (CW) exhibit significant carbonation reactivity in the order of RCPs > RCFs > RCAs. In a CO_2-rich environment, the unhydrated calcium silicate phases (C_2S and C_3S) and hydration products (portlandite, calcium silicate hydrate, and calcium aluminate hydrates) associated with residual cement paste react with CO_2 to form calcium carbonate ($CaCO_3$) and Si-rich gel. The accelerated carbonation of CW commonly takes place at atmospheric pressure and ambient temperatures, aiming to enhance feasibility for use in industrial applications. This process involves gas–liquid-solid reactions and can occur in either a dry environment with 50–80% humidity (semi-dry carbonation route) or a wet environment (aqueous carbonation route). In general, under ambient conditions, unhydrated clinker phases, hydrated products of residual cement paste in concrete waste fractions, as well as CO_2 gas, dissolve in the carbonation solution, and resulting products precipitate from the solution through two main kinetics steps [11–13]. A schematic of phase evaluation and pH changes during the carbonation of residual cement paste in concrete waste is shown in Fig. 32.1.

(1) *In the first kinetic step,* carbonation is limited by the availability of CO_2 and the dissolution rate of Ca-based components in concrete waste. The pH of the solution rises to 9–12 within few minutes due to the rapid dissolution of hydrates in cement paste ($Ca(OH)_2$, followed by ettringite, AFm phases and partial decalcification of the calcium silicate hydrate (C-S-H) phases) and CO_2 consumption. This pH increase is influenced by particle size and the residual cement paste content in concrete waste fractions [14–16]. It is important to note that $CaCO_3$

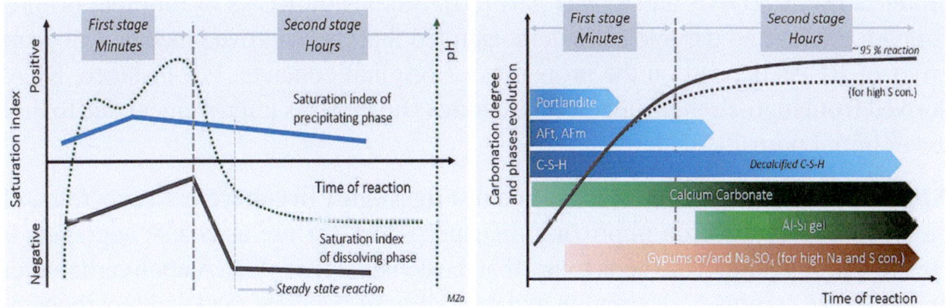

Fig. 32.1 A schematic of phase evolution and pH changes during the carbonation of residual hydrated cement paste in concrete waste [14]

nucleation and precipitation are critical at this stage, and delays in nucleation can cause a pH decrease.

(2) *In the second kinetic step,* the carbonation reaction is controlled by the dissolution rate of the hydrates and the transport of dissolved Ca ions to the surface of particles. As CO_2 concentration increases in the solution over time, the pH begins to drop due to the reduction in saturation indices. At this stage, only the larger cement-paste grains react since smaller particles have already fully carbonated during the first reaction step, leading to the precipitation of additional $CaCO_3$. During this step, the final carbonation of calcium silicate hydrate phases occurs, and all aluminum-bearing hydrates except hydrotalcite react with CO_2, forming Si-rich gel and $CaCO_3$.

32.3 Properties and Application of Carbonated CW Materials

32.3.1 Carbonated Recycled Coarse Aggregates

Physical and microstructural characteristics: Experimental studies have shown that carbonation reactions significantly refine the pore structure of RCAs, reducing the amount of capillary pores and total porosity by up to 30% [17]. This refinement leads to a 5–45% decrease in the water absorption capacity of RCAs. Additionally, the electrical conductivity and chloride permeability of RCAs can be reduced by up to 70% after 7 days of carbonation [18]. Notably, the reduction of porosity and densification of RCA microstructure occurs not only in the residual paste on the aggregate surface but also in the old ITZ section, which can reduce the likelihood of internal failure [19]. Moreover, the precipitation of $CaCO_3$ on the surface of carbonated RCAs contributes to improved mechanical properties, resulting in higher microhardness and a lower of crushing value. This phenomenon could be related to the higher elastic modulus of carbonated products ($CaCO_3$) compared to hydration products (i.e. $Ca(OH)_2$ and C-S-H gel) [20]. Studies on RCAs of different origins, strength grades and particle sizes demonstrated that the improved mechanical properties of RCAs depend on the properties of original concrete. For instance, RCAs derived from high-strength concrete structures showed less improvement due to their lower initial porosity [21].

Application of carbonated RCAs as coarse aggregate in concrete: Recent research on RCAs has focused on improving their properties for use as coarse aggregate in concrete production. It has been found that the densification of RCA upon carbonation enhances the compressive strength and durability of concrete made with carbonated RCAs compared to those made with non-carbonated RCAs [22] (see Fig. 32.2). Also, the strength loss of RCA-based is limited to 10% when 100% coarse aggregate

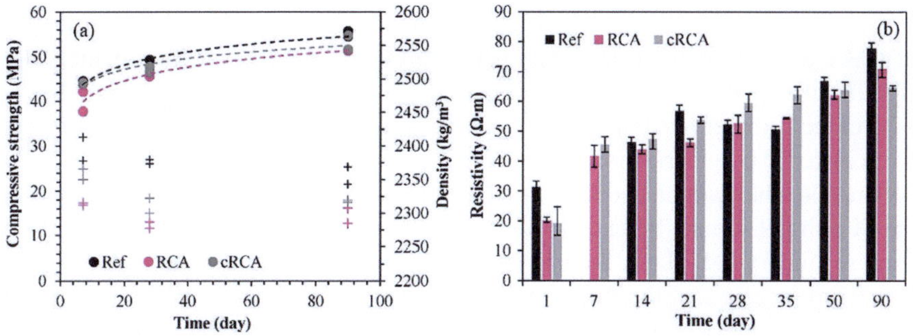

Fig. 32.2 (**a**) Compressive strength, density and (**b**) electrical resistivity of concrete with non-carbonated and carbonated RCA [22]

is replaced with carbonated RCAs, while its elastic modulus became closer to that normal concrete [22, 23].

The decrease in drying shrinkage by up to 25% and the improvement in water penetration and chloride resistance of concrete by up to 36% were other benefits of carbonation of RCAs when they were used as coarse aggregates in concrete [22, 24]. Research findings have demonstrated that carbonated RCAs could serve as internal curing agents due to their porous structure, thereby facilitating additional cement hydration in the new ITZ (i.e. the region between RCAs and new binder) by gradually releasing water [25]. Besides, the $CaCO_3$ precipitated on the surface of RCAs could act as nuclei for accelerating cement hydration, resulting in a 12% decrease in the porosity of the new ITZ in concrete. As a result, the bond between RCAs and the new cement paste was strengthened by up to 90% [26]. In summary, the densification of RCAs upon carbonation treatment and the improved quality of new ITZ are key factors enhancing the performance of RCA-based concrete.

32.3.2 Carbonated Recycled Concrete Fines

Properties of carbonated RCFs: Compared to RCAs, RCFs exhibited higher water absorption and carbonation capacity due to their finer particle size and greater residual hydrated cement paste. Water absorption in RCFs ranges from 4% to16%, depending on the amount of hydrated cement paste on the aggregate surface. However, it was found that carbonation reduces water absorption in RCFs by 3% to 60%, a more substantial decrease than in RCAs [3]. Interestingly, pre-soaking RCFs in limewater before carbonation further reduced the water absorption by 21% due to the formation of more $CaCO_3$ [27]. Additionally, the carbonation of RCFs in a wet environment (i.e. aqueous carbonation route) was more efficient than the dry carbonation route; so that, the aqueous carbonated RCFs showed similar water absorption and density value after 6 h of carbonation compared to 24 h of dry carbonation [28].

In addition to the densifying of RCF microstructure caused by the precipitation of $CaCO_3$, the formation of Si-rich gel with Q^3 and Q^4 silicon coordination also significantly increased the pozzolanic activity of RCFs. Research has shown that 7–17% of the silica tetrahedral in the residual hydrated cement paste on the RCF surface participated in polymerization, resulting in the formation of Si-rich gel (see Fig. 32.3). As a result, when carbonated RCFs are used as sand in mortar or concrete production, the Si-rich gel formed by carbonation reacts with $Ca(OH)_2$ in new cement paste to form secondary C-S-H gel, thereby increasing the pozzolanic activity of RCFs [29]. It is worth noting that the finer fractions of RCF (i.e. 0.3–0.6 mm), as well as the aqueous carbonation route, enhanced the pozzolanic reactivity due to the formation of more Si-rich gel [30]. Another significant approach for improving the properties of carbonated RCFs was to modify the polymorphs and morphologies of $CaCO_3$. For instance, the formation of whisker-like $CaCO_3$ during carbonation could act as micro-fibers, strengthening the bonding between carbonated RCF and new cement paste in concrete [19]. Moreover, the formation of spear-like $CaCO_3$ significantly increased the surface roughness of RCFs, facilitating the effective interlocking of carbonated RCFs into new mortar or concrete.

Application of carbonated RCFs as sand in construction materials: Recent research has shown that carbonating RCFs before use in concrete can mitigate the strength loss observed in RCF-based concrete [30]. This strength improvement could

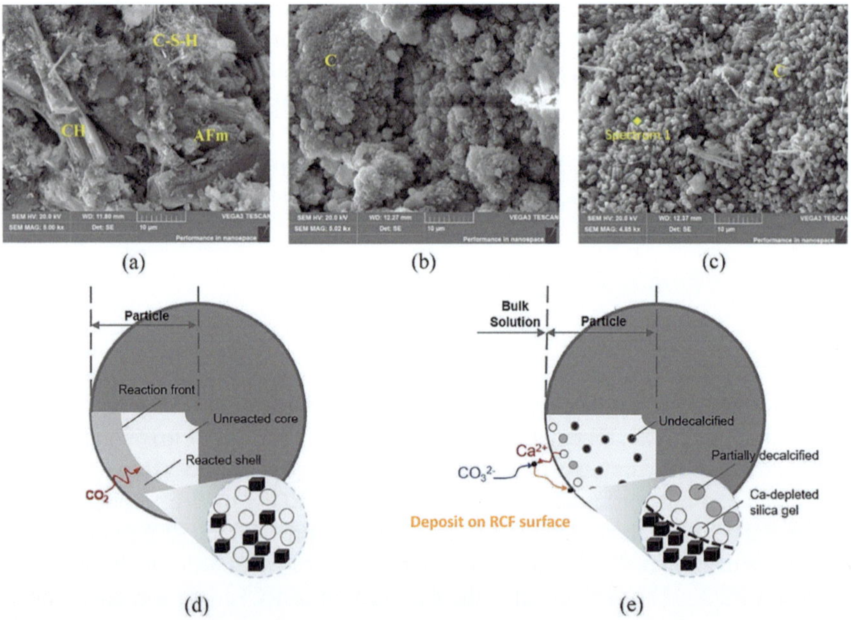

Fig. 32.3 SEM images and microstructural changes of: (**a**) non-carbonated RCF, (**b, d**) RCF carbonated under pressurized dry route, and (**c, e**) RCF carbonated under aqueous route [3]

be attributed to the nucleation effect of $CaCO_3$ and the pozzolanic reaction of the Si-rich gel precipitated on the surface of carbonated RCFs, which accelerated the cement hydration. As a result, concrete containing carbonated RCF achieved a comparable strength to that of concrete made with natural sand [3], while also experiencing reduced chloride permeability and the risk of steel corrosion [24]. However, due to the higher water demand of carbonated RCFs compared to natural sand, it was still necessary to add an extra amount of superplasticizer to adjust the workability of the concrete. Interestingly, using carbonated RCFs as fine aggregate in concrete was more effective in developing the strength of new ITZ than using RCAs as coarse aggregate, thus facilitating overall strength improvement [3]. From an economic perspective, the treatment of RCFs via accelerated carbonation was found to be a cost-effective method to improve the properties of weak recycled concrete aggregates (i.e. RCFs and RCAs) compared to the slurry coating modified method [31].

32.3.3 Carbonated Recycled Concrete Powders

Carbonation mechanisms and carbonated products: The recycled concrete powders (RCPs) composed of 80–90 wt.% hydrated cement paste. Hence, most research has focused on the carbonation of hydrated cement paste without considering the inert materials present in real RCPs. Studies have shown that RCPs can nearly fully carbonate in a wet environment, whereas the carbonation degree reaches only 40–60% with the semi-dry carbonation route [14]. Like RCAs and RCFs, the main carbonation products of RCPs are $CaCO_3$ and an amorphous alumina-silica gel, regardless of the reaction conditions (Fig. 32.4). At ambient temperature, calcite is the dominant carbonate product, whereas the presence of alkalis or sulfate ions in the carbonation solution promotes the formation of aragonite and vaterite polymorphs of $CaCO_3$ [32, 33]. In the carbonation process, the amorphous alumina-silica gel (with a high Si ratio) also forms from the carbonation of ettringite, AFm phases, and C-(A)-S-H phases. It is worth noting that the phase assemblage of carbonated RCPs can be influenced by the composition of the carbonation solution. For instance, high sulfate concentrations in the carbonation solution may reduce $CaCO_3$ formation due to gypsum precipitation, while high alkali concentration in the solution can counteract gypsum formation, increasing the carbonation potential.

Application of carbonated RCPs as cementitious materials: Carbonated RCPs, regardless of the carbonation method, could be successfully utilized as supplementary cementitious materials (SCMs) up to 20% of binder without compromising compressive strength [7–9]. The hydration mechanism of carbonated RCP-cements resembles that of ordinary Portland cement. In the blended cements, $CaCO_3$ in carbonated RCP acts as a nucleation site for C_3S hydration, accelerating the cement hydration. In addition, the amorphous (Al)-Si gel reacts rapidly with $Ca(OH)_2$ produced during the hydration of cement particles, leading to the formation of secondary C-(A)-S-H

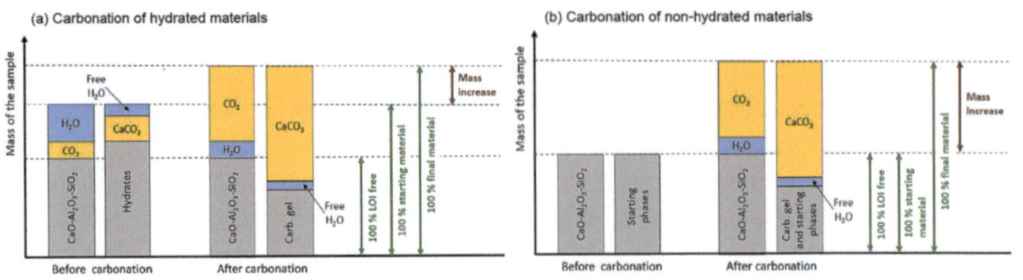

Fig. 32.4 Mass changes of (**a**) hydrated and (**b**) non-hydrated components in RCP after carbonation [14]

gel [34]. It was also observed that the incorporation of RCPs decreased the flowability and compressive strength of cement paste, with reduction becoming more obvious at higher RCP contents. Interestingly, carbonation treatment reduced the negative impact of RCPs on the fresh and mechanical properties of blended cements. Studies suggest that the blended cement containing carbonated RCPs could achieve a compressive strength of 43–61 MPa, depending on the water-to-binder ratio (0.3–0.4), which was notably 10–40% higher than those made with non-carbonated RCPs [7–9, 9, 34].

32.4 Challenges for Industrial Upscaling and Future Perspective

Semi-dry carbonation of CW can be conducted in CO_2 curing chambers where relative humidity and CO_2 concentrations are controlled. Studies have shown that adjusting relative humidity at 50–80% and CO_2 concentration above 20% is necessary to maximize CO_2 sequestration. However, the high energy consumption and long carbonation duration (28 days) pose significant challenges for industrial applications. Compared to semi-dry carbonation, the aqueous carbonation route offers higher efficiency, completing the process in 3–6 h. This method enhances CO_2 diffusion and calcium leaching rate for carbonation by homogenizing CW and saturating pores of CW during the reaction. Additionally, temperatures up to 100°C enhance the carbonation rate due to an increase in the leaching rate. However, the high-temperature requirement for aqueous carbonation complicates scaling up for real-world applications.

In addition to carbonation efficiency, the environmental and economic impacts of carbonation routes are pivotal in decision-making for upscaling. The financial benefits of re-circulating carbonated CW in the construction industry and reducing the costs of traditional recycling processes encourage the applicability of the carbonation process. However, the higher cost of producing carbonated products than the natural aggregates, attributed to the transportation distances, may reduce the financial benefits. Therefore, optimizing the location placement for of units within the construction industry is essential. Moreover, most studies have focused on the carbonation of CW using pure CO_2 gas, while the purification of CO_2 gas is an expensive process. Hence, the use of flue gas for CW carbonation should be considered to further optimize CO_2 sequestration and carbonated products in actual industrial applications.

32.5 Conclusions

Accelerated carbonation provides a sustainable approach to achieving total recycling of recycled concrete waste (CW) in C&D wastes, while also sequestrating CO_2. This process involves the carbonation of residual hydrated cement pastes present in three distinct fraction sizes sourced from CW: recycled coarse aggregates (RCAs), recycled concrete fines (RCFs), and recycled concrete powders (RCPs). The carbonation reactions typically occur in a moist or wet CO_2-rich environment at atmospheric pressure. Carbonating RCAs mainly consumes $Ca(OH)_2$ in the residual cement paste to form $CaCO_3$, resulting in a reduction in the porosity of aggregate by up to 30%, thus enhancing the mechanical strength and durability properties of RCA-based concrete. Notably, the involvement of C-S-H phases in finer particles of CW in the carbonation reaction not only improves their physical properties through the precipitation of $CaCO_3$ but also enhances the pozzolanic reactivity of the RCFs and RCPs by forming of Si-rich gel. As a result, the modification of RCF surfaces through the formation of needle-like and spear-like $CaCO_3$ benefits the bonding of RCF with the new cement paste in the mortar or concrete, while Si-rich gel enhances cement hydration in new ITZ. Moreover, incorporating carbonated RCPs up to 30% of cement or traditional SCMs in cementitious products holds promise for reclaiming CW and reducing global CO_2 emissions.

Acknowledgements Financial support from the European Union's Horizon Europe research and innovation programme under the Marie Skłodowska-Curie grant agreement No. 101066240 is greatly acknowledged.

References

1. G. Haber, S.A. Miller, V.M. John, J.L. Provis, A. Favier, A. Harvath, K.L. Scrivener, Environmental impacts and decarbonization strategies in the cement and concrete industries. Nat. Rev. Earth Environ. **1**, 559–573 (2020)
2. C.S. Poon, P. Shen, Y. Jiang, Z. Ma, D. Xuan, Total recycling of concrete waste using accelerated carbonation: a review. Cem. Concr. Res. **173**, 107284 (2023)
3. T. Zhang, M. Chen, Y. Wang, M. Zhang, Roles of carbonated recycled fines and aggregates in hydration, microstructure and mechanical properties of concrete: a critical review. Cem. Concr. Compos. **138**, 104994 (2023)
4. L. Jin, Z. Wang, T. Wu, P. Liu, P. Zhou, D. Zhu, X. Wang, Mesoscale-based mechanical parameters determination and compressive properties of fully recycled coarse aggregate concrete. J. Build. Eng. **90**, 109366 (2024)

5. Y. Chen, B. Zhan, L. Hong, B. Guo, C. Wang, H. Li, Q. Yu, New insights into the effect of residual paste content on the properties of recycled fine aggregate mortar and concrete. Constr. Build. Mater. **435**, 136744 (2024)
6. B. Cantero, M. Bravo, J. de Brito, I.F. Saez del Bosque, C. Medina, Mechanical behaviour of structural concrete with ground recycled concrete cement and mixed recycled aggregate. J. Clean. Prod. **275**, 122913 (2020)
7. H. Mehdizadeh, T.C. Ling, X. Cheng, S.Y. Pan, K.H. Mo, CO_2 treatment of hydrated cement powder: characterization and application consideration. J. Mater. Civ. Eng. **33**(4), 04021041–04021051 (2021)
8. H. Mehdizadeh, T.C. Ling, X. Cheng, K.H. Mo, Effect of particle size and CO_2 treatment of waste cement powder on properties of cement paste. Can. J. Civ. Eng. **48**(5), 522–531 (2021)
9. H. Mehdizadeh, X. Cheng, K.H. Mo, T.C. Ling, Upcycling of waste hydrated cement paste containing supplementary cementitious materials via CO_2 pre-treatment. J. Build. Eng. **52**, 104396 (2022)
10. H.J. Ho, A. Iizuka, E. Shibata, H. Tomita, K. Takano, T. Endo, CO_2 utilization via direct aqueous carbonation of synthesized concrete fines under atmospheric pressure. ACS Omega **5**, 15877–15890 (2020)
11. H.J. Ho, A. Iizuka, E. Shibata, H. Tomita, K. Takano, T. Endo, Utilization of CO_2 in direct aqueous carbonation of concrete fines generated from aggregate recycling: influences of the solid-liquid ration and CO_2 concentration. J. Cleaner Prod. **312**, 127832 (2021)
12. Y. Mao, P. He, S. Drissi, J. Zhang, X. Hu, C. Shi, Effect of conditions on wet carbonation products of recycled cement paste powder. Cem. Concr. Compos. **144**, 105307 (2023)
13. M. Zajac, J. Skibsted, P. Durdzinski, F. Bullerjahn, J. Skocek, M. Ben Haha, Kinetics of enforced carbonation of cement paste. Cem. Concr. Res. **131** 106013 (2020)
14. M. Zajac, I. Maruyama, A. Iizuka, J. Skibsted, Enforced carbonation of cementitious materials. Cem. Concr. Res. **174**, 107285 (2023)
15. Y. jiang, L. Peng, Z. Ma, J.X. Lu, P. Shen, C.S. Poon, Enhancing the treatment efficiency of recycled concrete fines with aqueous carbonation. Cem. Concr. Res. **174** 107338 (2023)
16. N. Li, C. Unluer, Enhancement of the wet carbonation of artificial recycled concrete aggregates in seawater. Cem. Concr. Res. **175**, 107387 (2024)
17. A. Gholizadeh-Vayghan, A. Bellinkx, R. Snellings, B. Vandoren, M. Quaghebeur, The effects of carbonation conditions on the physical and microstructural properties of recycled concrete coarse aggregates. Constr. Build. Mater. **257**, 119486 (2020)
18. B.J. Zhan, D.X. Xuan, W. Zeng, C.S. Poon, Carbonation treatment of recycled concrete aggregate: effect on transport properties and steel corrosion of recycled aggregate concrete. Cem. Concr. Compos. **104**, 103360 (2019)
19. Y. Leng, Y. Rui, S. Zhonghe, F. Dingqiang, W. Jinnan, Y. Yonghuan, L. Qiqing, H. Xiang, Development of an environmental ultra-high performance concrete (UHPC) incorporating carbonated recycled coarse aggregate. Constr. Build. Mater. **362**, 129657 (2023)
20. J. Fu, S. Kamali-Bernard, F. Bernard, M. Cornen, Comparison of mechanical properties of C-S-H and portlandite between nano-indentation experiments and a modeling approach using various simulation techniques, compos. Part B **151**, 127–138 (2018)
21. M. Sereng, A. Djerbi, O.O. Metalssi, P. Dangla, J.-M. Torrenti, Improvement of recycled aggregates properties by means of CO_2 uptake. Appl. Sci. **11**, 6571 (2021)
22. N. Russo, F. Lollini, Effect of carbonated recycled coarse aggregates on the mechanical and durability properties of concrete. J. Build. Eng. **51**, 104290 (2022)
23. J. Wang, J. Zhang, D. Cao, H. Dang, B. Ding, Comparison of recycled aggregate treatment methods on the performance of recycled concrete. Constr. Build. Mater. **234**, 117366 (2020)
24. C. Liang, H. Ma, Y. Pan, Z. Ma, Z. Duan, Z. He, Chloride permeability and the caused steel corrosion in the concrete with carbonated recycled aggregate. Constr. Build. Mater. **218**, 506–518 (2019)
25. J. Liu, K. Ma, J. Shen, Y. Xie, G. Long, Influence of CO_2 enhancement of recycled aggregate on microstructure of ITZs in recycled concrete. J. Build. Eng. **65**, 105805 (2023)

26. B.J. Zhan, D.X. Xuan, C.S. Poon, K.L. Scrivener, Characterization of interfacial transition zone in concrete prepared with carbonated modeled recycled concrete aggregates. Cem. Concr. Res. **136**, 106175 (2020)
27. G. Pan, M. Zhan, M. Fu, Y. Wang, X. Lu, Effect of CO_2 curing on demolition recycled fine aggregates enhanced by calcium hydroxide pre-soaking. Constr. Build. Mater. **154**, 810–818 (2017)
28. S. Liu, P. Shen, D. Xuan, L. Li, A. Sojobi, B. Zhan, C.S. Poon, A comparison of liquid-solid and gas-solid accelerated carbonation for enhancement of recycled concrete aggregate. Cem. Concr. Compos. **118**, 103988 (2021)
29. J. Suazo-Hernandez, V. Letelier, A. Suazo, M. Bustamante, B. Wenzel, J.M. Ortega, Assessment of mortar with fine recycled aggregates and nanoparticles of titanium dioxide under accelerated carbonation. Constr. Build. Mater. **431**, 136555 (2023)
30. Y. Jiang, L. Li, J.X. Lu, P. Shen, T.C. Ling, C.S. Poon, Mechanism of carbonating recycled concrete fines in aqueous environment: the particle size effect. Cem. Concr. Compos. **133**, 104655 (2022)
31. C. Shi, Z. Wu, Z. Cao, T.C. Ling, J. Zheng, Performance of mortar prepared with recycled concrete aggregate enhanced by CO2 and pozzolan slurry. Cem. Concr. Compos. **86**, 130–138 (2018)
32. M. Zajac, J. Skibsted, B. Lothenbach, F. Bullerjahn, J. Skocek, M. Ben Haha, Effect of sulfate on CO_2 binding efficiency of recycled alkaline materials. Cem. Concr. Res. **157** 106804 (2022)
33. M. Zajac, J. Skibsted, P. Durdzinski, M. Ben Haha, Effects of alkalis on products of enforced carbonation of cement paste. Constr. Build. Mater. **291** 123203 (2021)
34. M. Zajac, J. Skocek, P. Durdzinski, F. Bullerjahn, J. Skibsted, M. Ben Haha, Effect of carbonated cement paste on composite hydration and performance. Cem. Concr. Res. **134** 106090 (2020)
35. B. Lu, C. Shi, J. Zhang, J. Wang, Effects of carbonated hardened cement paste powder on hydration and microstructure of Portland cement. Constr. Build. Mater. **186**, 699–708 (2018)

Chapter 33
Analysis of Energy Systems. Recovering Nature-Based Solutions for Future Sustainable Development

Anya Montserrat Pulido Cavada, Federico de Isidro Gordejuela, and Roberto Alonso González-Lezcano

Abstract Nowadays, optimising the use of energy resources to reduce energy consumption is essential to stop the cycle of environmental pollution. The Anthropocene, the current geological era that is drowning in overconsumption of resources, where energy plays a key role in achieving the energy and climate goals that many political and legislative agendas have been setting since the first oil crisis, is a double-sided coin. While energy plays a key role in the development of the world economy and the growth and development of ecosystems, it has a major impact, posing a challenge in addressing global warming and the depletion of biodiversity. This chapter aims to examine energy systems in the face of increasing global energy needs, anticipating population and economic growth, and analysing resource availability and potential bottlenecks. It also proposes the earth as a solution based on nature and environmental protection within safe planetary biophysical limits that are already being exceeded.

33.1 Introduction

Efforts to address the environmental impacts that humans have created through pollution, deforestation and other environmentally damaging actions are just beginning. The last few decades have witnessed the major repercussions of dependence on and intensive use of energy systems; the increasing acidification of the oceans caused by the growing concentration of gases such as carbon dioxide, methane and nitrous

A. M. P. Cavada (✉) · F. de Isidro Gordejuela · R. A. González-Lezcano
Department of Architecture and Design, Higher Polytechnic School, San Pablo CEU University, Urbanización Montepríncipe, Alcorcón, Madrid 28668, Spain
e-mail: anya.pulidocavada@ceu.es

F. de Isidro Gordejuela
e-mail: isidro.eps@ceu.es

R. A. González-Lezcano
e-mail: rgonzalezcano@ceu.es

D. A. H. Hanaor (ed.), *Sustainable Materials for the Built Environment*, Springer Series in Materials Science 352, https://doi.org/10.1007/978-3-031-97818-0_33

oxide in the atmosphere is a clear reflection of the major impact of human activities on natural systems on a global scale through greenhouse gas (GHG) emissions. At the same time, to account for increasing demand for food, agricultural runoff has generated large 'dead zones' in some areas of the ocean, leading to massive losses of marine biodiversity [1].

The construction sector accounts for 40% of global resource consumption and, together with the transport sector, represents the main driver of global demand for materials, followed by the food and energy sectors; together they account for 90% of global demand for natural resources, and this share is increasing steadily. The growing demand for minerals demanded by the construction sector dominates the global growth of raw material extraction, which has doubled its development in the last 30 years [2]. The feared resource scarcity that is already on the horizon seems to be less critical than the continuous impacts generated by their extraction, transformation, use and disposal of derived waste at the end of their useful life [3]. There is growing international concern about developing strategies to prevent, or at least reduce, a supply chain gap and to achieve this goal it is necessary to provide tools to monitor the use of materials, energy, waste and greenhouse gas emissions from the production-consumption chain [4]. Urgent and large-scale action must be taken to curb and reduce the increasing concentrations of gases in the atmosphere by considering the multiple pathways linked to changes in energy systems [1]:

I. More effective management of human activities that interact with natural systems on a large scale; to this end, a balance must be achieved in the consumption of energy sources, promoting clean and sustainable energy sources, as well as a reduction in the use of other materials such as black soot, a by-product of incomplete combustion of organic materials with high carbon content.
II. Reduction of energy demand as a response to high energy requirements in sectors that are in continuous progress, such as the construction and transport sectors. Any improvement in efficiency reduces the amount of energy supply needed; good insulation of infrastructures would imply a reduction in the consumption of indoor air-conditioning installations and, consequently, of water heating and/or cooling.
III. Increased effectiveness in terms of energy conservation; low cost and high availability have led to imprudent use.

According to Renewables, the share of renewable energy included in all other power generation systems in 2023 will grow by 50%, with China being the region with the highest growth rate, commissioning as much photovoltaic power in 2023 as the rest of the world in 2022. Despite the continued growth of renewables in recent years, there is an urgent need to respond to the current paradigm: an unprecedented increase in global energy consumption in response to continued population growth which, in parallel, requires new infrastructure and higher energy demand. We are facing a snake biting its own tail, a cycle that responds to the accelerated demographic transition that took place since the first Industrial Revolution and that has been made possible by the massive use of hydrocarbons.

On 25 September 2015, the UN General Assembly took place, where more than 150 world leaders adopted 17 global goals as part of a sustainable development agenda. Within these 17 goals were 169 integrated targets covering economic, social and environmental spheres, all of which were to be achieved over the next 15 years. Now, with six years to go until the Agenda for Sustainable Development is achieved, are we any closer to meeting the proposed goals?

33.2 Analysis

The last 50 years have seen significant shifts in urban patterns [5], economic output, demographic indices [6] and individual income [7]. Such trends imply changes in natural resource consumption and environmental transformations at all scales, putting pressure on ecosystem health and biodiversity.

One of the elements to consider when referring to Energy Sustainability is to reduce GHGs from continued energy use, while providing appropriate energy supplies to meet economic, food, manufacturing and mobility needs without compromising future sustainable development. Lowering the costs and increasing the reliability and development of clean energy infrastructures will require technological advances if much of future demand is to be met from renewable energy rather than relying on stored resources, the replenishment rate of which is lower than their utilisation rate. To achieve this goal, it is essential to achieve a systematic reduction in the intensive demand for material and energy resources in high-consumption contexts. This can be applied to several sectors; firstly, changes in food diets can reduce the land area required by 5% by 2060 compared to the levels required in 2020, due to a reduction in high impact products—such as animal protein—and thus a reduction in food waste. Secondly, this change can be applied to the transport sector; encouraging the use of shared and active transport can lead to a 50% reduction in the stock needs of related materials, a 50% reduction in energy demand and a 60% reduction in GHG emissions. Neighborhoods characterised by a compact and balanced urban fabric and structure characterised by a continued use of recycled materials in construction and other circular economy measures can achieve reductions in building material stocks of up to 25% by 2060, leading to a reduction in energy demand and emissions of close to 30% compared to current trends [8].

The path to achieving sustainability on a global scale is narrowing with each passing day, due to the failure to deliver on most of the political commitments that have been made since the first Earth Summit in June 1972. Ensuring continued progress requires integrating economic and environmental policies that enhance sustainable production and consumption within the main sectors—food, construction, transport and energy—and that result in a significant reduction in material and energy supplies. Three scenarios to be considered that transcend the Sustainable Development Goals and the quest for energy efficient infrastructures are analysed below, to support the introduction of a building material that was at the heart of much of the international vernacular architecture of the century prior to the Common Era: earth.

33.2.1 *Anthropogenic Mass Versus Biomass*

Due to the technological and economic transformation process brought about by the Third Industrial Revolution in 1970, the global population has progressively increased by 1.5% annually, reaching 8 billion people today, as shown in Fig. 33.1. During this period, Asia and the Pacific have been the two regions with the highest recorded population growth rates, starting from a population of about 2 billion in 1970 and reaching about 4.5 billion in 2024; this figure represents more than 50% of the world's current population. However, population growth rates have shifted from an average of 1.9% at the end of the twentieth century to a decline of an estimated 0.8% per year between 2020 and 2024.

Paul Crutzen, a Dutch chemist who won the Nobel Prize for chemistry in 1995, popularised the term Anthropocene at the beginning of the twenty-first century, referring to a new geological era that would succeed the Holocene—the period since the last ice age—and begin because of the global impact of human activities on terrestrial ecosystems. Although the term "anthropocene"—from the Greek *anthropos*, meaning human, and kainos, meaning new—has been a contested term in the scientific community, it has been associated by many with the onset of the Third Industrial Revolution. The anthropogenic mass generated since the second half of the twentieth century has been accelerated by technological progress, overpopulation and consequently increased production and consumption. The continuous interaction of

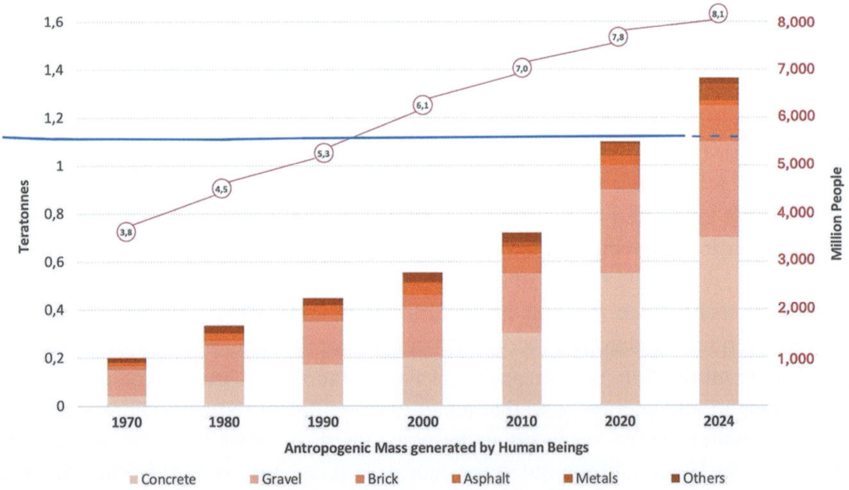

Fig. 33.1 Anthropogenic mass outweighs natural mass, exceeding 1, 1 teratonnes

these three factors has led to an unrestrained extraction and transformation of the raw materials available in nature, as well as to an uncontrolled expansion of agricultural land, urban networks and transport infrastructures and networks.

Energy Science and Technology must come together to address the challenges of ecosystem health and sustainability. It is indisputable that Energy is key to ecosystem functioning and key to the global economy; however, it is a double-sided coin. In 2018, Global Energy Needs were projected to increase by 60% in the face of projected population and economic growth by 2030 [9].

During the CAST (China Association for Science and Technology) Frontier Forum on Ecology and Energy in Beijing (China) in June 2018, the country of China as the world's largest major emerging economy was taken as an example, looking at the main challenges and opportunities it would face in the years to come. Considering that the demographic population grew by 58% in 2016, it was estimated for the coming years [10]:

- An increase of about 50% in food demand, which would lead to an increase in the continued use of nitrogen fertilisers by 200% and phosphorus by 80%.
- 19.7 billion tonnes of untreated rural wastewater per year. Among them, about 4128 tonnes of phosphorus and 82,560 tonnes of nitrogen discharged into watersheds [11]. This continued discharge results in agricultural pollution and reduced runoff, causing major pollution of watersheds in the dry season and consequently affecting Food Security.
- A production of more than 3.6 million tonnes of solid waste and around 900,000 tonnes of toxic landfill leachate per day. This would bring a great risk to more than 16% of agricultural land due to heavy metal contamination [10, 12], among which 36.43% are classified as "unsafe" [13].

The demand for energy from fossil fuels such as oil, natural gas and coal continues to increase in proportion to population growth and the consumption demand it demands. In contrast, Renewable Energies such as wind, hydro and solar photovoltaic represent a small fraction of the world's energy supply. Even if annual population growth rates can be maintained in the coming decades, it may take more than 20 years before renewable energy sources supply more than 20% of total electricity demand [1].

To achieve a reduction of energy resources, resource availability and potential bottlenecks must be analysed [10]. The geographically uneven distribution of resources is one of the main problems affecting energy production; a clear example of this is cobalt [14]. The rise of lithium-ion batteries used in electric vehicles has boosted demand for this material, with the highest concentration of deposits found in the Democratic Republic of Congo. If electric car production is to account for 20% of total vehicle production by 2025, it will require a doubling of lithium mining, which was around 140,000 tonnes in 2020 according to the US Geological Survey.

All environmental impacts are increasing; waste and emissions have increased in proportion to the increase in demand for materials. The extraction and transformation of natural resources is responsible for more than 55% of greenhouse gas emissions into the atmosphere and thus for approximately 40% of health impacts from airborne particulate pollutants—these include heavy metals, volatile chemicals, combustion by-products and solvents. If land use change is considered, environmental impacts and risks from biodiversity loss increase to 60%, with biomass contributing the most at 28%, followed by fossil fuels and metallic and non-metallic minerals.

33.2.2 Overconsumption by a Minority

Today, the competencies to deal with anthropogenic impacts and their derivatives have not yet been brought together on a global scale; we are still moving towards a climate emergency that is reaching catastrophic levels for current and future generations [15]. The worsening of this situation hinders global cooperation, leading to fragmentation in political, economic and food systems, making it possible to aggravate other storage systems (such as a nuclear catastrophe) [16].

For the most part, ecological impacts are driven by the luxury consumption of most of the northern hemisphere and certain populations in other parts of the globe, whose emissions may become more complex to decarbonise compared to the emissions emitted by those living with some level of sufficiency [17]; more importantly, despite the positive social outcomes that certain countries have achieved - considering factors such as average life expectancy and the Human Development Index—none have managed to stay below the planetary boundary [18]. Societies characterised by inequality, where the basic needs of the vast majority are insufficient, can be a cause of socio-economic instability and violent conflict, episodes that may be exacerbated by the resurgence of climate change [19, 20].

The correlation between inequality and climate collapse is clear; reducing inequality in the social, economic and political spheres is key to enabling emissions mitigation [21].

Taking as an archetype the work done by Goldemberg et al. [22] and successive estimates of the energy required to ensure an optimal standard of living on a regional and global scale, the study by Joel Millward-Hopkins [23] updates the aforementioned model by assessing an inventory of consumption, technological advances and knowledge of human needs, thereby providing a series of static estimates of global energy needs for the year 2050. According to this analysis, the near-term scenario for high-income and lower-middle income individuals is virtually identical, except for the 1% of the population represented by the billionaire elite, whose lifestyle-related ecological impact has a strong influence on future sustainable development because of their political power and investments in destructive industries [24, 25]. As a result, achieving an egalitarian quality of life, but with high material inequality close to current levels, could result in a doubling of energy consumption compared to an egalitarian world with inequalities based on needs alone. Estimating an 18% increase in global energy consumption by 2050 due to population growth, it will be necessary to consider ambitious technology deployment accompanied by lower population growth or, instead, a deployment of negative emissions technologies and practices (NETP) to complement emissions reductions (or both).

The problems caused by economic inequality and ecological collapse have had a greater impact since the Third Industrial Revolution, leading to the implementation of multiple sustainable development goals and the creation of various social movements and groups such as Fridays For Future and Occupy Wall Street. If we look at the expectations suggested by researchers for future urban development, the evolution is more than positive; however, because of this reality, social inequality is increasing, a condition that tends to accompany the collapse of civilisations. While resources are exploited to meet the needs of the average citizen, the poorer regions experience the immediate effects; meanwhile, the elite citizen relies on uninterrupted luxury consumption.

Due to the intensive exploitation of materials such as carbon, iron, steel and cement, and the growing demand and production of energy, global greenhouse gas emissions have been increasing considerably since 1970. According to an Emissions Database for Global Atmospheric Research (EDGAR) report conducted in 2023, in the last fifty years, GHG emissions have doubled from 20 billion tonnes in 1970 to about 43 billion tonnes in 2020, without considering their global warming potential. Furthermore, the production of carbon dioxide (CO2) represents the largest proportion of gases released into the atmosphere compared to other gases derived from the burning of fuels in various industries (Fig. 33.2), such as fluorinated refrigerant gases (F-gases) or oligogases, such as methane (CH_4) and nitrous oxide (N_2O). The latter two, together with CO_2, make up the main GHGs—Greenhouse Gases-, gases that concentrate in the atmosphere, trapping heat from the sun and preventing it from escaping into space, thus causing global warming.

The behaviour of CO_2 is completely different from the behaviour of oligogases; according to a study conducted by the Intergovernmental Panel on Climate Change

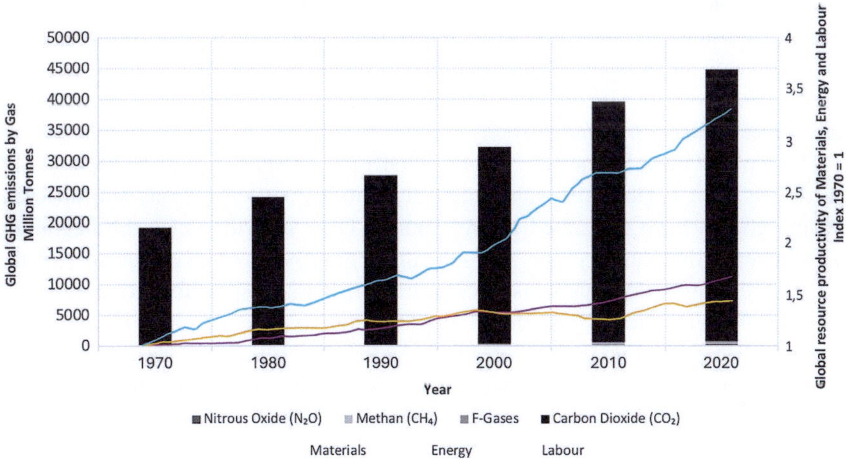

Fig. 33.2 Global GHG emissions by gas related to global resource productivity of materials, energy and labour from 1970 to 2020

(ipcc) in 2007, any moderate reduction in CO_2 emissions shows increasing concentrations due to the exchange processes characteristic of the carbon cycle in the climate system, such as the transfer of gases between the ocean and the atmosphere, and the chemical and biological processes that are integrated into it [26]. Considering the removal processes of various trace gases such as methane (CH_4), which is removed from the atmosphere by chemical processes, and nitrous oxide (N_2O), the largest percentage of which can be destroyed in the upper atmosphere by solar radiation, the atmospheric permanence of these gases can be relatively accurately determined, operating on a time scale of no more than 120 years. In contrast, defining the atmospheric lifetime of CO_2 is difficult.

From a study relating the residence time and removal processes of different gases in order to observe the evolution of concentrations from certain reductions—10, 30 and 50compared to the current level—it concludes that a small or moderate reduction of CO_2 emissions would not lead to stabilisation of its concentration, but would only reduce the rate of its growth over the next two decades, i.e. the current rate of carbon dioxide emission represents a higher rate than its rate of removal.

According to data from joint research between Oxfam and the Stockholm Environment Institute (SEI), the extreme inequality of carbon emissions in recent decades is closely linked to the average income level of the population. This study reflects how, during the years between 1990 and 2015 in which humanity doubled its concentration of dioxide in the atmosphere, an enormous population gap occurred between individuals with the most and the least income. Consequently, the poorest 50% of the population—represented by approximately 3.1 billion people—only generate around 7% of global emissions attributed to individual consumption, compared to the richest 1% of the world population—represented by 63 approximately millions of people—who generate 15% of accumulated emissions; that is, twice as much as the poorest half of the population. As Fig. 33.3 shows, the percentage of people living in high-income countries has decreased by around 7% between 2000 and 2024; This translates into a lower population growth rate in regions with higher per capita income. However, low-income regions have recorded faster population growth, with an increase rate of 2.6% annually.

The global Gross Domestic Product (GDP) per capita between 1970 and 2024 multiplied by 2.4; However, substantial disparities between regions and different income groups have generated uneven annual GDP growth. Since the beginning of the twenty-first century, the Asia–Pacific region has overtaken the rest of the world as the most dominant resource extractor, being responsible for more than half of global greenhouse gas emissions. Returning to the example of China as one of the largest emerging markets in recent years, according to a study carried out by the International Resource Panel in 2024, most of the materials used worldwide were extracted in middle-income countries-high. In 2020, 48.9 billion tons of natural resources—that is, 51% of the world total—were extracted in the Asia–Pacific region, where, in addition, countries like China with an upper-middle income have replaced high-income population as main emitters (Fig. 33.3). This reflects a change in the global energy system along with a saturation of urbanization and living standards in the group of middle-income countries. However, in per capita terms, high-income

Fig. 33.3 Total domestic material consumption (DMC) related to greenhouse gases (GHG) emissions by seven world regions and income groups in 1970, 2000 and 2020

countries continue to represent the largest group of greenhouse gas emitters, about seven times higher than low-income economies since 1970.

The population affected by greater economic inequality lives mostly in the country's most vulnerable to climate change. This extreme inequality and the social and economic poverty it generates causes a CO_2 emissions gap; In other words, the problem to be considered is not reduced to the imminent growth of the population, but to the level of income. People with daily incomes greater than $23—around 10% of the population—contribute 43% of CO2 emissions; On the contrary, those individuals with an income below $1.25 per day—around 18% of the population—produce 3% of CO_2 emissions [27, 28]. According to the data collected in the twelfth Global Wealth Report of 2021, prepared by Credit Suisse Research Institute, it is ensured that the total wealth of the richest adults in the world has multiplied almost four-fold, with an increase of 7.4% in world wealth. However, according to the NGO, the richest 5% of the world's population is responsible for a third of the total increase in emissions (Fig. 33.4) [29].

Global materials productivity, defined as the relationship between GDP and national consumption of materials, increased by 0.7% on average annually between 1970 and 2020. This improvement in productivity was favored due to various causes, such as the outsourcing of transformation processes to foreign countries and the change from the primary sector to tertiary sector activities. The material productivity gap in each region is more evident if we look at the income groups. In 1970, high-income countries had a productivity nine times higher than the productivity generated by low-income countries, with low and upper-middle income representing 20% of the average for high-income countries. By 2024, a proportion thirteen times greater than that mentioned is expected. It is a fact: the excessive consumption of a wealthy minority that increases over the years is inflaming the climate crisis and slowing down efforts to address global impacts [8].

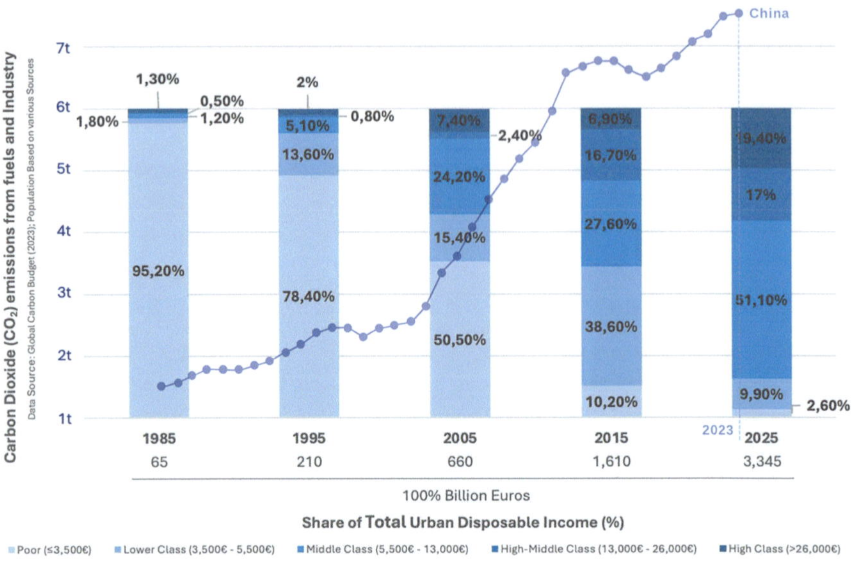

Fig. 33.4 Increase in CO_2 emissions from fuels and industry in proportion to the increase in urban disposable income in China from 1985 to 2025

The limits between stability, inequality and vulnerability to possible climate shocks have been little studied by the scientific community; No study to date has brought together the foundations that promote a healthy, resilient quality of life for humans and the planet, remaining under the ecological limits required to guarantee social security [18]. Without urgent and common action on a global scale to change the way resources are used, the extraction and transformation of material resources could increase by 60% by 2060 compared to the rates obtained in 2020; This translates into intensive exploitation of 100,000 to 160,000 million tons, exceeding the limits imposed by the Sustainable Development Goals to satisfy human needs.

The question that arises then is how to stay in the race towards "zero emissions"?

33.2.3 The City. Strategic Point of Action

Information technologies, communication and energy systems continue to increase as urbanizations expand; The urban energy network demanded by cities represented two-thirds of the world's energy in mid-2017, and this percentage is not hindered by the safe biophysical limits already exceeded in the first decade of the twenty-first century [30].

Among the main supply services that satisfy human needs, the following stand out:

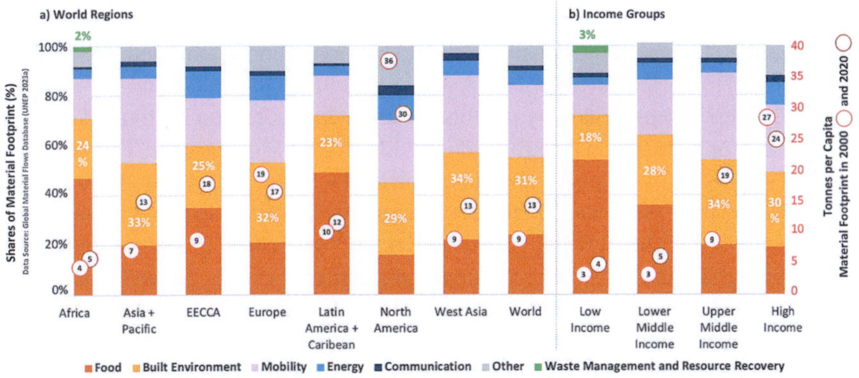

Fig. 33.5 Shares of material footprint related to material footprint by seven world regions and income groups in 2000 and 2020

1. Food
2. Built environment
3. Mobility
4. Energy
5. Communication
6. Waste Management and Resource Recovery

To provide the service to the population, all of them require an infrastructure and a series of specific materials to carry out their construction, maintenance and operation of said infrastructure. As can be seen in Fig. 33.5, during 2020, 59,000 million tons of materials were required to meet the demand of the construction, transportation and infrastructure sectors; This is equivalent to 55% of all material demand.

Considering the Material Footprint (MF) index—an indicator based on demand, reflecting the material consumption and investment needs of a country's capital regardless of its origin—the Asia and Pacific regions recorded an increase of 27% in global material footprint between 2000 and 2020, reaching more than half of the registered global footprint with a value of 56%. Although North America, along with Europe, minimized this index by 17% during the same period, the per capita footprint was almost triple the world average. In parallel, looking at groups by income index, the figures between the high-income and upper-middle-income classes in 2000 represented 51% and 34% respectively. In 2020 the figures were practically reversed (Fig. 33.5).

Based on Fig. 33.5, the construction sector, together with the transportation sector, is one of the sectors with the greatest potential to improve energy performance and achieve the energy and climate objectives that have guided numerous political and legislative agendas since the first crisis of the oil:

i. Sustainable Cities and Communities
ii. Production and Responsible Consumption
iii. Affordable and non-polluting energy
iv. Health & Wellness

Upper-middle-income countries are responsible for the largest proportion of materials demand directed at the construction and transportation sector, with 68% of the total material footprint. On the contrary, in groups with a low-middle and low income, this percentage is represented by the food sector, which represents a total of 54% of the demand for materials in the group of individuals with a low income. This perspective is reiterated at the regional level in countries such as Africa, Latin America, the Caucasus and the Caribbean, where material provision systems respond to social and cultural practices. The massive growth in demand for natural resources, the paralysis of the productivity of the use of materials and the benefits derived from them, are distributed unevenly among nations and populations. This unsustainable scale of prolonged and comprehensive use of materials makes it impossible at different scales to achieve benefits in the social, economic and environmental spheres. As the middle class expands and the industrialization process advances in response to this continuous growth, the corresponding increase in materials is evident, which brings with it an increase in the use of land and water, waste generation and production greenhouse gases; this situation intensifies the pressure on the natural ecosystem when it comes to regenerating resources and managing waste. This opens the doors to the search for more eco-efficient production systems and the implementation of greater green infrastructure [31].

The current problem is no longer focused on questioning the need for a transformation towards sustainable use of natural resources on a global scale; The concern lies in how to materialize this objective by ensuring decent standards of living at all scales within safe planetary limits [30].

Each challenge manifests itself at a local, regional and global scale, and must complement and support national policies to be able to integrate, first of all, sustainable goods and services, regulating the marketing practices that cause uncontrolled consumption; secondly, define means for a sustainable implementation of resources that complement global agendas to achieve the United Nations SDGs; and thirdly, promote the circular economy, applying production and consumption models that contribute to prolonging the useful life of materials and achieving more efficient production of goods and services through repair, reuse, renewal and recycling.

The transition towards a sustainable and efficient use of natural resources that does not lead to inequality and inefficiency requires the implementation of low-impact, high-performance supply systems. The problems are well known, but new approaches are required, and for this it is necessary to involve the public in the construction of new alternative models that ensure a future ecological civilization, raising awareness about climate change and its negative effects: scarcity of resources, pollution environmental degradation, degradation of ecosystem services and functions linked to the loss of habitats and species, loss of soil and agricultural production, and the scarcity of resources that hinders the supply of water, soil, energy and food, in a sustainable manner [10].

Between energy supply and demand trends, cities demand two-thirds of the world's energy. In 2017, nuclear and coal energy production in Germany fell by 51% (from 150 TWh to 73.6 TWh) and 22.5% (310 TWH to 240.1 TWh) compared to production in 1990, leading to an increase in renewables by 95% (from 10 TWh to 218.3 TWh). The great impact that future urbanization will have on energy consumption is indisputable, potentially causing an excess in plans for the introduction of renewable energies [32]; However, the high rate of consumption demanded by cities makes them become the point of greatest strategic interest in which to implement new alternatives that bring us closer to meeting the SDGs.

It is essential to curb fossil fuel production and investments in energy-intensive infrastructure if renewable solutions are to be promoted and future carbon lock-ins avoided. To do this, it must be considered that the construction sector represents 30% of national energy consumption and that, between the production, execution and operation of buildings and most of the materials used in architecture. In contemporary times, the sector is one of the largest emitters of GHG, as can be seen in Fig. 33.5. The large carbon footprint left by the modern city, added to other more indirect consequences such as inequality, crime and the heat island effect, raise alarm bells towards new future urban models. According to Marta Olazabal and María Ruíz de Gopegui, researchers at the Basque Center for Climate Change (BC3) in the magazine Landscape and Urban Planning, the current cities' plans for adaptation to climate change in most cities are large cities in the world will not achieve effective results in the medium and long term.

We are faced with a double situation: the need to confront the increase in population that brings with it an increase in socioeconomic and political inequality, faced with the obligation to reduce global warming and the decrease in biodiversity derived from human activities, which they respond to the first need. Since the end of the twentieth century, there has been a call for the use of new technologies and alternative practices that respond to the current need to build with decarbonized materials that reduce the embodied carbon of buildings. Increasing investment in solutions based on nature and environmental protection favors:

- The adaptation of urgent measures to combat climate change (SDG 13), reducing extreme weather events, protecting and restoring eco-systematic services, improving the management and treatment of solid waste that degrades the soil, etc.
- Sustainable management of forests and the fight against desertification, halting and reversing land degradation and loss of biodiversity (SDG 15)

In retrospect, vernacular architecture has left great clues about how the design of the future building stock should be approached. This period, characterized by a close relationship between man and the environment, is based on an architectural design that responds to the cultural, technological, environmental and economic dimension; These four dimensions lay the foundations for a model that is respectful of the envi ronment, reflecting the identity of a territory and a culture that collects and transforms the resources that the environment offers. However, this type of traditional infrastructure closely related to human ecology and due to its low environmental impact, has not been compatible with the ideology of progress that prevailed throughout the twentieth century, which is why its use has been declining, being unique and exclusively accessible to the very rich or the very poor. Such is the case of earth, a material that has been used worldwide throughout history due to its great accessibility and versatility, its minimal cost and the history of tradition it has [33]; It currently constitutes around 30% of the world's cultural heritage.

Currently, there are numerous research groups and architectural studies that are dedicated to the research, application and dissemination of knowledge related to this type of architecture and its development along with other local materials. Both developing countries, where economical, quick and easy-to-build housing models are required, and more developed countries—such as the United States, Australia, France and Germany—have focused their research on experimental projects for improvements in composition, additives, compositional and aesthetic concepts [34]; However, there is a continuous struggle regarding the introduction of earth as a construction material in large cities, dominated by materials such as concrete and steel, which are a reflection of modernity and the industrial progress that has been developing since the Third Industrial Revolution. Referring to the words of American poet Robert Frost, "Humanity has reached a fork in the road". It is clear that now we have to choose the road less traveled, because that will make the difference. In contrast, many ask: Why and how does the earth naturally incorporate high-tech properties for sustainable construction?

33.3 Results

Adopting a supply systems perspective can help rethink the way human needs are met, designing new urban models that meet sustainability objectives. There are more and more platforms that integrate and share improvements in resource efficiency, development of circular economy and how to make use of existing ones. Such coalitions help provide resources and knowledge for the design of capacity development plans through demonstration facilities; An example of this is the Global Alliance on Circular Economy and Resource Efficiency (GACERE), the African Circular Economy Alliance, the Latin American Circular Economy Alliance or the Platform to Accelerate the Circular Economy (PACE).

The Intergovernmental Scientific Policy Platform on Biological Diversity and Ecosystem Services (IPBES) is currently underway with the aim of, in cohesion with governments, academia and civil society, establishing a series of foundations that reinforce policy formulation to maximize the benefits of biodiversity and ecosystems for quality of life and health in an equitable manner. One of the main problems when it comes to 'recycling' a material after the end of its useful life is found in the chemical modification to which it has been subjected in its extraction and transformation process; Most conventional construction materials have been extracted from bedrock and geological deposits and, during their manufacturing process, have been chemically altered, hindering their ability to be recycled at low energy costs and making it impossible to return to the natural environment. The earth, on the other hand, adjusts to the principles of the circular economy, benefiting from its:

a. High Thermal Mass
b. Noise Control
c. Humidity control
d. Pest Control
e. Low Carbon footprint
f. Good structural support
g. Ease of Transportation and Automation of on-site manufacturing

The technical barriers that arise from the use of earth as a construction material have been arousing great interest on the part of the scientific community since 1990 and publications have multiplied since then. Other types of impediment categories, such as economic, organizational, political and social, have not yet been addressed in the scientific literature with such exhaustiveness; This may be one of the reasons why earth as a construction material remains a niche market despite benefiting from most of the attributes to combat the current and future climate and resource crises. Other impediments such as the inaccuracy of its composition in terms of adequate material content, the availability of land suitable for construction with earth and the impact on the availability of land that its development could cause are aspects to consider. However, it is clear the growing interest that has emerged in recent decades in relation to a search for energy balance in accordance with the trends of resource use in the built environment of the present and the future.

33.4 Conclusions

The construction industry is responsible for high energy consumption, especially associated with the use of heating and cooling in buildings in which interior thermal comfort is sought, which has been altered in recent decades due to extreme changes of temperature resulting from climate change. For years, power generation has benefited from non-renewable energy sources derived from fossil fuels (coal, oil and gas), which has caused a call for attention towards the development of new clean and renewable energy models and the implementation of new materials that minimize energy consumption in buildings thanks to their thermal storage capacity. During the last two decades, the various earth construction techniques have garnered great interest in research projects and scientific groups; All of this has optimized the modernization of the traditional techniques worked by the great masters who shaped the land, authors of constructions still present such as the ancient walled city of Shibam in Yemen or the citadel of Arg-é Bam, known for being the largest building in molded clay of the world. However, its introduction in large cities involves breaking down certain social, economic and political barriers, in addition to improving its industrialization and mechanization to optimize processes. Promoting the use of biodegradable materials such as soil will not only provide a social benefit but will also generate an economic benefit associated with the reduction of waste generated and operating costs, a decrease in energy demand and operating costs and therefore to reducing environmental impact. All this translates into sustainable construction.

33.5 Conflicts of Interest

The authors declare no conflict of interest.

References

1. S.M. Benson, F.M. Orr, Sustainability and energy conversions. MRS Bull. **33**, 297–302 (2008)
2. H. Sameer, S. Bringezu, Life cycle input indicators of material resource use for enhancing sustainability assessment schemes of buildings. J. Build. Eng. **21** (2019)
3. S. Bringezu, Possible target corridor for sustainable use of global material resources. Resources **4**, 25–54 (2015)
4. S.M. Benson, F.M. Orr, Sustainability and energy conversions, MRS bulletin, **33**(4), 297–302
5. L. Jiang, B.C. O'Neill, Global urbanization projections for the shared socioeconomic pathways. Glob. Environ. Chang. **42**, 193–199 (2017)
6. M. Lenzen, A. Geschke, J. West, J. Fry, A. Malik, S. Giljum et al., Implementing the material footprint to measure progress towards sustainable development goals 8 and 12. Nat Sustain. **5**, 157–166 (2022)
7. R. Dellink, J. Chateau, E. Lanzi, B. Magné, Long-term economic growth projections in the shared socioeconomic pathways. Glob. Environ. Chang. **42**, 200–214 (2017)
8. UNEP, International Resource Panel (IRP), Global Resources Outlook 2024
9. R. Warren, J. Price, E. Graham, N. Forstenhaeusler, J. Vanderwal, Report title: the projected effect on insects, vertebrates and plants of limiting global warming to 1.5°C rather than 2°C. Science **360**(6390), 791–795 (2018)
10. S.K. Chou, R. Costanza, P. Earis, K. Hubacek, B.L. Li, Y. Lu et al (2018) Priority areas at the frontiers of ecology and energy. Ecosyst. Health Sustain. 243–6
11. M. Zhou, H. Wang, J. Zhu, W. Chen, L. Wang, S. Liu et al., Cause-specific mortality for 240 causes in China during 1990–2013: a systematic subnational analysis for the global burden of disease study 2013. Lancet **387**, 251–272 (2016)
12. G. Zhang, C. Azorin-Molina, X. Wang, T.R. McVicar, J.A. Guijarro, A. Chappell et al., Rapid urbanization induced daily maximum wind speed 1 decline in metropolitan areas: a case study in the Yangtze. Urban Clim. **43**, 101147
13. E.C. Stephens, A.D. Jones, D. Parsons, Agricultural systems research and global food security in the 21st century: an overview and roadmap for future opportunities. Agric. Syst. **163**, 1–6 (2018)
14. E.A. Olivetti, G. Ceder, G.G. Gaustad, X. Fu, *Lithium-Ion Battery Supply Chain Considerations: Analysis of Potential Bottlenecks in Critical Metals* (Joule. Cell Press, 2017), pp. 229–43
15. C.J.A. Bradshaw, P.R. Ehrlich, A. Beattie, G. Ceballos, E. Crist, J. Diamond et al., Response: commentary: underestimating the challenges of avoiding a ghastly future. Front. Conserv. Sci. (2021)
16. C. Parthemore, F. Femia, C. Werrell, The global responsibility to prepare for intersecting climate and nuclear risks. Bull. Atm. Sci. **74**, 374–378 (2018)
17. L. Chancel, Global carbon inequality over 1990–2019. Nat Sustain. **5**, 931–938 (2022)
18. W.F. Lamb, J.K. Steinberger, Human well-being and climate change mitigation. Wiley Interdiscip. Rev. Clim. Change. (2017)

19. S.J. Beard, L. Holt, A. Tzachor, L. Kemp, S. Avin, P. Torres et al., Assessing climate change's contribution to global catastrophic risk. Futures **127**, 102673 (2021)
20. K.J. Mach, C.M. Kraan, W.N. Adger, H. Buhaug, M. Burke, J.D. Fearon et al., Climate as a risk factor for armed conflict. Nature **571**, 193–197 (2019)
21. N.D. Rao, J. Min, Less global inequality can improve climate outcomes. Wiley Interdiscip. Rev. Clim. Change. **9** (2018)
22. J. Goldemberg, T.B. Johansson, Basic needs and much more with one kilowatt per capita [Internet]. Available from: http://www.jstor.org/stable/4313148
23. J. Millward-Hopkins, Inequality can double the energy required to secure universal decent living. Nat Commun. **13** (2022)
24. M.G. Ceddia, The super-rich and cropland expansion via direct investments in agriculture. Nat Sustain. **3**, 312–318 (2020)
25. T. Wiedmann, M. Lenzen, L.T. Keyßer, J.K. Steinberger, Scientists' warning on affluence. Nat Commun. **11** (2020)
26. R.K. Pachauri, A. Reisinger, L. Bernstein, Intergovernmental panel on climate change. Core writing team, intergovernmental panel on climate change. Clim Chang (2007)
27. A.K. Jorgenson, S. Fiske, K. Hubacek, J. Li, T. McGovern, T. Rick et al., Social science perspectives on drivers of and responses to global climate change. Wiley Interdiscip Rev Clim Change. (2019)
28. K. Song, G. Baiocchi, K. Feng, K. Hubacek, L. Sun, Unequal household carbon footprints in the peak-and-decline pattern of U.S. greenhouse gas emissions. J Clean Prod. **368** (2022)
29. J. Lin, V. Treichel, W. Bank, Learning from China's rise to escape the middle-income trap : a new structural economics approach to Latin America [Internet]. (2012). Available from: http://econ.worldbank.org
30. J. Rockström, W. Steffen, K. Noone, A. Persson, F.S. Chapin III., E.F. Lambin et al., A safe operating space for humanity. Nature **461**, 472–475 (2009)
31. J.H. Spangenberg, S. Lorek, Sufficiency and consumer behaviour: From theory to policy. Energy Policy **129**, 1070–1079 (2019)
32. M. Wang, GREET Bioenergy Life Cycle Analysis and Key Issues for Woody Feedstocks (2014)
33. Miguel Tutor Vicente Director A. La recuperación de la tierra pisada en la arquitectura contemporánea (2015)
34. M. Goy, Construir con tierra: Todo lo que hay que saber de un material que se redescubre [Internet]. 2022 July. Available from: https://www.ellitoral.com/area-metropolitana/construccion-tierra-barro-tecnologia-btc-ingeniero-tecnologica-utn_0_lkGlCoGayF.html

Chapter 34
Current Perspectives, Challenges and Opportunities for Ash Utilisation and Pond Reclamation in Australia

Tushar Gupta, Alec Miller, Tim T. Werner, Xiaoling Wu, and Mohan Yellishetty

Abstract Coal ash mitigation and utilisation has been a major challenge in many countries with a greater dependence on coal power. The unutilised ash gets dumped in open ponds and dykes, wasting large amount of land and adversely impacting the local population, flora and fauna. This study presents the current scenario of ash dumping in Australian ash ponds and their impact on population and biodiversity. The locations and areas of ash ponds were mapped for all ash ponds in Australia using satellite image analysis. Statewide data for pond sizes were compared with their population, province area, and power plant capacities. Potential risks to local population and biodiversity arising from ash ponds were also indexed and mapped. Possible beneficial applications of fly ash are further discussed to reduce the adverse impact of ash dumping.

T. Gupta
Department of Mining Engineering, National Institute of Technology Rourkela, Rourkela, Odisha, India
e-mail: guptat@nitrkl.ac.in

A. Miller · X. Wu · M. Yellishetty (✉)
Department of Civil and Environmental Engineering, Monash University, Clayton, Australia
e-mail: mohan.yellishetty@monash.edu

A. Miller
e-mail: alec.j.miller@outlook.com

X. Wu
e-mail: Xiaoling.Wu@monash.edu

T. T. Werner
Department of Geography, Earth and Atmospheric Sciences, University of Melbourne, Carlton, VIC, Australia
e-mail: tim.werner@unimelb.edu.au

D. A. H. Hanaor (ed.), *Sustainable Materials for the Built Environment*, Springer Series in Materials Science 352, https://doi.org/10.1007/978-3-031-97818-0_34

34.1 Introduction

Coal has long served as a major global energy source. At present, ~ 27% of the world's total energy demand is met by coal [1]. In 2022 the world produced 8,318 million tonnes (Mt) of coal [2], out of which the major contributors were China (4237 Mt), India (893 Mt), the United States (535), Indonesia (622 Mt) and Australia (446 Mt) (IEA, 2022). While the proportional contribution of coal towards global energy needs may decrease due to the development of alternate energy sources, the total amount of coal produced and consumed globally is still growing, or is at least stable. Consequently, issues pertaining to coal waste management are also likely to persist. In this study, we explore the current scenario and options for coal fly ash utilisation in Australia. Fly ash has been described as one of Australia's biggest waste challenges [3], yet rarely does it enter the public consciousness.

Mitigating the large volumes of fly ash produced during the coal burning process is a major challenge. Annual amounts of fly ash generated worldwide is in excess of 900Mt per year [4]. This figure changes from country to country based on the ash content of the local coal. Australian coal has a relatively low ash content (10–15% [5]), resulting in the production of 11.4 Mt of fly ash in year 2021 (ADAA, 2022). In comparison, other countries like India produce coal with a higher ash content (30–50% [6]). However, in comparison with the major coal-producing nations, Australia produces the largest amount of ash per capita. In 2021, Australia produced approximately 460 kg of ash for every person, greater than much more populous countries such as China which produces 416 kg per person. The latest available fly ash production statistics for major coal-consuming countries are shown in Table 34.1.

There is a large variation in the type of coal burnt in the power plants within Australia. For instance, the lignite (brown coal) burnt in the Latrobe valley in Victoria exhibit different properties than the traditional black coal burnt in power plants located in other provinces such as New South Wales. The nature and amounts of fly ash generated also differ in these cases. Therefore, it is important to understand the physical and chemical properties of the generated ash.

Table 34.1 Ash production and utilisation of major coal producing countries

Country	Ash produced (Mt)	Percentage utilised (%)	Year	Ash produced per person (tonne)	References
Australia	11.4	46	2021	0.46	[5]
India	270.82	96	2021	0.13	[6]
China	550	70	2018	0.41	[7]
Russia	26.6	18.79	2010	0.19	[8]
USA	28	67.08	2021	0.08	[9]

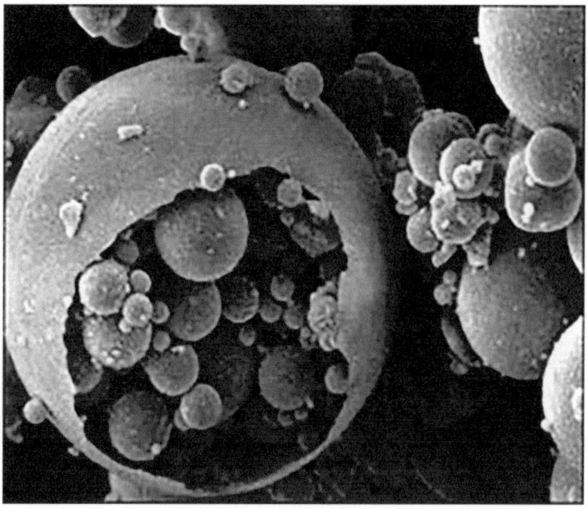

Fig. 34.1 SEM image of fly ash particles [10]

34.2 Physical and Chemical Properties of Fly Ash

Fly ash often appears in the form of a fine grey powder, but its colour may change based on its contents, specifically the amount of iron and free carbon. The chemical composition of fly ash is commonly based on the source and characteristics of the coal that is burnt in a power plant. Various chemical characterisation studies have shown that a major portion of fly ash is composed of Silica (SiO_2), Alumina (Al_2O_3), Ferric oxides (Fe_2O_3), and Lime (CaO). Some other mineral oxides such as the oxides of Magnesium (MgO), Sodium (Na_2O), Potassium (K_2O), Manganese (MnO), and Titanium (TiO_2) can be present based on the coal burnt. Fly ash mostly contains spherical glass-like particles of silica that are formed under a process of glassification which occurs during the combustion of coal in thermal power plants (Fig. 34.1). The size of these spherical particles can range from < 1 μm (μm) to > 100 μm, however in general, over 90% of the particles are < 75 μm. Chemical compositions of fly ash in some of countries are presented in Table 34.2.

34.3 Pozzolanic Nature of Fly Ash

One of the unique properties that makes fly ash a highly desirable coal combustion by-product is its pozzolanic nature. While mostly non-reactive in its original state, it can attain useful cementing abilities when lime is introduced from an external source. This is due to the presence of silica and alumina in fly ash which react with the added lime to form cement-like end products, thus giving strength to the material. The pozzolanic nature of fly ash (especially of low lime ASTM Class F fly ash) makes it a viable partial substitute for cement in the civil and construction industry.

Table 34.2 Chemical characteristics of fly ash from different countries

Fly ash source	Chemical composition (%)										Reference study
	SiO_2	Al_2O_3	CaO	Fe_2O_3	Na_2O	MgO	K_2O	SO_3	TiO_2	LOI	
Avon Lake, Ohio	55.07	28.61	1.97	6.22	0.38	1.08	0.00	0.19	–	1.82	[11]
Bandel, India	53.10	24.40	1.80	6.50	0.35	1.30	2.21	0.20	1.03	8.20	[12]
Kemerkoy, Turkey	26.40	10.00	33.80	3.40	0.15	1.72	1.00	–	0.43	–	[13]
Mea Moh, Thailand	38.70	20.80	16.60	15.30	1.30	1.30	2.10	2.60	0.25	0.80	[14]
Hebei, China	47.42	30.90	1.88	5.38	0.59	1.93	1.36	0.78	1.50	7.80	[15]
Svilajnac, Serbia	51.74	21.36	11.34	8.80	0.64	1.56	1.35	1.27	–	1.60	[16]
Class F, Spain	53.09	24.80	2.44	8.01	0.73	1.94	3.78	0.23	–	3.59	[17]
Nova Scotia, Canada	47.10	23.00	1.21	20.40	0.54	1.17	3.16	0.67	–	2.88	[18]
Tarong, Australia	65.90	28.89	0.06	0.38	0.05	0.15	0.26	0.03	1.97	1.24	[19]
Collie Australia	51.50	23.63	1.74	15.30	0.38	1.20	0.84	0.28	1.32	1.78	[20]

Several research studies and practical applications have proven the effectiveness of fly ash in construction industries. Table 34.3 provides a summary of the effects of fly ash on concrete when partially substituted in place of regular Ordinary Portland Cement.

34.4 Fly Ash Utilisation in Australia

Despite its abundance and a range of viable industrial uses, the uptake and utilisation of fly ash is poor. Though its utilisation has increased over the years, roughly 50% has been reported to remain as waste globally [4], with utilization rates ranging between 3 and 57% between countries. Australia has reported a fly ash utilisation rate of ~ 46% [5, 6].

A large portion of the fly ash utilised in Australia is used by cement and concrete industries as a partial substitute for cement. In 2021, almost 40% of beneficially utilised fly ash (fly utilising its pozzolanic nature) was used in cement and concrete

Table 34.3 Summary of some of the effects of fly ash addition on concrete

Aspect /property	Impact/change by fly ash addition	Reference(s)
Setting time	Initial and final setting time increases	[21–23]
Workability	Improves on the addition of fly ash, but up to a certain percentage addition	[24, 25]
Water requirement	Reduces due to fly ash addition	[26–28]
Bleeding	Most fly ash reduces bleeding in fresh concrete as compared to no addition	[29, 30]
Strength development	The rate of strength gain becomes slow. However, the later-stage strengths are similar to, or more than no fly ash concretes	[31–41]
Elasticity	Similar to impact on strength but to a lesser extent	[41–44]
Creep behaviour	Remains almost the same or reduces for moderate replacement percentages (0–25%). Higher percentages result in more creep	[32, 44–47]
Shrinkage	No significant change in shrinkage during moist curing, however, increases in case of dry curing	[44, 48–51]
Abrasion and erosion resistance	Similar strength properties. Observed to improve due to fly ash addition	[38, 41, 52, 53]

industries, while 58% was used in backfilling and stabilisation operations (Fig. 34.2). This contrasts with near 85% used for cement and concrete as recently as 2016.

34.5 Unutilised Fly Ash

Although efforts are being made to better utilise fly ash, a major portion of generated fly ash is still not utilised but is stored in form of ash ponds and ash dykes. ~ 53% of Australian ash was dumped in ash ponds and ash dykes in 2021, a proportion that has remained relatively constant in recent years.

Dumping fly ash in open fields requires substantial land use. In addition, fly ash dumping is a major source of air, water and land pollution in the locality which can be detrimental to health of local human population and biodiversity including the flora and fauna.

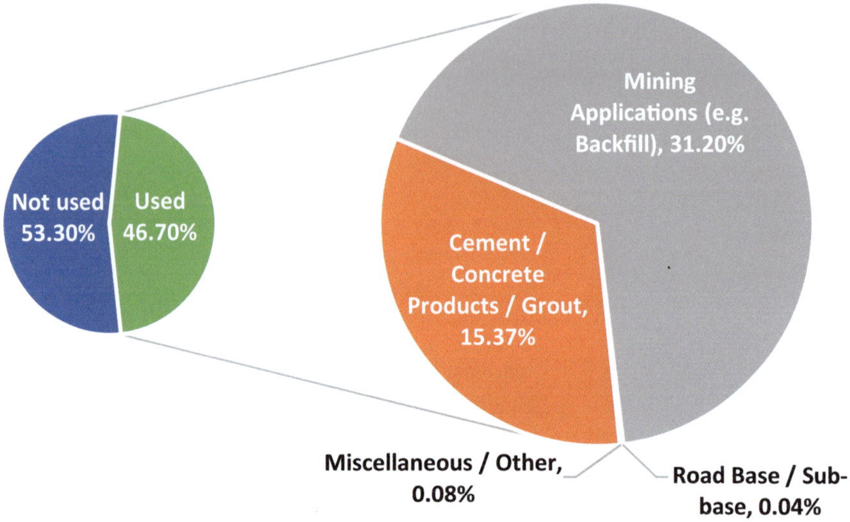

Fig. 34.2 Fly ash utilisation scenario in Australia (2021, data sourced from ADAA, 2022)

34.5.1 Land Wastage Due to Fly Ash Dumping

A satellite image analysis of major fly ash storage areas in Australia was carried out in 2022 to get a perspective of the extent of land wastage (Fig. 34.3). A list of major coal-fired power plants in Australia was obtained from web resources [54]. These details were cross-referenced and geo-tagged on the map and the associated ash ponds/dykes were located for each power plant (Fig. 34.4).

The area of these ash ponds was calculated manually using measurement tools available in Google Maps. Since the ash ponds are typically made in the vicinity of the coal power plants to avoid transportation costs and logistics, the area measurements were based on manual searching and demarcation of the ash pond boundaries. These demarcations were based on observation of some of the pond design characteristics such as stepped dykes, lagoons, water escape structures and ash embankments [55]. Additionally, the reclaimed and revegetated ash ponds are not considered in this analysis. A summary of all the power plants and the related ash pond area is provided in Table 34.4.

The analysis revealed that the amount of generated ash and corresponding area of ash ponds varies based on the type of coal being used. Sites using lignite or brown coal in power plants, such as Victoria, generate lower amounts of fly ash, which is reflected in the area of ash ponds. The area increases in other locations using black coal, with averages ranging from 0.32 km^2 in Victoria to 1.72 km^2 in New South Wales (Fig. 34.5). Due to remediation South Australia's ash pond footprint is now indicated as zero.

Fig. 34.3 Satellite image of former ash pond near North coal power plant, South Australia. This ash pond area is now under remediation

Excluding South Australia, Victoria has the lowest ash dumped area in relation to the population size, wasting 0.19 km^2 of land per million people. In comparison, the highest amount of land wasted area for fly ash dumping is in Queensland, which uses almost 1.49 km^2 of land area per million people (Fig. 34.6).

The dumped area per unit capacity of the coal-fired power plants is most significant in Queensland with almost 1.63 km^2 of ash dumped area per GW capacity at their coal power plants (Fig. 34.7).

New South Wales wastes the largest amount of area compared to the total area of the state, with more than 11 m^2 of land wasted per km^2 of total area. Since recent rehabilitation works, South Australia now wastes the least amount of land with 0 m^2 of ash pond area wasted per km^2 (Fig. 34.8).

34.5.2 Environmental Impact of Open Dumping of Fly Ash

Dumping of fly ash in these ponds and dykes not only wastes precious land resources, but also has serious effects on the local environment (Fig. 34.9). Dumping fly ash

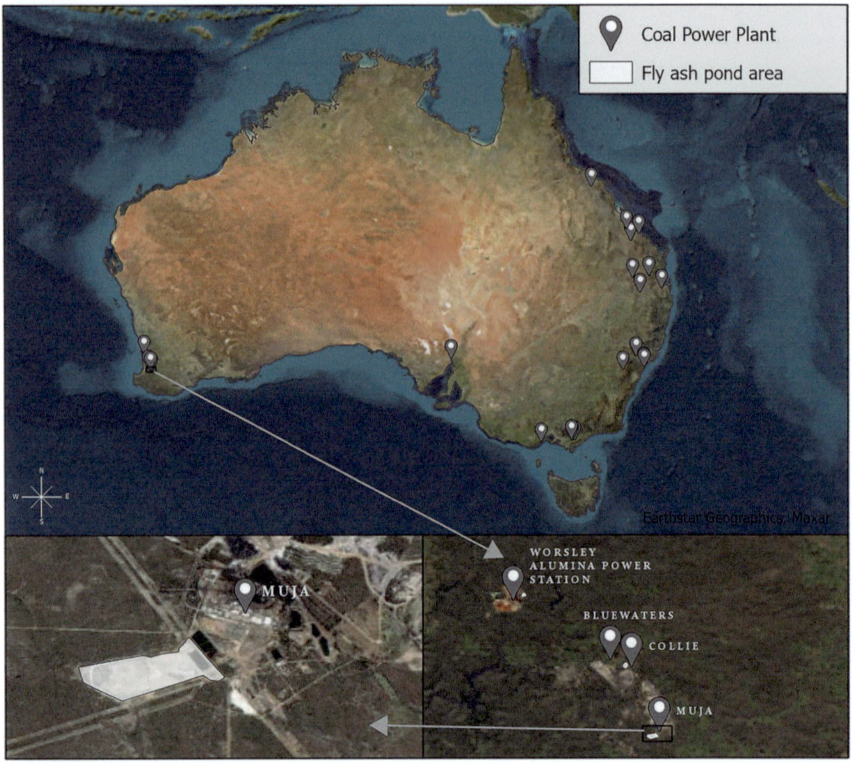

Fig. 34.4 Geo-locations of all coal power plants in Australia with ash pond area example

into ash ponds often requires slurry formation and pumping of the ash slurry to the ash pond. This water slowly seeps out of the pond but is contaminated with heavy metals and other toxic elements present in the ash itself.

This contaminated water is highly detrimental to local water bodies and the underground water table, which can make the local water unsuitable for drinking as well as being toxic to flora/fauna. This effect has been seen in many studies on local water quality near ash ponds. Chakrabarti 2005, and Mudd 1998 have performed detailed studies for the coal ash leaching in Latrobe valley ash ponds, Victoria, and demonstrated the presence of higher amounts of heavy metals during leaching [56, 57].

The water near Parichha Thermal Power Plant Jhansi, India, had excessive concentrations of Total Dissolved Solids (TDS) and high turbidity due to the leaching of ash-contaminated water [58]. In Nagpur, India, concentrations of elements like Lead, Copper, Zinc, Manganese, and Iron, increased near fly ash dumps, polluting local river bodies, and making the water unsuitable for drinking. This also resulted in bioaccumulation of chemicals in the local fish population [59]. Similar effects were seen in well water near Kolaghat thermal power station, West Bengal, India where

Table 34.4 Coal Ash pond for all coal power plants in Australia

Province	Power-Plant Name	Capacity (MWe)	Pond area (km^2)
Victoria	Loy Yang	3165	0.59
	Yallourn	1480	0.36
	Hazelwood	1470	0.22
	Morwell	170	0.36
	Anglesea	150	0.05
New South Wales	Eraring	2840	1.66
	Bayswater	2640	1.54
	Liddell	2000	2.74
	Mt Piper	1400	0.54
	Wallerawang	1000	0.53
	Vales Point	1320	3.36
Queensland	Gladstone	1680	2.85
	Callide	1600	1.72
	Stanwell	1460	0.4
	Tarong	1400	2.97
	Millmerran	880	0
	Kogan Creek	750	0.28
	Swanbank	480	0.49
	Tarong North	443	0
	Collinsville	190	0
South Australia	Northern	544	0
	Playford B	240	0
Western Australia	Muja	854	0.58
	Kwinana	420	0
	Bluewaters	416	0
	Collie	330	0.24
	Worsley Alumina Power Station	244	0

higher amounts of elements like Pb, As, Mn, Mo, Ba etc. were observed [60]. Hansen [61] demonstrated in a generic analysis of fly ash dumping that the land foot print of salinity (measure of dissolved ions) due to ash dumping was almost double the expected value.

Apart from the leached water polluting groundwater and the water-table, a major form of contamination produced by the fly ash is the particulate pollution. Dry ash dumping into the ponds is also a big source of local dust pollution because it can easily become airborne, and its fine particle size makes fly ash a hazardous air pollutant. In fact, prolonged inhalation of fly ash in the suspended fly ash prone areas can lead to diseases such as silicosis, and can aggravate the conditions of bronchitis,

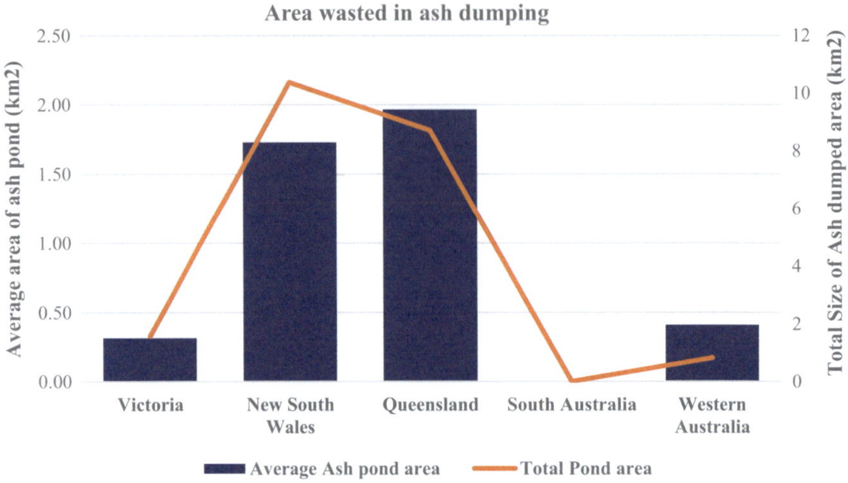

Fig. 34.5 Statewide average ash pond size and total ash dumped area

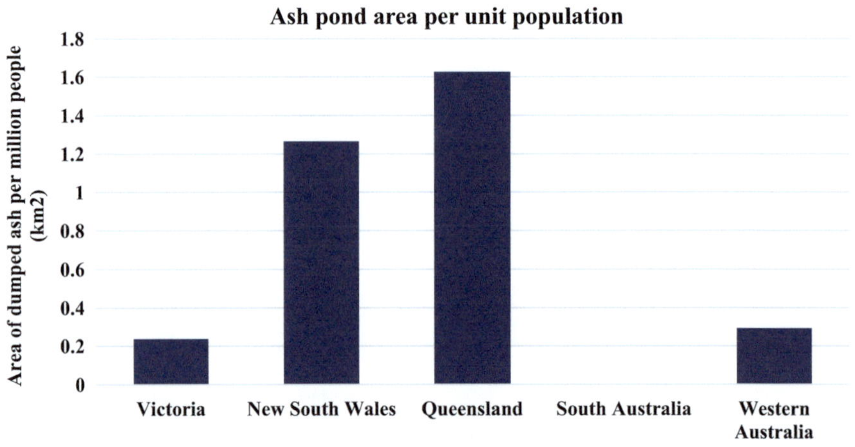

Fig. 34.6 Statewide ash dumped area per million people

asthma, and even lung cancer [62]. Pandey [63] conducted an extensive study on the nearby flora of the thermal power plant in Mirzapur, India, and found significant particulate settlement on the leaves of trees, thus reducing the health of trees, plants, and shrubs in the local region. Although fly ash is seen to improve the pH of soil, there are also concerns for the detrimental effects of fly ash on soils such as excessive increase of salinity in soil and high amounts of phytotoxic elements such as boron [64]. Studies have also shown that vegetation near fly ash dumping sites can develop bioaccumulation of heavy metals, which can then enter animal and human food

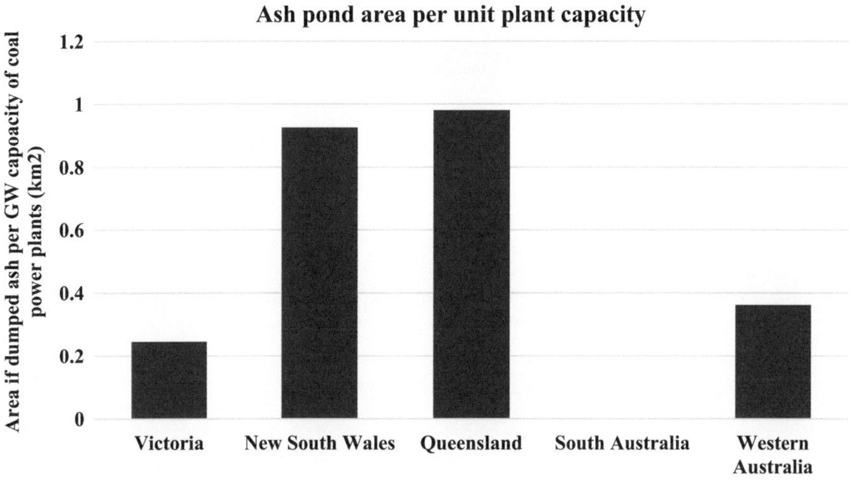

Fig. 34.7 Statewide ash dumped area per unit power plant capacity

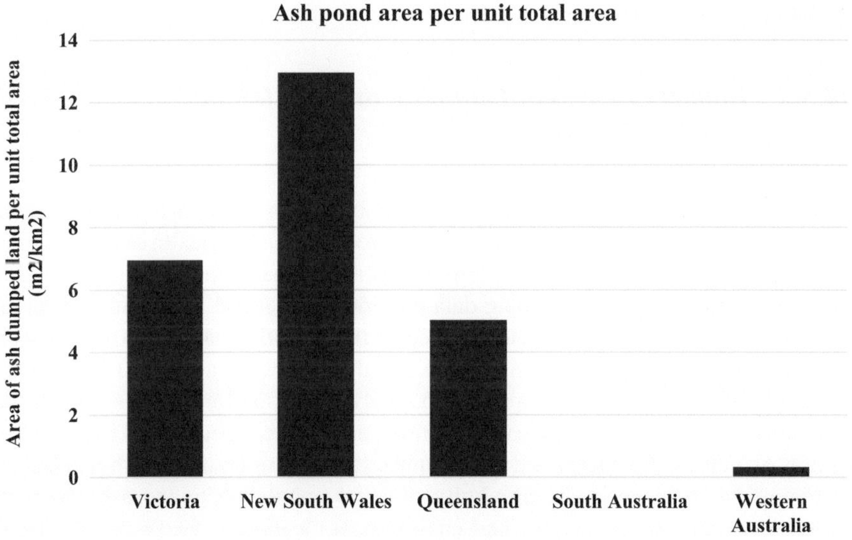

Fig. 34.8 Statewide ash dumped area per unit total province area

chains. Fly ash can also have genotoxic nature, and can cause DNA damage in human and plant cells [65].

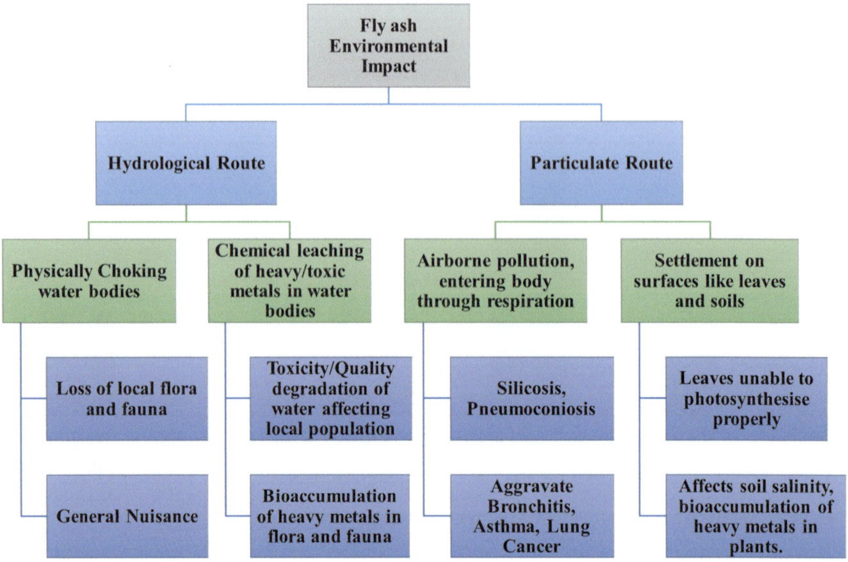

Fig. 34.9 Environmental impact of fly ash

34.5.3 Impact of Fly Ash Dumping on Population and Biodiversity

As discussed in the previous section, dumping of fly ash in ponds and dykes create environmental impacts that can directly or indirectly affect the health of local human population and can be detrimental to the flora and fauna of the region. Using GIS mapping and open-source mapping data provided by Federal and State Australian Government, a study was conducted to observe the potential risks of ash dumping on humans and animals alike. Considering a circular area with radius of 1 km around the ash ponds as the highest impact zone, analysis was performed to establish the proximity of ash ponds to the local population, endangered species, and biodiversity centres, which can be interpreted as the impact quantification due to ash ponds.

Figure 34.10 displays the proximity of local population to the ash ponds in Australia. Although legislations and environment clearances necessitate the location of power plants to be at regulated distance from local communities, many times this is not practical for distribution and logistical purposes [66]. As is clear from the Fig. 34.11, some ash ponds in Queensland, like those of Gladstone and Swanbank power plants, are in very close proximity to the local population centres as compared to other regions with lower population densities [67]. These locations as well as other in Western Australia and Victoria have reported increased incidences of particulate pollution and general nuisance due to ash in close proximity to the communities [66].

Australian biodiversity is a key component of Australia's national identity. It is not only an essential component of the lives and culture of indigenous Australians, but

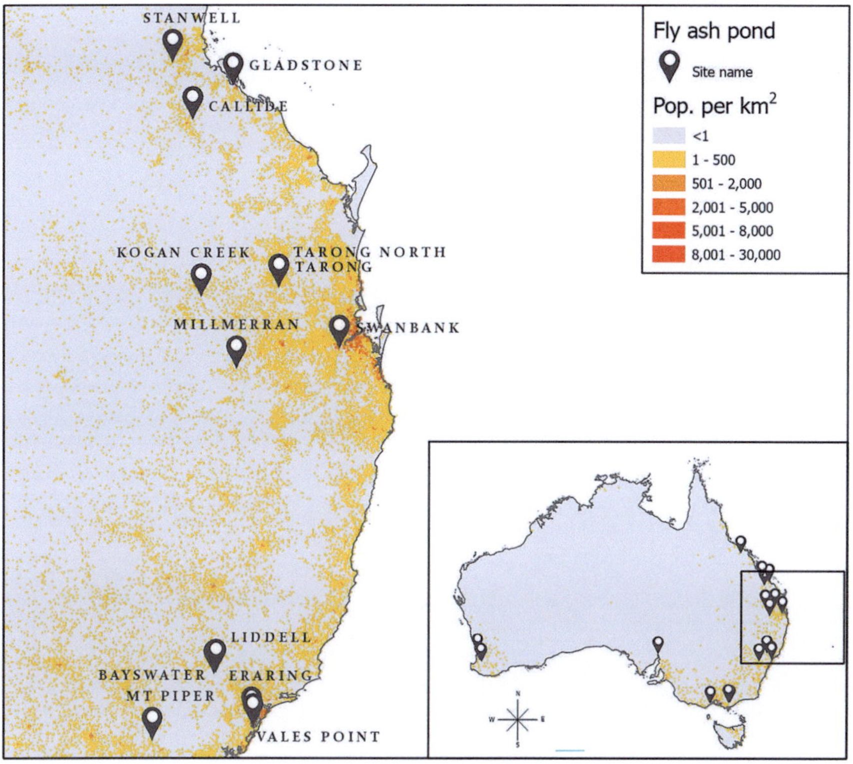

Fig. 34.10 Population proximity map for each fly ash pond

also supports health, environment and human wellbeing [68]. It was thus important to understand how the extensive ash dumping affects the biodiversity at various locations in Australia.

The Department of Climate, Change, Energy, the Environment and Water (DCCEEW) maintains records of Species of National Environmental Significance (SNES) under The Environment Protection and Biodiversity Conservation (EPBC) Act 1999 [69]. The SNES is a database of EPBC listed species which include threatened, migratory, marine and cetacean species that have known and predicted areas of occurrence, including areas of potential habitat. The DCCEEW [69] states that the data is indicative rather than definitive with the publicly available data to be used as a starting point for further investigation. The number of EPBC species that were located within the same area as the ash pond were studied using the available database.

Figure 34.11 indicates the number of Species of National Environmental Significance that are within the same area as the ash ponds with a range of between 20 and 114 species. A filter was used to overlay only "Critically Endangered" species with the range of overlay species with ash ponds being much smaller between 1 and

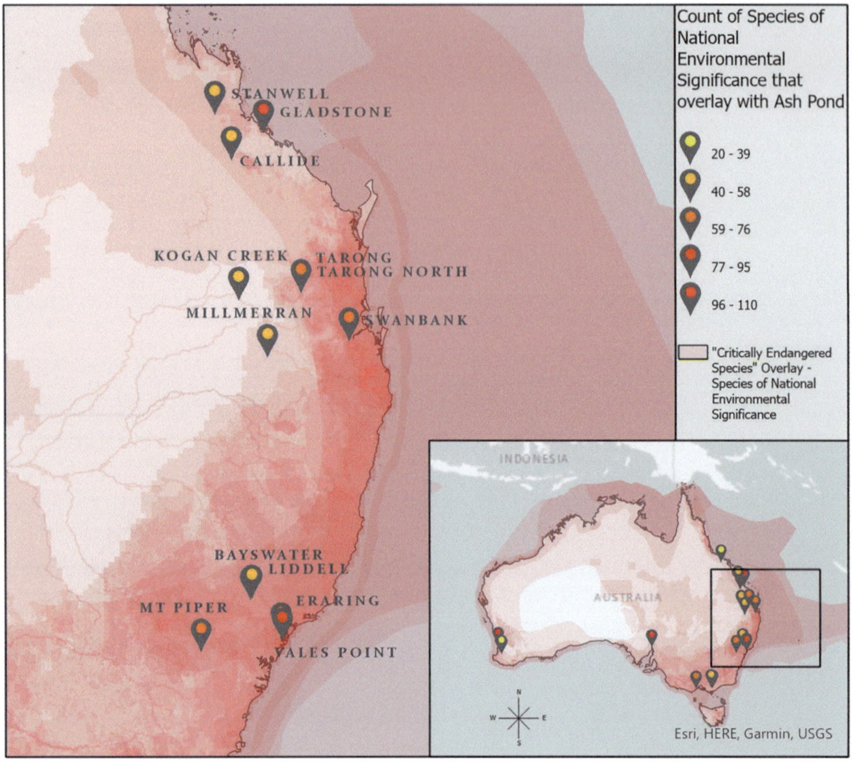

Fig. 34.11 Count of SNES that overlay with fly ash ponds

10 counts. Vales Point had the greatest count of Critically Endangered species with a count of 10, with the analysis showing that ash ponds that are closer to the ocean typically have a higher count of overlay with Critically Endangered species [69].

The local animal life is specially affected due to the potential of dust and pollution of water bodies near the ash ponds [66]. Sincere consideration and necessary efforts are therefore required to curtail the problems posed by the open dumping of fly ash.

34.6 Challenges for Fly Ash Utilisation

Fly ash utilisation has increased in many countries recently, however the problem of not utilising fly ash adequately still remains prevalent. There are many challenges that hinder the complete utilisation of fly ash today. Some of the major challenges are discussed here.

1. *Transportation cost*: A major hindrance in the use of fly ash is the transportation cost involved. The cost of fly ash itself is negligible as compared to other construction material such as bricks, cement or aggregate. Costs associated with using fly ash occur at the place of application because being a material with low bulk density, fly ash occupies more space for transportation therefore, increasing transportation and logistical costs. Traditionally, the coal-fired power plants, which are the source of fly ash generation, are located at considerable distance from major cities and communities for several socio-environmental reasons. However, this exacerbates the transportation costs as most of the fly ash utilisation is in civil construction industries near the cities and towns.
2. *Large quantities of produced fly ash*: The quantitative amount of the generated fly ash is also a major hindrance in its complete utilisation. As can be seen in Table 34.4, only a few coal power plants have complete ash mitigation which has resulted in no ash being dumped in the ash ponds. Contrary to this, most coal power plants have to dump some fly ash in the open dumps resulting in ash ponds that are more than 2.5 km^2 in some locations. These large quantities of fly ash are currently not financially feasible for a single field of utilisation such as in cement and construction industry, hindering complete utilisation of fly ash.
3. *Lack of sufficient research*: Although the potential applications for fly ash utilisation have been researched for more than a century, there is still a lack of significant research that can utilises the fly ash quantitatively and qualitatively. To ensure fly ash is suitable and safe for reuse research is also required to remove undesirable contaminants from the fly ash such as Pb and Cu that may be at concentrations above their states EPA regulations. As discussed before, fly ash possesses a pozzolanic nature which makes it a viable product in the construction industry. However, fly ash is still being used as landfills, in mine backfilling, and other similar operations where any other material could easily be used [5, 6]. This may be due to mandatory utilisation targets set by the local governments [70], however it still is under-utilising a high-potential industrial by-product. Thus, a lack of significant research exists that qualitatively investigates the utilisation of fly ash to improve its application potential.
4. *Lack of required regulations*: There are some government policies that can be changed in order to facilitate improved utilisation of fly ash. For instance, in some states fly ash is considered a waste material in Australia, regardless of the recycling process and concentration of comingled contaminants. This means that it is subject to additional environmental controls resulting in a market preference for fresh material as compared to the alternates such as fly ash [71]. Another example of the regulative hinderance is the difference in opinion between Australian federal and state government laws and regulations. Because each state follows its own Environment Protection Acts and operational regulations, fly ash can be classified as 'inert waste' in one state and 'prescribed waste' in another, even when it has the same chemical composition. Further, the management of fly ash ponds can differ between states with Victoria classifying fly ash ponds as a 'landfill' under the Victorian Environment Protection Act 2017 in comparison

to NSW classifying these as a 'dam' under the Dams Safety Act 1978 and the Protection of the Environment Operations Act 1997.

34.7 Potential Opportunities for Fly Ash Utilisation in Australia

In Australia 10.2 M tonnes of cement was produced and imported between 2020 and 2021 [72]. Although, the composition of cement and concrete can vary depending on the manufacturer and end-use requirements, fly ash can potentially replace 15–30% of cement in some of these products. This would range between 1.53 and 3.06 M tonnes of fly ash shown as a comparison to total cement in Fig. 34.12.

Currently, fly ash is most commonly utilised in the cement and concrete industry due to its pozzolanic nature. The historic utilisation of fly ash in the cement-concrete industry is summarised in Fig. 34.13.

In addition to the cement industry, there are various other fields of application that have been and are presently being researched. These not only improve the utilisation of fly ash, thus reducing the environmental impact of fly ash dumping, but also contribute to the economics of the utilisation. Some of these application areas and their related research studies are summarised in the Table 34.5.

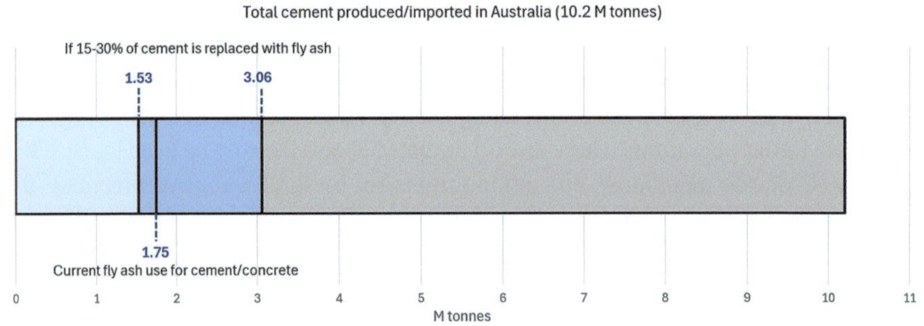

Fig. 34.12 Total cement produced/imported with marker indicating potential fly ash use range

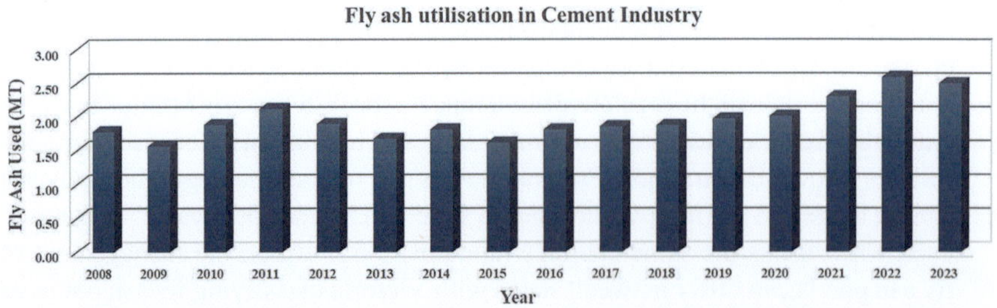

Fig. 34.13 Historic utilisation of fly ash in cement-concrete industry in Australia

Table 34.5 Various fields of utilisation of fly ash with related literature

Utilisation field	Broad use/benefit	Reference studies
Cement substitute in cement and concrete industry	As a partial replacement of cement, thus reducing cost of construction and reducing the carbon footprint during cement manufacturing	[12, 18, 19, 73–79]
Bricks and masonry	Can replace clay bricks and can be used in underground mine stoppings and dams	[73, 75]
Agriculture, soil beneficiation	Can improve soil pH, especially in acidic soil conditions. Artificial soils can be used for mine reclamation	[73, 80–86]
Stabilisation of soil slopes, roads, etc.	Uses pozzolanic nature to provide strength to slopes at marginal costs as compared to traditional cement like binders	[75, 79, 83, 87–92]
Ceramic and glass raw material	Acts as a good source of silica, which is major component in most types of fly ash	[73, 93–96]
Reclamation and backfilling	Acts as an effective filler due to its low density and particulate size	[97–101]
Metal recovery/extraction	A good source of metals, especially trace metals such as Ge, Ti and Ga. Specifically profitable where fly ash amounts are comparatively larger	[102], Al ([103, 104], Ti [105], Ge [106–109], Ga [110–113]
Adsorbents/filters/water treatment	Can act as an effective, low-cost filter capable of removing ionic, organic and inorganic impurities from waste water	[77, 114–121]
Zeolites	Molecular level filters. Can also act as efficient gas filter and many more industrial applications	[122–125]
Mesoporous Materials	Nano-pore materials with advanced application in chemical and petroleum industries	[126–129]
Cenospheres recovery	Silica spheres that form major part of some types of fly ash. Used in making synthetic foams and other micro composites	[130–132]
Polymeric fillers and composites	Enhances properties of polymers for specific applications	[73, 133–136]

Enhancing the utilisation of fly ash in some of these areas will also generate many job opportunities for local populations. Thus, increased utilisation of fly ash will further add to the economics and wellbeing of a community. However, overcoming the challenges towards complete fly ash utilisation is a necessary step to reach a stage of a zero-waste target in the coal power sector.

A possible way to deal with these challenges is to find applications for fly ash that are within close proximity of coal-fired power plant, therefore reducing the transportation cost. These sites would also need the capacity to utilise larger amounts of fly ash to its fullest potential. One of the likely venues for such utilisation is in opencast coal mines. These opencast mines experience heavy material movement throughout their working life due to large-scale production with extensive excavation of land area.

In most cases, coal-fired power plants are strategically located near coal mines to reduce the transportation cost of coal from the mines to the power plant. Conversely, transportation cost of fly ash will be reduced considerably if it can be directly utilised back in mines, thus addressing the first major challenge of fly ash utilisation. An opencast coal mine is also large in size and the potential for large scale utilisation of fly ash easily exists in such scenarios, thus addressing the second major challenge in fly ash utilisation.

Some of the research projects involving utilisation of fly ash in opencast coal mines have already been initiated successfully under the Resources Engineering division of the Department of Civil and Environmental Engineering, Monash University, Clayton, along with their national and international collaborators. These and other potential fly ash applications in coal mines are discussed in brief hereafter.

34.7.1 Fly Ash in Haul Road Construction

Haul roads are the lifelines of surface mining operation and a critical component for a smooth mine production cycle. However, due to mine economics and the dynamic nature of mining operation, these haul roads are made of blasted waste rock material (overburden) (Fig. 34.14). This makes them susceptible to breakdowns and failures, especially under heavy dump truck movement and watery conditions.

Utilising the pozzolanic nature of fly ash, a construction material was synthesised that could not only provide strength to the haul roads but was also economically and environmentally safe. The construction material synthesised here used high volumes of fly ash with polymeric additives, which made the composite competent enough to deal with the water-related structural challenges of the mine haul road. This construction material was rigorously tested in haul road simulation environments and was proved to be capable of sufficiently strengthening the haul roads, simultaneously improving its hydrological behaviour (Fig. 34.15). This drastically reduced the maintenance cost of haul roads and dumpers alike, simultaneously increasing the fly ash utilisation [138].

Fig. 34.14 A typical mine haul road made of overburden material [137]

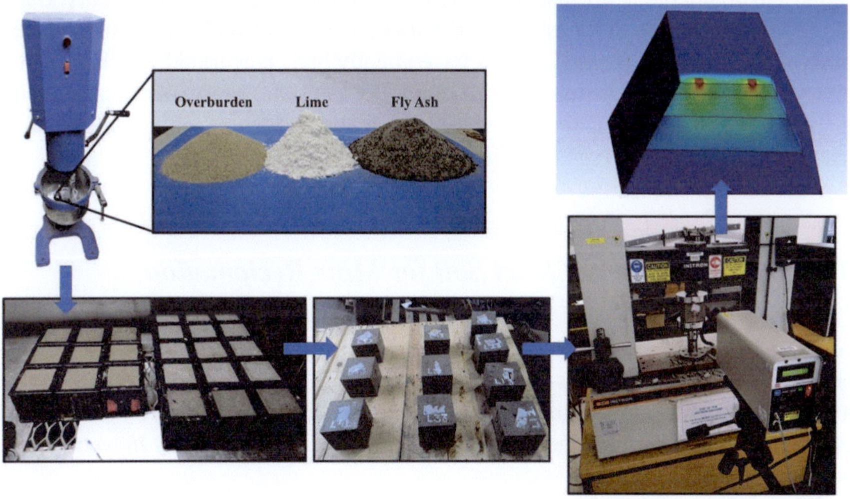

Fig. 34.15 Synthesis and testing of fly ash composite for application in haul road construction [138]

The experimental and numerical analysis conducted using the Latrobe Valley fly ash revealed that a strong cement-free fly ash based composite can be synthesized by categorically optimising the percentages of fly ash and the added lime. For the Latrobe Valley ash, this optimisation was achieved at 52.5% Fly ash, 27.5% Lime, 20% OB with the addition of 3% Gypsum by weight. Simulation results also revealed that the use of this composite in haul roads could safely reduce the surface layer thickness by 34% [138]. Further studies to universalize the approach of fly ash composite design optimisation, irrespective of the source of ash and other raw materials, are in progress.

34.7.2 Fly Ash in Waste Dump Stabilisation

In a surface mining operation, huge amounts of waste rock (overburden) have to be removed in order to reach the mineral. This overburden is usually stored in form of dumps, either inside the mine (internal dumps) or outside the mine (external dumps), based on the working method. With increasing technological advancement, higher amounts of overburden can be removed economically for mining out a unit of mineral, resulting in the need of higher and steeper dumps. To make sure these dumps remain stable and do not break down it is required to stabilise these dumps artificially. Using the pozzolanic nature of fly ash, these dumps can be stabilised by introducing layers of fly ash composite. Several studies have already been performed which prove the effectiveness of fly ash in dump stabilisation. Fly ash stabilisation of dumps not only helps protect the dump against failure, but also allows for steeper and taller dumps leading to improved storage capacities [83, 87, 139, 140].

34.7.3 Fly Ash in Artificial Soil for Mine Reclamation

A major problem faced in conducting mine rehabilitation of coal mines is the revegetation of overburden dumps and other mine sites. Apart from other factors such as a lack of nutrients and chemical wash-off during the long storage of overburden, a major reason for the infertility of overburden material is the acidity induced due to acid mine drainage (AMD).

The addition of fly ash with the overburden material increases its pH due to the alkaline nature of fly ash. The missing nutrients in the soil can be supplemented by the addition of compost obtained from local sewage plants (Fig. 34.16). The artificial soil thus formed has proven to be substantially effective in promoting vegetation in various laboratory-based experiments (Fig. 34.17) [82, 141].

Fig. 34.16 Pot experiments conducted for optimising artificial soil for mine reclamation [82]

Fig. 34.17 Vegetation studies conducted to test the effectiveness of artificial soil [141]

At present stage these artificial soils are being used for progressive mine rehabilitation in Latrobe valley's coal mines in collaboration with Monash University. These efforts will not only help in effective mine rehabilitation of present sites but also become blueprints for the rehabilitation of other mines around Australia and the world [142].

34.8 Conclusion

Substantial benefits are achievable if fly ash utilisation is prioritised by policymakers. With a significant resource of fly ash, alongside industry and academic expertise, Australia has the capacity to establish state-of-the-art research facilities for fly ash utilisation and mitigation. This will not only help Australia reach its target of a zero-waste power sector but will also act as a framework for multinational research, establishing Australia as the leader in this field. Focused efforts and research in this direction can result in a better environment, a thriving economy, and higher employment for the local communities.

References

1. BP, BP Statistical Review of World Energy 2022, British Petroleum (2022)
2. IEA, Coal Information 2022, International Energy Agency (2022)
3. B. Millington, in Coal ash has become one of Australia's biggest waste problems—and a solution is being ignored, News Article (ABC News, Online, 2019)
4. C. Heidrich, H.J. Feuerborn, A. Weir, Coal Combustion Products: A global Perspective, in *World of Coal Ash* (WOCA, Lexington, 2013)
5. ADAA, Annual Membership Survey Results, Ash Development Association of Australia (2022)
6. CEA, Fly ash generation at coal/lignite based thermal power stations and its utilization in the country for the year 2021–22, Central Electricity Authority (2022)
7. Y. Luo, Y. Wu, S. Ma, S. Zheng, Y. Zhang, P.K. Chu, Environ. Sci. Pollut. Res. **28**, 18727–18740 (2021)
8. I. Putilova, V. Putilov, Properties of coal ash in Russia, in *EUROCOALASH 2010* (Eurocoalash, Copenhagen, Denmark., 2010)
9. ACAA, in *Coal Ash Recycling Rate Increases Slightly in 2021; Use of Harvested Ash Grows Significantly* (American Coal Ash Association, online, 2021)
10. Coal Combustion Products, (Industrial Resource Council) http://www.industrialresourcescouncil.org/Materials/CoalCombustionProducts/tabid/363/Default.aspx. Accessed 18 Jan 2018
11. E. Diaz, E. Allouche, S. Eklund, Fuel **89**, 992–996 (2010)
12. A.K. Das, D. Ghosh, S. Maitra, A. Das, Int. Res. J. Eng. Technol. (IRJET) **2**, 2451–2455 (2015)
13. A. Ugurlu, Environ. Geol. **46**, 890–895 (2004)
14. P. Chindaprasirt, C. Jaturapitakkul, W. Chalee, U. Rattanasak, Waste Manage. **29**, 539–543 (2009)
15. J. Xu, Y. Zhou, Q. Chang, H. Qu, Mater. Lett. **60**, 820–822 (2006)
16. M. Komljenović, Z. Baščarević, V. Bradić, J. Hazard. Mater. **181**, 35–42 (2010)
17. A. Fernández-Jiménez, A. Palomo, Fuel **82**, 2259–2265 (2003)
18. G.M. Giaccio, V. Malhotra, in *Cement, Concrete and Aggregates*, vol. 10 (1988), pp. 88–95
19. H. Myadaraboina, M. Solikin, I. Patnaikuni, S. Setunge, Development of high volume fly ash concrete using ultra-fine fly ash, in *23rd Australasian Conference on the Mechanics of Structures and Materials (ACMSM23)* ed. by S. T. Smith (Byron Bay, Australia, 2014)
20. J. Temuujin, R. Williams, A. Van Riessen, J. Mater. Process. Technol. **209**, 5276–5280 (2009)
21. G. Carette, V. Malhotra, Can. J. Civ. Eng. **14**, 667–682 (1987)
22. EPRI, in *Investigation of High-Volume Fly Ash concrete Systems, Electrical Power Research Institute* (1993)
23. D. Montgomery, D. Hughes, R. Williams, Cem. Concr. Res. **11**, 591–603 (1981)

24. P. Owens, Concrete **13**, 21–26 (1979)
25. J.H. Brown, The strength and workability of concrete with PFA substitution, in *Proceedings International Symposium on the Use of PFA in Concrete* (University of Leeds, Leeds, West Yorkshire United Kingdom, 1982), pp. 151–161.
26. L. Minnick, W. Webster, E. Purdy, J. Mater. **6**, 163–187 (1971)
27. K. Stuart, D. Anderson, P. Cady, Cem. Concr. Res. **10**, 823–832 (1980)
28. R. Helmuth, in *Special Publication* vol. 91, (1986), pp. 723–740
29. B. Copeland, PFA concrete for hydraulic tunnels and shafts, Dinorwick pumped storage scheme—case history, in *International Symposium on the Use of PFA in Concrete*, edited by C. J. G., C. A. R., (Leeds, UK, 1982), pp. 323–343
30. B. Johnson, The use of fly ash in Cape Town RMC operations, in *5th International Conference on Alkali-Aggregate Reaction in Concrete* (Cape Town, South Africa, 1981)
31. J. William, P. Owens, The implications of a selected grade of United Kingdom pulverized fuel ash on the engineering design and use in structural concrete, in *International Symposium on the Use of PFA in Concrete*, ed. by C. J.G., and C. A.R., (Leeds, UK, 1982), pp. 301–313.
32. R.L. Yuan, J.E. Cook, Study of a class C fly ash concrete, in *Proceedings of 1st International Conference on the Use of Fly Ash, Silica Fume, Slag, and Other Mineral Byproducts in Concrete*, vol. 79 (American Concrete Institute, Montebello, PQ, 1983), pp. 307–320
33. F. Raba Jr, S. Smith, M. Mearing, Subbituminous fly ash utilization in concrete, in *Symposium on Fly Ash Incorporation on Hyderated Cement Systems* (Materials Research Society, Boston, 1981), pp. 296–306
34. S.H. Gebler, P. Klieger, Effect of fly ash on physical properties of concrete, in *2nd CANMET/ACI International Conference on the Use of Fly Ash, Silica Fume, Slag, and Natural Pozzolan in Concrete*, vol. 91 (American Concrete Institute, Madrid, Spain, 1986), pp. 1–50
35. P. Tikalsky, P. Carrasquillo, R. Carrasquillo, Mater. J. **85**, 505–511 (1988)
36. D. Ravina, Production and collection of fly ash for use in concrete, in *Symposium of Fly Ash Incorporation in Hydrated Cement Systems* (Materials Research Society, Boston, 1981), pp. 2–11
37. M. Monk, Mag. Concr. Res. **35**, 131–141 (1983)
38. A.M. Neville, *Properties of Concrete*, vol. 4 (Longman London, 1995)
39. M. Kokubu, S. Takano, I. Miura, R. Sugiki, Trans. Jpn. Soc. Civ. Eng. **1960**, c1–c10 (1960)
40. D. Ravina, Efficient concrete utilization of coarse and fine fly ash in precast concrete by incorporating thermal curing. J. Am. Concr. Inst. **78**, 194–200 (1981)
41. E.A. Abdun-Nur, in *Fly Ash in Concrete: An Evaluation* (Highway Research Board, U.S. Army Engineer Research and Development Center, 1961), p. 138
42. R.O. Lane, J. Best, Concr. Int. **4**, 81–92 (1982)
43. R.E. Davis, A review of pozzolanic materials and their use in concretes, in *Symposium on Use of Pozzolanic Materials in Mortars and Concretes* (ASTM International, 1950)
44. R.S. Ghosh, J. Timusk, Creep of fly ash concrete. J. Am. Concr. Inst. **78**, 351–357 (1981)
45. P. Bamforth, Proc. Inst. Civ. Eng. **69**, 777–800 (1980)
46. R. Lohtia, B. Nautiyal, O. Jain, J. Am. Concr. Inst. **73**, 469–472 (1976)
47. K. Nasser, A. Al-Manaseer, Shrinkage and creep of concrete containing 50 percent lignite fly ash at different stress-strength ratios, in *2nd CANMET/ACI International Conference on the Use of Fly Ash, Silica Fume, Slag, and Natural Pozzolans in Concrete*, vol. 91 (American Concrete Institute, Spain, 1986), pp. 433–448.
48. R.J. Elfert Jr, Bureau of Reclamation experiences with fly ash and other pozzolans in concrete, in *3rd International Ash Utilization Symposium*, (Bureau of Mines, Pittsburgh, PA, 1973), pp. 80–93
49. R. Yuan, J. Cook, Time dependent deformations of high strength fly ash concrete, in *Proceedings, of the International Symposium on the Use of PFA in Concrete*, vol. 1 (Leeds, UK, 1982), pp. 255–260
50. J. Munday, L. Ong, L. Wong, R. Dhir, Load independent movements in opc/pfa concrete, in *Proceedings, International Symposium on the use of PFA in concrete*, (Leeds UK, 1982), pp. 243–246

51. R.S. Ravindrarajah, C. Tam, Properties of concrete containing low-calcium fly ash under hot and humid climate, in *3rd CANMET/ACI International Conference on the Use of Fly Ash, Silica Fume, Slag, and Natural Pozzolans in Concrete*, vol. 114 (Norway, 1989), pp. 139–156
52. P.M. Carrasquillo, Durability of concrete containing fly ash for use in highway applications, in *Katharine and Bryant Mather International Conference on Concrete Durability*, vol. 100 (American Concrete Institute, Atlanta, GA, 1987), pp. 843–862
53. T.C. Liu, Maintenance and preservation of concrete structures. Report 3 (Abrasion-Erosion Resistance of Concrete, U.S. Army Corps of Engineers, 1980)
54. Global Energy Observatory, (Global Energy Observatory) http://globalenergyobservatory.org/. Accessed 20 Apr 2018
55. CEA, Guidelines on design, construction, O&M and annual certification of coal ash ponds (Government of India, Ministry of Power, Central Electricity Authority, 2022)
56. S. Chakrabarti, G. Mudd, J. Kodikara, Coupled atmopsheric-unsaturated flow modelling of leached ash disposal in the Latrobe valley, Australia, in *1st International Conference on Engineering for Waste Treatment* (Albi, France, 2005), p. 8
57. G.M. Mudd, J. Kodikara, Coal ash leachability: detailed field studies, in *Australian Institute of Energy (AIE) - 8TH Australian Coal Science Conference* (Sydney, NSW, 1998), pp. 357–362
58. S. Kanchan, V. Kumar, K.K. Yadav, N. Gupta, S. Arya, S. Sharma, Current World Environ. **10**, 572 (2015)
59. V. Pandey, Z. Usmani, A. Chandra, R.K. Mishra, V. Kumar, Chem. Ecol. **33**, 777–794 (2017)
60. A. Mandal, D. Sengupta, Curr. Sci. **88**, 617–624 (2005)
61. Y. Hansen, P.J. Notten, J.G. Petrie, Appl. Geochem. **17**, 1131–1141 (2002)
62. M. Dutta, in *Not So Inert After All! Fly Ash: An Environment and Health Perspective, Centre for Science and Environment* (1997)
63. S.N. Pandey, Water. Air Soil Pollut. **19**, 87–100 (1983)
64. R. Blissett, N. Rowson, Fuel **97**, 1–23 (2012)
65. R. Chakraborty, A. Mukherjee, Ecotoxicol. Environ. Saf. **72**, 838–842 (2009)
66. B. Lipski, N. Rivers, J. Whelan, in *Toxic and Terminal: How the Regulation of Coal-Fired Power Stations Fails Australian Communities* (2017)
67. ABS, *Regional Population, 2021–22 financial year, Australian Bureau of Statistics* (2023)
68. I. Cresswell, H. Murphy, in *Biodiversity, Department of the Environment and Energy* (2017)
69. DCCEEW, in *Species of National Environmental Significance, Department of Climate Change, Energy, the Environment and Water* (2023)
70. G.C. Chattopadhyay, in *Issues in Utilization of ash by Thermal Power Plants in the Country* (Indian Audit and Accounts department, 2015)
71. C. Heidrich, Ash Utilisation-an Australian perspective, in *Geopolymers 2002 International Conference, Melbourne, Australia, Siloxo* (2002)
72. CIF, 2022 Cement industry federation industry report (2022)
73. A. Dwivedi, M.K. Jain, Recent. Res. Sci. Technol. **6** (2014)
74. V. Kumar, P. Sharma, M. Mathur, Fly ash disposal: mission beyond 2000 AD, in *Conference on Fly ash disposal and Deposition: Beyond 2000AD*, (IIT, Kanpur, Kanpur, 2000), pp. 223–230
75. J. Alam, M. Akhtar, Int. J. Emerg. Trends Eng. Dev. **1**, 1–14 (2011)
76. R.E. Davis, R.W. Carlson, J.W. Kelly, H.E. Davis, Properties of cements and concretes containing fly ash. J. Proc. **33**, 577–612 (1937)
77. S. Dhadse, P. Kumari, L. Bhagia, J. Sci. Ind. Res. **67**, 11–18 (2008)
78. D.F. Aponte, M. Barra, E. Vàzquez, Constr. Build. Mater. **30**, 537–546 (2012)
79. A. González, R. Navia, N. Moreno, Waste Manage. Res. **27**, 976–987 (2009)
80. L. Giardini, Rivista di Ingegnaria Agraria **12**, 679–689 (1991)
81. D.W. Evans, K. Whysner, Beneficial reuse of CCR for site closure using mining reclamation technologies, in *World of Coal Ash*, (flyash.info, Lexington, KY (USA), 2017)
82. M. Yellishetty, V. Wong, M. Taylor, J. Li, Forming artificial soils from waste materials for mine site rehabilitation, in *EGU General Assembly Conference Abstracts* (2014), p. 16
83. T. Gupta, M. Yellishetty, Singh T., in *Proceedings of MPES 2015-Smart Innovation in Mining* (The Southern African Institute of Mining and Metallurgy Johannesburg, South Africa, 2015), pp. 997–1004.

84. V.C. Pandey, N. Singh, Agric., Ecosyst. Environ. **136**, 16–27 (2010)
85. V. Manoharan, I. Yunusa, P. Loganathan, R. Lawrie, C. Skilbeck, M. Burchett, B. Murray, D. Eamus, Fuel **89**, 3498–3504 (2010)
86. L.C. Ram, R.E. Masto, J. Environ. Manage. **91**, 603–617 (2010)
87. A.K. Gupta, B. Paul, Current World Environment **11**, 204 (2016)
88. S. Matsumoto, S. Ogata, H. Shimada, T. Sasaoka, G.J. Kusuma, R.S. Gautama, Adv. Mater. Sci. Eng. **2016** (2016)
89. B. Leelavathamma, N. Pandian, J. Test. Eval. **33**, 88–93 (2005)
90. N. Pandian, K. Krishna, A. Sridharan, J. Test. Eval. **29**, 220–226 (2001)
91. Z. Nalbantoğlu, Constr. Build. Mater. **18**, 377–381 (2004)
92. F. Zha, S. Liu, Y. Du, K. Cui, Nat. Hazards **47**, 509–523 (2008)
93. M. Erol, S. Küçükbayrak, A. Ersoy-Mericboyu, J. Hazard. Mater. **153**, 418–425 (2008)
94. R. Sokolar, L. Vodova, Ceram. Int. **37**, 2879–2885 (2011)
95. Y. He, W. Cheng, H. Cai, J. Hazard. Mater. **120**, 265–269 (2005)
96. F. Peng, K.-M. Liang, A.-M. Hu, Fuel **84**, 341–346 (2005)
97. E. Brännvall, J. Kumpiene, Environ Sci Process Impacts **18**, 11–21 (2016)
98. S. Choudhary, *B. Tech, National Institute of Technology Rourkela* (2013)
99. J.P. Capp, L.M. Adams, ACS Div. Fuel Chem. Preprints **15**, 26 (1971)
100. R.J. Hornberger, A.D. Dalberto, M.J. Menghini, T.C. Kania, B.E. Scheetz, D.J. Koury, *Coal ash beneficial use at mine sites in Pennsylvania,* in *World of Coal Ash* (flyash.info, Kentucky, USA, 2005)
101. V.K. Jha, M. Matsuda, M. Miyake, J. Ceram. Soc. Jpn. **116**, 167–175 (2008)
102. T. Okada, H. Tomikawa, Waste Manage. **33**, 605–614 (2013)
103. R. Matjie, J. Bunt, J. Van Heerden, Miner. Eng. **18**, 299–310 (2005)
104. G.-H. Bai, T. Wei, X.-G. Wang, J.-G. Qin, X. Peng, P.-C. Li, Trans. Nonferrous Metals Soc. China **20**, s169–s175 (2010)
105. Y. Shabtai, I. Mukmenev, J. Biotechnol. **51**, 209–217 (1996)
106. A. Hernández-Expósito, J. Chimenos, A. Fernández, O. Font, X. Querol, P. Coca, F.G. Peña, Chem. Eng. J. **118**, 69–75 (2006)
107. J. Marco-Lozar, D. Cazorla-Amorós, A. Linares-Solano, Carbon **45**, 2519–2528 (2007)
108. F.A. Torralvo, C. Fernández-Pereira, Miner. Eng. **24**, 35–41 (2011)
109. F. Arroyo, C. Fernández-Pereira, Ind. Eng. Chem. Res. **47**, 3186–3191 (2008)
110. R. Moskalyk, Miner. Eng. **16**, 921–929 (2003)
111. Z. Fang, H. Gesser, Hydrometallurgy **41**, 187–200 (1996)
112. O. Font, X. Querol, R. Juan, R. Casado, C.R. Ruiz, Á. López-Soler, P. Coca, F.G. Peña, J. Hazard. Mater. **139**, 413–423 (2007)
113. B. Gutiérrez, C. Pazos, J. Coca, Waste Manage. Res. **15**, 371–382 (1997)
114. D. Goswami, A.K. Das, J. Sci. Ind. Res. **65**, 77–79 (2006)
115. B. Dutta, J.K. Basu, S. DasGupta, Sep. Sci. Technol. **38**, 1345–1360 (2003)
116. D. Mohan, K.P. Singh, G. Singh, K. Kumar, Ind. Eng. Chem. Res. **41**, 3688–3695 (2002)
117. V. Hequet, P. Ricou, I. Lecuyer, P. Le Cloirec, Removal of Cu^{2+} and Zn^{2+} in aqueous solutions by sorption onto fly ash and fly ash mixtures, in *1999 International Ash Utilization Symposium* (1999), pp. 41–46
118. H. Cho, D. Oh, K. Kim, J. Hazard. Mater. **127**, 187–195 (2005)
119. I. Alinnor, Fuel **86**, 853–857 (2007)
120. S. Mohan, R. Gandhimathi, J. Hazard. Mater. **169**, 351–359 (2009)
121. G. Itskos, N. Koukouzas, C. Vasilatos, I. Megremi, A. Moutsatsou, J. Hazard. Mater. **183**, 787–792 (2010)
122. X. Querol, J.C. Umaña, F. Plana, A. Alastuey, A. Lopez-Soler, A. Medinaceli, A. Valero, M.J. Domingo, E. Garcia-Rojo, Fuel **80**, 857–865 (2001)
123. H. Holler, U. Wirsching, Fortschr. Mineral. **63**, 21–43 (1985)
124. G. Hollman, G. Steenbruggen, M. Janssen-Jurkovičová, Fuel **78**, 1225–1230 (1999)
125. X. Querol, N. Moreno, J.T. Umaña, A. Alastuey, E. Hernández, A. Lopez-Soler, F. Plana, Int. J. Coal Geol. **50**, 413–423 (2002)

126. P. Kumar, N. Mal, Y. Oumi, K. Yamana, T. Sano, J. Mater. Chem. **11**, 3285–3290 (2001)
127. M. Halina, S. Ramesh, M.A. Yarmo, R. Kamarudin, Mater. Chem. Phys. **101**, 344–351 (2007)
128. K. Hui, C.Y.H. Chao, J. Hazard. Mater. **137**, 401–409 (2006)
129. A.O. Dhokte, S.L. Khillare, M.K. Lande, B.R. Arbad, J. Ind. Eng. Chem. **17**, 742–746 (2011)
130. T. Hirajima, H. Petrus, Y. Oosako, M. Nonaka, K. Sasaki, T. Ando, Int. J. Miner. Process. **95**, 18–24 (2010)
131. G. Jegadeesan, S.R. Al-Abed, P. Pinto, Fuel **87**, 1887–1893 (2008)
132. H. Petrus, T. Hirajima, Y. Oosako, M. Nonaka, K. Sasaki, T. Ando, Int. J. Miner. Process. **98**, 15–23 (2011)
133. X. Guo, H. Shi, W.A. Dick, Cement Concr. Compos. **32**, 142–147 (2010)
134. P. Chindaprasirt, T. Chareerat, S. Hatanaka, T. Cao, J. Mater. Civ. Eng. **23**, 264–270 (2010)
135. D. Hardjito, S.E. Wallah, D.M. Sumajouw, B.V. Rangan, Mater. J. **101**, 467–472 (2004)
136. J.S. Van Deventer, J.L. Provis, P. Duxson, Miner. Eng. **29**, 89–104 (2012)
137. Robey P., Mining Photography,Portfolio, (2011)
138. T. Gupta, *Ph.D., Indian Institute of Technology Bombay* (2018)
139. T. Gupta, T. Singh, Int. J. Adv. Sci. Eng. Inf. Technol. **8**, 405–410 (2018)
140. T. Gupta, S.P. Pradhan, T.N. Singh, *A Critical Study on Feasibility of Fly Ash Utilization in Overburden Dumps of Opencast Coal Mines,* in *Seminar on Sustainable Development in Mineral & Earth Resources*, (The Indian Mining & Engineering Journal, New Delhi, 2014), pp. 149–153
141. M. Taylor, M. Yellishetty, B.C. Panther, Geotechnical and hydrogeological evaluation of artificial soils to remediate acid mine drainage and improve mine rehabilitation—an Australian case study, in *Mine Planning and Equipment Selection 2013*, ed. by C. Drebenstedt, R. Singhal, (Springer, Dresden, Germany, 2013), pp. 855–865
142. Artificial soils could be key in rehabilitation of Latrobe Valley coal mines, (phys.org) https://phys.org/news/2017-03-artificial-soils-key-latrobe-valley.html. Accessed 12 Oct 2023

Chapter 35
An Exploration of the Embodied Energy in Construction Materials to Support the Diversification of Greener Design

Koorosh Gharehbaghi, Naiyla Goverdhan, Ken Farnes, and Kong Fah Tee

Abstract The construction sector is a significant consumer of natural resources and a major contributor to energy consumption, leading to considerable environmental repercussions that have raised serious concerns. This study seeks to enhance the process of selecting building materials, given the adverse effects that many conventional construction materials impose on the environment. Certain materials utilized in building construction are associated with elevated levels of Embodied Energy (EE) and Carbon Dioxide (CO_2) emissions, necessitating the identification of more sustainable and environmentally friendly alternatives to foster a healthier future. Furthermore, this research includes a comparative analysis of three houses constructed with varied materials: a standard burnt brick house, a cement block house, and a mudbrick house. A mixed-methods approach was utilized to quantify each construction technique's EE and CO_2 emission. After thorough evaluation, three houses of similar floor area and exposed to identical climatic conditions were chosen for the study. The findings revealed that the mudbrick house (House 2) emerged as the most sustainable option, primarily since mudbrick generates no EE or CO_2 emissions. In contrast, materials such as cement blocks, steel, concrete, and aluminum were found to significantly contribute to environmental degradation and toxicity. This research advocates for a practical framework for the adoption of alternative building materials in residential construction, emphasizing the urgent need for decisive actions to combat climate change and promote sustainability.

K. Gharehbaghi (✉) · N. Goverdhan · K. Farnes
RMIT University, Melbourne, Australia
e-mail: koorosh.gharehbaghi@rmit.edu.au

K. F. Tee
Department of Civil and Environmental Engineering, King Fahd University of Petroleum and Minerals, Dhahran, Saudi Arabia

Interdisciplinary Research Center for Construction and Building Materials, KFUPM, Dhahran, Saudi Arabia

D. A. H. Hanaor (ed.), *Sustainable Materials for the Built Environment*, Springer Series in Materials Science 352, https://doi.org/10.1007/978-3-031-97818-0_35

35.1 Introduction and Background

The surge in residential construction activities in recent years has highlighted the environmental challenges linked to this sector. It is crucial to explore strategies for enhancing material selection to mitigate the adverse effects on the environment [1]. This concern is particularly pertinent given that building materials are among the largest consumers of energy and resources, thus contributing significantly to environmental degradation [2]. Research by [3] indicates that building materials account for 30–50% of global raw resource consumption, with approximately 40% of landfill waste in OECD countries originating from construction materials. This issue is increasingly critical in light of rising global temperatures and ozone layer depletion [4]. Annually, construction materials generate over a million tons of waste, exacerbating environmental challenges. Furthermore, the materials utilized in construction often possess prominent levels of Embodied Energy (EE), and their production processes are responsible for substantial Carbon Dioxide (CO_2) emissions.

The typical stages encompassed in a building's lifecycle consist of the extraction and processing of raw materials, the construction phase, the operational period of the building, and the demolition once the building has reached the end of its useful life. Furthermore, shifts in lifestyle and evolving demands can contribute to a reduction in a building's lifespan, potentially leading to increased landfill waste or recycling efforts as a result of renovations and demolitions. The substantial consumption of construction materials, along with the associated EE, has led to significant resource depletion on a global scale [5]. The choice of environmentally sustainable building materials and products is often shaped by the various stakeholders involved in the construction sector [6].

35.1.1 Research Objectives

The construction sector must prioritize environmental preservation and the establishment of a sustainable future for upcoming generations. The processes, methodologies, and materials employed within the industry must be innovatively developed to facilitate the global adoption of sustainable building practices, making this a fundamental consideration at every phase of construction [7]. Data from the Australian Bureau of Statistics [8] indicates a significant increase in the consumption of construction materials. This research aims to assist in the selection of building materials that minimize adverse environmental effects while also shedding light on the ecological consequences associated with various construction materials. Utilizing materials that can be reused or recycled is crucial in decreasing the demand for newly produced materials. Edil [9] posits that recycling not only mitigates environmental harm but also leads to substantial cost savings by reducing landfill expenses, creating economically viable alternatives, and lowering energy usage.

Furthermore, the energy expended during material extraction, processing, and transportation contributes significantly to pollution levels [10, 11]. Sourcing construction materials locally can prove advantageous by diminishing transportation costs and CO_2 emissions, thereby reducing greenhouse gas output.

The selection of materials in construction is influenced by numerous factors, including budget constraints and the availability of resources. However, the assessment of EE and CO_2 emissions presents significant challenges due to the complexity of the data involved, which encompasses aspects such as the geographical origin of materials, the technologies employed, and the methodologies utilized in manufacturing processes [12]. Consequently, it is crucial to comprehend the environmental ramifications associated with the choice of construction materials and to implement strategies aimed at mitigating these impacts. Two effective approaches for significantly lowering the environmental footprint of construction materials are recycling and reusing existing materials. The careful selection of materials during the design phase plays a vital role in minimizing environmental waste, as these practices not only decrease the demand for added resources but also lead to cost reductions and a decrease in the volume of new materials required for construction. Therefore, it is essential to consider factors such as the source of materials, their EE, and the CO_2 generated throughout the extraction, manufacturing, and construction processes. While these strategies are particularly relevant to the construction sector, they can also be adapted for use in other industries. Thoughtful material selection can contribute to improved quality of life and overall human well-being.

This study explores the benefits associated with the utilization of sustainable materials in the construction of residential buildings. As previously noted, construction materials are often characterized by significant levels of EE, and their production and disposal contribute to substantial CO_2 emissions. Identifying which materials exert the least environmental impact in contrast to those that are more detrimental is crucial. Furthermore, this research aims to provide a comprehensive overview and a deeper insight into how selecting construction materials and processes can mitigate environmental harm. By comparing three distinct types of houses, the study evaluates the materials employed, the associated EE, and the CO_2 produced. This analysis will reveal which building practices have minimal environmental impact and which are more harmful.

The construction industry must prioritize the careful selection of sustainable materials, as this practice can mitigate the adverse effects associated with materials that possess high EE. It is essential to evaluate materials throughout the entire construction process, encompassing stages such as extraction, sourcing, manufacturing, and, demolition, while also considering their potential for recycling and reuse. By establishing material selection strategies during the initial phases of construction, particularly in design and planning, it becomes possible to make environmentally beneficial choices that promote the use of sustainable materials. Consequently, this research focuses on the following questions:

- (RQ1) How can the environmental impact be reduced through better building materials usage?

- (RQ2) How can the control and decrease of waste within the construction industry be better addressed?
- (RQ3) How can the CO_2 footprint be reduced through increased material performance?

35.1.2 Research Gap

A comprehensive examination of existing literature and extensive research indicates that the process of selecting building materials, along with the environmental costs tied to these materials, lacks transparency. While sustainable materials offer potential environmental advantages, they often come at a higher price compared to conventional standardized options. The elevated costs associated with sustainable materials can be attributed to the more intensive processes required for their cultivation and production. Additionally, the principle of economies of scale plays a significant role in pricing; as demand for sustainable materials remains relatively low, their prices tend to stay elevated. Consequently, a rise in demand could lead to increased production levels, which in turn would result in a reduction of unit costs.

35.2 Literature Review and Background

Researchers have observed that significant amounts of CO_2 are released into the atmosphere throughout various stages of the construction life cycle [13–15]. The connection between emissions generated during material production and construction practices has been extensively examined, particularly concerning the choice of construction materials. Hu et al. [16], Costa et al. [17], Rossello-Batle [18] propose that the environmental footprint can be mitigated by substituting the most energy-demanding components of a building with more ecologically sustainable alternatives. Furthermore, studies conducted by [12, 19] highlight that construction activities are significantly contributing to the swift exhaustion of non-renewable resources.

The construction of buildings necessitates various resources that undergo both physical and chemical transformations throughout their lifecycle. The manufacturing processes associated with these products generate waste outputs that contribute to the release of harmful emissions into the environment. As urban populations continue to grow and cities expand, there is a corresponding rise in resource consumption, leading to an increased demand for housing and infrastructure. This heightened demand results in greater waste production, as well as elevated levels of discharge and CO_2 affecting land, water, and air quality. Wackernagel et al. [20], Nguyen [21] assert that sustainability involves living within the environmental regenerative capacity, a threshold that human activities have surpassed. This study investigates the EE associated with the production of building materials, encompassing their extraction, processing, manufacturing, and transportation. The integration of recycled and reused materials in construction practices is beneficial, as it helps to minimize waste generation and associated CO_2 emissions, thereby reducing the amount of waste sent to landfills. Additionally, implementing an effective construction waste management strategy is crucial for reducing costs and conserving resources. Finally, materials characterized as sustainable building materials are those that are environmentally friendly and generate minimal waste, including low-cost options that can be sourced locally to further decrease transportation emissions.

35.2.1 Embodied Energy (EE)

EE refers to the cumulative energy utilized throughout the various direct and indirect processes linked to a building's lifecycle, adhering to the cradle-to-grave framework [22]. This concept encompasses a wide range of activities, including the extraction of raw materials, their manufacturing, transportation to the construction site, the actual construction process, any refurbishment or replacement efforts, and the disposal of the building once it reaches the end of its useful life.

Buildings are constructed using a diverse array of materials, each of which requires energy at various stages, including production, utilization, and eventual dismantling [23]. This energy consumption encompasses several critical processes such as the extraction of raw materials, their transportation, manufacturing, assembly, installation, disassembly, deconstruction, and decomposition [24, 25]. The manufacturing and transportation of numerous building materials are particularly energy-intensive. In recent years, there has been a growing focus on the environmental implications of the building and construction sector. The production processes associated with building materials are significant contributors to greenhouse gas emissions, particularly CO_2 [26, 27].

The construction industry is recognized as one of the largest consumers of commercial energy, primarily derived from electricity and fossil fuel combustion. Urge-Vorsatz and Novikova [28] noted that in 2004, buildings accounted for 32% of global CO_2 emissions, with projections indicating that by 2030, this figure could rise to between 35 and 42%. Bansal et al. [29] suggest that energy consumption can be mitigated by utilizing materials with low EE, which are also cost-effective. The overall life cycle energy of a building comprises both EE and operational energy, with the former arising from the production, on-site construction, demolition, and disposal of building materials. In contrast, operational energy pertains to the management of the internal environment, which includes heating, cooling, lighting, and the operation of appliances [24, 30, 31] provide a detailed examination of the numerous benefits associated with green buildings in the context of high-rise construction, as illustrated in Fig. 35.1.

Figure 35.1 illustrates that the primary characteristics of green buildings encompass structural performance, constructability and maintenance, innovation and design processes, and materials and resources. These attributes are crucial for conducting Life Cycle Assessments (LCA) and understanding their associated energy dynamics.

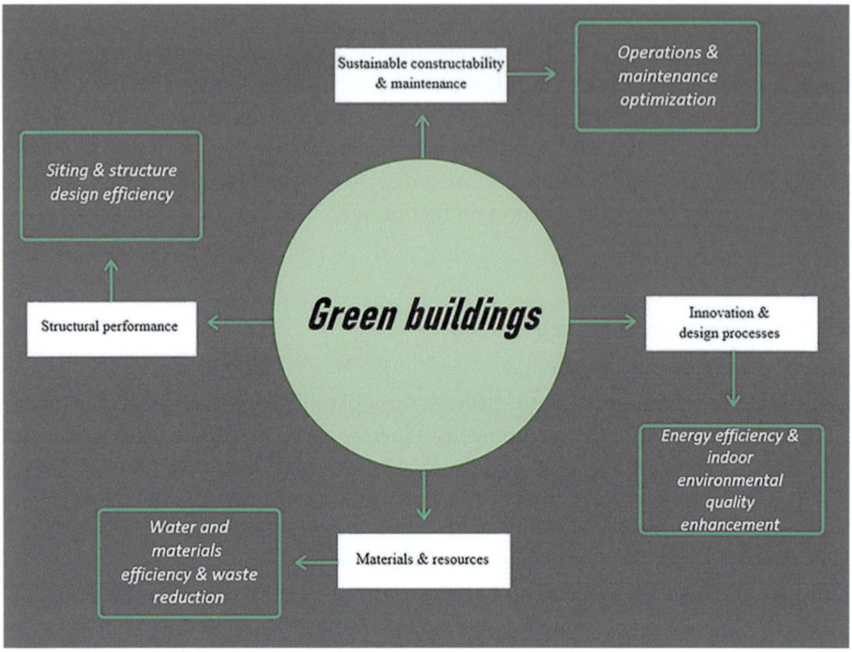

Fig. 35.1 The influences of green buildings for high-rise construction [31]

Conversely, Fig. 35.2 presents an overview of the energy associated with the building life cycle.

Various scholars have approached the concept of EE and its assessment from different perspectives. The analyses conducted by [24, 32, 33] characterize EE as the cumulative energy necessary for the construction of a building, which encompasses both the direct energy utilized during the construction and assembly phases, as well as the indirect energy associated with the production of the building's materials and components. Furthermore, [34] articulate EE as the total energy needed to deliver a product, encompassing both direct and indirect energy inputs throughout all upstream processes. A more comprehensive elucidation is offered by [35–39], who collectively describe EE as the sum of energy expended during the extraction and processing of raw materials, the transportation of these materials, the manufacturing of building components, and the energy consumed in various activities throughout the construction and eventual demolition of the structure, as referenced in [40].

Numerous research investigations including [41–43] have focused on the CO_2 emissions associated with the energy demands of producing and processing various building materials. Consequently, architects and designers globally are actively seeking strategies to mitigate the adverse environmental effects attributed to buildings, as highlighted by [44]. Adopting energy-efficient alternative building materials can significantly lower the EE within a structure, as noted by [26]. By distinguishing

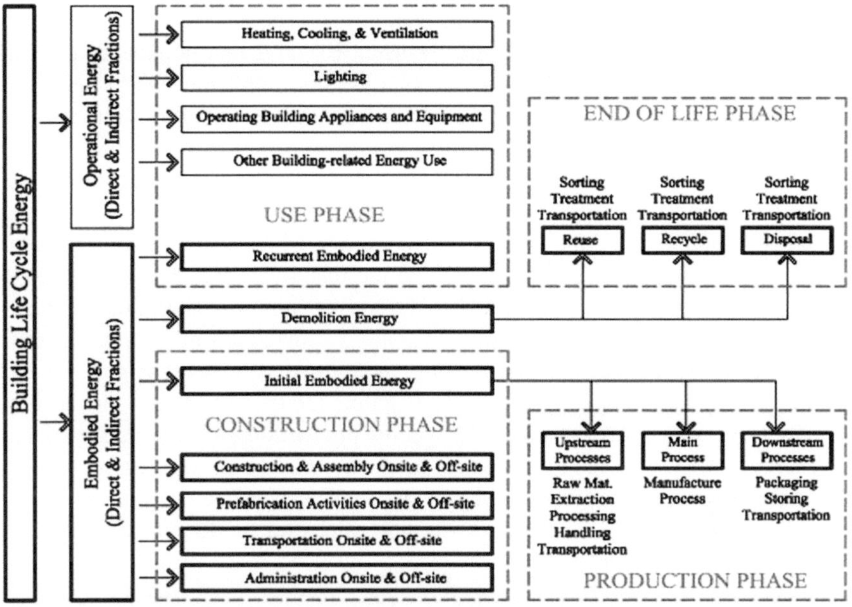

Fig. 35.2 Building life cycle energy [24]

between the EE expended during the construction phase and that used in the operation and maintenance phase, it becomes possible to pinpoint the most effective opportunities for reducing EE and, in turn, lessening harmful environmental impacts.

The construction sector is a significant consumer of energy and natural resources, accounting for approximately 40% of the materials that enter the global economy and contributing to 40–50% of worldwide greenhouse gas emissions, as reported by [45]. The energy consumption of buildings, influenced by the types and quantities of construction materials utilized, represents about 10–20% of the total energy expended throughout their construction and operational lifespan, according to [29, 46]. The diversity in building typologies leads to varying levels of EE, as different structures require differing amounts of construction materials. Natural resource-derived materials such as bricks, cement, aggregates, timber, glass, steel, and aluminum undergo multiple manufacturing processes that embed energy, as discussed by [47].

Moreover, EE encompasses the total energy invested in a product throughout its lifecycle, beginning with the extraction of raw materials and extending through transportation, refining, manufacturing, and assembly. The energy consumed during these stages can be quantified and assessed with relative ease; however, the indirect energy released, which includes the energy utilized by machinery that facilitates these processes, poses a greater challenge in terms of measurement. Direct energy is typically associated with on-site and off-site operations such as construction, prefabrication, administration, and transportation activities, as noted by [48]. Conversely, the energy associated with the production of building materials, as well as activities related to renovation, refurbishment, and demolition, is classified as indirect energy. This category encompasses initial EE, recurrent EE, and energy expended during demolition. As previously mentioned, EE signifies the cumulative energy necessary for the production of a product, spanning from the extraction of raw materials to the final product's readiness for use.

In the case of bricks, this process typically involves three key stages: first, the extraction of raw materials, which refers to the energy expended in mining clay and other essential materials; second, transportation, which accounts for the energy consumed in moving these raw materials to the manufacturing facility and subsequently transporting the finished bricks to the construction site; and third, the manufacturing process itself, which includes the energy utilized during various stages of brick production, such as crushing, mixing, molding, drying, and firing. The EE of bricks can vary based on several factors, including the type of brick—such as fired clay bricks versus concrete bricks—the specific production methods employed, and the energy sources utilized. For instance, fired clay bricks typically exhibit a higher EE compared to concrete bricks, primarily due to the elevated temperatures required during the firing process. Further, the average values for the EE of common brick types are:

- Fired clay bricks: approximately 2.5–3.5 MJ/kg.
- Concrete bricks: approximately 0.8–1.5 MJ/kg.

A significant approach to minimizing the EE of bricks involves firing them at lower temperatures. This method not only contributes to a reduction in CO_2 but also helps maintain the inherent properties of the raw materials, thereby enhancing production efficiency by optimizing kiln usage and potentially boosting overall output. The process of extracting raw materials, along with their manufacturing and transportation to construction sites, represents the initial EE invested in material production. In contrast, the energy expended on maintenance and refurbishment throughout the building's lifespan is categorized as recurrent EE. Additionally, the energy required for the deconstruction and disposal of materials at the end of a building's life cycle is referred to as demolition energy [24]. Various methodologies have been developed to accurately assess EE, including process analysis, input–output analysis, and hybrid analysis [49]. The careful selection and utilization of building materials in the construction sector can significantly diminish the overall EE associated with a project. Consequently, the reduction of EE has emerged as a critical concern for researchers, industry professionals, builders, and material manufacturers alike.

35.2.2 Recycled and Reusable Materials

Research consistently highlights recycling as a pivotal strategy for diminishing the overall consumption of energy and resources within society [50]. The construction sector, as noted by [51, 52], is increasingly incorporating recycled materials into its practices. This trend is further bolstered by selective demolition techniques, the integration of recycled materials, and the architectural design of structures that promote future recycling efforts [53]. Wagner [54] emphasizes that the construction industry is progressively utilizing raw materials at a higher rate than other sectors. The processes of reusing and recycling present a range of opportunities and challenges, which not only yield environmental benefits but also result in economic savings [55]. Consequently, the adoption of recycled materials in construction can significantly mitigate the demand for new materials, thereby conserving financial resources, natural assets, and EE [56].

Numerous nations have enacted regulations aimed at promoting the circular economy to enhance sustainable development and improve the recycling of construction waste. For instance, the European Union's Waste Framework Directive [57] established in 2008, set an ambitious target for member states to ensure that at least 70% of non-hazardous construction waste is reused or recycled by the year 2020. This directive applies to countries within the EU and the European Economic Area (EEA) [57]. In Australia, the transition towards a circular economy began with the implementation of the National Waste Policy in 2009 [58, 59]. The construction sector has significantly refined its operational practices to foster the development of sustainable and eco-friendly buildings. Construction waste encompasses a variety of materials, including timber, concrete, plasterboard, bricks, vegetation, rock, and soil, which are primarily produced during construction, demolition, or renovation activities. Notably, waste generated from these processes account for more than one-third of all landfill waste, with over 75% of landfill contents consisting of clean, excavated materials such as concrete, bricks, and timber that are suitable for recycling [60]. While solid waste-to-energy incinerators are frequently promoted as a rapid and effective method for managing increasing waste volumes while simultaneously generating energy, they also pose environmental challenges, including water pollution, unpleasant odors, noise, and vibrations, which can adversely affect nearby residential and commercial areas.

Furthermore, the incineration process generates hazardous byproducts, particularly fly ash and bottom ash, which necessitate meticulous management and disposal [61]. As a result, waste incinerators significantly contribute to the environmental challenges faced by the construction sector. These facilities emit a considerable volume of CO_2, surpassing the emissions produced by power plants that utilize coal, oil, or natural gas. In light of these concerns, scientists and scholars have undertaken extensive research to explore alternative strategies for alleviating the environmental impacts of waste generation and pollution. For instance, [62] proposes that the replacement of raw materials with recycled alternatives can conserve natural resources. Their investigation, which involved surveying companies regarding the advantages of utilizing recycled materials, revealed that the primary benefit was a 38% reduction in landfill waste, followed by improvements in quality at 33%, and cost savings at 27%.

A variety of construction materials are amenable to recycling, with studies indicating that both aluminum and steel can be entirely recycled, as noted by [63, 64]. This recycling process significantly mitigates the EE associated with these materials,

Table 35.1 Environmental profiles for a range of building materials (based on [118])

Material	EE (kWh/m^3)	Renewable source	Local extraction disruption	Indoor effects (toxicity)	Recycling potential
Brick	300	Non-renewable	High	Dust if exposed	Low-med
Timber	110	Renewable	Low	None	Med
Concrete	800	Non-renewable	High	Dust if exposed	Med
Cement	2860	Non-renewable	High	Dust if exposed	Med
Steel	103,000	Non-renewable	High	None	High
Aluminum	75,600	Non-renewable	High	None	High
Glass	23,000	Non-renewable	High	None	High

achieving reductions of 72% for steel and an impressive 95% for aluminum. Furthermore, while plasterboard poses environmental hazards by releasing toxic hydrogen sulfide (H2S) in landfills, it remains a recyclable material. Timber is particularly notable for its versatility, as it can be repurposed into particleboard or transformed into mulch for horticultural applications. Windows, too, can be reused or recycled for similar purposes. The recycling of glass results in a 20% decrease in EE, and the resultant recycled glass can serve as aggregate in the production of concrete, bricks, and tiles, as highlighted by [65]. Additionally, concrete can incorporate industrial by-products such as fly ash, slag, and silica fume, which are derived from coal-fired power generation, steel manufacturing, and silicon metal production, respectively [65]. The environmental impacts of commonly utilized materials in the construction sector are summarized in Table 35.1.

The government of New South Wales adopted the principles of reduce, reuse, and recycle (3R) as a foundational waste management strategy at Sydney Olympic Park, aiming to optimize outcomes [66]. Munn and Soebarto [65] highlighted that various reused and recycled materials were incorporated into concrete for several prominent Australian landmarks, including the Sydney Harbour Tunnel, the Sydney Parallel Runway, the Melbourne Crown Casino, and the white-water canoe course for the Sydney Olympics located in Penrith. Furthermore, they pointed out that the Sydney International Shooting Gallery, constructed for the 2000 Sydney Olympics, was made from reused timber, with 90% of its materials sourced from this category and the remaining 10% derived from plantation timber.

35.2.3 Construction Waste Management

Construction and Demolition (C&D) waste represents one of the most significant waste streams globally. This type of waste originates from various activities, including land excavation, civil and building construction, site clearance, roadwork, and refurbishment, all of which are integral to construction, renovation, and demolition processes [67]. Research conducted by [68–70] indicates that C&D waste constitutes 30–40% of total solid waste, a figure that has escalated due to the increasing frequency of C&D activities driven by urbanization and city redevelopment. The rapid urban growth has led to heightened consumption of non-renewable resources, raising significant environmental concerns [71]. In certain nations, such as Bangladesh, challenges persist, including low public awareness, inadequate law enforcement, and reliance on outdated technologies [72, 73]. Furthermore, deficiencies in communication, coordination, oversight, and regulatory frameworks, along with insufficient contractor awareness and poor project waste budgeting, contribute to the excessive generation of C&D waste. In the United States, it is reported that over one hundred million tons of C&D waste are produced annually, while the United Kingdom accounts for more than 50% of its C&D waste being sent to landfills [74], with an annual rejection of 70 million tons of C&D waste.

Furthermore, the yearly production of C&D waste in Hong Kong experienced a significant increase, having doubled between 1993 and 2004, reaching approximately 20 million tons by 2004 [75]. Numerous studies and industry analysts suggest that the volume of C&D waste continues to rise markedly each year, underscoring the critical need to comprehend the magnitude of the C&D waste generated by construction operations.

Effective management of C&D waste can transform it into a significant resource rather than a source of environmental concern. Various materials, including metals, mortar, concrete, bricks, plastics, timber, ceramics, and glass, have the potential for reuse, thereby allowing the C&D waste stream to generate revenue instead of contributing to pollution. According to [76], the field of C&D waste management is gaining traction as a vital discipline, with efficient management practices leading to decreased energy consumption, conservation of raw materials, and minimized transportation requirements. The worldwide C&D waste management industry is experiencing growth, creating economic prospects that appeal to entrepreneurs [77]. Although the ongoing growth of C&D waste management companies has not yet substantially reduced environmental repercussions, it is paving the way for enhanced waste management strategies. C&D waste can be divided into two main categories: inert waste, which consists of both soft and hard materials, and non-inert waste, which comprises residuals such as metals, wood, plastics, and glass. This categorization is depicted in the life cycle flowchart of various C&D wastes, as illustrated in Fig. 35.3.

The challenge of material waste in the construction industry is a pressing concern that requires immediate attention. To effectively manage waste throughout all stages of a construction project, the implementation of a robust waste management system is essential [78]. Shifting the construction sector towards a 'Zero waste' paradigm demands informed decision-making at every phase of the construction life cycle, as well as the development and enforcement of appropriate standards [4, 79]. Emphasizing the principles of reduce, reuse, and recycle (3R) during construction activities aims to decrease the amount of mixed construction waste sent to landfills, thereby promoting material waste reduction. Research by [80] underscores the considerable

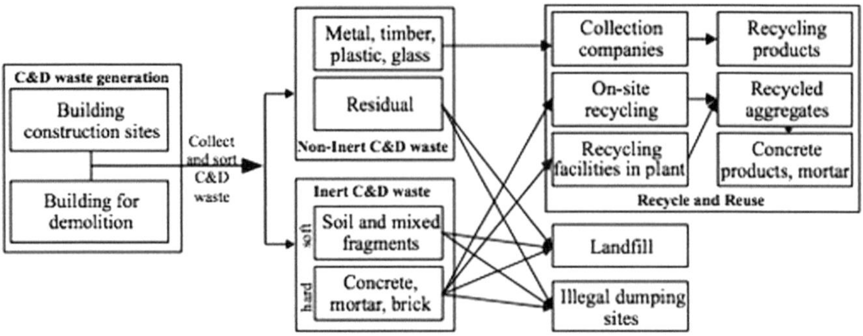

Fig. 35.3 Lifecycle flowchart of C&D waste [77]

environmental risks and socio-economic impacts linked to C&D waste in landfill sites.

The depletion of natural resources is accelerating, driven by the growing demands of infrastructure development and the diminishing availability of alternative resources. To address the environmental challenges posed by construction waste, the principles of reduce, reuse, recycle (3R) and reduce, reuse, recycle, and recover (4R) can be effectively integrated throughout the entire construction process. The concept of reduction emphasizes minimizing resource consumption at its origin, thereby curtailing unnecessary material usage, which can be applied from the initial planning stages to the final handover of projects by ensuring meticulous and precise material planning. Reuse involves repurposing the same materials in construction without altering their original function or composition, such as utilizing formwork multiple times or adapting materials for new applications, thereby extending their lifespan and reducing waste generation while benefiting the environment. This approach is particularly advantageous as it requires minimal processing and energy input. Recycling, on the other hand, entails reprocessing materials to create new products once they are no longer usable in their original form, which helps decrease the demand for virgin materials and mitigates the volume of waste directed to landfills. Finally, recovery pertains to salvaging discarded materials for alternative uses, allowing for their processing rather than outright disposal [81].

35.2.4 Sustainable Building Materials

A significant body of research has explored the challenges associated with material selection; however, these studies often neglect to establish a clear definition of 'sustainable building materials' [82]. Sustainable building materials are characterized by their natural origins and offer numerous benefits to users, including reduced maintenance requirements, enhanced energy efficiency, and improved health and comfort for occupants [83]. The selection of materials typically takes place during the design phase, where detailed specifications and working plans are formulated. Halliday [84] notes that various methodologies and tools have been created to facilitate optimal material selection based on their physical and mechanical properties. Research indicates that certain materials and resources are becoming increasingly scarce, necessitating careful management of the remaining supplies [84]. Reddy [85] argues that the development of sustainable alternative building technologies can promote energy conservation, reduce reliance on materials with high energy embeddedness, encourage environmentally friendly practices, minimize transportation needs, maximize the use of local resources and skills, incorporate industrial waste into building materials, and utilize recycling and renewable energy sources. By adhering to these principles, construction methods can evolve to become sustainable, thereby minimizing their negative impact on the environment.

Glass is widely recognized as a prevalent material in the construction industry, primarily utilized for its ability to facilitate daylight harvesting. However, its effectiveness is hindered by limitations in light management, inadequate thermal insulation, and a significant CO_2 footprint. The thermal properties of glass allow heat to transfer easily, leading to increased energy expenses, particularly as warmth escapes during colder months and enters during warmer periods. In contrast, transparent wood is emerging as a highly promising alternative to glass, offering substantial advantages for future applications. This innovative material not only demonstrates superior thermal insulation—boasting thermal conductivity that is at least three times lower than that of glass—but also presents a cost-effective solution for energy savings. Furthermore, transparent wood is derived from sustainable and renewable resources, contributing to lower CO_2 [86–88]. Research conducted by [88] has illustrated the potential of transparent wood composites as building materials that can effectively harness sunlight, thereby ensuring consistent and uniform indoor lighting while enhancing thermal insulation.

The production, utilization, and disposal of carpets raise considerable environmental and health issues. The majority of carpets are constructed from synthetic materials that necessitate substantial energy and water consumption, resulting in the generation of toxic waste during their production and lacking biodegradability. In contrast, carpets made from natural fibers possess the advantage of being biodegradable. For instance, wool carpets, particularly those crafted in India, are steeped in a rich Persian-style tradition. According to [89], the Indian carpet industry is hindered by a deficiency in innovation and reliance on antiquated technologies. This sector is highly fragmented and faces numerous challenges that require urgent attention. The manufacturing process consumes excessive water and employs chemical dyes, which pose significant risks to both the environment and human health, contributing to substantial waste that exacerbates landfill issues. In response to these concerns, a new standard for evaluating the environmental sustainability of carpets was established in November 2007 and received approval from the American National Standards Institute (ANSI). While hardwood and laminated flooring are aesthetically pleasing, they contribute to the depletion of natural timber resources. Conversely, bamboo flooring, derived from grass, offers superior hardness and durability compared to hardwood and laminated options, and it is also more sustainable due to the rapid regeneration of bamboo.

35.2.4.1 Locally Sourced Materials

Transportation plays a crucial role throughout the various phases of construction projects, encompassing the movement of goods and materials to and from the construction site, as well as site extraction, processing, and actual construction activities. An increase in transportation distance correlates with a rise in both EE and greenhouse gas emissions. By utilizing locally sourced materials, energy expenditure is significantly reduced, thereby fostering a more sustainable environment. Research indicates that employing locally sourced materials can lead to a reduction in energy consumption in buildings by as much as 215% and a decrease in transportation impacts by 453% [10]. Life Cycle Assessment (LCA) studies have demonstrated that buildings incorporating a high proportion of bio-based materials exhibit the least environmental impact [90]. Furthermore, [91] found that greenhouse gas emissions are highly variable based on the geographical origin of the materials. Locally grown crops used in construction not only consume less energy but also lower transportation energy costs [92]. The use of locally sourced materials can also expedite the construction timeline [93]. Commonly available low-cost raw materials include soil, clay, straw, and sand, whereas many materials prevalent in the construction sector, such as steel, cement, glass, aluminum, plastics, and bricks, are energy-intensive and typically sourced from greater distances, exacerbating their environmental impact through increased CO_2 emissions and energy consumption during manufacturing [94].

It is essential to reduce the consumption of energy resources and raw materials to create simple, energy-efficient, and sustainable alternatives in building construction. The construction sector faces significant challenges in embracing sustainable practices, primarily due to insufficient information, technological advancements, and evaluation methodologies. The gradual integration of sustainable building techniques is hindered by a reluctance to adopt modern technologies, the necessary changes in processes, and concerns regarding potential risks and unforeseen expenses, as noted by [95]. This resistance complicates the promotion of sustainable construction methods, as highlighted by [96]. Nevertheless, there is a growing trend towards utilizing locally sourced materials, which aligns with the principles of sustainable development. The use of heavy construction materials, such as aggregates and bricks, presents limitations in transportation capacity due to their weight, while lighter materials, like insulation and wind cowls, necessitate more delivery loads. Consequently, it is crucial to evaluate the average transportation distances based on both weight and volume. Research conducted by [97] indicates that sourcing materials locally can significantly lower CO_2 emissions without incurring additional costs, simply by opting for local suppliers. Therefore, prioritizing sustainability and adopting a 3R or 4R approach to materials is vital for achieving energy-efficient designs, although this endeavor can prove to be quite challenging.

35.2.4.2 Life Cycle Assessment (LCA)

The assessment of EE and CO_2 in buildings is typically conducted through a modified version of LCA, which serves as a framework for evaluating the environmental repercussions associated with the material, energy, and pollutant flows throughout a product's entire lifespan. According to the International Standard ISO 14044:2006, LCA encompasses four essential stages: 'Define goal and scope', 'Life cycle inventory analysis' (LCI), 'Impact Assessment', and 'Interpretation' [98]. Central to the LCA process is the functional unit, which must be quantifiable, relevant, and consistent, encompassing aspects such as building materials, energy systems, packaging, and transportation. Additionally, the ISO standards related to LCA are integrated within the ISO 14000 series, which outlines environmental management standards. This series includes (I) ISO 14040: Environmental Management—Life Cycle Assessment—Principles and Framework; (II) ISO 14044: Environmental Management—Life Cycle Assessment—Requirements and Guidelines. Other supportive ISO standards for LCA include ISO 14025: Environmental Labels and Declarations, ISO 14067: Greenhouse Gases, and ISO 14046: Environmental Management, among others. These standards are crucial as they provide the necessary guidelines and requirements for conducting LCA assessments, ensuring that evaluations are performed consistently and credibly.

Sustainable building materials are fundamental to the welfare of the environment and the LCA enables analysis and helps to select the best sustainable materials. Cole [99] reports that the use of the LCA approach enables the comparison of alternative materials, components, and services in buildings. In addition, based on [100] it allows for measuring the environmental impacts of buildings. In Europe and the United States, the LCA is being used for resource comparison, however, it has slowly been implemented and included in government policy, strategic planning, and product design [101]. The input and output flows of materials, energy, and pollutants to and from the environment life cycle perspective enable complex processes to be analyzed [40]. The process of a LCA is depicted in Fig. 35.4.

In addition to the specific LCA processes, the main principles of LCA help in (a) decision-making in selecting alternatives by measuring and assessing the environmental performance of materials or processes; and (b) analyzing the environmental performance of the system for potential improvements, hence modification, or designing a system to reduce the impact. This can occur for specific stages in the life cycle [40].

35.2.4.3 Energy Efficiency

For better building performance through energy efficiency, building design, system, and occupants need to be taken into consideration. It is found that globally buildings contribute to 20–30% of primary energy use and CO_2 emissions. According to [102], Fig. 35.5, demonstrates the separate contributions of individual factors in buildings to energy consumption.

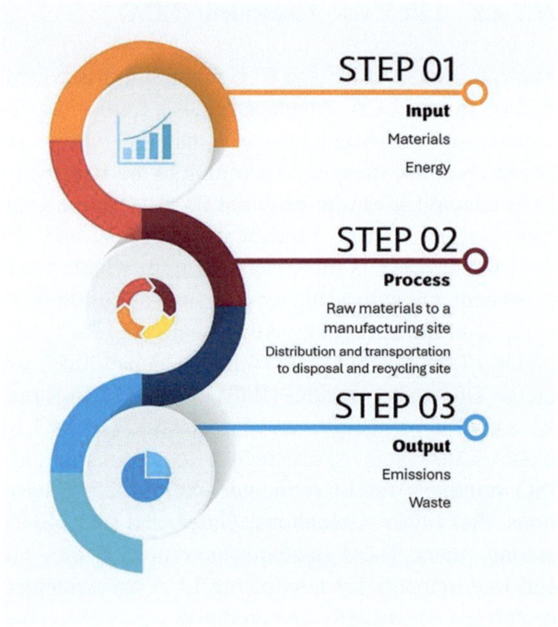

Fig. 35.4 Input and output flows of materials, energy, and pollutants from a life cycle perspective (based on [40])

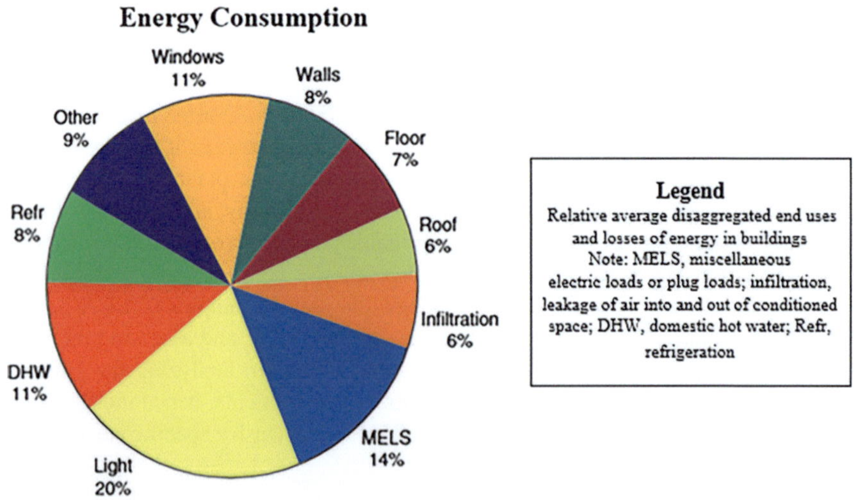

Fig. 35.5 The separate contributions of individual factors in buildings to energy consumption (data sourced from [88, 102])

Figure 35.5 demonstrates the relative average disaggregated end uses and losses of energy in buildings. While MELS represents miscellaneous electric loads or plug loads; Infiltration, on the other hand, shows leakage of air into and out of conditioned space. In comparison, DHW includes domestic hot water and refrigeration. If the building design is optimized efficiently this will allow to reduce the system capacity which will result in the minimization of the need for services. Therefore, in the early design stages when project planning occurs the important parameters that affect the building performance can include shape, orientation, window-to-wall ratio, interior layout, etc. [103].

35.3 Research Development and Process

This study investigates the benefits associated with the utilization of sustainable materials in the construction of residential buildings, addressing the growing concerns regarding the environmental repercussions of construction activities, which have escalated alongside the rise in both construction volume and the production and consumption of construction materials. Significant advancements have been made in techniques and practices aimed at mitigating environmental impacts through the careful selection of sustainable building materials and methodologies. The environmental consequences of construction encompass issues such as resource depletion, greenhouse gas emissions, solid waste generation, and climate change, with the construction sector being identified as one of the most significant contributors to these challenges. Analyzing the EE and CO_2 emissions of various materials will facilitate the identification of those that are most beneficial for the environment, as well as for human health and well-being.

The study employs an analytical framework that integrates both quantitative and qualitative methodologies, beginning with an extensive literature review from which relevant case studies will be identified and summarized. The quantitative aspect of the research will involve examining various scenarios where sustainable construction materials are evaluated against traditional building materials, to quantify the data to bolster the case for sustainable options. In contrast, the qualitative component will focus on an analysis of three houses of comparable size situated in similar climatic conditions but constructed using different methods. Furthermore, the research will assess the materials utilized in the selected case studies, employing logical reasoning and a comprehensive literature review to elucidate the advantages of sustainable construction materials over their conventional counterparts.

35.3.1 Research Methodology

The data-gathering process for this research is structured into two distinct segments. Initially, an extensive review of existing literature was performed to assess the environmental implications associated with various construction materials. This phase involved utilizing multiple search queries that incorporated terms such as 'comparative material selection' and 'buildings constructed with recycled or reused materials' to systematically identify and evaluate relevant scholarly works. The subsequent segment of the study comprised three case studies focusing on houses built from varied materials, all of which were comparable in size and situated within similar environmental contexts. Numerous researchers have published evaluations of individual or multiple building case studies, while others have synthesized these findings to provide a comprehensive overview of the embodied environmental impacts.

The literature review indicated a substantial body of research focused on the environmental implications of building materials. By synthesizing relevant findings from prior studies, this analysis aimed to enhance the understanding of material selection processes. Moreover, the research investigates the environmental consequences associated with the choice of construction materials by evaluating the EE and CO_2 emissions linked to various buildings. The central research question posed is, "In what ways can the environmental impact of construction be mitigated through improved selection of building materials?" The primary aim of this study was to perform a comparative analysis of residential structures constructed from several types of materials. Specifically, this research examined three distinct houses: one built with conventional standardized burnt bricks, another utilizing cement blocks, and a third constructed from soil cement, or mudbrick. This investigation relied on secondary sources, online databases, and web searches to gather comprehensive information regarding buildings made from recycled or reusable materials. Additionally, by employing search criteria such as "comparative material selection" and "buildings made with recycled/reused materials," the case studies yield pertinent information and data. Particularly, the selected houses for comparison are of similar dimensions and situated in comparable environmental contexts, thereby facilitating a reliable comparison of the gathered data.

The main data was collected through academic journals that display the release of EE and CO_2 emissions. Hence, Tables and Figures (graphs) were created to demonstrate which house construction method has the most severe impact on the environment. The Tables analyzed each house and were divided into different building elements such as the floor, roof, external walls, windows, doors, and finishes. This allowed the depiction of which house construction method has the lowest release of EE and CO_2 emissions as well as which material has the most consequence. Thus, it created a framework that will enable the makeup of sustainable houses in the future due to lower detrimental effects on the environment. Figure 35.6 represents the research methodology.

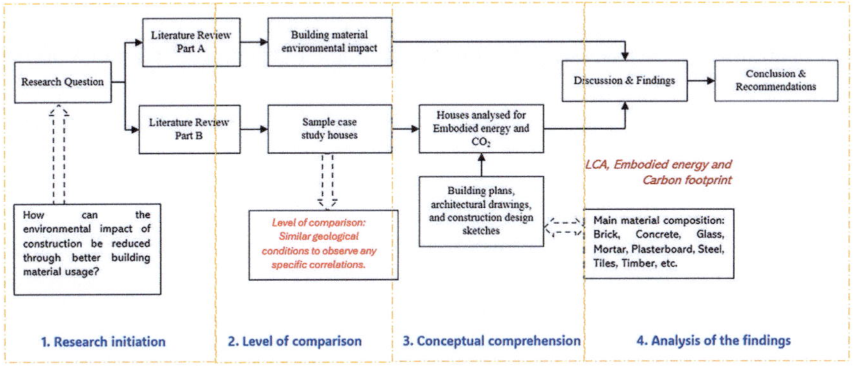

Fig. 35.6 The research methodology. *Source* Authors

35.3.2 Research Methods and Data Sources

Through vigorous research conducted through numerous sources of peer-reviewed academic literature, initial data was acquired. Further, as previously mentioned, the next step was the calculation of EE, and CO_2 emissions figures based on the material used for the buildings. The data attained was collected through a mixed-method approach of quantitative data and qualitative data with the use of secondary research. A quantitative case study offers ‘emphasis on numbers and figures in the collection and analysis of data’ [104] whereas qualitative research enables us to systemically measure variables and explore the concepts and experiences more in-depth. The quantitative data will be able to depict the total EE in megajoules (MJ) and the CO_2.

A substantial body of research has been undertaken through a variety of case studies that examine the advantages associated with the incorporation of recycled materials in construction. Specific findings from these studies reveal the environmental implications of utilizing materials such as concrete, brick, steel, and wood, as well as the significance of geographical distance. Additionally, the research is grounded in empirical evaluations that facilitate comparisons with other case studies featuring similar building dimensions, which employed new and standardized materials, leading to detrimental environmental consequences. The analysis has been meticulously conducted, underscoring the research outcomes from these case studies that demonstrate the superior environmental benefits of utilizing a considerable proportion of reused or recycled materials in the construction process.

35.3.3 Comparative Analysis

This research investigated three houses made with different construction materials. While House 1 is located in South Australia (Australia), Houses 2 and 3 are located in Cameroon, Africa. Although House 1 is on a different continent from Houses 2 and 3, all three houses possessed similar conditions. Furthermore, the three houses have similar building types (residential), building age (< 15 years old), structural similarities, comparable environmental conditions (climate and building characteristics), and geological conditions. Careful study was undertaken in choosing the right houses for the research and there were common denominators when comparing the building materials, this is depicted in Table 35.2.

Table 35.2 also highlights the overall materials composition of the three houses. As already noted, House 1 is located in South Australia, Australia, which is a typical house. It consists of 2 bedrooms with a floor size of 111m^2. This house is built with standardized materials which are the commonly used construction materials in the residential sector. Houses 2 and 3 are located in Cameroon, Africa, and are also considered typical and common houses. House 2 consists of 2 bedrooms with a floor size of 78.66 m^2 and is made of mudbrick. Similarly, House 3 is a 3-bedroom house with a 109.98 m^2 floor size which is built from cement blocks (hollow concrete blocks). All the houses are single-story and are approximately the same building

Table 35.2 Assessment of 3 case studies

House No.	Climate conditions	Dwelling size	Main material composition
House 1 (Standard build)	28.3 °C	Single story—111 m^2	Steel—2% Concrete—69% Timber—4% Brick—21% Glass—0.1% Tiles—0.4% Carpet—0.2% Plasterboard—1% Mortar—3%
House 2 (Mudbrick)	25.5 °C	Single story—78.66 m^2	Concrete—31% Timber—5% Substrate—0.4% Aluminum—0.2% Mortar—7% Brick (mud)—57% Glass—0.002% Tiles—0.1%
House 3 (Cement block: *Hollow Concrete Block*)	25.5 °C	Single story—109.98 m^2	Concrete—31% Timber—0.6% Substrate—0.4% Aluminum—0.2% Mortar—8.3% Cement block—60% Glass—0.002% Tiles—0.2%

Source Authors

Table 35.3 Data sources and analysis overview

Data	Source	Analysis purpose
Recyclable	[105–109]	Waste management/minimization, reusable materials
Sustainable building	[40, 110–112]	Resource efficiency, life-cycle assessment, design strategies
EE & CO_2 emissions	[113–115]	Processes associated with the production of buildings (manufacturing, transportation, production)

Source Authors

size. The environmental conditions between the two different countries on average are also relatively equivalent as it ranges between 25 and 29 °C.

Although the three houses were selected from various locations, they also had similar geological conditions. They were of the same size and located in similar environmental conditions thus providing data that can be confidentially compared concerning the amount of EE and CO_2 emissions being released. The examination of the Houses is summarized through the data synopsis depicted in Table 35.3.

35.4 Findings and Discussion

The comparative analysis of the building materials from three houses was based on the release of EE and CO_2 through the quantity of materials used. This research enables the construction industry to identify which building materials are less harmful to the environment. This research also helps to identify the most sustainable form of a house built within the same climate conditions and of similar dwelling sizes. The principal building materials used in each house are listed in Table 35.2 above, which shows the composition of each house and the percentage of the diverse types of materials used. Knowing the quantity of materials used assists in identifying which materials are more popular and have a larger impact, this will encourage the construction industry to look for other alternative materials.

35.4.1 Results

As shown in Tables 35.4 and 35.5 below, the three houses are compared by the amount of EE (MJ) and CO_2 emissions (kg CO_2) being released into the environment. House 2 (Mudbrick) is shown as having the lowest EE and CO_2 of 71,883 MJ and 7383 kg CO_2 (Refer to Tables 35.4 and 35.5). House 3 (Cement block: hollow concrete blocks) was recorded with the second lowest amount of EE at 189,045 MJ and CO_2 at 24,404 kg CO_2, and lastly House 1 (Standard build) recorded the highest amount of EE of 283,649 MJ and CO_2 at 25,582 kg CO_2. Tables 35.4 and 35.5 shown below exhibit the findings of the comparative analysis, while Figs. 35.8 and 35.9 display a simpler and more understandable version of Tables 35.4 and 35.5. Figures 35.7 and 35.8 are simplified to show the comparisons of each house of the amount of EE and CO_2 being produced.

Explicitly, a more comprehensive analysis is provided in the following sections, from floors to finishes.

Table 35.4 EE assessment of the houses

		EE (MJ)					
Element	Materials	QTY (kg)	House 1	QTY (kg)	House 2	QTY (kg)	House 3
Floor	Steel	740	19,011	–	–	–	–
	Concrete	45,408	65,716	20,856	18,978	27,600	25,116
	Timber	–	–	63	630	81	810
	Substrate	–	–	380	31	537	44
	Sub-total		***84,727***		***19,639***		***25,970***
Roof	Timber	1445	28,345	120	1247	144	1499
	Steel	370	49,338	–	–	–	–
	Plasterboard	654	8307	–	–	–	–
	Aluminum	–	–	184	28,458	243	37,664
	Sub-total	–	***85,990***		***29,705***		***39,163***
External walls	Brick (Standard)	13,781	48,708	–	–	–	–
	Mortar	1832	2652	8662	9615	10,065	11,172
	Timber	710	13,918	1260	1260	153	1530
	Brick walls (mud)	–	–	50,862	0	–	–
	Concrete	–	–	7200	7380	9600	9840
	Cement blocks (hollow concrete blocks)	–	–	–	–	72,200	96,026
	Sub-total		***65,278***		***18,255***		***118,568***
Windows & Doors	Timber	246	4829	294	3057	294	3057
	Glass	69	6558	2	33	3	48
	Sub-total	–	***11,387***		***3090***		***3105***
Finishes	Tiles	237	21,371	94	1194	176	2239
	Carpet	159	14,896	–	–	–	–
	Sub-total		***36,267***		***1194***		***2239***
Grand-total		65,651	**283,649**	89,977	**71,883**	121,096	**189,045**

Source Authors

Table 35.5 Embodied CO_2 assessment of the houses

Embodied CO_2 Emissions (Kg CO_2)							
Element	Materials	QTY (kg)	House 1	QTY (kg)	House 2	QTY (kg)	House 3
Floor	Steel	740	1758	–	–	–	–
	Concrete	45,408	5835	20,856	2732	27,600	3615
	Timber	–	–	63	44	81	57
	Substrate	–	–	380	2	537	3
	Sub-total		***7593***		***2778***		***3675***
Roof	Timber	1445	2687	120	106	144	128
	Steel	370	4533	–	–	–	–
	Plasterboard	654	737	–	–	–	–
	Aluminum	–	–	184	1511	243	2000
	Sub-total	–	***7957***		***1617***		***2128***
External walls	Brick (Standard)	13,781	4093	–		–	–
	Mortar	1832	235	8662	1481	10,065	1721
	Timber	710	1319	1260	89	153	18
	Brick walls (mud)	–	–	50,862	0	–	–
	Concrete	–	–	7200	1083	9600	1444
	Cement blocks (hollow concrete blocks)	–	–	–	–	72,200	15,017
	Sub-total		***5647***		***2653***		***18,200***
Windows and Doors	Timber	246	457	294	261	294	261
	Glass	69	569	2	2	3	3
	Sub-total	–	***1026***		***263***		***264***
Finishes	Tiles	237	1908	94	72	176	137
	Carpet	159	1451	–	–	–	–
	Sub-total		***3359***		***72***		***137***
Grand-total		65,651	**25,582**	89,977	**7383**	121,096	**24,404**

Source Authors

35.4.1.1 Floor

Table 35.4 shows that the floor element of the houses produces the second-highest EE. The main materials used in House 1 for the floors were steel and concrete, which resulted in a total of 84,727 MJ producing 7593 kg CO_2. House 3 used concrete, timber, and substrate for the floors which resulted in 25,970 MJ and 3675 kg CO_2. The results for House 3 were similar to the results for House 2 as they used the same

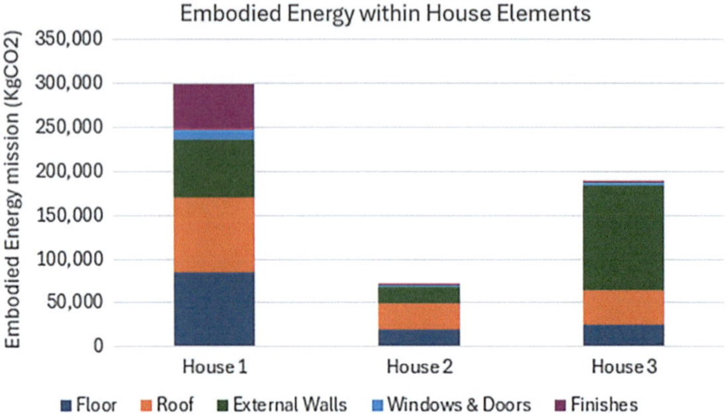

Fig. 35.7 Comparison of EE. *Source* Authors

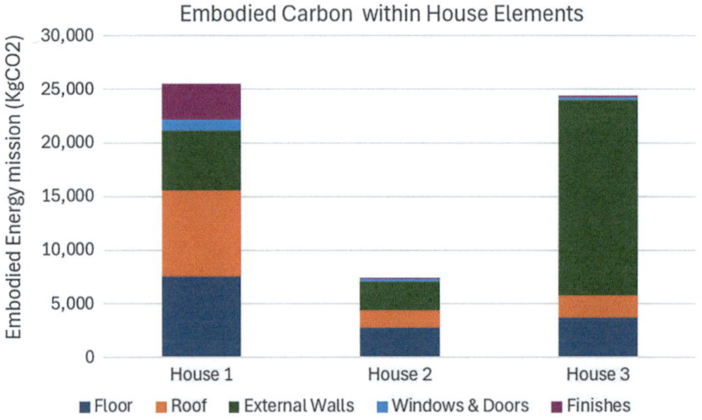

Fig. 35.8 Comparison of embodied CO_2. *Source* Authors

materials, however, house 2 was slightly smaller house by 31.32 m^2, and the EE and CO_2 were consequently slightly lower resulting in House 2 having a total EE of 19,630 MJ and 2778 kg CO_2.

35.4.1.2 Roof

Roofing is a major component of a house, it has a significant impact on the environment, hence more sustainable materials need to be selected. House 1 (Standard build) uses three main materials for the roofing, timber, steel, and plasterboard for the internal ceilings. The manufacturing of steel has a large negative impact on the environment. The iron ore used to produce steel is a non-renewable resource although steel itself is 100% recyclable. The EE for House 1 roofing is 85,990 MJ with CO_2

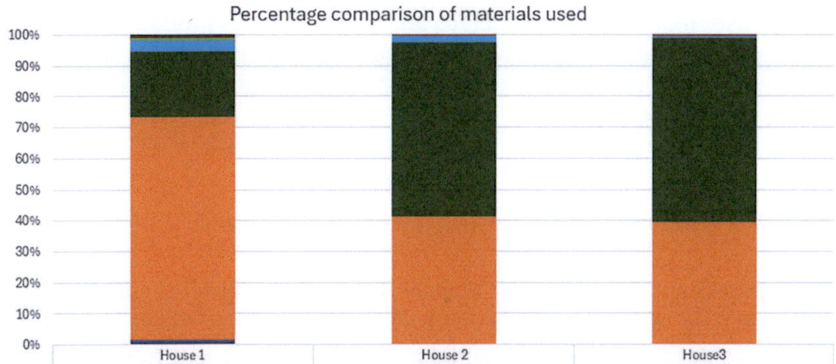

	House 1	House 2	House3
Carpet	0%	0%	0%
Tiles	0%	0%	0%
Glass	0%	0%	0%
Plasterboard	1%	0%	0%
Aluminum	0%	0%	0%
Timber	4%	2%	1%
Bricks & blocks	21%	57%	60%
Concrete	72%	41%	39%
Steel	2%	0%	0%

Fig. 35.9 Comparison of material composition. *Source* Authors

at 7957 kg CO_2. House 3 (Cement block: Cement blocks—hollow concrete blocks) has the second highest EE and CO_2 emissions. Further, Hollow CMUs were used. This was used for all the blocks: Bullnose blocks, Corner blocks, Lintel blocks, and Split-face blocks.

House 3 used timber and aluminum which resulted in an EE of 39,163 MJ and CO_2 of 2128 kg CO_2. House 2 (Mudbrick) recorded the lowest impact on the environment producing EE of 29,705 MJ and CO_2 at 1617 kg CO_2. The common roofing material for the three houses is timber, however, House 1 utilizes steel which has a larger amount of EE and CO_2 emissions, and House 2 utilizes aluminum which also has high EE and CO_2 in comparison to timber employed in House 2.

35.4.1.3 External Walls

The external walls of houses refer to the outer walls of the building. The three houses being compared have a similar floor size, ranging from $78m^2$ to $111m^2$. They are all single-story houses. As mentioned previously, the weather where the houses are located is warm, with an average of 27.5 °C. This demonstrates that the use of the materials for the external walls can be compared due to the same climate conditions. House 1 (Standard build) records the second-highest EE and CO_2 emissions. This house's external walls consist of standard brick, mortar, and timber resulting in a total EE of 65,278 MJ and CO_2 at 5647 kg CO_2. House 2 (Mudbrick) has the lowest EE of 18,255 MJ in the external walls, this is 100,313 MJ lower than the EE of House 3, it also has 2653 kg CO_2 of CO_2 which is 15,547 kg CO_2 lower than house 3.

House 3 uses mortar, timber, concrete, and Cement blocks (hollow concrete blocks), resulting in the external walls having the highest EE of 118,568 MJ and CO_2 emission at 18,200 kg CO_2. House 2 (Mudbrick) ranks number one with the lowest amount of EE and CO_2 being produced.

35.4.1.4 Windows and Doors

The windows and doors within the three different houses all shared the use of two main materials, which were timber and glass. All three houses have appropriately the same range of EE and CO_2 being produced through the timber frames of both doors and windows with House 1 (Standard build) recording an EE for the timber frames of both doors and windows at 4829 MJ, while Houses 2 and 3 recorded 3057 MJ. The CO_2 for House 1 was recorded at 457 kg CO_2 while Houses 2 and 3 recorded CO_2 of 261 kg CO_2. The three houses are of assorted designs, and the amount of glass in each house differs in size. As shown in Table 35.2, House 1 incorporates considerably more glass for windows in comparison to Houses 2 and 3. The total EE for House 1 is 11,387 MJ and the CO_2 are 1026 kg CO_2, House 2 (Mudbrick) records a total EE for windows and doors at 3105 MJ and CO_2 emission of 264 kg CO_2, while House 3 (Cement block) recorded total EE at 3090 MJ and CO_2 of 263 kg CO_2.

35.4.1.5 Finishes

The last element being compared with the three houses is the finishes. All three houses use tiles however, House 1 also uses carpet. House 1 in using two types of finishes increases the amount of EE and CO_2 emissions. This results in House 1 (Standard build) recording a total EE of 36,267 MJ and CO_2 at 3359 kg CO_2. Contrastingly, Houses 2 and 3 only use tiles, hence, releasing similar amounts of EE and CO_2 emissions. House 2 (Mudbrick) with the lowest amount of EE produced 1194 MJ whereas House 3 (Cement block) had 2239 MJ for CO_2 emissions, House 2 recorded the least amount at 72 kg CO_2 while House 3 recorded CO_2 of 137 kg CO_2.

35.4.2 Overall Material Composition Analysis

The material composition is derived from the amounts of the materials used in each house in kilograms (Tables 35.4 and 35.5), then further converted to the percentages as depicted through the assessment of the 3 case studies (Table 35.2). Figure 35.9 depicts the percentage breakdown of the materials that are being used in each house. This can demonstrate which materials can have the most impact on the environment or can be replaced with another sustainable material. This can be beneficial for

Table 35.6 Inclusive final analysis

Research Aims	Objectives	Research topic	Final Analysis
Examining the advantages of the use of sustainable building materials in construction	Conducting a comparative analysis on 3 houses and seeing the EE and CO_2 emissions	The comparison of the houses depicted which house has the least impact on the environment Hence, this relates to the research topic as it is about the '*reduction of environmental impact through improved construction material selection and processes*	**EE**: House 1—283,649 MJ House 2—71,883 MJ House 3—189,045 **CO_2 Emissions:** House 1—25,582 kg CO_2 House 2—7383 kg CO_2 House 3—24,404 kg CO_2 **Overall House 2** is the most sustainable building and the least harmful to the environment

Source Authors

future sustainable buildings as it demonstrates which materials to use more when also compared to Figs. 35.7 and 35.8, the relation between the amount used with EE and CO_2 emissions.

Figure 35.9 suggests that concrete (including concrete byproducts) formed a sizable portion of the material used within the three houses, with House 1 (Standard build) having 72%, House 2 (Mudbrick) having 41% and House 3 at 39%. It also depicts the use of brick was quite significant, standard bricks made up 21% of the material for house 1. Mudbricks made up 57% of the material used in House 2 (Mudbrick). Mudbricks are a sustainable building material that does not cause detrimental impacts on the environment. House 1 also displays the usage of steel, carpet, and plasterboard which can be harmful and can be easily avoided when planning and designing the house for sustainability. House 3 (Cement block) shows the lowest percentage of timber being used in comparison to the other houses.

Furthermore, Table 35.6 highlights the final analysis of the research in conjunction with the research aims, objectives, and research topic.

Table 35.6 above summarizes the research analysis conducted in this study and demonstrates how the research aims, objectives, and research topic were achieved. The results are discussed further in more depth in the following section.

35.4.3 Findings and Discussion

Attention has been drawn to the environmental issues associated with residential construction. This research aimed to analyze building materials to address potential ways of improving material selection to reduce the detrimental impacts on the environment by the construction industry. Based on the attributes of the materials that are used, it is possible to show the amount of EE and CO_2 that are released by the three houses and which of the materials being used can be beneficial. Comparing the houses, House 2 (Mudbrick) has the lowest EE and CO_2 being produced (Refer to Figs. 35.7 and 35.8), when converted to energy and CO_2 footprint per square meter (using Gross Floor Area (GFA) = 78.66m^2 from Table 35.2), the values are 913.84 MJ/m^2 and 93.90 kg CO_2/m^2. Additionally, for House 1 (using GFA of 111 m^2), the values are obtained: 2555.40 MJ/m^2 and 230.47 kg CO_2/m^2, and for House 3 (using GFA of 109.98 m^2) the values being: 1718.90 MJ/m^2 and 221.89 kg CO_2/m^2.

There have been several research studies comparing EE and CO_2 of different construction materials, results from other studies are discussed as a comparison. A study by [26] considered the EE of a variety of construction materials including clay brick masonry walls buildings with reinforced concrete structure (4210 MJ/m^2), burnt clay brickwork (2920 MJ/m^2), and soil–cement block houses (1610 MJ/m^2). Another study undertaken in India by [116], compared the EE of reinforced cement concrete (3702.3 MJ/m^2) and mudbrick houses (2298.8 MJ/m^2), see Table 35.7.

It is important to acknowledge that the studies in question exhibit limitations in compatibility due to the varying dimensions and locations of the houses analyzed (for further details, refer to Sect. 3.3). Nevertheless, these studies facilitate a comprehensive comparison. Reddy and Jagadish [26] found that a mudbrick house, constructed from compressed soil cement, possesses a lower EE compared to both burnt brick and reinforced concrete houses. Similarly, [116] reported that a mudbrick house demonstrates lower EE than a reinforced concrete house of comparable size. The findings of the current study corroborate the results of both [26, 116], revealing that House 2 (Mudbrick) has the least EE, while House 1 (Standard build—Burnt brick) exhibits the highest, followed by House 3 (Concrete block). As illustrated in Table 35.5, the CO_2 from House 1 (Standard build—Burnt brick) and House 3 (Concrete block) are at least 3.5 and 3.3 times greater than those from House 2 (Mudbrick), respectively. While it is crucial to evaluate the total EE and CO_2 for each house, it is equally essential to analyze the impact of the individual materials used in construction.

Table 35.7 Results summary from comparative studies

Structure type	This study	Reddy and Jagadish [26]	Chel and Tiwari [116]
Burnt brick (House 1)	2555.40 MJ/m^2	2920 MJ/m^2	×
Mudbrick (House 2)	913.84 MJ/m^2	1610 MJ/m^2	2298.8 MJ/m^2
Concrete block (House 3)	1718.90 MJ/m^2	×	×
Reinforced concrete	×	4210 MJ/m^2	3702.3 MJ/m^2
EPS 100 mm	×	×	×
EPS 150 mm	×	×	×

Source Authors

Significant reductions in EE and CO_2 within residential buildings can be realized by substituting high-energy and high-CO_2 construction materials with more sustainable alternatives. Research has identified specific materials that contribute the most to EE, as evidenced by the works of [38, 117, 118]. For instance, an analysis reveals that House 1, constructed with burnt brick, incorporates steel, while Houses 2 and 3, built with mud and concrete blocks respectively, do not utilize this material. By replacing steel with timber, the overall EE and environmental impact can be diminished. Additionally, aluminum, known for its substantial EE and CO_2 footprint, can also be substituted with more eco-friendly materials. Timber stands out as the most sustainable option due to its low EE and CO_2 emissions, along with its recyclability and renewability, as highlighted in Table 35.1. This has led to a renewed interest in timber as a preferred building material, particularly as a substitute for concrete and steel in modern construction practices, as noted by [119].

The use of natural, replenishable resources can be beneficial to the environment as they have fewer negative environmental impacts. Earthen construction materials, such as mudbricks, can have cost advantages as reported by [120], where the South African Rand (R) cost per unit of mudbrick is between R1.20-R1.50 ($0.11–0.14 AUD) while conventional industrially burnt bricks or cement blocks vary between R2.50-R3 ($0.23–0.28 AUD) per unit. Adegun and Adedji [120] also suggest that earthen materials have thermal benefits as they absorb heat slowly hence, during hot afternoons the interior remains cool. Moreover, the accumulated heat dissipates gradually at night during cold nights helping to maintain an even temperature within the house.

As previously mentioned, glass is a prevalent material in the construction industry; however, its use incurs significant economic and environmental costs. The data presented in Fig. 35.9 indicates that House 1, constructed with standard burnt brick, contains the highest proportion of glass. Transitioning to transparent wood could offer substantial cost savings, as it provides superior thermal insulation compared to glass, thereby reducing energy expenses. Transparent wood is derived from sustainable, renewable resources and is associated with lower CO_2 [86–88]. Conversely, the carpet utilized in House 1, as illustrated in Fig. 35.9, is not a sustainable choice, despite its common application in residential settings. Carpets are composed of synthetic materials derived from petroleum, rendering them unsustainable and significantly contributing to landfill waste. In 1999, Anderson, while chairing the President's Council for Sustainable Development, advocated for stricter penalties regarding carpet waste in landfills [121]. More sustainable alternatives to traditional carpeting include bamboo flooring, recycled timber, and natural linoleum, all of which are recyclable and sourced from renewable materials.

35.5 Conclusion and Recommendations

This research presents a comparative evaluation of the building materials used in three distinct houses, focusing on the total EE and CO_2 associated with each. The findings facilitate an understanding of which materials exert the most significant negative effects on the environment, thereby guiding the search for more sustainable alternatives in future construction projects. The study employed a quantitative approach to assess the three houses by integrating material characteristics with construction quantities. The analysis of EE and CO_2 emissions, as detailed in Tables 35.4 and 35.5, allows for a comparative evaluation of the environmental impacts of the various house designs. The central research inquiry revolved around identifying strategies to mitigate environmental impacts through improved selection of building materials. The study confirms that careful material selection can indeed lead to a reduction in environmental harm. Specifically, the incorporation of recyclable materials and judicious choices made during the planning and design stages can significantly lower both EE and CO_2 emissions. To ensure the reliability of the results, the three houses were selected based on comparable climatic conditions and similar sizes, thereby providing a robust framework for analysis.

The findings of the research indicate that House 2, constructed from mudbrick, demonstrates the highest level of sustainability when compared to the other housing options analyzed. This superior sustainability is attributed to the utilization of mudbricks, which are characterized by their zero EE and CO_2 emissions. In contrast, conventional materials such as burnt bricks and cement blocks significantly contribute to an environmentally detrimental and toxic atmosphere. Furthermore, the case studies reveal that the use of construction materials like concrete, steel, and aluminum should be minimized due to their elevated levels of EE and CO_2 emissions, particularly in comparison to more sustainable alternatives like timber and earthen materials. An effective building design should prioritize the incorporation of non-virgin materials that are sustainably sourced, ideally from local suppliers, to bolster local economies, mitigate high energy consumption, and reduce the environmental impact associated with the long-distance transport of materials. This research advocates for a more sustainable approach to material selection, emphasizing that the adoption of environmentally friendly materials can significantly diminish ecological footprints and foster healthier living conditions. The key considerations taken from this research include:

I. Increased renewable and recycled material selection, such as recycled Steel and timber reducing the need for more production which is energy intensive.
II. Increased use of low EE materials that can be sourced naturally such as Straw Bale, Hempcrete, etc.
III. Boosted use of locally sourced materials such as local Stone and aggregate to reduce the transportation impacts and support sustainable forestry practices.
IV. Increase the use of high thermal mass materials that can store and release thermal energy, reducing heating and cooling needs. More sustainable roofing materials

such as green and cool roofs covered with vegetation, provide insulation, absorb rainwater, and reduce the urban heat island effect.

V. Appropriate lifecycle considerations to increase durability, ease of maintenance, and End-of-Life recyclability. This in turn reduces landfill waste.

VI. The utilization of highly innovative materials such as Cross-Laminated Timber (CLT) which provides strong, sustainable properties and good thermal properties. Along with better insulation techniques such as using aerogel insulation which has highly effective insulators with low thickness.

The persistent reliance on standardized materials poses a significant risk to the future of sustainable communities, potentially hindering the attainment of global sustainable development objectives. Consequently, the construction sector must prioritize efforts aimed at minimizing EE and CO_2 emissions. A thorough examination of current practices is essential to identify and mitigate unnecessary contributions to these environmental impacts. To foster a sustainable future, it is crucial to implement rigorous strategies and interventions that address the primary contributors to climate change.

35.5.1 Recommendations

The findings of this study suggest several avenues for recommendations that could enhance the accuracy and precision of results. Given that the current research relies on secondary data, the incorporation of primary data could yield more reliable and trustworthy outcomes. In utilizing primary data, the researcher would engage in direct collection, thereby mitigating the potential biases associated with secondary data sources. Additionally, a comparative analysis of houses with identical floor areas situated within the same vicinity could facilitate a more equitable assessment, particularly when examining the diverse materials employed, such as standardized materials versus fully recyclable and earthen alternatives. This approach would also encompass an evaluation of the EE and CO_2 associated with the various material phases. Future investigations could benefit from a broader sample size of houses, which would enhance the accuracy of mean values, as the current analysis of three houses presents a limited margin of error. Furthermore, adopting a more qualitative methodology could enrich the research, involving human-centered approaches such as surveys and interviews to gather insights from residents regarding their experiences in sustainable versus standardized homes, as well as inquiries directed at builders concerning the manageability of varied materials. This multifaceted approach would provide a more comprehensive perspective, leading to more robust and dependable research outcomes.

References

1. K. Al-Kodmany, Sustainable high-rise buildings: toward resilient built environment. Front. Sustain. Cities **4** (2022).https://doi.org/10.3389/frsc.2022.782007
2. M. Klijn-Chevalerias, S. Javed, The Dutch approach for assessing and reducing environmental impacts of building materials. Build. Environ. **111**, 147–159 (2017). https://doi.org/10.1016/j.buildenv.2016.11.003
3. A. Walker-Morison, T. Grant, S. McAlister, *The Environmental Impact of Building Materials* (BEDP Environment Design Guide, 2007) pp. 1–9
4. S. Karumba, S. Sethuvenkatraman, V. Dedeoglu, R. Jurdak, S. Kanhere, Barriers to blockchain-based decentralised energy trading: a systematic review. Int. J. Sustain. Energ. **42**(1), 41–71 (2023). https://doi.org/10.1080/14786451.2023.2171417
5. S.S. Petkar, *Environmental Impact of Construction Materials and Practices* (National Institute of Construction Management and Research, 2014)
6. DPTI (Department of Planning, Transport and Infrastructure), *Environmentally Sustainable Building Materials—Selection* (2017). https://www.dit.sa.gov.au/__data/assets/pdf_file/0009/293688/Environmentally_Sustainable_Building_Materials.pdf
7. A. Warsame, A. Alasow, M. Salad, Energy consumption and economic growth nexus in Somalia: an empirical evidence from nonlinear ARDL technique. Int. J. Sustain. Energ. **43**, 1 (2024). https://doi.org/10.1080/14786451.2023.2287780
8. ABS, *Environment by Numbers: Selected Articles on Australia's Environment*. Australian Bureau of Statistics (2003). http://www.abs.gov.au/ausstats/abs@.nsf/mf/4617.0
9. T.B. Edil, Green highways: strategy for recycling materials for sustainable construction practices, in *7th International Congress on Advances in Civil Engineering* (2006), pp. 1–20
10. J.C. Morel, A. Mesbah, M. Oggero, P. Walker, Building houses with local materials: means to drastically reduce the environmental impact of construction. Build. Environ. **36**(10), 1119–1126 (2001). https://doi.org/10.1016/S0360-1323(00)00054-8
11. I. Aşchilean, N. Cobîrzan, A. Bolboaca, R. Boieru, R. Felseghi, Pairing solar power to sustainable energy storage solutions within a residential building: a case study. Int. J. Energy Res. **45**(10), 15495–15511 (2021). https://doi.org/10.1002/er.6982
12. L. Qarout, Reducing the environmental impacts of building materials: embodied energy analysis of a high-performance building. Thesis, The University of Wisconsin-Milwaukee (2017)
13. J. Navarro, M.J. González, Assessment of the decrease of CO_2 emissions in the construction field through the selection of materials: practical case study of three houses of low environmental impact. Build. Environ. **41**(7), 902–909 (2006). https://doi.org/10.1016/j.buildenv.2005.04.006
14. K. Gharehbaghi, M. Georgy, F. Rahmani, Composite high-rise structures: structural health monitoring (SHM) and case studies. Mater. Technol. Eng. 146–152 (2018)
15. W. Ferdous, Y. Bai, T.D. Ngo, A. Manalo, P. Mendis, New advancements, challenges and opportunities of multi-storey modular buildings—a state-of-the-art review. Eng. Struct. **183**, 883–893 (2019). https://doi.org/10.1016/j.engstruct.2019.01.061
16. R. Hu, K. Chen, W. Fang, L. Zheng, J. Xu, The technology-environment relationship revisited: Evidence from the impact of prefabrication on reducing construction waste. J. Clean. Prod. **341**, 130883 (2022). https://doi.org/10.1016/j.jclepro.2022.130883
17. S. Costa, M.S. Carvalho, C. Pimentel, C. Duarte, A systematic literature review and conceptual framework of construction industrialization. J. Constr. Eng. Manag. **149**(2) (2023). https://doi.org/10.1061/(ASCE)CO.1943-7862.0002410
18. B. Rossello-Batle, C. Ribas, A. Moià-Pol, V. Martínez-Moll, Saving potential for embodied energy and CO_2 emissions from building elements: a case study. J. Building Phys. **39**, 261–284 (2015). https://doi.org/10.1177/1744259114543982
19. W. Lu, W.M.W. Lee, F. Xue, J. Xu, Revisiting the effects of prefabrication on construction waste minimization: a quantitative study using bigger data, resources. Conserv. Recycl. **170**, 105579 (2021). https://doi.org/10.1016/j.resconrec.2021.105579

20. M. Wackernagel, N.B. Schulz, D. Demuling, A.C. Linares, M. Jenkins, V. Kapos, C. Monfreda, J. Loh, N. Myers, R. Norgaard, J. Randers, Tracking the ecological overshoot of the human economy. Proc. Natl. Acad. Sci. **99**(14), 9266–9271 (2002). https://doi.org/10.1073/pnas.142033699
21. L.H. Nguyen, The impact modeling of project management function behaviors on construction labor productivity. Int. J. Product. Perform. Manag. **71**(7), 2991–3008 (2022). https://doi.org/10.1108/IJPPM-09-2020-0478
22. C. Nwachukwu, E. Diemuodeke, T. Briggs, M. Ojapah, C. Okereke, K. Okedu, A. Kalam, Low/zero carbon technology diffusion and mapping for Nigeria's decarbonization. Int. J. Sustain. Energy **43**(1) (2024). https://doi.org/10.1080/14786451.2024.2317146
23. A. Podder, S. Pless, S. Rothgeb, N. Klammer, C. Blazek, J. Louis, K. Rashid, *Integrating Energy Efficiency Strategies with Industrialized Construction for our Clean Energy Future* (2020), https://www.osti.gov/biblio/1668862
24. M.K. Dixit, J.L. Fernandez-Solis, S. Lavy, C.H. Culp, Identification of parameters for embodied energy measurement. Energy Build. **42**(8), 1238–1247 (2010). https://doi.org/10.1016/j.enbuild.2010.02.016
25. L. Blackburne, K. Gharehbaghi, A. Hosseinian-Far, The knock-on effects of green buildings: high-rise construction design implications. Int. J. Struct. Integr. **13**(1), 57–77 (2022). https://doi.org/10.1108/IJSI-06-2021-0062
26. B.V.V. Reddy, K.S. Jagadish, Embodied energy of common and alternative building materials and technologies. Energy Build. **35**(2), 129–137 (2003). https://doi.org/10.1016/S0378-7788(01)00141-4
27. Y. Teng, K. Li, W. Pan, T. Ng, Reducing building life cycle carbon emissions through prefabrication: evidence from and gaps in empirical studies. Build. Environ. **132**, 125–136 (2018). https://doi.org/10.1016/j.buildenv.2018.01.026
28. D. Urge-Vorsatz, A. Novikova, Opportunities and costs of carbon dioxide mitigation in the worlds domestic sector, in *International Energy Efficiency in Domestic Appliances and Lighting Conference* (2006).https://doi.org/10.1016/j.enpol.2007.10.009
29. D. Bansal, R. Singh, R.L. Sawhney, Effect of construction materials on embodied energy and cost of buildings—a case study of residential houses in India up to 60 m of plinth area. Energy Build. **69**, 260–266 (2014). https://doi.org/10.1016/j.enbuild.2013.11.006
30. R.Z.C. Lim, D.T.-W. Looi, M.-T. Chen, H.-H. Tsang, J. Wilson, A component-based macro-mechanical model for inter-module connections in steel volumetric buildings. J. Constr. Steel Res. **207** (2023). https://doi.org/10.1016/j.jcsr.2023.107954
31. K. Gharehbaghi, K. Farnes, L. Kucharski, S. Fragomeni, The adaptability of evolving green high-rise construction: embodied energy dynamics in Australian high-rise buildings. Int. J. Sustain. Energ. **41**(10), 1383–1398 (2022). https://doi.org/10.1080/14786451.2022.2052291
32. A. Miller, Embodied energy—a life-cycle of transportation energy embodied in construction materials, in *Proceedings* of *the RICS Foundation Construction and Building Research Conference. COBRA 2001* (2001)
33. P. Crowther, Re-Valuing construction materials and components through design for disassembly, in *Unmaking Waste in Production and Consumption: Towards the Circular Economy* (Emerald Publishing Limited, 2018), pp. 309–321. https://doi.org/10.1108/978-1-78714-619-820181024
34. G.J. Treloar, P.E. Love, G.D. Holt, Using national input output data for embodied energy analysis of individual residential buildings. Constr. Manag. Econ. **19**(1), 49–61 (2001). https://doi.org/10.1080/014461901452076
35. L. Blackburne, K. Gharehbaghi, K. Farnes, O. Moore, M. Russo, Application of Confirmatory Factor Analysis (CFA) as the basis of the evaluation of the green building certification systems. J. Sci. Technol. Policy Manag. **14**(4), 696–712 (2023). https://doi.org/10.1108/JSTPM-04-2021-0066
36. B.E. Shari, S. Madougou, O.S. Ohunakin, P. Blechinger, Y. Moumouni, A. Ahmed, Y. Tukur, Exploring the dynamics of stakeholders' perspectives towards planning low-carbon energy transitions: a case of the Nigerian power sector. Int. J. Sustain. Energ. **42**, 209–235 (2023). https://doi.org/10.1080/14786451.2023.2186147

37. A.C. Sorensen, S. Jagtap, A. Warell, Material selection-a qualitative case study of five design consultancies, Vancouver, Canada, DS 87-1, in *Proceedings of the 21st International Conference on Engineering Design (ICED 17). Vol 1: Resource Sensitive Design, Design Research Applications and Case Studies, Vancouver, Canada*, 21–25 Aug 2017 (2017), pp. 439–448
38. B. Lawson, *Building Materials, Energy and the Environment: Towards Ecologically Sustainable Development*. Red Hill, ACT, Royal Australian Institute of Architects (1996)
39. R.J. Cole, P.C. Kernan, Life-cycle energy use in office buildings. Build. Environ. **31**(4), 307–317 (1996)
40. G.K.C. Ding, Life cycle assessment (LCA) of sustainable building materials: an overview. Eco-Effic. Constr. Build. Mater. 38–622014).https://doi.org/10.1533/9780857097729.1.38
41. M. Suzuki, T. Oka, K. Okada, The estimation of energy consumption and CO_2 emission due to housing construction in Japan. Energy Build. **22**, 165–169 (1995). https://doi.org/10.1016/0378-7788(95)00914-J
42. T. Oka, M. Suzuki, T. Konnya, The estimation of energy consumption and amount of pollutants due to the construction of buildings. Energy Build. **19**(4), 303–311 (1993). https://doi.org/10.1016/0378-7788(93)90016-N
43. P. Tiwari, J. Parikh, Cost of CO_2 reduction in building construction. Energy **20**(6), 531–547 (1995). https://doi.org/10.1016/0360-5442(94)00084-G
44. K. Gharehbaghi, M. Georgy, K.M. Robson, S. Wilkinson, K. Farnes, Versatility in sustainable building design (SBD) practices: an empirical study. Int. J. Build. Pathol. Adapt. **40**(5), 728–752 (2022)
45. M. Asif, T. Muneer, R. Kelley, Life cycle assessment: a case study of a dwelling home in Scotland. Build. Environ. **42**(3), 1391–1394 (2007). https://doi.org/10.1016/j.buildenv.2005.11.023
46. A. De Luca, L. Chen, K. Gharehbaghi, Sustainable utilization of recycled aggregates: robust construction and demolition waste reduction strategies. Int. J. Build. Pathol. Adapt. **39**(4), 666–682 (2021). https://doi.org/10.1108/IJBPA-04-2020-0029
47. K. Gharehbaghi, R. Chenery, Fiber Reinforced Concrete (FRC) for high rise construction: case studies. Mater. Sci. Eng. **272**, 223–229 (2017). https://doi.org/10.1088/1757-899X/272/1/012034
48. K. Gharibashvili, F. Rahmani, Viscoelastic materials: innovation and development countenance. Solid State Phenom. Key Eng. Mater. **266**, 165–171 (2017). https://doi.org/10.4028/www.scientific.net/SSP.266.165
49. R. Fay, G. Treloar, U.I. Raniga, Life-cycle energy analysis of buildings: a case study. Build. Res. Inf. **28**(1), 31–41 (2000). https://doi.org/10.1080/096132100369073
50. J. McIntosh, K. Gharehbaghi, K.F. Tee, A. De Luca, J. Taylor, Viability rubric of fibre-reinforced polymer composites for civil infrastructure applications. Infrastruct. Asset Manag. **11**(2), 78–87 (2024). https://doi.org/10.1680/jinam.23.00009
51. C. Thormark, Including recycling potential in energy use into the life cycle of buildings. Building Res. Inf. **28**(3), 176–183 (2000). https://doi.org/10.1080/096132100368948
52. V. Tatýrek, D. Měšťanová, Use of recycled materials in road construction. Bus. IT (Praha, On-line). **XI**, 54–60 (2021). https://doi.org/10.14311/bit.2021.01.06
53. K. Gharehbaghi, K. Tee, K. Farnes, L. Blackburne, Pertinency of geopolymer concrete for the Australian construction industry: a broader contextual appraisal, in *Cement and Concrete: Design, Performance and Structure,* (2023), pp. 153–190
54. B. Wagner, Developments of material flow cost accounting in Germany (2003)
55. A. Horvath, Construction materials and the environment. Annu. Rev. Environ. Resour. **29**, 181–204 (2004). https://doi.org/10.1146/annurev.energy.29.062403.102215
56. G.J. Treloar, H. Gupta, P.E. Love, B. Nguyen, An analysis of factors influencing waste minimisation and use of recycled materials for the construction of residential buildings. Manag. Environ. Qual. **14**(1) 2003. 10.https://doi.org/10.1108/14777830310460432.
57. Directive 2008/98/EC on waste and repealing certain Directives. L 312/3. Brussels, Belgium.: European Parliament and Council, http://eur-lex.europa.eu/LexUriServ/LexUriServ.do?uri=OJ:L:2008:312:0003:0030:en:PDF

58. COAG (Council of Australian Governments), *2018 National Waste Policy: Less Waste More Resources* (2018). https://www.environment.gov.au/protection/wasteresource-recovery/publications/national-waste-policy-2018
59. COAG (Council of Australian Governments), *National Waste Policy Action Plan* (2019). https://www.environment.gov.au/protection/waste/publications/national-wastepolicy-action-plan
60. Sustainability Victoria, *Dispose of Construction Waste* (2021). https://www.sustainability.vic.gov.au/recycling-and-reducing-waste/at-home/recycling-at-home/construction-waste
61. D. Mutz, D. Hengvevoss, C. Hugi, T. Gros, *Waste-to-Energy Options in Municipal Solid Waste Management* (2017). https://www.giz.de/en/downloads/GIZ_WasteToEnergy_Guidelines_2017.pdf
62. J. Bolden, T. Abu-Lebdeh, E. Fini, Utilization of recycled and waste materials in various construction applications. American Journal of Environmental Science. **9**(1), 14–24 (2013). https://doi.org/10.3844/ajessp.2013.14.24
63. Y. Xiao, M.A. Reuter, U.D.O. Boin, Aluminium recycling and environmental issues of salt slag treatment. J. Environ. Sci. Health **40**(10), 1861–1875 (2005). https://doi.org/10.1080/10934520500183824
64. Johnson (2023) Press Release: Johnson Controls launches a partnership with Nucor to recycle nearly 100% of scrap steel at major Johnson Controls facilities in the United States, Dow Jones Institutional News website.
65. S. Munn, V. Soebarto, The issues of using recycled materials in architecture, in *38th international conference of architectural science association ANZAScA "Contexts of architecture"*, Launceston, Tasmania (2004)
66. N. Campbell, Integrating ESD—the environmental legacy of Sydney Olympic Park, in *Environmental Symposium Old Parliament House*, Canberra, 14 Sept 2001 (2001)
67. L.Y. Shen, C.M. Tam, D. Drew, Mapping approach for examining waste management on construction sites. J. Constr. Eng. Manag. **130**(4), 472–481 (2004). https://doi.org/10.1061/(ASCE)0733-9364(2004)130:4(472)
68. A. Akhtar, A.K. Sarmah, Construction and demolition waste generation and properties of recycled aggregate concrete: a global perspective. J. Clean. Prod. **186**, 262–281 (2018). https://doi.org/10.1016/j.jclepro.2018.03.085
69. R. Jin, B. Li, T. Zhou, D. Wanatowski, P. Piroozfar, An empirical study of perceptions towards construction and demolition waste recycling and reuse in China. Resour. Conserv. Recycl. **126**, 86–98 (2017). https://doi.org/10.1016/j.resconrec.2017.07.034
70. W. Zhao, R.B. Leeftink, V.S. Rotter, Evaluation of the economic feasibility for the recycling of construction and demolition waste in China—the case of Chongqing. Resour. Conserv. Recycl. **54**(6), 377–389 (2010). https://doi.org/10.1016/j.resconrec.2009.09.003
71. S. Suranjan Salins, S.V.K. Reddy, S. Kumar, P.S. Nair, Application of biomass-based wood shaving packing in a liquid desiccant dehumidification system—a source of sustainable energy. Int. J. Sustain. Energy **43**(1) (2024). https://doi.org/10.1080/14786451.2023.2287783
72. M. Chowdhury, A. Upadhay, A. Briggs, M. Belal, An empirical analysis of green supply chain management practices in Bangladesh construction industry, Trondheim, Norway, in *European Operation Management Association (EurOMA) Conference* (2016), pp. 1–10
73. S. Harris, G. Ryan, M. Marino, K. Gharehbaghi, The green design of high-rise buildings: how is the construction industry evolving. Technological Sustainability **2**(1), 89–99 (2023). https://doi.org/10.1108/TECHS-04-2022-0017
74. J. Ferguson, N. Kermode, C.L. Nash, W.A. Sketch, R.P. Huxford, *Managing and minimising construction waste: a practical guide* (Institution of Civil Engineers, London, 1995), pp.1–60
75. C.S. Poon, Reducing construction waste. Waste Manage. **27**, 1715–1716 (2007). https://doi.org/10.1016/j.wasman.2007.08.013
76. W. Lu, H. Yuan, A framework for understanding waste management studies in construction. Waste Manage. **31**(6), 1252–1260 (2011). https://doi.org/10.1016/j.wasman.2011.01.018
77. R. Islam, T.H. Nazifa, A. Yuniarto, A.S. Uddin, S. Salmiati, S. Shahid, An empirical study of construction and demolition waste generation and implication of recycling. Waste Manage. **95**, 10–21 (2019). https://doi.org/10.1016/j.wasman.2019.05.049

78. K.R. Kareem, R.K. Pandey, Study of management and control of waste construction materials in civil construction project. Int. J. Eng. Adv. Technol. (IJEAT) **2**(3) (2013)
79. A.A. Gulghane, P.V. Khandve, Management for construction materials and control of construction waste in construction industry: a review. Int. J. Eng. Res. Appl. **5**(4), 59–64 (2015)
80. K. Cochran, T. Townsend, D. Reinhart, H. Heck, Estimation of regional building related C&D debris generation and composition: case study for Florida, US. Waste Manage. **27**(7), 921–931 (2007). https://doi.org/10.1016/j.wasman.2006.03.023
81. M. Kazerooni Sadi, A. Abdullah, M. Navazandeh Sajoudi, M.M. Kamal, F. Torshizi, R. Taherkhani, Reduce, reuse, recycle and recovery in sustainable construction waste management. Advanced Materials Research. **446**, 937–944 (2012). https://doi.org/10.4028/www.scientific.net/AMR.446-449.937
82. E. Franzoni, Materials selection for green buildings: which tools for engineers and architects? Procedia Engineering. **21**, 883–890 (2011)
83. G. Ding, The development of a multi-criteria approach for the measurement of sustainable performance for built projects and facilities, Sydney. Ph.D. Thesis, University of technology (2004). https://opus.lib.uts.edu.au/bitstream/10453/20191/7/02Whole.pdf
84. S. Halliday, *Sustainable Construction* (Butterworth Heinemann, London, 2008). https://doi.org/10.4324/9780080557168.
85. B.V. Reddy, Sustainable building technologies. Curr. Sci. **87**(7), 899–907 (2004)
86. A. Androff, *Transparent Wood Could Be the Window of the Future* (2020). https://www.usda.gov/media/blog/2020/10/01/transparent-wood-could-be-window-future
87. T. Li, M. Zhu, Z. Yang, J. Song, J. Dai, Y. Yao, W. Luo, G. Pastel, B. Yang, L. Hu, Wood composite as an energy efficient building material: guided sunlight transmittance and effective thermal insulation. Adv. Energy Mater. **6**(22) 2016. https://doi.org/10.1002/aenm.201601122
88. C. Montanari, Y. Ogawa, P. Olsén, L.A. Berglund, High performance, fully bio-based, and optically transparent wood biocomposites. Adv. Sci. **8** (2021).https://doi.org/10.1002/advs.202100559
89. N.B. Das, R.K. Sharma, A. Pandey, B.G. Narayanan, The existence of carpet industry in Bhadohi, India. Trends Text. Eng. Fash. Technol. **3**, 323–332 (2018). https://doi.org/10.31031/TTEFT.2018.03.000563
90. D. Peñaloza, M. Erlandsson, A. Falk, Exploring the climate impact effects of increased use of bio-based materials in buildings. Constr. Build. Mater. **125**, 219–226 (2016). https://doi.org/10.1016/j.conbuildmat.2016.08.041
91. M.U. Hossain, A. Sohail, T.S. Ng, Developing a GHG-based methodological approach to support the sourcing of sustainable construction materials and products. Resour. Conserv. Recycl. **145**, 160–169 (2019). https://doi.org/10.1016/j.resconrec.2019.02.030
92. M. Lawrence, Reducing the environmental impact of construction by using renewable materials. J. Renew. Mater. **3**(3), 163–174 (2015). https://doi.org/10.7569/JRM.2015.634105
93. W. Drozd, A. Leśniak, S. Zaworski, Construction time of three wall types made of locally sourced materials: a comparative study. Adv. Mater. Sci. Eng. 1–8 (2018).https://doi.org/10.1155/2018/2172575
94. I. Marchetti, E. Rego, Social impacts of energy resource planning: assessment methodology and case study. Int. J. Sustain. Energ. **42**(1), 331–350 (2023). https://doi.org/10.1080/14786451.2023.2191145
95. T. Häkkinen, K. Belloni, Barriers and drivers for sustainable building. Building research and information. Int. J. Res. Dev. Demonstr. **39**, 239–255 (2011). https://doi.org/10.1080/09613218.2011.561948
96. J. McKenna, S. Harris, K. Heinrich, T. Stewart, K. Gharehbaghi, The evaluation of green building's feasibility: comparative analysis across different geological conditions. Engineering **4**, 2034–2054 (2023). https://doi.org/10.3390/eng4030115
97. N. Lazarus, Construction materials report. Bio-Regional Development Group (2003), http://www.eurogypsum.org/wp-content/uploads/2015/05/N1019.pdf

98. M. Finkbeiner, A. Inaba, R. Tan, K. Christiansen, H.J. Klüppel, The new international standards for life cycle assessment: ISO 14040 and ISO 14044. Int. J. Life Cycle Assess **11**, 80–85 (2006). https://doi.org/10.1065/lca2006.02.002
99. R.J. Cole, Energy and greenhouse gas emissions associated with the construction of alternative structural systems. Build. Environ. **34**(3), 335–348 (1998). https://doi.org/10.1016/s0360-1323(98)00020-1
100. I. Page, Timber in government buildings-cost and environmental impact analysis. E408, Project No. QC, 5018. (2006).
101. C. Scheuer, G.A. Keoleian, P. Reppe, Life cycle energy and environmental performance of a new university building: modeling challenges and design implications. Energy Build. **35**, 1049–1064 (2003). https://doi.org/10.1016/S0378-7788(03)00066-5
102. R. Judkoff, Increasing building energy efficiency through advances in materials. MRS Bull. **33**(4), 449–454 (2008). https://doi.org/10.1557/mrs2008.88
103. W. Yu, B. Li, H. Jia, M. Zhang, D. Wang, Application of multi-objective genetic algorithm to optimize energy efficiency and thermal comfort in building design. Energy and Buildings **88**, 135–143 (2015). https://doi.org/10.1016/j.enbuild.2014.11.063
104. D. Eyisi, The usefulness of qualitative and quantitative approaches and methods in researching problem-solving ability in science education curriculum. J. Educ. Pract. **7**(15), 91–98 (2016)
105. G. Rounce, Quality, waste and cost considerations in architectural building design management. Int. J. Project Manage. **16**(2), 123–127 (1998). https://doi.org/10.1016/S0263-7863(97)00042-2
106. S. Coventry, B. Shorter, M. Kingsley, *Demonstrating Waste Minimisation benefits in Construction* (CIRIA, London, 2001)
107. R. Emmanuel, Estimating the environmental suitability of wall materials: preliminary results from Sri Lanka. Build. Environ. **39**(10), 1253–1261 (2004). https://doi.org/10.1016/j.buildenv.2004.02.012
108. N. Lawson, I. Douglas, S. Garvin, C. McGrath, D. Manning, J. Vetterlein, Recycling construction and demolition waste—a UK perspective. Environ. Manag. Health **12**(2), 146–157 (2001). https://doi.org/10.1108/09566160110389898
109. M. Osmani, J. Glass, A.D. Price, Architect's perspectives on construction waste reduction by design. Waste Manage. **28**(7), 1147–1158 (2007). https://doi.org/10.1016/j.wasman.2007.05.011
110. A. Invidiata, M. Lavagna, E. Ghisi, Selecting design strategies using multi-criteria decision making to improve the sustainability of buildings. Build. Environ. **139**, 58–68 (2018)
111. M.M. Khasreen, P.F. Banfill, G.F. Menzies, Life-cycle assessment and the environmental impact of buildings: a review. Sustainability **1**(3), 674–701 (2009). https://doi.org/10.3390/su1030674
112. I.Z. Bribian, A.V. Capilla, A.A. Uson, Life cycle assessment of building materials: Comparative analysis of energy and environmental impacts and evaluation of the eco-efficiency improvement potential. Build. Environ. **46**(5), 1133–1140 (2011). https://doi.org/10.1016/j.buildenv.2010.12.002
113. G.P. Hammond, C.I. Jones, Embodied energy and carbon in construction materials. Proc. Inst. Civ. Eng.-Energy **161**(2), 87–98 (2008). https://doi.org/10.1680/ener.2008.161.2.87
114. L.F. Cabeza, C. Barreneche, L. Mireo, J.M. Morera, E. Bartoli, A.I. Fernandez, Low carbon and low embodied energy materials in buildings: a review. Renew. Sustain. Energy Rev. **23**, 536–542 (2013). https://doi.org/10.1016/j.rser.2013.03.017
115. A.A. Morini, M.J. Ribeiro, D. Hotza, Early-stage materials selection based on embodied energy and carbon footprint. Mater. Des. **178**, 107861 (2019). https://doi.org/10.1016/j.matdes.2019.107861
116. A. Chel, G. Tiwari, Thermal performance and embodied energy analysis of a passive house-case study of vault roof mud-house in India. Appl. Energy **86**, 1956–1969 (2009). https://doi.org/10.1016/j.apenergy.2008.12.033
117. K.A. Ali, M.I. Ahmad, Y. Yusup, Issues, impacts, and mitigations of carbon dioxide emissions in the building sector. Sustainability **12**(7427), 1–11 (2020). https://doi.org/10.3390/SU12187427

118. D. Harris, A quantitative approach to the assessment of the environmental impact of building materials. Build. Environ. **34**(6), 751–758 (1999). https://doi.org/10.1016/S0360-1323(98)00058-4
119. R.H. Crawford, X. Cadorel, A framework for assessing the environmental benefits of mass timber construction. Procedia Eng. **196**, 838–846 (2017). https://doi.org/10.1016/j.proeng.2017.08.015
120. O.B. Adegun, Y.M.D. Adedji, Review of economic and environmental benefits of earthen materials for housing in Africa. Front. Arch. Res. **6**(4), 19–528 (2017). https://doi.org/10.1016/j.foar.2017.08.003
121. A. Luqmani, M. Leach, D. Jesson, Factors behind sustainable business innovation: The case of a global carpet manufacturing company. Environ. Innov. Soc. Trans. **24**, 94–105 (2017). https://doi.org/10.1016/j.eist.2016.10.007